Springer **M***onographs* in **M***athematics*

For further volumes:
www.springer.com/series/3733

Kenji Iohara • Yoshiyuki Koga

Representation Theory of the Virasoro Algebra

Kenji Iohara
Université Claude Bernard Lyon 1
Institut Camille Jordan
43 Boulevard du 11 Novembre 1918
69622 Villeurbanne Cedex
France
iohara@math.univ-lyon1.fr

Yoshiyuki Koga
University of Fukui
Department of Applied Physics
Faculty of Engineering
3-9-1 Bunkyo
910-8507 Fukui
Japan
koga@quantum.apphy.u-fukui.ac.jp

ISSN 1439-7382
ISBN 978-1-4471-2609-6 ISBN 978-0-85729-160-8 (eBook)
DOI 10.1007/978-0-85729-160-8
Springer London Dordrecht Heidelberg New York

British Library Cataloguing in Publication Data
A catalogue record for this book is available from the British Library

Mathematics Subject Classification (2010): 17B10, 17B50, 17B55, 17B67, 17B68, 17B69, 17B70, 81R10

Cover design: VTEX, Vilnius

Printed on acid-free paper

Springer is part of Springer Science+Business Media (www.springer.com)

Dedicated to Professor Michio Jimbo on the occasion of his sixtieth birthday.

Preface

The Virasoro algebra is an infinite dimensional Lie algebra which has appeared in several context. For example, E. Cartan [C] in 1909 classified simple infinite dimensional linearly compact Lie algebras over $\mathbb{C}$ which contains the Lie algebra $\overline{W}_m$ of all formal vector fields in m determinates. Around 1940, the Lie algebra of derivations of the group algebra of $\mathbb{Z}/p\mathbb{Z}$ over a field of characteristic $p > 2$ was studied by colleagues of E. Witt, e.g., H. Zassenhaus [Za] showed that this Lie algebra is simple and H.J. Chang [Ch] studied its representations. All these works treated the so-called **Witt algebra**, or the centreless Virasoro algebra. It was only in 1966 that its non-trivial central extension first appeared in the work of R. Block [Bl] over a field of positive characteristic in classifying certain class of Lie algebras. In 1968, I. M. Gelfand and D. B. Fuchs [GF] determined the cohomology ring of the Lie algebra of the vector fields on the circle.

In 1970, the Virasoro algebra appeared in the article [Vir] of M. Virasoro on his work on string theory, hence the name **Virasoro algebra**. See, e.g., [Mand] for an account of this period. A great of impact was made in 1984 by A. A. Belavin, A. M. Polyakov and A. B. Zamolodchikov [BPZ1], [BPZ2] where they treated the phenomena at critical points in 2-dimensional statistical mechanical models.

In the 1980s, the representation theory of the Virasoro algebra was intensively studied. In particular, the structure of its Verma modules and its Fock modules were completely determined by B. Feigin and D. Fuchs [FeFu4]. For a further beautiful historical description, see [GR].

The aim of this book is to describe some fundamental facts about the representation theory of the Virasoro algebra in a self-contained manner. The topics covered in this book are the structure of Verma modules and Fock modules, the classification of (unitarisable) Harish-Chandra modules, tilting equivalence, and the rational vertex operator algebras associated to the so-called BPZ series representations. A detailed description of the contents of this book will be given in the introduction.

In 1995, the first author was proposed by his supervisor M. Jimbo to write a survey about known important facts about the representation theory of the Virasoro algebra as his PhD thesis, since there confusion in the literature as the theory was developed both in mathematics and in physics at the same time. After the PhD thesis of the first author, which had nothing to do with the proposition, he had occasion to take part in a joint project with the second author. Around 2000, we decided to realise the proposition and finally it took us nearly 10 years.

Acknowledgements

During the preparation of the manuscript, we have visited several institutes such as Fields Institute, Korean Institute for Advanced Study, Mathematische Forschungsinstitut Oberwolfach, Max-Planck-Institut für Mathematik, Mathematical Sciences Research Institute, and Research Institute of Mathematical Sciences. It is a pleasure to thank these institutions and the colleagues who inspired us. In particular, we would like to thank B. Feigin, M. Jimbo, F. Malikov, O. Mathieu and M. Miyamoto for their discussions and helpful suggestions. The second author is partly supported by JSPS Grant-in-Aid for Scientific Research.

Last but not least, the first author would like to thank his wife Yuko Iohara and the second author to thank his wife Mokako Koga for their constant encouragement without which this work would not have been accomplished.

Lyon, France
Fukui, Japan
July 2010.

Kenji Iohara
Yoshiyuki Koga

Contents

Introduction

The aim of this book is to describe some fundamental facts about the representation theory of the Virasoro algebra. We also collect some non-standard but basic facts used to describe its representations.

Here is a detailed outline of the contents of the main parts of the book.

Chapter 1 We introduce the Virasoro algebra Vir as the universal central extension of the Witt algebra. This Lie algebra is a $\mathbb{Z}$-graded Lie algebra and it is appropriate to work within the framework of what we call Q-graded Lie algebras. Hence, we also present a categorical setup for representations of Q-graded Lie algebras and some of their properties such as local composition series, (co)homologies, and Berstein-Gelfand-Gelfand duality.

Chapter 2 We explain the proof of the conjecture of V. Kac which says that any simple $\mathbb{Z}$-graded Vir-module with finite multiplicities is either a highest weight module, a lowest weight module, or the module of type $t^{\lambda}\mathbb{C}[t,t^{-1}](dt)^{\mu}$. The proof is given by the reduction to positive characteristic cases. Hence, for the reader who is not familiar with Lie p-algebras and their representations, we make a brief survey of the general theory of Lie p-algebras in Appendix B that are used in the proof. This chapter can be read independently of other chapters.

Chapter 3 The representation theory of the Virasoro algebra Vir behaves like a rank 2 Lie algebra with triangular decomposition and the Jantzen filtration is an effective tool to analyze Verma modules over these algebras. Here, we present the Jantzen filtration and explain some of their properties. We also generalise this filtration to analyze Fock modules over Vir in Chapter 8. The generalisation given here first appeared in this book.

Chapter 4 The so-called Kac determinant formula of Verma modules over Vir is given with its proof in the language of vertex algebras. This result plays a crucial role when we analyze the Verma modules in Chapters 5 and 6. The determinant formulae of the Vir-module maps from a Verma module to the Fock module (with the same highest weight) and from a Fock module to the contragredient dual of the Verma module (with the same highest weight) play an important role to analyze Fock modules in Chapter 8 and they are also given in this chapter.

Chapters 5 and 6 The structure of Verma modules is analyzed in detail. Starting from the classification of the highest weights, the structure of the Jantzen filtration of Verma modules is completely determined from which the Bernstein–Gelfand–Gelfand type resolution follows. As a simple corollary, the characters of the all irreducible highest weight modules over Vir are given. In particular, the characters of minimal series representations, with fixed central charge, forms a vector-valued $SL(2, \mathbb{Z})$-modular form and its modular transformations are also calculated.

Chapter 7 The tilting equivalence for a certain class of $\mathbb{Z}$-graded Lie algebras is explained. In particular, for the Virasoro algebra, this explains the structural duality between Verma modules with highest weights (c, h) and $(26 - c, 1 - h)$ which can be observed by analyzing Verma modules. This chapter can be read independently of other chapters.

Chapter 8 The structure of Fock modules is analyzed in detail as an application of our generalised Jantzen filtration. Two other topics are also presented: i) singular vectors of Fock modules in terms of Jack symmetric polynomials, and ii) the relation between semi-infinite forms and the bosonic Fock modules. This chapter uses some results from Chapters 5 and 6.

Chapter 9 The rationality of the vertex operator algebras associated to minimal series representations is given as an application of the results obtained in Chapters 5 and 6. The fusion algebra associated to such a vertex operator algebra is also given. One of the beautiful and important characterisations of the BPZ series representations is presented with two appendices which provide some necessary background.

Chapters 10 and 11 We show that certain irreducible highest weight Vir-modules are unitarisable with the aid of unitarisability of integrable highest weight $\hat{\mathfrak{sl}}_2$-modules. The complete classification of the unitarisable Harish-Chandra modules is given as the goal of these two chapters.

There are also three appendices for the reader's convenience:

Appendix A Some facts from homological algebras are recalled. Most of them are given without proof.

Appendix B The general theory of Lie p-algebras, in particular, of completely solvable Lie p-algebras is recalled. A proof of some facts are given.

Appendix C We collect some facts about the rationality of vertex operator algebras. A generalisation to vertex operator superalgebras is briefly discussed.

Now, we briefly discuss the difference between this book and two other references: one by V. G. Kac and A. K. Raina [KR] and the other by L. Guieu and C. Roger [GR].

In [KR], the authors treated the so-called Kac determinant of Verma modules, that is explained in Chapter 4. They also showed that the condition on the highest weight given by D. Freidan, Z. Qiu and S. Shenker [FQS1], [FQS2] for a highest weight module to be unitarisable is sufficient. In this book, we have shown that, indeed, the above condition is necessary and sufficient for highest weight modules in Chapters 10 and 11.

In [GR], the authors mainly discussed some algebraic and geometric aspects of the Virasoro algebra itself and not its representations. Hence, the contents of this book are rather complementary.

Chapter 1
Preliminary

In this chapter, we will collect some fundamental objects in the representation theory of the Virasoro algebra.

In Section 1.1, we will define the Virasoro algebra as the universal central extension of the Witt algebra, i.e., the Lie algebra which consists of the derivations of the Laurent polynomial ring with one variable.

In Section 1.2, we will introduce a class of Lie algebras with triangular decomposition and an anti-involution, called *Q-graded Lie algebras* [RW1]. In fact, many important Lie algebras, e.g., the Virasoro algebra and a Kac–Moody algebra, are Q-graded Lie algebras. We will develop a general theory on representations over a Q-graded Lie algebra, such as categories of modules, highest weight modules, Verma modules, contragredient duals and so on.

In Section 1.3, we will define Lie algebra homology and cohomology in terms of derived functors. Moreover, we will state some formulae related with extensions of modules and duality, e.g., Frobenius reciprocity.

In Section 1.4, we will explain the so-called Berstein-Gelfand-Gelfand duality.

In the appendix of this chapter, we will prove the propositions stated in Subsection 1.1.1. An alternative proof, in terms of standard (co)complex, of a proposition proved in Section 1.3 will be also given.

Until the end of this chapter, we assume that $\mathbb{K}$ is a field whose characteristic is zero.

1.1 Virasoro Algebra

We introduce the Virasoro algebra as the universal central extension of the Witt algebra. For the reader's convenience, we first recall basic properties of (universal) central extensions. (For the details and the proofs, see § 1.A.)

K. Iohara, Y. Koga, *Representation Theory of the Virasoro Algebra*,
Springer Monographs in Mathematics, DOI 10.1007/978-0-85729-160-8_1,

1.1.1 Universal Central Extension

Let $\mathfrak{a}$ and $\mathfrak{u}$ be Lie algebras over $\mathbb{K}$, and let V be a $\mathbb{K}$-vector space. Here, we regard V as a commutative Lie algebra. A short exact sequence

$$0 \longrightarrow V \xrightarrow{\iota} \mathfrak{u} \xrightarrow{\alpha} \mathfrak{a} \longrightarrow 0$$

is called a **central extension** of $\mathfrak{a}$ if

$$[\iota(V), \mathfrak{u}] = \{0\}.$$

V is called the **kernel** of the central extension. We sometimes refer to $\mathfrak{u}$ as a central extension of $\mathfrak{a}$.

Definition 1.1 *A central extension*

$$0 \longrightarrow V \xrightarrow{\iota} \mathfrak{u} \xrightarrow{\alpha} \mathfrak{a} \longrightarrow 0$$

of $\mathfrak{a}$ *is called a* **universal central extension** *if*

1. $\mathfrak{u}$ *is perfect, i.e.,* $\mathfrak{u} = [\mathfrak{u}, \mathfrak{u}]$,
2. *for any central extension* $\beta : \mathfrak{b} \to \mathfrak{a}$, *there exists* $\gamma : \mathfrak{u} \to \mathfrak{b}$ *such that the following diagram commutes:*

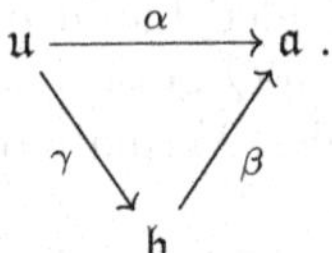

Remark that uniqueness of γ in Definition 1.1 follows from perfectness of $\mathfrak{u}$. Indeed, let $\gamma' : \mathfrak{u} \to \mathfrak{b}$ be another homomorphism such that $\alpha = \beta \circ \gamma'$. Then, for x, $y \in \mathfrak{u}$, we have

$$\begin{aligned}(\gamma - \gamma')([x, y]) &= [\gamma(x), \gamma(y)] - [\gamma'(x), \gamma'(y)] \\ &= [\gamma(x) - \gamma'(x), \gamma(y)] + [\gamma'(x), \gamma(y) - \gamma'(y)] \\ &= 0,\end{aligned}$$

since $\gamma(x) - \gamma'(x)$, $\gamma(y) - \gamma'(y) \in \mathrm{Ker}\alpha = \iota(V)$. Hence, $\gamma = \gamma'$ on $[\mathfrak{u}, \mathfrak{u}]$.

The following proposition holds:

Proposition 1.1 ([Gar]) *1. A Lie algebra* $\mathfrak{a}$ *admits a universal central extension if and only if* $\mathfrak{a}$ *is perfect.*
2. A universal central extension is unique up to an isomorphism of Lie algebras.

The kernel of the universal central extension of $\mathfrak{a}$ can be described by means of the second homology group $H_2(\mathfrak{a}, \mathbb{K}) := Z_2(\mathfrak{a}, \mathbb{K})/B_2(\mathfrak{a}, \mathbb{K})$ (cf. § A.3):

$$Z_2(\mathfrak{a},\mathbb{K}) := \left\{ \sum_i x_i \wedge y_i \,\middle|\, x_i, y_i \in \mathfrak{a},\ \sum_i [x_i, y_i] = 0 \right\},$$
$$B_2(\mathfrak{a},\mathbb{K}) := \{ x \wedge [y,z] + y \wedge [z,x] + z \wedge [x,y] | x,y,z \in \mathfrak{a} \}.$$

Proposition 1.2 *Suppose that a Lie algebra $\mathfrak{a}$ is perfect. Let $\mathfrak{c}$ be the kernel of the universal central extension of $\mathfrak{a}$. Then, we have*

$$\mathfrak{c} \simeq H_2(\mathfrak{a},\mathbb{K}).$$

Next, we recall the following relation between extensions and Lie algebra cohomology. For a $\mathbb{K}$-vector space V, we regard V as a trivial $\mathfrak{a}$-module. The second cohomology group $H^2(\mathfrak{a},V)$ is defined by $H^2(\mathfrak{a},V) := Z^2(\mathfrak{a},V)/B^2(\mathfrak{a},V)$ (cf. § A.3), where

$$Z^2(\mathfrak{a},V) := \left\{ f : \mathfrak{a}\times\mathfrak{a} \to V \,\middle|\, \begin{array}{l} \text{(i) } f(x,y) = -f(y,x), \\ \text{(ii) } f(x,[y,z]) + f(z,[x,y]) + f(y,[z,x]) = 0, \\ \qquad (\forall\, x,y,z \in \mathfrak{a}) \end{array} \right\}$$
$$B^2(\mathfrak{a},V) := \{ f : \mathfrak{a}\times\mathfrak{a} \to V \,|\, f(x,y) = g([x,y]),\ \exists g : \mathfrak{a} \to V \text{ (linear)} \}.$$

Proposition 1.3 *There exists a one-to-one correspondence between $H^2(\mathfrak{a},V)$ and set of equivalence classes of the central extensions of $\mathfrak{a}$ by V.*

1.1.2 Witt Algebra and its Universal Central Extension

The **Witt algebra** $\mathcal{D}$ is the Lie algebra which consists of derivations on the Laurent polynomial ring $\mathbb{K}[z,z^{-1}]$. In fact, $\mathcal{D}$ is given by

$$\mathcal{D} = \mathbb{K}[z,z^{-1}]\frac{d}{dz}. \tag{1.1}$$

We set $d_n := -z^{n+1}\frac{d}{dz}$, then

$$\mathcal{D} = \bigoplus_{n\in\mathbb{Z}} \mathbb{K} d_n,$$

and these generators satisfy the commutation relations:

$$[d_m, d_n] = (m-n) d_{m+n}.$$

The Virasoro algebra is the universal central extension of $\mathcal{D}$. Since $\mathcal{D}$ is perfect, there exists the universal central extension of $\mathcal{D}$ and it is described by the second (co)homology group (Propositions 1.1, 1.2 and 1.3).

Proposition 1.4 *1.* $\dim H_2(\mathcal{D},\mathbb{K}) = 1$.
2. $\dim H^2(\mathcal{D},\mathbb{K}) = 1$.

Proof. We show the first assertion. One can show that the set

$$\left\{ \begin{array}{l|l} (m-n)d_{m+n}\wedge d_0 - (m+n)d_m \wedge d_n & m+n\neq 0 \wedge m>n \\ d_k \wedge d_{-k} - kd_1 \wedge d_{-1} & k>1 \end{array} \right\}$$

forms a $\mathbb{K}$-basis of $Z_2(\mathcal{D},\mathbb{K})$.

On the other hand, $B_2(\mathcal{D},\mathbb{K})$ is spanned by the elements of the form

$$(m-n)d_l \wedge d_{m+n} + (l-m)d_n \wedge d_{l+m} + (n-l)d_m \wedge d_{n+l}. \tag{1.2}$$

By setting $l := 0$ in (1.2), we obtain

$$(m-n)d_{m+n}\wedge d_0 - (m+n)d_m \wedge d_n \equiv 0 \quad \text{in} \quad H_2(\mathcal{D},\mathbb{K}).$$

In the case $l+m+n=0$, (1.2) can be written as

$$\begin{aligned} &(l+2m)d_l \wedge d_{-l} + (l-m)d_{-l-m}\wedge d_{l+m} + (-2l-m)d_m \wedge d_{-m} \\ &= (m-l)v_{l+m} - (2l+m)v_m + (l+2m)v_l, \end{aligned}$$

where we set $v_k := d_k \wedge d_{-k} - kd_1 \wedge d_{-1}$. In particular, by setting $m=1$, we obtain

$$(1-l)v_{l+1} + (l+2)v_l \equiv 0 \quad \text{in} \quad H_2(\mathcal{D},\mathbb{K}).$$

Hence, we see that $\dim H_2(\mathcal{D},\mathbb{K}) = 1$.

We show the second assertion. Notice that for any $f \in Z^2(\mathcal{D},\mathbb{K})$ there exists $\tilde{f} \in Z^2(\mathcal{D},\mathbb{K})$ such that

$$f - \tilde{f} \in B^2(\mathcal{D},\mathbb{K}) \quad \text{and} \quad \tilde{f}(d_0, x) = 0 \ (\forall\, x \in \mathcal{D}).$$

Indeed, if we define $g_f : \mathcal{D} \to \mathbb{K}$ by

$$g_f(d_n) := \begin{cases} f(d_0,d_n)/n & \text{if } n \neq 0 \\ 0 & \text{if } n = 0 \end{cases},$$

then

$$\tilde{f}(x,y) := f(x,y) + g_f([x,y])$$

satisfies the above condition. Hence, for any $f + B^2(\mathcal{D},\mathbb{K}) \in H^2(\mathcal{D},\mathbb{K})$, we can take its representative f such that $f(d_0,x) = 0$ holds for any $x \in \mathcal{D}$.

On the other hand, since $f \in Z^2(\mathcal{D},\mathbb{K})$, we have

$$f(d_l,[d_m,d_n]) + f(d_n,[d_l,d_m]) + f(d_m,[d_n,d_l]) = 0,$$

and thus

$$(m-n)f(d_l, d_{m+n}) + (l-m)f(d_n, d_{l+m}) + (n-l)f(d_m, d_{n+l}) = 0. \quad (1.3)$$

We first set $l := 0$ in (1.3). Then, we have

$$(m+n)f(d_m, d_n) = 0,$$

since $f(d_0, x) = 0$. This implies that $f(d_m, d_n)$ can be written as

$$f(d_m, d_n) = \delta_{m+n,0} f_m$$

for some $\{f_m \in \mathbb{C} | m \in \mathbb{Z}\}$ such that $f_{-m} = -f_m$.

Next, by setting $m := 1$ and $n := -l-1$ in (1.3), we get

$$(l+2)f_l - (l-1)f_{l+1} - (2l+1)f_1 = 0. \quad (1.4)$$

Hence, $\{f_l | l \in \mathbb{Z}\}$ satisfies (1.4) and $f_{-l} = -f_l$. It is easy to see that the space of the solutions of these linear recursion relations is at most 2-dimensional and both $f_l = l^3$ and $f_l = l$ are solutions. Moreover, $f \in B^2(\mathcal{D}, \mathbb{K})$ if and only if $f_l = Kl$ for some $K \in \mathbb{K}$. We have proved the second assertion. □

By the above lemma, the Lie algebra Vir defined below is the universal central extension of $\mathcal{D}$.

Definition 1.2 *The* **Virasoro algebra**

$$\mathrm{Vir} := \bigoplus_{n \in \mathbb{Z}} \mathbb{K} L_n \oplus \mathbb{K} C$$

is the Lie algebra which satisfies the following commutation relations:

$$[L_m, L_n] = (m-n)L_{m+n} + \frac{1}{12}(m^3 - m)\delta_{m+n,0} C,$$
$$[\mathrm{Vir}, C] = \{0\},$$

where $\delta_{i,j}$ denotes the Kronecker delta.

1.2 Q-graded Lie Algebra

A Q-graded Lie algebra which is defined in this section is a generalisation of several important Lie algebras such as the Virasoro algebra. Some categories and their important objects, highest weight modules, Verma modules and the contragredient dual of modules are also introduced in this section.

1.2.1 Γ-graded Vector Spaces

Let Γ be an abelian group. Let $\mathrm{Vect}_{\mathbb{K}}^{\Gamma}$ be the category of $\mathbb{K}$-vector spaces defined as follows:

Definition 1.3 *1. A Γ-graded $\mathbb{K}$-vector space $V = \bigoplus_{\gamma\in\Gamma} V^{\gamma}$ is an object of $\mathrm{Vect}_{\mathbb{K}}^{\Gamma}$ if and only if $\{\gamma \in \Gamma | V^{\gamma} \neq \{0\}\}$ is at most countable.*
2. For Γ-graded $\mathbb{K}$-vector spaces V and W,

$$\mathrm{Hom}_{\mathrm{Vect}_{\mathbb{K}}^{\Gamma}}(V, W) := \{f \in \mathrm{Hom}_{\mathbb{K}}(V, W) | f(V^{\gamma}) \subset W^{\gamma}\ (\forall \gamma \in \Gamma)\}.$$

For simplicity, we set $\mathcal{P}(V) := \{\gamma \in \Gamma | V^{\gamma} \neq \{0\}\}$, and denote $\mathrm{Hom}_{\mathrm{Vect}_{\mathbb{K}}^{\Gamma}}(V, W)$ by $\mathrm{Hom}_{\mathbb{K}}^{\Gamma}(V, W)$.

Next, we introduce bifunctors $(\cdot) \otimes_{\mathbb{K}} (\cdot)$ and $\overline{\mathrm{Hom}}_{\mathbb{K}}(\cdot, \cdot)$ on the category $\mathrm{Vect}_{\mathbb{K}}^{\Gamma}$. For $V = \bigoplus_{\gamma\in\Gamma} V^{\gamma}$ and $W = \bigoplus_{\gamma\in\Gamma} W^{\gamma} \in \mathrm{Ob}(\mathrm{Vect}_{\mathbb{K}}^{\Gamma})$, we define $V \otimes_{\mathbb{K}} W$ and $\overline{\mathrm{Hom}}_{\mathbb{K}}(V, W)$ as follows:

$$V \otimes_{\mathbb{K}} W := \bigoplus_{\gamma\in\Gamma} (V \otimes_{\mathbb{K}} W)^{\gamma},$$

where $(V \otimes_{\mathbb{K}} W)^{\gamma} := \bigoplus_{\alpha\in\Gamma} V^{\alpha} \otimes_{\mathbb{K}} W^{\gamma-\alpha}$, and

$$\overline{\mathrm{Hom}}_{\mathbb{K}}(V, W) := \bigoplus_{\gamma\in\Gamma} \overline{\mathrm{Hom}}_{\mathbb{K}}(V, W)^{\gamma},$$

where $\overline{\mathrm{Hom}}_{\mathbb{K}}(V, W)^{\gamma} := \prod_{\alpha\in\Gamma} \mathrm{Hom}_{\mathbb{K}}(V^{\alpha}, W^{\gamma+\alpha})$.

Lemma 1.1. *For U, V and $W \in \mathrm{Ob}(\mathrm{Vect}_{\mathbb{K}}^{\Gamma})$, there exists an isomorphism*

$$\mathrm{Hom}_{\mathbb{K}}^{\Gamma}(U \otimes_{\mathbb{K}} V, W) \simeq \mathrm{Hom}_{\mathbb{K}}^{\Gamma}(U, \overline{\mathrm{Hom}}_{\mathbb{K}}(V, W)) \tag{1.5}$$

of $\mathbb{K}$-vector spaces.

Proof. By definition, we have

$$\begin{aligned}
\mathrm{Hom}_{\mathbb{K}}^{\Gamma}(U \otimes_{\mathbb{K}} V, W) &\simeq \prod_{\gamma} \mathrm{Hom}_{\mathbb{K}}((U \otimes V)^{\gamma}, W^{\gamma}) \\
&\simeq \prod_{\gamma} \prod_{\beta} \mathrm{Hom}_{\mathbb{K}}(U^{\beta} \otimes_{\mathbb{K}} V^{\gamma-\beta}, W^{\gamma}) \\
&\simeq \prod_{\beta} \prod_{\gamma} \mathrm{Hom}_{\mathbb{K}}(U^{\beta}, \mathrm{Hom}_{\mathbb{K}}(V^{\gamma-\beta}, W^{\gamma})) \\
&\simeq \prod_{\beta} \mathrm{Hom}_{\mathbb{K}}(U^{\beta}, \overline{\mathrm{Hom}}_{\mathbb{K}}(V, W)^{\beta}) \\
&\simeq \mathrm{Hom}_{\mathbb{K}}^{\Gamma}(U, \overline{\mathrm{Hom}}_{\mathbb{K}}(V, W)).
\end{aligned}$$

□

Moreover, we define a functor $D : \mathrm{Vect}_{\mathbb{K}}^{\Gamma} \to \mathrm{Vect}_{\mathbb{K}}^{\Gamma}$ as follows: For $V \in \mathrm{Ob}(\mathrm{Vect}_{\mathbb{K}}^{\Gamma})$, we set

$$D(V) := \bigoplus_{\gamma\in\Gamma} D(V)^{\gamma} \quad (D(V)^{\gamma} := V^{-\gamma}).$$

We introduce two dualising functors on $\mathrm{Vect}_{\mathbb{K}}^{\Gamma}$.

Definition 1.4 *Let V and W be objects of $\mathrm{Vect}_{\mathbb{K}}^{\Gamma}$, and let f be a morphism from V to W.*

1. *We define $V^{\sharp+} \in \mathrm{Ob}(\mathrm{Vect}_{\mathbb{K}}^{\Gamma})$ by $V^{\sharp+} := D(\overline{\mathrm{Hom}_{\mathbb{K}}}(V, \mathbb{K}^0))$, and $f^{\sharp+} \in \mathrm{Hom}_{\mathbb{K}}^{\Gamma}(W^{\sharp+}, V^{\sharp+})$ by the transpose of f.*
2. *We define $V^{\sharp-} \in \mathrm{Ob}(\mathrm{Vect}_{\mathbb{K}}^{\Gamma})$ by $V^{\sharp-} := \overline{\mathrm{Hom}_{\mathbb{K}}}(V, \mathbb{K}^0)$ and $f^{\sharp-} \in \mathrm{Hom}_{\mathbb{K}}^{\Gamma}(W^{\sharp-}, V^{\sharp-})$ by the transpose of f.*

In the sequel, for a $\mathbb{K}$-vector space V, let us denote $\mathrm{Hom}_{\mathbb{K}}(V, \mathbb{K})$ by V^*.

Remark 1.1 *Strictly speaking, the transpose ${}^t f$ of $f \in \mathrm{Hom}_{\mathbb{K}}^{\Gamma}(V, W)$ is an element of $\mathrm{Hom}_{\mathbb{K}}(W^*, V^*)$. By noticing that ${}^t f(W^{\sharp\pm}) \subset V^{\sharp\pm}$, the above $f^{\sharp\pm}$ is just the map ${}^t f|_{W^{\sharp\pm}}$.*

Finally, we define a **Γ-graded Lie algebra** as follows:

Definition 1.5 *A Lie algebra $\mathfrak{g} = \bigoplus_{\alpha\in\Gamma} \mathfrak{g}^{\alpha} \in \mathrm{Ob}(\mathrm{Vect}_{\mathbb{K}}^{\Gamma})$ is called Γ-graded, if it satisfies*

$$[\mathfrak{g}^{\alpha}, \mathfrak{g}^{\beta}] \subset \mathfrak{g}^{\alpha+\beta} \quad (\forall \alpha, \beta \in \Gamma).$$

1.2.2 Definitions

Let Q be a free abelian group of finite rank, say r, and let $\mathfrak{g}$ be a Lie algebra with a commutative subalgebra $\mathfrak{h}$.

Definition 1.6 *We say that a pair $(\mathfrak{g}, \mathfrak{h})$ is a **Q-graded Lie algebra** if it satisfies the following conditions.*

C0. A Lie algebra $\mathfrak{g} = \bigoplus_{\alpha\in Q} \mathfrak{g}^{\alpha}$ is Q-graded, $\mathfrak{h} = \mathfrak{g}^0$, and $\{\alpha \in Q | \mathfrak{g}^{\alpha} \neq \{0\}\}$ generates Q.

C1. For any $\alpha \in Q$ such that $\mathfrak{g}^{\alpha} \neq \{0\}$, there exists a unique $\lambda_{\alpha} \in \mathfrak{h}^*$ such that

$$[h, x] = \lambda_{\alpha}(h)x \quad (\forall h \in \mathfrak{h},\ \forall x \in \mathfrak{g}^{\alpha}).$$

C2. For any $\alpha \in Q$, $\dim \mathfrak{g}^{\alpha} < \infty$.

C3. There exists a basis $\{\alpha_i | i = 1, 2, \cdots, r\}$ of Q such that for any $\alpha \in Q$ with $\mathfrak{g}^{\alpha} \neq \{0\}$, one has

$$\alpha \in \sum_{i=1}^{r} \mathbb{Z}_{\geq 0}\alpha_i \ \text{ or } \ \alpha \in \sum_{i=1}^{r} \mathbb{Z}_{\leq 0}\alpha_i.$$

For $\alpha = \sum_i m_i \alpha_i \in Q$, we set $\mathrm{ht}\alpha := \sum_i m_i$. We say that a Lie subalgebra $\mathfrak{a}$ of a Q-graded Lie algebra $(\mathfrak{g}, \mathfrak{h})$ is Q-graded if $\mathfrak{a} = \bigoplus_{\alpha \in Q} \mathfrak{a}^\alpha$, where $\mathfrak{a}^\alpha := \mathfrak{a} \cap \mathfrak{g}^\alpha$.

In the sequel, let π_Q be the homomorphism defined by

$$Q \longrightarrow \mathfrak{h}^*; \quad \alpha \longmapsto \lambda_\alpha. \tag{1.6}$$

Remark 1.2 *Let $P^\vee := \mathrm{Hom}_{\mathbb{Z}}(Q, \mathbb{Z})$ be the dual lattice of Q and $\langle \cdot, \cdot \rangle : P^\vee \times Q \to \mathbb{Z}$ the dual pairing. We set $\mathfrak{d} := \mathbb{K} \otimes_{\mathbb{Z}} P^\vee$. We extend the Q-graded Lie algebra structure of $\mathfrak{g}$ to $\mathfrak{g}^e := \mathfrak{g} \oplus \mathfrak{d}$ (the direct sum of $\mathbb{K}$-vector spaces) as follows:*

$$(\mathfrak{g}^e)^\alpha := \begin{cases} \mathfrak{g}^\alpha & \alpha \neq 0 \\ \mathfrak{h} \oplus \mathfrak{d} & \alpha = 0 \end{cases},$$

and

$$[(g, d), (g', d')] := ([g, g'] + \langle d, \beta \rangle g' - \langle d', \alpha \rangle g, 0),$$

where $g \in \mathfrak{g}^\alpha$, $g' \in \mathfrak{g}^\beta$ and $d, d' \in \mathfrak{d}$. Let $\mathfrak{h}^e$ be the commutative subalgebra $\mathfrak{h} \oplus \mathfrak{d}$ of $\mathfrak{g}^e$. We define $\iota : Q \hookrightarrow (\mathfrak{h}^e)^*$ by

$$\iota(\alpha)((h, d)) = \lambda_\alpha(h) + \langle d, \alpha \rangle \quad (\forall (h, d) \in \mathfrak{h} \oplus \mathfrak{d}).$$

Then, $\mathfrak{g} = \bigoplus_{\alpha \in Q} \mathfrak{g}^\alpha$ is the simultaneous eigenspace decomposition of $\mathfrak{g} \subset \mathfrak{g}^e$ with respect to the adjoint action of $\mathfrak{h}^e$, i.e.,

$$\mathfrak{g}^\alpha = \{ g \in \mathfrak{g} \,|\, [h, g] = \iota(\alpha)(h) \ (\forall h \in \mathfrak{h}^e) \}.$$

The condition **C3** implies that a Q-graded Lie algebra admits a *triangular decomposition.* If we set $Q^+ := \sum_{i=1}^r \mathbb{Z}_{\geq 0} \alpha_i$ and

$$\mathfrak{g}^\pm := \bigoplus_{\pm \alpha \in Q^+ \setminus \{0\}} \mathfrak{g}^\alpha,$$

then we have $\mathfrak{g} = \mathfrak{g}^- \oplus \mathfrak{h} \oplus \mathfrak{g}^+$. For later use, we set $\mathfrak{g}^{\geq} := \mathfrak{h} \oplus \mathfrak{g}^+$ and $\mathfrak{g}^{\leq} := \mathfrak{g}^- \oplus \mathfrak{h}$.

Let $\mathfrak{a}$ be a Lie algebra over $\mathbb{K}$. Throughout this book, we denote the universal enveloping algebra of $\mathfrak{a}$ by $U(\mathfrak{a})$.

A linear map $\sigma : \mathfrak{a} \to \mathfrak{a}$ is called an **anti-involution** of $\mathfrak{a}$, if it satisfies $\sigma([x, y]) = [\sigma(y), \sigma(x)]$ $(\forall x, y \in \mathfrak{a})$ and $\sigma^2 = \mathrm{id}$. σ naturally extends to an anti-involution of the algebra $U(\mathfrak{a})$, and it is denoted by the same symbol σ.

The restriction of σ to $\mathfrak{a}$ is called an anti-involution of $\mathfrak{a}$, and is denoted by the same symbol σ.

Definition 1.7 *Let σ be an anti-involution of a Q-graded Lie algebra $(\mathfrak{g}, \mathfrak{h})$. We call σ a **Q-graded anti-involution**, if $\sigma(\mathfrak{g}^\alpha) \subset \mathfrak{g}^{-\alpha}$ for any $\alpha \in Q \setminus \{0\}$ and $\sigma|_{\mathfrak{h}} = \mathrm{id}_{\mathfrak{h}}$.*

1.2.3 Examples

Here, we provide three examples of the Q-graded Lie algebras, the *Virasoro algebra, affine Lie algebras* and a *Heisenberg Lie algebra.*

1. The **Virasoro algebra**: Let $\mathrm{Vir} := \bigoplus_{n\in\mathbb{Z}} \mathbb{K}L_n \oplus \mathbb{K}C$ be the Virasoro algebra (cf. Definition 1.2), and let $Q := \mathbb{Z}\alpha$ be a free abelian group of rank one. Setting $\mathfrak{h} := \mathbb{K}L_0 \oplus \mathbb{K}C$ and

$$\mathrm{Vir}^{n\alpha} := \begin{cases} \mathbb{K}L_n & \text{if } n \neq 0 \\ \mathfrak{h} & \text{if } n = 0 \end{cases},$$

we have a Q-gradation $\mathrm{Vir} = \bigoplus_{\beta\in Q} \mathrm{Vir}^{\beta}$. Let σ be the anti-involution defined by

$$\sigma(L_n) = L_{-n}, \quad \sigma(C) = C.$$

Then, $(\mathrm{Vir}, \mathfrak{h})$ is a Q-graded Lie algebra with the Q-graded anti-involution σ.

2. **Affine Lie algebras**: Let $\bar{\mathfrak{g}}$ be a simple finite dimensional Lie algebra over $\mathbb{C}$. Let $\bar{\mathfrak{h}}$ be a Cartan subalgebra of $\bar{\mathfrak{g}}$, and let $(\ ,\ )$ be a non-degenerate invariant bilinear form on $\bar{\mathfrak{g}}$. We set

$$\mathfrak{g} := \bar{\mathfrak{g}} \otimes \mathbb{C}[t, t^{-1}] \oplus \mathbb{C}K \oplus \mathbb{C}d,$$

and define the bracket on $\mathfrak{g}$ by

$$[x \otimes t^m, y \otimes t^n] := [x, y] \otimes t^{m+n} + m\delta_{m+n,0}(x, y)K,$$
$$[K, \mathfrak{g}] := \{0\}, \quad [d, x \otimes t^m] := mx \otimes t^m,$$

then $\mathfrak{g}$ is a Lie algebra, called an (*untwisted*) *affine Lie algebra.*
To describe the Q-graded Lie algebra structure of $\mathfrak{g}$, we first introduce some notation. Let $\bar{\Delta}$ be the set of the roots of $\bar{\mathfrak{g}}$ with respect to $\bar{\mathfrak{h}}$, and let $\bar{\mathfrak{g}}^{\beta}$ be the root space of $\bar{\mathfrak{g}}$ with root $\beta \in \bar{\Delta}$. Hence, $\bar{\mathfrak{g}} = \bar{\mathfrak{h}} \oplus \bigoplus_{\beta\in\bar{\Delta}} \bar{\mathfrak{g}}^{\beta}$. In the sequel, we fix a set of the simple roots $\bar{\Pi} = \{\alpha_i | i = 1, 2, \cdots, r\}$ of $\bar{\mathfrak{g}}$. Here, we denote the highest root by θ. Further, let $\bar{\sigma} : \bar{\mathfrak{g}} \to \bar{\mathfrak{g}}$ be an anti-involution of $\bar{\mathfrak{g}}$ such that $\bar{\sigma}|_{\bar{\mathfrak{h}}} = \mathrm{id}_{\bar{\mathfrak{h}}}$ and $\bar{\sigma}(\bar{\mathfrak{g}}^{\beta}) = \bar{\mathfrak{g}}^{-\beta}$ $(\forall \beta \in \bar{\Delta})$.
We put

$$\mathfrak{h} := \bar{\mathfrak{h}} \otimes 1 \oplus \mathbb{C}K \oplus \mathbb{C}d,$$

and regard $\bar{\Delta} \subset \mathfrak{h}^*$ via $\beta(h \otimes 1) := \beta(h)$, $\beta(K) := 0$ and $\beta(d) := 0$ $(\beta \in \bar{\Delta})$. Let $\delta \in \mathfrak{h}^*$ such that $\delta(\bar{\mathfrak{h}} \otimes 1) := \{0\}$, $\delta(K) := 0$ and $\delta(d) := 1$. We set

$$Q := \mathbb{Z}\bar{\Delta} \oplus \mathbb{Z}\delta \ (\subset \mathfrak{h}^*).$$

For each $\alpha \in Q$, we set

$$\mathfrak{g}^{\alpha} := \begin{cases} \bar{\mathfrak{g}}^{\beta} \otimes t^n & \text{if } \alpha = \beta + n\delta \ (\exists \beta \in \bar{\Delta},\ n \in \mathbb{Z}) \\ \bar{\mathfrak{h}} \otimes t^n & \text{if } \alpha = n\delta \ (n \in \mathbb{Z} \setminus \{0\}) \\ \mathfrak{h} & \text{if } \alpha = 0 \\ \{0\} & \text{otherwise} \end{cases}.$$

Then, $\mathfrak{g} = \bigoplus_{\alpha \in Q} \mathfrak{g}^{\alpha}$ is the root space decomposition with respect to $\mathfrak{h}$, and $\Pi := \bar{\Pi} \cup \{\alpha_0\}$ ($\alpha_0 := \delta - \theta$) is a $\mathbb{Z}$-basis of Q which satisfies **C3** in Definition 1.6. Hence, $(\mathfrak{g}, \mathfrak{h})$ is a Q-graded Lie algebra. We define a linear map $\sigma : \mathfrak{g} \to \mathfrak{g}$ by

$$\sigma(x \otimes t^n) := \bar{\sigma}(x) \otimes t^{-n} \ (x \in \bar{\mathfrak{g}}), \quad \sigma(K) = K, \quad \sigma(d) = d.$$

Then, σ is a Q-graded anti-involution of $(\mathfrak{g}, \mathfrak{h})$.

Remark 1.3 *Let us introduce a Q-graded Lie algebra structure on*

$$\mathfrak{g}' := [\mathfrak{g}, \mathfrak{g}] = \bar{\mathfrak{g}} \otimes \mathbb{C}[t, t^{-1}] \oplus \mathbb{C}K.$$

We set $\mathfrak{h}' := \mathfrak{h} \cap \mathfrak{g}'$. Then, $(\mathfrak{g}', \mathfrak{h}')$ is a Q-graded Lie algebra with Q-gradation

$$(\mathfrak{g}')^{\alpha} := \begin{cases} \mathfrak{g}^{\alpha} & \alpha \neq 0 \\ \mathfrak{h}' & \alpha = 0 \end{cases},$$

which admits the Q-graded anti-involution $\sigma|_{\mathfrak{g}'}$. In this case, the map $\pi_Q : Q \to (\mathfrak{h}')^*$ is not injective. An irreducible highest weight $\mathfrak{g}$-module is always irreducible as $\mathfrak{g}'$-module (cf. [Kac4]). The reader should notice that each weight subspace of the irreducible highest weight $\mathfrak{g}$-module with respect to the $\mathfrak{h}'$-action is in general not finite dimensional.

3. The **Heisenberg Lie algebra** of rank one: Let

$$\mathcal{H} := \bigoplus_{n \in \mathbb{Z}} \mathbb{K}a_n \oplus \mathbb{K}K_{\mathcal{H}}$$

be the Lie algebra with commutation relations

$$[a_m, a_n] := m\delta_{m+n,0}K_{\mathcal{H}}, \quad [K_{\mathcal{H}}, \mathcal{H}] = \{0\}.$$

Let $Q := \mathbb{Z}\alpha_{\mathcal{H}}$ be a free abelian group of rank one. We set $\mathfrak{h} := \mathbb{K}a_0 \oplus \mathbb{K}K_{\mathcal{H}}$ and

$$\mathcal{H}^{n\alpha_{\mathcal{H}}} := \begin{cases} \mathbb{K}a_n & \text{if } n \neq 0 \\ \mathfrak{h} & \text{if } n = 0 \end{cases}$$

for $n \in \mathbb{Z}$. Then, $\mathcal{H} = \bigoplus_{\beta \in Q} \mathcal{H}^{\beta}$ is Q-gradation of $\mathcal{H}$, and $(\mathcal{H}, \mathfrak{h})$ is a Q-graded Lie algebra. $\mathcal{H}$ admits the Q-graded anti-involution $\sigma_{\mathcal{H}}$ defined by

$$\sigma_{\mathcal{H}}(a_n) = a_{-n}, \quad \sigma_{\mathcal{H}}(K_{\mathcal{H}}) = K_{\mathcal{H}}.$$

Remark that, in this case, the map π_Q defined in (1.6) is trivial, hence it is not injective.

1.2.4 Categories of $(\mathfrak{g}, \mathfrak{h})$-modules

Before defining categories of $(\mathfrak{g}, \mathfrak{h})$-modules, we first introduce the notion of Γ-graded modules over a Γ-graded Lie algebra $\mathfrak{g}$.

Suppose that Γ is an abelian group and $\mathfrak{g} = \bigoplus_{\alpha\in\Gamma} \mathfrak{g}^\alpha$ is a Γ-graded Lie algebra.

Definition 1.8 *A $\mathfrak{g}$-module M is called Γ-graded if $M = \bigoplus_{\alpha\in\Gamma} M^\alpha \in \mathrm{Ob}(\mathrm{Vect}_{\mathbb{K}}^{\Gamma})$, and it satisfies*

$$\mathfrak{g}^\alpha.M^\beta \subset M^{\alpha+\beta} \quad (\forall\alpha, \beta \in \Gamma).$$

For Γ-graded $\mathfrak{g}$-modules M and N, we set

$$\mathrm{Hom}_{\mathfrak{g}}^{\Gamma}(M, N) := \{f \in \mathrm{Hom}_{\mathbb{K}}^{\Gamma}(M, N) | f(x.v) = x.f(v) \ (x \in \mathfrak{g}, \ v \in M)\}.$$

Definition 1.9 *Let $\mathrm{Mod}_{\mathfrak{g}}^{\Gamma}$ be the category of all Γ-graded $\mathfrak{g}$-modules whose morphisms are given by*

$$\mathrm{Hom}_{\mathrm{Mod}_{\mathfrak{g}}^{\Gamma}}(M, N) := \mathrm{Hom}_{\mathfrak{g}}^{\Gamma}(M, N)$$

for M, $N \in \mathrm{Ob}(\mathrm{Mod}_{\mathfrak{g}}^{\Gamma})$.

Next, suppose that Q is a free abelian group of finite rank and $(\mathfrak{g}, \mathfrak{h})$ is a Q-graded Lie algebra. Let π_Q be the map defined by (1.6).

To consider modules over a Q-graded Lie algebra $(\mathfrak{g}, \mathfrak{h})$ even in the case where π_Q is not injective, here, we introduce categories which are generalisations of $\mathcal{C}_{(\mathfrak{g},\mathfrak{h})}$ in [RW1] and $\mathcal{O}$ in [BGG1].

In the sequel, we fix a homomorphism of free abelian group

$$\iota : \mathrm{Im}\pi_Q \longrightarrow Q$$

such that $\pi_Q \circ \iota = \mathrm{id}$. Then, we have $\mathrm{Im}\iota \cap \mathrm{Ker}\pi_Q = \{0\}$ and thus $Q = \mathrm{Im}\iota \oplus \mathrm{Ker}\pi_Q$.

For simplicity, we set $G := Q/\mathrm{Im}\iota$. Let

$$\mathbf{p} : Q \longrightarrow G \tag{1.7}$$

be the canonical projection. Then, we have the following isomorphism

$$Q \xrightarrow{\sim} G \oplus \mathrm{Im}\pi_Q; \quad \alpha \longmapsto (\mathbf{p}(\alpha), \pi_Q(\alpha)). \tag{1.8}$$

Definition 1.10 *1. An $\mathfrak{h}$-module M is called **$\mathfrak{h}$-diagonalisable**, if*

$$M = \bigoplus_{\lambda \in \mathfrak{h}^*} M_\lambda,$$

where $M_\lambda := \{v \in M | h.v = \lambda(h)v \ (\forall h \in \mathfrak{h})\}$ for $\lambda \in \mathfrak{h}^*$.
*2. An $\mathfrak{h}$-diagonalisable module M is called **$\mathfrak{h}$-semi-simple**, if*

$$\dim M_\lambda < \infty \quad \forall \lambda \in \mathfrak{h}^*.$$

Definition 1.11 *A $\mathfrak{g}$-module is said to be a* **$(\mathfrak{g}, \mathfrak{h})$-module**, *if it is $\mathfrak{h}$-diagonalisable.*

We regard $\mathfrak{g}$ as $G \times \mathfrak{h}^*$-graded Lie algebra via the isomorphism (1.8), i.e., $\mathfrak{g} = \bigoplus_{(\gamma,\lambda) \in G \times \mathfrak{h}^*} \mathfrak{g}_\lambda^\gamma$ and

$$\mathfrak{g}_\lambda^\gamma = \begin{cases} \mathfrak{g}^\alpha & \text{if } \exists \alpha \in Q \text{ s.t. } \gamma = \mathbf{p}(\alpha),\ \lambda = \pi_Q(\alpha) \\ \{0\} & \text{otherwise} \end{cases}.$$

For a $G \times \mathfrak{h}^*$-graded $(\mathfrak{g}, \mathfrak{h})$-module M, we denote its $(\alpha, \lambda) \in G \times \mathfrak{h}^*$ component by M_λ^α. Although, $G \times \mathfrak{h}^*$-graded structure depends on the choice of the map ι, we omit the symbol ι in the notations for simplicity.

In the sequel, for a $G \times \mathfrak{h}^*$-graded $(\mathfrak{g}, \mathfrak{h})$-module $M = \bigoplus_{(\alpha,\lambda) \in G \times \mathfrak{h}^*} M_\lambda^\alpha$, we set $M^\alpha := \bigoplus_{\lambda \in \mathfrak{h}^*} M_\lambda^\alpha$ for each $\alpha \in G$, and regard $M = \bigoplus_{\alpha \in G} M^\alpha$ as a G-graded module.

We next introduce categories of $G \times \mathfrak{h}^*$-graded $(\mathfrak{g}, \mathfrak{h})$-modules. First, we define the category $\mathcal{C}^\iota_{(\mathfrak{g},\mathfrak{h})}$.

Definition 1.12 *Let $\mathcal{C}^\iota_{(\mathfrak{g},\mathfrak{h})}$ be the category of $G \times \mathfrak{h}^*$-graded $(\mathfrak{g}, \mathfrak{h})$-modules defined as follows:*

1. *M is an object of $\mathcal{C}^\iota_{(\mathfrak{g},\mathfrak{h})}$ if and only if M is a $G \times \mathfrak{h}^*$-graded $(\mathfrak{g}, \mathfrak{h})$-module.*
2. *For $M, N \in \mathrm{Ob}(\mathcal{C}^\iota_{(\mathfrak{g},\mathfrak{h})})$,*

$$\mathrm{Hom}_{\mathcal{C}^\iota_{(\mathfrak{g},\mathfrak{h})}}(M, N) := \mathrm{Hom}_{\mathfrak{g}}^G(M, N).$$

We have

Proposition 1.5 *$\mathcal{C}^\iota_{(\mathfrak{g},\mathfrak{h})}$ is an abelian category.*

Remark 1.4 *In general, a submodule of an object of $\mathcal{C}^\iota_{(\mathfrak{g},\mathfrak{h})}$ is not necessarily an object of $\mathcal{C}^\iota_{(\mathfrak{g},\mathfrak{h})}$. In the case where π_Q is injective, the linear independence of the condition* **C3** *ensures that any submodule of an object of $\mathcal{C}^\iota_{(\mathfrak{g},\mathfrak{h})}$ is also an object of $\mathcal{C}^\iota_{(\mathfrak{g},\mathfrak{h})}$.*

Second, we introduce the subcategory of $\mathcal{C}^{\iota}_{(\mathfrak{g},\mathfrak{h})}$ which consists of all $\mathfrak{h}$-semi-simple modules.

Definition 1.13 *The category $\mathcal{C}^{\iota}_{\mathrm{adm}}$ is the full subcategory of $\mathcal{C}^{\iota}_{(\mathfrak{g},\mathfrak{h})}$ whose objects consist of $M \in \mathrm{Ob}(\mathcal{C}^{\iota}_{(\mathfrak{g},\mathfrak{h})})$ such that M^{α} is $\mathfrak{h}$-semi-simple for any $\alpha \in G$. We call an object of $\mathcal{C}^{\iota}_{\mathrm{adm}}$ an admissible $(\mathfrak{g},\mathfrak{h})$-module.*

Remark 1.5 *In the case where π_Q in not injective, an object of $\mathcal{C}^{\iota}_{\mathrm{adm}}$ is in general not $\mathfrak{h}$-semi-simple. For example, although a Fock module over the Heisenberg Lie algebra $\mathcal{H}$ (i.e., Verma module over $\mathcal{H}$) is not $\mathfrak{h}$-semi-simple, it is an object of $\mathcal{C}^{\iota}_{\mathrm{adm}}$ (cf. § 4.1.1).*

Third, we define a category $\mathcal{O}^{\iota}$. For $(\alpha,\lambda) \in G \times \mathfrak{h}^*$, set

$$D(\alpha,\lambda) := \{(\beta,\mu) \in G \times \mathfrak{h}^* | \beta = \alpha - \mathbf{p}(\gamma),\ \mu = \lambda - \pi_Q(\gamma)\ (\gamma \in Q^+)\}.$$

Definition 1.14 *The category $\mathcal{O}^{\iota}$ is the full subcategory of $\mathcal{C}^{\iota}_{\mathrm{adm}}$ whose objects consist of $M \in \mathrm{Ob}(\mathcal{C}^{\iota}_{\mathrm{adm}})$ with the following properties: There exist finitely many $(\beta_i,\lambda_i) \in G \times \mathfrak{h}^*$ such that*

$$\mathcal{P}(M) \subset \bigcup_i D(\beta_i,\lambda_i).$$

In the case where π_Q is injective, we have $G = \{0\}$ and $\iota = \mathrm{id}$. Hence, in this case, we sometimes omit the symbol ι and $\alpha \in G$ in the notations, namely, we abbreviate $\mathcal{C}^{\iota}_{(\mathfrak{g},\mathfrak{h})}$, $\mathcal{C}^{\iota}_{\mathrm{adm}}$, $\mathcal{O}^{\iota}$ and the weight subspace M^{α}_{λ} to $\mathcal{C}_{(\mathfrak{g},\mathfrak{h})}$, $\mathcal{C}_{\mathrm{adm}}$, $\mathcal{O}$ and M_{λ} for simplicity. Remark that the categories $\mathcal{C}_{(\mathfrak{g},\mathfrak{h})}$ and $\mathcal{O}$ are nothing but the category of $\mathfrak{h}$-diagonalisable $\mathfrak{g}$-modules introduced in [RW1] and the so-called BGG (Bernstein–Gelfand–Gelfand) category introduced in [BGG1] respectively.

Here, it should be noted that in the case where π_Q is not injective, a $G \times \mathfrak{h}^*$-graded $\mathfrak{g}$-module which has no non-trivial proper $G \times \mathfrak{h}^*$-graded submodule is not necessarily a simple $\mathfrak{g}$-module. This observation leads us to

Definition 1.15 *Let M be an object of $\mathcal{C}^{\iota}_{(\mathfrak{g},\mathfrak{h})}$.*

1. *M is called a* ***simple graded $\mathfrak{g}$-module*** *if M has no non-trivial $G \times \mathfrak{h}^*$-graded submodule.*
2. *M is called a* ***graded simple $\mathfrak{g}$-module*** *if M has no non-trivial submodule.*

1.2.5 Some Objects of the Category $\mathcal{C}^{\iota}_{(\mathfrak{g},\mathfrak{h})}$

In this subsection, let $(\mathfrak{g},\mathfrak{h})$ be a Q-graded Lie algebra. Here, we introduce important objects of the category $\mathcal{C}^{\iota}_{(\mathfrak{g},\mathfrak{h})}$ called *highest weight modules* and *lowest weight modules*. We also classify simple objects in $\mathrm{Ob}(\mathcal{O}^{\iota})$.

Definition 1.16 *Suppose that* $M \in \mathrm{Ob}(\mathcal{C}^{\iota}_{(\mathfrak{g},\mathfrak{h})})$ *and* $(\alpha, \lambda) \in G \times \mathfrak{h}^*$. M *is called a* **highest weight module** *with highest weight* (α, λ), *if there exists a non-zero vector* $v \in M^{\alpha}_{\lambda}$ *such that*

1. $x.v = 0$ *for any* $x \in \mathfrak{g}^+$,
2. $U(\mathfrak{g}^-).v = M$.

The vector v *is called a* ***highest weight vector*** *of* M.

Remark that a highest weight module is always an object of $\mathcal{O}^{\iota}$.

Next, we introduce highest weight modules with some universal property, called Verma modules. For $(\alpha, \lambda) \in G \times \mathfrak{h}^*$, let

$$\mathbb{K}^{\alpha}_{\lambda} := \mathbb{K}\mathbf{1}^{\alpha}_{\lambda} \tag{1.9}$$

be the one-dimensional $\mathfrak{g}^{\geq}$-module defined by

1. $\mathbb{K}^{\alpha}_{\lambda}$ is a G-graded $\mathbb{K}$-vector space with
$$(\mathbb{K}^{\alpha}_{\lambda})^{\beta} = \begin{cases} \{0\} & \text{if } \beta \neq \alpha \\ \mathbb{K}^{\alpha}_{\lambda} & \text{if } \beta = \alpha \end{cases},$$
2. $h.\mathbf{1}^{\alpha}_{\lambda} := \lambda(h)\mathbf{1}^{\alpha}_{\lambda}$ for $h \in \mathfrak{h}$,
3. $x.\mathbf{1}^{\alpha}_{\lambda} := 0$ for $x \in \mathfrak{g}^+$.

Definition 1.17 *For* $(\alpha, \lambda) \in G \times \mathfrak{h}^*$, *we set*

$$M(\alpha, \lambda) := \mathrm{Ind}^{\mathfrak{g}}_{\mathfrak{g}^{\geq}} \mathbb{K}^{\alpha}_{\lambda} \left(= U(\mathfrak{g}) \otimes_{U(\mathfrak{g}^{\geq})} \mathbb{K}^{\alpha}_{\lambda}\right),$$

and call it the **Verma module** *with highest weight* (α, λ).

The Verma module $M(\alpha, \lambda)$ is a highest weight module with highest weight vector $1 \otimes \mathbf{1}^{\alpha}_{\lambda}$. Verma modules enjoy the following properties:

Proposition 1.6 *1.* **(Universal property)** *For any highest weight module* M *with highest weight* $(\alpha, \lambda) \in G \times \mathfrak{h}^*$, *there exists a surjective homomorphism* $\phi : M(\alpha, \lambda) \to M$.

2. The Verma module $M(\alpha, \lambda)$ *has a unique maximal proper* $G \times \mathfrak{h}^*$*-graded submodule* $J(\alpha, \lambda) \in \mathrm{Ob}(\mathcal{O}^{\iota})$, *i.e.,* $M(\alpha, \lambda)/J(\alpha, \lambda)$ *is a simple graded* $\mathfrak{g}$*-module. Moreover,* $J(\alpha, \lambda)$ *is the maximal proper submodule of* $M(\alpha, \lambda)$, *i.e.,* $M(\alpha, \lambda)/J(\alpha, \lambda)$ *is a graded simple* $\mathfrak{g}$*-module.*

Proof. The first statement follows by definition. We show the second one. It is easy to see that there exists a unique maximal proper $G \times \mathfrak{h}^*$-graded submodule $J(\alpha, \lambda)$ of $M(\alpha, \lambda)$. We show that $J(\alpha, \lambda)$ is the maximal proper submodule. We assume that there exists a proper submodule J' of $M(\alpha, \lambda)$ such that $J' \supsetneq J(\alpha, \lambda)$ and lead to a contradiction.

For $v \in M(\alpha, \lambda)$, we express $v = \sum_i v^{\beta_i}_{\mu_i}$, where $\{(\beta_i, \mu_i)\}$ are distinct and $v^{\beta_i}_{\mu_i} \in M(\alpha, \lambda)^{\beta_i}_{\mu_i}$. Since $(\beta_i, \mu_i) \in D(\alpha, \lambda)$ for each i, there exists $\gamma_i \in Q^+$ such that $\beta_i = \alpha - \mathbf{p}(\gamma_i)$ and $\mu_i = \lambda - \pi_Q(\gamma_i)$. Here, we set $\mathrm{ht}v := \sum_i \mathrm{ht}\gamma_i$.

Let us take a non-zero vector $v = \sum_{i=1}^{N} v_{\mu_i}^{\beta_i} \in J' \setminus J(\alpha, \lambda)$ such that $\mathrm{ht}v$ is minimal. By the minimality of $\mathrm{ht}v$, we have $\mathfrak{g}^+.v = \{0\}$. Since $\mathfrak{g}^+$ is $G \times \mathfrak{h}^*$-graded, $\mathfrak{g}^+.v_{\mu_i}^{\beta_i} = \{0\}$ holds for any i. Hence, for i such that $(\beta_i, \mu_i) \neq (\alpha, \lambda)$, $U(\mathfrak{g})v_{\mu_i}^{\beta_i}$ is a proper $G \times \mathfrak{h}^*$-graded submodule of $M(\alpha, \lambda)$, and thus, $v_{\mu_i}^{\beta_i} \in J(\alpha, \lambda)$. Notice that if $v_{\mu_i}^{\beta_i} \in J(\alpha, \lambda)$, then $v - v_{\mu_i}^{\beta_i} \in J' \setminus J(\alpha, \lambda)$ since $J(\alpha, \lambda) \subset J'$. Since $\mathrm{ht}(v - v_{\mu_i}^{\beta_i}) < \mathrm{ht}v$, we conclude that $N = 1$ and $(\beta_1, \mu_1) = (\alpha, \lambda)$. This implies that J' contains a highest weight vector of $M(\alpha, \lambda)$. Hence, we have $J' = M(\alpha, \lambda)$. This is a contradiction. □

As a corollary of this proposition, we have

Corollary 1.1 *Any highest weight module has the unique maximal proper $G \times \mathfrak{h}^*$-graded submodule, and it is a maximal proper submodule.*

Proof. Let M be a highest weight module with highest weight (α, λ), and let $\phi : M(\alpha, \lambda) \to M$ be a surjection given by the proposition. Then, $\phi(J(\alpha, \lambda))$ is the unique maximal proper $G \times \mathfrak{h}^*$-graded submodule of M. □

We will explain another important property of Verma modules in § 1.4. We set

$$L(\alpha, \lambda) := M(\alpha, \lambda)/J(\alpha, \lambda).$$

It is obvious that the module $L(\alpha, \lambda)$ is an **irreducible highest weight module** with highest weight (α, λ), and thus, $L(\alpha, \lambda) \in \mathrm{Ob}(\mathcal{O}^\iota)$. Moreover, we have

Lemma 1.2. $\{L(\alpha, \lambda) | (\alpha, \lambda) \in G \times \mathfrak{h}^*\}$ *exhaust the simple objects of the category $\mathcal{O}^\iota$.*

Proof. To prove this lemma, we introduce a partial order on $G \times \mathfrak{h}^*$ as follows:

$$\begin{aligned} &(\beta_1, \lambda_1) < (\beta_2, \lambda_2) \\ \Leftrightarrow \quad &\exists \gamma \in Q^+ \text{ s.t. } \beta_2 - \beta_1 = \mathbf{p}(\gamma),\ \lambda_2 - \lambda_1 = \pi_Q(\gamma). \end{aligned} \tag{1.10}$$

Suppose that M is an irreducible module in $\mathrm{Ob}(\mathcal{O}^\iota)$. Since M is irreducible, for a maximal element (β, μ) of $\mathcal{P}(M)$ and $v \in M_\mu^\beta \setminus \{0\}$, we have $M = U(\mathfrak{g}).v$. Moreover, by the maximality of (β, μ), we have $M = U(\mathfrak{g}^-).v$. Hence, M is a highest weight module with highest weight (β, μ), and thus, the lemma holds. □

Finally, we define lowest weight modules and introduce lowest weight Verma modules.

Definition 1.18 *We say that $M \in \mathrm{Ob}(\mathcal{C}^\iota_{(\mathfrak{g},\mathfrak{h})})$ is a* ***lowest weight module*** *with lowest weight $(\alpha, \lambda) \in G \times \mathfrak{h}^*$, if there exists $v \in M_\lambda^\alpha \setminus \{0\}$ such that*

1. *$x.v = 0$ for any $x \in \mathfrak{g}^-$,*
2. *$U(\mathfrak{g}^+).v = M$.*

*The vector v is called a **lowest weight vector** of M.*

Definition 1.19 *For $(\alpha,\lambda) \in G \times \mathfrak{h}^*$, let $\mathbb{K}_\lambda^{\alpha:-} = \mathbb{K}\mathbf{1}_\lambda^{\alpha:-}$ be the one-dimensional $\mathfrak{g}^{\leq}$-module defined by*

1. *$\mathbb{K}_\lambda^{\alpha:-}$ is a G-graded $\mathbb{K}$-vector space with*

$$(\mathbb{K}_\lambda^{\alpha:-})^\beta = \begin{cases} \{0\} & \text{if } \beta \neq \alpha \\ \mathbb{K}_\lambda^{\alpha:-} & \text{if } \beta = \alpha \end{cases},$$

2. *$h.\mathbf{1}_\lambda^{\alpha:-} = \lambda(h)\mathbf{1}_\lambda^{\alpha:-}$ for $h \in \mathfrak{h}$, and*
3. *$x.\mathbf{1}_\lambda^{\alpha:-} = 0$ for any $x \in \mathfrak{g}^-$.*

We set

$$M^-(\alpha,\lambda) := \mathrm{Ind}_{\mathfrak{g}^{\leq}}^{\mathfrak{g}} \mathbb{K}_\lambda^{\alpha:-},$$

*and call it the **lowest weight Verma module** with lowest weight (α,λ).*

One can similarly show that $M^-(\alpha,\lambda)$ has the unique maximal proper $G\times\mathfrak{h}^*$-graded submodule $J^-(\alpha,\lambda)$, which is also a maximal proper submodule of $M^-(\alpha,\lambda)$. We set $L^-(\alpha,\lambda) := M^-(\alpha,\lambda)/J^-(\alpha,\lambda)$.

Remark 1.6 *$M^-(\alpha,\lambda)$ and $L^-(\alpha,\lambda)$ are in general not objects of the category $\mathcal{O}^\iota$, but are objects of $\mathcal{C}^\iota_{\mathrm{adm}}$.*

In the case where $\pi_Q : Q \to \mathfrak{h}^*$ is injective, i.e., ι is the identity, we abbreviate $M(\alpha,\lambda)$, $L(\alpha,\lambda)$, $M^-(\alpha,\lambda)$ and $L^-(\alpha,\lambda)$ to $M(\lambda)$, $L(\lambda)$, $M^-(\lambda)$ and $L^-(\lambda)$ for simplicity.

1.2.6 Simple Objects of the Category $\mathcal{C}_{\mathrm{adm}}$ (the Virasoro Case)

In this subsection, we classify simple objects of the category $\mathcal{C}_{\mathrm{adm}}$ in the case of the Virasoro algebra. (For the proof, see the next chapter.)

We first introduce irreducible Vir-modules called *intermediate series*. For $a, b \in \mathbb{K}$, let $V_{a,b} := \bigoplus_{n\in\mathbb{Z}} \mathbb{K}v_n$ be the $\mathbb{Z}$-graded Vir-module defined by

$$\begin{aligned} L_s.v_n &= (as+b-n)v_{n+s}, \\ C.v_n &= 0. \end{aligned} \tag{1.11}$$

By definition, $V_{a,b} \in \mathrm{Ob}(\mathcal{C}_{\mathrm{adm}})$.

Remark 1.7 *For $\lambda \in \mathbb{Z}_{\leq 0}$ and $\mu \in \mathbb{K}$, let $t^\mu\mathbb{K}[t,t^{-1}]dt^{-\lambda}$ be the module over the Witt algebra $\mathcal{D}$ (1.1) defined by*

$$(f(t)\frac{d}{dt}).\phi(t)dt^{-\lambda} := \{f(t)\phi'(t) - \lambda f'(t)\phi(t)\}\, dt^{-\lambda},$$

where $f(t)\frac{d}{dt} \in \mathcal{D}$, $\phi(t) \in t^{\mu}\mathbb{K}[t,t^{-1}]$ *and* $(\cdot)'$ *denotes the derivative with respect to* t. *If* $a \in \mathbb{Z}_{\leq 0}$ *and* $b \in \mathbb{K}$, *then*

$$V_{a,b} \simeq t^{a-b}\mathbb{K}[t,t^{-1}]dt^{-a} \quad (v_n \mapsto t^{a-b+n}dt^{-a})$$

as $\mathrm{Vir}/\mathbb{K}C(\simeq \mathcal{D})$*-modules.*

Then, the following hold:

Proposition 1.7 *1. If* $a \neq 0, -1$ *or* $b \notin \mathbb{Z}$, *then* $V_{a,b}$ *is an irreducible* Vir-*module.*
2. If $a = 0$ *and* $b \in \mathbb{Z}$, *then there exists a submodule* V *of* $V_{a,b}$ *such that* $V \simeq \mathbb{K}$ *and* $V_{a,b}/V$ *is irreducible.*
3. If $a = -1$ *and* $b \in \mathbb{Z}$, *then there exists a submodule* V *of* $V_{a,b}$ *such that* $V_{a,b}/V \simeq \mathbb{K}\frac{dt}{t}$ *and* V *is irreducible.*

Proof. We will prove this proposition in a more general setting, i.e., where the characteristic of $\mathbb{K}$ is not necessarily zero. See Proposition 2.1. □

Definition 1.20 *The irreducible modules* $V_{a,b}$, $V_{a,b}/V$ *and* V *given in the above proposition are called the* **intermediate series** *of the Virasoro algebra.*

Theorem 1.1 ([Mat2]) *The intermediate series, the irreducible highest weight modules and the irreducible lowest weight modules exhaust the* ***Harish-Chandra modules*** *over the Virasoro algebra, i.e., simple objects of the category* $\mathcal{C}_{\mathrm{adm}}$.

This theorem will be proved in Chapter 2.

Proposition 1.7 reveals the structure of the intermediate series. From now on, we will mainly investigate the structure of highest weight modules, in particular, Verma modules. The modules $V_{a,b}$ will appear in § 8.5 to construct fermionic Fock modules.

1.2.7 Dualising Functors

Let $(\mathfrak{g}, \mathfrak{h})$ be a Q-graded Lie algebra. Let $a : U(\mathfrak{g}) \to U(\mathfrak{g})$ be the antipode of the standard Hopf algebra structure on $U(\mathfrak{g})$, i.e., the anti-automorphism defined by $a(x) := -x$ $(x \in \mathfrak{g})$.

We introduce the antipode dual of an object of $\mathcal{C}^{\iota}_{(\mathfrak{g},\mathfrak{h})}$.

Definition 1.21 *We define the functor* $(\cdot)^{\sharp a} : \mathcal{C}^{\iota}_{(\mathfrak{g},\mathfrak{h})} \to \mathcal{C}^{\iota}_{(\mathfrak{g},\mathfrak{h})}$ *as follows:*

1. For $M \in \mathrm{Ob}(\mathcal{C}^{\iota}_{(\mathfrak{g},\mathfrak{h})})$, *we set* $M^{\sharp a} := M^{\sharp -}$ *and regard it as* $G \times \mathfrak{h}^*$*-graded* $(\mathfrak{g}, \mathfrak{h})$*-module via*

$$(x.\varphi)(v) := \varphi(a(x).v) \quad (\varphi \in M^{\sharp a},\ v \in M,\ x \in U(\mathfrak{g})). \tag{1.12}$$

2. For $f \in \mathrm{Hom}_{\mathcal{C}^{\iota}_{(\mathfrak{g},\mathfrak{h})}}(M,N)$*, we let* $f^{\sharp a} := f^{\sharp -}$*.*

The module $M^{\sharp a}$ *is called the* ***antipode dual*** *of* M*.*

We have

Lemma 1.3. *1.* $(\cdot)^{\sharp a}$ *is contravariant and is exact.*
2. $L(\alpha,\lambda)^{\sharp a} \simeq L^{-}(-\alpha,-\lambda)$ *for* $(\alpha,\lambda) \in G \times \mathfrak{h}^*$*.*
3. The category $\mathcal{C}^{\iota}_{\mathrm{adm}}$ *is stable under taking the antipode dual.*

We next suppose that $(\mathfrak{g},\mathfrak{h})$ has a Q-graded anti-involution σ. Let us introduce the contragredient dual of an object of $\mathcal{C}^{\iota}_{(\mathfrak{g},\mathfrak{h})}$.

Definition 1.22 *We define the functor* $(\cdot)^c : \mathcal{C}^{\iota}_{(\mathfrak{g},\mathfrak{h})} \to \mathcal{C}^{\iota}_{(\mathfrak{g},\mathfrak{h})}$ *as follows:*

1. For $M \in \mathrm{Ob}(\mathcal{C}^{\iota}_{(\mathfrak{g},\mathfrak{h})})$*, we set* $M^c := M^{\sharp +}$ *and regard it as* $G \times \mathfrak{h}^*$*-graded* $(\mathfrak{g},\mathfrak{h})$*-module via*

$$(x.\varphi)(v) := \varphi(\sigma(x).v) \quad (\varphi \in M^c,\ v \in M,\ x \in U(\mathfrak{g})). \tag{1.13}$$

2. For $f \in \mathrm{Hom}_{\mathcal{C}^{\iota}_{(\mathfrak{g},\mathfrak{h})}}(M,N)$*, we let* $f^c := f^{\sharp +}$*.*

The module M^c *is called the* ***contragredient dual*** *of* M*.*

Then, one can check the following lemma.

Lemma 1.4. *1.* $(\cdot)^c$ *is contravariant and is exact.*
2. $L(\alpha,\lambda)^c \simeq L(\alpha,\lambda)$ *for* $(\alpha,\lambda) \in G \times \mathfrak{h}^*$*.*
3. The categories $\mathcal{C}^{\iota}_{\mathrm{adm}}$ *and* $\mathcal{O}^{\iota}$ *are stable under taking the contragredient dual.*

1.2.8 Local Composition Series and Formal Character

Throughout this subsection, we assume that $(\mathfrak{g},\mathfrak{h})$ is a Q-graded Lie algebra unless otherwise stated. In the following chapters, our main ingredients are objects of the category $\mathcal{O}^{\iota}$. In general, an object of the category $\mathcal{O}^{\iota}$ does not necessarily have a composition series of finite length. Hence, we have to consider a 'local' version of a composition series. Here, we recall the local composition series of an object of $\mathcal{O}^{\iota}$ and its formal character.

We first show the existence of local composition series.

Proposition 1.8 *For any* $V \in \mathrm{Ob}(\mathcal{O}^{\iota})$ *and* $(\alpha,\lambda) \in G \times \mathfrak{h}^*$*, there exists a finite filtration*

$$V = V_t \supset V_{t-1} \supset \cdots \supset V_1 \supset V_0 = \{0\}$$

of V *by a sequence of submodules, and a subset* $J \subset \{1,2,\cdots,t\}$ *such that*

(i) if $j \in J$*, then* $V_j/V_{j-1} \simeq L(\alpha_j,\lambda_j)$ *for some* $(\alpha_j,\lambda_j) \geq (\alpha,\lambda)$*,*
(ii) if $j \notin J$*, then* $(V_j/V_{j-1})^{\beta}_{\mu} = \{0\}$ *for any* $(\beta,\mu) \geq (\alpha,\lambda)$*,*

where the order $\leq$ on $G \times \mathfrak{h}^*$ is defined in (1.10).

Proof. The proof given here is essentially the same as the one in [DGK]. For $(\alpha, \lambda) \in G \times \mathfrak{h}^*$ and $V = \bigoplus_{(\beta,\mu)\in G\times\mathfrak{h}^*} V_\mu^\beta \in \mathrm{Ob}(\mathcal{O}^\iota)$, we set

$$a(V, (\alpha, \lambda)) := \sum_{(\beta,\mu)\geq(\alpha,\lambda)} \dim V_\mu^\beta,$$

and show this proposition by induction on $a(V, (\alpha, \lambda))$.

In the case where $a(V, (\alpha, \lambda)) = 0$, the statement holds by choosing $\{0\} = V_0 \subset V_1 = V$ as the filtration. We assume that $a(V, (\alpha, \lambda)) > 0$. We take a maximal element (β, μ) of $\mathcal{P}(V)$. Let $v \in V_\mu^\beta \backslash \{0\}$ and set $W := U(\mathfrak{g}).v$. Then, W is a highest weight module with highest weight (β, μ). By Corollary 1.1, there exists the unique maximal proper $G \times \mathfrak{h}^*$-graded submodule W' of W, which is, in fact, a maximal proper submodule. We have

1. $\{0\} \subset W' \subset W \subset V$,
2. $W'/W \simeq L(\beta, \mu)$.

Since $a(W', (\alpha, \lambda)) < a(V, (\alpha, \lambda))$ and $a(V/W, (\alpha, \lambda)) < a(V, (\alpha, \lambda))$, we obtain a filtration of V which satisfies the conditions of the proposition by combining a filtration of W' with the pull back of a filtration of V/W with respect to the map $V \twoheadrightarrow V/W$. □

Any such filtration obtained in Proposition 1.8 is called a **local composition series** of V at (α, λ).

Here, it should be remarked that Proposition 1.8 does not ensure the existence of a local composition series $V = V_t \supset V_{t-1} \supset \cdots \supset V_1 \supset V_0 = \{0\}$ such that V_{t-1} is a maximal proper $G \times \mathfrak{h}^*$-graded submodule of V. In the case where V is finitely generated, the following lemma implies that there exists a local composition series $V = V_t \supset V_{t-1} \supset \cdots \supset V_1 \supset V_0 = \{0\}$ of V such that V_{t-1} is a maximal proper $G \times \mathfrak{h}^*$-graded submodule of V.

Lemma 1.5. *Suppose that $M \in \mathrm{Ob}(\mathcal{O}^\iota)$ is a finitely generated $(\mathfrak{g}, \mathfrak{h})$-module. For any (not necessarily finitely generated) proper $G \times \mathfrak{h}^*$-graded submodule $M' \in \mathrm{Ob}(\mathcal{O}^\iota)$ of M, there exists a maximal proper $G \times \mathfrak{h}^*$-graded submodule $N \in \mathrm{Ob}(\mathcal{O}^\iota)$ of M such that $M' \subset N$.*

Proof. Let $\mathcal{V}$ be the set of $G \times \mathfrak{h}^*$-graded (not necessarily finitely generated) proper submodules V of M such that $M' \subset V$. Then, $\mathcal{V}$ is a partially ordered set via inclusion. Hence, we show that $\mathcal{V}$ is an inductive set. Let $\{V_i\} \subset \mathcal{V}$ be a totally ordered subset. We suppose that $\bigcup_i V_i \notin \mathcal{V}$, i.e., $\bigcup_i V_i = M$, and lead to a contradiction. Let us take a set of homogeneous generators $\{x_1, x_2, \cdots, x_n\}$ of M. If $\bigcup_i V_i = M$, then there exists i such that $x_k \in V_i$ for any $1 \leq k \leq n$, since n is finite. This contradicts $V_i \in \mathcal{V}$. Hence, $\mathcal{V}$ is an inductive set. By Zorn's lemma, $\mathcal{V}$ has a maximal element N. □

Lemma 1.5 also holds for finitely generated $(\mathfrak{g}, \mathfrak{h})$-modules which are objects of $\mathcal{C}^{\iota}_{(\mathfrak{g},\mathfrak{h})}$ or $\mathcal{C}^{\iota}_{\mathrm{adm}}$.

By definition, for (α, λ) and $(\beta, \mu) \in G \times \mathfrak{h}^*$ such that $(\alpha, \lambda) \geq (\beta, \mu)$, a local composition series at (β, μ) is also a local composition series at (α, λ). On the other hand, although a local composition series at (α, λ) is not necessarily a local composition series at (β, μ), there exists a 'refinement' of a local composition series at (α, λ), which is a local composition series at (β, μ). Before making the statement precise, we define a refinement of a sequence of submodules.

Definition 1.23 *Let V be an object of $\mathcal{O}^{\iota}$ and let*

$$V \supset V_t \supset V_{t-1} \supset \cdots \supset V_1 \supset V_0 = \{0\} \tag{1.14}$$

be a sequence of $G \times \mathfrak{h}^*$-graded submodules of V (not necessarily local composition series of V). We say that a sequence

$$V = V'_t \supset V'_{t-1} \supset \cdots \supset V'_1 \supset V'_0 = \{0\}$$

of $G \times \mathfrak{h}^*$-graded submodules of V is a **refinement** *of (1.14) if for any $1 \leq i \leq s$, there exists j such that $V_i = V'_j$.*

We have

Lemma 1.6. *Let V be an object of $\mathcal{O}^{\iota}$, and let*

$$V = V_s \supset V_{s-1} \supset \cdots \supset V_1 \supset V_0 = \{0\} \tag{1.15}$$

be a local composition series of V at (α, λ).

1. *Let*
$$V = V'_t \supset V'_{t-1} \supset \cdots \supset V'_1 \supset V'_0 = \{0\} \tag{1.16}$$
be a refinement of (1.15). Then, (1.16) is a local composition series of V at $(\alpha, \lambda) \in G \times \mathfrak{h}^*$.
2. *Suppose that $(\beta, \mu) \in G \times \mathfrak{h}^*$ satisfy $(\beta, \mu) \leq (\alpha, \lambda)$. Then, there exists a local composition series*
$$V = V''_t \supset V''_{t-1} \supset \cdots \supset V''_1 \supset V''_0 = \{0\} \tag{1.17}$$
at (β, μ) which is a refinement of (1.15).

Proof. The first statement follows by definition. We show the second one. Since V_i/V_{i-1} has a local composition series at (α, λ), by taking the pull back of the series under the canonical map $V_i \twoheadrightarrow V_i/V_{i-1}$ and combining these series, we obtain a local composition series of V at (α, λ). □

We next show that under a mild condition two local composition series of $V \in \mathrm{Ob}(\mathcal{O}^{\iota})$ have a common refinement in the following sense.

Definition 1.24 *1. We say that two sequences*

$$V = M_s \supset M_{s-1} \supset \cdots \supset M_1 \supset M_0 = \{0\},$$
$$V = M'_t \supset M'_{t-1} \supset \cdots \supset M'_1 \supset M'_0 = \{0\}$$

of $G \times \mathfrak{h}^*$-graded submodules of $V \in \mathrm{Ob}(\mathcal{O}^\iota)$ are ***equivalent*** *if $s = t$ and there exists a bijection ϕ on $\{1, 2, \cdots, s\}$ such that*

$$M_j/M_{j-1} \simeq M'_{\phi(j)}/M'_{\phi(j-1)} \quad (\forall j \in \{1, 2, \cdots, s\}).$$

2. We say that two local composition series

$$V = N_s \supset N_{s-1} \supset \cdots \supset N_1 \supset N_0 = \{0\},$$
$$V = N'_t \supset N'_{t-1} \supset \cdots \supset N'_1 \supset N'_0 = \{0\}$$

of V at (α, λ) are ***equivalent as local composition series*** *if there exists a bijection ψ between the sets $J \subset \{1, 2, \cdots, s\}$ and $J' \subset \{1, 2, \cdots, t\}$ given in Proposition 1.8 such that*

$$N_j/N_{j-1} \simeq N'_{\psi(j)}/N'_{\psi(j-1)} \quad (\forall j \in J)$$

as $G \times \mathfrak{h}^*$-graded $(\mathfrak{g}, \mathfrak{h})$-module.

Remark 1.8 *The local composition series (1.16) and (1.17) are equivalent to (1.15) as local composition series at (α, λ).*

Proposition 1.9 *Suppose that $V \in \mathrm{Ob}(\mathcal{O}^\iota)$. Let*

$$V = M_s \supset M_{s-1} \supset \cdots \supset M_1 \supset M_0 = \{0\}, \tag{1.18}$$
$$V = N_t \supset N_{t-1} \supset \cdots \supset N_1 \supset N_0 = \{0\}, \tag{1.19}$$

be local composition series of V at (α, λ) and $(\beta, \mu) \in G \times \mathfrak{h}^*$ satisfying $D(\alpha, \lambda) \cap D(\beta, \mu) \neq \emptyset$. Then, there exists $(\gamma, \nu) \in D(\alpha, \lambda) \cap D(\beta, \mu)$ and local composition series of V at (γ, ν) which is equivalent to (1.18) and (1.19) as local composition series at (α, λ) and (β, μ).

We can prove this proposition by an argument similar to Schreier's refinement theorem. We need a preliminary lemma.

Lemma 1.7. *Let W_1 and W'_1 be $G \times \mathfrak{h}^*$-graded submodules of a $G \times \mathfrak{h}^*$-graded $(\mathfrak{g}, \mathfrak{h})$-module V. Let W_2 and W'_2 be $G \times \mathfrak{h}^*$-graded submodules of W_1 and W'_1 respectively. Then, the following isomorphism holds.*

$$\frac{W_2 + (W_1 \cap W'_1)}{W_2 + (W_1 \cap W'_2)} \simeq \frac{W'_2 + (W_1 \cap W'_1)}{W'_2 + (W_2 \cap W'_1)}. \tag{1.20}$$

Proof. One can easily check that each side of (1.20) is isomorphic to

$$\frac{(W_1 \cap W_1')}{(W_1 \cap W_2') + (W_1' \cap W_2)}. \qquad \square$$

Proof of Proposition 1.9. For each $i \in \{1, 2, \cdots, s\}$ and $j \in \{1, 2, \cdots, t\}$, we set

$$M_i^j := M_i + (M_{i+1} \cap N_j),$$
$$N_j^i := N_j + (N_{j+1} \cap M_i).$$

Since $M_i^0 = M_{i-1}^t = M_i$ and $N_j^0 = N_{j-1}^s = N_j$, we have refinements

$$\begin{aligned} V =& M_{s-1}^t \supset M_{s-1}^{t-1} \supset \cdots \supset M_{s-1}^1 \supset M_{s-1}^0 = M_{s-2}^t \supset \cdots \\ & \cdots \supset M_1^0 = M_0^t \supset M_0^{t-1} \supset \cdots \supset M_0^1 \supset M_0^0 = \{0\} \end{aligned} \tag{1.21}$$

of the sequence (1.18), and

$$\begin{aligned} V =& N_{t-1}^s \supset N_{t-1}^{s-1} \supset \cdots \supset N_{t-1}^1 \supset N_{t-1}^0 = N_{t-2}^s \supset \cdots \\ & \cdots \supset N_1^0 = N_0^s \supset N_0^{s-1} \supset \cdots \supset N_0^1 \supset N_0^0 = \{0\} \end{aligned} \tag{1.22}$$

of the sequence (1.19). By Lemma 1.7, we have

$$M_{i-1}^j / M_{i-1}^{j-1} \simeq N_{j-1}^i / N_{j-1}^{i-1},$$

and thus, the sequences (1.21) and (1.22) are equivalent (in the sense of Definition 1.24. 1). Here, notice that a sequence of $G \times \mathfrak{h}^*$-graded submodules of V which is equivalent to a local composition series of V at (α, λ) is also a local composition series of V at (α, λ). Hence, by Lemma 1.6. 1, the sequence (1.21) is a local composition series not only at (α, λ) but also at (β, μ). By Lemma 1.6. 2, there exists a local composition series of V at (γ, ν), which is a refinement of (1.21). By Remark 1.8, this local composition series is equivalent to (1.18) and (1.19) as local composition series at (α, λ) and (β, μ) respectively. □

The following proposition ensures that the multiplicity of V at $L(\alpha, \lambda)$ is well-defined for any $V \in \mathrm{Ob}(\mathcal{O}^\iota)$. For $(\alpha, \lambda) \in G \times \mathfrak{h}^*$, we fix an element (β, μ) such that $(\beta, \mu) \leq (\alpha, \lambda)$, and take a filtration of V as above. Then, by Proposition 1.9, we have

Proposition 1.10 *The number*

$$\sharp\{j \in J | V_j / V_{j-1} \simeq L(\alpha, \lambda)\}$$

does not depend on the choice of (β, μ) and the filtration.

The number $\sharp\{j \in J | V_j / V_{j-1} \simeq L(\alpha, \lambda)\}$ is called the **multiplicity** of V at $L(\alpha, \lambda)$. We denote it by $[V : L(\alpha, \lambda)]$.

Next, we define formal characters for objects in the category $\mathcal{O}^\iota$. Let $\mathcal{E}$ be the $\mathbb{K}$-algebra which consists of the elements of the form

$$\sum_{(\alpha,\lambda)\in G\times\mathfrak{h}^*} c_{(\alpha,\lambda)}e(\alpha,\lambda),$$

where $c_{(\alpha,\lambda)} \in \mathbb{K}$ and there exist finitely many $\alpha_1,\cdots,\alpha_m \in G$ and $\lambda_1,\cdots,\lambda_m \in \mathfrak{h}^*$ such that $c_{(\alpha,\lambda)} = 0$ if

$$(\alpha,\lambda) \notin \bigcup_i D(\alpha_i,\lambda_i).$$

The ring structure of $\mathcal{E}$ is given as follows. We set $e(\alpha,\lambda)e(\beta,\mu) := e(\alpha+\beta,\lambda+\mu)$, and extend to $\mathcal{E}$ by linearity.

Definition 1.25 *For $V \in \mathrm{Ob}(\mathcal{O}^\iota)$, we set*

$$\mathrm{ch}\, V := \sum_{(\alpha,\lambda)\in G\times\mathfrak{h}^*} (\dim V_\lambda^\alpha)e(\alpha,\lambda)$$

and call it the **formal character** *of V.*

Notice that, by Lemma 1.2, for any $V \in \mathrm{Ob}(\mathcal{O}^\iota)$, $\mathrm{ch}\, V$ is expressed as a linear combination of $\{\mathrm{ch}\, L(\alpha,\lambda) | (\alpha,\lambda) \in G\times\mathfrak{h}^*\}$. By definition, we have

Proposition 1.11 ([DGK]) *For $V \in \mathrm{Ob}(\mathcal{O}^\iota)$, we have*

$$\mathrm{ch}\, V = \sum_{(\alpha,\lambda)\in G\times\mathfrak{h}^*} [V : L(\alpha,\lambda)]\,\mathrm{ch}\, L(\alpha,\lambda).$$

Finally, we present some properties of multiplicity.

Lemma 1.8. *For each $(\alpha,\lambda) \in G\times\mathfrak{h}^*$, $[\cdot : L(\alpha,\lambda)]$ is additive, i.e., for any exact sequence*

$$0 \longrightarrow V_1 \longrightarrow V_2 \longrightarrow V_3 \longrightarrow 0$$

in the category $\mathcal{O}^\iota$,

$$[V_1 : L(\alpha,\lambda)] + [V_3 : L(\alpha,\lambda)] = [V_2 : L(\alpha,\lambda)]$$

holds.

The following lemma is a simple but useful application of Proposition 1.8.

Lemma 1.9 ([KT]). *Suppose that $V \in \mathrm{Ob}(\mathcal{O}^\iota)$ and $(\alpha,\lambda) \in G\times\mathfrak{h}^*$. Then, we have*

$$\dim \mathrm{Hom}_{\mathcal{O}^\iota}(M(\alpha,\lambda), V) \le [V : L(\alpha,\lambda)].$$

Moreover, if $(\mathfrak{g},\mathfrak{h})$ is a Q-graded Lie algebra with a Q-graded anti-involuion, then

$$\dim \mathrm{Hom}_{\mathcal{O}^\iota}(V, M(\alpha,\lambda)^c) \le [V : L(\alpha,\lambda)].$$

Proof. Let us take $(\beta,\mu) \in G\times\mathfrak{h}^*$ such that $(\beta,\mu) \le (\alpha,\lambda)$, and a local composition series $V = V_t \supset V_{t-1} \supset \cdots \supset V_1 \supset V_0 = \{0\}$ at (β,μ) with a subset J of $\{1,2,\cdots,t\}$ given as in Proposition 1.8. If $j \in J$ then there exists $(\beta_j, \mu_j) \in G \times \mathfrak{h}^*$ such that $(\beta_j, \mu_j) \geq (\beta, \mu)$ and the sequence $0 \to V_{j-1} \to V_j \to L(\beta_j, \mu_j) \to 0$ is exact. From this short exact sequence, we obtain the left exact sequence

$$\begin{aligned} 0 \to \mathrm{Hom}_{\mathcal{O}^{\iota}}(M(\alpha,\lambda), V_{j-1}) &\to \mathrm{Hom}_{\mathcal{O}^{\iota}}(M(\alpha,\lambda), V_j) \\ &\to \mathrm{Hom}_{\mathcal{O}^{\iota}}(M(\alpha,\lambda), L(\beta_j,\mu_j)). \end{aligned}$$

Hence, we have

$$\begin{aligned} \dim \mathrm{Hom}_{\mathcal{O}^{\iota}}(M(\alpha,\lambda), V_j) \leq & \dim \mathrm{Hom}_{\mathcal{O}^{\iota}}(M(\alpha,\lambda), V_{j-1}) \\ &+ \dim \mathrm{Hom}_{\mathcal{O}^{\iota}}(M(\alpha,\lambda), L(\beta_j,\mu_j)). \end{aligned}$$

In the case where $j \notin J$, since $\mathrm{Hom}_{\mathcal{O}^{\iota}}(M(\alpha,\lambda), V_j/V_{j-1}) = \{0\}$, we have

$$\dim \mathrm{Hom}_{\mathcal{O}^{\iota}}(M(\alpha,\lambda), V_j) = \dim \mathrm{Hom}_{\mathcal{O}^{\iota}}(M(\alpha,\lambda), V_{j-1})$$

by a similar argument. Since

$$\dim \mathrm{Hom}_{\mathcal{O}^{\iota}}(M(\alpha,\lambda), L(\beta_j,\lambda_j)) = \begin{cases} 0 & \text{if } (\beta_j,\mu_j) \neq (\alpha,\lambda) \\ 1 & \text{if } (\beta_j,\mu_j) = (\alpha,\lambda) \end{cases},$$

consequently, we have

$$\dim \mathrm{Hom}_{\mathcal{O}^{\iota}}(M(\alpha,\lambda), V) \leq [V : L(\alpha,\lambda)].$$

Hence, the first inequality holds. By taking the contragredient dual, we obtain the second inequality. □

1.3 (Co)homology of a Q-graded Lie Algebra

We have two different ways to define Lie algebra (co)homology. One is the definition using the so-called *Chevalley–Eilenberg (co)complex* (cf. § A.3), and the other is the definition as derived functors of the (co)invariant functors. Here, we define *(co)homology groups of Q-graded Lie algebra* $(\mathfrak{g}, \mathfrak{h})$ *with coefficients in a* $(\mathfrak{g}, \mathfrak{h})$*-module* M by means of derived functors, and state their properties.

1.3.1 Preliminaries

In this section, we use the following notation. Let $(\mathfrak{g}, \mathfrak{h})$ be a Q-graded Lie algebra over $\mathbb{K}$. Note that, for some statements, it is necessary that $(\mathfrak{g}, \mathfrak{h})$ is a Q-graded Lie algebra with a Q-graded anti-involution.

For a Q-graded Lie subalgebra $\mathfrak{a}$ of $(\mathfrak{g}, \mathfrak{h})$ and $G \times \mathfrak{h}^*$-graded (left) $\mathfrak{a}$-modules M and N, we denote the subspace

$$\left\{ f \in \mathrm{Hom}_{\mathbb{K}}^{G\times\mathfrak{h}^*}(M,N) \middle| f(a.m) = a.f(m) \ (a \in \mathfrak{a}, \ m \in M) \right\}$$

by $\mathrm{Hom}_{\mathfrak{a}}^{G\times\mathfrak{h}^*}(M,N)$. We further set

$$\overline{\mathrm{Hom}}_{\mathfrak{a}}(M,N) := \{ f \in \overline{\mathrm{Hom}}_{\mathbb{K}}(M,N) | f(a.m) = a.f(m) \ (a \in \mathfrak{a}, \ m \in M)\}.$$

Note that in the case where M has a right $\mathfrak{g}$-module structure which commutes with the left $\mathfrak{a}$-action, $\overline{\mathrm{Hom}}_{\mathfrak{a}}(M,N)$ is stable under the $\mathfrak{g}$-action

$$(x.f)(m) := f(mx) \quad (x \in \mathfrak{g}, \ f \in \overline{\mathrm{Hom}}_{\mathfrak{a}}(M,N), m \in M). \tag{1.23}$$

Hence, we sometimes regard $\overline{\mathrm{Hom}}_{\mathfrak{a}}(M,N)$ as a left $\mathfrak{g}$-module via this action.

Throughout this section, let $\mathfrak{a}$ be a Q-graded Lie subalgebra of $(\mathfrak{g}, \mathfrak{h})$, which contains $\mathfrak{h}$. (Notice that if $\mathfrak{a}$ is a Q-graded Lie subalgebra of $(\mathfrak{g}, \mathfrak{h})$, then $\mathfrak{a}+\mathfrak{h}$ is also a Q-graded Lie subalgebra.) Let $\mathcal{C}^{\iota}_{(\mathfrak{a},\mathfrak{h})}$ be the category of $G\times\mathfrak{h}^*$-graded $(\mathfrak{a}, \mathfrak{h})$-modules.

For $M \in \mathrm{Ob}(\mathcal{C}^{\iota}_{(\mathfrak{a},\mathfrak{h})})$, we set

$$\mathrm{Ind}_{\mathfrak{a}}^{\mathfrak{g}} M := U(\mathfrak{g}) \otimes_{U(\mathfrak{a})} M,$$

and regard it as a left $\mathfrak{g}$-module via

$$x.(y \otimes n) := (xy) \otimes m \quad (x \in \mathfrak{g}, \ y \in U(\mathfrak{g}), \ m \in M). \tag{1.24}$$

For $M \in \mathrm{Ob}(\mathcal{C}^{\iota}_{(\mathfrak{g},\mathfrak{h})})$, we regard M as an $\mathfrak{a}$-module by forgetting the unnecessary action, and denote the $\mathfrak{a}$-module by $\mathrm{Res}_{\mathfrak{a}}^{\mathfrak{g}} M$. By definition, we have

1. $\mathrm{Ind}_{\mathfrak{a}}^{\mathfrak{g}} V \in \mathrm{Ob}(\mathcal{C}^{\iota}_{(\mathfrak{g},\mathfrak{h})})$ for any $V \in \mathrm{Ob}(\mathcal{C}^{\iota}_{(\mathfrak{a},\mathfrak{h})})$.
2. $\mathrm{Res}_{\mathfrak{a}}^{\mathfrak{g}} W \in \mathrm{Ob}(\mathcal{C}^{\iota}_{(\mathfrak{a},\mathfrak{h})})$ for any $W \in \mathrm{Ob}(\mathcal{C}^{\iota}_{(\mathfrak{g},\mathfrak{h})})$.

Hence, $\mathrm{Ind}_{\mathfrak{a}}^{\mathfrak{g}}(\cdot)$ and $\mathrm{Res}_{\mathfrak{a}}^{\mathfrak{g}}(\cdot)$ define functors between these categories, i.e.,

$$\mathrm{Ind}_{\mathfrak{a}}^{\mathfrak{g}}(\cdot) : \mathcal{C}^{\iota}_{(\mathfrak{a},\mathfrak{h})} \longrightarrow \mathcal{C}^{\iota}_{(\mathfrak{g},\mathfrak{h})},$$
$$\mathrm{Res}_{\mathfrak{a}}^{\mathfrak{g}}(\cdot) : \mathcal{C}^{\iota}_{(\mathfrak{g},\mathfrak{h})} \longrightarrow \mathcal{C}^{\iota}_{(\mathfrak{a},\mathfrak{h})}.$$

1.3.2 Frobenius Reciprocity

We first show a preliminary lemma. Let $\mathfrak{a}$ and $\mathfrak{b}$ be Q-graded Lie subalgebras of $(\mathfrak{g}, \mathfrak{h})$, which contain $\mathfrak{h}$. Suppose that $M \in \mathrm{Ob}(\mathcal{C}^{\iota}_{(\mathfrak{a},\mathfrak{h})})$ and $N \in \mathrm{Ob}(\mathcal{C}^{\iota}_{(\mathfrak{b},\mathfrak{h})})$.

Lemma 1.10. *The following isomorphism of $\mathbb{K}$-vector spaces holds:*

$$\mathrm{Hom}_{\mathfrak{b}}^{G\times\mathfrak{h}^*}(U(\mathfrak{g}) \otimes_{U(\mathfrak{a})} M, N) \simeq \mathrm{Hom}_{\mathfrak{a}}^{G\times\mathfrak{h}^*}(M, \overline{\mathrm{Hom}}_{\mathfrak{b}}(U(\mathfrak{g}), N)). \tag{1.25}$$

Proof. The left-hand side (resp. the right-hand side) of (1.25) is nothing but $\overline{\mathrm{Hom}}_{\mathfrak{b}}(U(\mathfrak{g}) \otimes_{U(\mathfrak{a})} M, N)_0^0$ (resp. $\overline{\mathrm{Hom}}_{\mathfrak{a}}(M, \overline{\mathrm{Hom}}_{\mathfrak{b}}(U(\mathfrak{g}), N))_0^0$). Hence, it is enough to show that

$$\overline{\mathrm{Hom}}_{\mathfrak{b}}(U(\mathfrak{g}) \otimes_{U(\mathfrak{a})} M, N) \simeq \overline{\mathrm{Hom}}_{\mathfrak{a}}(M, \overline{\mathrm{Hom}}_{\mathfrak{b}}(U(\mathfrak{g}), N)). \tag{1.26}$$

Let Ψ and Φ be the following maps:

$$\Psi : \overline{\mathrm{Hom}}_{\mathfrak{b}}(U(\mathfrak{g}) \otimes_{U(\mathfrak{a})} M, N) \longrightarrow \overline{\mathrm{Hom}}_{\mathfrak{a}}(M, \overline{\mathrm{Hom}}_{\mathfrak{b}}(U(\mathfrak{g}), N)),$$
$$\Phi : \overline{\mathrm{Hom}}_{\mathfrak{a}}(M, \overline{\mathrm{Hom}}_{\mathfrak{b}}(U(\mathfrak{g}), N)) \longrightarrow \overline{\mathrm{Hom}}_{\mathfrak{b}}(U(\mathfrak{g}) \otimes_{U(\mathfrak{a})} M, N)$$

defined by

$$(\Psi(f)(m))(x) := f(x \otimes m)$$

where $f \in \overline{\mathrm{Hom}}_{\mathfrak{b}}(U(\mathfrak{g}) \otimes_{U(\mathfrak{a})} M, N)$, $m \in M$ and $x \in U(\mathfrak{g})$, and

$$\Phi(p)(x \otimes m) := p(m)(x),$$

where $p \in \overline{\mathrm{Hom}}_{\mathfrak{a}}(M, \overline{\mathrm{Hom}}_{\mathfrak{b}}(U(\mathfrak{g}), N))$, $m \in M$ and $x \in U(\mathfrak{g})$.

In fact, these maps are well-defined.

<u>Map Ψ</u> We check that

1. $\Psi(f)(a.m) = a.\Psi(f)(m)$ $(\forall a \in \mathfrak{a})$,
2. $\Psi(f)(m)(b.x) = b.(\Psi(f)(m)(x))$ $(\forall b \in \mathfrak{b})$.

For the first formula, we have

$$\begin{aligned}\Psi(f)(a.m)(x) &= f(x \otimes (a.m)) = f((xa) \otimes m)\\ &= \Psi(f)(m)(x.a) = a.(\Psi(f)(m))(x).\end{aligned}$$

For the second formula, we have

$$\begin{aligned}\Psi(f)(m)(b.x) &= f((b.x) \otimes m) = f(b.(x \otimes m))\\ &b.(f(x \otimes m)) = b.(\Psi(f)(m)(x)).\end{aligned}$$

Hence, Ψ is well-defined.

Map Φ We check that

1. $\Phi(p)(b.(x \otimes m)) = b.(\Phi(p)(x \otimes m))$ $(\forall b \in \mathfrak{b})$,
2. $\Phi(p)((x.a) \otimes m) = \Phi(p)(x \otimes (a.m))$ $(\forall a \in \mathfrak{a})$.

For the first formula, we have

$$\begin{aligned}\Phi(p)(b.(x \otimes m)) &= \Phi(p)((b.x) \otimes m) = p(m)(b.x) \\ &= b.(p(m)(x)) = b.(\Phi(p)(x \otimes m)).\end{aligned}$$

For the second formula, we have

$$\begin{aligned}\Phi(p)((x.a) \otimes m) &= p(m)(x.a) = (a.p(m))(x) \\ &= p(a.m)(x) = \Phi(p)(x \otimes (a.m)).\end{aligned}$$

Hence, Φ is well-defined.

By definition, $\Psi \circ \Phi$ and $\Phi \circ \Psi$ are the identity maps. Hence, we have proved the isomorphism (1.26), and thus the lemma follows. □

As a corollary, we show a graded version of *Frobenius reciprocity.*

Lemma 1.11. *For $M \in \mathrm{Ob}(\mathcal{C}^{\iota}_{(\mathfrak{a},\mathfrak{h})})$ and $N \in \mathrm{Ob}(\mathcal{C}^{\iota}_{(\mathfrak{g},\mathfrak{h})})$, we have*

$$\mathrm{Hom}_{\mathfrak{g}}^{G\times\mathfrak{h}^*}(\mathrm{Ind}_{\mathfrak{a}}^{\mathfrak{g}} M, N) \simeq \mathrm{Hom}_{\mathfrak{a}}^{G\times\mathfrak{h}^*}(M, \mathrm{Res}_{\mathfrak{a}}^{\mathfrak{g}} N).$$

Proof. We first notice an isomorphism

$$\overline{\mathrm{Hom}}_{\mathfrak{g}}(U(\mathfrak{g}), N) \simeq N \tag{1.27}$$

as $\mathfrak{g}$-modules. Indeed, for each $n \in N$, we define $f_n \in \overline{\mathrm{Hom}}_{\mathfrak{g}}(U(\mathfrak{g}), N)$ by

$$f_n(x) := xn \quad (x \in U(\mathfrak{g})).$$

Then, by (1.23), the map from $N \to \overline{\mathrm{Hom}}_{\mathfrak{g}}(U(\mathfrak{g}), N)$ $(n \mapsto f_n)$ is a homomorphism of $\mathfrak{g}$-modules, and clearly it is a bijection. Hence, by taking $\mathfrak{b} = \mathfrak{g}$ in Lemma 1.10, we have

$$\begin{aligned}\mathrm{Hom}_{\mathfrak{g}}^{G\times\mathfrak{h}^*}(\mathrm{Ind}_{\mathfrak{a}}^{\mathfrak{g}} M, N) &\simeq \mathrm{Hom}_{\mathfrak{a}}^{G\times\mathfrak{h}^*}(M, \overline{\mathrm{Hom}}_{\mathfrak{g}}(U(\mathfrak{g}), N)) \\ &\simeq \mathrm{Hom}_{\mathfrak{a}}^{G\times\mathfrak{h}^*}(M, N) \\ &\simeq \mathrm{Hom}_{\mathfrak{a}}^{G\times\mathfrak{h}^*}(M, \mathrm{Res}_{\mathfrak{a}}^{\mathfrak{g}} N).\end{aligned}$$

Now, we obtain the lemma. □

1.3.3 Definitions

In this subsection, a Q-graded Lie subalgebra $\mathfrak{p}$ of $(\mathfrak{g}, \mathfrak{h})$ does not necessarily contain the subalgebra $\mathfrak{h}$. For $M \in \mathrm{Ob}(\mathcal{C}^{\iota}_{(\mathfrak{g},\mathfrak{h})})$, we set

$$G_{\mathfrak{p}}(M) := M/\mathfrak{p}M,$$
$$F_{\mathfrak{p}}(M) := M^{\mathfrak{p}}(:= \{m \in M | a.m = 0 \ (\forall a \in \mathfrak{p})\}).$$

By definition, $G_{\mathfrak{p}}(M)$ and $F_{\mathfrak{p}}(M)$ are objects of $\mathrm{Vect}_{\mathbb{K}}^{G\times\mathfrak{h}^*}$. Hence, $F_{\mathfrak{p}}$ and $G_{\mathfrak{p}}$ define functors from $\mathcal{C}^{\iota}_{(\mathfrak{g},\mathfrak{h})}$ to $\mathrm{Vect}_{\mathbb{K}}^{G\times\mathfrak{h}^*}$. We call $G_{\mathfrak{p}}$ and $F_{\mathfrak{p}}$ the **coinvariant functor** and **invariant functor** respectively.

For $\alpha \in G$, let $\mathbb{K}^{\alpha}$ be the trivial representation of $\mathfrak{p}$ with G-gradation given by

$$(\mathbb{K}^{\alpha})^{\beta} = \begin{cases} \mathbb{K}^{\alpha} & \beta = \alpha \\ \{0\} & \beta \neq \alpha \end{cases}.$$

Noticing that the above functors can be written as

$$G_{\mathfrak{p}}(M) = \mathbb{K}^0 \otimes_{\mathfrak{p}} M, \quad F_{\mathfrak{p}}(M) = \overline{\mathrm{Hom}}_{\mathfrak{p}}(\mathbb{K}^0, M),$$

and we obtain

Lemma 1.12. *1. $G_{\mathfrak{p}}$ is a covariant and right exact functor.*
2. $F_{\mathfrak{p}}$ is a covariant and left exact functor.

Remark 1.9 *The functors $G_{\mathfrak{p}}$ and $F_{\mathfrak{p}}$ are not exact. Let $\mathfrak{p} = \mathbb{K}e$ be a one-dimensional Q-graded Lie algebra with trivial Q-gradation, i.e., $\mathfrak{p}^0 = \mathfrak{p}$, and let $V := \mathbb{K}v_1 \oplus \mathbb{K}v_2$ be the two-dimensional $\mathfrak{p}$-module defined by $e.v_1 = 0$ and $e.v_2 := v_1$.*

$W := \mathbb{K}v_1$ is a $\mathfrak{p}$-submodule of V. Then, we have the exact sequence

$$0 \longrightarrow W \longrightarrow V \longrightarrow V/W \longrightarrow 0$$

of $\mathfrak{p}$-modules. Since

$$\mathfrak{p}.W = \{0\}, \quad \mathfrak{p}.V = W, \quad \mathfrak{p}.(V/W) = \{0\},$$
$$W^{\mathfrak{p}} = \mathbb{K}v_1, \quad V^{\mathfrak{p}} = \mathbb{K}v_1, \quad (V/W)^{\mathfrak{p}} = \mathbb{K}(v_2 + W),$$

$G_{\mathfrak{p}}(W) \to G_{\mathfrak{p}}(V)$ is not injective and $F_{\mathfrak{p}}(V) \to F_{\mathfrak{p}}(V/W)$ is not surjective.

To define the (co)homology group as derived functors of (co)invariant functors, we show the following proposition.

Proposition 1.12 *1. $\mathcal{C}^{\iota}_{(\mathfrak{g},\mathfrak{h})}$ has enough projectives and injectives (cf. [RW1]).*
2. $\mathcal{O}^{\iota}$ has enough projective $U(\mathfrak{g}^-)$-modules and injective $U(\mathfrak{g}^+)$-modules.

Proof. First, we show that $\mathcal{C}^{\iota}_{(\mathfrak{g},\mathfrak{h})}$ has enough projectives. For any $M \in \mathrm{Ob}(\mathcal{C}^{\iota}_{(\mathfrak{g},\mathfrak{h})})$, we have to prove that there exists a projective $P \in \mathrm{Ob}(\mathcal{C}^{\iota}_{(\mathfrak{g},\mathfrak{h})})$ such that $P \to M \to 0$ is exact. We set

$$\tilde{M} := \mathrm{Ind}^{\mathfrak{g}}_{\mathfrak{h}}(\mathrm{Res}^{\mathfrak{g}}_{\mathfrak{h}} M) = U(\mathfrak{g}) \otimes_{U(\mathfrak{h})} \mathrm{Res}^{\mathfrak{g}}_{\mathfrak{h}} M.$$

We first prove that $\tilde{M}$ is a projective $U(\mathfrak{g})$-module. It is enough to show that $\mathrm{Hom}^{G\times\mathfrak{h}^*}_{\mathfrak{g}}(\tilde{M}, \cdot)$ is exact. Lemma 1.11 implies that

$$\mathrm{Hom}^{G\times\mathfrak{h}^*}_{\mathfrak{g}}(\tilde{M}, N) \simeq \mathrm{Hom}^{G\times\mathfrak{h}^*}_{\mathfrak{h}}(\mathrm{Res}^{\mathfrak{g}}_{\mathfrak{h}} M, \mathrm{Res}^{\mathfrak{g}}_{\mathfrak{h}} N)$$

for any $N \in \mathrm{Ob}(\mathcal{C}^{\iota}_{(\mathfrak{g},\mathfrak{h})})$. On the other hand, $\mathrm{Hom}^{G\times\mathfrak{h}^*}_{\mathfrak{h}}(\mathrm{Res}^{\mathfrak{g}}_{\mathfrak{h}} M, \cdot)$ is exact, since any objects of $\mathcal{C}^{\iota}_{(\mathfrak{g},\mathfrak{h})}$ are $G \times \mathfrak{h}^*$-graded. Hence, $\tilde{M}$ is projective. Moreover, $\tilde{M} \to M$ such that $x \otimes v \mapsto x.v$ gives a surjective $\mathfrak{g}$-homomorphism. Hence, we can take $\tilde{M}$ as a projective P, and we have proved that $\mathcal{C}^{\iota}_{(\mathfrak{g},\mathfrak{h})}$ has enough projectives.

Next, we show that $\mathcal{C}^{\iota}_{(\mathfrak{g},\mathfrak{h})}$ has enough injectives, i.e., for any $M \in \mathrm{Ob}(\mathcal{C}^{\iota}_{(\mathfrak{g},\mathfrak{h})})$, there exists an injective $I \in \mathrm{Ob}(\mathcal{C}^{\iota}_{(\mathfrak{g},\mathfrak{h})})$ such that $0 \to M \to I$ is exact. Let P be a projective $U(\mathfrak{g})$-module such that $P \to M^{\sharp a} \to 0$ is exact. Then, $I := P^{\sharp a}$ is an injective $U(\mathfrak{g})$-module such that $0 \to M \to I$ is exact.

We show the second statement. Let $M(0,0)$ be the Verma module with highest weight $(0,0) \in G \times \mathfrak{h}^*$. Since $L(0,0)$ is isomorphic to the trivial $\mathfrak{g}$-module $\mathbb{K}^0_0$, there exists a surjective homomorphism $M(0,0) \twoheadrightarrow \mathbb{K}^0_0$. By tensoring M with this sequence, we have

$$M(0,0) \otimes M \twoheadrightarrow \mathbb{K}^0_0 \otimes M \simeq M.$$

Since $M(0,0)$ is a $U(\mathfrak{g}^-)$-free module, by Corollary A.1 in § A.3.3 we see that $M(0,0) \otimes M$ is $U(\mathfrak{g}^-)$-free, i.e., $U(\mathfrak{g}^-)$-projective. Hence, $\mathcal{O}^{\iota}$ has enough projective $U(\mathfrak{g}^-)$-modules.

Finally, we show that $\mathcal{O}^{\iota}$ has enough injective $U(\mathfrak{g}^+)$-modules. Let $M^-(0,0)$ be the lowest weight module with lowest weight $(0,0) \in G \times \mathfrak{h}^*$. Here, for V and $W \in \mathrm{Ob}(\mathcal{C}^{\iota}_{(\mathfrak{g},\mathfrak{h})})$, we regard $\overline{\mathrm{Hom}_{\mathbb{K}}}(V, W)$ as a left $\mathfrak{g}$-module via

$$(x.f)(v) := x.(f(v)) - f(x.v) \quad (x \in \mathfrak{g},\ f \in \overline{\mathrm{Hom}_{\mathbb{K}}}(V, W),\ v \in V).$$

Applying the contravariant exact functor $\overline{\mathrm{Hom}_{\mathbb{K}}}(\cdot, M)$ $(M \in \mathrm{Ob}(\mathcal{C}^{\iota}_{(\mathfrak{g},\mathfrak{h})}))$ to $M^-(0,0) \twoheadrightarrow \mathbb{K}^0_0$, we have

$$M \simeq \overline{\mathrm{Hom}_{\mathbb{K}}}(\mathbb{K}^0_0, M) \hookrightarrow \overline{\mathrm{Hom}_{\mathbb{K}}}(M^-(0,0), M).$$

Hence, it suffices to see that $\overline{\mathrm{Hom}_{\mathbb{K}}}(M^-(0,0), M)$ is an injective $U(\mathfrak{g}^+)$-module. First, we show that for $N \in \mathrm{Ob}(\mathcal{C}^{\iota}_{(\mathfrak{g},\mathfrak{h})})$, the isomorphism of $\mathbb{K}$-vector spaces

$$\mathrm{Hom}_{\mathbb{K}}^{G\times\mathfrak{h}^*}(N,\overline{\mathrm{Hom}}_{\mathbb{K}}(M^-(0,0),M)) \simeq \mathrm{Hom}_{\mathbb{K}}^{G\times\mathfrak{h}^*}(N\otimes_{\mathbb{K}}M^-(0,0),M), \quad (1.28)$$

given in Lemma 1.1 is, in fact, an isomorphism of $\mathfrak{g}$-modules, where $N \otimes_{\mathbb{K}} M^-(0,0)$ is the tensor product of $\mathfrak{g}$-modules. One can check that the isomorphism is explicitly given by

$$\Phi(f)(n \otimes m) := f(n)(m)$$

for $f \in \mathrm{Hom}_{\mathbb{K}}^{G\times\mathfrak{h}^*}(N,\overline{\mathrm{Hom}}_{\mathbb{K}}(M^-(0,0),M))$, $n \in N$ and $m \in M^-(0,0)$. For $x \in \mathfrak{g}$, we have

$$\begin{aligned}\Phi(x.f)(n \otimes m) &= ((x.f)(n))(m)\\ &= (x.(f(n)) - f(x.n))(m)\\ &= x.(f(n)(m)) - f(n)(x.m) - f(x.n)(m)\\ &= x.(\Phi(f)(n \otimes m)) - \Phi(f)(n \otimes x.m + x.n \otimes m)\\ &= (x.\Phi(f))(n \otimes m),\end{aligned}$$

and thus (1.28) is an isomorphism of $\mathfrak{g}$-modules. Taking $\mathfrak{g}^+$-invariants of both sides of (1.28), we obtain

$$\mathrm{Hom}_{\mathfrak{g}^+}^{G\times\mathfrak{h}^*}(N,\overline{\mathrm{Hom}}_{\mathbb{K}}(M^-(0,0),M)) \simeq \mathrm{Hom}_{\mathfrak{g}^+}^{G\times\mathfrak{h}^*}(N \otimes_{\mathbb{K}} M^-(0,0),M).$$

Let $\mathfrak{F} : \mathcal{C}^{\iota}_{(\mathfrak{g},\mathfrak{h})} \to \mathrm{Vect}_{\mathbb{K}}^{G\times\mathfrak{h}^*}$ be the forgetful functor. Since, Corollary A.1 implies the following isomorphism of $\mathfrak{g}^+$-modules

$$N \otimes_{\mathbb{K}} M^-(0,0) \simeq M^-(0,0) \otimes_{\mathbb{K}} N \simeq U(\mathfrak{g}^+) \otimes_{\mathbb{K}} \mathfrak{F}N,$$

where $\mathfrak{g}^+$ acts on $U(\mathfrak{g}^+) \otimes_{\mathbb{K}} \mathfrak{F}N$ via the left multiplication, we have

$$\mathrm{Hom}_{\mathfrak{g}^+}^{G\times\mathfrak{h}^*}(N \otimes_{\mathbb{K}} M^-(0,0),M) \simeq \mathrm{Hom}_{\mathfrak{g}^+}^{G\times\mathfrak{h}^*}(U(\mathfrak{g}^+) \otimes_{\mathbb{K}} \mathfrak{F}N,M).$$

Moreover, the map

$$\Psi : \mathrm{Hom}_{\mathfrak{g}^+}^{G\times\mathfrak{h}^*}(U(\mathfrak{g}^+) \otimes_{\mathbb{K}} \mathfrak{F}N,M) \longrightarrow \mathrm{Hom}_{\mathbb{K}}^{G\times\mathfrak{h}^*}(\mathfrak{F}N,\mathfrak{F}M)$$

defined by

$$\Psi(f)(n) := f(1 \otimes n) \quad (f \in \mathrm{Hom}_{\mathfrak{g}^+}^{G\times\mathfrak{h}^*}(U(\mathfrak{g}^+) \otimes_{\mathbb{K}} \mathfrak{F}N,M),\ n \in \mathfrak{F}N)$$

is an isomorphism. Since any object of $\mathcal{C}^{\iota}_{(\mathfrak{g},\mathfrak{h})}$ is $G\times\mathfrak{h}^*$-graded, $\mathrm{Hom}_{\mathbb{K}}^{G\times\mathfrak{h}^*}(\cdot,\mathfrak{F}M)$ is exact, and thus, $\mathrm{Hom}_{\mathbb{K}}(M^-(0,0),M)$ is an injective $U(\mathfrak{g}^+)$-module. Hence, $\mathcal{O}^{\iota}$ has enough injective $U(\mathfrak{g}^+)$-modules. Now, we have completed the proof of the proposition. □

We define the homology group $H_n(\mathfrak{g},M)$ and the cohomology group $H^n(\mathfrak{g},M)$ as follows:

Definition 1.26 *For $n \in \mathbb{Z}_{\geq 0}$, we set*

$$H_n(\mathfrak{g}, M) := L_n G_{\mathfrak{g}}(M),$$
$$H^n(\mathfrak{g}, M) := R^n F_{\mathfrak{g}}(M).$$

Remark that by Lemma 1.12 and Lemma A.2, we have

$$H_0(\mathfrak{g}, M) = G_{\mathfrak{g}}(M),$$
$$H^0(\mathfrak{g}, M) = F_{\mathfrak{g}}(M).$$

1.3.4 Some Properties

We state some fundamental properties of homology and cohomology groups of a Q-graded Lie algebra. Using Frobenius reciprocity, we show *Shapiro's lemma.*

Let $\mathfrak{a}$ be a Q-graded Lie subalgebra of $(\mathfrak{g}, \mathfrak{h})$ which contains $\mathfrak{h}$.

Lemma 1.13. *For $V \in \mathrm{Ob}(\mathcal{C}^{\iota}_{(\mathfrak{a},\mathfrak{h})})$ and $W \in \mathrm{Ob}(\mathcal{C}^{\iota}_{(\mathfrak{g},\mathfrak{h})})$, we have*

$$\mathrm{Ext}^n_{\mathcal{C}^{\iota}_{(\mathfrak{g},\mathfrak{h})}}(\mathrm{Ind}^{\mathfrak{g}}_{\mathfrak{a}} V, W) \simeq \mathrm{Ext}^n_{\mathcal{C}^{\iota}_{(\mathfrak{a},\mathfrak{h})}}(V, \mathrm{Res}^{\mathfrak{g}}_{\mathfrak{a}} W).$$

Proof. Here we set

$$F^W_{\mathfrak{g}} := \mathrm{Hom}^{G\times\mathfrak{h}^*}_{\mathfrak{g}}(\cdot, W) : \mathcal{C}^{\iota}_{(\mathfrak{g},\mathfrak{h})} \longrightarrow \mathrm{Vect}^{G\times\mathfrak{h}^*}_{\mathbb{K}},$$
$$F^{W'}_{\mathfrak{a}} := \mathrm{Hom}^{G\times\mathfrak{h}^*}_{\mathfrak{a}}(\cdot, W') : \mathcal{C}^{\iota}_{(\mathfrak{a},\mathfrak{h})} \longrightarrow \mathrm{Vect}^{G\times\mathfrak{h}^*}_{\mathbb{K}}$$

for $W \in \mathrm{Ob}(\mathcal{C}^{\iota}_{(\mathfrak{g},\mathfrak{h})})$ and $W' \in \mathrm{Ob}(\mathcal{C}^{\iota}_{(\mathfrak{a},\mathfrak{h})})$. To prove this lemma, we notice the following:

(i) $\mathrm{Ind}^{\mathfrak{g}}_{\mathfrak{a}}(\cdot) = U(\mathfrak{g}) \otimes_{U(\mathfrak{a})} (\cdot)$ is covariant exact, since $U(\mathfrak{g})$ is a free right $U(\mathfrak{a})$-module. Hence, the induction functor maps $U(\mathfrak{a})$-projectives to $U(\mathfrak{g})$-projectives.
(ii) $F^W_{\mathfrak{g}} \circ \mathrm{Ind}^{\mathfrak{g}}_{\mathfrak{a}} = F^{W'}_{\mathfrak{a}}$ ($W' := \mathrm{Res}^{\mathfrak{g}}_{\mathfrak{a}} W$) by Frobenius reciprocity.

Hence, we have

$$\begin{aligned}
\mathrm{Ext}^n_{\mathcal{C}^{\iota}_{(\mathfrak{g},\mathfrak{h})}}(\mathrm{Ind}^{\mathfrak{g}}_{\mathfrak{a}} V, W) &\simeq (R^n F^W_{\mathfrak{a}}) \circ \mathrm{Ind}^{\mathfrak{g}}_{\mathfrak{a}}(V) \\
&\simeq R^n \left(F^W_{\mathfrak{a}} \circ \mathrm{Ind}^{\mathfrak{g}}_{\mathfrak{a}}\right)(V) && \text{(i)+Proposition A.2} \\
&\simeq R^n F^{W'}_{\mathfrak{a}}(V) && \text{(ii)} \\
&\simeq \mathrm{Ext}^n_{\mathcal{C}^{\iota}_{(\mathfrak{a},\mathfrak{h})}}(V, \mathrm{Res}^{\mathfrak{g}}_{\mathfrak{a}} W).
\end{aligned}$$

□

Let us prove two useful propositions.

Proposition 1.13 ([RW1]) *For $V \in \mathrm{Ob}(\mathcal{C}^\iota_{(\mathfrak{g},\mathfrak{h})})$ and $(\alpha,\nu) \in G \times \mathfrak{h}^*$, we have*

$$\mathrm{Ext}^n_{\mathcal{C}^\iota_{(\mathfrak{g},\mathfrak{h})}}(M(\alpha,\nu),V) \simeq \mathrm{Hom}_{\mathfrak{h}}^{G\times\mathfrak{h}^*}(\mathbb{K}^\alpha_\nu, H^n(\mathfrak{g}^+, \mathrm{Res}^{\mathfrak{g}}_{\mathfrak{g}^\geq} V)).$$

In particular, for $n = 0$,

$$\mathrm{Hom}_{\mathfrak{g}}^{G\times\mathfrak{h}^*}(M(\alpha,\nu),V) \simeq \mathrm{Hom}_{\mathfrak{h}}^{G\times\mathfrak{h}^*}(\mathbb{K}^\alpha_\nu, (\mathrm{Res}^{\mathfrak{g}}_{\mathfrak{g}^\geq} V)^{\mathfrak{g}^+}).$$

Proof. For simplicity, we set

$$F_{\mathfrak{g}^+} := \mathrm{Hom}_{\mathfrak{g}^+}^{G\times\mathfrak{h}^*}(\mathbb{K}^0,\cdot),$$
$$F_{\mathfrak{h}}^{(\alpha,\nu)} := \mathrm{Hom}_{\mathfrak{h}}^{G\times\mathfrak{h}^*}(\mathbb{K}^\alpha_\nu,\cdot),$$

where $\mathbb{K}^0$ denotes the trivial representation of $\mathfrak{g}^+$. A key of our proof is the following fact:

$$\mathrm{Hom}_{\mathcal{C}^\iota_{(\mathfrak{g}^\geq,\mathfrak{h})}}(\mathbb{K}^\alpha_\nu, W) = \{F_{\mathfrak{h}}^{(\alpha,\nu)} \circ F_{\mathfrak{g}^+}\}(W) \qquad (1.29)$$

holds for any $W \in \mathrm{Ob}(\mathcal{C}^\iota_{(\mathfrak{g}^\geq,\mathfrak{h})})$. From Shapiro's lemma, we obtain

$$\begin{aligned}\mathrm{Ext}^n_{\mathcal{C}^\iota_{(\mathfrak{g},\mathfrak{h})}}(M(\alpha,\nu),V) &\simeq \mathrm{Ext}^n_{\mathcal{C}^\iota_{(\mathfrak{g}^\geq,\mathfrak{h})}}(\mathbb{K}^\alpha_\nu, \mathrm{Res}^{\mathfrak{g}}_{\mathfrak{g}^\geq} V)\\ &= R^n\mathrm{Hom}_{\mathcal{C}^\iota_{(\mathfrak{g}^\geq,\mathfrak{h})}}(\mathbb{K}^\alpha_\nu,\cdot)(\mathrm{Res}^{\mathfrak{g}}_{\mathfrak{g}^\geq} V).\end{aligned}$$

Since $F_{\mathfrak{h}}^{(\alpha,\nu)}$ is covariant exact, (1.29) and Proposition A.2 imply that

$$\begin{aligned}R^n\mathrm{Hom}_{\mathfrak{g}^\geq}^{G\times\mathfrak{h}^*}(\mathbb{K}^\alpha_\nu,\cdot)(\mathrm{Res}^{\mathfrak{g}}_{\mathfrak{g}^\geq} V) &\simeq R^n\left(F_{\mathfrak{h}}^{(\alpha,\nu)} \circ F_{\mathfrak{g}^+}\right)(\mathrm{Res}^{\mathfrak{g}}_{\mathfrak{g}^\geq} V)\\ &\simeq F_{\mathfrak{h}}^{(\alpha,\nu)} \circ (R^n F_{\mathfrak{g}^+})(\mathrm{Res}^{\mathfrak{g}}_{\mathfrak{g}^\geq} V)\\ &= \mathrm{Hom}_{\mathfrak{h}}^{G\times\mathfrak{h}^*}(\mathbb{K}^\alpha_\nu, H^n(\mathfrak{g}^+, \mathrm{Res}^{\mathfrak{g}}_{\mathfrak{g}^\geq} V)).\end{aligned}$$ □

Here, we assume that $(\mathfrak{g},\mathfrak{h})$ is a Q-graded Lie algebra with a Q-graded anti-involution σ.

Proposition 1.14 ([DGK], [Liu]) *Suppose that $M \in \mathrm{Ob}(\mathcal{O}^\iota)$. For each $n \in \mathbb{Z}_{\geq 0}$, the following isomorphism*

$$H^n(\mathfrak{g}^+, M^c) \simeq H_n(\mathfrak{g}^-, M)^{\sharp+}$$

of $G \times \mathfrak{h}^*$-graded $\mathbb{K}$-vector space holds.

Proof. For simplicity, we denote the functors $\overline{\mathrm{Hom}}_{\mathfrak{g}^+}(\mathbb{K}^0,\cdot)$ and $\mathbb{K}^0 \otimes_{\mathfrak{g}^-} (\cdot)$ by F_+ and G^- respectively. We first show that

$$F_+ \circ (\cdot)^c = (\cdot)^{\sharp+} \circ G^-, \qquad (1.30)$$

i.e., $\overline{\mathrm{Hom}}_{\mathfrak{g}^+}(\mathbb{K}^0, M^c) = (\mathbb{K}^0 \otimes_{\mathfrak{g}^-} M)^{\sharp+}$ for $M \in \mathrm{Ob}(\mathcal{O}^\iota)$. We introduce

$$\begin{aligned} &\Psi : \overline{\mathrm{Hom}}_{\mathfrak{g}^+}(\mathbb{K}^0, M^c) \longrightarrow (\mathbb{K}^0 \otimes_{\mathfrak{g}^-} M)^{\sharp+}, \\ &\Phi : (\mathbb{K}^0 \otimes_{\mathfrak{g}^-} M)^{\sharp+} \longrightarrow \overline{\mathrm{Hom}}_{\mathfrak{g}^+}(\mathbb{K}^0, M^c) \end{aligned}$$

as follows: For $k \in \mathbb{K}^0$ and $m \in M$,

$$\begin{aligned} &\Psi(f)(k \otimes m) := f(k)(m) \quad f \in \overline{\mathrm{Hom}}_{\mathfrak{g}^+}(\mathbb{K}^0, M^c), \\ &(\Phi(g)(k))(m) := g(k \otimes m) \; g \in (\mathbb{K}^0 \otimes_{\mathfrak{g}^-} M)^{\sharp+}. \end{aligned}$$

Let us check that Ψ and Φ are well-defined. Since $\sigma(\mathfrak{g}^-) = \mathfrak{g}^+$, we have

$$\Psi(f)(k \otimes x.m) = f(k)(x.m) = (\sigma(x).f(k))(m) = 0,$$

for $x \in \mathfrak{g}^-$ and $f \in \overline{\mathrm{Hom}}_{\mathfrak{g}^+}(\mathbb{K}^0, M^c)$. Furthermore, $\Psi(f)(k \otimes m) = 0$ for $k \otimes m \in (\mathbb{K} \otimes_{\mathfrak{g}^-} M)^\beta_\mu$ except for finitely many $(\beta, \mu) \in G \times \mathfrak{h}^*$, since $f(k) \in M^c$. Hence, $\Psi(f) \in (\mathbb{K} \otimes_{\mathfrak{g}^-} M)^{\sharp+}$ and Ψ is well-defined. On the other hand, we have

$$(y.\Phi(g)(k))(m) = (\Phi(g)(k))(\sigma(y).m) = g(k \otimes \sigma(y).m) = 0$$

for $y \in \mathfrak{g}^+$, and $(\Phi(g)(k))(m) = 0$ for $m \in M^\beta_\mu$ except for finitely many $(\beta, \mu) \in G \times \mathfrak{h}^*$. Hence, $\Phi(g) \in \overline{\mathrm{Hom}}_{\mathfrak{g}^+}(\mathbb{K}^0, M^c)$ and Φ is well-defined. By definition, one can easily check that $\Phi \circ \Psi = \mathrm{id}$ and $\Psi \circ \Phi = \mathrm{id}$, and thus (1.30) is proved.

Since the functors $(\cdot)^c$ and $(\cdot)^{\sharp+}$ are contravariant exact, we have

$$\begin{aligned} H^n(\mathfrak{g}^+, M^c) &= (R^n F_+) \circ (\cdot)^c(M) \\ &\simeq R^n(F_+ \circ (\cdot)^c)(M) && \text{Proposition A.2} \\ &\simeq R^n((\cdot)^{\sharp+} \circ G^-)(M) && (1.30) \\ &\simeq (\cdot)^{\sharp+} \circ (L_n G^-)(M) && \text{Proposition A.2} \\ &= H_n(\mathfrak{g}^-, M)^{\sharp+}. \end{aligned}$$

Therefore, we complete the proof. □

1.4 Bernstein–Gelfand–Gelfand Duality

In this section, we state the so-called Bernstein–Gelfand–Gelfand duality. Here, we assume that the map $\pi_Q : Q \to \mathfrak{h}^*$ defined in (1.6) is injective, and hence, any submodules of an object of the category $\mathcal{O}^\iota$ are also objects of $\mathcal{O}^\iota$. Thus, we abbreviate $\mathcal{O}^\iota$, $M(\alpha, \lambda)$, $L(\alpha, \lambda)$ etc., to $\mathcal{O}$, $M(\lambda)$, $L(\lambda)$ etc. for simplicity.

1.4.1 Preliminaries

In this subsection, we introduce the notion of Verma composition series for objects of the category $\mathcal{O}$, and state some properties.

Definition 1.27 *We say that $M \in \mathrm{Ob}\mathcal{O}$ has a* **Verma composition series** *(VCS for short) of length l if there exists a filtration*

$$M = M_0 \supset M_1 \supset \cdots \supset M_l \supset M_{l+1} = \{0\} \tag{1.31}$$

of $\mathfrak{g}$-modules such that

$$M_i/M_{i+1} \simeq M(\mu_i) \quad (\exists \mu_i \in \mathfrak{h}^*).$$

For an object $M \in \mathrm{Ob}\mathcal{O}$ with a VCS of the form (1.31), we set

$$[M : M(\mu)] := \sharp\{i | \mu_i = \mu\}.$$

This is well-defined. Indeed, if M has a VCS (1.31), then

$$\mathrm{ch}\, M = \sum_{i=1}^{l} \mathrm{ch}\, M(\mu_i) = \sum_{\mu \in \mathfrak{h}^*} [M : M(\mu)]\, \mathrm{ch}\, M(\mu).$$

Since $\{\mathrm{ch}\, M(\mu) | \mu \in \mathfrak{h}^*\}$ are linearly independent, $[M : M(\mu)]$ does not depend on the choice of a VCS.

We give some properties of VCS.

Lemma 1.14. *Suppose that $M \in \mathrm{Ob}\mathcal{O}$ has a VCS of length l. For a maximal element $\mu \in \mathcal{P}(M)$ and $v \in M^{\mu} \setminus \{0\}$, we set $M' := U(\mathfrak{g}).v$. Then,*

1. *$M' \simeq M(\mu)$,*
2. *M/M' has a VCS of length $l - 1$.*

Proof. We show this lemma by induction on l. In the case $l = 0$, we have nothing to prove. We suppose that $l > 0$. Let

$$M = M_0 \supset M_1 \supset \cdots \supset M_l \supset M_{l+1} = \{0\}$$

be a VCS of M. If $v \in M_1$, then the lemma holds by induction hypothesis. Hence, we may suppose that $v \notin M_1$. Since $M_0/M_1 \simeq M(\mu_0)$ for some $\mu_0 \in \mathfrak{h}^*$, by the maximality of μ, we have $\mu = \mu_0$ and $v + M_1$ is a highest weight vector of $M(\mu)$. On the other hand, by the universality of $M(\mu)$, there exists

$$M(\mu) \to M' \quad (1 \otimes \mathbf{1}_{\mu} \mapsto v).$$

Hence, an exact sequence

$$0 \longrightarrow M_1 \longrightarrow M_0 \longrightarrow M(\mu) \longrightarrow 0$$

splits, i.e., $M_0 \simeq M_1 \oplus M(\mu)$. Thus, the lemma follows. □

By the above lemma, we have

Proposition 1.15 *Let M and N be objects of $\mathcal{O}$ such that $M \oplus N$ has a VCS. Then, both M and N have VCSs.*

Proof. Suppose that $M \oplus N$ has a VCS of length l. We prove this proposition by induction on l. The case $l = 0$ follows by definition. In the case $l > 0$, let us take a maximal element $\mu \in \mathcal{P}(M \oplus N)$. We may assume that $\mu \in \mathcal{P}(M)$ without loss of generality. For $v \in M^{\mu} \setminus \{0\}$, we set $M' := U(\mathfrak{g}).v$. Then, by Lemma 1.14,

$$(M \oplus N)/M' \simeq (M/M') \oplus N$$

has a VCS of length $l - 1$. By induction hypothesis, we see that M/M' and N have VCSs. By pulling back a VCS of M/M' via the canonical projection $M \twoheadrightarrow M/M'$, the proposition follows. □

1.4.2 Truncated Category

For each $\Lambda \in \mathfrak{h}^*$, we introduce a full subcategory $\mathcal{C}(\Lambda)$ of $\mathcal{O}$. We set

$$\mathcal{P}_{\Lambda}^{-} := \{\Lambda - \alpha | \alpha \in Q^+\}.$$

Definition 1.28 *For $\Lambda \in \mathfrak{h}^*$, let $\mathcal{P}_{\Lambda}^{-}$ be the full subcategory of $\mathcal{O}$ consisting of these objects satisfying*

$$\mathcal{P}(M) \subset \mathcal{P}_{\Lambda}^{-}.$$

By Lemma 1.5 and Remark 1.4, we have

Lemma 1.15. *Let $M \in \mathrm{Ob}\mathcal{C}(\Lambda)$ be a finitely generated $\mathfrak{g}$-module. For any proper submodule M' of M, there exists a maximal proper submodule $N \in \mathrm{Ob}\mathcal{C}(\Lambda)$ of M such that $M' \subset N$.*

Remark that N is not necessarily finitely generated.

Proposition 1.16 *For any finitely generated $\mathfrak{g}$-module $M \in \mathrm{Ob}\mathcal{C}(\Lambda)$, there exists finite number of indecomposable modules $M^{(k)} \in \mathrm{Ob}\mathcal{C}(\Lambda)$ of finite type such that*

$$M \simeq \bigoplus_k M^{(k)}.$$

Proof. For finitely generated $\mathfrak{g}$-module $M \in \mathrm{Ob}\mathcal{C}(\Lambda)$, we set

$$\mathrm{rk}M := \inf\left\{ \sum_i \dim M_{\mu_i} \,\middle|\, \mu_i \in \mathcal{P}(M) \text{ and } \bigoplus_i M_{\mu_i} \text{ generates } M \right\}.$$

Since M is finitely generated, rk$M < \infty$. We show the existence of $M^{(k)}$s by induction on rk$M$.

Suppose that rk$M = 1$, i.e., there exists $\mu \in \mathfrak{h}^*$ such that $\dim M_\mu = 1$ and M is generated by M_μ. Then, M is indecomposable. Indeed, if $M = M_1 \oplus M_2$, then $(M_1)_\mu \neq \{0\}$ and $(M_2)_\mu \neq \{0\}$, i.e., $\dim M_\mu \geq 2$.

Suppose that the proposition holds for any $N \in \mathrm{Ob}\mathcal{C}(\Lambda)$ such that rk$N \leq l$. If $M \in \mathrm{Ob}\mathcal{C}(\Lambda)$ such that rk$M = l+1$ is not indecomposable, then $M = M_1 \oplus M_2$. Since rk$M_1 \leq l$ and rk$M_2 \leq l$, M_1 and M_2 decompose into indecomposable modules by the induction hypothesis. Hence, the first part of this proposition has been proved.

Assuming that M is generated by $\bigoplus_i M_{\mu_i}$, $M^{(k)}$ is generated by $(\bigoplus_i M_{\mu_i}) \cap M^{(k)}$. Hence, $M^{(k)}$ is of finite type. □

1.4.3 Projective Objects

In this subsection, we introduce a projective object $P(\mu)$, which plays an important role in the proof of the duality theorem (Theorem 1.2).

From now on, we suppose that $\mu \in \mathcal{P}^-_\Lambda$. Let $M^-(\mu)$ be the lowest weight Verma module with lowest weight μ (Definition 1.19). For simplicity, in this subsection, we set

$$N(\mu) := \bigoplus_{\nu \in \mathcal{P}(M^-(\mu)) \setminus \mathcal{P}^-_\Lambda} M^-(\mu)_\nu,$$

and regard it as a $\mathfrak{g}^{\geq}$-module in a natural way. We further set

$$W(\mu) := M^-(\mu)/N(\mu).$$

For each $\mu \in \mathcal{P}^-_\Lambda$, we define $P(\mu)$ by

$$P(\mu) := U(\mathfrak{g}) \otimes_{U(\mathfrak{g}^{\geq})} W(\mu).$$

By definition, $P(\mu) \in \mathrm{Ob}\mathcal{C}(\Lambda)$, and it is a finitely generated $\mathfrak{g}$-module since $\dim W(\mu) < \infty$.

Proposition 1.17 *For any $M \in \mathrm{Ob}\mathcal{C}(\Lambda)$, there exists the following isomorphism of $\mathbb{K}$-vector spaces:*

$$\mathrm{Hom}_{\mathfrak{g}}(P(\mu), M) \simeq \mathrm{Hom}_{\mathfrak{h}}(\mathbb{K}_\mu, \mathrm{Res}^{\mathfrak{g}}_{\mathfrak{h}} M).$$

Proof. By Frobenius reciprocity (Lemma 1.11), we have

$$\mathrm{Hom}_{\mathfrak{g}}(P(\mu), M) \simeq \mathrm{Hom}_{\mathfrak{g}^{\geq}}(W(\mu), \mathrm{Res}^{\mathfrak{g}}_{\mathfrak{g}^{\geq}} M).$$

From $0 \to N(\mu) \to M^-(\mu) \to W(\mu) \to 0$, we obtain

$$0 \longrightarrow \mathrm{Hom}_{\mathfrak{g}_\geq}(W(\mu), \mathrm{Res}^{\mathfrak{g}}_{\mathfrak{g}_\geq} M) \longrightarrow \mathrm{Hom}_{\mathfrak{g}_\geq}(M^-(\mu), \mathrm{Res}^{\mathfrak{g}}_{\mathfrak{g}_\geq} M)$$
$$\longrightarrow \mathrm{Hom}_{\mathfrak{g}_\geq}(N(\mu), \mathrm{Res}^{\mathfrak{g}}_{\mathfrak{g}_\geq} M).$$

Since $\mathrm{Hom}_{\mathfrak{g}_\geq}(N(\mu), \mathrm{Res}^{\mathfrak{g}}_{\mathfrak{g}_\geq} M) = \{0\}$ by the definition of $N(\mu)$, we get

$$\mathrm{Hom}_{\mathfrak{g}_\geq}(W(\mu), \mathrm{Res}^{\mathfrak{g}}_{\mathfrak{g}_\geq} M) \simeq \mathrm{Hom}_{\mathfrak{g}_\geq}(M^-(\mu), \mathrm{Res}^{\mathfrak{g}}_{\mathfrak{g}_\geq} M).$$

Moreover, by Frobenius reciprocity, we have

$$\mathrm{Hom}_{\mathfrak{g}_\geq}(M^-(\mu), \mathrm{Res}^{\mathfrak{g}}_{\mathfrak{g}_\geq} M) \simeq \mathrm{Hom}_{\mathfrak{h}}(\mathbb{K}_\mu, \mathrm{Res}^{\mathfrak{g}}_{\mathfrak{h}} M).$$ □

Since $\mathrm{Hom}_{\mathfrak{g}}(P(\mu), \cdot)$ is exact by this proposition, we have

Corollary 1.2 *$P(\mu)$ is a projective object in the category $\mathcal{C}(\Lambda)$.*

Moreover, by the above proposition, we obtain

Corollary 1.3 *For any finitely generated $\mathfrak{g}$-module $M \in \mathrm{Ob}\mathcal{C}(\Lambda)$, there exist $\mu_1, \cdots, \mu_l \in \mathcal{P}_\Lambda^-$ such that the following surjection exists:*

$$\bigoplus_{i=1}^{l} P(\mu_i) \longrightarrow M.$$

Proof. Let $\{m_1, \cdots, m_l\}$ be a set of weight vectors which generates M. Suppose that

$$m_i \in M_{\mu_i} \setminus \{0\} \quad (\mu_i \in \mathcal{P}_\Lambda^-).$$

Proposition 1.17 implies that for each i, there exists a map $f_i : P(\mu_i) \to M$ such that

$$f_i : 1 \otimes (v_{\mu_i}^- + N(\mu_i)) \mapsto m_i,$$

where $v_{\mu_i}^-$ is a highest weight vector of $M^-(\mu_i)$. $f := \sum_i f_i$ gives the desired surjection. □

Moreover, the following lemma holds:

Lemma 1.16. *1. For any $\mu \in \mathcal{P}_\Lambda^-$, there exist finitely generated $\mathfrak{g}$-modules $P(\mu)^{(1)}, \cdots, P(\mu)^{(l)}$, which are indecomposable and projective in $\mathcal{C}(\Lambda)$, such that*

$$P(\mu) \simeq \bigoplus_{i=1}^{l} P(\mu)^{(i)}.$$

2. For any finitely generated $\mathfrak{g}$-module M, which is indecomposable and projective in $\mathcal{C}(\Lambda)$, there exist $\mu \in \mathcal{P}_\Lambda^-$ and i such that $M \simeq P(\mu)^{(i)}$.

Proof. Since a direct summand of a projective module is also projective, the first statement follows from Proposition 1.16. Since M is indecomposable, by Corollary 1.3, there exists $P(\mu) \twoheadrightarrow M$. Since M is projective, M is a direct summand of $P(\mu)$. Hence, the second assertion follows. □

Proposition 1.18 *1. $P(\mu)$ has a VCS.*
2. For any $\lambda, \mu \in \mathcal{P}_{\Lambda}^{-}$, we have

$$[P(\lambda) : M(\mu)] = \dim \operatorname{Hom}_{\mathfrak{g}}(P(\lambda), M(\mu)).$$

Proof. Let $\{w_i | 1 \leq i \leq l\}$ be a basis of $W(\lambda)$. We may assume that w_i is a weight vector of weight λ_i and

$$\lambda_i - \lambda_j \in Q_+ \Rightarrow i \geq j$$

holds for any i and j. Hence, by setting

$$F_k W(\lambda) := \bigoplus_{i \geq k} \mathbb{K} w_i,$$

$\{F_k W(\lambda) | 1 \leq k \leq l+1\}$ defines a decreasing filtration of $\mathfrak{g}^{\geq}$-modules. We set

$$F_k P(\lambda) := U(\mathfrak{g}) \otimes_{U(\mathfrak{g}^{\geq})} F_k W(\lambda).$$

Then, it induces a VCS of $P(\lambda)$. Indeed, we have

$$F_k P(\lambda)/F_{k+1} P(\lambda) \simeq U(\mathfrak{g}) \otimes_{U(\mathfrak{g}^{\geq})} \{F_k W(\lambda)/F_{k+1} W(\lambda)\} \simeq M(\lambda).$$

We show the next statement. By construction, we have

$$[P(\lambda) : M(\mu)] = \dim W(\lambda)_{\mu}.$$

On the other hand, Proposition 1.17 implies that

$$\dim \operatorname{Hom}_{\mathfrak{g}}(P(\lambda), M(\mu)) = \dim M(\mu)_{\lambda}.$$

Since $\dim W(\lambda)_{\mu} = \dim M^{-}(\lambda)_{\mu} = \dim M(\mu)_{\lambda}$, the second statement holds. □

Hence, we have

Corollary 1.4 *Any finitely generated indecomposable projective $\mathfrak{g}$-module in $\mathcal{C}(\Lambda)$ has a VCS.*

1.4.4 Indecomposable Projective Objects

In this subsection, we show the existence and the uniqueness of a projective cover $I(\mu)$ of $L(\mu)$. By definition $I(\mu)$ is projective and indecomposable, and there exists a surjection $I(\mu) \twoheadrightarrow L(\mu)$.

We start with the next lemma.

Lemma 1.17. *For any finitely generated $\mathfrak{g}$-module M which is indecomposable projective in $\mathcal{C}(\Lambda)$, there uniquely exists a maximal proper submodule N of M.*

Proof. By Lemma 1.15, the existence of a maximal proper submodule N of M follows. Hence, for any two maximal proper submodules N_1 and N_2, we show that $N_1 = N_2$. Suppose that $N_1 \neq N_2$. By their maximality, we see that $N_1 + N_2 = M$. Hence, there exists a surjection

$$\phi : N_1 \oplus N_2 \longrightarrow M \quad ((n_1, n_2) \mapsto n_1 + n_2).$$

Since M is projective, there exists $\psi : M \to N_1 \oplus N_2$ such that $\phi \circ \psi = \mathrm{id}$. For $i = 1, 2$, let $\pi_i : N_1 \oplus N_2 \to N_i$ be the canonical projection, and let $\iota_i : N_i \hookrightarrow M$ be the inclusion. We set $\psi_i := \iota_i \circ \pi_i \circ \psi : M \to N_i \hookrightarrow M$. By definition, $\psi_1 + \psi_2 = \mathrm{id}$. This implies that $\psi_1 \circ \psi_2 = \psi_2 \circ \psi_1$.

In the sequel, we show that both ψ_1 and ψ_2 are nilpotent. We set

$$I_1 := \bigcap_{k \in \mathbb{Z}_{>0}} \mathrm{Im}\psi_1^k, \quad K_1 := \bigcup_{k \in \mathbb{Z}_{>0}} \mathrm{Ker}\psi_1^k.$$

We show that

$$I_1 \oplus K_1 = M. \tag{1.32}$$

By definition, $I_1 \cap K_1 = \{0\}$. Moreover, since ψ_1 preserves each weight subspace, for each $\nu \in \mathcal{P}(M)$, there exists $k \in \mathbb{Z}_{>0}$ such that

$$I_1 \cap M_\nu = (\mathrm{Im}\psi_1^k) \cap M_\nu, \quad K_1 \cap M_\nu = (\mathrm{Ker}\psi_1^k) \cap M_\nu.$$

Since we have

$$\dim\{(\mathrm{Im}\psi_1^k) \cap M_\nu\} + \dim\{(\mathrm{Ker}\psi_1^k) \cap M_\nu\} = \dim M_\nu,$$

(1.32) holds. Then, since M is indecomposable and $K_1 \neq \{0\}$ by assumption, we have $I_1 = \{0\}$. Since ψ_1 preserves each weight subspace and M is finitely generated, we see that there exists $n_1 \in \mathbb{Z}_{>0}$ such that $\psi_1^{n_1} = 0$, i.e., ψ_1 is nilpotent.

In the same way, one can show that so is ψ_2. Hence, $\psi_1 + \psi_2$ is nilpotent. This is a contradiction. □

Proposition 1.19 *For any $\mu \in \mathcal{P}_\Lambda^-$, there uniquely exists a $\mathfrak{g}$-module $I(\mu)$ such that*

1. *there exists $I(\mu) \twoheadrightarrow L(\mu)$,*
2. *$I(\mu)$ is finitely generated,*
3. *$I(\mu)$ is indecomposable and projective in $\mathcal{C}(\Lambda)$.*

Moreover, the set $\{I(\mu) | \mu \in \mathcal{P}_\Lambda^-\}$ exhausts the finitely generated $\mathfrak{g}$-modules which are indecomposable and projective in $\mathcal{C}(\Lambda)$.

Proof. Since $L(\mu) \in \mathrm{Ob}\mathcal{C}(\Lambda)$ is finitely generated and irreducible, by Corollary 1.3, there exists a surjective homomorphism from an indecomposable component of $P(\mu)$ to $L(\mu)$. Hence, it suffices to see that for finitely generated indecomposable $\mathfrak{g}$-modules I_1 and I_2 which are projective in $\mathcal{C}(\Lambda)$ with surjections $\pi_i : I_i \twoheadrightarrow L(\lambda)$, $I_1 \simeq I_2$ as $\mathfrak{g}$-module. Since I_1 is projective, there exists $h : I_1 \to I_2$ such that $\pi_2 \circ h = \pi_1$.

We first show that h is a surjection of $\mathfrak{g}$-modules. We assume that h is not surjective. Then, the unique maximal proper submodule of I_2 contains $\mathrm{Im}h$. On the other hand, by the above lemma, $\mathrm{Ker}\pi_2$ coincides with the maximal proper submodule. Hence, $\pi_2 \circ h = 0$, and this is a contradiction. Hence, $h : I_1 \to I_2$ is surjective. Since I_2 is projective, I_2 is a direct summand of I_1. Since I_1 is indecomposable, $I_1 \simeq I_2$.

The rest of this proposition follows from Lemma 1.16. □

Corollary 1.5 *For any $\mu \in \mathcal{P}_\Lambda^-$,*

$$P(\mu) = \bigoplus_{\lambda \leq \mu} I(\lambda)^{\oplus m_{\lambda,\mu}},$$

where $m_{\lambda,\mu} \in \mathbb{Z}_{\geq 0}$ and $m_{\mu,\mu} = 1$.

Proof. By Lemma 1.16 and Proposition 1.19, we have

$$P(\mu) = \bigoplus_{\lambda \in \mathcal{P}_\Lambda^-} I(\lambda)^{\oplus m_{\lambda,\mu}}.$$

Moreover, by Proposition 1.19,

$$m_{\lambda,\mu} = \dim \mathrm{Hom}_{\mathfrak{g}}(P(\mu), L(\lambda)).$$

Since we have $\mathrm{Hom}_{\mathfrak{g}}(P(\mu), L(\lambda)) \simeq \mathrm{Hom}_{\mathfrak{h}}(\mathbb{K}_\mu, \mathrm{Res}^{\mathfrak{g}}_{\mathfrak{h}} L(\lambda))$ by Proposition 1.17, we see that $m_{\lambda,\mu} = 0$ for $\mu \not\leq \lambda$ and $m_{\lambda,\lambda} = 1$. □

The following is a key of the proof of the duality theorem given in the next subsection.

Proposition 1.20 *For any $M \in \mathrm{Ob}\mathcal{C}(\Lambda)$ and $\mu \in \mathfrak{h}^*$, we have*

$$[M : L(\mu)] = \dim \mathrm{Hom}_{\mathfrak{g}}(I(\mu), M).$$

Proof. We take a local composition series

$$M = M_t \supset M_{t-1} \supset \cdots \supset M_1 \supset M_0 = \{0\},$$

such that there exists $J \subset \{1, 2, \cdots, t\}$ satisfying

1. if $j \in J$, then $M_j/M_{j-1} \simeq L(\mu_j)$ for some $\mu_j \geq \mu$,
2. if $j \notin J$, then $(M_j/M_{j-1})_\nu = \{0\}$ for any $\nu \geq \mu$.

Since $I(\mu)$ is projective, we see that

$$\dim \mathrm{Hom}_{\mathfrak{g}}(I(\mu), M) = \sum_j \dim \mathrm{Hom}_{\mathfrak{g}}(I(\mu), M_j/M_{j+1}).$$

Since, by the above proposition,

$$\dim \mathrm{Hom}_{\mathfrak{g}}(I(\mu), L(\lambda)) = \delta_{\mu,\lambda},$$

we have

$$\dim \mathrm{Hom}_{\mathfrak{g}}(I(\mu), M_j/M_{j+1}) = \begin{cases} 1 & \text{if } j \in J \wedge \mu_j = \mu \\ 0 & \text{otherwise} \end{cases}.$$

Hence, we have completed the proof. □

1.4.5 Duality Theorem

As an application of some results obtained in this section, we show the following main theorem of this section which was originally obtained by I. N. Bernstein, I. M. Gelfand and S. I. Gelfand [BGG1] for a finite dimensional simple Lie algebra.

Theorem 1.2 *For any μ, $\lambda \in \mathcal{P}_\Lambda^-$, we have*

$$[I(\mu) : M(\lambda)] = [M(\lambda) : L(\mu)].$$

Proof. By Proposition 1.20, it is enough to see that

$$[I(\mu) : M(\lambda)] = \dim \mathrm{Hom}_{\mathfrak{g}}(I(\mu), M(\lambda)).$$

This follows from Corollary 1.5 and Proposition 1.18. □

1.5 Bibliographical Notes and Comments

In 1948, C. Chevalley and S. Eilenberg [CE] showed that the isomorphic classes of central extensions of a Lie algebra can be parameterised by the second cohomology. For the Lie algebra of smooth vector fields on the circle $\mathrm{Vect}S^1$, I. M. Gelfand and D. B. Fuchs [GF] showed that the cohomology ring $H^*(\mathrm{Vect}S^1)$ is generated by two generators, one is of degree two and the other is of degree three. In 1980, H. Garland [Gar] proved that the kernel of the universal central extension of a Lie algebra is just the second homology group. Thus, by a more or less well-known Proposition 1.14, the kernel of the universal central extension of VectS^1 is one dimensional. It seems that the central extension of the Witt algebra over a field $\mathbb{K}$ of characteristic $p > 0$, which is defined as the derivation algebra of $\mathbb{K}[\mathbb{Z}/p\mathbb{Z}]$, was first discovered by R. E. Block [Bl] in 1966.

In 1982, A. Rocha-Caridi and N. R. Wallach [RW1] introduced a nice class of infinite dimensional Lie algebras called Q-graded Lie algebras. Some fundamental tools of representation theory of Q-graded Lie algebras were developed, e.g., in [RW1] and [DGK]. Here, we have generalised the notion of the Q-graded Lie algebra and have stated their properties which will be used in later chapters.

The reader should be careful that, in the infinite dimensional case, there are two different definitions of the cohomology group. One allows all sort of cocyles and the other allows only those of compact support. For the interested reader, we suggest to compare our treatment with the one in [DGK] and [Liu].

1.A Appendix: Proof of Propositions 1.1, 1.2 and 1.3

PROOF OF PROPOSITION 1.3. Notice that for each central extension $0 \to V \to \mathfrak{a}_1 \to \mathfrak{a} \to 0$, one may associate a 2-cocycle $F \in Z^2(\mathfrak{a}, V)$. Indeed, for $x, y \in \mathfrak{a}$, if we set

$$F(x,y) := [(x,0),(y,0)] - ([x,y],0) \in \mathfrak{a}_1,$$

then we have $F(x,y) \in V$ and F satisfies the 2-cocycle conditions.

Conversely, for each $f \in Z^2(\mathfrak{a}, V)$, one can define a central extension

$$0 \longrightarrow V \longrightarrow \mathfrak{a}_f \longrightarrow \mathfrak{a} \longrightarrow 0,$$

by

$$[(x,v),(y,w)]_f := ([x,y], f(x,y)),$$

where $x, y \in \mathfrak{a}$ and $v, w \in V$.

Let f and g be elements of $Z^2(\mathfrak{a}, V)$ such that $f - g \in B^2(\mathfrak{a}, V)$, i.e., $(f-g)(x,y) = h([x,y])$, where $h : \mathfrak{a} \to V$ is some $\mathbb{K}$-linear map. Now, we prove that the extensions defined by f and g are equivalent. Let us define $\Phi : \mathfrak{a}_f \to \mathfrak{a}_g$ by

$$\Phi((x,v)) := (x, v - h(x)).$$

It is clear that Φ is bijective. We check that Φ is a homomorphism of Lie algebras. We have

$$\begin{aligned}[\Phi((x,v)),\Phi((y,w))]_g &= [(x,v-h(x)),(y,w-h(y))]_g\\ &= ([x,y],g(x,y))\\ &= ([x,y],f(x,y)-h([x,y]))\\ &= \Phi\left(([x,y],f(x,y))\right)\\ &= \Phi([(x,v),(y,w)]_f).\end{aligned}$$

Next, we show that for $f,g \in Z^2(\mathfrak{a},V)$ such that the central extensions $\mathfrak{a}_f \to \mathfrak{a}$ and $\mathfrak{a}_g \to \mathfrak{a}$ are equivalent, we have $f-g \in B^2(\mathfrak{a},V)$. Let Φ be a homomorphism of Lie algebras such that

$$\begin{array}{ccccccccc} 0 & \longrightarrow & V & \longrightarrow & \mathfrak{a}_f & \longrightarrow & \mathfrak{a} & \longrightarrow & 0\\ & & \| & & \Phi\downarrow & & \| & & \\ 0 & \longrightarrow & V & \longrightarrow & \mathfrak{a}_g & \longrightarrow & \mathfrak{a} & \longrightarrow & 0\end{array}$$

commutes. We can express $\Phi(x,v) = (x,v-h(x))$ for some $\mathbb{K}$-linear map $h:\mathfrak{a}\to V$. Then, we have

$$\begin{aligned}\Phi([(x,v),(y,w)]_f) &= \Phi(([x,y],f(x,y)))\\ &= ([x,y],f(x,y)-h([x,y])),\\ [\Phi((x,v)),\Phi((y,w))]_g &= [(x,v-h(x)),(y,w-h(y))]_g\\ &= ([x,y],g(x,y)),\end{aligned}$$

and thus, $(f-g)(x,y) = h([x,y])$, i.e., $f-g \in B^2(\mathfrak{a},V)$. We have completed the proof. □

Proof of Proposition 1.1. Suppose that $\alpha : \mathfrak{u} \to \mathfrak{a}$ is the universal central extension. By definition, $\mathfrak{u}$ is perfect, and hence,

$$\mathfrak{a} = \alpha(\mathfrak{u}) = \alpha([\mathfrak{u},\mathfrak{u}]) = [\alpha(\mathfrak{u}),\alpha(\mathfrak{u})] = [\mathfrak{a},\mathfrak{a}].$$

Next, we suppose that $\mathfrak{a}$ is perfect. We set

$$\begin{aligned}W' &:= \bigwedge^2 \mathfrak{a} = (\mathfrak{a}\otimes\mathfrak{a})/\langle x\otimes y + y\otimes x \mid x,y\in\mathfrak{a}\rangle_{\mathbb{K}},\\ I &:= B_2(\mathfrak{a},k) = \langle x\wedge[y,z] - [x,y]\wedge z - y\wedge[x,z] \mid x,y,z\in\mathfrak{a}\rangle_{\mathbb{K}}\end{aligned}$$

and $W := W'/I$. Let $\omega : W' \to W$ be the canonical projection. By definition, $\omega \in Z^2(\mathfrak{a},W)$. We consider the central extension

$$0 \longrightarrow W \longrightarrow \mathfrak{a}_\omega \longrightarrow \mathfrak{a} \longrightarrow 0,$$

defined by ω. Using this central extension, we construct the universal central extension of $\mathfrak{a}$. Let V be an arbitrary $\mathbb{K}$-vector space and $f \in Z^2(\mathfrak{a},V)$. Since $f(x,y) = -f(y,x)$, we have a $\mathbb{K}$-linear map

$$\psi' : W \longrightarrow V \quad \text{such that} \quad \omega(x,y) \mapsto f(x,y).$$

We define $\phi' : \mathfrak{a}_\omega \to \mathfrak{a}_f$ by

$$\phi'((x,u)) := (x, \psi'(u)).$$

Then, it is clear that the diagram

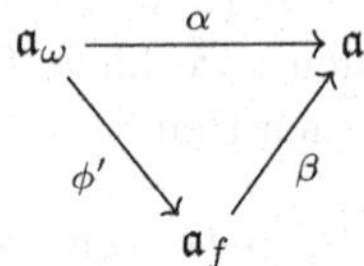

commutes. Now, let us set

$$\hat{\mathfrak{a}} := [\mathfrak{a}_\omega, \mathfrak{a}_\omega].$$

Since $\mathfrak{a}$ is perfect, it follows that $\hat{\mathfrak{a}} + W = \mathfrak{a}_\omega$. This implies that $\hat{\mathfrak{a}}$ is perfect since

$$\hat{\mathfrak{a}} = [\hat{\mathfrak{a}} + W, \hat{\mathfrak{a}} + W] = [\hat{\mathfrak{a}}, \hat{\mathfrak{a}}].$$

Furthermore, if we set

$$\mathfrak{c} := W \cap \hat{\mathfrak{a}},$$

then we have a central extension

$$0 \longrightarrow \mathfrak{c} \longrightarrow \hat{\mathfrak{a}} \longrightarrow \mathfrak{a} \longrightarrow 0$$

such that $\hat{\mathfrak{a}}$ is perfect. Now, if we define ϕ as the restriction of ϕ' to the subalgebra $\hat{\mathfrak{a}}$, then the following diagram commutes:

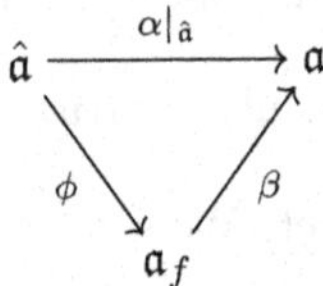

Therefore, $\hat{\mathfrak{a}} \to \mathfrak{a}$ is the universal central extension and the proof is completed. □

As a corollary, we obtain Proposition 1.2. Indeed, from the proof of Proposition 1.1, we see that

$$\mathfrak{c} = \left\{ \sum_i \alpha(x_i, y_i) \,\middle|\, \sum_i x_i \wedge y_i \in Z_2(\mathfrak{a}, k) \right\} \Big/ B_2(\mathfrak{a}, k).$$

This leads to Proposition 1.2.

1.B Appendix: Alternative Proof of Proposition 1.14

In § 1.3, we introduced Lie algebra homology and cohomology as derived functors. On the other hand, one can define them by means of the standard (co)complex (see § A.3). Here, using the second definition of Lie algebra (co)homology and the Koszul complex, we give an alternative proof of Proposition 1.14.

Here, we suppose that $M \in \mathrm{Ob}(\mathcal{O})$. It is enough to prove that

$$H^n(\mathfrak{g}^+, M^c) \simeq H_n(\mathfrak{g}^-, M)^{\sharp +}. \tag{1.33}$$

First, we introduce the following three complexes C_1, C_2 and C_3. Let

$$C_1 : \cdots \to \Lambda^n \mathfrak{g}^- \otimes M \xrightarrow{\partial_n} \Lambda^{n-1}\mathfrak{g}^- \otimes M \to \cdots \to \Lambda^0 \mathfrak{g}^- \otimes M \to 0$$

be the standard complex of the $\mathfrak{g}^-$-module M. By taking the contragredient dual of C_1, we obtain

$$C_2 : \cdots \leftarrow (\Lambda^{n+1}\mathfrak{g}^- \otimes M)^{\sharp +} \xleftarrow{d^n} (\Lambda^n \mathfrak{g}^- \otimes M)^{\sharp +} \leftarrow \cdots \leftarrow (\Lambda^0 \mathfrak{g}^- \otimes M)^{\sharp +} \leftarrow 0,$$

where $d^n(f) = f \circ \partial_{n+1}$. Let

$$C_3 : \cdots \leftarrow \mathrm{Hom}_{\mathbb{K}}(\Lambda^{n+1}\mathfrak{g}^+, M^c) \xleftarrow{\partial^n} \mathrm{Hom}_{\mathbb{K}}(\Lambda^n \mathfrak{g}^+, M^c) \leftarrow$$
$$\cdots \leftarrow \mathrm{Hom}_{\mathbb{K}}(\Lambda^0 \mathfrak{g}^+, M^c) \leftarrow 0$$

be the cocomplex defined in § A.3.

By definition, we have

$$\begin{matrix} H_n(\mathfrak{g}^-, M) = H^n(C_1) \\ H^n(\mathfrak{g}^+, M^c) = H^n(C_3) \end{matrix} \quad (\forall n \in \mathbb{Z}).$$

To show (1.33), we first prove the lemma below:

Lemma 1.18.

$$H^n(C_3) \simeq H^n(C_2) \quad (\forall n \in \mathbb{Z}).$$

Proof. We define $\Phi_n : \mathrm{Hom}_{\mathbb{K}}(\Lambda^n \mathfrak{g}^+, M^c) \to (\Lambda^n \mathfrak{g}^- \otimes M)^{\sharp +}$ by

$$\Phi_n(f)(x_1 \wedge \cdots \wedge x_n \otimes m) := f(\omega(x_1) \wedge \cdots \wedge \omega(x_n))(m),$$

where $f \in \mathrm{Hom}_{\mathbb{K}}(\Lambda^n \mathfrak{g}^+, M^c)$, $x_i \in \mathfrak{g}^-$ $(1 \leq i \leq n)$, $m \in M$ and $\omega := -\sigma : \mathfrak{g} \to \mathfrak{g}$ (an involution of $\mathfrak{g}$).

If $f \in \mathrm{Ker}\Phi_n$, then $f(\omega(x_1) \wedge \cdots \wedge \omega(x_n))(m) = 0$ for any $x_i \in \mathfrak{g}^-$ and $m \in M$. Hence, $f(\omega(x_1) \wedge \cdots \wedge \omega(x_n)) = 0$ for any $x_i \in \mathfrak{g}^-$. This means that $f = 0$, and thus Φ_n is injective.

For $g \in (\Lambda^n \mathfrak{g}^- \otimes M)^{\sharp +}$, we define $f_g \in \mathrm{Hom}_{\mathbb{K}}(\Lambda^n \mathfrak{g}^+, M^c)$ by

$$f_g(y_1 \wedge \cdots \wedge y_n)(m) := g(\omega(y_1) \wedge \cdots \wedge \omega(y_n) \otimes m) \quad (y_i \in \mathfrak{g}^+,\ m \in M).$$

By definition, $\Phi_n(f_g) = g$. Hence, Φ_n is surjective. Moreover, one can directly check that $\Phi_{n+1} \circ \partial^n = d^n \circ \Phi_n$. Hence, we have proved the lemma. □

On the other hand, since $(\cdot)^{\sharp+}$ is contravariant exact, the following isomorphism holds:

$$H^n(C_2) \simeq H_n(C_1)^{\sharp+} \quad (\forall n \in \mathbb{Z}).$$

Hence, we have proved the isomorphism (1.33).

Chapter 2
Classification of Harish-Chandra Modules

In this chapter, we will prove a theorem of O. Mathieu [Mat2] saying that any simple $\mathbb{Z}$-graded Vir-module of finite type is either a highest weight module, a lowest weight module, or a simple subquotient of the module of type $V_{a,b}$ introduced in Chapter 1. (See Theorem 2.1, for detail.) This was a conjecture of V. G. Kac [Kac3].

First, we will classify irreducible modules in the case of positive characteristic, and will prove the results in the characteristic zero case by the semi-continuity principle.

In Section 2.1, we will recall some basic notion, and will state the main results in a precise form. The rest of the sections are devoted to the proof of the main results. In Section 2.2, we will recall basic facts about the 'partial Lie algebras' and their 'modules' with detailed proof. In Section 2.3, we will prove some facts about $\mathbb{Z}$-graded Lie algebras, and will prove that the dimensions of any simple $\mathbb{Z}$-graded Vir-module without highest nor lowest degree are uniformly bounded. In Section 2.4, we will study representations of Lie p-algebras $W(m)$, quotients of the Witt algebra in characteristic $p \neq 2, 3$. Finally, in Section 2.5, after recalling some facts about Dedekind rings, we will prove the main theorem.

Through this chapter, an associative algebra is not necessarily unital. When an algebra has to be unital, we always indicate it.

2.1 Main Result

2.1.1 Notations and Conventions

Let $\mathbb{K}$ be a field. For an abelian group G, we say that a Lie algebra $\mathfrak{g} = \bigoplus_{\pi \in G} \mathfrak{g}_\pi$ over $\mathbb{K}$ is **G-graded** if it satisfies

$$[\mathfrak{g}_\pi, \mathfrak{g}_{\pi'}] \subset \mathfrak{g}_{\pi+\pi'} \quad (\forall \pi, \pi' \in G).$$

K. Iohara, Y. Koga, *Representation Theory of the Virasoro Algebra*, Springer Monographs in Mathematics, DOI 10.1007/978-0-85729-160-8_2,

Remark that, in this chapter, the condition $\dim \mathfrak{g}_\pi < \infty$ is not necessarily assumed. Hence, a G-graded Lie algebra with $G = Q$ does not mean a Q-graded Lie algebra in Definition 1.6. A module $M = \bigoplus_{\pi \in G} M_\pi$ over a G-graded Lie algebra $\mathfrak{g}$ is called **G-graded** if $\mathfrak{g}_\pi . M_{\pi'} \subset M_{\pi+\pi'}$ for any $\pi, \pi' \in G$. $\mathfrak{g}$ (resp. M) is said to be **finite** if $\dim \mathfrak{g}_\pi < \infty$ (resp. $\dim M_\pi < \infty$) for any $\pi \in G$.

Definition 2.1 *Let $\mathfrak{g} = \bigoplus_{\pi \in G} \mathfrak{g}_\pi$ be a G-graded Lie algebra, and let $M = \bigoplus_{\pi \in G} M_\pi$ be a G-graded $\mathfrak{g}$-module.*

1. *M is called a **simple G-graded $\mathfrak{g}$-module** if M has no non-trivial G-graded submodule.*
2. *M is called a **G-graded simple $\mathfrak{g}$-module** if M has no non-trivial submodule.*

For simplicity, we often omit G in the terminology.

In this chapter, we mainly deal with the cases $G = \mathbb{Z}$ and $G = \mathbb{Z}/N\mathbb{Z}$. Here, we introduce some notations for $\mathbb{Z}$-graded Lie algebras. For a $\mathbb{Z}$-graded $\mathbb{K}$-vector space $M = \bigoplus_{n \in \mathbb{Z}} M_n$ and an integer a, we set

$$M^{\geq a} := \bigoplus_{n \geq a} M_n, \quad M^{\leq a} := \bigoplus_{n \leq a} M_n.$$

For simplicity, we denote $M^{\geq 1}$, $M^{\leq 1}$, $M^{\geq 0}$ and $M^{\leq 0}$ by M^+, M^-, $M^{\geq}$ and $M^{\leq}$ respectively. A $\mathbb{Z}$-graded Lie algebra $\mathfrak{g} = \bigoplus_{n \in \mathbb{Z}} \mathfrak{g}_n$ has a triangular decomposition $\mathfrak{g} = \mathfrak{g}^- \oplus \mathfrak{g}_0 \oplus \mathfrak{g}^+$.

2.1.2 Definitions

Let $\mathbb{K}$ be a field of characteristic $p \neq 2, 3$. Similarly to the characteristic zero case, the Virasoro algebra over $\mathbb{K}$ is, by definition,

$$\mathrm{Vir}_{\mathbb{K}} := \bigoplus_{n \in \mathbb{Z}} \mathbb{K} L_n \oplus \mathbb{K} C$$

as vector space satisfying

$$[L_m, L_n] = (m-n) L_{m+n} + \frac{1}{12}(m^3 - m)\delta_{m+n,0} C,$$
$$[C, \mathrm{Vir}_{\mathbb{K}}] = \{0\}.$$

We set $\mathfrak{h} := \mathbb{K} L_0 \oplus \mathbb{K} C$.

We recall the definition of Harish-Chandra modules over $\mathrm{Vir}_{\mathbb{K}}$. In this definition, we suppose that the characteristic of $\mathbb{K}$ is zero.

Definition 2.2 *Let M be an absolutely simple module over* $\mathrm{Vir}_{\mathbb{K}}$. *$M$ is called a Harish-Chandra module over* $\mathrm{Vir}_{\mathbb{K}}$ *if M is $\mathfrak{h}$-diagonalisable and any weight subspaces are finite dimensional.*

To state the classification theorem of Harish-Chandra modules over $\mathrm{Vir}_{\mathbb{K}}$, we recall the intermediate series of the Virasoro algebra. In the sequel, we regard $\mathbb{Z}/p\mathbb{Z} \subset \mathbb{K}$. For $n \in \mathbb{Z}$, we often regard n as an element of $\mathbb{Z}/p\mathbb{Z} \subset \mathbb{K}$ via the canonical map $\mathbb{Z} \to \mathbb{Z}/p\mathbb{Z}$. For $a, b \in \mathbb{K}$, let

$$V_{a,b} = \bigoplus_{n \in \mathbb{Z}} \mathbb{K} v_n$$

be a $\mathbb{Z}$-graded $\mathrm{Vir}_{\mathbb{K}}$-module defined by

$$\begin{aligned} L_s.v_n &= (as + b - n)v_{n+s}, \\ C.v_n &= 0. \end{aligned}$$

Proposition 2.1 *1. If $a \neq 0, -1$ or $b \notin \mathbb{Z}/p\mathbb{Z}$, then $V_{a,b}$ is simple graded.*
2. If $a = 0$ and $b \in \mathbb{Z}/p\mathbb{Z}$, then there exists a submodule V of $V_{a,b}$ such that the quotient module $V_{a,b}/V$ is simple graded.
3. If $a = -1$ and $b \in \mathbb{Z}/p\mathbb{Z}$, then there exists a simple graded submodule V of $V_{a,b}$.

Proof. Suppose that $V_{a,b}$ is not simple graded, i.e., there exists a non-trivial proper graded submodule $M = \bigoplus_{n \in \mathbb{Z}} M_n$ of $V_{a,b}$, where $M_n \subset \mathbb{K} v_n$. Since each graded subspace of $V_{a,b}$ is one-dimensional, there exists $u \in \mathbb{Z}$ such that

$$M_u \neq \{0\} \text{ and } \{\ M_{u+1} = \{0\} \text{ or } M_{u-1} = \{0\}\ \}.$$

In this proof, we only consider the case $M_{u-1} = \{0\}$, since the other case can be similarly treated.

Notice that

$$L_{-1}.v_u = (-a + b - u)v_{u-1} \in M_{u-1} = \{0\}.$$

Hence, $-a + b - u = 0$, and thus

$$L_s.v_u = (s+1)av_{u+s}. \tag{2.1}$$

We consider the following cases.

<u>$a = 0$</u> Since $-a + b - u = 0$ in $\mathbb{K}$, $b = u \in \mathbb{Z}/p\mathbb{Z}$. In this case, we have

$$L_{s'}.v_n = (u - n)v_{n+s'}. \tag{2.2}$$

Hence,

$$V := \bigoplus_{\substack{n \in \mathbb{Z} \\ n = b \text{ in } \mathbb{Z}/p\mathbb{Z}}} \mathbb{K} v_n$$

is a direct sum of trivial $\mathrm{Vir}_{\mathbb{K}}$-modules. By (2.2), we see that $V_{a,b}/V$ is simple graded.

$\underline{a \neq 0}$ By (2.1), if $s \not\equiv -1 \pmod p$, then $v_{u+s} \in M$. Let us check whether $v_{u+s} \in M$ or not for $s \equiv -1 \pmod p$. Since $-a+b-u=0$ in $\mathbb{K}$, we have

$$L_{s'}.v_n = \{a(s'+1)+(u-n)\}v_{n+s'}. \tag{2.3}$$

In particular, if $s'+1 \equiv u-n \pmod p$, then

$$L_{s'}.v_n = (a+1)(u-n)v_{n+s'}. \tag{2.4}$$

$\underline{a \neq -1}$ One can find integers $n, s' \in \mathbb{Z}$ such that $n \not\equiv u \pmod p$ and $n+s' \equiv u-1 \pmod p$. Hence, $v_n \in M$ for any $n \in \mathbb{Z}$. This is a contradiction, since M is a proper submodule. Hence, $V_{a,b}$ is simple graded.
$\underline{a=-1}$ By (2.3) and (2.4),

$$V := \bigoplus_{\substack{n\in\mathbb{Z} \\ n\neq b \text{ in } \mathbb{Z}/p\mathbb{Z}}} \mathbb{K}v_n.$$

is a simple graded submodule of $V_{a,b}$. □

For each $a, b \in \mathbb{K}$, we set

$$V'_{a,b} := \begin{cases} V_{a,b} & (a \neq 0, -1 \wedge b \notin \mathbb{Z}/p\mathbb{Z}) \\ V_{a,b}/V & (a = 0 \wedge b \in \mathbb{Z}/p\mathbb{Z}) \\ V & (a = -1 \wedge b \in \mathbb{Z}/p\mathbb{Z}) \end{cases} \tag{2.5}$$

where $V_{a,b}$ and V are as in the above proposition.

Definition 2.3 *The irreducible representations $V'_{a,b}$ $(a, b \in \mathbb{K})$ over* $\mathrm{Vir}_{\mathbb{K}}$ *are called the* ***intermediate series****.*

The following is the main result of this chapter.

Theorem 2.1 *Let V be a Harish-Chandra module over* $\mathrm{Vir}_{\mathbb{K}}$*, where the base field $\mathbb{K}$ is an algebraically closed field of characteristic zero. Then, V is isomorphic to an irreducible highest weight module, an irreducible lowest weight module or one of the intermediate series.*

2.2 Partial Lie Algebras

A partial Lie algebra introduced in [Mat3] plays an essential role in the proof of Theorem 2.1. In this section, we recall its definition and state fundamental properties.

2.2.1 Definition and Main Theorems

First, we introduce the notion of partial Lie algebras and their modules. Let (d, e) be a pair of integers such that $d \le 0 \le e$. Let

$$\Gamma := \bigoplus_{d \le i \le e} \Gamma_i$$

be a graded $\mathbb{K}$-vector space. Throughout this section, we always assume that Γ is finite dimensional.

Definition 2.4 *We say that Γ is a* **partial Lie algebra** *of size (d, e), if there exists a bilinear map*

$$[\cdot, \cdot] : \Gamma \times \Gamma \longrightarrow \Gamma$$

with the following properties:

1. *for i and j such that $d \le i, j, i + j \le e$,*
$$[\Gamma_i, \Gamma_j] \subset \Gamma_{i+j},$$
2. *for i and j such that $d \le i, j, i + j \le e$,*
$$[x_i, x_j] + [x_j, x_i] = 0 \quad (x_i \in \Gamma_i,\ x_j \in \Gamma_j),$$
3. *for i, j and k such that $d \le i, j, k, i + j, j + k, k + i, i + j + k \le e$,*
$$[x_i, [x_j, x_k]] + [x_k, [x_i, x_j]] + [x_j, [x_k, x_i]] = 0$$
$$(x_i \in \Gamma_i,\ x_j \in \Gamma_j,\ x_k \in \Gamma_k).$$

For a partial Lie algebra $\Gamma = \bigoplus_{d \le i \le e} \Gamma_i$, we set

$$\Gamma^- := \bigoplus_{d \le i < 0} \Gamma_i, \quad \Gamma^+ := \bigoplus_{0 < i \le e} \Gamma_i,$$

and regard them as partial Lie algebras of size $(d, -1)$ and $(1, e)$ respectively. Then, we have a triangular decomposition $\Gamma = \Gamma^- \oplus \Gamma_0 \oplus \Gamma^+$.

For a $\mathbb{Z}$-graded Lie algebra $\mathcal{L} = \bigoplus_{i \in \mathbb{Z}} \mathcal{L}_i$, we set

$$\mathrm{Par}_d^e \mathcal{L} := \bigoplus_{i=d}^{e} \mathcal{L}_i.$$

Then, $\mathrm{Par}_d^e \mathcal{L}$ is naturally equipped with a partial Lie algebra structure. We call it the **partial part** of $\mathcal{L}$ of size (d, e).

Let Γ and Γ' be partial Lie algebras of size (d, e). A linear map $\phi : \Gamma \to \Gamma'$ is called a **homomorphism** of partial Lie algebras if the following hold:

1. ϕ is a homomorphism of graded vectors spaces, i.e., for any $d \le i \le e$,

$$\phi(\Gamma_i) \subset \Gamma'_i,$$

2. for i and j such that $d \le i, j, i+j \le e$,

$$\phi([x_i, x_j]) = [\phi(x_i), \phi(x_j)] \quad (x_i \in \Gamma_i,\ x_j \in \Gamma_j).$$

A bijective homomorphism of partial Lie algebras is called an **isomorphism**.

Let I be a graded subspace of a partial Lie algebra Γ, i.e., $I = \bigoplus_{d \le i \le e} I_i$, where $I_i := I \cap \Gamma_i$. I is called a **subalgebra** (resp. an **ideal**) of Γ, if it satisfies $[I_i, I_j] \subset I_{i+j}$ (resp. $[I_i, \Gamma_j] \subset I_{i+j}$) for any i and j such that $d \le i, j, i+j \le e$.

Next, we introduce a partial module over a partial Lie algebra.

Definition 2.5 *A pair (V, ρ) of a graded vector space $V = \bigoplus_{d \le i \le e} V_i$ and a linear map $\rho : \Gamma \to \mathrm{End} V$ is called a* **partial module** *of Γ if it satisfies*

1. *for i and j such that $d \le i, j, i+j \le e$,*

$$\rho(x_i)v_j \in V_{i+j} \quad (x_i \in \Gamma_i,\ v_j \in V_j),$$

2. *for i, j and k such that $d \le i, j, k, i+j, j+k, k+i, i+j+k \le e$,*

$$(\rho(x_i)\rho(x_j) - \rho(x_j)\rho(x_i))v_k = \rho([x_i, x_j])v_k \quad (x_i \in \Gamma_i,\ x_j \in \Gamma_j,\ v_k \in V_k).$$

For a $\mathbb{Z}$-graded module $M = \bigoplus_{i \in \mathbb{Z}} M_i$ over a $\mathbb{Z}$-graded Lie algebra $\mathcal{L}$, we define the **partial part** $\mathrm{Par}_d^e M$ of M by

$$\mathrm{Par}_d^e M := \bigoplus_{d \le i \le e} M_i,$$

and regard it as a partial module over $\mathrm{Par}_d^e \mathcal{L}$ in a natural way.

Let Γ be a partial Lie algebra of size (d, e), $(V \rho)$ and (V', ρ') be partial Γ-modules. A homomorphism $\phi : (V, \rho) \to (V', \rho')$ of graded vector spaces is a **homomorphism** of partial Γ-modules if

$$\phi(x_i.v_j) = x_i.\phi(v_j) \quad (x_i \in \Gamma_i,\ v_j \in V_j),$$

for any i and j such that $d \le i, j, i+j \le e$. A bijective homomorphism of partial modules is called an **isomorphism**. A partial module is said to be **simple** if there is no non-trivial partial submodule.

The following two theorems are the main results of this section.

Theorem 2.2 *Let Γ be a partial Lie algebra of size (d, e).*

1. *There exists a unique $\mathbb{Z}$-graded Lie algebra $\mathcal{L}_{\max}(\Gamma)$ which satisfies the following:*

   a. *$\mathrm{Par}_d^e \mathcal{L}_{\max}(\Gamma) \simeq \Gamma$ as partial Lie algebra.*

b. *For any* $\mathbb{Z}$*-graded Lie algebra* $\mathfrak{g}$ *and a partial Lie algebra homomorphism* $\phi : \Gamma \to \mathrm{Par}_d^e \mathfrak{g}$*, there exists a unique Lie algebra homomorphism* $\Phi : \mathcal{L}_{\max}(\Gamma) \to \mathfrak{g}$ *whose restriction to* Γ *is* ϕ*.*

c. *As* $\mathbb{Z}$*-graded* $\mathbb{K}$*-vector space,*

$$\mathcal{L}_{\max}(\Gamma) \simeq \mathcal{L}_{\max}(\Gamma^-) \oplus \Gamma_0 \oplus \mathcal{L}_{\max}(\Gamma^+). \tag{2.6}$$

2. *There exists a unique* $\mathbb{Z}$*-graded Lie algebra* $\mathcal{L}_{\min}(\Gamma)$ *which satisfies the following:*

   a. $\mathrm{Par}_d^e \mathcal{L}_{\min}(\Gamma) \simeq \Gamma$ *as partial Lie algebra.*

   b. *For any* $\mathbb{Z}$*-graded Lie algebra* $\mathfrak{g}$ *such that* $\mathfrak{g}$ *is generated by* $\mathrm{Par}_d^e \mathfrak{g}$ *and a surjective homomorphism* $\psi : \mathrm{Par}_d^e \mathfrak{g} \twoheadrightarrow \Gamma$*, there exists a unique homomorphism* $\Psi : \mathfrak{g} \to \mathcal{L}_{\min}(\Gamma)$ *whose restriction to* $\mathrm{Par}_d^e \mathfrak{g}$ *is* ψ*.*

Theorem 2.3 *Let* Γ *be a partial Lie algebra of size* (d, e)*, and let* V *be a partial module over* Γ*. Let* $\mathcal{L} := \mathcal{L}_{\max}(\Gamma)$ *be the* $\mathbb{Z}$*-graded Lie algebra given by Theorem 2.2. Then, we have*

1. *There exists a unique* $\mathbb{Z}$*-graded* $\mathcal{L}$*-module* $M_{\max}(V)$ *such that*

   a. $\mathrm{Par}_d^e M_{\max}(V) \simeq V$ *as partial* Γ*-module.*

   b. *For any* $\mathbb{Z}$*-graded* $\mathcal{L}$*-module* M *and a homomorphism* $\phi : V \to \mathrm{Par}_d^e M$*, there uniquely exists a homomorphism of* $\mathcal{L}$*-modules* $\Phi : M_{\max}(V) \to M$ *such that its restriction to the partial part coincides with* ϕ*.*

2. *There exists a unique* $\mathbb{Z}$*-graded* $\mathcal{L}$*-module* $M_{\min}(V)$ *such that*

   a. $\mathrm{Par}_d^e M_{\min}(V) \simeq V$ *as partial* Γ*-module.*

   b. *For any* $\mathbb{Z}$*-graded* $\mathcal{L}$*-module* M *generated by its partial part as* $\mathcal{L}$*-module and a surjective homomorphism* $\psi : \mathrm{Par}_d^e M \to V$*, there uniquely exists a homomorphism of* $\mathcal{L}$*-modules* $\Psi : M \to M_{\min}(V)$ *such that its restriction to the partial part coincides with* ψ*.*

We prove these theorems in the following subsections.

2.2.2 *Proof of Theorem 2.2*

In this section, we construct $\mathcal{L}_{\max}(\Gamma)$ as a quotient of the free Lie algebra on Γ by the ideal generated by the relations of Γ. Moreover, we show that $\mathcal{L}_{\min}(\Gamma)$ can be realised as a certain quotient of $\mathcal{L}_{\max}(\Gamma)$. Remark that we essentially follow arguments used to construct Kac–Moody algebras in Chapter 1 of [Kac4].

2.2.2.1 Notation

We first introduce some notation for free Lie algebras. Let V be a vector space over $\mathbb{K}$, and let

$$\mathcal{T}(V) = \bigoplus_{n \in \mathbb{Z}_{\geq 0}} \mathcal{T}^n(V) \qquad \mathcal{T}^n(V) := \begin{cases} V \otimes V \otimes \cdots \otimes V \ (n \text{ copies}) & n \geq 1 \\ \mathbb{K} & n = 0 \end{cases}$$

be the tensor algebra on V. We regard the associative algebra $\mathcal{T}(V)$ as a Lie algebra in the natural way. The free Lie algebra $\mathcal{F}(V)$ on V is the Lie subalgebra of $\mathcal{T}(V)$ generated by V. By definition, we have

Lemma 2.1. *Suppose that V is a module over a Lie algebra $\mathfrak{a}$. Then, $\mathcal{F}(V)$ is an $\mathfrak{a}$-submodule of $\mathcal{T}(V)$.*

In the case where $V = \bigoplus_{i \in \mathbb{Z}} V_i$ is a $\mathbb{Z}$-graded vector space, $\mathcal{F}(V)$ naturally inherits a $\mathbb{Z}$-graded Lie algebra structure. We denote this $\mathbb{Z}$-gradation by

$$\mathcal{F}(V) = \bigoplus_{i \in \mathbb{Z}} \mathcal{F}(V)_i.$$

2.2.2.2 Construction of $\mathcal{G}(\Gamma)$

In the following, let Γ be a partial Lie algebra. Here, we introduce a $\mathbb{Z}$-graded Lie algebra $\mathcal{G}(\Gamma)$ which plays a role similar to those of $\tilde{\mathfrak{g}}(A)$ in Chapter 1 of [Kac4].

Let $\mathcal{F}(\Gamma)$ be the free Lie algebra on Γ, and let $\mathcal{J}(\Gamma)$ be the ideal of $\mathcal{F}(\Gamma)$ generated by

$$\{u \otimes v - v \otimes u - [u, v]_\Gamma | u \in \Gamma_i,\ v \in \Gamma_j,\ d \leq i, j, i + j \leq e,\ ij \leq 0\}, \quad (2.7)$$

where $[\ ,\ ]_\Gamma$ denotes the partial Lie bracket on Γ. We set

$$\mathcal{G}(\Gamma) := \mathcal{F}(\Gamma)/\mathcal{J}(\Gamma). \quad (2.8)$$

By definition, $\mathcal{G}(\Gamma)$ is a $\mathbb{Z}$-graded Lie algebra. We denote its triangular decomposition by $\mathcal{G}(\Gamma) = \mathcal{G}(\Gamma)^- \oplus \mathcal{G}(\Gamma)_0 \oplus \mathcal{G}(\Gamma)^+$.

We first show the following proposition (cf. Theorem 1.2. in [Kac4]):

Proposition 2.2 *There exists an isomorphism of $\mathbb{Z}$-graded $\mathbb{K}$-vector spaces:*

$$\mathcal{G}(\Gamma) \simeq \mathcal{F}(\Gamma^-) \oplus \Gamma_0 \oplus \mathcal{F}(\Gamma^+). \quad (2.9)$$

Moreover, $\mathcal{G}(\Gamma)^\pm \simeq \mathcal{F}(\Gamma^\pm)$ and $\mathcal{G}(\Gamma)_0 \simeq \Gamma_0$ as $\mathbb{Z}$-graded Lie algebra.

To prove this proposition, we have to introduce an appropriate Lie algebra structure on the direct sum in the right-hand side of (2.9). Here, we show the following general statement.

Proposition 2.3 *Let X_0 be a Lie algebra and let $X^{\pm}$ be X_0-modules. We set $X := X^- \oplus X_0 \oplus X^+$ and*

$$\mathcal{X} := \mathcal{F}(X^-) \oplus X_0 \oplus \mathcal{F}(X^+).$$

Let $\phi : X^+ \otimes_{\mathbb{K}} X^- \to X$ be a homomorphism of X_0-modules. Then, there exists a Lie algebra structure $[\cdot,\cdot]_{\mathcal{X}}$ on $\mathcal{X}$ which satisfies the following properties:

1. *$\mathcal{F}(X^{\pm})$ and X_0 are Lie subalgebras of $\mathcal{X}$.*
2. *For any $x \in X_0$ and $u \in \mathcal{F}(X^{\pm})$, $[x,u]_{\mathcal{X}} = x.u$, where the action of X_0 on $\mathcal{F}(X^{\pm})$ in the right-hand side is given by Lemma 2.1.*
3. *For any $u \in X^+$ and $v \in X^-$, $[u,v]_{\mathcal{X}} = \phi(u \otimes v)$ holds.*

Proof. We set $X^{\geq} := X_0 \oplus X^+$ and $X^{\leq} := X^- \oplus X_0$ for simplicity, and introduce a bilinear map $\phi_X : X^{\geq} \times X^{\leq} \to X$ as follows:

$$\begin{aligned} \phi_X(u,v) &:= \phi(u \otimes v), & \phi_X(w,w') &:= [w,w']_{X_0}, \\ \phi_X(u,w) &:= -w.u, & \phi_X(w,v) &:= w.v, \end{aligned} \tag{2.10}$$

where $u \in X^+$, $v \in X^-$ and $w, w' \in X_0$ and $[\cdot,\cdot]_{X_0}$ denotes the Lie bracket on X_0. Using this notation, we introduce a $\mathbb{K}$-associative algebra $\mathcal{U}(X)$ as follows: Let $\mathcal{K}(X)$ be the two-sided ideal of the tensor algebra $\mathcal{T}(X)$ on X generated by

$$\{x \otimes y - y \otimes x - \phi_X(x,y) | x \in X^{\geq},\ y \in X^{\leq}\}.$$

We set

$$\mathcal{U}(X) := \mathcal{T}(X)/\mathcal{K}(X).$$

In the sequel, we naturally regard $\mathcal{U}(X)$ as Lie algebra and realise the Lie algebra $\mathcal{X}$ as a Lie subalgebra of $\mathcal{U}(X)$.

Let $\mathcal{D}(X)$ (resp. $\mathcal{D}(X^{\pm})$ and $\mathcal{D}(X_0)$) be the Lie subalgebra of $\mathcal{U}(X)$ generated by X (resp. $X^{\pm}$ and X_0). We denote the Lie bracket on $\mathcal{D}(X)$ by $[\cdot,\cdot]_{\mathcal{D}}$. By definition, the following commutation relations hold:

$$[x,y]_{\mathcal{D}} = \phi_X(x,y), \quad (\forall x \in X^{\geq}, \forall y \in X^{\leq}). \tag{2.11}$$

By using them, we can show that

Lemma 2.2. *There exists an isomorphism of $\mathbb{K}$-vector spaces:*

$$\mathcal{D}(X) \simeq \mathcal{D}(X^-) \oplus \mathcal{D}(X_0) \oplus \mathcal{D}(X^+).$$

Proof. For simplicity, we set $\widetilde{\mathcal{D}}(X) := \mathcal{D}(X^-) \oplus \mathcal{D}(X_0) \oplus \mathcal{D}(X^+)$. Notice that $X \subset \widetilde{\mathcal{D}}(X) \subset \mathcal{D}(X)$ and $\mathcal{D}(X)$ is generated by X. Hence, we prove that $\widetilde{\mathcal{D}}(X)$ is a Lie subalgebra of $\mathcal{D}(X)$, i.e.,

$$[X, \widetilde{\mathcal{D}}(X)]_{\mathcal{D}} \subset \widetilde{\mathcal{D}}(X).$$

By the commutation relations (2.11), $[X^{\pm}, \mathcal{D}(X_0)]_{\mathcal{D}} \subset X^{\pm}$ and $[X_0, \mathcal{D}(X^{\pm})] \subset \mathcal{D}(X^{\pm})$ hold. Hence, it suffices to show that

$$[X^+, \mathcal{D}(X^-)] \subset X^{\geq} \oplus \mathcal{D}(X^-), \quad [X^-, \mathcal{D}(X^+)] \subset X^{\leq} \oplus \mathcal{D}(X^+).$$

Here, we prove the first inclusion since the second one can be proved similarly.

By definition, $\mathcal{D}(X^-)$ is spanned by elements of the form

$$x = [x_m, [x_{m-1}, [\cdots [x_2, x_1]_{\mathcal{D}}]_{\mathcal{D}} \cdots]_{\mathcal{D}} \quad (x_i \in X^-).$$

By induction on m, one can show that

$$[y, x] \in X^{\geq} \oplus \mathcal{D}(X^-) \quad (\forall y \in X^+).$$

Hence, we have the first inclusion, and thus, the lemma holds. □

On the other hand, by an argument similar to the proof of the Poincaré–Birkhoff–Witt theorem in [Jac], we can show that

Lemma 2.3. *There exists an isomorphism of $\mathbb{K}$-vector spaces:*

$$\mathcal{U}(X) \simeq \mathcal{T}(X^-) \otimes_{\mathbb{K}} U(X_0) \otimes_{\mathbb{K}} \mathcal{T}(X^+),$$

where $U(X_0)$ denotes the universal enveloping algebra of X_0.

Proof. We can prove this lemma in a way similar to the proof of the PBW theorem in Chapter V Section 2 of [Jac]. Hence, we only indicate an outline of the proof.

Let $\{x_i | i \in I^{\pm}\}$ (resp. $\{x_i | i \in I^0\}$) be $\mathbb{K}$-bases of $X^{\pm}$ (resp. X_0), and let $\leq$ be a total order on I^0. We set $I := I^- \sqcup I^0 \sqcup I^+$. Then, $\{x_i | i \in I\}$ forms a $\mathbb{K}$-basis of X.

Let us construct a $\mathbb{K}$-basis of $\mathcal{U}(X)$ by using $\{x_i | i \in I\}$. For each $i, j \in I$, we set

$$\eta(i,j) := \begin{cases} 1 & \text{if} \begin{cases} i \in I^+ \wedge j \in I^0 \\ i \in I^+ \wedge j \in I^- \\ i, j \in I^0 \wedge i > j \\ i \in I^0 \wedge j \in I^- \end{cases} \\ 0 & \text{otherwise} \end{cases}.$$

Remark that for a sequence $(i_1, \cdots, i_n)$ consisting of elements of I,

$$\begin{aligned} &\eta(i_k, i_{k+1}) = 0 \quad (\forall k = 1, 2, \cdots n-1) \\ \Leftrightarrow &\begin{cases} i_1, \cdots, i_{s-1} \in I^- \\ i_s, \cdots, i_{t-1} \in I^0 \wedge i_s \leq \cdots \leq i_{t-1} \\ i_t, \cdots, i_n \in I^+ \end{cases} \quad (\exists s,\ t,\ 1 \leq s \leq t \leq n+1). \end{aligned} \tag{2.12}$$

One can easily check that $\mathcal{U}(X)$ is spanned by the vectors of the form

$$x_{i_1} \otimes \cdots \otimes x_{i_n} + \mathcal{K}(X) \quad (n \in \mathbb{Z}_{\geq 0},\ \eta(i_k, i_{k+1}) = 0\ (\forall k)), \tag{2.13}$$

where in the case of $n = 0$, we regard $x_{i_1} \otimes \cdots \otimes x_{i_n} = 1$. Moreover, the following lemma ensures that these vectors are linearly independent. Let $S(X_0)$ be the symmetric algebra on X_0.

Lemma 2.4. *For each $n \in \mathbb{Z}_{>0}$, there exists a linear map*

$$\sigma_n : \mathcal{T}^{\leq n}(X) \to \mathcal{T}(X^-) \otimes_{\mathbb{K}} S(X_0) \otimes_{\mathbb{K}} \mathcal{T}(X^+)$$

which satisfies the following conditions:

1. *if $\eta(i_k, i_{k+1}) = 0$ for any k, then*

$$\sigma_n(x_{i_1} \otimes \cdots \otimes x_{i_n}) = (x_{i_1} \otimes \cdots \otimes x_{i_{s-1}}) \otimes (x_{i_s} \cdots x_{i_{t-1}}) \otimes (x_{i_t} \otimes \cdots \otimes x_{i_n}),$$

where s and t are given as in (2.12).

2. *if $\eta(i_k, i_{k+1}) = 1$ for some k, then*

$$\begin{aligned} &\sigma_n(x_{i_1} \otimes \cdots \otimes x_{i_n}) - \sigma_n(x_{i_1} \otimes \cdots \otimes x_{i_{k+1}} \otimes x_{i_k} \otimes \cdots \otimes x_{i_n}) \\ &= \sigma_n(x_{i_1} \otimes \cdots \otimes \phi_X(x_{i_k}, x_{i_{k+1}}) \otimes \cdots \otimes x_{i_n}), \end{aligned}$$

where $\phi_X : X^{\geq} \times X^{\leq} \to X$ is the bilinear map defined in (2.10).

Proof. Using the Jacobi identity of X_0 and the facts that $X^{\pm}$ are X_0-modules and $\phi : X^+ \otimes X^- \to X$ is an X_0-module map, one can show this lemma by induction on n. □

The map σ_n induces a linear map

$$\begin{aligned} &\bar{\sigma}_n : (\mathcal{T}^{\leq n}(X) + \mathcal{K}(X))/\mathcal{K}(X) \longrightarrow \mathcal{T}(X^-) \otimes_{\mathbb{K}} S(X_0) \otimes_{\mathbb{K}} \mathcal{T}(X^+), \\ &\quad \bar{\sigma}_n(x_{i_1} \otimes \cdots \otimes x_{i_n} + \mathcal{K}(X)) \\ &\quad = (x_{i_1} \otimes \cdots \otimes x_{i_{s-1}}) \otimes (x_{i_s} \cdots x_{i_{t-1}}) \otimes (x_{i_t} \otimes \cdots \otimes x_{i_n}). \end{aligned}$$

Hence, the vectors (2.13) are linearly independent, and thus, Lemma 2.3 holds. □

As a corollary of this lemma, we have

$$\mathcal{D}(X^{\pm}) \simeq \mathcal{F}(X^{\pm}), \quad \mathcal{D}(X_0) \simeq X_0,$$

and hence, $\mathcal{D}(X) \simeq \mathcal{X}$ as $\mathbb{K}$-vector space. Through this isomorphism, we obtain the required Lie algebra structure on $\mathcal{X}$. Thus, we have proved Proposition 2.3. □

Next, we introduce a Lie algebra structure on the direct sum in the right-hand side of (2.9). For simplicity, we denote the direct sum by $\tilde{\mathcal{G}}(\Gamma)$. By Proposition 2.3, there exists a Lie bracket $[\cdot,\cdot]_{\tilde{\mathcal{G}}}$ on $\tilde{\mathcal{G}}(\Gamma)$ which satisfies

$$[x, y]_{\tilde{\mathcal{G}}} = [x, y]_{\Gamma} \quad (x \in \Gamma_i,\ y \in \Gamma_j) \tag{2.14}$$

for any integers i and j such that $d \leq i, j, i+j \leq e$ and $ij \leq 0$. Hence, there exists a surjective homomorphism of Lie algebras:

$$i_{\mathcal{G}} : \mathcal{G}(\Gamma) \longrightarrow \tilde{\mathcal{G}}(\Gamma). \tag{2.15}$$

Now, Proposition 2.2 follows from the next proposition:

Lemma 2.5. *The homomorphism (2.15) is bijective.*

Proof. By an argument similar to the proof of Lemma 2.2, one can show that $\mathcal{G}(\Gamma)^{\pm}$ and $\mathcal{G}(\Gamma)_0$ are generated by $\Gamma^{\pm}$ and Γ_0 respectively. Hence, the universality of the free Lie algebras $\mathcal{F}(\Gamma^{\pm})$ implies that $i_{\mathcal{G}}$ is injective. □

2.2.2.3 Construction of $\mathcal{L}(\Gamma)$

Here, we introduce a $\mathbb{Z}$-graded Lie algebra $\mathcal{L}(\Gamma)$ which is the quotient of $\mathcal{F}(\Gamma)$ by the ideal generated by the relations of Γ. We also describe the triangular decomposition of $\mathcal{L}(\Gamma)$ by using that of $\mathcal{G}(\Gamma)$, and in the next subsubsection, we check that $\mathcal{L}(\Gamma)$ is equipped with the properties required for $\mathcal{L}_{\max}(\Gamma)$.

Let $\mathcal{I}(\Gamma)$ be the ideal of $\mathcal{F}(\Gamma)$ generated by

$$\{u \otimes v - v \otimes u - [u, v]_{\Gamma} | u \in \Gamma_i,\ v \in \Gamma_j,\ d \leq i, j, i+j \leq e\}, \tag{2.16}$$

and let $\mathcal{L}(\Gamma)$ be the Lie algebra defined by

$$\mathcal{L}(\Gamma) := \mathcal{F}(\Gamma)/\mathcal{I}(\Gamma). \tag{2.17}$$

By definition, $\mathcal{J}(\Gamma) \subset \mathcal{I}(\Gamma)$, and thus, there exists a canonical projection:

$$\pi_{\mathcal{G}} : \mathcal{G}(\Gamma) \twoheadrightarrow \mathcal{L}(\Gamma).$$

Using this map, we describe the triangular decomposition of $\mathcal{L}(\Gamma)$ explicitly.

To describe $\mathrm{Ker}\pi_{\mathcal{G}}$, we introduce some notation. By replacing Γ with $\Gamma^{\pm}$, we define the ideals $\mathcal{I}(\Gamma^{\pm})$ of $\mathcal{F}(\Gamma^{\pm})$ and set $\mathcal{L}(\Gamma^{\pm}) := \mathcal{F}(\Gamma^{\pm})/\mathcal{I}(\Gamma^{\pm})$. Moreover, via the isomorphism (2.9), we regard $\mathcal{I}(\Gamma^{\pm}) \subset \mathcal{F}(\Gamma^{\pm}) \subset \mathcal{G}(\Gamma)$. Then, we have

Lemma 2.6. *$\mathcal{I}(\Gamma^{\pm})$ are ideals of the Lie algebra $\mathcal{G}(\Gamma)$.*

Proof. For simplicity, we set $\mathcal{I}^{\pm} := \mathcal{I}(\Gamma^{\pm})$. Since $\mathcal{G}(\Gamma)$ is generated by Γ, it is enough to show that $[\Gamma, \mathcal{I}^{\pm}] \subset \mathcal{I}^{\pm}$. By using commutation relations of $\mathcal{G}(\Gamma)$, $[\Gamma_0, \mathcal{I}^{\pm}] \subset \mathcal{I}^{\pm}$ holds. Hence, we show that $[\Gamma^{+}, \mathcal{I}^{-}] \subset \mathcal{I}^{-}$ and $[\Gamma^{-}, \mathcal{I}^{+}] \subset \mathcal{I}^{+}$. We prove the first inclusion, since the second one can be proved similarly.

We denote the linear span of

$$\{u \otimes v - v \otimes u - [u, v]_{\Gamma} | u \in \Gamma_i,\ v \in \Gamma_j,\ d \leq i, j, i+j \leq -1\}$$

by S^-, where $S^- \subset \mathcal{F}(\Gamma^-) \subset \mathcal{G}(\Gamma)$. Then, $\mathcal{I}^-$ is spanned by elements of the form

$$y = [y_m, [y_{m-1}, \cdots, [y_1, s] \cdots] \quad (y_i \in \mathcal{F}(\Gamma^-),\ s \in S^-).$$

Using the Jacobi identity, we have

$$[\Gamma^+, S^-] \subset S^-.$$

Moreover, one can show that

$$[x, y] \in \mathcal{I}^- \quad (\forall x \in \Gamma_k\ (1 \leq k \leq e))$$

by induction on m. Hence, the lemma holds. □

By this lemma, $\mathcal{I}(\Gamma^+) \oplus \mathcal{I}(\Gamma^-)$ is an ideal of $\mathcal{G}(\Gamma)$. Moreover, it coincides with $\mathrm{Ker}\pi_{\mathcal{G}}$, namely, the following holds:

Proposition 2.4 *The following isomorphism of $\mathbb{Z}$-graded Lie algebras holds:*

$$\mathcal{L}(\Gamma) \simeq \mathcal{G}(\Gamma)/(\mathcal{I}(\Gamma^-) \oplus \mathcal{I}(\Gamma^+)). \tag{2.18}$$

Hence, as $\mathbb{Z}$-graded vector space,

$$\mathcal{L}(\Gamma) \simeq \mathcal{L}(\Gamma^-) \oplus \Gamma_0 \oplus \mathcal{L}(\Gamma^+). \tag{2.19}$$

Proof. For simplicity, we denote the right-hand side of (2.18) by $\widetilde{\mathcal{L}}(\Gamma)$. The inclusion $\mathcal{I}(\Gamma^+) + \mathcal{I}(\Gamma^-) \subset \mathcal{I}(\Gamma)/\mathcal{J}(\Gamma)$ in $\mathcal{G}(\Gamma)$ implies that $\pi_{\mathcal{G}} : \mathcal{G}(\Gamma) \twoheadrightarrow \mathcal{L}(\Gamma)$ factors as follows:

$$\begin{array}{ccc} \mathcal{G}(\Gamma) & \xrightarrow{\pi_{\mathcal{G}}} & \mathcal{L}(\Gamma)\,. \\ & \searrow \quad \nearrow_{\bar{\pi}} & \\ & \widetilde{\mathcal{L}}(\Gamma) & \end{array}$$

On the other hand, the kernel of the composition $\mathcal{F}(\Gamma) \twoheadrightarrow \mathcal{G}(\Gamma) \twoheadrightarrow \tilde{\mathcal{L}}(\Gamma)$ of canonical projections is an ideal of $\mathcal{F}(\Gamma)$ which contains

$$\{u \otimes v - v \otimes u - [u, v]_\Gamma | u \in \Gamma_i,\ v \in \Gamma_j,\ d \leq i, j, i + j \leq e\}.$$

Hence, this composition factors as follows:

$$\begin{array}{ccccc} \mathcal{F}(\Gamma) & \longrightarrow & \mathcal{G}(\Gamma) & \longrightarrow & \widetilde{\mathcal{L}}(\Gamma)\,. \\ & \searrow & & \nearrow_{\bar{\psi}} & \\ & & \mathcal{L}(\Gamma) & & \end{array}$$

By definition, we have

$$\bar{\psi} \circ \bar{\pi}|_{\mathrm{Par}_d^e \widetilde{\mathcal{L}}(\Gamma)} = \mathrm{id}_{\mathrm{Par}_d^e \widetilde{\mathcal{L}}(\Gamma)}, \quad \bar{\pi} \circ \bar{\psi}|_{\mathrm{Par}_d^e \mathcal{L}(\Gamma)} = \mathrm{id}_{\mathrm{Par}_d^e \mathcal{L}(\Gamma)}.$$

Hence, $\bar{\pi}$ and $\bar{\psi}$ are isomorphisms, and thus, the proposition follows. □

2.2.2.4 Proof of Theorem 2.2.1

We complete the proof of the first statement of Theorem 2.2. By construction, $\mathcal{L}(\Gamma)$ enjoys the universal property required for $\mathcal{L}_{\max}(\Gamma)$. Moreover, by Proposition 2.4, the triangular decomposition (2.19) holds. Hence, it suffices to show that $\mathrm{Par}_d^e\mathcal{L}(\Gamma) \simeq \Gamma$. The following lemma is a key of the proof.

Lemma 2.7. $\mathrm{Par}_1^e\mathcal{L}(\Gamma^+) \simeq \Gamma^+$ *and* $\mathrm{Par}_d^{-1}\mathcal{L}(\Gamma^-) \simeq \Gamma^-$ *hold.*

Proof. Here, we show this lemma for Γ^+, since the case of Γ^- can be shown similarly. In this case, $\mathrm{Par}_1^e\mathcal{F}(\Gamma^+)$ is spanned by elements of the form

$$x = [x_m, [x_{m-1}, \cdots, [x_2, x_1]_{\mathcal{F}} \cdots]_{\mathcal{F}} \quad (x_i \in \Gamma_{n_i},\ 0 < n_i \leq e,\ \sum_{i=1}^m n_i \leq e), \tag{2.20}$$

where $[\cdot,\cdot]_{\mathcal{F}}$ denotes the Lie bracket on $\mathcal{F}(\Gamma^+)$. Hence, there exists a partial Lie algebra homomorphism

$$\psi : \mathrm{Par}_1^e\mathcal{F}(\Gamma^+) \longrightarrow \Gamma^+$$

which sends x of the form (2.20) to

$$[x_m, [x_{m-1}, \cdots, [x_2, x_1]_\Gamma \cdots]_\Gamma,$$

where $[\cdot,\cdot]_\Gamma$ is the partial Lie bracket on Γ^+. Moreover, by the definition of $\mathcal{I}(\Gamma^+)$,

$$\mathrm{Par}_1^e\mathcal{I}(\Gamma^+) \subset \mathrm{Ker}\psi.$$

Hence, we have a homomorphism $\bar{\psi} : \mathrm{Par}_1^e\mathcal{L}(\Gamma^+) \to \Gamma^+$.

On the other hand, there is a homomorphism of partial Lie algebra

$$\phi : \Gamma^+ \longrightarrow \mathrm{Par}_1^e\mathcal{L}(\Gamma^+); \quad x \longmapsto x + \mathrm{Par}_1^e\mathcal{I}(\Gamma^+),$$

which satisfies $\bar{\psi} \circ \phi = \mathrm{id}_{\Gamma^+}$ and $\phi \circ \bar{\psi} = \mathrm{id}_{\mathrm{Par}_1^e\mathcal{L}(\Gamma^+)}$. Thus, we have proved Lemma 2.7. □

Combining this lemma with the triangular decomposition (2.19), we obtain $\mathrm{Par}_d^e\mathcal{L}(\Gamma) \simeq \Gamma$. Now, we have completed the proof of Theorem 2.2.1.

2.2.2.5 Proof of Theorem 2.2.2

Here, we construct $\mathcal{L}_{\min}(\Gamma)$ and prove Theorem 2.2.2. Let $\mathcal{M}(\Gamma)$ be the maximal $\mathbb{Z}$-graded ideal of $\mathcal{L}(\Gamma)$ such that $\mathcal{M}(\Gamma) \cap \Gamma = \{0\}$.

We show that $\mathcal{L}(\Gamma)/\mathcal{M}(\Gamma)$ gives the Lie algebra $\mathcal{L}_{\min}(\Gamma)$. In fact, the first property

$$\mathrm{Par}_d^e(\mathcal{L}(\Gamma)/\mathcal{M}(\Gamma)) \simeq \Gamma$$

follows by definition. Hence, we show the second property (the universal property).

Let $\mathfrak{g}$ be a $\mathbb{Z}$-graded Lie algebra and let $\psi : \mathrm{Par}_d^e\mathfrak{g} \to \Gamma$ be a surjective homomorphism of partial Lie algebras. We show that there exists a homomorphism of $\mathbb{Z}$-graded Lie algebras $\mathfrak{g} \to \mathcal{L}(\Gamma)/\mathcal{M}(\Gamma)$ whose restriction to the partial part coincides with ψ.

By Theorem 2.2.1, there exist Lie algebra homomorphisms

$$\Psi_1 : \mathcal{L}(\mathrm{Par}_d^e\mathfrak{g}) \to \mathcal{L}(\Gamma) \quad \text{and} \quad \Psi_2 : \mathcal{L}(\mathrm{Par}_d^e\mathfrak{g}) \to \mathfrak{g}.$$

Note that Ψ_1 is surjective, since $\mathcal{L}(\Gamma)$ is generated by Γ. Hence, Ψ_1 maps an ideal of $\mathcal{L}(\mathrm{Par}_d^e\mathfrak{g})$ to that of $\mathcal{L}(\Gamma)$. Since $\Psi_2|_{\mathrm{Par}_d^e\mathfrak{g}} = \mathrm{id}_{\mathrm{Par}_d^e\mathfrak{g}}$, we have $\Psi_1(\mathrm{Ker}\Psi_2) \cap \Gamma = \{0\}$, and thus, $\Psi_1(\mathrm{Ker}\Psi_2) \subset \mathcal{M}(\Gamma)$. Hence, the composition $\mathcal{L}(\mathrm{Par}_d^e\mathfrak{g}) \to \mathcal{L}(\Gamma) \twoheadrightarrow \mathcal{L}(\Gamma)/\mathcal{M}(\Gamma)$ factors as

$$\begin{array}{ccccc} \mathcal{L}(\mathrm{Par}_d^e\mathfrak{g}) & \longrightarrow & \mathcal{L}(\Gamma) & \longrightarrow & \mathcal{L}(\Gamma)/\mathcal{M}(\Gamma) \\ & \searrow & & \nearrow & \\ & & \mathfrak{g} & & \end{array},$$

where the restriction of the homomorphism $\mathfrak{g} \to \mathcal{L}(\Gamma)/\mathcal{M}(\Gamma)$ to its partial part coincides with ψ.

2.2.3 Proof of Theorem 2.3

To show Theorem 2.3, we introduce an $\mathcal{L}_{\max}(\Gamma)$-module $M(V)$ associated with a partial Γ-module V and show that it enjoys the properties required for $M_{\max}(V)$.

2.2.3.1 Construction of $M(V)$

To construct $M(V)$, we first introduce the semi-direct product of partial Lie algebra and its partial module. For a partial Lie algebra Γ and its partial module V the semi-direct product $\Gamma \ltimes V$ is defined as follows: We set $\Gamma \ltimes V := \Gamma \oplus V$ (the direct sum as vector space), and define a bilinear operation $[\ ,\ ]$ on $\Gamma \ltimes V$ by

1. $[x_i, x_j] := [x_i, x_j]_\Gamma$ ($x_i \in \Gamma_i$, $x_j \in \Gamma_j$)
2. $[x_i, y_j] := x_i.y_j$ ($x_i \in \Gamma_i$, $y_j \in V_j$),
3. $[y_i, y_j] := 0$ ($y_i \in V_i$, $y_j \in V_j$),

where $[\ ,\ ]_\Gamma$ is the partial Lie bracket of Γ, and $i, j \in \mathbb{Z}$ satisfy $d \le i, j, i+j \le e$. By definition, $\Gamma \ltimes V$ is a partial Lie algebra of size (d, e). We should remind the reader that the semi-direct product of a $\mathbb{Z}$-graded Lie algebra $\mathcal{L}$ and its $\mathbb{Z}$-graded module M is defined by setting $d := -\infty$ and $e := \infty$ formally.

For simplicity, we set $\mathcal{L} := \mathcal{L}_{\max}(\Gamma)$, $\tilde{\Gamma} := \Gamma \ltimes V$ and $\tilde{\mathcal{L}} := \mathcal{L}_{\max}(\tilde{\Gamma})$. Let $\phi_1 : \tilde{\Gamma} \twoheadrightarrow \Gamma$ be the canonical projection. Theorem 2.2 implies that there uniquely exists a homomorphism of $\mathbb{Z}$-graded Lie algebras

$$\Phi_1 : \tilde{\mathcal{L}} \to \mathcal{L} \tag{2.21}$$

such that $\Phi_1|_{\mathrm{Par}_d^e \tilde{\mathcal{L}}} = \phi_1$. We set

$$\mathcal{K} := \mathrm{Ker}\Phi_1.$$

Remark that $\mathcal{K}$ is a $\mathbb{Z}$-graded Lie subalgebra of $\tilde{\mathcal{L}}$.

On the other hand, by Theorem 2.2, the inclusion map $\Gamma \hookrightarrow \tilde{\Gamma}$ induces a homomorphism of $\mathbb{Z}$-graded Lie algebras from $\mathcal{L}$ to $\tilde{\mathcal{L}}$. We regard $\mathcal{K}$ as $\mathcal{L}$-module via the homomorphism $\mathcal{L} \to \tilde{\mathcal{L}}$. Moreover, $[\mathcal{K}, \mathcal{K}]$ is an $\mathcal{L}$-submodule of $\mathcal{K}$.

We define the $\mathcal{L}$-module $M(V)$ by

$$M(V) := \mathcal{K}/[\mathcal{K}, \mathcal{K}]. \tag{2.22}$$

In the following, we check that $M(V)$ satisfies the conditions for $M_{\max}(V)$.

2.2.3.2 Proof of Theorem 2.3.1

We first show that $\mathrm{Par}_d^e M(V) \simeq V$. Since $\mathrm{Par}_d^e \mathcal{K} \simeq V$ by definition, we have to show that $\mathrm{Par}_d^e [\mathcal{K}, \mathcal{K}] \simeq \{0\}$. This fact follows from the following lemma:

Lemma 2.8.

$$[\mathcal{K}, \mathcal{K}] = [\mathcal{K}^+, \mathcal{K}^+] \oplus [\mathcal{K}^-, \mathcal{K}^-].$$

Proof. We first show the following lemma:

Lemma 2.9. *The positive part $\mathcal{K}^+$ (resp. the negative part $\mathcal{K}^-$) of $\mathcal{K}$ coincides with the ideal of $\tilde{\mathcal{L}}^+$ (resp. $\tilde{\mathcal{L}}^-$) generated by V^+ (resp. V^-).*

Proof. Let $\tilde{\mathcal{K}}^+$ (resp. $\tilde{\mathcal{K}}^-$) be the ideal of $\tilde{\mathcal{L}}^+$ (resp. $\tilde{\mathcal{L}}^-$) generated by V^+ (resp. V^-). We set $\tilde{\mathcal{K}} := \tilde{\mathcal{K}}^- \oplus V_0 \oplus \tilde{\mathcal{K}}^+$ and show that $\mathcal{K} = \tilde{\mathcal{K}}$. Since $V \subset \tilde{\mathcal{K}}$, it is enough to show that $\tilde{\mathcal{K}}$ is stable under the adjoint action of $\tilde{\Gamma}$.

By an argument similar to the proof of Lemma 2.2, one can prove that

$$[\tilde{\Gamma}, \mathcal{K}^+] \subset \mathcal{K}^+ \oplus V^{\le}, \quad [\tilde{\Gamma}, \mathcal{K}^-] \subset \mathcal{K}^- \oplus V^{\ge}. \tag{2.23}$$

Hence, $\mathcal{K} = \tilde{\mathcal{K}}$, and thus, the lemma holds. □

Second, we show the following two facts:

$$[V, \mathcal{K}^{\pm}] \subset [\mathcal{K}^{\pm}, \mathcal{K}^{\pm}], \tag{2.24}$$

$$[\tilde{\Gamma}, [\mathcal{K}^{\pm}, \mathcal{K}^{\pm}]] \subset [\mathcal{K}^{\pm}, \mathcal{K}^{\pm}]. \tag{2.25}$$

By the Jacobi identity, we have $[\tilde{\Gamma}, [\mathcal{K}^{\pm}, \mathcal{K}^{\pm}]] \subset [[\tilde{\Gamma}, \mathcal{K}^{\pm}], \mathcal{K}^{\pm}]$. Hence, the second fact follows from the first one and (2.23) and we show the first fact. Here, we prove $[V, \mathcal{K}^-] \subset [\mathcal{K}^-, \mathcal{K}^-]$. It is enough to show that $[V^{\geq}, \mathcal{K}^-] \subset [\mathcal{K}^-, \mathcal{K}^-]$.

By the above lemma, $\mathcal{K}^-$ is spanned by elements of the form

$$y := [y_m, [y_{m-1}, [\cdots, [y_1, v] \cdots] \quad (y_i \in \tilde{\Gamma}^-,\ v \in V^-). \tag{2.26}$$

Noticing this fact, by induction on m, one can show that

$$[u, y] \in [\mathcal{K}^-, \mathcal{K}^-] \quad (\forall u \in V^{\geq}),$$

and thus, (2.24) holds.

We show the lemma. Since $\mathcal{K} = \mathcal{K}^- \oplus V_0 \oplus \mathcal{K}^+$, it suffices to show

$$[\mathcal{K}^-, \mathcal{K}^+] \subset [\mathcal{K}^-, \mathcal{K}^-] \oplus [\mathcal{K}^+, \mathcal{K}^+]. \tag{2.27}$$

Suppose that $y \in \mathcal{K}^-$ is of the form (2.26) and $x \in \mathcal{K}^+$. Using the Jacobi identity, (2.23), (2.24) and (2.25), one can check

$$[x, y] \in [\mathcal{K}^-, \mathcal{K}^-] \oplus [\mathcal{K}^+, \mathcal{K}^+]$$

by induction on m. Hence, (2.27) holds. We have completed the proof. □

This lemma implies that

$$\mathrm{Par}_1^e[\mathcal{K}^+, \mathcal{K}^+] = \{0\}, \quad \mathrm{Par}_d^{-1}[\mathcal{K}^-, \mathcal{K}^-] = \{0\},$$

since $[V_i, V_j] = \{0\}$ if $d \leq i, j, i + j \leq e$. Hence, we have $\mathrm{Par}_d^e[\mathcal{K}, \mathcal{K}] \simeq \{0\}$.

Next, we state the universal property of $M(V)$. Let M be a $\mathbb{Z}$-graded $\mathcal{L}$-module. Suppose that there exists a homomorphism of partial Γ-modules $\phi : V \to \mathrm{Par}_d^e M$.

Proposition 2.5 *There exists a unique homomorphism of $\mathcal{L}$-modules $M(V) \to M$ whose restriction to the partial part V coincides with ϕ.*

Proof. Since $\mathrm{Par}_d^e(\mathcal{L} \ltimes M) = \Gamma \ltimes \mathrm{Par}_d^e M$, the following homomorphism of partial Lie algebras exists:

$$\phi_2 : \tilde{\Gamma} \longrightarrow \mathrm{Par}_d^e(\mathcal{L} \ltimes M); \quad (x, v) \longmapsto (x, \phi(v)) \quad (x \in \Gamma,\ v \in V).$$

Theorem 2.2 implies that there exists a unique homomorphism of $\mathbb{Z}$-graded Lie algebras $\Phi_2 : \tilde{\mathcal{L}} \longrightarrow \mathcal{L} \ltimes M$ such that $\Phi_2|_{\tilde{\Gamma}} = \phi_2$.

Let $\Phi_1 : \tilde{\mathcal{L}} \to \mathcal{L}$ be the homomorphism (2.21), and let $\Phi_3 : \mathcal{L} \ltimes M \twoheadrightarrow \mathcal{L}$ be the canonical projection. By definition, we have $\Phi_1|_{\tilde{\Gamma}} = \Phi_3 \circ \Phi_2|_{\tilde{\Gamma}}$, and thus, the following diagram commutes:

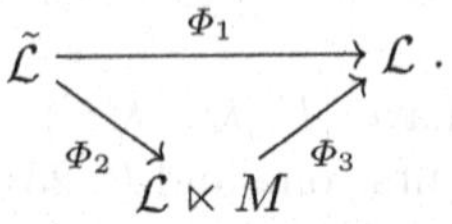

$$\begin{array}{ccc} \tilde{\mathcal{L}} & \xrightarrow{\Phi_1} & \mathcal{L}\,. \\ & {\scriptstyle \Phi_2}\searrow \quad \nearrow{\scriptstyle \Phi_3} & \\ & \mathcal{L} \ltimes M & \end{array}$$

Hence, $\Phi_2(\mathcal{K}) \subset M$ holds, since $\mathcal{K} = \mathrm{Ker}\Phi_1$ and $\mathrm{Ker}\Phi_3 = M$.

Here, we show that $\Phi_2|_{\mathcal{K}} : \mathcal{K} \to M$ is a homomorphism of $\mathcal{L}$-modules. Let $\Psi : \mathcal{L} \to \tilde{\mathcal{L}}$ be the homomorphism induced from the inclusion $\Gamma \hookrightarrow \tilde{\Gamma}$. Since $\Phi_1 \circ \Psi = \mathrm{id}_{\mathcal{L}}$, we see that $\Phi_2 \circ \Psi(x) = (x, 0) \in \mathcal{L} \ltimes M$ for $x \in \mathcal{L}$. Since the $\mathcal{L}$-module structure of $\mathcal{K}$ is given by Ψ, we have

$$\Phi_2(x.y) = \Phi_2([\Psi(x), y]) = [\Phi_2 \circ \Psi(x), \Phi_2(y)] = x.\Phi_2(y),$$

for $x \in \mathcal{L}$ and $y \in \mathcal{K}$, and thus, $\Phi_2|_{\mathcal{K}}$ is an $\mathcal{L}$-module homomorphism.

Moreover, we have $\Phi_2([\mathcal{K}, \mathcal{K}]) = \{0\}$, since M is a commutative subalgebra of $\mathcal{L} \ltimes M$. Hence, $\Phi_2|_{\mathcal{K}}$ induces an $\mathcal{L}$-module homomorphism $\Phi : M(V) \to M$. By definition, $\Phi|_V = \phi$.

Finally, we show the uniqueness of the homomorphism Φ, namely, if $\phi = 0$, then $\Phi = 0$. In fact, if $\phi = 0$, then $\Phi_2(\tilde{\mathcal{L}}) \subset \mathcal{L}$. Noticing that $\Phi_3|_{\mathcal{L}} = \mathrm{id}_{\mathcal{L}}$, we have $\mathrm{Ker}\Phi_1 = \mathrm{Ker}\Phi_2$. Hence, $\Phi_2|_{\mathcal{K}} = 0$, and thus, $\Phi = 0$. □

2.2.3.3 Proof of Theorem 2.3.2

We construct $M_{\min}(V)$ as a quotient of $M(V)$. Let $J(V)$ be the $\mathbb{Z}$-graded maximal proper submodule of $M(V)$ such that $J(V) \cap V = \{0\}$. Then, one can show that the quotient module $M(V)/J(V)$ satisfies the conditions for $M_{\min}(V)$ in Theorem 2.3.2 in a way similar to § 2.2.2.5.

2.3 $\mathbb{Z}$-graded Lie Algebras

In this section, we collect some properties of $\mathbb{Z}$-graded Lie algebras and $\mathbb{Z}$-graded modules, which are necessary for the proof of Theorem 2.1. Through this section, let $\mathfrak{g} = \bigoplus_{n \in \mathbb{Z}} \mathfrak{g}_n$ be a $\mathbb{Z}$-graded Lie algebra over $\mathbb{K}$, and let $M = \bigoplus_{n \in \mathbb{Z}} M_n$ be a $\mathbb{Z}$-graded $\mathfrak{g}$-module.

2.3.1 $\mathbb{Z}$-graded Modules

In this subsection, we give a necessary condition for which there exist highest or lowest degree of a $\mathbb{Z}$-graded module. As an application, we show that the dimensions of homogeneous component of a simple graded $\mathrm{Vir}_{\mathbb{K}}$-module without highest or lowest degree are uniformly bounded.

Lemma 2.10. *Suppose that there exist $d, e \in \mathbb{Z}$ $(d \leq 0 \leq e)$ such that $\mathfrak{g}$ is generated by its partial part $\Gamma := \mathrm{Par}_d^e \mathfrak{g}$. If there exist $a, b \in \mathbb{Z}$ $(a \leq b)$ such that M is generated by $\mathrm{Par}_a^b M$ as $\mathfrak{g}$-modules, then*

1. *for any $s \geq b$, $\mathfrak{g}^+$-module $M^{\geq s}$ is generated by $\mathrm{Par}_s^{s+e} M$,*
2. *for any $t \leq a$, $\mathfrak{g}^-$-module $M^{\leq t}$ is generated by $\mathrm{Par}_{t+d}^t M$.*

Proof. We first notice that, by Theorem 2.2, $\mathfrak{g}^{\pm}$ are generated by $\Gamma^{\pm}$ respectively. Since M is generated by $\mathrm{Par}_a^b M$, M is spanned by the elements of the form

$$x_k x_{k-1} \cdots x_1 y z m$$

where $x_i \in \Gamma^+$, $y \in U(\mathfrak{g}_0)$, $z \in U(\mathfrak{g}^-)$ and $m \in \mathrm{Par}_a^b M$ are homogeneous elements. Suppose that

$$x_k x_{k-1} \cdots x_1 y z m \in M^{\geq s}.$$

In the case $s > b$, we have $k > 0$. Since $1 \leq$ degree of $x_i \leq e$ is satisfied for each i, there exists k' $(1 \leq k' \leq k)$ such that

$$x_{k'} \cdots x_1 y z m \in \mathrm{Par}_s^{s+e} M.$$

Hence, $M^{\geq s}$ is generated by $\mathrm{Par}_s^{s+e} M$. On the other hand, in the case $s = b$, if $k = 0$, then the assertion holds by definition. If $k > 0$, then it follows as above. The other statement for $M^{\leq t}$ can be proved similarly. □

By using Lemma 2.10, we have

Proposition 2.6 *Suppose that a $\mathbb{Z}$-graded Lie algebra $\mathfrak{g}$ and a $\mathbb{Z}$-graded $\mathfrak{g}$-module M satisfy the following conditions:*

$\mathfrak{g}$ is finite (i.e., $\dim \mathfrak{g}_n < \infty$ for any $n \in \mathbb{Z}$), finitely generated, and $[\mathfrak{g}^-, \mathfrak{g}^{\geq n}] = \mathfrak{g}$ for any $n \in \mathbb{Z}_{>0}$, and
M is finite (i.e., $\dim M_n < \infty$ for any $n \in \mathbb{Z}$), simple graded, and there exist $s \in \mathbb{Z}$ and $v \in M \setminus \{0\}$ such that $\mathfrak{g}^{\geq s}.v = \{0\}$.

Then, for some $k \in \mathbb{Z}$, $M^{\geq k} = \{0\}$.

Proof. We may assume that v is a homogeneous element without loss of generality.

We define a subspace N of M by

$$N := \{w \in M | \mathfrak{g}^{\geq l}.w = \{0\} \text{ for some } l\}.$$

We first show that N is a graded submodule of M. For $w \in N$, let k be a positive integer such that $\mathfrak{g}^{\geq k}.w = \{0\}$. Since for any $g \in \mathfrak{g}_l$, we have $[\mathfrak{g}^{\geq k-l}, g] \subset \mathfrak{g}^{\geq k}$, the following holds:

$$\mathfrak{g}^{\geq k-l}.(g.w) \subset [\mathfrak{g}^{\geq k-l}, g].w + g.\mathfrak{g}^{\geq k-l}.w = \{0\}.$$

Hence, $g.w \in N$ for any $g \in \mathfrak{g}$. Moreover, N is graded by definition. Since $v \in N \neq \{0\}$ and M is simple graded, we have

$$N = M. \tag{2.28}$$

Next, we show that M^+ is a finitely generated $\mathfrak{g}^+$-module. We may assume that $M_a \neq \{0\}$ for some $a \in \mathbb{Z}_{>0}$. Since M is simple graded, M is generated by M_a. On the other hand, since $\mathfrak{g}$ is finitely generated, there exist integers d, e $(d \leq 0 \leq e)$ such that $\mathfrak{g}$ is generated by $\Gamma := \mathrm{Par}_d^e \mathfrak{g}$. By Lemma 2.10, $\mathfrak{g}^+$-module $M^{\geq a}$ is generated by $\mathrm{Par}_a^{a+e} M$. Hence, M^+ is generated by $\mathrm{Par}_1^{a+e} M$. Since M is finite, $\mathrm{Par}_1^{a+e} M$ is finite dimensional. Hence, the $\mathfrak{g}^+$-module M^+ is finitely generated.

Let X be a set of generators of the $\mathfrak{g}^+$-module M^+. We may assume that the cardinality of X is finite. Since $M^+ = N^+$ by (2.28), there exists $s \in \mathbb{Z}_{>0}$ such that

$$\mathfrak{g}^{\geq s}.X = \{0\}.$$

Hence, one can easily show that

$$\mathfrak{g}^{\geq s}.M^+ = \{0\},$$

since $\mathfrak{g}^{\geq s}$ is an ideal of $\mathfrak{g}^+$.

By the assumption that $\mathfrak{g}$ is finite and $[\mathfrak{g}^-, \mathfrak{g}^{\geq n}] = \mathfrak{g}$, we have

$$\mathfrak{g}^+ \subset \mathfrak{g}^{\geq s} + [\mathrm{Par}_{-t}^{-1}\mathfrak{g}, \mathfrak{g}^{\geq s}]$$

for some $t \in \mathbb{Z}_{>0}$. Hence, we obtain

$$\begin{aligned}
\mathfrak{g}^+.M^{\geq t} &\subset \{\mathfrak{g}^{\geq s} + [\mathrm{Par}_{-t}^{-1}\mathfrak{g}, \mathfrak{g}^{\geq s}]\}.M^{\geq t} \\
&= \mathfrak{g}^{\geq s}(\mathrm{Par}_{-t}^{-1}\mathfrak{g}).M^{\geq t} \\
&\subset \mathfrak{g}^{\geq s}.M^+ \\
&= \{0\}.
\end{aligned}$$

Taking $k \in \mathbb{Z}_{>0}$ such that $k \geq t$ and $M_k \neq \{0\}$, we see that $U(\mathfrak{g}).M_k$ is a non-zero graded submodule of M such that $U(\mathfrak{g}).M_k \subset M^{\leq k}$. Since M is simple graded, we have $M^{>k} = \{0\}$. □

As a corollary of Proposition 2.6, we have the following proposition on simple graded modules over the Virasoro algebra.

Proposition 2.7 *Suppose that the characteristic of $\mathbb{K}$ is zero. Let M be a simple graded* $\mathrm{Vir}_{\mathbb{K}}$*-module without highest or lowest degree. Then, the dimensions of the homogeneous components of M are uniformly bounded.*

To show this proposition, a preliminary lemma is necessary.

Lemma 2.11. *For each positive integer $n \in \mathbb{Z}_{>0}$, let $\mathfrak{s}_n$ be a subalgebra of* $\mathrm{Vir}_{\mathbb{K}}^{+}$ *generated by $\{L_n, L_{n+1}\}$. Then, the codimension of $\mathfrak{s}_n$ in* $\mathrm{Vir}_{\mathbb{K}}^{+}$ *is finite.*

Proof. For each positive integer m, we have

$$L_m \in \mathfrak{s}_n \text{ if } \exists \alpha, \beta \in \mathbb{Z}_{>0} \text{ such that } m = \alpha n + \beta(n+1).$$

Notice that if m satisfies $nk < m < (n+1)k$, then

$$m = \{(n+1)k - m\}n + (m - nk)(n+1).$$

Moreover, we have

$$\begin{aligned}
&\text{in the case } m = nk, \text{ if } k > n+1, \text{ then } m = (k-n-1)n + n(n+1),\\
&\text{in the case } m = (n+1)k, \text{ if } k > n, \text{ then } m = (n+1)n + (k-n)(n+1).
\end{aligned}$$

Hence, we obtain

$$L_m \in \mathfrak{s}_n$$

for any $m \in \mathbb{Z}_{>0}$ such that $m > n(n+1)$, and thus, the codimension of $\mathfrak{s}_n$ in $\mathrm{Vir}_{\mathbb{K}}^{+}$ is finite. □

Proof of Proposition 2.7. We first show that

$$\{\dim M_{-n} | n \in \mathbb{Z}_{>0}\}$$

are uniformly bounded.

By the above lemma, there exists $k \in \mathbb{Z}_{>0}$ such that $\mathrm{Vir}_{\mathbb{K}}^{\geq k} \subset \mathfrak{s}_n$. By Proposition 2.6, if M does not have highest or lowest degree, then $\mathrm{Vir}_{\mathbb{K}}^{\geq k}.v \neq \{0\}$ for any $v \in M$, and thus

$$M^{\mathfrak{s}_n} \subset M^{\mathrm{Vir}_{\mathbb{K}}^{\geq k}} = \{0\}.$$

Hence, we have

$$\mathrm{Ker}\rho(L_n) \cap \mathrm{Ker}\rho(L_{n+1}) = \{0\}, \tag{2.29}$$

where $\rho : \mathrm{Vir}_{\mathbb{K}} \to \mathrm{End}M$.

On the other hand, we have

$$\begin{aligned}
\dim M_0 &\geq \dim \mathrm{Im}\rho(L_n)|_{M_{-n}} = \dim M_{-n} - \dim \mathrm{Ker}\rho(L_n)|_{M_{-n}},\\
\dim M_1 &\geq \dim \mathrm{Im}\rho(L_{n+1})|_{M_{-n}} = \dim M_{-n} - \dim \mathrm{Ker}\rho(L_{n+1})|_{M_{-n}}
\end{aligned}$$

and thus,

$$\dim M_0 + \dim M_1 \geq 2\dim M_{-n} - \dim \mathrm{Ker}\rho(L_n)|_{M_{-n}} \\ - \dim \mathrm{Ker}\rho(L_{n+1})|_{M_{-n}}.$$

Here, (2.29) implies that

$$\begin{aligned} &\dim \mathrm{Ker}\rho(L_n)|_{M_{-n}} + \dim \mathrm{Ker}\rho(L_{n+1})|_{M_{-n}} \\ &= \dim \left(\mathrm{Ker}\rho(L_n)|_{M_{-n}} \oplus \mathrm{Ker}\rho(L_{n+1})|_{M_{-n}}\right) \\ &\leq \dim M_{-n}. \end{aligned}$$

Therefore, we see that

$$\dim M_{-n} \leq \dim M_0 + \dim M_1.$$

One can similarly check that

$$\dim M_n \leq \dim M_0 + \dim M_{-1}$$

holds for any $n \in \mathbb{Z}_{>0}$. Now, we have completed the proof. □

2.3.2 Correspondence between Simple $\mathbb{Z}$-graded Modules and Simple $\mathbb{Z}/N\mathbb{Z}$-graded Modules

Let $\mathfrak{g} = \bigoplus_{n\in\mathbb{Z}} \mathfrak{g}_n$ be a $\mathbb{Z}$-graded Lie algebra over the field $\mathbb{K}$. We first introduce some notation. For a $\mathbb{Z}$-graded vector space $V = \bigoplus_{n\in\mathbb{Z}} V_n$ and $m \in \mathbb{Z}$, we set

$$\mathrm{End}^m V := \{f \in \mathrm{End}V | f(V_n) \subset V_{n+m}\ (\forall n \in \mathbb{Z})\}.$$

In the case where V is a $\mathbb{Z}$-graded $\mathfrak{g}$-module, we further set

$$\mathrm{End}^m_{\mathfrak{g}} V := \mathrm{End}^m V \cap \mathrm{End}_{\mathfrak{g}} V.$$

For an integer N, one can naturally regard $\mathfrak{g}$ as a $\mathbb{Z}/N\mathbb{Z}$-graded Lie algebra, i.e.,

$$\mathfrak{g} = \bigoplus_{\alpha\in\mathbb{Z}/N\mathbb{Z}} \mathfrak{g}_\alpha, \quad \left(\mathfrak{g}_\alpha := \bigoplus_{\substack{n\in\mathbb{Z} \\ \alpha=n+N\mathbb{Z}}} \mathfrak{g}_n\right).$$

Let $\mathbf{M} = \bigoplus_{\alpha\in\mathbb{Z}/N\mathbb{Z}} \mathbf{M}_\alpha$ be a $\mathbb{Z}/N\mathbb{Z}$-graded $\mathfrak{g}$-module. For each $n \in \mathbb{Z}$, we set

$$\tilde{\mathbf{M}} = \bigoplus_{n\in\mathbb{Z}} \tilde{\mathbf{M}}_n, \quad \left(\tilde{\mathbf{M}}_n := \mathbf{M}_{n+N\mathbb{Z}}\right),$$

and regard $\tilde{\mathbf{M}}$ as a $\mathbb{Z}$-graded $\mathfrak{g}$-module in a natural way.

Definition 2.6 *A simple $\mathbb{Z}/N\mathbb{Z}$-graded $\mathfrak{g}$-module $\mathbf{M}$ is called* **relevant** *if $\tilde{\mathbf{M}}$ is a simple $\mathbb{Z}$-graded $\mathfrak{g}$-module.*

For $\lambda \in \mathbb{K}$, we define $\theta_\lambda \in \mathrm{End}^N \tilde{\mathbf{M}}$ by

$$\theta_\lambda(m) := \lambda m \in \tilde{\mathbf{M}}_{n+N} \quad (\forall m \in \tilde{\mathbf{M}}_n).$$

Remark 2.1 *1. $\theta_\lambda \in \mathrm{End}_{\mathfrak{g}} \tilde{\mathbf{M}}$,*
2. $\tilde{\mathbf{M}}$ is not $\mathbb{Z}$-graded simple, since $\mathrm{Im}(\mathrm{id}_{\tilde{\mathbf{M}}} - \theta_\lambda)$ is a non-trivial proper submodule for $\lambda \neq 0$.
3. $\tilde{\mathbf{M}}/\mathrm{Im}(\mathrm{id}_{\tilde{\mathbf{M}}} - \theta_{\lambda_1}) \simeq \tilde{\mathbf{M}}/\mathrm{Im}(\mathrm{id}_{\tilde{\mathbf{M}}} - \theta_{\lambda_2})$ if and only if $\lambda_1 = \lambda_2$.
4. If $\lambda = 1$, then $\mathbf{M} \simeq \tilde{\mathbf{M}}/\mathrm{Im}(\mathrm{id}_{\tilde{\mathbf{M}}} - \theta_\lambda)$.

From now on, we assume that the base field $\mathbb{K}$ is an algebraically closed field.

Proposition 2.8 *Suppose that a $\mathfrak{g}$-module $M = \bigoplus_{n\in\mathbb{Z}} M_n$ is finite simple $\mathbb{Z}$-graded and not $\mathbb{Z}$-graded simple. Then, there exists a positive integer N and an invertible homomorphism $\theta \in \mathrm{End}_{\mathfrak{g}}^N M$ such that*

$$\mathrm{End}_{\mathfrak{g}} M = \mathbb{K}[\theta, \theta^{-1}].$$

θ is called a **generating endomorphism** *of M.*

Proof. We divide the proof into two steps.

Step I: We show that $\mathrm{End}_{\mathfrak{g}}^m M \neq \{0\}$ for some non-zero integer m. For $v \in M$ such that

$$v = v_{k_1} + v_{k_2} + \cdots + v_{k_s} \quad (v_{k_i} \in M_{k_i} \setminus \{0\}),$$

where $k_i \in \mathbb{Z}$ $(1 \le i \le s)$ and $k_1 < k_2 < \cdots < k_s$, we set

$$\ell(v) := k_s - k_1.$$

Note that if v is a homogeneous vector, then $\ell(v) = 0$.

Since M is not $\mathbb{Z}$-graded simple, there exits a non-trivial proper submodule M' of M. We set

$$N_0 := \min\{\ell(v) | v \in M' \setminus \{0\}\}.$$

Since M is simple $\mathbb{Z}$-graded, M' does not contain homogeneous vectors. Hence, we see $N_0 > 0$. We fix $w \in M' \setminus \{0\}$, which attains N_0. Suppose that

$$w = w_{k_1} + w_{k_2} + \cdots + w_{k_s} \quad (w_{k_i} \in M_{k_i} \setminus \{0\}),$$

where $k_i \in \mathbb{Z}$ $(1 \le i \le s)$ and $k_1 < k_2 < \cdots < k_s$. Notice that $N_0 = k_s - k_1$. Since M is simple $\mathbb{Z}$-graded, we have $U(\mathfrak{g}).w_{k_1} = M$. Hence, we define $f \in \mathrm{End} M$ by

$$f(x.w_{k_1}) := x.w_{k_2} \quad (x \in U(\mathfrak{g})).$$

Indeed, one can check that f is well-defined as follows: It is enough to see that

$$x.w_{k_1} = y.w_{k_1} \Rightarrow x.w_{k_2} = y.w_{k_2}$$

for any $x, y \in U(\mathfrak{g})$. We may assume that x and y are homogeneous. Since

$$(x-y).w = (x-y).w_{k_2} + \cdots + (x-y).w_{k_s} \in M'$$

and $k_s - k_2 < N_0$, from the assumption on N_0, we see that $(x-y).w = 0$. Since x and y are homogeneous, we have

$$x.w_{k_2} = y.w_{k_2}.$$

Hence, f is well-defined. By definition, $f \notin \mathbb{K}\mathrm{id}_M$. Thus, we see that $\mathrm{End}_{\mathfrak{g}} M \neq \mathbb{K}\mathrm{id}_M$.

Step II: Remark that, in general,

$$\bigoplus_{m\in\mathbb{Z}} \mathrm{End}^m M \subsetneq \mathrm{End} M.$$

Nevertheless, the following lemma holds:

Lemma 2.12.

$$\mathrm{End}_{\mathfrak{g}} M = \bigoplus_{m\in\mathbb{Z}} \mathrm{End}_{\mathfrak{g}}^m M.$$

Proof. The inclusion $\supset$ is clear. We show $\subset$.

We first show that for $f \in \mathrm{End}_{\mathfrak{g}} M$,

$$f \in \bigoplus_{m\in\mathbb{Z}} \mathrm{End}^m M.$$

Any f can be expressed as

$$f = \sum_{i\in\mathbb{Z}} f_i \quad (f_i \in \mathrm{End}^i M), \tag{2.30}$$

where the sum in the right-hand side is not necessarily finite. Let us take a homogeneous vector $v \in M$ such that $f(v) \neq 0$. It follows from $f(v) \in M = \bigoplus_{i\in\mathbb{Z}} M_i$ that $f_i(v) = 0$ for all but a finite number of $i \in \mathbb{Z}$. Since $f \in \mathrm{End}_{\mathfrak{g}} M$, we have

$$f(x.v) = \sum_{i\in\mathbb{Z}} x.f_i(v). \tag{2.31}$$

Since $U(\mathfrak{g}).v = M$, we conclude that the sum (2.30) is finite.

Moreover, if $x \in U(\mathfrak{g})$ is a homogeneous element, then (2.31) implies that $f_i(x.v) = x.f_i(v)$ for any $i \in \mathbb{Z}$. Hence, $f_i \in \mathrm{End}_{\mathfrak{g}}^i M$ for any i, i.e.,

$$f \in \bigoplus_{m \in \mathbb{Z}} \mathrm{End}_{\mathfrak{g}}^m M.$$

We notice that any $f \in \mathrm{End}_{\mathfrak{g}}^m M \setminus \{0\}$ is invertible. Indeed, since $\mathrm{Ker} f$ and $\mathrm{Im} f$ are $\mathbb{Z}$-graded submodules of M, $\mathrm{Ker} f = \{0\}$ and $\mathrm{Im} f = M$.

We set

$$N := \min\{m \in \mathbb{Z}_{>0} | \mathrm{End}_{\mathfrak{g}}^m M \neq \{0\}\},$$

and fix $\theta \in \mathrm{End}_{\mathfrak{g}}^N M \setminus \{0\}$. Notice that θ is invertible by the above fact. Let us show that

$$\mathrm{End}_{\mathfrak{g}}^m M = \begin{cases} \{0\} & (m \neq kN \text{ for all } k \in \mathbb{Z}) \\ \mathbb{K}\theta^k & (m = kN \text{ for some } k \in \mathbb{Z}) \end{cases}.$$

First, we show the case $m = 0$. Suppose that $f \in \mathrm{End}_{\mathfrak{g}}^0 M$. Since M is finite, we have $\dim M_0 < \infty$. Recall that $\mathbb{K}$ is algebraically closed. Hence, there exists an eigenvalue $\lambda \in \mathbb{K}$ of $f|_{M_0}$. Since $f - \lambda \mathrm{id}_M$ is not invertible, the above fact implies $f = \lambda \mathrm{id}_M$. Thus, $\mathrm{End}_{\mathfrak{g}}^0 = \mathbb{K}\mathrm{id}_M$. The rest of the assertions follows from the minimality of N. □

We complete the proof of Proposition 2.8. □

Remark 2.2 *It follows from the proof of Proposition 2.8 that for a finite simple $\mathbb{Z}$-graded $\mathfrak{g}$-module M, if $\mathrm{End}_{\mathfrak{g}} M \neq \mathbb{K}\mathrm{id}_M$, then M is not $\mathbb{Z}$-graded simple.*

Lemma 2.13. *Let M be a finite simple $\mathbb{Z}$-graded and not $\mathbb{Z}$-graded simple $\mathfrak{g}$-module, and let θ be a generating endomorphism of M. We set*

$$\mathbf{M}_\theta := M/\mathrm{Im}(\mathrm{id}_M - \theta).$$

Then, $\mathbf{M}_\theta$ is a finite simple $\mathbb{Z}/N\mathbb{Z}$-graded relevant $\mathfrak{g}$-module.

Proof. One can check that $\mathbf{M}_\theta$ is finite and simple $\mathbb{Z}/N\mathbb{Z}$-graded. Hence, we show that $\mathbf{M}_\theta$ is relevant. Notice that $M \simeq \tilde{\mathbf{M}}_\theta$, since

$$M_n \ni x \mapsto x + \mathrm{Im}(\mathrm{id}_M - \theta) \in (\tilde{\mathbf{M}}_\theta)_n$$

gives an isomorphism of $\mathbb{Z}$-graded $\mathfrak{g}$-modules. Hence, $\tilde{\mathbf{M}}_\theta$ is simple $\mathbb{Z}$-graded and thus, $\mathbf{M}_\theta$ is relevant. □

Remark 2.3 *Let M be as above. Then, we have*

1. *for any generating endomorphisms θ_i $(i = 1, 2)$ of M (by Proposition 2.8, $\theta_1 \propto \theta_2$),*

$$\mathbf{M}_{\theta_1} \simeq \mathbf{M}_{\theta_2} \Leftrightarrow \theta_1 = \theta_2,$$

2. *for any generating endomorphisms θ of M,*

$$\tilde{\mathbf{M}}_\theta \simeq M$$

as $\mathbb{Z}$-graded $\mathfrak{g}$-module.

Let $\mathcal{M}$ be the set of the pairs (M, θ) such that M is a finite simple $\mathbb{Z}$-graded and not $\mathbb{Z}$-graded simple $\mathfrak{g}$-module, and θ is a generating endomorphism of M. We define an equivalence relation $\sim$ on $\mathcal{M}$ by

$$(M,\theta) \sim (M',\theta') \Leftrightarrow \exists f : M \to M' : \text{an isomorphism of } \mathfrak{g}\text{-modules such that}$$
$$\text{(i)}\ \ f(M_n) \subset M'_n \quad (\forall n \in \mathbb{Z}),$$
$$\text{(ii)}\ \ f \circ \theta = \theta' \circ f.$$

Set $\bar{\mathcal{M}} := \mathcal{M}/\sim$. We further denote the set of the isomorphism classes of finite simple $\mathbb{Z}/N\mathbb{Z}$-graded relevant $\mathfrak{g}$-modules by $\mathcal{N}$. Then, we have

Proposition 2.9 *There exists a bijective correspondence between the sets $\bar{\mathcal{M}}$ and $\mathcal{N}$ which sends an equivalence class represented by (M, θ) to an isomorphism class represented by $\mathbf{M}_\theta := M/\mathrm{Im}(\mathrm{id}_M - \theta)$.*

Proof. By Lemma 2.13, $\mathbf{M}_\theta$ is a finite simple $\mathbb{Z}/N\mathbb{Z}$-graded relevant $\mathfrak{g}$-module. On the other hand, one can show that the correspondence defined from the map $\mathbf{M} \mapsto (\tilde{\mathbf{M}}, \theta)$, where θ is a generating endomorphism of $\tilde{\mathbf{M}}$, gives the inverse of the correspondence by Remark 2.1. □

2.3.3 R-forms

Let R be a subring of the base field $\mathbb{K}$. In this subsection, we state a lemma on R-forms of a $\mathbb{Z}$-graded Lie algebra, a partial Lie algebra and their modules.

For a finite $\mathbb{Z}$-graded vector space $M = \bigoplus_{n\in\mathbb{Z}} M_n$, let M_R be a $\mathbb{Z}$-graded R-submodule of M such that

$$M_R = \bigoplus_{n\in\mathbb{Z}} (M_R)_n, \quad (M_R)_n := M_R \cap M_n.$$

M_R is called an R-**form** of M if

1. $M = \mathbb{K} \otimes_R M_R$,
2. for each $n \in \mathbb{Z}$, $(M_R)_n$ is a finitely generated R-submodule of M_n.

For a finite $\mathbb{Z}$-graded Lie algebra $\mathfrak{g}$, an R-form $\mathfrak{g}_R$ of $\mathfrak{g}$ as a $\mathbb{Z}$-graded vector space is called an R-form of $\mathfrak{g}$ if it is an R-Lie subalgebra of $\mathfrak{g}$. Similarly, for a finite $\mathbb{Z}$-graded $\mathfrak{g}$-module M, an R-form M_R of M is an R-form of M as a $\mathbb{Z}$-graded vector space and a $\mathfrak{g}_R$-submodule of M. For a partial Lie algebra and its partial module, their R-forms are defined similarly.

Lemma 2.14. *Let $\mathfrak{g}$ be a finite $\mathbb{Z}$-graded Lie algebra generated by its partial part $\Gamma := \mathrm{Par}^e_d\mathfrak{g}$, and let M be a finite $\mathbb{Z}$-graded $\mathfrak{g}$-module. Set $V := \mathrm{Par}^e_d M$.*

For an R-form Γ_R of Γ, and an R-form V_R of V, let $\mathfrak{g}_R$ be an R-Lie subalgebra generated by Γ_R, and let M_R be a $\mathfrak{g}_R$-submodule of M generated by V_R. Then, $\mathfrak{g}_R$ (resp. M_R) is an R-form of $\mathfrak{g}$ (resp. M) with a partial part Γ_R (resp. V_R).

Proof. For simplicity, we set $U^\pm := U(\mathfrak{g}^\pm)$. Let $U_R^\pm$ be R-subalgebras of $U^\pm$ generated by $\Gamma_R^\pm$. Then, one can show that $(U_R^\pm)_n$ is a finitely generated R-module and $\mathbb{K}\otimes_R U_R^\pm = U^\pm$. Hence, $U_R^\pm$ is an R-form of $U^\pm$. Let us introduce filtrations $\{F_n U_R^\pm | n \in \mathbb{Z}_{>0}\}$ as follows:

$$F_1 U_R^\pm := \Gamma_R^\pm \oplus R1, \quad F_n U_R^\pm := \Gamma_R^\pm F_{n-1} U_R^\pm + F_{n-1} U_R^\pm \quad (n \geq 2).$$

We set

$$\mathfrak{g}_R := U_R^-.\Gamma_R^- \oplus (\Gamma_R)_0 \oplus U_R^+.\Gamma_R^+ \subset \mathfrak{g},$$

where $U_R^\pm$ acts on $\mathfrak{g}$ via the adjoint action. By induction on n in $F_n U_R^\pm$, one can show that $\mathfrak{g}_R$ is an R-Lie subalgebra of $\mathfrak{g}$. By construction, for each $n \in \mathbb{Z}$, $(\mathfrak{g}_R)_n$ is a finitely generated R-module and $\mathbb{K} \otimes_R \mathfrak{g}_R = \mathfrak{g}$. Moreover, by definition, $\mathrm{Par}_d^e \mathfrak{g}_R = \Gamma_R$. Hence, $\mathfrak{g}_R$ is an R-form which satisfies the conditions in this lemma.

Similarly, if we set

$$M_R := U_R^-.V_R^- \oplus (V_R)_0 \oplus U_R^+.V_R^+ \subset M,$$

then M_R is the desired R-form of M. □

2.4 Lie p-algebra $W(m)$

For the classification of the Harish-Chandra modules over the Virasoro algebra, we use the representations over a Lie p-algebra $W(m)$. In this section, we estimate the dimension of irreducible representations over $W(m)$.

Through this section, let $\mathbb{K}$ be a field whose characteristic is $p > 3$, unless otherwise stated.

2.4.1 Definitions

Let m be a positive integer. We first introduce the Lie p-algebra $W(m)$. Let $\mathbb{K}[t]$ be a polynomial ring in a variable t. We set

$$\begin{aligned} W(m) &:= \mathrm{Der}\,(\mathbb{K}[t]/(t^{pm})) \\ &= \bigoplus_{i=-1}^{pm-2} \mathbb{K}e_i \qquad (e_i := -t^{i+1}\frac{d}{dt}). \end{aligned}$$

For simplicity, we set $e_i := 0$ for $i > pm - 2$. Since

$$(\mathrm{ad}e_i)^p e_j = \prod_{k=-1}^{p-2} (-ik - j)e_{pi+j} = \begin{cases} -je_{pi+j} & (i \equiv 0 \pmod p) \\ 0 & (i \not\equiv 0 \pmod p) \end{cases},$$

by Proposition B.2, $W(m)$ is a Lie p-algebra with the pth power operation given by

$$e_i^{[p]} := \begin{cases} e_{pi} & (i \equiv 0 \pmod p) \\ 0 & (i \not\equiv 0 \pmod p) \end{cases}. \tag{2.32}$$

For study of representations over $W(m)$, we introduce a p-subalgebra and ideals of $W(m)$. We set

$$B(m) := \bigoplus_{i=0}^{pm-2} \mathbb{K}e_i.$$

Then, it is a completely solvable Lie p-algebra (see Definition B.6). Indeed, by setting

$$B(m)_k := \bigoplus_{i=pm-1-k}^{pm-2} \mathbb{K}e_i$$

for $1 \leq k \leq pm - 1$, we have a chain

$$\{0\} =: B(m)_0 \subset B(m)_1 \subset B(m)_2 \subset \cdots \subset B(m)_{pm-1} = B(m) \tag{2.33}$$

of ideals of $B(m)$ such that $\dim B(m)_k = k$. Further, we have

Lemma 2.15. *$B(m)_k$ is a p-ideal of $B(m)$.*

Proof. It suffices to show that

$$\left(\sum_{j=pm-1-k}^{pm-2} c_j e_j\right)^{[p]} \in B(m)_k$$

for any $c_j \in \mathbb{K}$. By Lemma B.3, for any $x, y \in B(m)_k$,

$$s_i(x, y) \in B(m)_k \quad (i = 1, 2, \cdots, p).$$

Hence, combining this fact with (2.32), we obtain the lemma. □

Next, we set $I(m) := B(m)_p \subset W(m)$. By definition, we have

$$[e_{-1}, I(m)] \subset I(m).$$

In addition, Lemma 2.15 implies that $I(m)$ is stable under pth power operation. Hence, $I(m)$ is a p-ideal of $W(m)$. Notice that the pth power of an element of $I(m)$ for $m \geq 2$ is trivial, i.e.,

Lemma 2.16. *Suppose that $m \geq 2$. Then,*

$$x^{[p]} = 0 \quad (\forall x \in I(m)).$$

Proof. By Lemma B.3,

$$s_i(x, y) = 0 \quad (i = 1, 2, \cdots, p),$$

since $[x, y] = 0$ for any $x, y \in I(m)$. Hence, from the Lie p-algebra structure (2.32), we obtain the lemma. □

2.4.2 Preliminaries

Until the end of Lemma 2.18, the base field $\mathbb{K}$ is not necessarily of positive characteristic.

Let V be a finite dimensional vector space over $\mathbb{K}$, and let S be a subset of End$V$ closed under $[\cdot, \cdot]$, i.e., for any $x, y \in S$, $[x, y] := xy - yx \in S$. Here, let us denote by $\langle S \rangle$ the associative subalgebra of End$V$ generated by S. We first show the following theorem due to N. Jacobson.

Theorem 2.4 ([Jac] Chapter *II*) *Suppose that any elements of S are nilpotent. Then, $\mathfrak{S} := \langle S \rangle$ is nilpotent, i.e., there exists $k \in \mathbb{Z}_{>0}$ such that $\mathfrak{S}^k = \{0\}$.*

To prove this theorem, we need the following lemma:

Lemma 2.17. *Suppose that a subset T of S is closed under $[\cdot, \cdot]$. Set $\mathfrak{T} := \langle T \rangle$.*

1. If $x \in S$ satisfies $[T, x] \subset \mathfrak{T}$, then

$$x\mathfrak{T} \subset \mathfrak{T}x + \mathfrak{T}.$$

2. Suppose that $\mathfrak{T}$ is nilpotent and $\mathfrak{T} \subsetneq \mathfrak{S}$. Then, there exists $x \in S$ such that $x \notin \mathfrak{T}$ and $[T, x] \subset \mathfrak{T}$.

Proof. The first assertion follows from $[T, x] \subset \mathfrak{T}$. Hence, we show the second one.

We assume that $[T, x] \not\subset \mathfrak{T}$ for any $x \in S \setminus \mathfrak{T}$, and lead to a contradiction. Let us fix $x \in S \setminus \mathfrak{T}$. By the above assumption, there exists $t_1 \in T$ such that

$$[t_1, x] \notin \mathfrak{T}.$$

Moreover, since $[t_1, x] \in S \setminus \mathfrak{T}$, there exists $t_2 \in T$ such that

$$[t_2, [t_1, x]] \notin \mathfrak{T}.$$

Hence, for any integer i, there exists $t_i \in T$ such that

$$[t_i, [t_{i-1}, \cdots, [t_2, [t_1, x]] \cdots] \notin \mathfrak{T}.$$

On the other hand, since $\mathfrak{T}$ is nilpotent, there exists $k \in \mathbb{Z}_{>0}$ such that $\mathfrak{T}^k = \{0\}$. Then, we have

$$(\mathrm{ad}T)^{2k} = \{0\} \quad \text{in } \mathrm{End}V.$$

This is a contradiction. □

Proof of Theorem 2.4. We show this theorem by induction on $\dim V$. In the case where $\dim V = 0$ or $\mathfrak{S} = \{0\}$, the theorem follows by definition. Let us assume that $\dim V > 0$ and $\mathfrak{S} \neq \{0\}$.

We set

$$\Omega := \{S' | S' \subset S,\ [S', S'] \subset S',\ \langle S' \rangle : \text{nilpotent}\}.$$

Let T be an element of Ω such that $\dim\langle T \rangle$ is maximal. To show this theorem, it is enough to see that $\mathfrak{S} = \langle T \rangle$. For simplicity, we set $\mathfrak{T} := \langle T \rangle$.

First, we see that $\mathfrak{T} \neq \{0\}$. Indeed, for a non-zero element x of S, if we set

$$X := \langle \{x\} \rangle \cap S,$$

then X is closed under $[\cdot, \cdot]$, and we have

$$\langle X \rangle = \langle \{x\} \rangle = \sum_{i \geq 1} \mathbb{K} x^i.$$

Hence, $X \in \Omega$ and $\dim\langle X \rangle > 0$.

Second, we set $W := \mathfrak{T}.V$. Notice that $W \subsetneq V$, since $\mathfrak{T}$ is nilpotent. Here, we further set

$$S' := \{x \in S | x.W \subset W\}.$$

By definition, $T \subset S'$. We show that $S' \in \Omega$. We may regard $S' \subset \mathrm{End}W$, and consider an associative subalgebra $\mathfrak{S}_1'$ of $\mathrm{End}W$ generated by S'. We also regard $S' \subset \mathrm{End}(V/W)$, and consider an associative subalgebra $\mathfrak{S}_2'$ of $\mathrm{End}(V/W)$ generated by S'. Notice that $\dim W, \dim(V/W) < \dim V$. Hence, by induction hypothesis, both $\mathfrak{S}_1'$ and $\mathfrak{S}_2'$ are nilpotent, i.e., there exist positive integers k_1 and k_2 such that

$$(\mathfrak{S}_1')^{k_1} = \{0\} \quad \text{and} \quad (\mathfrak{S}_2')^{k_2} = \{0\}.$$

This means that, if we set $\mathfrak{S}' := \langle S' \rangle \subset \mathrm{End}V$, then

$$(\mathfrak{S}')^{k_2}.V \subset W \text{ and } (\mathfrak{S}')^{k_1}.W = \{0\}.$$

Hence, $(\mathfrak{S}')^{k_1+k_2} = \{0\}$, i.e., $\mathfrak{S}'$ is nilpotent. Since S' is closed under $[\cdot,\cdot]$ by definition, we have $S' \in \Omega$.

Third, we assume $\mathfrak{T} \subsetneq \mathfrak{S}$, and lead to a contradiction. By Lemma 2.17.2, there exists $x \in S$ such that $x \notin \mathfrak{T}$ and $[T, x] \subset \mathfrak{T}$. Hence, by Lemma 2.17.1, we get

$$x.W = x\mathfrak{T}.V \subset (\mathfrak{T}x + \mathfrak{T}).V \subset W.$$

Hence, $x \in S'$. Since $x \notin \mathfrak{T}$, we have

$$\dim \mathfrak{S}' \geq \dim \mathfrak{T} + 1.$$

This contradicts the assumption that $\dim \mathfrak{T}$ is maximal. □

Lemma 2.18. *Let $\mathfrak{a}$ be a Lie algebra over an algebraically closed field $\mathbb{K}$, and let V be an irreducible $\mathfrak{a}$-module. Suppose that for any $x \in [\mathfrak{a}, \mathfrak{a}]$, $\rho(x)$ is nilpotent, where $\rho : \mathfrak{a} \to \mathrm{End}V$. Then,*

$$\dim V = 1.$$

Proof. If we set

$$S := \{\rho(x) | x \in [\mathfrak{a}, \mathfrak{a}]\},$$

then S satisfies the conditions in Theorem 2.4. Hence, $\mathfrak{S} := \langle S \rangle$ is a nilpotent associative subalgebra of $\mathrm{End}V$. Hence,

$$W := \{v \in V | \rho([\mathfrak{a}, \mathfrak{a}]).v = \{0\}\}$$

satisfies $W \neq \{0\}$. Moreover, W is an $\mathfrak{a}$-submodule of V. Since V is irreducible, we have $V = W$. This means that V is an irreducible $(\mathfrak{a}/[\mathfrak{a}, \mathfrak{a}])$-module. Since $(\mathfrak{a}/[\mathfrak{a}, \mathfrak{a}])$ is an abelian Lie algebra and $\mathbb{K}$ is algebraically closed, we conclude that $\dim V = 1$. □

From now on, we assume that $\mathbb{K}$ is a field whose characteristic is $p > 0$ again. Let $\mathfrak{g}$ be a Lie p-algebra over $\mathbb{K}$ with a pth power operation $(\cdot)^{[p]}$. For $x \in \mathfrak{g}$, we say that x is *p-nilpotent* if there exist $k \in \mathbb{Z}_{>0}$ such that

$$x^{[p^k]} = 0,$$

where for $n \in \mathbb{Z}_{>0}$ we set

$$x^{[p^n]} := ((\cdots(x \underbrace{^{[p]})^{[p]})^{[p]} \ldots)^{[p]}}_{n \text{ times}}.$$

Lemma 2.19. *Let M (and $\rho : \mathfrak{g} \to \mathrm{End}M$) be a $\mathfrak{g}$-module with central character $\chi \in \mathfrak{g}^*$. For a p-nilpotent element $x \in \mathfrak{g}$, if*

$$\chi(x^{[p^n]}) = 0$$

for any $n \in \mathbb{Z}_{\geq 0}$, then $\rho(x)$ is nilpotent. In particular, if any $x \in \mathfrak{g}$ are p-nilpotent and M has the trivial central character, then $\rho(x)$ is nilpotent for any $x \in \mathfrak{g}$.

Proof. For $x \in \mathfrak{g}$, let k be an integer such that $x^{[p^k]} = 0$. Note that $x^{[p^{s+1}]}$ commutes with $x^{[p^s]}$, since

$$[x^{[p^{s+1}]}, x^{[p^s]}] = \mathrm{ad}(x^{[p^s]})^p(x^{[p^s]}) = 0.$$

Hence, we have

$$\begin{aligned}
&\sum_{s=0}^{k-1}\{(x^{[p^s]})^p - x^{[p^{s+1}]}\}^{p^{k-1-s}} \\
&= \sum_{s=0}^{k-1}\{(x^{[p^s]})^{p^{k-s}} - (x^{[p^{s+1}]})^{p^{k-1-s}}\} \\
&= x^{p^k} - x^{[p^k]} \\
&= x^{p^k}.
\end{aligned}$$

On the other hand, since $\chi(x^{[p^s]}) = 0$ for any $s \in \mathbb{Z}_{\geq 0}$,

$$\rho(x^{[p^s]})^p - \rho(x^{[p^{s+1}]}) = \chi(x^{[p^s]})^p = 0.$$

Hence, $\rho(x)^{p^k} = 0$, i.e., $\rho(x)$ is nilpotent. □

Lemma 2.20. *Let $\mathfrak{g}$ be a finite dimensional Lie p-algebra, $\mathfrak{h}$ be a p-subalgebra of $\mathfrak{g}$ and $\mathfrak{k}$ be a p-ideal of $\mathfrak{h}$ such that any $x \in \mathfrak{k}$ are p-nilpotent. Let M be an irreducible $\mathfrak{g}$-module with the central character $\chi \in \mathfrak{g}^*$ such that*

$$\chi(\mathfrak{k}) = \{0\}.$$

Moreover, assume that there exist $x, y \in \mathfrak{g}$ which satisfy the following conditions:

1. *$\mathfrak{g} = \mathbb{K}x \oplus \mathfrak{h}$,*
2. *$y \in \mathfrak{k}$ and $[x,y], [x,[x,y]] \in \mathfrak{h}$,*
3. *$\rho([x,y])$ is invertible where $\rho : \mathfrak{g} \to \mathrm{End}M$.*

Then, for any irreducible $\mathfrak{h}$-submodule M' of M, the following holds:

$$\dim M = p \dim M'.$$

Proof. Since M is an irreducible $\mathfrak{g}$-module, by Proposition B.4, there exists a surjective map

$$\mathrm{Ind}_{\mathfrak{h}}^{\mathfrak{g}}(M'; \chi) \twoheadrightarrow M.$$

Further, since $\dim \mathfrak{h} = \dim \mathfrak{g} - 1$, we have $\dim \mathrm{Ind}_{\mathfrak{h}}^{\mathfrak{g}}(M'; \chi) = p \dim M'$. Hence, the inequality

$$\dim M \leq p \dim M' \tag{2.34}$$

holds. Here, we assume that $\dim M < p \dim M'$ and leads to a contradiction.

We first show that

$$\mathfrak{k}.M' = \{0\}. \tag{2.35}$$

By Lemma 2.19, for any $z \in \mathfrak{k}$, $\rho(z)$ is nilpotent. Hence, by Theorem 2.4,

$$\{\rho(z)|_{M'} | z \in \mathfrak{k}\}$$

generates a nilpotent subalgebra of EndM'. This means that $\mathfrak{k}.M'$ is a proper $\mathfrak{h}$-submodule of M'. Since M' is an irreducible $\mathfrak{h}$-module, we obtain (2.35).

By the assumption $\dim M < p \dim M'$, there exist an integer n $(0 < n < p)$ and $u_0, \cdots, u_n \in M' \subset M$ such that

$$\sum_{s=0}^{n} x^s.u_s = 0. \tag{2.36}$$

We fix the minimal integer n such that (2.36) holds. Notice that

$$yx^s = x^s y + \sum_{i=1}^{s} \binom{s}{i} x^{s-i} \underbrace{[\cdots[y,x],x]\cdots x]}_{i \text{ times}}.$$

Since $[x,[x,y]] \in \mathfrak{h}$ and $\sum_{i=0}^{s} x^i M'$ is $\mathfrak{h}$-invariant for $0 \leq s < p$, we have

$$\underbrace{[\cdots[y,x],x]\cdots x]}_{s \text{ times}}.M' \subset \{(\mathrm{ad}x)^{s-2}\mathfrak{h}\}.M' \subset \sum_{i=0}^{s-2} x^i M'$$

for $s \geq 2$. Hence, there exist $u'_s \in M'$ $(0 \leq s \leq n)$ such that

$$\begin{aligned} 0 = y.(\sum_{s=0}^{n} x^s.u_s) &= x^n y u_n + x^{n-1}(y.u_{n-1} + n[y,x].u_n) + \sum_{s=0}^{n-2} x^s.u'_s \\ &= x^{n-1} n[y,x].u_n + \sum_{s=0}^{n-2} x^s.u'_s, \end{aligned}$$

since $y.u_n = y.u_{n-1} = 0$ by (2.35). It follows from the choice of n that $n[y,x].u_n = 0$. Since $[x,y]$ is invertible on M, we obtain $u_n = 0$. This contradicts the choice of n. Now, we have completed the proof. □

2.4.3 Irreducible Representations of $W(m)$ $(m \geq 2)$

In this subsection, we assume that $\mathbb{K}$ is an algebraically closed field. The main result of this subsection is the following theorem:

Theorem 2.5 *Let M be an irreducible faithful representation of $W(m)$. Then,*

$$\dim M \geq p^{\frac{1}{2}(m-1)(p-1)}.$$

For the proof, we need a preliminary lemma. First, we notice that $I(m).M \neq \{0\}$, since M is a faithful representation, and that $I(m).M$ is a submodule of M, since $I(m)$ is an ideal of $W(m)$. Hence, we see that $I(m).M = M$. This implies that there exists an irreducible $B(m)$-subquotient M' of M such that $I(m).M' \neq \{0\}$. Hence, we have

$$I(m).M' = M'. \tag{2.37}$$

To prove Theorem 2.5, here, we show

$$\dim M' \geq p^{\frac{1}{2}(m-1)(p-1)}.$$

Let $\chi \in B(m)^*$ be the central character of M'. Theorem B.4 ensures that there exist $f \in B(m)^*$ and the Vergne polarisation $\mathfrak{p}$ of $B(m)$ at f constructed from the chain (2.33) such that the induced representation

$$\mathrm{Ind}_{\mathfrak{p}}^{B(m)}(\mathbb{K}_f;\chi)$$

is isomorphic to M'. We have

Lemma 2.21. *1. There exists j $(pm-p-1 \leq j \leq pm-2)$ such that $\chi(e_j) \neq 0$.*
2. $f = \chi$ on $I(m)$.

Proof. By the definition of central character, for $j \geq pm-p-1$

$$\chi(e_j) = 0 \Leftrightarrow e_j^p.M' = \{0\},$$

since $e_j^{[p]} = 0$. We assume that $\chi(e_j) = 0$ for $pm-p-1 \leq j \leq pm-2$, and lead to a contradiction. We set

$$S := \bigcup_{j=pm-p-1}^{pm-2} \mathbb{K}e_j.$$

Then, S satisfies the conditions in Theorem 2.4. Hence, $\langle S\rangle$ is nilpotent, Since $I(m) \subset \langle S\rangle$ by definition, there exists $k \in \mathbb{Z}_{>0}$ such that

$$I(m)^k.M' = \{0\}.$$

This contradicts (2.37). The first statement follows.

To show the second statement, we notice that $I(m) = B(m)_p$ and

$$[B(m)_p, B(m)_p] = \{0\}.$$

By definition,

$$\mathfrak{p} \supset \mathfrak{c}_{B(m)_p}(f|_{B(m)_p}) = B(m)_p = I(m).$$

On the other hand, as stated in Remark B.2,

$$f(x) - f(x^{[p]})^{\frac{1}{p}} = \chi(x)$$

holds for any $x \in \mathfrak{p}$. Hence, by Lemma 2.16, the second statement follows. □

PROOF OF THEOREM 2.5. We set

$$j := \max\{i | f(e_i) \neq 0\}.$$

By the above lemma, we see that

$$j \geq pm - p - 1.$$

Hence, considering the matrix expression of the form df with respect to the basis $\{e_0, e_1, \cdots, e_{pm-2}\} \subset B(m)$, we see that

$$\mathrm{rank} df \geq \mathrm{rank} de_j^*.$$

Further, considering the matrix expression of the form de_j^*, we have

$$\begin{aligned}
\mathrm{rank} de_j^* &= \sharp\{(k,l) | de_j^*(e_k, e_l) \neq 0\} \\
&= \sharp\{(k,l) | (e_k.e_j^*)(e_l) \neq 0\} \\
&= \sharp\{k | e_k.e_j^* \neq 0\},
\end{aligned}$$

where $B(m)$ acts on $B(m)^*$ via the coadjoint action, i.e.,

$$e_k.e_j^* = (j - 2k)e_{j-k}^*.$$

Hence, we see that

$$\begin{aligned}
\mathrm{rank} de_j^* &= (j+1) - \sharp\{k | 0 \leq k \leq j \wedge j - 2k \equiv 0 \pmod{p}\} \\
&\geq (pm - p - 1) + 1 - (m-1) \\
&= (p-1)(m-1).
\end{aligned}$$

Therefore, by Theorem B.4, we obtain

$$\dim M \geq \dim M' \geq p^{\frac{1}{2}(m-1)(p-1)}.$$

□

2.4.4 Irreducible Representations of W(1)

The purpose of this subsection is also to give an effective estimate of dimension for faithful and irreducible $W(1)$-modules. In this subsection, we assume that $\mathbb{K}$ is algebraically closed.

To study representations over $W(1)$, it is convenient to use the following new basis of $W(1)$: Recall that $W(1) = \mathrm{Der}(\mathbb{K}[t]/(t^p))$. We set

$$\ell_s := -(1+t)^{s+1}\frac{d}{dt} \quad (s = 0, 1, \cdots, p-1).$$

Since $(1+t)^p = 1$ in $\mathbb{K}[t]/(t^p)$, we can regard $\{\ell_s\}$ as elements indexed over $\mathbb{Z}/p\mathbb{Z}$. By definition, we have

$$W(1) = \bigoplus_{s\in\mathbb{Z}/p\mathbb{Z}} \mathbb{K}\ell_s.$$

The above basis elements satisfy the following commutation relations:

$$[\ell_r, \ell_s] = (r-s)\ell_{r+s},$$

where $r, s \in \mathbb{Z}/p\mathbb{Z} \subset \mathbb{K}$.

For each $a, b \in \mathbb{K}$, we introduce a $W(1)$-module $M_{a,b}$. Set

$$M_{a,b} := \bigoplus_{n\in\mathbb{Z}/p\mathbb{Z}} \mathbb{K}v_n, \tag{2.38}$$

and regard $M_{a,b}$ as a $W(1)$-module via

$$\ell_s.v_n := (as+b-n)v_{n+s} \quad (s, n \in \mathbb{Z}/p\mathbb{Z}).$$

It should be noted that any submodule of $M_{a,b}$ is ℓ_0-diagonalisable. Hence, by an argument similar to the proof of Proposition 2.1, we obtain

Lemma 2.22. *1. If $a \neq 0, -1$ or $b \notin \mathbb{Z}/p\mathbb{Z}$, then $M_{a,b}$ is irreducible.*

2. If $a = 0$ and $b \in \mathbb{Z}/p\mathbb{Z}$, then $M_{0,b}$ contains the trivial representation $\mathbb{K}v_b$ as a submodule, and the quotient module

$$M_{0,b}/\mathbb{K}v_b$$

is irreducible.

3. If $a = -1$ and $b \in \mathbb{Z}/p\mathbb{Z}$, then

$$\bigoplus_{n\neq b} \mathbb{K}v_n$$

is an irreducible submodule of $M_{a,b}$.

In the sequel, for $a, b \in \mathbb{K}$, we set

$$M'_{a,b} := \begin{cases} M_{a,b} & (a \neq 0, -1 \wedge b \notin \mathbb{Z}/p\mathbb{Z}) \\ M_{0,b}/\mathbb{K}v_b & (a = 0 \wedge b \in \mathbb{Z}/p\mathbb{Z}) \\ \bigoplus_{n \neq b} \mathbb{K}v_n & (a = -1 \wedge b \in \mathbb{Z}/p\mathbb{Z}) \end{cases} . \tag{2.39}$$

The main result of this subsection is

Theorem 2.6 ([Ch]) *Let M be an irreducible representation of $W(1)$. Then, one of the following holds:*

1. $\dim M \geq p^2$,
2. $\dim M < p^2$ *and*
$$M \simeq M'_{a,b} \ \text{ or } \ M \simeq \mathbb{K}.$$

In particular, if M is faithful, then

$$\dim M \geq p^2 \quad \text{or} \quad \dim M = p, \ p-1.$$

Proof (cf. [St]). Let $\chi \in W(1)^*$ be the central character of M. We divide the proof into the following three cases:

Case I: $\chi(e_{p-2}) \neq 0$.
Case II: There exists j $(2 \leq j \leq p-2)$ such that $\chi(e_k) = 0$ for all $k \geq j$ and $\chi(e_{j-1}) \neq 0$.
Case III: $\chi(e_k) = 0$ for all $k \geq 1$.

In the following, we show that $\dim M \geq p^2$ in Cases I and II and $\dim M < p^2$ in Case III.

Case I Let M' be an irreducible $B(1)$-submodule of M. Set $\chi' := \chi|_{B(1)} \in B(1)^*$. Then, χ' is the central character of M'. By Theorem B.4, there exist $f \in B(1)^*$ and the Vergne polarisation $\mathfrak{p}$ of $B(1)$ at f constructed from the chain (2.33) such that

$$M' \simeq \mathrm{Ind}_{\mathfrak{p}}^{B(1)}(\mathbb{K}_f; \chi').$$

We first show that

$$f(e_{p-2}) = \chi(e_{p-2}). \tag{2.40}$$

Notice that $B(1)_1 = \mathbb{K}e_{p-2}$ and $[B(1)_1, B(1)_1] = \{0\}$. By the definition of the Vergne polarisation, we see that

$$\mathfrak{p} \supset \mathfrak{c}_{B(1)_1}(f|_{B(1)_1}) = B(1)_1.$$

Hence, by Remark B.2 and $e_{p-2}^{[p]} = 0$, we obtain (2.40). Hence, by the assumption of Case I, we see that $f(e_{p-2}) \neq 0$. Considering the matrix expression of df with respect to the base $\{e_0, e_1, \cdots, e_{p-2}\}$, we see that

$$\mathrm{rank} df \geq \mathrm{rank} de_{p-2}^*.$$

One can directly check that rank$de^*_{p-2} = p-1$. By Theorem B.4, we obtain

$$\dim M' \geq p^{\frac{1}{2}(p-1)}.$$

Since $p > 3$, we have the conclusion $\dim M' \geq p^2$.

<u>Case II</u> We use Lemma 2.20. We set

$$\mathfrak{g} := W(1), \quad \mathfrak{h} := B(1), \quad \mathfrak{k} := \bigoplus_{l \geq j} \mathbb{K}e_l,$$

and $x := e_{-1}$, $y := e_j$. Indeed, since $[x, y] = -(j+1)e_{j-1}$ and $e^p_{j-1} = \chi(e_{j-1})^p \mathrm{id}_M$ on M, we see that $[x, y]$ is invertible on M. Applying Lemma 2.20, we see that for any irreducible $B(1)$-submodule M' of M,

$$\dim M = p \dim M'. \tag{2.41}$$

We take $f \in B(1)^*$ and the Vergne polarisation $\mathfrak{p}$ of $B(1)$ at f constructed from the chain (2.33) such that

$$M' \simeq \mathrm{Ind}_{\mathfrak{p}}^{B(1)}(\mathbb{K}_f; \chi')$$

(see Theorem B.4). In this case,

$$df \neq 0 \tag{2.42}$$

holds. In fact, if $df = 0$, then

$$\mathfrak{p} \supset \mathfrak{c}_{B(1)}(f) = B(1).$$

Hence, by Remark B.2, $f(e_i) = \chi(e_i)$ for $i \geq 1$. Moreover, $df = 0$ implies $f(e_i) = 0$ for $i \geq 1$. This contradicts the assumption of Case II. Hence, (2.42) holds. Hence, by Theorem B.4, we have

$$\dim M' \geq p.$$

Combining this estimation with (2.41), we obtain the conclusion $\dim M \geq p^2$.

<u>Case III</u> We set

$$B'(1) := [B(1), B(1)] = \bigoplus_{i=1}^{p-2} \mathbb{K}e_i.$$

From the assumption of Case III, we see that χ vanishes on $B'(1)$. By definition, one can show that $x^{[p]} = 0$ for any $x \in B'(1)$. Hence, by Lemma 2.19, any $x \in B'(1)$ are nilpotent on M, and thus, by Lemma 2.18, any irreducible $B(1)$-submodules are one-dimensional. Hence, there exists a non-zero element $u \in M$ such that $\mathbb{K}u$ is an irreducible $B(1)$-submodule of M. Since e_i $(i \geq 1)$

are nilpotent on M, there exists $\lambda \in \mathbb{K}$ such that

$$e_i.u = \begin{cases} \lambda u & (i=0) \\ 0 & (i \geq 1) \end{cases}.$$

If we set

$$N := \mathrm{Ind}_{B(1)}^{W(1)}(\mathbb{K}u; \chi),$$

then there exists a surjective homomorphism of $W(1)$-modules $N \twoheadrightarrow M$, since M is irreducible.

For this $W(1)$-module N, we have

Lemma 2.23.

$$N \simeq M_{\lambda-1,\lambda-1}, \tag{2.43}$$

where $M_{a,b}$ is defined in (2.38).

Proof. Note that

$$N = \bigoplus_{i=0}^{p-1} \mathbb{K}(e_{-1})^i \otimes u.$$

By using

$$e_j(e_{-1})^i = (e_{-1})^i e_j + \sum_{k=1}^{i} \binom{i}{k} (e_{-1})^{i-k} [\cdots [[e_j, \underbrace{e_{-1}], e_{-1}] \cdots, e_{-1}}_{k \text{ times}}],$$

for $j \geq 1$ we have

$$\begin{aligned} e_j.(e_{-1})^i \otimes u &= \binom{i}{j} (e_{-1})^{i-j}(j+1)! e_0 \otimes u \\ &\quad + \binom{i}{j+1} (e_{-1})^{i-j-1}(j+1)! e_{-1} \otimes u \\ &= \frac{i!}{(i-j)!}\{(j+1)\lambda + (i-j)\}(e_{-1})^{i-j} \otimes u. \end{aligned}$$

Notice that this formula still holds for $j = 0, -1$. By setting

$$v_i := \frac{1}{(p-1-i)!}(e_{-1})^{p-1-i} \otimes u \quad (0 \leq i \leq p-1),$$

$\{v_i | 0 \leq i \leq p-1\}$ forms a basis of N. Moreover, by direct computation, one can check that

$$e_j.v_i = \{(\lambda-1)j + (\lambda-1) - i\} v_{i+j}$$

for any $0 \leq i, j \leq p-1$. Moreover, for $0 \leq i \leq p-1$, we set

$$u_i := \sum_{k=0}^{i} \binom{i}{k} v_k.$$

Identifying the index set of $\{u_i\}$ with $\mathbb{Z}/p\mathbb{Z}$, we have

$$\ell_j.u_i = \{(\lambda - 1)j + (\lambda - 1) - i\}u_{i+j}$$

by direct computation. Hence, the isomorphism (2.43) has been proved. □

Therefore, Case III of Theorem 2.6 follows from Lemma 2.22. Now, we have completed the proof. □

2.4.5 $\mathbb{Z}/N\mathbb{Z}$-graded Modules over $\mathrm{Vir}_{\mathbb{K}}$

In this subsection, let $\mathbb{K}$ be an algebraically closed field of *positive* characteristic $p \neq 2, 3$.

Let $\mathcal{D}_{\mathbb{K}} := \mathbb{K}[t, t^{-1}]\frac{d}{dt}$ be the Lie algebra with commutation relation

$$[f_1(t)\frac{d}{dt}, f_2(t)\frac{d}{dt}] = (f_1(t)f_2'(t) - f_1'(t)f_2(t))\frac{d}{dt}.$$

For $g(t) \in \mathbb{K}[t]$ such that $g(0) \neq 0$, we set

$$I(g(t)) := g(t)^p\mathbb{K}[t, t^{-1}]\frac{d}{dt} \subset \mathcal{D}_{\mathbb{K}}. \tag{2.44}$$

Since $\frac{d}{dt}g(t)^p = 0$, $I(g(t))$ is an ideal of $\mathcal{D}_{\mathbb{K}}$. Moreover, since $\mathbb{K}$ is algebraically closed, there exist $\alpha_1, \alpha_2, \cdots, \alpha_s \in \mathbb{K} \setminus \{0\}$ and $m_1, m_2, \cdots, m_s \in \mathbb{Z}_{>0}$ such that

$$g(t) = (t - \alpha_1)^{m_1}(t - \alpha_2)^{m_2} \cdots (t - \alpha_s)^{m_s}.$$

Now, $(t - \alpha)^{mp} = t^{mp} - \alpha^{mp}$ implies

$$\begin{aligned}\mathbb{K}[t, t^{-1}]\frac{d}{dt} \Big/ (t - \alpha)^{mp}\mathbb{K}[t, t^{-1}]\frac{d}{dt} &\simeq (\mathbb{K}[t]/(t^{mp} - \alpha^{mp}))\,\frac{d}{dt} \\ &\simeq \mathrm{Der}\,(\mathbb{K}[t - \alpha]/((t - \alpha)^{mp})) \\ &\simeq W(m).\end{aligned}$$

Hence, we have

$$\mathcal{D}_{\mathbb{K}}/I(g(t)) \simeq W(m_1) \oplus W(m_2) \oplus \cdots \oplus W(m_s).$$

Here, let us denote the canonical projection $\mathrm{Vir}_{\mathbb{K}} \twoheadrightarrow \mathcal{D}_{\mathbb{K}}$ by π. For $g(t) \in \mathbb{K}[t]$ such that $g(0) \neq 0$, we set

$$\mathcal{I}(g(t)) := \pi^{-1}(I(g(t))).$$

Lemma 2.24. *1. For any ideal I of $\mathcal{D}_{\mathbb{K}}$, there exists a polynomial $g(t) \in \mathbb{K}[t]$ such that $I = I(g(t))$.*
2. For any ideal $\mathcal{I}$ of $\mathrm{Vir}_{\mathbb{K}}$*, $C \in \mathcal{I}$ holds.*

Proof. For an ideal I of $\mathcal{D}_{\mathbb{K}}$, we set

$$J_I := \left\{ g(t) \in \mathbb{K}[t,t^{-1}] \,\middle|\, g(t)\frac{d}{dt} \in I \right\}.$$

We first show that J_I is an ideal of $\mathbb{K}[t,t^{-1}]$. For $g(t) \in \mathbb{K}[t,t^{-1}]$, we express it in the form

$$g(t) = \sum_{i=0}^{p-1} g_i(t^p)t^i \qquad (g_i(t) \in \mathbb{K}[t,t^{-1}]).$$

For $g(t) \in J_I$ and $a(t) \in \mathbb{K}[t,t^{-1}]$, $a(t)g'(t) - a'(t)g(t) \in J_I$ holds by definition. By taking $a(t) = t$, we have $tg'(t) \in J_I$. Hence,

$$\sum_{i=0}^{p-1} i^s g_i(t^p)t^i \in J_I \qquad (s \geq 0),$$

and thus, $g_i(t^p)t^i \in J_I$ for any i. For $a(t) = t^m$ $(m \in \mathbb{Z})$ and $g(t) = \tilde{g}(t^p)t^n \in \mathbb{K}[t,t^{-1}]$, we have $a(t)g'(t) - a'(t)g(t) = (n-m)\tilde{g}(t^p)t^{m+n-1}$. Specialising it appropriately, at most twice, we see that $g_i(t^p)t^j \in J_I$ for any integer j. Hence, $g(t)t^j \in J_I$, and thus, J_I is an ideal of $\mathbb{K}[t,t^{-1}]$. Moreover, since J_I is generated by elements of the form $\tilde{g}(t^p)$ for $\tilde{g}(t) \in \mathbb{K}[t,t^{-1}]$, we have

$$J_I = (\tilde{g}(t^p))$$

for some $\tilde{g}(t^p) \in \mathbb{K}[t,t^{-1}]$. Since there exists $g(t) \in \mathbb{K}[t,t^{-1}]$ such that $\tilde{g}(t^p) = g(t)^p$, we conclude that $I = I(g(t))$.

Next, we show the second statement. Suppose that there exists a non-trivial ideal $\mathcal{I}$ of $\mathrm{Vir}_{\mathbb{K}}$ such that $C \notin \mathcal{I}$. Let ω be the non-trivial 2-cocycle of $\mathcal{D}_{\mathbb{K}}$ defined by

$$\omega(L_m, L_n) := \delta_{m+n,0}\frac{m^3 - m}{12},$$

where we set $L_m := -t^{m+1}\frac{d}{dt} \in \mathcal{D}_{\mathbb{K}}$ by abuse of notation. Because of the well-definedness of the commutation relations of $\mathrm{Vir}_{\mathbb{K}}/\mathcal{I}$, we see that

$$\omega(I, \mathcal{D}_{\mathbb{K}}) = \{0\},$$

where $I := \pi(\mathcal{I})$. Hence, I is a subset of the radical of ω which is given by $\bigoplus_{n \in p\mathbb{Z}} \{\mathbb{K}L_{n-1} \oplus \mathbb{K}L_n \oplus \mathbb{K}L_{n+1}\}$. By the first statement, we deduce that $I = \{0\}$. This is a contradiction. □

Notice that for any integer N, $\mathrm{Vir}_{\mathbb{K}}$ is naturally $\mathbb{Z}/N\mathbb{Z}$-graded.

Lemma 2.25. *Let N be a positive integer such that there exist integers u and v satisfying $N = vp^u$, $u \geq 1$ and $(p, v) = 1$. Let $\mathcal{I}$ be a non-trivial and non-central $\mathbb{Z}/N\mathbb{Z}$-graded ideal of $\mathrm{Vir}_{\mathbb{K}}$. Set $\mathcal{G} := \mathrm{Vir}_{\mathbb{K}}/\mathcal{I}$. Let M be a simple and faithful $\mathcal{G}$-module. Then, the following three inequalities hold:*

A

$$\dim M \geq p^{\frac{1}{2}v(p-1)(p^{u-1}-1)},$$

B *if $u = 1$, then*

$$\dim M \geq (p-1)^v,$$

C *if $N = p$ and $\dim M \neq p, p-1$, then*

$$\dim M \geq p^2.$$

Proof. By Lemma 2.24, there exists a non-constant polynomial $g(t)$ with $g(0) \neq 0$ such that $\mathcal{I} = \mathcal{I}(g(t))$. Since $\mathcal{I}$ is $\mathbb{Z}/N\mathbb{Z}$-graded, we may assume that $g(t)^p \in \mathbb{K}[t^N]$, and thus, $g(t) \in \mathbb{K}[t^{vp^{u-1}}]$. Hence, there exists $h(t) \in \mathbb{K}[t^v]$ such that $g(t) = h(t)^{p^{u-1}}$. The set of the roots of $h(t) = 0$ is stable under the multiplication of vth roots of unity. Hence, $g(t) = 0$ has at least v roots with multiplicity equal to or greater than p^{u-1}. This implies that

$$\mathcal{G} \simeq \bigoplus_i W(m_i),$$

for $\{m_i\}$ such that $\sharp\{i | m_i \geq p^{u-1}\} \geq v$. Since M is a faithful $\mathcal{G}$-module, the inequality in **A** follows from Theorem 2.5, and the inequalities in **B** and **C** follow from Theorem 2.6. □

Proposition 2.10 *Let k be a finite field, and let $\bar{k}$ be its algebraic closure. Let $\mathfrak{g} = \bigoplus_{n \in \mathbb{Z}} \mathfrak{g}_n$ be a finitely generated finite $\mathbb{Z}$-graded Lie algebra over $\bar{k}$. Let $M = \bigoplus_{n \in \mathbb{Z}} M_n$ be a simple $\mathbb{Z}$-graded $\mathfrak{g}$ module which satisfies $\dim M = \infty$ and $\dim M_n$ are uniformly bounded, i.e., there exists $d > 0$ such that $\dim M_n < d$ for any $n \in \mathbb{Z}$. Then, M is not graded simple.*

Proof. Since $\mathfrak{g}$ is finitely generated, there exists $s \in \mathbb{Z}_{>0}$ such that $\mathfrak{g}$ is generated by the partial part $\Gamma := \mathrm{Par}_{-s}^{s}\mathfrak{g}$. We set $\mathcal{L} := \mathcal{L}_{\max}(\Gamma)$. By Theorem 2.2, there exists a surjective homomorphism

$$\mathcal{L} \twoheadrightarrow \mathfrak{g}.$$

In the sequel, we regard M as $\mathcal{L}$ module via the surjection. Since M is a simple $\mathbb{Z}$-graded $\mathfrak{g}$-module, it is a simple $\mathbb{Z}$-graded $\mathcal{L}$-module.

We may assume that $M_0 \neq \{0\}$ without loss of generality. We set $\sigma := \mathrm{Par}_{-s}^{s}M$ and regard it as Γ-module. Then, M is generated by σ as an $\mathcal{L}$-module, since M is simple graded. By Theorem 2.3, there is a surjection

$$M \twoheadrightarrow M_{\min}(\sigma).$$

We fix a basis $\{x_1, \cdots, x_a\}$ ($a := \dim \Gamma$) of Γ and $\{v_1, \cdots, v_b\}$ ($b := \dim \sigma$) of σ. Let $\{c_{i,j}^l | 1 \leq i, j, l \leq a\}$ and $\{d_{i,j}^l | 1 \leq i \leq a,\ 1 \leq j, l \leq b\}$ be the structure constants

$$[x_i, x_j] = \sum_{l=1}^{a} c_{i,j}^l x_l, \quad x_i.v_j = \sum_{l=1}^{b} d_{i,j}^l v_l.$$

Let K be the extension field $k(c_{i,j}^l, d_{i,j}^l)$. Notice that K is a finite field. We set

$$\Gamma_K := \bigoplus K x_i, \quad \sigma_K := \bigoplus K v_i,$$

$\mathcal{L}_K := \mathcal{L}_{\max}(\Gamma_K)$ and $M_K := M_{\min}(\sigma_K)$. Then,

$$\Gamma \simeq \bar{k} \otimes \Gamma_K, \quad \sigma \simeq \bar{k} \otimes \sigma_K, \quad \mathcal{L} \simeq \bar{k} \otimes \mathcal{L}_K, \quad M \simeq \bar{k} \otimes M_K.$$

In particular,

$$\dim(M_K)_n = \dim M_n < d \quad (\forall n \in \mathbb{Z}).$$

Moreover, it is easy to see that M_K is simple $\mathbb{Z}$-graded.

For each $m \in \mathbb{Z}$, setting

$$\sigma(m)_K := \bigoplus_{|m-n| \leq s} (M_K)_n,$$

we naturally regard it as partial Γ-module. Here, it should be noted that the cardinality of the equivalence classes of the partial Γ-module $\tau = \bigoplus_{|n| \leq s} \tau_n$ such that $\dim \tau_n \leq d$ for any n is finite, since K is a finite field. Moreover, $\sigma(m)_K \neq \{0\}$ for infinitely many integers m. Therefore, there exist $m_1, m_2 \in \mathbb{Z}$ ($m_1 \neq m_2$) such that

$$\sigma(m_1)_K \simeq \sigma(m_2)_K.$$

Here, we allow any degree shift for an isomorphism of partial Lie algebras. This implies that there exists a generating homomorphism

$$\theta \in \mathrm{End}_{\mathcal{L}_K}(M_K)_{m_1 - m_2}.$$

By Remark 2.2, M_K is not graded simple, and thus, M is not graded simple. □

2.5 Proof of the Classification of Harish-Chandra Modules

We complete the proof of the classification of Harish-Chandra modules over the Virasoro algebra.

2.5.1 Structure of Simple $\mathbb{Z}$-graded Modules

In this subsection, we show a proposition on structures of finite simple $\mathbb{Z}$-graded modules over the Virasoro algebra in positive characteristic.

Proposition 2.11 *Suppose that $\mathbb{K} = \bar{\mathbb{F}}_p$ for $p > 0$ such that $p \neq 2, 3$. Let $M = \bigoplus_{n \in \mathbb{Z}} M_n$ be a finite simple $\mathbb{Z}$-graded* $\mathrm{Vir}_{\mathbb{K}}$*-module such that*

$$\dim M_n < \frac{1}{2p}(p-1)^2 \quad (\forall n \in \mathbb{Z}).$$

Then, there exist $a, b \in \mathbb{K}$ such that the following isomorphism holds:

$$M \simeq V'_{a,b}$$

where $V'_{a,b}$ are defined as in (2.5).

Proof. By Proposition 2.10, M is not graded simple. Hence, by Proposition 2.8, there exists a generating homomorphism $\theta \in \mathrm{End}_{\mathrm{Vir}_{\mathbb{K}}}(M)$ of degree $N > 0$. Here, we use the following notation:

$$\mathbf{M} := M/(\mathrm{id}_M - \theta)M.$$

Let $\mathcal{I}$ be the kernel of $\mathrm{Vir}_{\mathbb{K}} \to \mathrm{End}_{\mathbb{K}}(\mathbf{M})$. Then, one can show that $\mathcal{I}$ is a non-trivial and non-central $\mathbb{Z}/N\mathbb{Z}$-graded ideal. Indeed, if $\mathcal{I}$ is central, then $\dim(\mathrm{Vir}_{\mathbb{K}}/\mathcal{I}) = \infty$. On the other hand, since $\mathbf{M}$ is a faithful $(\mathrm{Vir}_{\mathbb{K}}/\mathcal{I})$-module, $\mathrm{Vir}_{\mathbb{K}}/\mathcal{I} \hookrightarrow \mathrm{End}_{\mathbb{K}}(\mathbf{M})$. But this contradicts $\dim_{\mathbb{K}} \mathbf{M} < \infty$. One can also show that $\mathcal{I}$ is a $\mathbb{Z}/N\mathbb{Z}$-graded ideal by definition.

Let u and v be integers such that $N = vp^u$ and $(v, p) = 1$. Since θ commutes with L_0, we see that $p|N$. Hence, $u \geq 1$. By assumption, we have

$$\dim \mathbf{M} < N\frac{1}{2p}(p-1)^2 = \frac{1}{2}vp^{u-1}(p-1)^2.$$

By Lemma 2.25. **A**, we have

$$p^{\frac{1}{2}v(p-1)(p^{u-1}-1)} < \frac{1}{2}vp^{u-1}(p-1)^2.$$

Hence, $u = 1$. By Lemma 2.25. **B**, we have

$$(p-1)^v < \frac{1}{2}v(p-1)^2.$$

Hence, $v = 1$ and $N = p$. By Lemma 2.25. **C**, we have

$$\dim \mathbf{M} = p,\ p-1.$$

By Lemma 2.24, there exists $m_1, \cdots, m_s$ such that

$$\mathrm{Vir}_{\mathbb{K}}/\mathcal{I} \simeq \bigoplus_{i=1}^{s} W(m_i).$$

Hence, by Theorem 2.5, we obtain $s = 1$ and $m_1 = 1$, and thus

$$\mathrm{Vir}_{\mathbb{K}}/\mathcal{I} \simeq W(1).$$

By Theorem 2.6,

$$\mathbf{M} \simeq M'_{a,b}$$

for some $a, b \in \mathbb{K}$. Hence, by Proposition 2.9, we obtain

$$M \simeq V'_{a,b}.$$

□

2.5.2 Semi-continuity Principle

We recall the definition and some properties of Dedekind domain, which we use in the proof of Theorem 2.1. For definitions and results on commutative algebra, which we omit in this subsection, see [AtM] or [Matsu].

The aim of this section is the following proposition:

Proposition 2.12 *Let K be an algebraic number field, i.e., a finite algebraic extension of $\mathbb{Q}$, and let R' be the ring of integers of K, i.e., the integral closure of $\mathbb{Z}$ in K. Let R be a localisation of R' such that $R \neq K$, and let k be a residue field of R. Let M be a finite dimensional K-vector space, and let M_R be an R-form of M. We set $M_k := k \otimes_R M_R$. Then, we have*

$$\dim M = \dim M_k.$$

This proposition is usually refered to as the **semi-continuity principle**.

Recall the definition of Dedekind domain (see, e.g., [AtM] Chapter 9).

Definition 2.7 *An integral domain A is called a Dedekind domain if*

1. *A is Noetherian,*
2. *every non-zero prime ideal is a maximal ideal of A, and*
3. *A is integrally closed.*

An important example is given by the next proposition:

Proposition 2.13 *The ring of integers in an algebraic number field is a Dedekind domain.*

Proof. Theorem 9.5 in [AtM]. □

To show Proposition 2.12, we recall two facts from commutative algebras:

Lemma 2.26. *Suppose that A is a Dedekind domain. Then, for any non-zero prime ideal $\mathfrak{p}$ of A, $A_{\mathfrak{p}}$ is a principal ideal domain.*

Proof. Proposition 9.2 and Theorem 9.3 in [AtM]. □

Lemma 2.27. *Let A be an integral domain, and let M be an A-module. M is torsion free if and only if $M_{\mathfrak{p}}$ is a torsion free $A_{\mathfrak{p}}$-module for any prime ideal $\mathfrak{p}$.*

Proof. Exercise 13 in Chapter 3 of [AtM]. □

PROOF OF PROPOSITION 2.12. First, we notice that M_R is torsion free. Hence, by Lemma 2.27, for any non-zero prime ideal $\mathfrak{p}$ of R, $(M_R)_{\mathfrak{p}}$ is a torsion free $R_{\mathfrak{p}}$-module. On the other hand, it follows from $R \subset R_{\mathfrak{p}} \subset K$ that $(M_R)_{\mathfrak{p}}$ is an $R_{\mathfrak{p}}$-form of M. Since by Lemma 2.26, $R_{\mathfrak{p}}$ is a principal ideal domain, we see that $(M_R)_{\mathfrak{p}}$ is a free $R_{\mathfrak{p}}$-module of rank $\dim M$. Hence, we have

$$\dim M_k = \dim M.$$ □

2.5.3 Proof of Theorem 2.1

To complete the proof of the classification theorem, we introduce some notation. We set

$$\Gamma := \bigoplus_{|i| \leq 2} \mathbb{K}L_i \oplus \mathbb{K}C \subset \mathrm{Vir}_{\mathbb{K}},$$

and regard it as a partial Lie algebra.

We denote the kernel of the map $\mathcal{L}_{\max}(\Gamma) \to \mathrm{Vir}_{\mathbb{K}}$ by $\mathcal{I}$.

Let $V = \bigoplus_{|i| \leq 2} V_i$ be a graded $\mathbb{K}$-vector space. Let $\mathcal{V}(V, \Gamma)$ be the variety of partial Γ-module structures on V. It is a closed subvariety of $(\mathrm{End}V)^{\dim \Gamma}$. A partial Γ-module σ can be regarded as an element of $\mathcal{V}(V, \Gamma)$, and we will often do so.

For a positive integer f, let $\mathcal{V}(V, \Gamma, \mathcal{I}, f)$ be the subvariety of $\mathcal{V}(V, \Gamma)$ defined by

$$\begin{aligned} &\sigma \in \mathcal{V}(V, \Gamma, \mathcal{I}, f) \\ &\Leftrightarrow \sigma : \text{simple},\ \mathcal{I}.M_{\min}(\sigma) = \{0\},\ \dim M_{\min}(\sigma)_n \leq f\ (\forall n \in \mathbb{Z}). \end{aligned}$$

By definition, $\mathcal{V}(V, \Gamma, \mathcal{I}, f)$ is a locally closed subvariety.

Let $\mathcal{D} = (D_{-2}, D_{-1}, D_0, D_1, D_2)$ be a quintuple of non-negative integers such that $D_i \leq f$ for $-2 \leq i \leq 2$. Let $V_{\mathcal{D}}$ be a graded $\mathbb{K}$-vector space $\bigoplus_{|i|\leq 2} V_i$ such that $\dim V_i = D_i$. We set

$$\mathcal{V}(f) := \bigcup_{\mathcal{D}} \mathcal{V}(V_{\mathcal{D}}, \Gamma, \mathcal{I}, f).$$

Moreover, let $\mathcal{X}$ be the subvariety of $\mathcal{V}(f)$ which consists of $\mathrm{Par}_{-2}^{2} V'_{a,b}$, where $V'_{a,b}$ is the irreducible $\mathrm{Vir}_{\mathbb{K}}$-module defined as in (2.5).

By Proposition 2.7, for any Harish-Chandra module

$$M = \bigoplus_{n\in\mathbb{Z}} M_n$$

without highest or lowest degree, there exists a positive integer f such that $\dim M_n \leq f$ for any $n \in \mathbb{Z}$. Hence, we have

$$\mathrm{Par}_{-2}^{2} M \in \mathcal{V}(f).$$

Hence, to show the theorem, it suffices to prove that $\mathcal{X} = \mathcal{V}(f)$.

It should be noted that the varieties $\mathcal{V}(V, \Gamma)$, $\mathcal{V}(V, \Gamma, \mathcal{I}, f)$, $\mathcal{V}(f)$ and $\mathcal{X}$ are defined over $\mathbb{Q}$. Hence, we may assume that the base field $\mathbb{K}$ is $\overline{\mathbb{Q}}$ without loss of generality.

In the sequel, suppose that $\sigma \in \mathcal{V}(f)$ is defined on some localisation R of the ring of integers of a number field.

Remark 2.4 *Let K be the field of the fractions of R. By assumption, there exists a K-basis $\{v_i\}$ of σ such that the structure constants $(c_{i,j}^{l})$ defined by*

$$L_i.v_j = \sum_l x_{i,j}^{l} v_l, \quad C.v_j = \sum_l y_j^l v_l$$

satisfy $(x_{i,j}^{l}, y_j^l) \subset R$. Hence, $\sigma_K := \bigoplus_j K v_j$ is a partial Γ_K-module, where $\Gamma_K := \bigoplus_{|i|\leq 2} KL_i \oplus KC$. Moreover, $\dim_{\mathbb{K}} \sigma_{\mathbb{K}} = \dim_K \sigma_K$ holds.

By localising again, if necessary, we may assume that all prime numbers p such that

$$\frac{1}{2p}(p-1)^2 \leq f$$

are invertible in R. Moreover,

Lemma 2.28. *By localising R appropriately, if necessary, we may assume that for any residue field k, $\sigma_k := \sigma_R \otimes_R k$ is absolutely simple.*

Proof. For simplicity, we put $r := \dim \sigma$. Here, we denote the set of $r \times r$ matrices whose elements belong to a ring A by $\mathrm{Mat}_r(A)$. To show

σ_k is absolutely simple, it is enough to check that the image of the map $U(\mathcal{L}_{\max}(\Gamma_k)) \to \mathrm{End}_k(\sigma_k)$ is isomorphic to $\mathrm{Mat}_r(k)$.

Since σ is absolutely simple, the map $\varphi : U(\mathcal{L}_{\max}(\Gamma)) \to \mathrm{End}_{\mathbb{K}}(\sigma)$ is surjective, i.e., $\mathrm{Im}\varphi \simeq \mathrm{Mat}_r(\mathbb{K})$. Hence, for any $1 \leq i, j \leq r$, there exists an element $X \in U(\mathcal{L}_{\max}(\Gamma_R))$ such that $\varphi(X) \in RE_{i,j}$, where $E_{i,j}$ denotes the matrix unit. Hence, by taking an appropriate localisation R' of R, the image $\varphi(U(\mathcal{L}_{\max}(\Gamma_{R'})))$ is isomorphic to $\mathrm{Mat}_r(R')$ where $\sigma_{R'} := \sigma_R \otimes R'$. Therefore, $\sigma_{k'}$ is absolutely simple, where k' is any residue field of R'. □

From now on, we denote the localisation R' by R for simplicity.

Remark 2.5 *Since* $\dim \sigma < \infty$, *we may assume that for infinitely many prime integers* $q \in \mathbb{Z}$, q *is a prime element of* R. *We use this assumption at the end of this proof.*

Let M_R be the $\mathcal{L}_{\max}(\Gamma_R)$-submodule of $M := M_{\min}(\sigma)$ generated by σ_R. By Lemma 2.14, M_R is an R-form of M. Set

$$M'_k := M_R \otimes k,$$

where k is a residue field of R. Let N_k be a graded maximal proper submodule of M'_k. We set

$$M_k := M'_k / N_k.$$

Since σ_k is a simple partial module and generates M'_k as $\mathcal{L}_{\max}(\Gamma_k)$-module, we have $\sigma_k \cap N_k = \{0\}$. Hence,

$$\mathrm{Par}^2_{-2} M_k \simeq \sigma_k.$$

Moreover, since M_k is generated by σ_k, by Theorem 2.3, there exists a surjection $M_k \twoheadrightarrow M_{\min}(\sigma_k)$. Since M_k is simple $\mathbb{Z}$-graded, we have

$$M_k \simeq M_{\min}(\sigma_k).$$

By Proposition 2.12 and Remark 2.4, we see that

$$\dim(M_k)_n \leq f.$$

Since the residue field k is finite, and the characteristic p of k satisfies

$$f < \frac{1}{2p}(p-1)^2,$$

it follows from Proposition 2.11 that

$$M_k \otimes \bar{k} \simeq V'_{a,b:\bar{k}}$$

where $V'_{a,b:\bar{k}}$ denotes the irreducible $\mathrm{Vir}_{\bar{k}}$-module defined as in (2.5), and to avoid confusion, we specify the ground field $\bar{k}$.

In the following, we divide the proof into the following cases:

1. $a \neq 0, -1$ or $b \notin \mathbb{Z}/p\mathbb{Z}$,
2. $a = 0, -1$ and $b \in \mathbb{Z}/p\mathbb{Z}$,

and here, we prove the theorem in the first case only, since the other case can be similarly treated.

Case: $a \neq 0, -1 \vee b \notin \mathbb{Z}/p\mathbb{Z}$ If we set $\sigma_{\bar{k}} := \sigma_k \otimes_k \bar{k}$, then

$$\dim(\sigma_{\bar{k}})_n = 1$$

for any $-2 \leq n \leq 2$. Since, by Proposition 2.12

$$\dim(\sigma)_n = \dim(\sigma_{\bar{k}})_n$$

holds, we have

$$\dim(\sigma)_n = 1 \quad (-2 \leq n \leq 2).$$

Let us fix $v_0 \in (V_{a,b:\mathbb{K}})_0$. Notice that for any $-2 \leq n \leq 2$, there exists $x_n \in U(\mathrm{Vir}_R)$ $(-2 \leq n \leq 2)$ such that $x_n.v_0 \in (V_{a,b:\mathbb{K}})_n \setminus \{0\}$. By using the elements $x_{-2}, x_{-1}, \cdots, x_2$, we define an R-form of the partial module $\tau := \mathrm{Par}_{-2}^2 V_{a,b}$ by

$$\tau_R := \sum_{n=-2}^{2} R x_i v_0,$$

and set $\tau_{\bar{k}} := \tau_R \otimes_R \bar{k}$. Then, by the isomorphism $M_k \otimes_k \bar{k} \simeq V_{a,b:\bar{k}}$, the following isomorphism holds:

$$\tau_{\bar{k}} \simeq \sigma_{\bar{k}}. \tag{2.45}$$

Let w_0 be an element of $(\sigma_R)_0$ such that $(\sigma_R)_0 = R w_0$. (Note that $\mathrm{rank}(\sigma_R)_0 = 1$, since $\dim(\sigma)_0 = 1$.) We define an R-linear map

$$\varphi_R : \tau_R \longrightarrow \sigma_R$$

by $\varphi_R(x_i.v_0) := x_i.w_0$. Let $\varphi_{\bar{k}} : \tau_{\bar{k}} \longrightarrow \sigma_{\bar{k}}$ be the $\bar{k}$-linear map induced from φ_R. Since both $\tau_{\bar{k}}$ and $\sigma_{\bar{k}}$ are simple and isomorphic to each other, we see that $\varphi_{\bar{k}}$ gives an isomorphism of $\mathrm{Vir}_{\bar{k}}$-modules. Hence, for any $v \in \tau_R$ and $-2 \leq n \leq 2$,

$$L_n \varphi_R(v) - \varphi_R(L_n.v) \in \left(\bigcap_{\mathfrak{m} \in \text{m-Spec}(R)} \mathfrak{m} \right) \sigma_R,$$

where we denote the set of the maximal ideals of R by m-Spec(R). Hence, the following lemma implies that φ_R commutes with the action of L_n for $-2 \leq n \leq 2$.

Lemma 2.29.

$$\bigcap_{\mathfrak{m}\in \text{m-Spec}(R)} \mathfrak{m} = \{0\}.$$

Proof. By Remark 2.5, for infinitely many prime integers $q \in \mathbb{Z}$, $(q) \in \text{Spec}R$. Since R is a Dedekind domain, $(q) \in \text{m-Spec}(R)$. This implies the result. □

Thus, we can extend φ_R to the $\mathbb{K}$-linear map

$$\varphi : \tau \longrightarrow \sigma.$$

Then, since $\dim(\sigma)_n = 1$ for any $-2 \le n \le 2$, φ gives an isomorphism of partial module over $\Gamma = \bigoplus_{n=-2}^{2} \mathbb{K}L_n \oplus \mathbb{K}C$. Therefore, we have proved that $\sigma \in \mathcal{X}$ for any $\sigma \in \mathcal{V}(f)$. The inverse inclusion $\mathcal{X} \subset \mathcal{V}(f)$ holds by definition, and thus, we have completed the proof of the classification theorem.

2.6 Bibliographical Notes and Comments

In 1981, V. G. Kac [Kac3] proposed some open problems with conjectures. Among them, the problem of classifying Harish-Chandra modules over the Virasoro algebra is the theme of this chapter. Let us briefly recall its history and some relevant topics.

In 1985, I. Kaplansky and L. J. Santharoubane [KaSa] classified the Harish-Chandra modules all of whose weight multiplicities are 1, which is summarised in the appendix of this chapter. Later, in 1988, V. Chari and A. Pressley [CP2] classified the unitarisable Harish-Chandra modules. In 1991, C. Martin and A. Piard [MaP] proved that any indecomposable $\mathbb{Z}$-graded Vir-module with bounded weight multiplicities contains a $\mathbb{Z}$-graded submodule whose weight multiplicities are less than or equal to 1. Thus, in particular, their results together with Proposition 2.7 gives another proof of the conjecture mentioned above. All of these results are obtained only by purely characteristic zero arguments but involve complicated computations. The complete proof of the conjecture obtained by O. Mathieu [Mat2], as explained in this book, relies on completely different methods, *partial Lie theory* and representation theory in positive characteristic.

In this direction, a more general result was obtained by J. Germoni [Ger]. In 2001, he showed that every block in the category of $\mathbb{Z}$-graded Vir-modules whose weight multiplicities are bounded is wild. Moreover, he also proved that the category of finite-length extensions of irreducible highest (resp. lowest) weight Vir-modules is also wild.

Let us explain the original usage of the *partial Lie theory*. In [Kac3], V. G. Kac conjectured that any infinite dimensional simple $\mathbb{Z}$-graded Lie algebras with finite growth should be isomorphic to either 1) a loop algebra (including a twisted algebra), 2) a Cartan type Lie algebra or 3) the Virasoro algebra. In 1986, O. Mathieu [Mat1] proved this conjecture in the case when the growth is less than or equal to one, and in 1992, again, O. Mathieu [Mat3] gave a complete affirmative answer to this conjecture. '*Partial Lie theory*' was the key technique in his proof of the classification problem.

Finally, let us make a brief remark on the classification of Harish-Chandra modules over an infinite dimensional simple graded Lie algebra of finite growth. The classification of the Harish-Chandra modules over a (untwisted-)loop algebra was obtained by V. Chari [Char] and that for a twisted loop algebra was obtained by V. Chari and A. Pressley [CP1]. The classification of Harish-Chandra modules over a Cartan type Lie algebra of rank ≥ 2 was obtained by I. A. Kostrikin [Kostr], and the only rank 1 case, i.e., what is called of type W_1, was treated by O. Mathieu [Mat2]. Therefore, the classification of Harish-Chandra modules over all of the infinite-dimensional simple $\mathbb{Z}$-graded Lie algebras is complete.

2.A Appendix: Indecomposable $\mathbb{Z}$-graded Vir-Modules with Weight Multiplicities 1

Here, we state the classification of the indecomposable $\mathbb{Z}$-graded modules with weight multiplicities 1 following [KaSa] and [Ka].

2.A.1 Definition of $A(\alpha)$ and $B(\beta)$

For $\alpha, \beta \in \mathbb{C}P^1$, the Vir-modules $A(\alpha)$ and $B(\beta)$ are defined as follows:

1. $A(\alpha) = \bigoplus_{n\in\mathbb{Z}} \mathbb{C}w_n;\qquad C.w_n = 0,$

   i) $\alpha \in \mathbb{C}$,

$$L_i.w_j := (-i-j)w_{i+j} \qquad (j \neq 0),$$
$$L_i.w_0 := -i(1+(i+1)\alpha)w_i.$$

   ii) $\alpha = \infty$,

$$L_i.w_j := (-i-j)w_{i+j} \qquad (j \neq 0),$$
$$L_i.w_0 := -i(i+1)w_i.$$

2. $B(\beta) = \bigoplus_{n\in\mathbb{Z}} \mathbb{C}w_n;\qquad C.w_n = 0,$

   i) $\beta \in \mathbb{C}$,

$$L_i.w_j := -jw_{j+i} \qquad (j \neq -i),$$
$$L_i.w_{-i} := i(1+(i+1)\beta)w_0.$$

ii) $\beta = \infty$,

$$L_i.w_j := -jw_{j+i} \qquad (j \neq -i),$$
$$L_i.w_{-i} := i(i+1)w_0.$$

Notice that in [KaSa], the case $\alpha = \beta = \infty$ was missing owing to an error in the calculation of, what they call, the inverted module of $A(\alpha)$ and $B(\beta)$. These modules are clearly indecomposable. In fact, the next lemma holds:

Lemma 2.30. *For $\alpha, \beta \in \mathbb{C}P^1$,*

1. *$A(\alpha)$ is always reducible and its submodule $\bigoplus_{n\neq 0}\mathbb{C}w_n$ is isomorphic to the non-trivial proper submodule of $V_{-1,0} \cong A(0)$.*
2. *$B(\beta)$ is always reducible and its irreducible quotient $B(\beta)/\mathbb{C}w_0$ is isomorphic to the irreducible quotient of $V_{0,0} \cong B(0)$.*

Remark 1.7 shows the geometric nature of these modules. Indeed, under an isomorphism $V_{a,b} \cong t^{a-b}\mathbb{C}[t,t^{-1}](dt)^{-a}$, we have

$$V_{-1,0}/\bigoplus_{n\neq 0}\mathbb{C}w_n \cong \mathbb{C}\frac{dt}{t}, \qquad V_{0,0} \cong \mathbb{C}[t,t^{-1}] \quad (w_0 \longmapsto 1).$$

See also Proposition 1.7.

2.A.2 Classification Theorem

In the previous subsection, we have seen that $\dim \mathrm{Ext}^1_{\mathcal{C}_{\mathrm{adm}}}(\mathbb{C}1, \mathbb{C}[t,t^{-1}]/\mathbb{C}1)$ and $\dim \mathrm{Ext}^1_{\mathcal{C}_{\mathrm{adm}}}(\mathbb{C}[t,t^{-1}]/\mathbb{C}1, \mathbb{C}1)$ are at least 2. In fact, by the theorem due to I. Kaplansky and L. J. Santharoubane [KaSa] which is stated below, it follows that

$$\mathrm{Ext}^1_{\mathcal{C}_{\mathrm{adm}}}(\mathbb{C}1, \mathbb{C}[t,t^{-1}]/\mathbb{C}1) \cong \mathbb{C}^2, \qquad \mathrm{Ext}^1_{\mathcal{C}_{\mathrm{adm}}}(\mathbb{C}[t,t^{-1}]/\mathbb{C}1, \mathbb{C}1) \cong \mathbb{C}^2,$$

and each isomorphism class is represented by $A(\alpha)$ (resp. $B(\beta)$). Indeed, they have proved the next stronger statement:

Theorem 2.7 ([KaSa]) *Any $\mathbb{Z}$-graded indecomposable* Vir*-module whose weight multiplicities are* 1 *is one of the following modules:*

$$V_{a,b}\ (a,b \in \mathbb{C}), \qquad A(\alpha)\ (\alpha \in \mathbb{C}P^1), \qquad B(\beta)\ (\beta \in \mathbb{C}P^1).$$

Their proof is based on straightforward calculations. For its proof see [KaSa] together with [Ka] where the errors in the proof of the first lemma which states that, for a $\mathbb{Z}$-graded Vir-module $V = \bigoplus_{n\in\mathbb{Z}}\mathbb{C}w_n$,

if $L_{\pm 1}.w_j \neq 0$ $(\forall\, j \in \mathbb{Z})$, then $\exists\, a, b \in \mathbb{C}$ such that $V \cong V_{a,b}$,

was corrected.

We remark that the modules $A(\alpha)$ and $B(\beta)$ ($\alpha, \beta \in \mathbb{C}P^1$) have the following dualities:

1. The antipode-dual:

$$A(\alpha)^{\sharp a} \cong B(\alpha), \qquad B(\beta)^{\sharp a} \cong A(\beta).$$

2. The contragredient dual:

$$A(\alpha)^c \cong B\left(-\frac{\alpha}{2\alpha+1}\right), \qquad B(\beta)^c \cong A\left(-\frac{\beta}{2\beta+1}\right).$$

Chapter 3
The Jantzen Filtration

The Jantzen filtration is a very useful tool in the representation theory of Lie algebras. For example, the structure of Verma modules over a rank two simple Lie algebra (or a Kac–Moody algebra) is completely determined by means of the Jantzen filtration [Ja1].

In the case of the Virasoro algebra, the situation is similar to those in the cases of rank two simple Lie algebras, i.e., the Jantzen filtration reveals the structure of Verma modules. However, the original Jantzen filtration is not necessarily applicable to other Vir-modules, e.g. Fock modules. Hence a generalisation of the Jantzen filtration was proposed by B. L. Feigin and D. B. Fuchs in order to study the Vir-modules defined on the space of semi-infinite forms [FeFu4].

Here we will generalise the Jantzen filtration following [FeFu4]. (In this book, we call this generalisation the *Jantzen filtration à la Feigin and Fuchs.*) Recall that the original Jantzen filtration is constructed based on the contravariant form of a Verma module. On the other hand, the contravariant form can be regarded as a map from the Verma module to its contragredient dual. From this view point, we will generalise the Jantzen filtration based on an arbitrary homomorphism between two modules both of which have the same character.

In Section 3.1, we will introduce the Shapovalov form on $U(\mathfrak{g})$ of a Q-graded Lie algebra $\mathfrak{g}$ with a Q-graded anti-involution σ and a contravariant form on a $\mathfrak{g}$-module. These give us motivation for the definition, though here we will define the Jantzen filtration in a more general setting (see Subsection 3.1.3). In Section 3.2, we will recall the original definition of the Jantzen filtration and its properties. In Section 3.3, we will define the Jantzen filtration à la Feigin and Fuchs in an algebraic setting similar to the previous section. In Section 3.4, we will give a geometric interpretation of the filtration given in the previous section. In the last section, we will discuss the Jantzen filtration of some quotients of Verma modules.

Throughout this chapter, let $\mathbb{K}$ be an arbitrary field of characteristic 0 and we assume that a $\mathbb{K}$-algebra is associative and commutative with the unit.

K. Iohara, Y. Koga, *Representation Theory of the Virasoro Algebra*, Springer Monographs in Mathematics, DOI 10.1007/978-0-85729-160-8_3,

3.1 Motivation

In this section, let $\mathfrak{g}$ be a Q-graded Lie algebra over $\mathbb{K}$ with a Q-graded anti-involution σ (see § 1.2.2 for definition). In order to explain the motivation for the Jantzen filtration, we introduce the Shapovalov form of $U(\mathfrak{g})$ and contravariant forms on $\mathfrak{g}$-modules. The reader who is familiar with this subject may skip this section.

3.1.1 Shapovalov Forms

First, we remark that the Poincaré–Birkhoff–Witt theorem gives us the following decomposition as a vector space:

$$U(\mathfrak{g}) = U(\mathfrak{h}) \oplus \{\mathfrak{g}^- U(\mathfrak{g}) + U(\mathfrak{g})\mathfrak{g}^+\}.$$

Let $S(\mathfrak{h})$ be the symmetric algebra of $\mathfrak{h}$. Since $\mathfrak{h}$ is commutative, we have the following linear map

$$\pi : U(\mathfrak{g}) \longrightarrow U(\mathfrak{h}) \simeq S(\mathfrak{h}),$$

where $U(\mathfrak{g}) \longrightarrow U(\mathfrak{h})$ is the projection with respect to the above decomposition.

Definition 3.1 *The bilinear form*

$$F : U(\mathfrak{g}) \times U(\mathfrak{g}) \longrightarrow S(\mathfrak{h})$$

defined by

$$F(x, y) := \pi(\sigma(x)y) \quad (x, y \in U(\mathfrak{g}))$$

is called the **Shapovalov form** *of* $\mathfrak{g}$.

The universal enveloping algebra $U(\mathfrak{g})$ has the following decomposition:

$$U(\mathfrak{g}) = \bigoplus_{\beta \in Q} U(\mathfrak{g})_\beta,$$
$$U(\mathfrak{g})_\beta := \{x \in U(\mathfrak{g}) | [h, x] = \beta(h)x \ (\forall h \in \mathfrak{h})\}.$$

We define $U(\mathfrak{g}^\pm)_\beta$, $U(\mathfrak{g}^\geq)_\beta$ and $U(\mathfrak{g}^\leq)_\beta$ similarly.

We have

Proposition 3.1 *1. F is symmetric.*
2. F is contravariant, i.e., $F(zx, y) = F(x, \sigma(z)y)$ for any $x, y, z \in U(\mathfrak{g})$.
3. For $\beta_1, \beta_2 \in Q$ such that $\beta_1 \neq \beta_2$,

$$F(x, y) = 0 \quad (\forall x \in U(\mathfrak{g})_{\beta_1},\ \forall y \in U(\mathfrak{g})_{\beta_2}).$$

Proof. These assertions are proved by using $\sigma|_{\mathfrak{h}} = \mathrm{id}_{\mathfrak{h}}$. □

Next, we introduce the Shapovalov determinant. For each $\gamma \in Q^+$, we fix a basis $\{X_i | i \in I\}$ of $U(\mathfrak{g}^-)_{-\gamma}$.

Definition 3.2 *We set*

$$D_\gamma := \det\left(F(X_i, X_j)\right)_{i,j \in I} \in S(\mathfrak{h}),$$

and call it a **Shapovalov determinant** *of* $\mathfrak{g}$.

Needless to say, D_γ is determined up to a non-zero scalar depending on the choice of a basis.

3.1.2 Contravariant Forms

The Jantzen filtration of a $\mathfrak{g}$-module M is introduced based on a *contravariant form* on M. We define a contravariant form on a $\mathfrak{g}$-module.

Definition 3.3 *For a* $\mathfrak{g}$*-module* M*, a bilinear form*

$$\langle \cdot, \cdot \rangle : M \times M \longrightarrow \mathbb{K}$$

is said to be **contravariant** *if it satisfies*

$$\langle g.x, y \rangle = \langle x, \sigma(g).y \rangle \tag{3.1}$$

for any $g \in U(\mathfrak{g})$ *and* $x, y \in M$.

In the case where M is a highest weight module, the following proposition holds:

Proposition 3.2 *Any highest weight module has a unique, up to a constant factor, contravariant bilinear form.*

To show this proposition, we first consider the case where M is a Verma module.

Proposition 3.3 *The Verma module* $M(\lambda)$ *has a unique contravariant form* $\langle\ ,\ \rangle_\lambda$*, which satisfies*

$$\langle 1 \otimes \mathbf{1}_\lambda, 1 \otimes \mathbf{1}_\lambda \rangle_\lambda = 1,$$

where $1 \otimes \mathbf{1}_\lambda$ *is defined in § 1.2.5.*

Since the uniqueness of a contravariant form on $M(\lambda)$ follows from the property (3.1), it is enough to show the existence.

For X, $Y \in U(\mathfrak{g})$, we set

$$\langle X \otimes \mathbf{1}_\lambda, Y \otimes \mathbf{1}_\lambda \rangle_\lambda := F(X, Y)(\lambda)$$

by using $F(X,Y) \in S(\mathfrak{h}) \simeq \mathbb{K}[\mathfrak{h}^*]$. The form $\langle\ ,\ \rangle_\lambda$ is well-defined by the next lemma which follows from the Poincaré–Birkhoff–Witt theorem:

Lemma 3.1. *The universal enveloping algebra $U(\mathfrak{g})$ has the following decomposition:*

$$U(\mathfrak{g}) = U(\mathfrak{g})\{\mathfrak{g}^+ + \sum_{h\in\mathfrak{h}} \mathbb{K}(h - \lambda(h))\} \oplus U(\mathfrak{g}^-).$$

Proof of Proposition 3.3. For $X, Y \in U(\mathfrak{g})$, by Lemma 3.1,

$$X = X_1 + X_2, \quad Y = Y_1 + Y_2, \tag{3.2}$$

where

$$X_1, Y_1 \in U(\mathfrak{g}^-), \quad X_2, Y_2 \in U(\mathfrak{g})\{\mathfrak{g}^+ + \sum_{h\in\mathfrak{h}} \mathbb{K}(h - \lambda(h))\}.$$

Since $F(X_2, Z)(\lambda) = 0 = F(Z, Y_2)(\lambda)$ for any $Z \in U(\mathfrak{g})$, we have

$$F(X,Y)(\lambda) = F(X_1, Y_1)(\lambda).$$

Hence, the form $\langle\ ,\ \rangle_\lambda$ is well-defined. The contravariance $\langle\ ,\ \rangle_\lambda$ follows from Proposition 3.1. □

Proposition 3.1 implies that $\langle\ ,\ \rangle_\lambda$ is symmetric and

$$\langle\ ,\ \rangle_\lambda|_{M(\lambda)_\mu \times M(\lambda)_\nu} = 0, \tag{3.3}$$

for any $\mu, \nu \in \mathfrak{h}^*$ such that $\mu \neq \nu$.

By the contravariance of $\langle\ ,\ \rangle_\lambda$, the following proposition holds:

Proposition 3.4 *For any $\lambda \in \mathfrak{h}^*$, $\mathrm{rad}\langle\ ,\ \rangle_\lambda$ coincides with the maximal proper submodule $J(\lambda)$ of $M(\lambda)$, where*

$$\mathrm{rad}\langle\ ,\ \rangle_\lambda := \{u \in M(\lambda) | \langle v, u\rangle_\lambda = 0\ (\forall v \in M(\lambda))\}.$$

Proof of Proposition 3.2. By Proposition 3.3 and Proposition 3.4, a highest weight module with highest weight λ has the contravariant form induced from $\langle\ ,\ \rangle_\lambda$. □

Let us introduce some notation. For each $\lambda \in \mathfrak{h}^*$ and $\gamma \in Q^+$, we put

$$\det(\lambda)_\gamma := \det\left(\langle X_i \otimes \mathbf{1}_\lambda, X_j \otimes \mathbf{1}_\lambda\rangle_\lambda\right)_{i,j\in I},$$

where $\{X_i | i \in I\}$ is a basis of $U(\mathfrak{g}^-)_{-\gamma}$.

By definition, we have

$$\det(\lambda)_\gamma = D_\gamma(\lambda).$$

3.1.3 What is the Jantzen Filtration?

The Jantzen filtration of a Verma module is a decreasing filtration of $\mathfrak{g}$-submodules of the Verma module. Hence, we fix $\lambda \in \mathfrak{h}^*$ such that $M(\lambda)$ is reducible. How can one construct such a filtration?

Intuitively, the Janzten filtration

$$M(\lambda) \supset M(\lambda)(1) \supset M(\lambda)(2) \supset \cdots$$

of $M(\lambda)$ is introduced as follows. We fix a curve $C \subset \mathfrak{h}^*$ containing λ and consider a perturbation of λ along the curve C. To define the Jantzen filtration, we assume that

$$M(\mu) \text{ is irreducible for any } \mu \in C \setminus \{\lambda\}. \tag{3.4}$$

Under this assumption, we introduce a filtration on $M(\lambda)$ by counting the degree of 'degeneracy' of the contravariant forms $\langle\ ,\ \rangle_\mu$ $(\mu \in C)$ as $\mu \to \lambda$. Indeed, the Jantzen filtration satisfies $M(\lambda)(1) = J(\lambda)$ and thus, it gives us much information on the structure of Verma modules.

Next, let us algebraise the above construction of the Jantzen filtration. The idea is to use Verma modules over the polynomial ring $\mathbb{K}[t]$, where t corresponds to a local parameter of C at λ.

Let $\mathfrak{g}_{\mathbb{K}(t)}$ be the Lie algebra $\mathfrak{g} \otimes_{\mathbb{K}} \mathbb{K}(t)$ over the field $\mathbb{K}(t)$, the quotient fields of $\mathbb{K}[t]$, and let $\mathfrak{g}_{\mathbb{K}[t]}$ be the Lie algebra $\mathfrak{g} \otimes_{\mathbb{K}} \mathbb{K}[t]$ over the ring $\mathbb{K}[t]$. We denote the canonical map $\mathbb{K}[t] \twoheadrightarrow \mathbb{K}[t]/t\mathbb{K}[t] \simeq \mathbb{K}$ by ϕ, and the reduction functor from $\mathrm{Mod}_{\mathbb{K}[t]}$ to $\mathrm{Vect}_{\mathbb{K}}$ which sends $M \mapsto M \otimes_{\mathbb{K}[t]} (\mathbb{K}[t]/t\mathbb{K}[t])$ by the same notation ϕ.

Let $M_{\mathbb{K}(t)}(\lambda(t))$ be the Verma module over $\mathfrak{g}_{\mathbb{K}(t)}$ with highest weight $\lambda(t) \in \mathfrak{h}^* \otimes_{\mathbb{K}} \mathbb{K}[t]$. We fix $\lambda(t)$ such that $\lambda = \phi(\lambda(t))$. Moreover, we suppose that $\lambda(t)$ satisfies

$$M_{\mathbb{K}(t)}(\lambda(t)) \text{ is irreducible.} \tag{3.5}$$

Let $M_{\mathbb{K}[t]}(\lambda(t))$ be the $\mathbb{K}[t]$-sublattice of $M_{\mathbb{K}(t)}(\lambda(t))$ generated by $1 \otimes \mathbf{1}_{\lambda(t)} \in M_{\mathbb{K}(t)}(\lambda(t))$ as $\mathfrak{g}_{\mathbb{K}[t]}$-module. Note that it satisfies

$$\langle M_{\mathbb{K}[t]}(\lambda(t)), M_{\mathbb{K}[t]}(\lambda(t))\rangle_{\lambda(t)} \subset \mathbb{K}[t].$$

For $k \in \mathbb{Z}_{>0}$, we set

$$\begin{aligned} &M_{\mathbb{K}[t]}(\lambda(t))(k) \\ :=&\{v \in M_{\mathbb{K}[t]}(\lambda(t)) | \langle v, w\rangle_{\lambda(t)} \in t^k\mathbb{K}[t] \ (\forall w \in M_{\mathbb{K}[t]}(\lambda(t)))\}, \end{aligned}$$

and define the Jantzen filtration $\{M(\lambda)(k)\}_{k \in \mathbb{Z}_{\geq 0}}$ on $M(\lambda)$ by setting

$$M(\lambda)(k) := \mathrm{Im}\{M_{\mathbb{K}[t]}(\lambda(t))(k) \overset{\iota}{\hookrightarrow} M_{\mathbb{K}[t]}(\lambda(t)) \overset{\phi}{\twoheadrightarrow} M(\lambda)\},$$

where ι stands for a natural inclusion.

Remark that important properties of the Jantzen filtration in the above setting follow from the fact that $\mathbb{K}[t]$ is a principal ideal domain. Moreover, taking the property (3.3) into account, one may restrict to each weight subspace of $M_{\mathbb{K}[t]}(\lambda(t))$, which is a $\mathbb{K}[t]$-free module of finite rank. Hence, in the following subsections, we formulate the Jantzen filtration in terms of free modules of finite rank over a principal ideal domain.

3.2 The original Jantzen Filtration

We recall the original version of the Jantzen filtration [Ja1]. The reader who is familiar with this subject may skip this section.

3.2.1 Integral Form

Throughout this subsection, let $\mathcal{R}$ be a $\mathbb{K}$-algebra. For a $\mathbb{K}$-vector space V, we set $V_{\mathcal{R}} := V \otimes_{\mathbb{K}} \mathcal{R}$, and regard V as a subset of $V_{\mathcal{R}}$ via the map $v \mapsto v \otimes 1$. In the case where V is a Lie algebra or an associative algebra, we regard $V_{\mathcal{R}}$ as a Lie algebra or an associative algebra over $\mathcal{R}$ in a natural way.

Let $(\mathfrak{g}, \mathfrak{h})$ be a Q-graded Lie algebra over $\mathbb{K}$ with a Q-graded anti-involution σ. Here, following [Ja1], we recall a natural setting to work on the representation theory of $\mathfrak{g}_{\mathcal{R}}$.

$\mathfrak{g}_{\mathcal{R}}$ has the following triangular decomposition:

$$\mathfrak{g}_{\mathcal{R}} = \mathfrak{g}_{\mathcal{R}}^- \oplus \mathfrak{h}_{\mathcal{R}} \oplus \mathfrak{g}_{\mathcal{R}}^+,$$

where $\mathfrak{g}_{\mathcal{R}}^{\pm} := (\mathfrak{g}^{\pm})_{\mathcal{R}}$. For simplicity, we set $\mathfrak{g}_{\mathcal{R}}^{\gtrless} := (\mathfrak{g}^{\gtrless})_{\mathcal{R}}$ and $\mathfrak{h}_{\mathcal{R}}^* := (\mathfrak{h}^*)_{\mathcal{R}}$.

Remark 3.1 *One has the following isomorphisms:*

1. $\mathfrak{h}_{\mathcal{R}}^* \simeq \mathrm{Hom}_{\mathcal{R}}(\mathfrak{h}_{\mathcal{R}}, \mathcal{R})$,
2. $U(\mathfrak{g})_{\mathcal{R}} \simeq U_{\mathcal{R}}(\mathfrak{g}_{\mathcal{R}})$, *where the right-hand side is the universal enveloping algebra of an $\mathcal{R}$-Lie algebra $\mathfrak{g}_{\mathcal{R}}$.*

Indeed, the first isomorphism follows by definition, and the second one follows from the universality of universal enveloping algebras.

First, we introduce weight submodules of $\mathfrak{g}_{\mathcal{R}}$-modules and formal characters, and state some properties. Let $\tilde{M}$ be a $\mathfrak{g}_{\mathcal{R}}$-module. For $\tilde{\lambda} \in \mathfrak{h}_{\mathcal{R}}^*$, we set

$$\tilde{M}_{\tilde{\lambda}} := \{m \in \tilde{M} | H.m = \tilde{\lambda}(H)m \ (\forall H \in \mathfrak{h}_{\mathcal{R}})\}.$$

Note that if $\tilde{M}$ is a torsion-free $\mathcal{R}$-module then $\tilde{M}_{\tilde{\lambda}} = \tilde{M}_{\tilde{\mu}} \neq \{0\}$ implies $\tilde{\lambda} = \tilde{\mu}$. For $\tilde{\lambda}$ and $\tilde{\mu} \in \mathfrak{h}_{\mathcal{R}}^*$, we define

$$\tilde{\lambda} \geqslant \tilde{\mu} \Leftrightarrow \tilde{\lambda} - \tilde{\mu} \in Q^+,$$

where Q^+ was defined in § 1.2.2, and set $D(\tilde{\lambda}) := \{\tilde{\mu} \in \mathfrak{h}^*_{\mathcal{R}} | \tilde{\mu} \leqslant \tilde{\lambda}\}$. Let $\tilde{\mathcal{E}}$ be the $\mathbb{Z}$-algebra consisting of the elements of the form

$$\sum_{i=1}^{n} \sum_{\tilde{\mu} \in D(\tilde{\lambda}_i)} c_{\tilde{\mu}} e(\tilde{\mu}) \quad (n \in \mathbb{Z}_{>0},\ \tilde{\lambda}_i \in \mathfrak{h}^*_{\mathcal{R}},\ c_{\tilde{\mu}} \in \mathbb{Z}),$$

where the algebra structure on $\tilde{\mathcal{E}}$ is defined by $e(\tilde{\lambda})e(\tilde{\mu}) = e(\tilde{\lambda} + \tilde{\mu})$. For a $\mathfrak{g}_{\mathcal{R}}$-module $\tilde{M}$ such that $\tilde{M} = \bigoplus_{\lambda \in \mathfrak{h}^*_{\mathcal{R}}} \tilde{M}_{\tilde{\lambda}}$, where $\tilde{M}_{\tilde{\lambda}}$ is an $\mathcal{R}$-free module of finite rank for each $\tilde{\lambda}$, we define the formal character of $\tilde{M}$ as follows:

$$\operatorname{ch} \tilde{M} := \sum_{\tilde{\lambda} \in \mathfrak{h}^*_{\mathcal{R}}} (\operatorname{rk}_{\mathcal{R}} \tilde{M}_{\tilde{\lambda}}) e(\tilde{\lambda}) \in \tilde{\mathcal{E}}.$$

For a $\mathfrak{g}$-module M, we naturally regard $M_{\mathcal{R}}$ as a $\mathfrak{g}_{\mathcal{R}}$-module. Let M be a $\mathfrak{h}$-diagonalisable $\mathfrak{g}$-module (see Definition 1.10). Then, we have $\operatorname{ch} M_{\mathcal{R}} = \operatorname{ch} M$.

Next, we introduce Verma modules over $\mathfrak{g}_{\mathcal{R}}$. For $\tilde{\lambda} \in \mathfrak{h}^*_{\mathcal{R}}$, we define the $\mathfrak{g}^{\geqslant}_{\mathcal{R}}$-module structure on the $\mathcal{R}$-free module $\mathcal{R}_{\tilde{\lambda}} := \mathcal{R}\mathbf{1}_{\tilde{\lambda}}$ by

1. $h.\mathbf{1}_{\tilde{\lambda}} = \tilde{\lambda}(h)\mathbf{1}_{\tilde{\lambda}}$ for $h \in \mathfrak{h}_{\mathcal{R}}$,
2. $x.\mathbf{1}_{\tilde{\lambda}} = 0$ for $x \in \mathfrak{g}^+_{\mathcal{R}}$.

For $\tilde{\lambda} \in \mathfrak{h}^*_{\mathcal{R}}$, we set

$$M_{\mathcal{R}}(\tilde{\lambda}) := U_{\mathcal{R}}(\mathfrak{g}_{\mathcal{R}}) \otimes_{U_{\mathcal{R}}(\mathfrak{g}^{\geqslant}_{\mathcal{R}})} \mathcal{R}_{\tilde{\lambda}}, \tag{3.6}$$

and call it the Verma module over $\mathfrak{g}_{\mathcal{R}}$ with highest weight $\tilde{\lambda}$. It has the following decomposition:

$$M_{\mathcal{R}}(\tilde{\lambda}) = \bigoplus_{\gamma \in Q^+} M_{\mathcal{R}}(\tilde{\lambda})_{\tilde{\lambda} - \gamma}.$$

Note that, if $\lambda \in \mathfrak{h}^* \subset \mathfrak{h}^*_{\mathcal{R}}$, then the following isomorphism of $\mathfrak{g}_{\mathcal{R}}$-modules holds:

$$M_{\mathcal{R}}(\lambda) \simeq M(\lambda)_{\mathcal{R}}.$$

We set $v_{\tilde{\lambda}} := 1 \otimes \mathbf{1}_{\tilde{\lambda}} \in M_{\mathcal{R}}(\tilde{\lambda})$. Then, by an argument similar to Proposition 3.3, one can show that there uniquely exists a contravariant form $\langle\ ,\ \rangle_{\tilde{\lambda}}$ on $M_{\mathcal{R}}(\tilde{\lambda})$ such that $\langle v_{\tilde{\lambda}}, v_{\tilde{\lambda}} \rangle_{\tilde{\lambda}} = 1$.

Let $\mathcal{A}$ be a $\mathbb{K}$-algebra, and let $\phi : \mathcal{R} \to \mathcal{A}$ be a homonorphism of $\mathbb{K}$-algebras. In the sequel, we regard $\mathcal{A}$-modules as $\mathcal{R}$-modules via the homomorphism ϕ.

For a $\mathbb{K}$-vector space V, let V_ϕ be the map defined by

$$V_{\mathcal{R}} \longrightarrow V_{\mathcal{A}};\quad v \otimes a \mapsto v \otimes \phi(a) \quad (v \in V,\ a \in \mathcal{R}).$$

By definition, this is a homomorphism of $\mathcal{R}$-modules. Moreover, if V is an associative algebra (resp. a Lie algebra), then V_ϕ is a homomorphism of associative algebras (resp. Lie algebras). For $\tilde{\lambda} \in \mathfrak{h}^*_{\mathcal{R}}$, we set $\tilde{\lambda}' := (\mathfrak{h}^*)_\phi(\tilde{\lambda}) \in \mathcal{A}$. Note that ϕ defines a homomorphism of $\mathfrak{g}^{\geq}_{\mathcal{R}}$-modules

$$\mathcal{R}_{\tilde{\lambda}} \longrightarrow \mathcal{A}_{\tilde{\lambda}'}; \quad a.\mathbf{1}_{\tilde{\lambda}} \mapsto \phi(a).\mathbf{1}_{\tilde{\lambda}'} \quad (a \in \mathcal{R}).$$

Since $U(\mathfrak{g})_\phi : U(\mathfrak{g})_{\mathcal{R}} \to U(\mathfrak{g})_{\mathcal{A}}$ is a homomorphism of right $\mathfrak{g}^{\geq}_{\mathcal{R}}$-modules, the map

$$M(\tilde{\lambda})_\phi : M_{\mathcal{R}}(\tilde{\lambda}) \longrightarrow M_{\mathcal{A}}(\tilde{\lambda}'); \quad u.v_{\tilde{\lambda}} \mapsto (U(\mathfrak{g})_\phi u).v_{\tilde{\lambda}'} \quad (a \in U(\mathfrak{g}_{\mathcal{R}}))$$

is well-defined. Moreover, since $U(\mathfrak{g})_\phi$ is a homomorphism of left $\mathfrak{g}_{\mathcal{R}}$-modules, $M(\tilde{\lambda})_\phi$ is a homomorphism of $\mathfrak{g}_{\mathcal{R}}$-modules, where $\mathfrak{g}_{\mathcal{R}}$-module structure on $M_{\mathcal{A}}(\tilde{\lambda}')$ is defined via the map $\mathfrak{g}_\phi : \mathfrak{g}_{\mathcal{R}} \to \mathfrak{g}_{\mathcal{A}}$. Note that, by definition, the following formula holds:

$$\phi(\langle v, v' \rangle_{\tilde{\lambda}}) = \langle M(\tilde{\lambda})_\phi v, M(\tilde{\lambda})_\phi v' \rangle_{\tilde{\lambda}'} \quad (v,\ v' \in M_{\mathcal{R}}(\tilde{\lambda})).$$

From now on, we denote $\mathfrak{g}_\phi$, $\mathfrak{g}^{\geq}_\phi$, $\mathfrak{h}^*_\phi$, $U(\mathfrak{g})_\phi$, $M(\tilde{\lambda})_\phi$ etc by the same notation ϕ for simplicity.

Let $K/\mathbb{K}$ be an extension of fields. By Proposition 3.4, for $\tilde{\lambda} \in \mathfrak{h}^*_K$, the Verma module $M_K(\tilde{\lambda})$ ($\tilde{\lambda} \in \mathfrak{h}^*_K$) has a unique maximal proper submodule $J_K(\tilde{\lambda}) = \mathrm{rad}\langle\ ,\ \rangle_{\tilde{\lambda}}$. If $\lambda \in \mathfrak{h}^* \subset \mathfrak{h}^*_K$, then the irreducible quotient $L_K(\lambda) := M_K(\lambda)/J_K(\lambda)$ satisfies

$$L_K(\lambda) = L(\lambda) \otimes_{\mathbb{K}} K,$$

since a contravariant form on $M(\lambda)$ can be naturally extended to the contravariant form on $M_K(\lambda)$.

In general, a highest weight module over $\mathfrak{g}_{\mathcal{R}}$ is defined as follows:

Definition 3.4 *A $\mathfrak{g}_{\mathcal{R}}$-module $\tilde{M}$ is called a highest weight module with highest weight $\tilde{\lambda} \in \mathfrak{h}^*_{\mathcal{R}}$, if*

1. *$\tilde{M}_{\tilde{\lambda}}$ is an $\mathcal{R}$-free module,*
2. *there exists a non-zero element $v \in \tilde{M}_{\tilde{\lambda}}$ such that*

$$\mathfrak{g}^+_{\mathcal{R}}.v = \{0\}, \quad \tilde{M} = U_{\mathcal{R}}(\mathfrak{g}^-_{\mathcal{R}})v, \quad \tilde{M}_{\tilde{\lambda}} = \mathcal{R}v.$$

Let $\phi : \mathcal{R} \to K$ be a homomorphism of $\mathbb{K}$-algebras. We regard K as an $\mathcal{R}$-module via ϕ. Then, $\tilde{M} \otimes_\phi K (\simeq \tilde{M}/(\mathrm{Ker}\phi)\tilde{M})$ is a highest weight $\mathfrak{g}$-module with highest weight $\phi(\lambda)$. Moreover, if $\tilde{M}_{\tilde{\mu}}$ is an $\mathcal{R}$-free module for any $\tilde{\mu}$ such that $\tilde{M}_{\tilde{\mu}} \neq \{0\}$, then the following formula holds:

$$\mathrm{ch}(\tilde{M} \otimes_\phi K) = e(\phi(\lambda) - \lambda)\, \mathrm{ch}\, \tilde{M}.$$

From now on, we assume that $\mathcal{R}$ is an integral domain. Let $Q(\mathcal{R})$ be the quotient field of $\mathcal{R}$.

Take $\tilde{\lambda} \in \mathfrak{h}^*_{\mathcal{R}} \subset \mathfrak{h}^*_{Q(\mathcal{R})}$. Let $L_{Q(\mathcal{R})}(\tilde{\lambda})$ be the irreducible highest weight module with highest weight $\tilde{\lambda}$. Set $\bar{v}_{\tilde{\lambda}} := v_{\tilde{\lambda}} + J_{Q(\mathcal{R})}(\tilde{\lambda}) \in L_{Q(\mathcal{R})}(\tilde{\lambda})$ and

$$\tilde{M} := U_{\mathcal{R}}(\mathfrak{g}_{\mathcal{R}}).\bar{v}_{\tilde{\lambda}}.$$

Since $\mathfrak{g}^+_{\mathcal{R}}.\bar{v}_{\tilde{\lambda}} = \{0\}$ and $U_{\mathcal{R}}(\mathfrak{h}_{\mathcal{R}}).\bar{v}_{\tilde{\lambda}} = \mathcal{R}\bar{v}_{\tilde{\lambda}}$, we obtain

$$\tilde{M} = U_{\mathcal{R}}(\mathfrak{g}^-_{\mathcal{R}}).\bar{v}_{\tilde{\lambda}}$$

and it is a highest weight module with highest weight $\tilde{\lambda}$. Moreover, $\tilde{M}$ is torsion-free as $\mathcal{R}$-module and $\tilde{M}_{\tilde{\mu}}$ is a finitely generated $\mathcal{R}$-module for each $\tilde{\mu} \in \mathfrak{h}^*_{\mathcal{R}}$ since $\tilde{M} \subset M_{Q(\mathcal{R})}(\tilde{\lambda})$. Hence, if $\mathcal{R}$ is a principal ideal domain, then each $\mathcal{R}$-submodule $\tilde{M}_{\tilde{\mu}}$ is an $\mathcal{R}$-free module. In particular, we have

$$L_{Q(\mathcal{R})}(\tilde{\lambda})_{\tilde{\mu}} = Q(\mathcal{R})\tilde{M}_{\tilde{\mu}}$$

for any $\tilde{\mu} \in \mathfrak{h}^*_{\mathcal{R}}$, from which we conclude $\operatorname{ch} \tilde{M} = \operatorname{ch} L_{Q(\mathcal{R})}(\tilde{\lambda})$.

3.2.2 Definitions

Throughout this subsection, let $\mathcal{R}$ be a $\mathbb{K}$-algebra and a principal ideal domain. We fix a prime element $t \in \mathcal{R}$ and denote the residue field $\mathcal{R}/t\mathcal{R}$ by K. Let $\phi : \mathcal{R} \twoheadrightarrow K$ be the canonical map, and regard K as an $\mathcal{R}$-module via this map. Here, we denote the functor $M \longmapsto M \otimes_{\mathcal{R}} K$ from the category $\mathrm{Mod}_{\mathcal{R}}$ of $\mathcal{R}$-modules to the category Vect_K of K-vector spaces by the same notation ϕ. Note that, for a morphism f in $\mathrm{Mod}_{\mathcal{R}}$, the morphism $\phi(f)$ in Vect_K is defined by $f \otimes \mathrm{id}_K$. In the sequel, we call ϕ a reduction functor.

Let $\tilde{M}$ be an $\mathcal{R}$-free module of rank $r \in \mathbb{Z}_{>0}$ with a *non-degenerate* symmetric bilinear form

$$(\cdot,\cdot)_{\tilde{M}} : \tilde{M} \times \tilde{M} \longrightarrow \mathcal{R}.$$

We set $M := \phi\tilde{M}$. The Jantzen filtration is defined as a filtration on this K-vector space M. Note that the K-vector space M admits a symmetric bilinear form induced from $(\cdot,\cdot)_{\tilde{M}}$, defined by

$$(\phi v_1, \phi v_2) := \phi((v_1, v_2)_{\tilde{M}}) \quad (v_1, v_2 \in \tilde{M}).$$

Here, by abuse of langauage, we set $\phi v := v \otimes 1 \in \phi\tilde{N}$ for $\tilde{N} \in \mathrm{Ob}(\mathrm{Mod}_{\mathcal{R}})$. In the case when $\tilde{N}$ is $\mathcal{R}$-free, there exists a $\mathbb{K}$-vector space V such that $\tilde{N} \cong V \otimes_{\mathbb{K}} \mathcal{R}$. In particular, ϕ defined here can be identified with V_ϕ defined in the previous subsection.

The original Jantzen filtration is defined as follows:

Definition 3.5 ([Ja1]) *For $m \in \mathbb{Z}_{\geq 0}$, we set*

$$\tilde{M}(m) := \{v \in \tilde{M} \mid (v, \tilde{M})_{\tilde{M}} \subset t^m \mathcal{R}\ \}.$$

Let $\iota_m : \tilde{M}(m) \hookrightarrow \tilde{M}$ be a natural embedding. Then

$$M(m) := \mathrm{Im}\phi(\iota_m)$$

defines a filtration of the K-vector space M

$$M = M(0) \supset M(1) \supset M(2) \supset \cdots.$$

This filtration is called the **Jantzen filtration***.*

We remark that, for $v \in \tilde{M}(m)$, one has $\phi(\iota_m)(\phi v) = \phi\iota_m(v)$.

Remark 3.2 *Since the reduction functor is in general not left exact, it does not necessarily preserve kernels and the space $M(m)$ does not necessarily coincide with $\phi\tilde{M}(m)$.*

3.2.3 Basic Properties

In general, the Jantzen filtration $\{M(m)\}$ of M enjoy the following properties.

Proposition 3.5 *1.*

$$\bigcap_{m \in \mathbb{Z}_{\geq 0}} M(m) = \{0\}.$$

2. $M(1) = \mathrm{rad}\ (\cdot,\cdot)$, *where* $\mathrm{rad}\ (\cdot,\cdot) = \{v \in M | (v, M) = \{0\}\}$.
3. There exists a symmetric bilinear form $(\cdot,\cdot)_m$ on $M(m)$ such that

$$\mathrm{rad}\ (\cdot,\cdot)_m = M(m+1).$$

To show this proposition, we recall the following well-known lemma:

Lemma 3.2. *Any submodule $\tilde{N}$ of $\tilde{M}$ is $\mathcal{R}$-free of rank s less than or equal to r. Moreover, there exists an $\mathcal{R}$-free basis $\{v_1, \cdots, v_r\}$ of $\tilde{M}$ such that*

$$\tilde{N} = \bigoplus_{i=1}^{s} \mathcal{R} a_i v_i$$

for some $a_i \in \mathcal{R}$ $(i = 1, \cdots, s)$.

Lemma 3.2 allows us the following description:

Lemma 3.3. *1. Let us set* $\tilde{M}^* := \mathrm{Hom}_{\mathcal{R}}(\tilde{M}, \mathcal{R})$*. Then, there exist*

$$\{e_1^*, e_2^*, \cdots, e_r^*\} : \mathcal{R}\text{-free basis of } \tilde{M}^*,$$
$$\{f_1, f_2, \cdots, f_r\} : \mathcal{R}\text{-free basis of } \tilde{M}$$

and elements $\{a_1, a_2, \cdots, a_r\} \subset \mathcal{R}$ *which satisfy*

$$(f_i, \cdot)_{\tilde{M}} = a_i e_i^*$$

for $i = 1, 2, \cdots, r$.
2. For the above $\mathcal{R}$*-free bases, we have*

$$\tilde{M}(m) = \bigoplus_{i;\nu_t(a_i)<m} \mathcal{R}t^{m-\nu_t(a_i)} f_i \oplus \bigoplus_{i;\nu_t(a_i)\geq m} \mathcal{R}f_i.$$

Proof. If we set $\tilde{N} := \{(v, \cdot)_{\tilde{M}} \mid v \in \tilde{M}\}$, then $\tilde{N}$ is an $\mathcal{R}$-free submodule of $\tilde{M}^*$ of rank r, since the form $(\cdot, \cdot)_{\tilde{M}}$ is non-degenerate on $\tilde{M}$. By Lemma 3.2, there exists an $\mathcal{R}$-free basis $\{e_1^*, e_2^*, \cdots, e_r^*\}$ of $\tilde{M}^*$ and $a_1, a_2, \cdots, a_r \in \mathcal{R}$ such that

$$\tilde{N} = \bigoplus_{i=1}^{r} \mathcal{R}a_i e_i^*.$$

Taking $f_i \in \tilde{M}$ satisfying $(f_i, \cdot)_{\tilde{M}} = a_i e_i^*$, we have proved the first statement. The second statement is an immediate consequence of the first one. □

Proof of Proposition 3.5. The first assertion of the proposition is an immediate consequence of Lemma 3.3. The second assertion of the proposition is obvious from the definition of the bilinear form $(\cdot, \cdot)$. Hence we show the last assertion. To do it, let us define the bilinear form $(\cdot, \cdot)_m$ on $M(m)$ as follows:

$$(\phi v_1, \phi v_2)_m := \phi(t^{-m}(v_1, v_2)_{\tilde{M}}) \quad (v_1, v_2 \in \tilde{M}(m)).$$

Indeed, this bilinear form is well-defined, since

$$\begin{aligned} t^{-m}(v, \tilde{M}(m))_{\tilde{M}} &\subset t^{-m}(t\tilde{M}, \tilde{M}(m))_{\tilde{M}} \\ &= t^{-m+1}(\tilde{M}, \tilde{M}(m))_{\tilde{M}} \\ &\subset t^{m-m+1}\mathcal{R} \end{aligned}$$

for $v \in \tilde{M}(m) \cap t\tilde{M}$.

Thus, to complete the proof, it is enough to verify that the radical of $(\cdot, \cdot)_m$ coincides with $M(m+1)$. First, we show that $M(m+1) \subset \mathrm{rad}\ (\cdot, \cdot)_m$. For $v \in \tilde{M}(m+1)$, we have

$$t^{-m}(v, \tilde{M}(m))_{\tilde{M}} \subset t^{-m}(v, \tilde{M})_{\tilde{M}} \subset t^{m-m+1}\mathcal{R}.$$

Hence, $\phi v \in \mathrm{rad}\,(\cdot,\cdot)_m$. The opposite inclusion follows from Lemma 3.3. 2. □

Recall that the Verma module $M(\lambda)$ over a Q-graded Lie algebra with a Q-graded anti-involution possesses a contravariant bilinear form. One can define the Jantzen filtration $\{M(\lambda)(l)\}$ of $M(\lambda)$ by the contravariant form. The next corollary follows from Propositions 3.4 and 3.5:

Corollary 3.1 $M(\lambda)(1)$ *coincides with the maximal proper submodule of* $M(\lambda)$.

3.2.4 Character Sum

Next, we discuss the *character sum*,

$$\sum_{m=1}^{\infty} \dim_K M(m)$$

which is one of the important properties of the Jantzen filtration. This quantity can be expressed in terms of the valuation of the determinant of the form $(\cdot,\cdot)_{\tilde{M}}$ on $\tilde{M}$ as follows. For a fixed $\mathcal{R}$-free basis $\{e_1, \cdots, e_r\}$ of $\tilde{M}$, set

$$\tilde{D} := \det\left((e_i, e_j)_{\tilde{M}}\right)_{1\le i,j\le r}.$$

$\tilde{D}$ is determined up to multiplication of the square of a unit of $\mathcal{R}$, since any two $\mathcal{R}$-free bases of $\tilde{M}$ can be transformed to each other by an element of $\mathrm{GL}_r(\mathcal{R})$. $\tilde{D}$ is called the *discriminant* of the form $(\cdot,\cdot)_{\tilde{M}}$.

As an application of Lemma 3.3, we obtain

Proposition 3.6

$$\nu_t(\tilde{D}) = \sum_{m=1}^{\infty} \dim_K M(m).$$

Proof. Let $\{e_1^*, e_2^*, \cdots, e_r^*\}$ be the $\mathcal{R}$-free basis of $\tilde{M}^*$ given in Lemma 3.3 and let $\{e_1, e_2, \cdots, e_r\}$ be the $\mathcal{R}$-dual basis. Further, let $\{f_1, f_2, \cdots, f_r\}$ be the $\mathcal{R}$-free bases of $\tilde{M}$ such that

$$(f_i, \cdot)_{\tilde{M}} = a_i e_i^*.$$

By using these $\mathcal{R}$-free bases, we can express $\nu_t(\tilde{D})$ in terms of $\nu_t(a_i)$. We notice that there exists $X := (x_{i,j}) \in \mathrm{GL}_r(\mathcal{R})$ such that $e_i = \sum_{j=1}^{r} x_{j,i} f_j$, since both $\{e_1, e_2, \cdots, e_r\}$ and $\{f_1, f_2, \cdots, f_r\}$ are $\mathcal{R}$-free bases of $\tilde{M}$.

Then, we have

$$\begin{aligned}\tilde{D} &= \det\left((e_i, e_j)_{\tilde{M}}\right)_{1\le i,j\le r}\\ &= \det\left((e_i, \sum_{k=1}^{r} x_{k,j} f_k)_{\tilde{M}}\right)_{1\le i,j\le r}\\ &= \prod_{i=1}^{r} a_i \times \det X,\end{aligned}$$

since

$$(f_i, e_j) = a_i e_i^*(e_j) = \delta_{i,j} a_i.$$

Therefore, we obtain

$$\nu_t(\tilde{D}) = \sum_{i=1}^{r} \nu_t(a_i),$$

since $X \in \mathrm{GL}_r(\mathcal{R})$, $\det X$ is an unit of $\mathcal{R}$.

On the other hand, the right-hand side of the proposition can be computed from the fact that

$$M(m) = \bigoplus_{i;\nu_t(a_i)\ge m} K\phi f_i,$$

which is also an immediate consequence of Lemma 3.3. We have

$$\begin{aligned}\sum_{m=1}^{\infty} \dim_K M(m) &= \sum_{m=1}^{\infty} \sharp\{i; \nu_t(a_i) \ge m\}\\ &= \sum_{i=1}^{r} \nu_t(a_i)\\ &= \nu_t(\tilde{D}).\end{aligned}$$

Thus, we have completed the proof. □

Finally, let us remark on the relation between the original Jantzen filtration and its generalisation constructed in the next section.

The original Jantzen filtration is constructed based on the bilinear form $(\cdot,\cdot)_{\tilde{M}}$ on $\tilde{M}$. On the other hand,

$$\tilde{M} \longrightarrow \tilde{M}^* \text{ such that } m \mapsto (m,\cdot)_{\tilde{M}} \tag{3.7}$$

induces an isomorphism between $Q(\mathcal{R})$-vector spaces $\tilde{M}_{Q(\mathcal{R})}$ and $\tilde{M}^*_{Q(\mathcal{R})}$. The original Jantzen filtration can be regarded as filtration based on the morphism (3.7).

In the next section, we will generalise the original Jantzen filtration starting from an arbitrary morphism $\tilde{S} : \tilde{V} \to \tilde{W}$ of $\mathcal{R}$-free modules of the same rank, which naturally extends to an isomorphism $\tilde{V}_{Q(\mathcal{R})} \to \tilde{W}_{Q(\mathcal{R})}$ of $Q(\mathcal{R})$-vector spaces.

3.3 The Jantzen Filtration à la Feigin and Fuchs I

In this section, we introduce a generalisation of the Jantzen filtration.

3.3.1 Definitions and Properties

For the reader's convenience, we recall some notation for commutative rings. Let $\mathcal{R}$ be a $\mathbb{K}$-algebra, which is a principal integral domain. We denote the quotient field of $\mathcal{R}$ by $Q(\mathcal{R})$. We fix a prime element $t \in \mathcal{R}$, and set $K := \mathcal{R}/t\mathcal{R}$. Let $\nu_t : \mathcal{R} \to \mathbb{Z}_{\geq 0} \cup \{\infty\}$ be the t-adic valuation, and let ϕ be the reduction functor from the category of $\mathcal{R}$-modules to the category of K-vector spaces defined in § 3.2.1.

The following notation is necessary to define the filtration: Let $\tilde{V}$ and $\tilde{W}$ be $\mathcal{R}$-free modules of the same rank, say $r \in \mathbb{Z}_{>0}$. Let $\tilde{S} : \tilde{V} \to \tilde{W}$ be a homomorphism of $\mathcal{R}$-modules whose extension $\tilde{S} \otimes \mathrm{id}_{Q(\mathcal{R})} : \tilde{V}_{Q(\mathcal{R})} \to \tilde{W}_{Q(\mathcal{R})}$ is an isomorphism of $Q(\mathcal{R})$-vector spaces. We denote the extension $\tilde{S} \otimes \mathrm{id}_{Q(\mathcal{R})}$ by $\tilde{S}_{Q(\mathcal{R})}$. Moreover, we set $V := \phi\tilde{V}$, $W := \phi\tilde{W}$ and $S := \phi(\tilde{S}) : V \to W$.

Definition 3.6 *Suppose that $m \in \mathbb{Z}_{\geq 0}$. A K-subspace $V(m)$ of V is defined by*

$$\begin{aligned}\tilde{V}(m) &:= \tilde{S}^{-1}(\mathrm{Im}\tilde{S} \cap t^m \tilde{W}),\\ V(m) &:= \mathrm{Im}\phi(\iota_m^V),\end{aligned}$$

where $\iota_m^V : \tilde{V}(m) \hookrightarrow \tilde{V}$ is a natural embedding. The quotient space $W(m)$ of W is defined by

$$\begin{aligned}\widetilde{IK}(m) &:= t^{-m}(\mathrm{Im}\tilde{S} \cap t^m \tilde{W}),\\ IK(m) &:= \mathrm{Im}\phi(\iota_m^{IK}),\\ W(m) &:= \begin{cases} W/IK(m-1) & m > 0\\ W & m = 0\end{cases},\end{aligned}$$

where $\iota_m^{IK} : \widetilde{IK}(m) \hookrightarrow \tilde{W}$ is a natural embedding. We denote the projection $W \to W(m)$ by π_m.

Note that we use the notation $IK(m)$ to denote the image $\mathrm{Im}\phi(\iota_m^{IK})$ which is the kernel of $W \to W(m+1)$ at the same time (not Iohara and Koga!).

By the definition, we have the following filtration:

$$V = V(0) \supset V(1) \supset V(2) \supset \cdots,$$

which is called the **Jantzen filtration**, and one can check

$$\bigcap_{m=1}^{\infty} V(m) = \{0\}, \tag{3.8}$$

in a way similar to the proof of Proposition 3.5. On the other hand, there exists a sequence of canonical projections

$$W = W(0) \twoheadrightarrow W(1) \twoheadrightarrow W(2) \twoheadrightarrow \cdots.$$

Here and after, we call this sequence the **Jantzen cofiltration**. The relation between the filtration $\{V(m)\}$ and the original one is the following:

Remark 3.3 *If* $\tilde{V} = \tilde{M}$, $\tilde{W} = \tilde{M}^*$ *and*

$$\tilde{S} : \tilde{M} \longrightarrow \tilde{M}^* \quad \text{such that} \quad v \mapsto (v, \cdot)_{\tilde{M}},$$

then the new filtration coincides with the original Jantzen filtration.

Next, we introduce 'higher derivatives' $S^{(m)}$ of the map $S : V \to W$.

Definition-Lemma 3.1 *Suppose that* $m \in \mathbb{Z}_{>0}$. *For* $u \in V(m)$, *take* $\tilde{u} \in \tilde{V}(m)$ *such that* $u = \phi(\iota_m^V)(\phi\tilde{u})$. *Then, the element*

$$\pi_m \circ \phi(\iota_m^{IK}) \left(\phi t^{-m} \tilde{S}(\tilde{u})\right)$$

does not depend on the choice of $\tilde{u}$. *Hence, the map*

$$S^{(m)} : V(m) \to W(m), \quad u \longmapsto \pi_m \circ \phi(\iota_m^{IK}) \left(\phi t^{-m} \tilde{S}(\tilde{u})\right)$$

is well-defined and is called the m*th derivative.*

Proof. It suffices to check that

$$\pi_m \circ \phi(\iota_m^{IK}) \left(\phi t^{-m} \tilde{S}(\tilde{u})\right) = 0$$

for any $\tilde{u} \in \tilde{V}(m) \cap t\tilde{V}$. To show this, we notice the following fact:

$$\tilde{V}(m) \cap t\tilde{V} = t\tilde{V}(m-1),$$

which can be proved easily. Hence, there exists $\tilde{v} \in \tilde{V}(m-1)$ such that $\tilde{u} = t\tilde{v}$, thus we have

$$t^{-m}\tilde{S}(\tilde{u}) = t^{-m+1}\tilde{S}(\tilde{v}) \in \widetilde{IK}(m-1),$$

i.e.,

$$\phi(\iota_m^{IK}) \left(\phi t^{-m} \tilde{S}(\tilde{u})\right) \in IK(m-1).$$

□

For simplicity, let us set $S^{(0)} := S$. The mth step of the Jantzen (co)-filtration $V(m)$ and $W(m)$ is characterised by $S^{(m-1)} : V(m-1) \to W(m-1)$. In fact, we have

Proposition 3.7 *For any $m \in \mathbb{Z}_{>0}$, the following hold:*

1. $V(m) = \mathrm{Ker}S^{(m-1)}$,
2. $W(m) = \mathrm{Coker}S^{(m-1)}$.

Proof. First, let us prove that $V(m) = \mathrm{Ker}S^{(m-1)}$. Since $V(m) \subset \mathrm{Ker}S^{(m-1)}$ is clear by definition, we show that $V(m) \supset \mathrm{Ker}S^{(m-1)}$. We take $u \in \mathrm{Ker}S^{(m-1)}$. There exists $\tilde{u} \in \tilde{V}(m-1)$ such that $\phi(\iota^{V}_{m-1})(\phi\tilde{u}) = u$. It follows from $u \in \mathrm{Ker}S^{(m-1)}$ that

$$\phi(\iota^{IK}_{m-1})\left(\phi t^{-m+1}\tilde{S}(\tilde{u})\right) \in IK(m-2),$$

which implies

$$t^{-m+1}\tilde{S}(\tilde{u}) \in \widetilde{IK}(m-2) + t\tilde{W},$$

i.e.,

$$\tilde{S}(\tilde{u}) \in \left\{t(t^{m-2}\tilde{W} \cap \mathrm{Im}\tilde{S}) + t^{m}\tilde{W} \cap \mathrm{Im}\tilde{S}\right\}.$$

Using the fact that $\tilde{S}$ is an isomorphism, we obtain

$$\tilde{u} \in t\tilde{V}(m-2) + \tilde{V}(m).$$

Hence $u \in V(m)$ and thus $V(m) \supset \mathrm{Ker}S^{(m-1)}$ is proved.

Second, let us prove that $W(m) = \mathrm{Coker}S^{(m-1)}$. By definition, it is enough to show that

$$\mathrm{Im}S^{(m-1)} = \pi_{m-1}(IK(m-1)).$$

This follows from the definition of $S^{(m-1)}$. □

3.3.2 Character Sum

Similarly to the original version, the character sum of $\{V(m)|m \in \mathbb{Z}_{>0}\}$ can be expressed in terms of the valuation of the determinant of $\tilde{S}$. Let us denote the determinant of the map $\tilde{S}$ by $\tilde{D}$, namely, for $\mathcal{R}$-free bases $\{v_1, \cdots, v_r\}$ and $\{w_1, \cdots, w_r\}$ of $\tilde{V}$ and $\tilde{W}$ such that

$$\tilde{S}(v_i) = \sum_{j=1}^{r} s_{i,j}w_j,$$

we set

$$\tilde{D} := \det(s_{i,j})_{1\le i,j\le r}.$$

As was mentioned before, this determinant is determined up to multiplication of a unit of $\mathcal{R}$. We have

Proposition 3.8

$$\nu_t(\tilde{D}) = \sum_{m=1}^{\infty} \dim_K V(m).$$

To show this proposition, we use the following lemma.

Lemma 3.4. *1. There exist*

$$\{f_1, f_2, \cdots, f_r\} : \mathcal{R}\text{-free basis of } \tilde{V},$$
$$\{e_1, e_2, \cdots, e_r\} : \mathcal{R}\text{-free basis of } \tilde{W}$$

and $\{a_1, a_2, \cdots, a_r\} \subset \mathcal{R}$ *which satisfy*

$$\tilde{S}(f_i) = a_i e_i$$

for $i = 1, 2, \cdots, r$.
2. For the above $\mathcal{R}$*-free bases, the following hold:*

$$\tilde{V}(m) = \bigoplus_{i;\nu_t(a_i)<m} \mathcal{R}t^{m-\nu_t(a_i)} f_i \oplus \bigoplus_{i;\nu_t(a_i)\geq m} \mathcal{R}f_i,$$

$$\widetilde{IK}(m) = \bigoplus_{i;\nu_t(a_i)<m} \mathcal{R}t^{-\nu_t(a_i)} a_i e_i \oplus \bigoplus_{i;\nu_t(a_i)\geq m} \mathcal{R}t^{-m} a_i e_i.$$

Proof. The first assertion easily follows from Lemma 3.2. The second assertion is a direct consequence of the first one. □

Proof of Proposition 3.8. Taking $\mathcal{R}$-free bases of $\tilde{V}$ and $\tilde{W}$ as in the above lemma, we can express the both sides of the proposition in terms of $\nu_t(a_i)$. Indeed, we have

$$\nu_t(\tilde{D}) = \sum_{i=1}^{r} \nu_t(a_i),$$

since $\tilde{S}(f_i) = a_i e_i$ $(i = 1, 2, \cdots, r)$. On the other hand, we see that

$$V(m) = \bigoplus_{i;\nu_t(a_i)\geq m} K\phi f_i.$$

Thus, we have

$$\begin{aligned}\sum_{m=1}^{\infty} \dim_K V(m) &= \sum_{m=1}^{\infty} \sharp\{i | \nu_t(a_i) \geq m\} \\ &= \sum_{i=1}^{r} \nu_t(a_i) \\ &= \nu_t(\tilde{D}).\end{aligned}$$

□

3.3.3 Duality

The Jantzen filtration defined in this section has the following duality. Let

$$\tilde{V}^* := \mathrm{Hom}_{\mathcal{R}}(\tilde{V}, \mathcal{R}), \quad \tilde{W}^* := \mathrm{Hom}_{\mathcal{R}}(\tilde{W}, \mathcal{R})$$

be the dual $\mathcal{R}$-modules of $\tilde{V}$ and $\tilde{W}$. Let

$${}^t\tilde{S} : \tilde{W}^* \longrightarrow \tilde{V}^*$$

be the transpose of the map $\tilde{S}$. Moreover, we set

$$V^* := \phi\tilde{V}^*, \quad W^* := \phi\tilde{W}^*, \quad {}^tS := \phi({}^t\tilde{S}^*).$$

Under this setting, we define the Jantzen filtration $\{W^*(m)\}$ of W^*, the Jantzen cofiltration $\{V^*(m)\}$ of V^* and the higher derivatives

$${}^tS^{(m)} : W^*(m) \longrightarrow V^*(m)$$

as in Definition 3.6 and Definition-Lemma 3.1.

Then, we have the following proposition.

Proposition 3.9 *For $m \in \mathbb{Z}_{\geq 0}$, the following hold:*

1. $V^*(m) \simeq V(m)^*$,
2. $W^*(m) \simeq W(m)^*$,

where $V(m)^*$ and $W(m)^*$ are the dual K-vector spaces of the mth step of the Jantzen (co)-filtration $V(m)$ and $W(m)$.

Proof. Since $\tilde{V}$ is an $\mathcal{R}$-free module, for $m = 0$, the statement holds. For $m > 0$, we set

$$\begin{aligned}\bar{V}_m &:= \{v^* \in V^* | v^*(V(m)) = \{0\}\}, \\ \bar{W}_m &:= \{w^* \in W^* | w^*(IK(m-1)) = \{0\}\}.\end{aligned}$$

By definition, we have

$$V(m)^* \simeq V^*/\bar{V}_m, \quad W(m)^* \simeq \bar{W}_m.$$

Hence, it suffices to see that

$$\bar{V}_m = IK^*(m-1), \quad \bar{W}_m = W^*(m),$$

where $IK^*(m)$ is defined by

$$\widetilde{IK}^*(m) := t^{-m}(\mathrm{Im}^t\tilde{S} \cap t^m\tilde{V}^*),$$

$$\iota_m^{IK*} : \widetilde{IK}^*(m) \hookrightarrow \tilde{V}^* \quad \text{embedding}$$

$$IK^*(m) := \mathrm{Im}\phi(\iota_m^{IK*}).$$

Taking the $\mathcal{R}$-dual basis of the $\mathcal{R}$-free basis in Lemma 3.4, the explicit expressions of $\widetilde{IK}^*(m-1)$ and $\tilde{W}^*(m)$ prove the proposition. □

3.4 The Jantzen Filtration à la Feigin and Fuchs II

In this section, we briefly explain a geometric interpretation of the Jantzen filtration à la Feigin and Fuchs.

3.4.1 Notation

The geometric version of the Jantzen filtration is constructed by means of vector bundles and sheaves of sections on them. Before defining the filtration, we introduce some notation for vector bundles and sheaves.

Let X be an algebraic variety over the field K, and let $(\mathbb{V}, \pi_{\mathbb{V}})$ be a vector bundle on X, where $\pi_{\mathbb{V}} : \mathbb{V} \to X$ is the canonical projection. For $x \in X$, we denote the fibre $\pi_{\mathbb{V}}^{-1}(x)$ of $\mathbb{V}$ by $\mathbb{V}_x$. For a sheaf $\mathcal{F}$ on X and an open subset $U \subset X$, we also denote the sections of $\mathcal{F}$ over U and the stalk of $\mathcal{F}$ at $P \in X$ by $\mathcal{F}(U)$ and $\mathcal{F}_P$ respectively. Let $\mathcal{O}_X$ be the structure sheaf of X. For $\mathcal{O}_X$-modules $\mathcal{F}$ and $\mathcal{G}$, let $\mathrm{Hom}_{\mathcal{O}_X}(\mathcal{F}, \mathcal{G})$ be the set of morphisms of $\mathcal{O}_X$-modules. $\varphi \in \mathrm{Hom}_{\mathcal{O}_X}(\mathcal{F}, \mathcal{G})$ induces an $\mathcal{O}_{X,P}$-map $\mathcal{F}_P \to \mathcal{G}_P$ on the stalks, which we denote by φ_P.

3.4.2 Definitions and Properties

The geometric version of the Jantzen filtration is constructed from two vector bundles with a common base space and a map between them. To be precise, the setting is as follows:

Let $(\mathbb{V}, \pi_{\mathbb{V}})$ and $(\mathbb{W}, \pi_{\mathbb{W}})$ be vector bundles on X of the same rank, say r. Let $f : \mathbb{V} \to \mathbb{W}$ be a morphism of vector bundles, i.e., the following diagram commutes

$$\begin{array}{ccc} \mathbb{V} & \xrightarrow{f} & \mathbb{W}, \\ & {\scriptstyle\pi_{\mathbb{V}}}\searrow \quad \swarrow{\scriptstyle\pi_{\mathbb{W}}} & \\ & X & \end{array}$$

and its induced morphism at each fibre is a linear map.

Until the end of this section, we fix a point $P \in X$ and a curve $C \subset X$ containing the point P. We assume that

Assumption 1: C is *regular* at P.

Let $\mathcal{O}_C$ be the sheaf of regular functions on C. By the above assumption, $\mathcal{O}_{C,P}$ is a discrete valuation ring. We denote its valuation $\mathcal{O}_{C,P} \to \mathbb{Z}_{\geq 0} \cup \{\infty\}$ by ν. By assumption, $\mathcal{O}_{C,P}$ is a discrete valuation ring with its unique maximal ideal $\mathfrak{m}_{C,P} = (t)$ $(t \in \mathcal{O}_{C,P})$.

Let $(\mathbb{V}_C, \pi_{\mathbb{V},C})$, $(\mathbb{W}_C, \pi_{\mathbb{W},C})$ and $f_C : \mathbb{V}_C \to \mathbb{W}_C$ be the restrictions of $\mathbb{V}$, $\mathbb{W}$ and f to the curve C, and let $\mathcal{V}_C$ and $\mathcal{W}_C$ be the sheaves of sections of $\mathbb{V}_C$ and $\mathbb{W}_C$ respectively. For any open set $U \subset C$, we define homomorphisms $\tilde{f}_C \in \operatorname{Hom}_{\mathcal{O}_C}(\mathcal{V}_C, \mathcal{W}_C)$ and $\tilde{f}_{C,P} \in \operatorname{Hom}_{\mathcal{O}_{C,P}}(\mathcal{V}_{C,P}, \mathcal{W}_{C,P})$ as follows:

$$\left.\begin{aligned} (\tilde{f}_C(U)(s))(x) &:= f_C(s(x)), \\ \tilde{f}_{C,P} &:= \tilde{f}_C|_P, \end{aligned}\right. \qquad (x \in U,\ s \in \mathcal{V}_C(U)).$$

For simplicity, we assume the following:

Assumption 2: $\operatorname{Im}\tilde{f}_{C,P}$ is of full rank, i.e., r.

Under the above setting, we introduce the Jantzen filtration à la Feigin and Fuchs (geometric version) associated to the quintuple $(\mathbb{V}, \mathbb{W}, f, C, P)$, which gives a filtration $\{\mathbb{V}_{C,P}(m)\}$ of $\mathbb{V}_P$ and cofiltration $\{\mathbb{W}_{C,P}(m)\}$ of $\mathbb{W}_P$.

Definition 3.7 *Suppose that $m \in \mathbb{Z}_{\geq 0}$. We define a subspace $\mathbb{V}_{C,P}(m)$ of $\mathbb{V}_P$ by*

$$\begin{aligned} \mathcal{V}_{C,P}(m) &:= \tilde{f}_{C,P}^{-1}(\mathfrak{m}_{C,P}^m \mathcal{W}_{C,P} \cap \operatorname{Im}\tilde{f}_{C,P}), \\ \mathbb{V}_{C,P}(m) &:= \{\tilde{v}(P) | \tilde{v} \in \mathcal{V}_{C,P}(m)\}. \end{aligned}$$

Further, we define a quotient space $\mathbb{W}_{C,P}(m)$ of $\mathbb{W}_P$ by

$$\mathcal{IK}_{C,P}(m) := t^{-m}(\mathfrak{m}_{C,P}^m \mathcal{W}_{C,P} \cap \operatorname{Im}\tilde{f}_{C,P}),$$
$$\mathbb{IK}_{C,P}(m) := \{\tilde{v}(P) | \tilde{v} \in \mathcal{IK}_{C,P}(m)\},$$
$$\mathbb{W}_{C,P}(m) := \begin{cases} \mathbb{W}_P / \mathbb{IK}_P(m-1) & m > 0 \\ \mathbb{W}_P & m = 0 \end{cases}.$$

Let $\pi_m : \mathbb{W}_P \to \mathbb{W}_{C,P}(m)$ be the canonical projection.

This Jantzen filtration à la Feigin and Fuchs (geometric version) admits the same properties as those of the algebraic version in the previous section. For the reader's convenience, we will list these properties without proof. (The proofs of them are quite similar to those of the corresponding statements in the algebraic version.)

First, we construct a map $f_{C,P}^{(m)} : \mathbb{V}_{C,P}(m) \to \mathbb{W}_{C,P}(m)$ as a 'higher derivative' of $f_P : \mathbb{V}_P \to \mathbb{W}_P$ along the curve C.

Definition-Lemma 3.2 *Suppose that $m \in \mathbb{Z}_{\geq 0}$. For $u \in V(m)$, we take $\tilde{u} \in \mathcal{V}_{C,P}(m)$ such that $u = \tilde{u}(P)$. Then,*

$$\pi_m((t^{-m}\tilde{f}_{C,P}(\tilde{u}))(P))$$

does not depend on the choice of $\tilde{u}$. Therefore, the map

$$f_{C,P}^{(m)} : \mathbb{V}_{C,P}(m) \to \mathbb{W}_{C,P}(m), \quad u \longmapsto \pi_m((t^{-m}\tilde{f}_{C,P}(\tilde{u}))(P))$$

is well-defined. In particular, we have $f_{C,P}^{(0)} = f_P$.

Second, we state a characterisation of $\mathbb{V}_{C,P}(m)$ and $\mathbb{W}_{C,P}(m)$ by means of the map $f_{C,P}^{(m-1)}$.

Proposition 3.10 *For any $m \in \mathbb{Z}_{>0}$, the following hold:*

1. $\mathbb{V}_{C,P}(m) = \operatorname{Ker} f_{C,P}^{(m-1)}$,
2. $\mathbb{W}_{C,P}(m) = \operatorname{Coker} f_{C,P}^{(m-1)}$.

Let us fix a point $P \in X$ and curves $C, C' \subset X$. Assume that both C and C' satisfy **Assumptions 1** and **2**. They define, à priori, two different filtrations and cofiltrations $\{\mathbb{V}_{C,P}(m)\}$ and $\{\mathbb{V}_{C',P}(m)\}$ (resp. $\{\mathbb{W}_{C,P}(m)\}$ and $\{\mathbb{W}_{C',P}(m)\}$). Comparing these two filtrations and cofiltrations, we arrived at the following:

Conjecture 1 *For $m \in \mathbb{Z}_{\geq 0}$, we have*

$$\mathbb{V}_{C,P}(m) = \mathbb{V}_{C',P}(m), \qquad \mathbb{W}_{C,P}(m) = \mathbb{W}_{C',P}(m).$$

Third, we state the duality of our Jantzen filtration. Let $\mathbb{V}^\vee$ and $\mathbb{W}^\vee$ be the dual vector bundles of $\mathbb{V}$ and $\mathbb{W}$ respectively, i.e., their fibres satisfy $(\mathbb{V}^\vee)_x = (\mathbb{V}_x)^*$ and $(\mathbb{W}^\vee)_x = (\mathbb{W}_x)^*$ for any $x \in X$. Let

$$^t f : \mathbb{W}^\vee \longrightarrow \mathbb{V}^\vee$$

be the transpose of $f : \mathbb{V} \to \mathbb{W}$, i.e., $^t f|_{(\mathbb{W}^\vee)_x} = {}^t(f|_{\mathbb{V}_x})$. Let us take the point $P \in X$ and the curve C as before, and denote the restrictions by $\mathbb{V}_C^\vee$, $\mathbb{W}_C^\vee$, $\mathcal{V}_C^\vee$ and $\mathcal{W}_C^\vee$. By **Assumption 2**, we see that

$$^t \tilde{f}_{C,P} : \mathcal{W}_{C,P}^\vee \longrightarrow \mathcal{V}_{C,P}^\vee$$

is of full rank.

We can define the Jantzen (co)filtration $\{\mathbb{W}_{C,P}^\vee(m)\}$, $\{\mathbb{V}_{C,P}^\vee(m)\}$ associated to the quintuple $(\mathbb{W}^\vee, \mathbb{V}^\vee, {}^t f, C, P)$ as in Definition 3.7 and Definition-Lemma 3.2. Then, we have the following duality:

Proposition 3.11 *For $m \in \mathbb{Z}_{>0}$, we have*

1. $\mathbb{W}_{C,P}^\vee(m) \simeq \mathbb{W}_{C,P}(m)^*$,
2. $\mathbb{V}_{C,P}^\vee(m) \simeq \mathbb{V}_{C,P}(m)^*$.

3.5 The Jantzen Filtration of Quotient Modules

The Jantzen filtration of quotients of Verma modules was first considered in [Ja1], and it played an important role in [RW3] to determine the structure of Verma modules over the Witt algebra. Although one can formulate the Jantzen filtration of quotient modules over a general Q-graded Lie algebra with a Q-graded anti-involution, it is not only technically cumbersome, but also it may obscure the essence. Hence, we do not go into the technical details, instead we describe the construction intuitively. A concrete application will be given in § 5.6. The informed reader may generalise the construction for a general Q-graded Lie algebra with a Q-graded anti-involution.

Let $\mathfrak{g}$ be a Q-graded Lie algebra with a Q-graded anti-involution σ, and let $M(\lambda)$ be the Verma module with highest weight λ. Suppose that $M(\lambda)$ is reducible. Let $v_{\gamma,\lambda} \in (M(\lambda)_{\lambda-\gamma})^{\mathfrak{g}_+}$ $(\gamma \in Q^+)$ be a non-zero singular vector. Here, we are going to consider the Jantzen filtration of the following:

$$\overline{M}(\lambda) := M(\lambda)/U(\mathfrak{g}).v_{\gamma,\lambda}.$$

Let $\Phi \in S(\mathfrak{h}) \simeq \mathbb{K}[\mathfrak{h}^*]$ be an irreducible component of the determinant D_γ (Definition 3.2) which satisfies $\Phi(\lambda) = 0$ for the above $\lambda \in \mathfrak{h}^*$. In order to define the Jantzen filtration of $\overline{M}(\lambda)$, let us consider the curve $C \subset \mathfrak{h}^*$ defined by $\Phi = 0$, and perturb λ along the curve C.

We assume that

I. there exists a neighbourhood $U_1 \subset C$ of λ such that a non-zero singular vector $v_{\gamma,\mu} \in (M(\mu)_{\mu-\gamma})^{\mathfrak{g}_+}$ $(\mu \in U_1)$ which satisfies $v_{\gamma,\mu} \to v_{\gamma,\lambda}$ as $\mu \to \lambda$ exists.

Remark that by Proposition 3.4 the contravariant form on $M(\mu)$ ($\mu \in U_1$) induces a contravariant form on

$$\overline{M}(\mu) := M(\mu)/U(\mathfrak{g}).v_{\gamma,\mu}.$$

Under the following assumption:

II. there exists a neighbourhood $U_2 \subset C$ of λ such that $\overline{M}(\mu)$ is irreducible for any $\mu \in U_2 \setminus \{\lambda\}$,

we define the filtration

$$\overline{M}(\lambda) \supset \overline{M}(\lambda)(1) \supset \overline{M}(\lambda)(2) \supset \cdots$$

of $\overline{M}(\lambda)$, in the same way as in § 3.2.

We expect that the character sum of the filtration $\{\overline{M}(\lambda)(k)\}$ can be described by using the character sums of $M(\lambda)$ and $M(\lambda - \gamma)$. However, for $\mu \in C$, since $M(\mu)$ is always reducible and $M(\mu - \gamma)$ is not necessarily irreducible, one cannot naively define the Jantzen filtrations of these modules.

Hence, we further assume

III. there exists a neighbourhood $U_3 \subset C$ of λ such that $M(\mu - \gamma)$ is irreducible for any $\mu \in U_3 \setminus \{0\}$.

Moreover, we assume

IV. the existence of a neighbourhood $U_4 \subset C$ of λ with the following property: for each $\mu \in U_4 \setminus \{\lambda\}$, there exists a neighbourhood U_μ of μ such that $U_\mu \cap C = \{\mu\}$ and $M(\nu)$ is irreducible for any $\nu \in U_\mu \setminus \{\mu\}$.

Under these assumptions, considering the limits $\nu \to \mu$ and $\mu \to \lambda$ in that order, one can compare the degeneration of the contravariant forms on $\overline{M}(\lambda)$, $M(\lambda)$ and $M(\lambda - \gamma)$. In fact, under some technical assumptions, we are able to describe the character sum of $\overline{M}(\lambda)$ explicitly by means of character sums of $M(\lambda)$ and $M(\lambda - \gamma)$.

3.6 Bibliographical Notes and Comments

In 1977, J. C. Jantzen [Ja1] introduced the so-called Jantzen filtration in his study of Verma modules over a semi-simple finite dimensional Lie algebra. This technique is very useful, in particular, when one tries to study the structure of Verma modules over a rank 2 Q-graded Lie algebra with a Q-graded anti-involution (see, e.g., [RW2]). Its generalisation was also considered in several contexts, e.g., [FeFu4], [Ja1], [RW3] and so on.

Jantzen's approach [Ja1, Ja3] is algebraic, and is used to study not only Verma modules but also Weyl modules. A. Rocha-Caridi and N. R. Wallach [RW3] modified the original definition in terms of C^∞-language to apply it to the Lie algebra of vector fields on the circle, i.e., the Witt algebra. Their modification also applies to a quotient of a Verma module. After this work, B. Feigin and D. B. Fuchs [FeFu4] further generalised this filtration to analyze the modules of semi-infinite wedges over the Virasoro algebra.

Our approach in § 3.3 is based on [FeFu4], but is formulated in an algebraic language. Our intuitive explanation of the Jantzen filtration of quotient modules is also based on the idea of [RW3].

Chapter 4
Determinant Formulae

Determinants of contravariant forms on Verma modules over a semisimple Lie algebra were calculated by N. N. Shapovalov [Sh], and the result was essentially used by J. C. Jantzen in order to study the structure of Verma modules [Ja1]. In the case of the Virasoro algebra, to reveal the structures of Verma and Fock modules by means of Jantzen filtration, it is also important to compute their determinants.

To calculate the determinants of contravariant forms on Verma modules, following [Ro], we will use screening currents that appear in the conformal field theory. On the other hand, in the case of Fock modules, we will compute determinants of homomorphisms from Verma modules to Fock modules and from Fock modules to the contragredient dual of Verma modules. These determinants were calculated by A. Tsuchiya and Y. Kanie [TK2]. Here, we will simplify their proof.

In Section 4.1, we will introduce a Heisenberg Lie algebra and their Fock modules. Two vertex (super)algebra structures are recalled. In Section 4.2, we will study some properties of Fock modules as Vir-module to calculate the determinants. In Section 4.3, we will construct screening currents, and give a sufficient condition so that compositions of screening currents are non-trivial. In Section 4.4 we will calculate the determinants of Verma modules, and in Section 4.5, we will determine those of Fock modules.

From this chapter, we will work over $\mathbb{C}$, the field of complex numbers.

4.1 Vertex (Super)algebra Structures associated to Bosonic Fock Modules

In this section, we define *Fock modules* over the Virasoro algebra. Further, we give some isomorphisms between Fock modules and their duality with respect to the contragredient dual.

K. Iohara, Y. Koga, *Representation Theory of the Virasoro Algebra*, Springer Monographs in Mathematics, DOI 10.1007/978-0-85729-160-8_4,

4.1.1 Definitions and Notation

Fock modules over the Virasoro algebra are defined by using the *Heisenberg Lie algebra of rank one* defined in § 1.2.3. The Heisenberg Lie algebra

$$\mathcal{H} := \bigoplus_{n\in\mathbb{Z}} \mathbb{C}a_n \oplus \mathbb{C}K_{\mathcal{H}}$$

is the Lie algebra whose commutation relations are given by

$$[a_m, a_n] := m\delta_{m+n,0}K_{\mathcal{H}}, \quad [\mathcal{H}, K_{\mathcal{H}}] = \{0\}.$$

Recall that $(\mathcal{H}, \mathcal{H}^0)$ is a Q-graded Lie algebra with $Q := \mathbb{Z}\alpha_{\mathcal{H}}$ (§ 1.2.3), and

$$\mathcal{H} = \bigoplus_{\beta\in Q} \mathcal{H}^\beta, \quad \mathcal{H}^{n\alpha_{\mathcal{H}}} := \begin{cases} \mathbb{C}a_n & n \neq 0 \\ \mathbb{C}a_0 \oplus \mathbb{C}K_{\mathcal{H}} & n = 0 \end{cases}.$$

Since the map π_Q defined in (1.6) is trivial for the Heisenberg Lie algebra, $G = Q$ by the definition (1.7) and the map $\mathbf{p} : Q \to G$ is the identity. Recall that $\mathcal{H}^\pm := \bigoplus_{\pm\beta\in Q^+\setminus\{0\}} \mathcal{H}^\beta$, where $Q^+ := \mathbb{Z}_{\geq 0}\alpha_{\mathcal{H}}$, and $\mathcal{H}^{\geq} := \mathcal{H}^0 \oplus \mathcal{H}^+$, $\mathcal{H}^{\leq} := \mathcal{H}^- \oplus \mathcal{H}^0$.

For $\eta \in \mathbb{C}$, let $\mathbb{C}_\eta := \mathbb{C}\mathbf{1}_\eta$ be the one-dimensional $\mathcal{H}^{\geq}$-module given by

1. $a_n.\mathbf{1}_\eta = \eta\delta_{n,0}\mathbf{1}_\eta$ $(n \in \mathbb{Z}_{\geq 0})$,
2. $K_{\mathcal{H}}.\mathbf{1}_\eta = \mathbf{1}_\eta$,
3. G-gradation: for $\beta \in G$,

$$(\mathbb{C}_\eta)^\beta = \begin{cases} \mathbb{C}_\eta & (\beta = 0) \\ \{0\} & (\beta \neq 0) \end{cases}.$$

For simplicity, we denote by $\mathcal{F}^\eta$ the Verma module over $\mathcal{H}$ with highest weight $(0, (\eta, 1)) \in G \times (\mathcal{H}^0)^*$, where $(\eta, 1)(a_0) := \eta$ and $(\eta, 1)(K_{\mathcal{H}}) := 1$, namely,

$$\mathcal{F}^\eta = \mathrm{Ind}_{\mathcal{H}^{\geq}}^{\mathcal{H}} \mathbb{C}_\eta.$$

$\mathcal{F}^\eta$ is often called the **(bosonic) Fock module** over $\mathcal{H}$. We denote a highest weight vector $1 \otimes \mathbf{1}_\eta$ by $|\eta\rangle$.

Remark that $\mathcal{F}^\eta$ is an object of the category $\mathcal{O}^\iota$ defined in § 1.2.4, but it is not $(\mathcal{H}^0)^*$-semi-simple in the sense of Definition 1.10, since $(\mathcal{F}^\eta)_{(\eta,1)} = \mathcal{F}^\eta$ is infinite dimensional.

4.1.2 Vertex Operator Algebra $\mathcal{F}^0$

Here, we briefly recall a vertex operator algebra structure on $\mathcal{F}^0$ and its modules.

A $\mathbb{Z}$-graded vertex algebra structure on $\mathcal{F}^0$ is defined as follows:

1. (**Gradation**) For $n_1 \leq n_2 \leq \cdots \leq n_m < 0$, set
$$\deg(a_{n_1}a_{n_2}\cdots a_{n_m}.|0\rangle) := -\sum_{i=1}^{m} n_i.$$
2. (**Vacuum vector**) $|0\rangle$.
3. (**Translation operator**) Set $T|0\rangle := 0$ and $[T, a_n] := -na_{n-1}$.
4. (**Vertex operators**) Set $Y(|0\rangle, z) := \mathrm{id}$ and
$$Y(a_{-1}.|0\rangle, z) := a(z) := \sum_{n\in\mathbb{Z}} a_n z^{-n-1}.$$

By the strong reconstruction theorem (cf. Theorem C.1), these data define a $\mathbb{Z}$-graded vertex algebra structure on $\mathcal{F}^0$.

The vertex algebra $\mathcal{F}^0$ has a one-parameter family of conformal vectors, namely, for $\lambda \in \mathbb{C}$, we set

$$\omega_\lambda := \left(\frac{1}{2}a_{-1}^2 + \lambda a_{-2}\right).|0\rangle. \tag{4.1}$$

Then, ω_λ is a conformal vector with central charge

$$c_\lambda := 1 - 12\lambda^2. \tag{4.2}$$

Indeed, setting

$$T_\lambda(z) := Y(\omega_\lambda, z) = \sum_{n\in\mathbb{Z}} L_n^\lambda z^{-n-2}, \tag{4.3}$$

one can check that this field satisfies the following OPE (cf. § C.1.4):

$$T_\lambda(z)T_\lambda(w) \sim \frac{\frac{1}{2}c_\lambda}{(z-w)^4}\mathrm{id} + \frac{2}{(z-w)^2}T_\lambda(w) + \frac{1}{z-w}\partial T_\lambda(w).$$

In particular, $L_{-1}^\lambda = T$ and L_0^λ is the degree operator. Since each graded subspace of $\mathcal{F}^0$ is of finite dimension, the $\mathbb{Z}$-graded vertex algebra $\mathcal{F}^0$ equipped with the conformal vector ω_λ becomes a vertex operator algebra.

For any $\eta \in \mathbb{C}$, it is easy to see that $\mathcal{F}^\eta$ is an irreducible $\mathcal{H}$-module. Moreover, it has an $(\mathcal{F}^0, \omega_\lambda)$-module structure by letting

$$Y^\eta(a_{-1}.|0\rangle, z) := a(z) \in (\mathrm{End}\mathcal{F}^\eta)[[z^{\pm 1}]].$$

4.1.3 Vertex Operator Superalgebra $V_{\sqrt{N}\mathbb{Z}}$

In this subsection, we briefly recall a vertex operator superalgebra structure on $V_{\sqrt{N}\mathbb{Z}} := \bigoplus_{\eta \in \sqrt{N}\mathbb{Z}} \mathcal{F}^{\eta}$, for $N \in \mathbb{Z}_{>0}$.

For $\mu \in \sqrt{N}\mathbb{Z}$, let $e^{\mu q} \in \mathrm{End}_{\mathcal{H}^-}(V_{\sqrt{N}\mathbb{Z}})$ be the shift operator $\mathcal{F}^{\eta} \longrightarrow \mathcal{F}^{\eta+\mu}$ defined by

$$e^{\mu q}.|\eta\rangle := |\eta + \mu\rangle.$$

A vertex superalgebra structure on $V_{\sqrt{N}\mathbb{Z}}$ is defined as follows:

1. (**$\mathbb{Z}/2\mathbb{Z}$-gradation**)

$$V^{\bar{0}} := \bigoplus_{\substack{\eta \in \sqrt{N}\mathbb{Z} \\ \eta^2 \equiv 0(2)}} \mathcal{F}^{\eta}, \qquad V^{\bar{1}} := \bigoplus_{\substack{\eta \in \sqrt{N}\mathbb{Z} \\ \eta^2 \equiv 1(2)}} \mathcal{F}^{\eta}.$$

2. (**Vacuum vector**) $|0\rangle$.
3. (**Translation operator**) $T := \sum_{n \geq 0} a_{-n-1} a_n$, in particular, one has $T.|\mu\rangle = \mu a_{-1}|\mu\rangle$.
4. (**Vertex operators**) Set $Y(|0\rangle, z) := \mathrm{id}$, $Y(a_{-1}.|0\rangle, z) := a(z)$. For $\mu \in \sqrt{N}\mathbb{Z}$, set

$$Y(|\mu\rangle, z) := e^{\mu q} z^{\mu a_0} \exp\left(\mu \sum_{n>0} \frac{a_{-n}}{n} z^n\right) \exp\left(-\mu \sum_{n>0} \frac{a_n}{n} z^{-n}\right).$$

   Note that, when we restrict $Y(|\mu\rangle, z)$ to $\mathcal{F}^{\eta}$, the factor $z^{\mu a_0}$ becomes $z^{\mu\eta}$, so that the operator $Y(|\mu\rangle, z)$ is well-defined on $V_{\sqrt{N}\mathbb{Z}}$.

By the strong reconstruction theorem (cf. Theorem C.1) for vertex superalgebras, these data define a vertex superalgebra structure on $V_{\sqrt{N}\mathbb{Z}}$. In particular, when N is a positive even integer, $V_{\sqrt{N}\mathbb{Z}}$ becomes purely even, i.e., an ordinary vertex algebra, and this is the only case we treat in this chapter. Later, in Chapter 8, we will treat $V_{\mathbb{Z}}$. For simplicity, we set $V_{\lambda}(z) := Y(|\lambda\rangle, z)$.

Let us explain how the super-structure appears in $V_{\sqrt{N}\mathbb{Z}}$. One can check that

$$V_{\eta}(z)V_{\mu}(w) = (z-w)^{\eta\mu} {}^{\circ}_{\circ} V_{\eta}(z)V_{\mu}(w) {}^{\circ}_{\circ},$$

where we set

$$\begin{aligned} {}^{\circ}_{\circ} V_{\eta}(z)V_{\mu}(w) {}^{\circ}_{\circ} := & e^{(\eta+\mu)q} z^{\eta a_0} w^{\mu a_0} \\ & \times \exp\left(\sum_{n>0} \frac{a_{-n}}{n}(\eta z^n + \mu w^n)\right) \exp\left(-\sum_{n>0} \frac{a_n}{n}(\eta z^{-n} + \mu w^{-n})\right). \end{aligned}$$

This shows that for a sufficiently big $M \in \mathbb{Z}_{\geq 0}$, it follows that

$$(z-w)^M \left(V_{\eta}(z)V_{\mu}(w) - (-1)^{\eta\mu} V_{\mu}(w)V_{\eta}(z)\right) = 0,$$

namely, $V_\eta(z)$ and $V_\mu(w)$ are local in the super-setting.

In the rest of this chapter, we consider only the case when N is even.

Lemma 4.1. *Suppose that N is even. The conformal vector (4.1) defines a $\mathbb{Z}$-graded structure on $V_{\sqrt{N}\mathbb{Z}}$ if and only if $\lambda \in \frac{1}{\sqrt{N}}\mathbb{Z}$. In particular, in this case, $(V_{\sqrt{N}\mathbb{Z}}, \omega_\lambda)$ becomes a vertex operator algebra.*

This can be easily verified by noting $L_0^\lambda|\eta\rangle = h_\lambda^\eta|\eta\rangle$, where we set

$$h_\lambda^\eta = \frac{1}{2}\eta(\eta - 2\lambda). \tag{4.4}$$

In 1993, C. Dong [Do] showed the following theorem:

Theorem 4.1 *Suppose that N is even.*

1. *$V_{\sqrt{N}\mathbb{Z}}$ is a rational vertex operator algebra.*
2. *Any simple $V_{\sqrt{N}\mathbb{Z}}$-module is of the form $V_{\frac{i}{\sqrt{N}}+\sqrt{N}\mathbb{Z}}$ with $0 \le i < N$, where we set*

$$V_{\frac{i}{\sqrt{N}}+\sqrt{N}\mathbb{Z}} := \bigoplus_{\eta \in \frac{i}{\sqrt{N}}+\sqrt{N}\mathbb{Z}} \mathcal{F}^\eta.$$

4.2 Isomorphisms among Fock Modules

In this section, we study isomorphisms among bosonic Fock modules as Vir-modules that are induced from isomorphisms as $\mathcal{H}$-modules. Here and after, regarding $\mathcal{F}^\eta$ as Vir-module via $L_n \mapsto L_n^\lambda$ and $C \mapsto c_\lambda \mathrm{id}$, we denote it by $\mathcal{F}_\lambda^\eta$.

4.2.1 Notation

For such an $(\mathcal{H}, \mathcal{H}^0)$-module M that the set of the weights $\mathcal{P}(M)$ satisfies $\mathcal{P}(M) \subset G \times \{\lambda\}$ for some $\lambda \in (\mathcal{H}^0)^*$, we often denote the $G \times (\mathcal{H}^0)^*$-graded component M_λ^β by M^β for simplicity. (See § 1.2.4.) Using this notation, we have

$$\mathcal{F}^\eta = \bigoplus_{n \in \mathbb{Z}_{\geq 0}} (\mathcal{F}^\eta)^{-n\alpha_{\mathcal{H}}}.$$

Moreover, each graded component $(\mathcal{F}^\eta)^{-n\alpha_{\mathcal{H}}}$ can be described as follows: Let $\mathcal{P}_n$ be the set of the partitions of $n \in \mathbb{Z}_{>0}$. For $\mathbb{I} = (1^{r_1}2^{r_2}\cdots n^{r_n}) \in \mathcal{P}_n$, we set

$$a_{\mathbb{I}} := a_{-n}^{r_n} \cdots a_{-2}^{r_2} a_{-1}^{r_1} \in U(\mathcal{H}). \tag{4.5}$$

Then, we have

$$(\mathcal{F}^\eta)^{-n\alpha_{\mathcal{H}}} = \begin{cases} \bigoplus_{\mathbb{I}\in\mathcal{P}_n} \mathbb{C}a_{\mathbb{I}}|\eta\rangle & n\in\mathbb{Z}_{>0} \\ \mathbb{C}|\eta\rangle & n=0 \end{cases}. \tag{4.6}$$

Here, we also introduce some notation for $(\mathrm{Vir}, \mathrm{Vir}^0)$-modules. Recall that in this case, $G=\{0\}$ and π_Q is injective. Let $\alpha\in Q$ be the $\mathbb{Z}$-basis of Q such that $\alpha(C)=0$, $\alpha(L_0)=-1$ (§ 1.2.3), and let Λ be the element of $(\mathrm{Vir}^0)^*$ such that $\Lambda(C)=1$ and $\Lambda(L_0)=0$. Then, $(\mathrm{Vir}^0)^*=\mathbb{C}\Lambda\oplus\mathbb{C}\alpha$.

When the centre C acts on a $(\mathrm{Vir}, \mathrm{Vir}^0)$-module M as $c\,\mathrm{id}_M$, we refer to c as the **central charge** of M. For a $(\mathrm{Vir}, \mathrm{Vir}^0)$-module M with central charge c, we denote $M_{c\Lambda-h\alpha}$ by M_h. Thus,

$$M_h=\{v\in M|L_0.v=hv\},$$

and $M=\bigoplus_{h\in\mathbb{C}} M_h$. Remark that as vector space,

$$(\mathcal{F}^\eta_\lambda)_{h^\eta_\lambda+n}=(\mathcal{F}^\eta_\lambda)^{-n\alpha_{\mathcal{H}}}$$

holds for $n\in\mathbb{Z}_{\geq 0}$.

4.2.2 Isomorphisms arising from Automorphisms of $\mathcal{H}$

For $\lambda,\kappa\in\mathbb{C}$, we set

$$T_{\lambda,\kappa}(z):=\frac{1}{2}{}^\circ_\circ a(z)^2{}^\circ_\circ+(\lambda\partial_z+\kappa z^{-1})a(z)+\frac{1}{2}\kappa(\kappa-2\lambda)z^{-2}, \tag{4.7}$$

and

$$T_{\lambda,\kappa}(z)=\sum_{n\in\mathbb{Z}} L^{\lambda,\kappa}_n z^{-n-2}.$$

By direct computation, we have

Proposition 4.1 *1. The Virasoro algebra* Vir *acts on the space* $\mathcal{F}^\eta$ *via*

$$L_n\mapsto L^{\lambda,\kappa}_n \quad \textit{and} \quad C\mapsto c_\lambda \mathrm{id}_{\mathcal{F}^\eta},$$

2. $L^{\lambda,\kappa}_0.(1\otimes \mathbf{1}_\eta)=h^{\eta+\kappa}_\lambda(1\otimes\mathbf{1}_\eta)$,
where c_λ *(resp.* h^η_λ*) is defined in (4.2) (resp. (4.4)).*

When we regard the space $\mathcal{F}^\eta$ as Vir-module via the above action, we denote it by $\mathcal{F}^\eta_{\lambda,\kappa}$. By definition, we have $T_{\lambda,0}(z)=T_\lambda(z)$ and $\mathcal{F}^\eta_{\lambda,0}=\mathcal{F}^\eta_\lambda$.

Let π^η be the homomorphism $\mathcal{H}\to\mathrm{End}(\mathcal{F}^\eta)$.

Lemma 4.2. *1. The map* $i_{\mathcal{H}}:\mathcal{H}\to\mathcal{H}$ *defined by*

$$a_n \mapsto -a_n, \quad K_{\mathcal{H}} \mapsto K_{\mathcal{H}}$$

is an automorphism of $\mathcal{H}$. This induces an $\mathcal{H}$-isomorphism

$$(\mathcal{F}^\eta, \pi^\eta \circ i_{\mathcal{H}}) \simeq (\mathcal{F}^{-\eta}, \pi^{-\eta}).$$

2. For each $\kappa \in \mathbb{C}$, the map $\phi_\kappa : \mathcal{H} \to \mathcal{H}$ defined by

$$a_n \mapsto a_n + \delta_{n,0}\, \kappa K_{\mathcal{H}}, \quad K_{\mathcal{H}} \mapsto K_{\mathcal{H}}$$

is an automorphism of $\mathcal{H}$. This induces an $\mathcal{H}$-isomorphism

$$(\mathcal{F}^\eta, \pi^\eta \circ \phi_\kappa) \simeq (\mathcal{F}^{\eta+\kappa}, \pi^{\eta+\kappa}).$$

Proof. Direct verification. □

These isomorphisms induce Vir-isomorphisms between Fock modules.

Proposition 4.2 *As* Vir*-modules, the following hold:*

1. $\mathcal{F}^\eta_{\lambda,\kappa} \simeq \mathcal{F}^{-\eta}_{-\lambda,-\kappa}$,
2. $\mathcal{F}^\eta_{\lambda,\kappa} \simeq \mathcal{F}^{\eta+\kappa}_{\lambda,0}$.

Proof. Here, we lift the isomorphisms $i_{\mathcal{H}}$ and ϕ_κ to automorphisms on $U(\text{Vir})$. By definition, for any $u \in \mathcal{F}^\eta$ we have

$$i_{\mathcal{H}}(T_{\lambda,\kappa}(z)).u = T_{-\lambda,-\kappa}(z).u,$$
$$\phi_\kappa(T_{\lambda,0}(z)).u = T_{\lambda,\kappa}(z).u.$$

Hence, 1 and 2 of Lemma 4.2 imply the first and the second statements respectively. □

By virtue of Proposition 4.2, we can assume that $\kappa = 0$ without loss of generality. Hence, in the sequel we fix $\kappa = 0$ unless otherwise stated.

4.2.3 Isomorphisms related to the Contragredient Dual

Recall that $\mathcal{H}$ has the Q-graded anti-involution defined by $\sigma_{\mathcal{H}}(a_n) = a_{-n}$ and $\sigma_{\mathcal{H}}(K_{\mathcal{H}}) = K_{\mathcal{H}}$ (§ 1.2.3). Let $(\mathcal{F}^\eta)^c$ be the contragredient dual of the $\mathcal{H}$-module $\mathcal{F}^\eta$ (§ 1.2.7). Here, we give some isomorphisms of $\mathcal{F}^\eta$ related with the contragrediant dual.

Let us first introduce some notation. For $\mathbb{I} = (1^{r_1}2^{r_2}\cdots n^{r_n}) \in \mathcal{P}_n$, we put

$$|\mathbb{I}| := \sum_{k=1}^{n} r_k. \tag{4.8}$$

For $m \in \mathbb{Z}_{>0}$ and $\mathbb{I} := (1^{r_1}2^{r_2}\cdots n^{r_n})$, we set

$$\mathbb{I} \pm (k^m) := (1^{r_1}2^{r_2}\cdots k^{r_k \pm m}\cdots n^{r_n}), \tag{4.9}$$

where $\mathbb{I} - (k^m)$ is defined for $\mathbb{I}$ such that $r_k \geq m$.

Lemma 4.3. *$(\mathcal{F}^\eta)^c \simeq \mathcal{F}^\eta$ as $\mathcal{H}$-modules.*

Proof. Let $\{\phi_{\mathbb{I}} \mid \mathbb{I} \in \mathcal{P}_n\}$ be the basis of $\{(\mathcal{F}^\eta)^{-n\alpha_{\mathcal{H}}}\}^*$ defined by

$$\phi_{\mathbb{I}}(a_{\mathbb{J}} \otimes \mathbf{1}_\eta) := \delta_{\mathbb{I},\mathbb{J}}[\mathbb{I}],$$

where we set

$$[\mathbb{I}] := \prod_{k=1}^{n} r_k! k^{r_k}$$

for $\mathbb{I} = (1^{r_1}2^{r_2}\cdots n^{r_n}) \in \mathcal{P}_n$.

From the following explicit form of the action:

$$a_k.a_{\mathbb{I}} \otimes \mathbf{1}_\eta = \begin{cases} k r_k a_{\mathbb{I}-(k^1)} \otimes \mathbf{1}_\eta & (k>0) \\ \eta a_{\mathbb{I}} \otimes \mathbf{1}_\eta & (k=0) \\ a_{\mathbb{I}+(k^1)} \otimes \mathbf{1}_\eta & (k<0) \end{cases},$$

where the notation $\mathbb{I} \pm (k^1)$ is defined as in (4.9) and $a_{\mathbb{I}-(k^1)} \otimes \mathbf{1}_\eta$ is regarded as 0 if $r_k = 0$, we see that $\phi_{\mathbb{I}} \mapsto a_{\mathbb{I}} \otimes \mathbf{1}_\eta$ is an isomorphism of $\mathcal{H}$-modules. □

From this isomorphism of the lemma, we obtain duality of Fock modules over the Virasoro algebra. Let $(\mathcal{F}^\eta_\lambda)^c$ be the contragredient dual of $\mathcal{F}^\eta_\lambda$ as Vir-modules, where the Q-graded anti-involution σ is defined by $\sigma(L_n) = L_{-n}$ and $\sigma(C) = C$ (§ 1.2.3).

From Lemma 4.3, we obtain

Proposition 4.3 *For any $\kappa \in \mathbb{C}$, $(\mathcal{F}^\eta_{\lambda,\kappa})^c \simeq \mathcal{F}^\eta_{-\lambda,\kappa-2\lambda}$ as* Vir*-modules.*

Proof. For $\phi \in (\mathcal{F}^\eta_{\lambda,\kappa})^c$ and $u \in \mathcal{F}^\eta_{\lambda,\kappa}$, we have

$$\begin{aligned}(L_n.\phi)(u) =& \phi(L_{-n}.u) \\ =& \phi(L^{\lambda,\kappa}_{-n}.u) \\ =& (L^{-\lambda,\kappa-2\lambda}_n.\phi)(u).\end{aligned}$$

Note that the first and the third lines follow from the contravariance of the anti-involution σ and $\sigma_{\mathcal{H}}$ respectively, and the second line follows from the Vir-module structure on $\mathcal{F}^\eta_{\lambda,\kappa}$. Hence the Vir-module structure of $(\mathcal{F}^\eta_{\lambda,\kappa})^c$ is given by

$$L_n.\phi = L^{-\lambda,\kappa-2\lambda}_n.\phi,$$

and we obtain the desired conclusion. □

Corollary 4.1 *$(\mathcal{F}^\eta_\lambda)^c \simeq \mathcal{F}^{2\lambda-\eta}_\lambda$ as* Vir*-modules.*

Proof. Proposition 4.2 and 4.3 imply that

$$(\mathcal{F}^\eta_\lambda)^c = (\mathcal{F}^\eta_{\lambda,0})^c \simeq \mathcal{F}^\eta_{-\lambda,-2\lambda} \simeq \mathcal{F}^{\eta-2\lambda}_{-\lambda} \simeq \mathcal{F}^{2\lambda-\eta}_{\lambda}. \qquad \square$$

4.3 Intertwining Operators

In this section, we obtain a sufficient condition for a Fourier coefficient of the vertex operator $V_\mu(z)$ to be an intertwining operator of Vir-modules. As a corollary, we obtain a non-trivial singular vector of $\mathcal{F}^\eta_\lambda$ which will play a crucial role in the next section.

Here, we work in the framework of the vertex operator algebra $(V_{\sqrt{N}\mathbb{Z}}, \omega_\lambda)$ with $N \in 2\mathbb{Z}_{>0}$ and $\lambda \in \frac{1}{\sqrt{N}}\mathbb{Z}$ (cf. Lemma 4.1).

4.3.1 Vertex Operator $V_\mu(z)$

In this subsection, we obtain a necessary and sufficient condition for the zero-mode $V_\mu(0)$ of the vertex operator

$$V_\mu(z) = \sum_{n\in\mathbb{Z}} V_\mu(n) z^{-n-1}$$

to be an intertwining operator of Vir-modules.

By direct computation, one can check the following OPE:

$$T_\lambda(z)V_\mu(w) \sim \frac{h^\mu_\lambda}{(z-w)^2}V_\mu(w) + \frac{1}{z-w}\partial V_\mu(w),$$

where h^μ_λ is defined in (4.4). Hence, by Proposition C.2, we obtain

Lemma 4.4. *For each $n \in \mathbb{Z}$, we have*

$$[L_n, V_\mu(w)] = w^n \left\{ w\frac{\partial}{\partial w} + h^\mu_\lambda(n+1)\right\} V_\mu(w).$$

As a corollary, we obtain the following condition:

Lemma 4.5. *The Fourier coefficient $V_\mu(m) : \mathcal{F}^\eta_\lambda \to \mathcal{F}^{\eta+\mu}_\lambda$ gives a* Vir-*homomorphism if and only if $m = 0$ and*

$$\lambda = \frac{1}{2}\mu - \frac{1}{\mu}.$$

Proof. By Lemma 4.4, we have

$$[L_n, V_\mu(m)] = ((n+1)(h_\lambda^\mu - 1) - m)V_\mu(m+n) \quad (m, n \in \mathbb{Z}).$$

Hence, the lemma follows from (4.4). □

The intertwining operator $\Sigma_\mu := V_\mu(0)$ is called a **screening operator**.

Now, we study non-trivial Vir-homomorphisms of the form

$$(\Sigma_{\sqrt{N}})^n : \mathcal{F}_\lambda^{\eta - n\sqrt{N}} \to \mathcal{F}_\lambda^\eta \quad (n \in \mathbb{Z}_{>0}).$$

From now on, we fix

$$\lambda = \frac{1}{2}\sqrt{N} - \frac{1}{\sqrt{N}} = \frac{N-2}{2\sqrt{N}},$$

and give a sufficient condition so that

$$(\Sigma_{\sqrt{N}})^n.|\eta - n\sqrt{N}\rangle \neq 0. \tag{4.10}$$

4.3.2 Criterion for Non-Triviality

The goal of this section is the following proposition:

Proposition 4.4 *Suppose that $n \in \mathbb{Z}_{>0}$, $N = 2t$ for an odd prime number t and $\eta \in \sqrt{N}\mathbb{Z}$. Then, we have*

$$(\Sigma_{\sqrt{N}})^n.|\eta - n\sqrt{N}\rangle \neq 0,$$

if $t > n$ and $(n+1)t - (u+1) \geq 0$, where $u := \sqrt{N}\eta$.

For the proof of this proposition, we start from the following lemma:

Lemma 4.6. *Let us take $z_i \in \mathbb{C}$ $(1 \leq i \leq n)$ satisfying $|z_1| > |z_2| > \cdots > |z_n| > 0$. We have*

$$\begin{aligned} &V_{\sqrt{N}}(z_1)V_{\sqrt{N}}(z_2)\cdots V_{\sqrt{N}}(z_n).|\eta - n\sqrt{N}\rangle \\ &= \prod_{1\leq i<j\leq n}(z_i - z_j)^N \prod_{i=1}^n z_i^{\sqrt{N}(\eta - n\sqrt{N})} \prod_{i=1}^n \exp\left(\sqrt{N}\sum_{k>0}\frac{a_{-k}}{k}z_i^k\right).|\eta\rangle. \end{aligned} \tag{4.11}$$

Proof. We obtain this lemma from the following formula:

$$
\begin{aligned}
&V_{\sqrt{N}}(z_1)V_{\sqrt{N}}(z_2)\cdots V_{\sqrt{N}}(z_n)\\
&= \prod_{1\le i<j\le n}(z_i-z_j)^N \prod_{i=1}^{n}\exp\left(\sqrt{N}\sum_{k>0}\frac{a_{-k}}{k}z_i^k\right)e^{n\sqrt{N}q}\\
&\quad\times\prod_{i=1}^{n}\exp\left(-\sqrt{N}\sum_{k>0}\frac{a_k}{k}z_i^{-k}\right)\prod_{i=1}^{n}z_i^{\sqrt{N}a_0}
\end{aligned}
$$

for $|z_1|>|z_2|>\cdots>|z_n|$. □

To look at the right-hand side of (4.11) in more detail, for $j\in\mathbb{Z}_{\ge 0}$, we introduce polynomials $f_{-j}\in\mathbb{C}[a_{-k}\ (k\in\mathbb{Z}_{>0})]$ by

$$
\sum_{j\in\mathbb{Z}_{\ge0}} f_{-j}z^j=\exp\left(\sqrt{N}\sum_{k>0}\frac{a_{-k}}{k}z^k\right).
$$

Note that the polynomials $\{f_{-j}\mid j\in\mathbb{Z}_{\ge0}\}$ are the elementary Schur polynomials with variables $\{\sqrt{N}a_{-k}/k\mid k\in\mathbb{Z}_{>0}\}$.

Lemma 4.7. *The set of polynomials* $\{f_{-j}\mid j\in\mathbb{Z}_{\ge0}\}$ *forms a transcendental basis of the polynomial ring* $\mathbb{C}[a_{-k}\ (k\in\mathbb{Z}_{>0})]$.

We set

$$
\sum_{\nu}c_\nu z^\nu:=\prod_{1\le i<j\le n}(z_i-z_j)^N,
$$

where ν denotes the multi-index $(\nu_1,\cdots,\nu_n)\in(\mathbb{Z}_{\ge0})^n$ and we set $z^\nu:=\prod_{i=1}^n z_i^{\nu_i}$.

The following lemma is a key step of our proof.

Lemma 4.8. *For* $(m_1,m_2,\cdots,m_n)\in\mathbb{Z}^n$, *we have*

$$
V_{\sqrt{N}}(m_1)V_{\sqrt{N}}(m_2)\cdots V_{\sqrt{N}}(m_n).|\eta-n\sqrt{N}\rangle\neq 0, \tag{4.12}
$$

if there exists $(s_1,\cdots,s_n)\in(\mathbb{Z}_{\ge0})^n$ *such that* $\sum_{\nu}c_\nu\neq0$, *where the sum runs over the set*

$$
\{(-\sqrt{N}\eta+nN-m_1-1-s_{\sigma^{-1}(1)},\cdots,-\sqrt{N}\eta+nN-m_n-1-s_{\sigma^{-1}(n)})\mid\sigma\in\mathfrak{S}_n\}
$$

without multiplicity.

Proof. From (4.11), we obtain

$$
\begin{aligned}
&V_{\sqrt{N}}(z_1)V_{\sqrt{N}}(z_2)\cdots V_{\sqrt{N}}(z_n).|\eta-n\sqrt{N}\rangle\\
&=\sum_{\nu}\sum_{(j_1,\cdots,j_n)\in(\mathbb{Z}_{\ge0})^n}c_\nu\left(\prod_{i=1}^n z_i^{\nu_i+\sqrt{N}(\eta-n\sqrt{N})+j_i}\right)\prod_{i=1}^n f_{-j_i}.|\eta\rangle.
\end{aligned}\tag{4.13}
$$

Considering the coefficients of $z_1^{-m_1-1}z_2^{-m_2-1}\cdots z_n^{-m_n-1}$ in (4.13), we see that

$\sum_\nu c_\nu$ equals the coefficient of $f_{-s_1}f_{-s_2}\cdots f_{-s_n}.|\eta\rangle$ in

$$V_{\sqrt{N}}(m_1)V_{\sqrt{N}}(m_2)\cdots V_{\sqrt{N}}(m_n).|\eta - n\sqrt{N}\rangle, \tag{4.14}$$

since $\{f_{-j} \mid j \in \mathbb{Z}_{\geq 0}\}$ forms a transcendental basis. □

PROOF OF PROPOSITION 4.4. Taking $\tilde{\nu} \in (\mathbb{Z}_{\geq 0})^n$ as

$$\tilde{\nu} := (\tilde{\nu}_1, \cdots, \tilde{\nu}_n), \quad \tilde{\nu}_i := t(n-1) \ (1 \leq i \leq n),$$

we first show that if $t > n$, then $c_{\tilde{\nu}} \neq 0$. Let

$$\Delta(z_1, \cdots, z_n) := \prod_{1\leq i<j\leq n} (z_i - z_j)$$

be the Vandermonde determinant. We see that

$$\Delta(z_1, \cdots, z_n)^{2t} \equiv \Delta(z_1^t, \cdots, z_n^t)^2 \pmod{t\,\mathbb{Z}[z_1, \cdots, z_n]},$$

since, for an odd prime integer t, we have

$$(z_i - z_j)^t \equiv z_i^t - z_j^t \qquad (\mathrm{mod}\ t\,\mathbb{Z}[z_1, \cdots, z_n]).$$

Moreover, the coefficient of $z_1^{t(n-1)}z_2^{t(n-1)}\cdots z_n^{t(n-1)}$ in $\Delta(z_1^t, \cdots, z_n^t)^2$ is equal to $(-1)^{\frac{n(n-1)}{2}}n!$, since

$$\begin{aligned}\Delta(z_1^t, \cdots, z_n^t)^2 = &\sum_{\sigma\in\mathfrak{S}_n} \mathrm{sgn}(\sigma)\, z_{\sigma(n)}^{t(n-1)}\cdots z_{\sigma(2)}^t z_{\sigma(1)}^0 \\ &\times \sum_{\tau\in\mathfrak{S}_n} \mathrm{sgn}(\tau)\, z_{\tau(n)}^0\cdots z_{\tau(2)}^{t(n-2)} z_{\tau(1)}^{t(n-1)}\, (-1)^{\frac{n(n-1)}{2}}.\end{aligned}$$

We see that the coefficient of $z_1^{t(n-1)}z_2^{t(n-1)}\cdots z_n^{t(n-1)}$ in $\Delta(z_1, \cdots, z_n)^{2t}$ coincides with $(-1)^{\frac{n(n-1)}{2}}n! \pmod{t}$, and hence, $c_{\tilde{\nu}} \neq 0$ if $t > n$.

Here, we note that

$$(n+1)t - (u+1) \geq 0 \Leftrightarrow -\sqrt{N}\eta + nN - 1 \geq t(n-1).$$

Hence, choosing $(s_1, \cdots, s_n) \in (\mathbb{Z}_{\geq 0})^n$ as

$$s_i := (-\sqrt{N}\eta + nN - 1) - t(n-1) \qquad 1 \leq i \leq n,$$

we have $\sum_\nu c_\nu = c_{\tilde{\nu}} \neq 0$. Now, Proposition 4.4 follows from Lemma 4.8. □

4.4 Determinants of Verma Modules

Using the homomorphism $(\Sigma_{\sqrt{N}})^n$ studied in the previous section, we compute the determinant of Verma modules over the Virasoro algebra.

4.4.1 Definitions and Formulae

Here, we summarise basic properties of contravariant forms on Verma modules (cf. § 3.1.2), and state the determinant formula.

Using the notation introduced in § 4.1.1, we have

$$M(c,h) = \bigoplus_{n\in\mathbb{Z}_{\geq 0}} M(c,h)_{h+n}.$$

We sometimes denote the highest weight vector $1 \otimes \mathbf{1}_{c,h}$ of $M(c,h)$ by $v_{c,h}$. Hence $M(c,h)_h = \mathbb{C}v_{c,h}$. We say that a vector $v \in M(c,h)$ is of **level** n if $v \in M(c,h)_{h+n}$.

There exists a unique bilinear form

$$\langle \cdot, \cdot \rangle_{c,h} : M(c,h) \times M(c,h) \longrightarrow \mathbb{C}$$

such that

1. $\langle v_{c,h}, v_{c,h} \rangle_{c,h} = 1$,
2. for $x \in U(\mathrm{Vir})$ and $u, v \in M(c,h)$,
$$\langle x.v, w \rangle_{c,h} = \langle v, \sigma(x).w \rangle_{c,h} \tag{4.15}$$
where σ is the anti-involution of Vir.

By the contravariance, one has

1. $\langle \cdot, \cdot \rangle_{z,h}|_{M(c,h)_{h_1} \times M(c,h)_{h_2}} = 0$ if $h_1 \neq h_2$,
2. $\mathrm{rad}\langle \cdot, \cdot \rangle_{c,h}$ coincides with the maximal proper submodule $J(c,h)$ of $M(c,h)$.

Definition 4.1 *For $n \in \mathbb{Z}_{\geq 0}$, we define $\det(c,h)_n$ as the discriminant of the bilinear form*

$$\langle \cdot, \cdot \rangle_{c,h;n} := \langle \cdot, \cdot \rangle_{c,h}|_{M(c,h)_{h+n} \times M(c,h)_{h+n}}.$$

It should be noticed that

$$\det(c,h)_n \in \mathbb{C}[c,h]$$

by definition. The following is the main theorem of this section:

Theorem 4.2 *For $n \in \mathbb{Z}_{>0}$, we have*

$$\det(c,h)_n \propto \prod_{\substack{r,s\in\mathbb{Z}_{>0}\\ r\geq s\\ 1\leq rs\leq n}} \Phi_{r,s}(c,h)^{p(n-rs)}, \tag{4.16}$$

where

$$\Phi_{r,s}(c,h) := \begin{cases} \left\{h+\dfrac{1}{24}(r^2-1)(c-13)+\dfrac{1}{2}(rs-1)\right\} & \\ \times\left\{h+\dfrac{1}{24}(s^2-1)(c-13)+\dfrac{1}{2}(rs-1)\right\} & \text{if } r\neq s \\ +\dfrac{1}{16}(r^2-s^2)^2 & \\ h+\dfrac{1}{24}(r^2-1)(c-13)+\dfrac{1}{2}(r^2-1) & \text{if } r=s \end{cases}$$

and $p(n)$ denotes the partition number of n.

4.4.2 Proof of Theorem 4.2

For the proof of the theorem, we first introduce some notation. For $\mathbb{I} = (1^{r_1}2^{r_2}\cdots n^{r_n})$, we set

$$e_{\mathbb{I}} := L_{-n}^{r_n}\cdots L_{-2}^{r_2}L_{-1}^{r_1} \in U(\mathrm{Vir}). \tag{4.17}$$

forms a basis of the weight subspaceThen, $\{e_{\mathbb{I}}.v_{c,h}|\mathbb{I}\in\mathcal{P}_n\}$ forms a basis of the weight subspace $M(c,h)_{h+n}$.

Before carrying out the proof of Theorem 4.2, let us explain the strategy of our proof. As the first step, we estimate the degree of $\det(c,h)_n$ as a polynomial of h (Lemma 4.10). As the second step, exploiting homomorphisms $(\Sigma_{\sqrt{N}})^n$, we write down as many factors of the determinant as possible (Lemma 4.11). Comparing the estimation of the h-degree of the determinant, it turns out that the factors in the second step exhaust those of the determinant.

We carry out the first step. For notation, see (4.8).

Lemma 4.9. *For $\mathbb{I}$, $\mathbb{J}\in\mathcal{P}_n$,*

1. *h-$\deg\langle e_{\mathbb{I}}.v_{c,h}, e_{\mathbb{J}}.v_{c,h}\rangle_{c,h} \leq \min\{|\mathbb{I}|,|\mathbb{J}|\}$,*
2. *h-$\deg\langle e_{\mathbb{I}}.v_{c,h}, e_{\mathbb{J}}.v_{c,h}\rangle_{c,h} \leq |\mathbb{I}|-1$, if $|\mathbb{I}|=|\mathbb{J}|$ and $\mathbb{I}\neq\mathbb{J}$,*
3. *h-$\deg\langle e_{\mathbb{I}}.v_{c,h}, e_{\mathbb{I}}.v_{c,h}\rangle_{c,h} = |\mathbb{I}|$.*

Proof. Since the contravariant form $\langle\cdot,\cdot\rangle_{c,h}$ is symmetric, we may assume that $|\mathbb{I}|\leq|\mathbb{J}|$ without loss of generality. Let us prove the first assertion by induction on $|\mathbb{I}|$. Using the commutation relations of Vir, we have

$$L_j.e_{\mathbb{I}}v_{c,h} = \sum_{\mathbb{I}'\in\mathcal{P}_{n-j}} c_{\mathbb{I}'}e_{\mathbb{I}'}v_{c,h} \quad (j\in\mathbb{Z}_{>0}), \tag{4.18}$$

where $c_{\mathbb{I}'} \in \mathbb{C}[c,h]$ such that h-$\deg c_{\mathbb{I}'} \leq 1$. From the property (4.15), the first assertion is an immediate consequence of (4.18). The rest of the statements can be proved by induction on $|\mathbb{I}|$. □

Let $\mathbb{I}_1, \cdots, \mathbb{I}_{p(n)}$ be the distinct elements of $\mathcal{P}_n$. We have

$$\det(c,h)_n \propto \det\left(\langle e_{\mathbb{I}_i} v_{c,h}, e_{\mathbb{I}_j} v_{c,h}\rangle_{c,h}\right)_{1\leq i,j\leq p(n)}$$
$$= \sum_{\sigma\in\mathfrak{S}_{p(n)}} \operatorname{sgn}\sigma \prod_{i=1}^{p(n)} \langle e_{\mathbb{I}_i} v_{c,h}, e_{\mathbb{I}_{\sigma(i)}} v_{c,h}\rangle_{c,h}.$$

The above lemma says that the diagonal part attains the maximal degree of $\det(c,h)_n$. Hence, we see that

$$h\text{-}\deg\det(c,h)_n = \sum_{\mathbb{I}\in\mathcal{P}_n} |\mathbb{I}|.$$

Moreover, the right-hand side can be written as follows:

Lemma 4.10.

$$\sum_{\mathbb{I}\in\mathcal{P}_n} |\mathbb{I}| = \sum_{\substack{r,s\in\mathbb{Z}_{>0}\\ 1\leq rs\leq n}} p(n-sr).$$

Proof. The proof is direct calculation. Indeed, we have

$$\sum_{N\in\mathbb{Z}_{>0}}\left(\sum_{\mathbb{I}\in\mathcal{P}_N}|\mathbb{I}|\right)q^N = \sum_{N\in\mathbb{Z}_{>0}}\left(\sum_{(1^{r_1}\cdots N^{r_N})\in\mathcal{P}_N}\sum_{k=1}^{N} r_k q^{\sum_i ir_i}\right)$$
$$= \sum_{r_1,r_2,\cdots,\in\mathbb{Z}_{\geq 0}}\sum_{k\in\mathbb{Z}_{>0}} r_k q^{kr_k}\prod_{i\neq k} q^{ir_i}$$
$$= \sum_{k\in\mathbb{Z}_{>0}}\frac{q^k}{1-q^k}\prod_{i\in\mathbb{Z}_{>0}}(1-q^i)^{-1}$$
$$= \sum_{k\in\mathbb{Z}_{>0}}\left(\sum_{s\in\mathbb{Z}_{>0}} q^{ks}\right)\sum_{j\in\mathbb{Z}_{>0}} p(j)q^j$$
$$= \sum_{N\in\mathbb{Z}_{>0}}\left(\sum_{\substack{r,s>0\\ 1\leq rs\leq N}} p(N-rs)\right)q^N.$$

Thus, we obtain

$$\sum_{\mathbb{I}\in\mathcal{P}_N}|\mathbb{I}| = \sum_{\substack{r,s>0\\ 1\leq rs\leq N}} p(N-rs).$$ □

On the other hand, in the right-hand side of Theorem 4.2, we have

$$h\text{-}\deg \prod_{\substack{r,s\in\mathbb{Z}_{>0}\\ r\geq s\\ 1\leq rs\leq n}} \Phi_{r,s}(c,h)^{p(n-rs)} = \sum_{\substack{r,s>0\\ 1\leq rs\leq n}} p(n-rs),$$

since

$$h\text{-}\deg\Phi_{r,s}(c,h) = \begin{cases} 2 & r\neq s \\ 1 & r=s \end{cases}.$$

Therefore, the maximal h-degrees of both sides of (4.16) coincide, and we complete the first step.

Let us carrying out the second step, i.e., to find factors of the determinant $\det(c,h)_n$.

The singular vectors given by Proposition 4.4 provide the following vanishing locus of the determinant:

Lemma 4.11. *Suppose that*

$$c = 1 - 6\frac{(t-1)^2}{t}, \quad h = \frac{(rt-s)^2-(t-1)^2}{4t},$$

for $r,s\in\mathbb{Z}_{>0}$ and an odd prime number t such that $t>r$. Then, we have

$$\det(c,h)_{rs} = 0.$$

Proof. By setting

$$N := 2t, \quad \lambda := \frac{t-1}{\sqrt{2t}}, \quad \eta := \frac{(r+1)t-(s+1)}{\sqrt{2t}} \quad \text{and } n := r,$$

Lemma 4.5 and Proposition 4.4 imply that

$$(\Sigma_{\sqrt{N}})^n.|\eta - n\sqrt{N}\rangle \in (\mathcal{F}_\lambda^\eta)^{\mathrm{Vir}^+}\setminus\{0\},$$

since

$$\lambda = \frac{N-2}{2\sqrt{N}}, \quad \sqrt{N}\eta\in\mathbb{Z}, \quad (n+1)t-(\sqrt{N}\eta+1)\geq 0.$$

Furthermore, by Lemma 4.5 we have

$$L_0(\Sigma_{\sqrt{N}})^n.|\eta-n\sqrt{N}\rangle = (h+rs)(\Sigma_{\sqrt{N}})^n.|\eta-n\sqrt{N}\rangle,$$

since

$$\frac{1}{2}(\eta-n\sqrt{N})(\eta-n\sqrt{N}-2\lambda) = h_\lambda^\eta + rs = h+rs.$$

Using these facts, we can show the lemma in the following manner:

Let $\mathcal{G}_\lambda^\eta$ be the Vir-submodule of $\mathcal{F}_\lambda^\eta$ generated by $|\eta\rangle$, i.e., $\mathcal{G}_\lambda^\eta := U(\mathrm{Vir}).|\eta\rangle$. Here, we notice that $\mathcal{G}_\lambda^\eta$ is a highest weight module, and hence, there exists a non-trivial surjective homomorphism $M(c,h) \to \mathcal{G}^{\eta}_{\lambda}$. Let us denote the surjection by ϕ.

There are two possible cases.

1. $(\Sigma_{\sqrt{N}})^n.|\eta - n\sqrt{N}\rangle \in \mathcal{G}^{\eta}_{\lambda}$,
2. $(\Sigma_{\sqrt{N}})^n.|\eta - n\sqrt{N}\rangle \notin \mathcal{G}^{\eta}_{\lambda}$.

In the first case, we have

$$M(c,h)_{h+rs} \cap \mathrm{rad}\langle\cdot,\cdot\rangle_{c,h} \neq \{0\},$$

since there exists a vector $v \in \mathrm{rad}\langle\cdot,\cdot\rangle_{c,h}$ such that $\phi(v) = (\Sigma_{\sqrt{N}})^n.|\eta - n\sqrt{N}\rangle$. Then, from Proposition 3.4, we obtain $\det(c,h)_{rs} = 0$. In the second case, we get

$$M(c,h)_{h+rs} \cap \mathrm{Ker}\phi \neq \{0\}.$$

Since $\mathrm{Ker}\phi$ is a proper submodule of $M(c,h)$, we have $\det(c,h)_{rs} = 0$. Now, we have arrived at the conclusion. □

For $r, s \in \mathbb{Z}_{>0}$ we introduce the curve $\mathcal{V}_{r,s}$ in $\mathbb{C}^2$ by

$$\mathcal{V}_{r,s} := \{(c,h) \in \mathbb{C}^2 \mid \Phi_{r,s}(c,h) = 0\}. \tag{4.19}$$

We have

Lemma 4.12. *The determinant* $\det(c,h)_{rs}$ *vanishes on the curve* $\mathcal{V}_{r,s}$.

Proof. It follows from Lemma 4.11 that $\det(c,h)_{rs} = 0$ on infinitely many points of $\mathcal{V}_{r,s}$. Since $\det(c,h)_{rs} \in \mathbb{C}[c,h]$, $\det(c,h)_{rs} = 0$ on the Zariski closure of the set of these points. Moreover, by the irreducibility of the curve $\mathcal{V}_{r,s}$, the closure coincides with $\mathcal{V}_{r,s}$. Hence, the lemma is proved. □

Using Lemma 4.12 and embeddings between Verma modules, we can obtain sufficiently many factors of the determinant. To do this, the following two preliminary lemmas are necessary:

Lemma 4.13. *Let* V *be a finite dimensional vector space over* $\mathbb{C}$, *and let* $A(t)$ *be an element of* $\mathrm{End}(V) \otimes_{\mathbb{C}} \mathbb{C}[[t]]$. *If* $\dim \mathrm{Ker} A(0) = n$, *then we have* $t^n | \det A(t)$.

Proof. We choose a basis $\{e_1, \cdots, e_{\dim V}\}$ of V such that $\{e_1, \cdots, e_{\dim V}\}$ forms a basis of $\mathrm{Ker}A(0)$. With respect to the above basis, the matrix elements of the kth row of $A(t)$ are divided by t for any k such that $1 \leq k \leq n$. Hence, the lemma is obvious. □

Lemma 4.14. *Fix a central charge* $c \in \mathbb{C}$. *Suppose that* $\det(c,h_0)_n = 0$ *for some* $h_0 \in \mathbb{C}$ *and* $n \in \mathbb{Z}_{>0}$. *Let* k *be the minimal integer such that* $\det(c,h_0)_k = 0$. *Then,* $\det(c,h)_n$ *is divided by* $(h - h_0)^{p(n-k)}$.

Proof. To show this lemma, we state the following simple but important property of Verma modules: For any singular vector $v_l \in M(c,h)_{h+l}$,

$$U(\mathrm{Vir}).v_l \simeq M(c, h+l). \tag{4.20}$$

Note that $M(c,h_0)$ has a singular vector v_k of level k, since $\det(c,h_0)_k = 0$ and $\det(c,h_0)_l \neq 0$ for $l < k$. From (4.20), we see that

$$\dim(U(\mathrm{Vir}).v_k \cap M(c,h_0)_{h+n}) = p(n-k).$$

Now, Lemma 4.13 implies this lemma. □

Combining Lemma 4.12 with the above lemma, we carry out the second step of the proof of Theorem 4.2. For $r, s \in \mathbb{Z}_{>0}$, we set

$$\begin{aligned} h^{\pm}_{r,s}(c) := &-\frac{1}{48}(r^2+s^2-2)(c-13) - \frac{1}{2}(rs-1) \\ &\pm \frac{1}{48}(r^2-s^2)\sqrt{(1-c)(25-c)} \end{aligned}$$

and

$$\tilde{\Phi}_{r,s}(c,h) := \begin{cases} (h-h^+_{r,s}(c))(h-h^-_{r,s}(c)) & (r \neq s) \\ (h-h^+_{r,r}(c)) & (r = s) \end{cases}.$$

Then, the theorem is equivalent to

$$\det(c,h)_n \propto \prod_{\substack{r,s\in\mathbb{Z}_{>0} \\ r\geq s \\ 1\leq rs\leq n}} \tilde{\Phi}_{r,s}(c,h)^{p(n-rs)}, \tag{4.21}$$

since $\tilde{\Phi}_{r,s}(c,h) = \Phi_{r,s}(c,h)$. Lemmas 4.12 and 4.14 imply that the determinant $\det(c,h)_n$ is divided by

$$\prod_{\substack{r,s\in\mathbb{Z}_{>0} \\ r\geq s \\ 1\leq rs\leq n}} \tilde{\Phi}_{r,s}(c,h)^{p(n-rs)}.$$

Combining Lemma 4.10, we have (4.21). Hence, we complete the proof of Theorem 4.2.

4.5 Determinants of Fock Modules

In this section, we compute the determinants of two Vir-homomorphisms; from Verma modules to Fock modules and from Fock modules to the contragredient dual of Verma modules. Note that a key of our computation of the determinants is duality of Fock modules in the first section.

4.5.1 Definitions and Formulae

Let us first introduce the determinants of Fock modules to be discussed here. We consider a Vir-homomorphism

$$\Gamma_{\lambda,\eta} : M(c_\lambda, h_\lambda^\eta) \longrightarrow \mathcal{F}_\lambda^\eta$$

which sends the highest weight vector $v_{c_\lambda, h_\lambda^\eta}$ to the vector $|\eta\rangle$. Moreover, let

$$\mathrm{L}^{\lambda,\eta} : \mathcal{F}_\lambda^\eta \longrightarrow M(c_\lambda, h_\lambda^\eta)^c$$

be the Vir-homomorphism which is defined by the composition

$$\mathcal{F}_\lambda^\eta \xrightarrow{\sim} (\mathcal{F}_{-\lambda}^{\eta-2\lambda})^c \xrightarrow{{}^t\Gamma_{-\lambda,\eta-2\lambda}} M(c_{-\lambda}, h_{-\lambda}^{\eta-2\lambda})^c = M(c_\lambda, h_\lambda^\eta)^c,$$

where the first map is the isomorphism in Corollary 4.1 and the third equality comes from $c_{-\lambda} = c_\lambda$ and $h_{-\lambda}^{\eta-2\lambda} = h_\lambda^\eta$. For simplicity, until the end of this section, we set

$$c := c_\lambda, \quad h := h_\lambda^\eta.$$

Similarly to the case of Verma modules, we restrict the maps $\Gamma_{\lambda,\eta}$ and $\mathrm{L}^{\lambda,\eta}$ to each weight subspace and define their determinants. Let

$$\mathcal{F}_\lambda^\eta = \bigoplus_{n \in \mathbb{Z}_{\geq 0}} (\mathcal{F}_\lambda^\eta)_{h+n},$$

$$M(c,h)^c = \bigoplus_{n \in \mathbb{Z}_{\geq 0}} M(c,h)_{h+n}^c,$$

be the weight space decompositions of $\mathcal{F}_\lambda^\eta$ and $M(c,h)^c$ respectively, where

$$(\mathcal{F}_\lambda^\eta)_{h+n} := \{u \in \mathcal{F}_\lambda^\eta | L_0.u = (h+n)u\},$$
$$M(c,h)_{h+n}^c := \{v \in M(c,h)^c | L_0.v = (h+n)v\}.$$

Recall that

$$(\mathcal{F}_\lambda^\eta)_{h+n} = (\mathcal{F}^\eta)^{-n\alpha_\mathcal{H}},$$

where the subspace $(\mathcal{F}^\eta)^{-n\alpha_\mathcal{H}}$ is defined in (4.6).

We set

$$(\Gamma_{\lambda,\eta})_n := \Gamma_{\lambda,\eta}|_{M(c,h)_{h+n}} : M(c,h)_{h+n} \longrightarrow (\mathcal{F}_\lambda^\eta)_{h+n},$$

$$(\mathrm{L}^{\lambda,\eta})_n := \mathrm{L}^{\lambda,\eta}|_{(\mathcal{F}_\lambda^\eta)_{h+n}} : (\mathcal{F}_\lambda^\eta)_{h+n} \longrightarrow M(c,h)_{h+n}^c.$$

Here, we fix bases of the weight subspaces $M(c,h)_{h+n}$, $(\mathcal{F}_\lambda^\eta)_{h+n}$ and $M(c,h)_{h+n}^c$ as

$$\{e_{\mathbb{I}}.v_{c,h} \mid \mathbb{I} \in \mathcal{P}_n\}, \tag{4.22}$$
$$\{a_{\mathbb{I}}.|\eta\rangle \mid \mathbb{I} \in \mathcal{P}_n\}, \tag{4.23}$$
$$\{(e_{\mathbb{I}}.v_{c,h})^* \mid \mathbb{I} \in \mathcal{P}_n\}, \tag{4.24}$$

where for $\mathbb{I} := (1^{r_1}2^{r_2}\cdots n^{r_n})$,

$$e_{\mathbb{I}} := L_{-n}^{r_n}\cdots L_{-2}^{r_2}L_{-1}^{r_1}, \quad a_{\mathbb{I}} := a_{-n}^{r_n}\cdots a_{-2}^{r_2}a_{-1}^{r_1},$$

and (4.24) is the dual basis of (4.22). Further, we introduce the matrices

$$C_n(\lambda,\eta) := (C_n(\lambda,\eta)_{\mathbb{I},\mathbb{J}})_{\mathbb{I},\mathbb{J}\in\mathcal{P}_n}, \quad C'_n(\lambda,\eta) := (C'_n(\lambda,\eta)_{\mathbb{I},\mathbb{J}})_{\mathbb{I},\mathbb{J}\in\mathcal{P}_n}$$

by

$$\Gamma_{\lambda,\eta}(e_{\mathbb{I}}.v_{c,h}) = \sum_{\mathbb{J}\in\mathcal{P}_n} C_n(\lambda,\eta)_{\mathbb{J},\mathbb{I}}\ a_{\mathbb{J}}.|\eta\rangle,$$
$$\mathrm{L}^{\lambda,\eta}(a_{\mathbb{I}}.|\eta\rangle) = \sum_{\mathbb{J}\in\mathcal{P}_n} C'_n(\lambda,\eta)_{\mathbb{J},\mathbb{I}}\ (e_{\mathbb{J}}.v_{c,h})^*,$$

and we define the determinants $\det(\Gamma_{\lambda,\eta})_n$ and $\det(\mathrm{L}^{\lambda,\eta})_n$ by

$$\det(\Gamma_{\lambda,\eta})_n := \det C_n(\lambda,\eta),$$
$$\det(\mathrm{L}^{\lambda,\eta})_n := \det C'_n(\lambda,\eta).$$

Note that these determinants are independent of the choice of bases up to a scalar.

Now, we are ready to give the main statement of this section.

Theorem 4.3 *For $n \in \mathbb{Z}_{>0}$, the following hold:*

1.

$$\det(\Gamma_{\lambda,\eta})_n \propto \prod_{\substack{r,s\in\mathbb{Z}_{>0}\\ 1\le rs\le n}} \Psi^+_{r,s}(\lambda,\eta)^{p(n-rs)}, \tag{4.25}$$

2.

$$\det(\mathrm{L}^{\lambda,\eta})_n \propto \prod_{\substack{r,s\in\mathbb{Z}_{>0}\\ 1\le rs\le n}} \Psi^-_{r,s}(\lambda,\eta)^{p(n-rs)}, \tag{4.26}$$

where

$$\Psi^{\pm}_{r,s}(\lambda,\eta) := (\eta-\lambda) \pm \left(\frac{1}{2}\lambda_+ r + \frac{1}{2}\lambda_- s\right),$$

and $\lambda_\pm := \lambda \pm \sqrt{\lambda^2+2}$ is the solutions of $\lambda = \frac{1}{2}\mu - \frac{1}{\mu}$ which appear in Lemma 4.5.

We remark that the right-hand sides of (4.25) and (4.26) are elements of $\mathbb{C}[\lambda,\eta]$, because if $r \neq s$,

$$\Psi^{\pm}_{r,s}(\lambda,\eta)\Psi^{\pm}_{s,r}(\lambda,\eta) = (\eta-\lambda)^2 \pm 2\lambda(r+s) + (\lambda^2+1)rs - \frac{1}{2}(r^2+s^2)$$

and if $r = s$,

$$\Psi^{\pm}_{r,r}(\lambda,\eta) = (\eta-\lambda) \pm \lambda r.$$

4.5.2 Proof of Theorem 4.3

Let us start the proof with the following lemma, which says that, to show the theorem, it is enough to look at the determinant $\det(\mathrm{L}^{\lambda,\eta})_n$.

Lemma 4.15.

$$\det(\Gamma_{\lambda,\eta})_n \propto \det(\mathrm{L}^{-\lambda,\eta-2\lambda})_n.$$

Proof. By definition of the map $\mathrm{L}^{\lambda,\eta}$ the lemma follows. □

Here, we regard the contravariant form $\langle\cdot,\cdot\rangle_{c,h}$ as a homomorphism from $M(c,h)$ to $M(c,h)^c$ in the following way:

$$\begin{array}{ccc} M(c,h) & \longrightarrow & M(c,h)^c, \\ v & \mapsto & \langle v,\cdot\rangle_{c,h}. \end{array}$$

Then, the following diagram commutes:

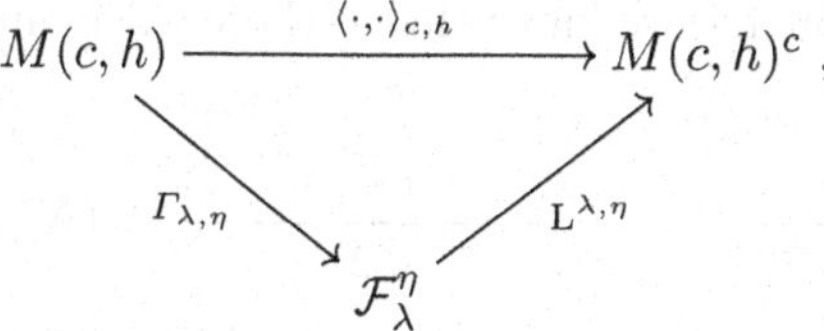

since a homomorphism from $M(c,h)$ to $M(c,h)^c$ is unique up to a scalar. This commutative diagram implies the lemma below.

Lemma 4.16.

$$\det(c,h)_n \propto \det(\Gamma_{\lambda,\eta})_n \times \det(\mathrm{L}^{\lambda,\eta})_n.$$

We have already computed $\det(c,h)_n$, and by Lemma 4.15 the second factor in the right-hand side can be computed from the first factor.

In fact, to prove Theorem 4.3, we find sufficiently many factors of $\det(\mathrm{L}^{\lambda,\eta})_n$ by using the homomorphism $(\Sigma_{\sqrt{N}})^n$ in § 4.3, and show that such factors exhaust factors of the determinant by Lemma 4.16.

For the proof of Theorem 4.3, we need preliminary lemmas. The first one is a slight modification of Lemma 4.13.

Lemma 4.17. *Let V and W be finite dimensional vector spaces of the same dimension k, and let $\{v_i\}$ and $\{w_i\}$ be bases of V and W respectively. For $A(t) \in \mathrm{Hom}_{\mathbb{C}}(V, W) \otimes_{\mathbb{C}} \mathbb{C}[[t]]$, we define a $k \times k$ matrix $M(t) := (m_{i,j}(t))$ by*

$$A(t)v_i = \sum_{j=1}^{k} m_{j,i}(t) w_j.$$

If $\dim \mathrm{Ker} A(0) = n$, *then* $\det M(t)$ *is divisible by* t^n.

Lemma 4.18. *For any highest weight $(c, h) \in \mathbb{C}^2$ and $n \in \mathbb{Z}_{>0}$, $M(c,h)^c_{h+n} \setminus \{0\}$ has no singular vector.*

Proof. The dual statement is that $M(c,h)_{h+n} \setminus \{0\}$ has no cosingular vector, which is clear, since $M(c,h)$ is generated from its highest weight vector. □

Using the above two lemmas, we show the following lemma, which is a key step of the proof of Theorem 4.3.

Lemma 4.19. *For $n \in \mathbb{Z}_{>0}$ and $r, s \in \mathbb{Z}_{>0}$ such that $rs \leq n$, we have*

1. $\det(\mathrm{L}^{\lambda,\eta})_n$ *is divisible by* $\left\{\Psi^-_{r,s}(\lambda,\eta)\Psi^-_{s,r}(\lambda,\eta)\right\}^{p(n-rs)}$, *if* $r \neq s$,
2. $\det(\mathrm{L}^{\lambda,\eta})_n$ *is divisible by* $\Psi^-_{r,r}(\lambda,\eta)^{p(n-r^2)}$, *if* $r = s$.

Proof. By Proposition 4.4, we know that if there exists an odd prime number $t > r$ such that

$$\lambda = \frac{t-1}{\sqrt{2t}}, \quad \eta = \frac{(r+1)t - (s+1)}{\sqrt{2t}}, \quad (N = 2t),$$

then

$$(\Sigma_{\sqrt{N}})^n . |\eta - n\sqrt{N}\rangle \in \{(\mathcal{F}^\eta_\lambda)_{h+rs}\}^{\mathrm{Vir}^+} \setminus \{0\}.$$

For such λ, η and t, we set

$$\mathcal{G}^t_{r,s} := U(\mathrm{Vir}).(\Sigma_{\sqrt{N}})^r |\eta - r\sqrt{N}\rangle.$$

Lemma 4.18 implies that

$$(\Sigma_{\sqrt{N}})^r |\eta - r\sqrt{N}\rangle \in \mathrm{Ker}\mathrm{L}^{\lambda,\eta},$$

and thus

$$\mathcal{G}^t_{r,s} \subset \mathrm{Ker}\mathrm{L}^{\lambda,\eta}.$$

Here, we notice the fact that for each r and s,

$$\det(c, h + rs)_m \neq 0 \quad (\forall\, m \leq n - rs)$$

hold except for finitely many t ($c := c_\lambda$, $h := h^\eta_\lambda$ for the above λ, η). Hence, there exist infinitely many t, which satisfy

$$\dim\{\mathcal{G}^t_{r,s} \cap (\mathcal{F}^\eta_\lambda)_{h+n}\} = p(n - rs).$$

Since the determinant $\det(\mathrm{L}^{\lambda,\eta})_n$ is algebraic with respect to the variables λ and η, the conclusion follows from Lemma 4.17. □

As a consequence of this lemma, we have

Corollary 4.2 *The determinant* $\det(\mathrm{L}^{\lambda,\eta})_n$ *is divisible by the factor*

$$\prod_{\substack{r,s\in\mathbb{Z}_{>0}\\ 1\leq rs\leq n}} \Psi^-_{r,s}(\lambda,\eta)^{p(n-rs)}.$$

Finally, we complete the proof of Theorem 4.3. Combining Lemma 4.15 and Corollary 4.2 and the fact that

$$\Psi^+_{r,s}(\lambda,\eta) = \Psi^-_{r,s}(-\lambda,\eta - 2\lambda),$$

we see that the determinant $\det(\Gamma_{\lambda,\eta})_n$ is divisible by

$$\prod_{\substack{r,s\in\mathbb{Z}_{>0}\\ 1\leq rs\leq n}} \Psi^+_{r,s}(\lambda,\eta)^{p(n-rs)}.$$

On the other hand, by direct computation, we have

$$\Phi_{r,s}(c,h) = \begin{cases} \dfrac{1}{4}\Psi^+_{r,s}(\lambda,\eta)\Psi^+_{s,r}(\lambda,\eta)\Psi^-_{r,s}(\lambda,\eta)\Psi^-_{s,r}(\lambda,\eta) & (r \neq s) \\ \\ \dfrac{1}{2}\Psi^+_{r,r}(\lambda,\eta)\Psi^-_{r,r}(\lambda,\eta) & (r = s) \end{cases}. \tag{4.27}$$

Hence, we obtain

$$\det(c,h)_n \propto \prod_{\substack{r,s\in\mathbb{Z}_{>0}\\ 1\leq rs\leq n}} \Psi^+_{r,s}(\lambda,\eta)^{p(n-rs)}\Psi^-_{r,s}(\lambda,\eta)^{p(n-rs)}.$$

Combining Lemmas 4.15 and 4.16, we see that

$$\det(\mathrm{L}^{\lambda,\eta})_n \propto \prod_{\substack{r,s\in\mathbb{Z}_{>0}\\ 1\leq rs\leq n}} \Psi^-_{r,s}(\lambda,\eta)^{p(n-rs)}.$$

Therefore, we complete the proof.

4.6 Bibliographical Notes and Comments

The formula given in Theorem 4.2, what physicists call the Kac determinant, was first stated in [Kac1] without proof. Its proof was given by several authors in different ways. B. Feigin and D. B. Fuchs [FeFu1] proved it through their study of morphisms between Verma modules and modules of semi-infinite wedges. A. Tsuchiya and Y. Kanie [TK2] proved it by studying morphisms between Verma modules and bosonic Fock modules. Since a module of semi-infinite wedges and a bosonic Fock module with appropriate parameters are isomorphic (see § 8.5), these two proofs are essentially the same. The latter approach was slightly simplified by A. Rocha-Caridi [Ro]. The proof by V. G. Kac and M. Wakimoto [KW1] is given as an application of the coset construction (see Chapter 10).

Theorem 4.3 was proved by A. Tsuchiya and Y. Kanie [TK2] and by E. Frenkel [Fr] by another method. Note that this theorem was also proved by B. Feigin and D. B. Fuchs [FeFu1], although the result itself is not stated explicitly.

Here, we proved these two theorems, based on the approaches due to [Ro] and [TK2]. Our proof of Theorem 4.3 is slightly simplified from the original argument given in [TK2].

Chapter 5
Verma Modules I: Preliminaries

In this and the next chapter, we will study Verma modules over the Virasoro algebra, and will reveal the structure of the Jantzen filtration of Verma modules. This chapter is a preliminary part for the structure theorem of Verma modules developed in the next chapter. Namely, first, we will classify highest weights. Second, we will show the uniqueness of singular vectors and the existence of Shapovalov elements. Third, we will construct embedding diagrams of Verma modules (at least partially). Finally, using the classification of highest weights, we will compute character sums of Jantzen filtration of Verma modules and of some quotient modules. For some special classes of highest weights, which are important in mathematical and theoretical physics, see Section 5.1.5.

5.1 Classification of Highest Weights

Throughout this chapter, let $\mathfrak{g}$ be the Virasoro algebra, and let $\mathfrak{h}$ be the subalgebra $\mathbb{C}L_0 \oplus \mathbb{C}C$. We identify $\mathfrak{h}^*$ with $\mathbb{C}^2$ as $\lambda = (c, h)$ if $\lambda(C) = c$ and $\lambda(L_0) = h$.

Let $\Phi_{\alpha,\beta}(c, h)$ be a factor of the determinant $\det(c, h)_n$ in Theorem 4.2. For each highest weight $(c, h) \in \mathbb{C}^2$, we put

$$\tilde{D}(c, h) := \{(\alpha, \beta) \in (\mathbb{Z}_{>0})^2 | \alpha \geq \beta \wedge \Phi_{\alpha,\beta}(c, h) = 0\}. \tag{5.1}$$

Further, we set

$$D(c, h) := \{\alpha\beta | (\alpha, \beta) \in \tilde{D}(c, h)\}, \tag{5.2}$$

and for $n \in D(c, h)$, we set

$$a(n) := \sharp\{(\alpha, \beta) \in \tilde{D}(c, h) | \alpha\beta = n\}. \tag{5.3}$$

K. Iohara, Y. Koga, *Representation Theory of the Virasoro Algebra*, Springer Monographs in Mathematics, DOI 10.1007/978-0-85729-160-8_5,

In this section, we classify the highest weights (c, h), and for each (c, h) we describe $D(c, h)$ and $a(n)$ $(n \in D(c, h))$ explicitly.

Here, we introduce some notation for later use. For $t \in \mathbb{C}\backslash\{0\}$ and $\alpha, \beta \in \mathbb{Z}$, we set

$$\begin{aligned} c(t) &:= 13 - 6(t + t^{-1}), \\ h_{\alpha,\beta}(t) &:= \frac{1}{4}(\alpha^2 - 1)t - \frac{1}{2}(\alpha\beta - 1) + \frac{1}{4}(\beta^2 - 1)t^{-1}. \end{aligned} \tag{5.4}$$

Note that $(c(t), h_{\alpha,\beta}(t))$ parameterises the curve $\mathcal{V}_{\alpha,\beta} := \{(c, h)|\Phi_{\alpha,\beta}(c, h) = 0\}$.

5.1.1 Strategy of Classification

First we introduce useful parameterisation of highest weight (c, h) due to [FeFu4]. For $P, Q \in \mathbb{C} \setminus \{0\}$, $m \in \mathbb{C}$, we set

$$c_{P,Q} := 13 - 6(\frac{P}{Q} + \frac{Q}{P}), \quad h_{P,Q:m} := \frac{m^2 - (P - Q)^2}{4PQ}. \tag{5.5}$$

Then, it is easy to see that

Lemma 5.1. *1. For any $(c, h) \in \mathfrak{h}^*$, there exists $P, Q \in \mathbb{C} \setminus \{0\}$ and $m \in \mathbb{C}$, such that $(c, h) = (c_{P,Q}, h_{P,Q:m})$.*
2. $(c_{P,Q}, h_{P,Q:m}) = (c_{P',Q'}, h_{P',Q':m'})$, if and only if

$$[P' : Q' : m'] = [P : Q : \pm m] \quad \text{or} \quad [P' : Q' : m'] = [Q : P : \pm m],$$

where $[P : Q : m] \in \mathbb{C}P^2$.

By using this parameterisation, $\Phi_{\alpha,\beta}(c, h)$ factors as follows:

$$\Phi_{\alpha,\beta}(c_{P,Q}, h_{P,Q:m}) = \begin{cases} \frac{1}{(4PQ)^2}(m - P\alpha + Q\beta)(m + P\alpha - Q\beta) \\ \qquad \times(m - Q\alpha + P\beta)(m + Q\alpha - P\beta) & (\alpha \neq \beta) \\ \frac{1}{4PQ}(m - P\alpha + Q\alpha)(m + P\alpha - Q\alpha) & (\alpha = \beta) \end{cases}. \tag{5.6}$$

Hence, the set $\tilde{D}(c, h)$ is described in terms of integral points on the four lines $P\alpha - Q\beta = \pm m$ and $Q\alpha - P\beta = \pm m$. Indeed, we have

$$\begin{aligned} &\tilde{D}(c_{P,Q}, h_{P,Q:m}) \\ &= \{(\alpha, \beta) \in (\mathbb{Z}_{>0})^2 | \alpha \geq \beta, (P\alpha - Q\beta = \pm m) \vee (Q\alpha - P\beta = \pm m)\}. \end{aligned} \tag{5.7}$$

As an immediate consequence of (5.7), we obtain the following lemma.

Lemma 5.2. *Suppose that* $\tilde{D}(c_{P,Q}, h_{P,Q:m}) \neq \emptyset$.

1. *If the slopes* $\frac{Q}{P}$ *and* $\frac{P}{Q}$ *of the lines are not rational, then*

$$\sharp\tilde{D}(c_{P,Q}, h_{P,Q:m}) = 1.$$

2. *If the slopes of the lines are rational, then*

$$\sharp\tilde{D}(c_{P,Q}, h_{P,Q:m}) = \infty \;\Leftrightarrow\; Q/P > 0.$$

Motivated by this lemma, we introduce the following classification **Class *VIR*** of highest weights (c, h):

Class V (Vacant): $\tilde{D}(c,h) = \emptyset$,
Class I (Irrational): $\tilde{D}(c,h) \neq \emptyset$ and $Q/P \notin \mathbb{Q}$,
Class R (Rational): $\tilde{D}(c,h) \neq \emptyset$ and $Q/P \in \mathbb{Q} \setminus \{0\}$.

Remark 5.1 *In the cases of* **Class** V *(resp.* **Class** I*), it is not difficult to calculate character sums and to construct embedding diagrams.*

From now on, we restrict to the case of **Class** R. We define subclasses of **Class** R by

Class R^+: $Q/P \in \mathbb{Q}_{>0}$,
Class R^-: $Q/P \in \mathbb{Q}_{<0}$.

In **Class** R, we may assume that $P, Q \in \mathbb{Z}$ and $(P, Q) = 1$ without loss of generality. Further, if $P, Q \in \mathbb{Z}$ and $m \notin \mathbb{Z}$, then $\tilde{D}(c, h) = \emptyset$. Hence, until the end of this section, we suppose that

1. $P \in \mathbb{Z}_{>0}$, $Q \in \mathbb{Z}_{>0}$ and $m \in \mathbb{Z}$ in **Class** R^+,
2. $P \in \mathbb{Z}_{>0}$, $Q \in \mathbb{Z}_{<0}$ and $m \in \mathbb{Z}$ in **Class** R^-.

Remark 5.2 *For a highest weight* (c, h) *which belongs to* **Class** $R^{\pm}$, *the graphs of the lines* $P\alpha - Q\beta = \pm m$ *and* $Q\alpha - P\beta = \pm m$ *are as follows: If* $Q/P \neq \pm 1$ *and* $m \neq 0$, *then the graphs are given as in Figure 5.1. If* $Q/P = \pm 1$ *or* $m = 0$, *then the lines degenerate into two or one line, and their graphs are given as in Figure 5.2 and Figure 5.3.*

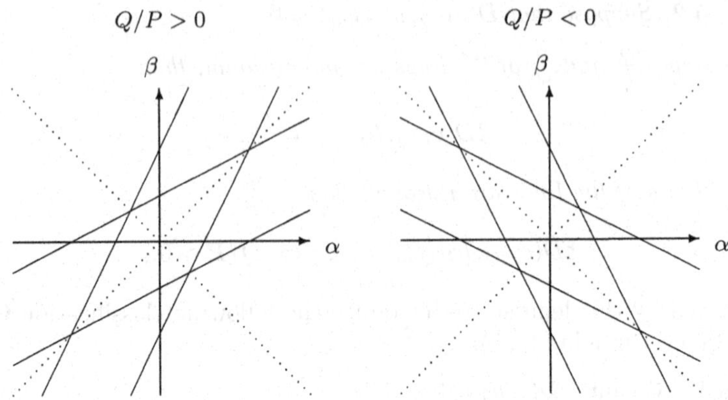

Fig. 5.1 $Q/P \neq \pm 1$ and $m \neq 0$

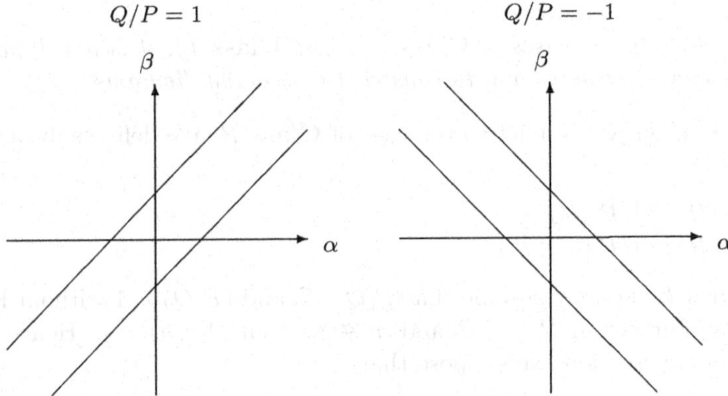

Fig. 5.2 $Q/P = \pm 1$ and $m \neq 0$

Note that if $Q/P < 0$ *and* $m = 0$*, then* $\tilde{D}(c, h) = \emptyset$*. Hence in this case,* (c, h) *belongs to* **Class** V.

Motivated by the above remark, we divide **Class** $R^{\pm}$ into the following types:

Type I: $Q/P \neq \pm 1$ and $m \neq 0$.
Type II: $Q/P = \pm 1$ and $m \neq 0$.
Type III: $m = 0$ (in this case $Q/P > 0$).

The strategy of our classification of highest weights of **Class** $R^{\pm}$ is as follows:

Step 1 For each (c, h), let us fix one of the four lines $P\alpha - Q\beta = \pm m$ and $Q\alpha - P\beta = \pm m$, and denote it by $\ell_{c,h}$. We give a bijection between $\tilde{D}(c, h)$ and a set of integral points on $\ell_{c,h}$. (By virtue of this bijection, we can describe $D(c, h)$ and $a(n)$ $(n \in D(c, h))$ by means of integral points on $\ell_{c,h}$.)

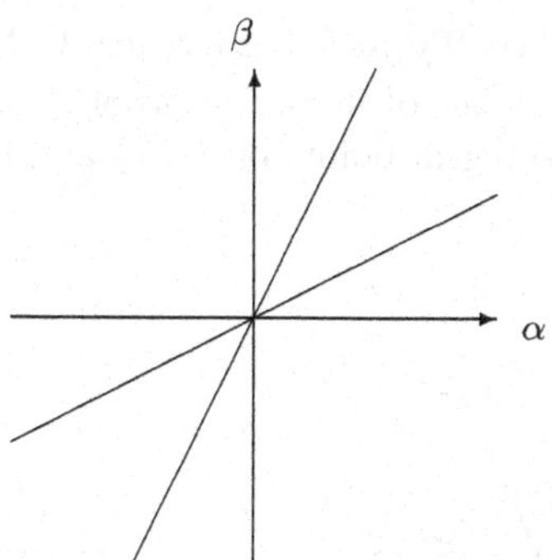

Fig. 5.3 $Q/P > 0$ and $m = 0$

Step 2 We write down the integral points on $\ell_{c,h}$.
Step 3 For each pair (P, Q), using the list of integral points, we describe $D(c, h)$ and $a(n)$ ($n \in D(c, h)$).

5.1.2 Bijection between $\tilde{D}(c, h)$ and Integral Points on $\ell_{c,h}$

Here, we carry out **Step** 1 in the above strategy.

Lemma 5.3. *1. There exists a one-to-one correspondence between $\tilde{D}(c, h)$ and the following set of integral points on $\ell_{c,h}$:*

Class $R^{\pm}$ *of* **Type** *I and* **Class** R^{-} *of* **Type** *II*:

$$\{(\alpha, \beta) \in \mathbb{Z}^2 \cap \ell_{c,h} | \ \alpha\beta > 0\}, \tag{5.8}$$

Class R^{+} *of* **Types** *II and III*:

$$\{(\alpha, \beta) \in (\mathbb{Z}_{>0})^2 \cap \ell_{c,h}\}. \tag{5.9}$$

2. The above correspondence preserves the product of the first and the second coordinates, i.e., if $(\alpha, \beta) \mapsto (\alpha', \beta')$ under the correspondence, then $\alpha\beta = \alpha'\beta'$. Hence, $D(c, h)$ is described as follows:

Class $R^{\pm}$ *of* **Type** *I and* **Class** R^{-} *of* **Type** *II*:

$$D(c, h) = \{\alpha\beta | (\alpha, \beta) \in \mathbb{Z}^2 \cap \ell_{c,h} \wedge \alpha\beta > 0\}. \tag{5.10}$$

Class R^{+} *of* **Types** *II and III*:

$$D(c,h) = \{\alpha\beta | (\alpha,\beta) \in (\mathbb{Z}_{>0})^2 \cap \ell_{c,h}\}. \tag{5.11}$$

PROOF FOR **Class** R^+ OF **Type** I. Let us take the line $\ell_{c,h}$ as in the following figure. (For the other choice of lines, the proof is similar.) Note that $\tilde{D}(c,h)$ is given by the set of integral points on ℓ_1, ℓ_2 and ℓ_3.

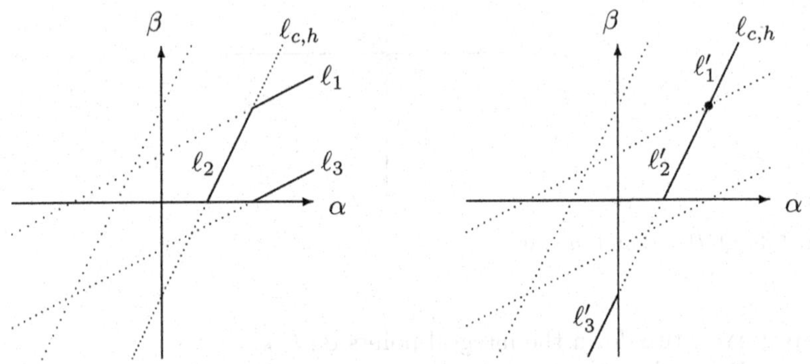

Considering three subsets ℓ_1', ℓ_2' and ℓ_3' of the line $\ell_{c,h}$ as above, we have the following bijection which maps integral points of ℓ_i to those of ℓ_i' $(i = 1,2,3)$:

$$\begin{aligned} \ell_1 \ni (\alpha,\beta) &\mapsto (\beta,\alpha) \in \ell_1', \\ \ell_2 \ni (\alpha,\beta) &\mapsto (\alpha,\beta) \in \ell_2', \\ \ell_3 \ni (\alpha,\beta) &\mapsto (-\beta,-\alpha) \in \ell_3'. \end{aligned}$$

It is obvious that this bijection preserves the value $\alpha\beta$. Now, the lemma for **Class** R^+ of **Type** I follows from the above figures. □

PROOF FOR THE OTHER CASES. First, we consider the case of **Class** R^- of **Type** I. The set $\tilde{D}(c,h)$ is given by the set of integral points on ℓ_1 and ℓ_2 in the figure below. Here, let us take $\ell_{c,h}$ as follows (for the other choice of lines, the proof is similar):

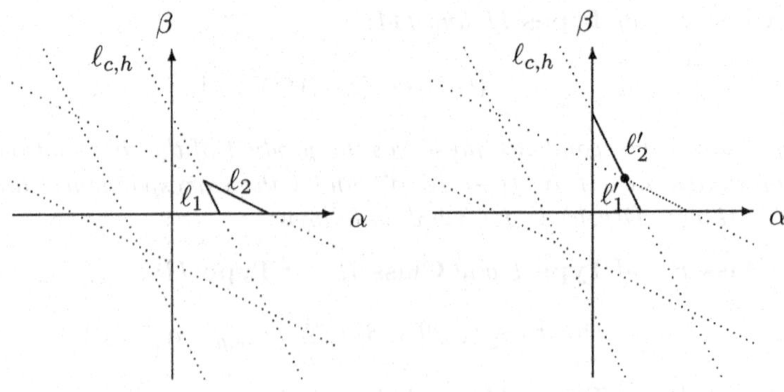

We take ℓ_1' and ℓ_2' as in the figure. Then, the bijection between $\tilde{D}(c,h)$ and the set of integral points on $\ell_1' \sqcup \ell_2' \subset \ell_{c,h}$ is given by

$$\ell_1 \ni (\alpha, \beta) \mapsto (\alpha, \beta) \in \ell_1',$$
$$\ell_2 \ni (\alpha, \beta) \mapsto (\beta, \alpha) \in \ell_2'.$$

This bijection preserves the product $\alpha\beta$. Hence, the rest of the part for **Class** R^- of **Type** I follows.

Next, we consider **Class** $R^\pm$ of **Types** II and III: In these cases, the set $\tilde{D}(c,h)$ is given by the integral points of the following rays or segments, and thus the lemma follows.

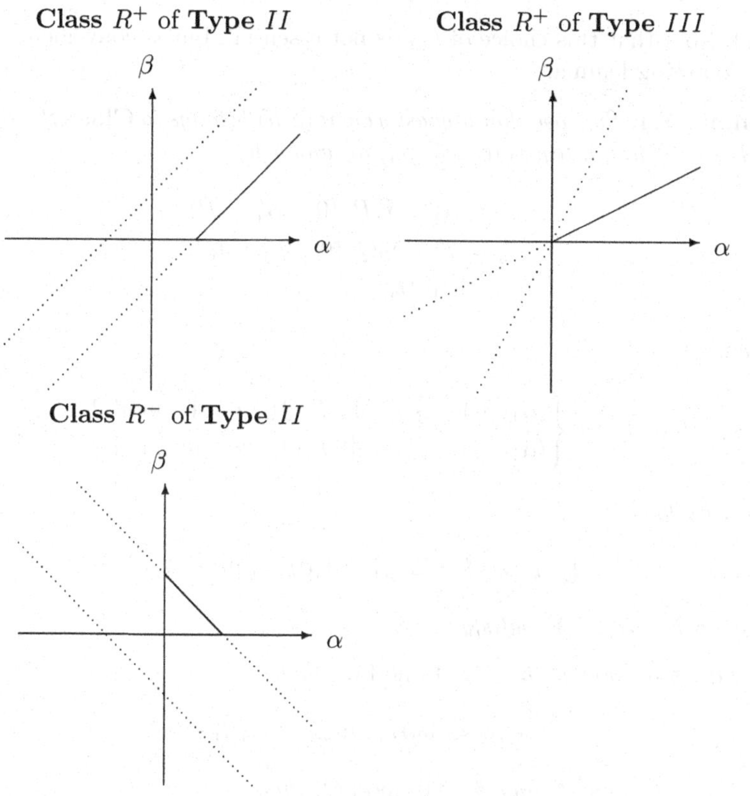

□

5.1.3 List of Integral Points of $\ell_{c,h}$

In this subsection, we enumerate the integral points on $\ell_{c,h}$, i.e., **Step** 2 in the strategy. The results are given in § 5.A of this chapter.

Class R^+ of Type *I*: Let $\{(\alpha_k, \beta_k)|k \in \mathbb{Z}_{>0}\}$ be the set of integral points of (5.8). Here, we suppose that

$$\alpha_1\beta_1 = \min_{k\in\mathbb{Z}}\{\alpha_k\beta_k\}. \tag{5.12}$$

To write down the integral points, we fix $\ell_{c,h}$ as

$$\ell_{c,h} : Q\alpha - P\beta = m,$$

where we choose the signature of m so that

$$\alpha_1 > 0 \text{ and } \beta_1 > 0$$

hold. Note that this choice of $\ell_{c,h}$ is not essential, but is convenient to state the following lemma.

Lemma 5.4. *Suppose that highest weight (c, h) belongs to* **Class** R^+ *of* **Type** *I. Let us define integers α_1', β_1', δ_α, δ_β and δ by*

$$\begin{aligned}
&\alpha_1 = \alpha_1' + \delta_\alpha P \ \ (0 \le \alpha_1' < P),\\
&\beta_1 = \beta_1' + \delta_\beta Q \ \ (0 \le \beta_1' < Q),\\
&\delta := \max\{\delta_\alpha, \delta_\beta\}.
\end{aligned}$$

If we set

$$(\tilde{\alpha}_i, \tilde{\beta}_i) := \begin{cases} (\alpha_1, \beta_1) + \frac{1}{2}(i-1)(P, Q) & i \equiv 1 \mod 2 \\ (\alpha_1, \beta_1) - (\frac{1}{2}i + \delta)(P, Q) & i \equiv 0 \mod 2 \end{cases}, \tag{5.13}$$

then, we have

$$\{(\alpha_k, \beta_k)|k \in \mathbb{Z}_{>0}\} = \{(\tilde{\alpha}_k, \tilde{\beta}_k)|k \in \mathbb{Z}_{>0}\}.$$

Moreover, $\{(\tilde{\alpha}_k, \tilde{\beta}_k)\}$ satisfy

1. *if $\tilde{\alpha}_1 \not\equiv 0 \mod P$ or $\tilde{\beta}_1 \not\equiv 0 \mod Q$, then*

$$\tilde{\alpha}_1\tilde{\beta}_1 < \tilde{\alpha}_2\tilde{\beta}_2 < \tilde{\alpha}_3\tilde{\beta}_3 < \tilde{\alpha}_4\tilde{\beta}_4 < \cdots,$$

2. *if $\tilde{\alpha}_1 \equiv 0 \mod P$ and $\tilde{\beta}_1 \equiv 0 \mod Q$, then*

$$\tilde{\alpha}_1\tilde{\beta}_1 = \tilde{\alpha}_2\tilde{\beta}_2 < \tilde{\alpha}_3\tilde{\beta}_3 = \tilde{\alpha}_4\tilde{\beta}_4 < \cdots.$$

Proof. In the (α, β)-plane, the integral points $\{(\tilde{\alpha}_k, \tilde{\beta}_k)\}$ are described as follows:

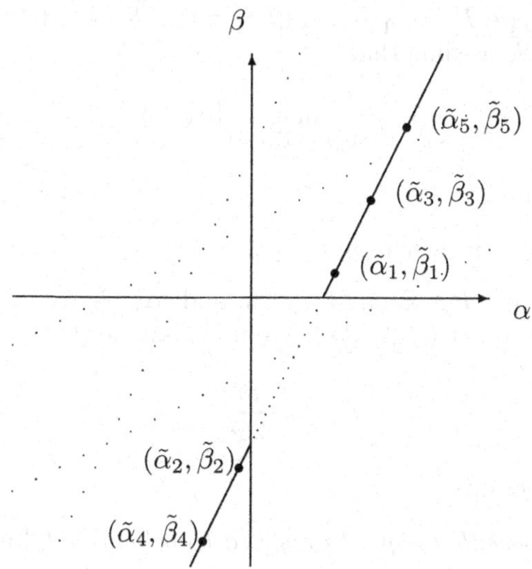

From this figure, it is obvious that $\{(\tilde{\alpha}_k, \tilde{\beta}_k)|k \in \mathbb{Z}_{>0}\}$ coincides with $\{(\alpha_k, \beta_k)|k \in \mathbb{Z}_{>0}\}$. The inequalities of the lemma are consequences of the assumption (5.12), i.e., $\tilde{\alpha}_k\tilde{\beta}_k \geq \alpha_1\beta_1$ holds for any k. Hence, the lemma has been proved. □

Class R^+ **of Types** *II* **and** *III*: Let $\{(\alpha_k, \beta_k)|k \in \mathbb{Z}_{>0}\}$ be the set of integral points of (5.9). We assume that

$$\alpha_1\beta_1 = \min_{k\in\mathbb{Z}_{>0}} \{\alpha_k\beta_k\}. \tag{5.14}$$

We fix $\ell_{c,h}$ as follows:

$\ell_{c,h} : Q\alpha - P\beta = m$, where $m \in \mathbb{Z}_{\leq 0}$.

Then, we have

Lemma 5.5. *Suppose that highest weight* (c, h) *belongs to* **Class** R^+ *of* ***Type*** *II or III. If we set*

$$(\tilde{\alpha}_i, \tilde{\beta}_i) = (\alpha_1, \beta_1) + (i - 1)(P, Q), \tag{5.15}$$

then, we have

$$\{(\alpha_k, \beta_k)|k \in \mathbb{Z}_{>0}\} = \{(\tilde{\alpha}_k, \tilde{\beta}_k)|k \in \mathbb{Z}_{>0}\}.$$

Moreover, $\{(\tilde{\alpha}_k, \tilde{\beta}_k)\}$ *satisfy*

$$\tilde{\alpha}_1\tilde{\beta}_1 < \tilde{\alpha}_2\tilde{\beta}_2 < \tilde{\alpha}_3\tilde{\beta}_3 < \tilde{\alpha}_4\tilde{\beta}_4 < \cdots .$$

Proof. The proof is similar to the case of **Class** R^+ of **Type** I. □

<u>**Class** R^- of **Type** I</u>: Let $\{(\alpha_k, \beta_k) | 1 \leq k \leq \sharp\tilde{D}(c,h)\}$ be the set of integral points (5.8). We assume that

$$\alpha_1\beta_1 = \max_{1\leq k\leq \sharp\tilde{D}(c,h)} \{\alpha_k\beta_k\}. \tag{5.16}$$

Here, we fix $\ell_{c,h}$ as

$\ell_{c,h} : -Q\alpha + P\beta = m$, where $m \in \mathbb{Z}_{>0}$.

Note that in this case, $P \in \mathbb{Z}_{>0}$, $Q \in \mathbb{Z}_{<0}$, and (α_1, β_1) is a nearest integral point on $\ell_{c,h}$ to the point $(-\frac{m}{2Q}, \frac{m}{2P})$. We may assume that

$$\alpha_1 \leq -\frac{m}{2Q}$$

without loss of generality.

Lemma 5.6. *Suppose that highest weight* (c,h) *belongs to* **Class** R^- *of* ***Type*** *I. If we set*

$$(\tilde{\alpha}_i, \tilde{\beta}_i) = \begin{cases} (\alpha_1, \beta_1) + \frac{1}{2}i(P,Q) & i \equiv 0 \mod 2 \\ (\alpha_1, \beta_1) - \frac{1}{2}(i-1)(P,Q) & i \equiv 1 \mod 2 \end{cases}, \tag{5.17}$$

then we have

$$\{(\alpha_k, \beta_k) | 1 \leq k \leq \sharp\tilde{D}(c,h)\} = \{(\tilde{\alpha}_k, \tilde{\beta}_k) | 1 \leq k \leq \sharp\tilde{D}(c,h)\}.$$

Moreover, $\{(\tilde{\alpha}_k, \tilde{\beta}_k)\}$ *satisfy*

1. *if* $\tilde{\alpha}_1 \neq -\frac{m}{2Q}$ *and* $\frac{1}{2}(\tilde{\alpha}_1 + \tilde{\alpha}_2) \neq -\frac{m}{2Q}$, *then*

$$\tilde{\alpha}_1\tilde{\beta}_1 > \tilde{\alpha}_2\tilde{\beta}_2 > \tilde{\alpha}_3\tilde{\beta}_3 > \tilde{\alpha}_4\tilde{\beta}_4 > \cdots,$$

2. *if* $\tilde{\alpha}_1 \neq -\frac{m}{2Q}$ *and* $\frac{1}{2}(\tilde{\alpha}_1 + \tilde{\alpha}_2) = -\frac{m}{2Q}$, *then*

$$\tilde{\alpha}_1\tilde{\beta}_1 = \tilde{\alpha}_2\tilde{\beta}_2 > \tilde{\alpha}_3\tilde{\beta}_3 = \tilde{\alpha}_4\tilde{\beta}_4 > \cdots,$$

3. *if* $\tilde{\alpha}_1 = -\frac{m}{2Q}$, *then*

$$\tilde{\alpha}_1\tilde{\beta}_1 > \tilde{\alpha}_2\tilde{\beta}_2 = \tilde{\alpha}_3\tilde{\beta}_3 > \tilde{\alpha}_4\tilde{\beta}_4 = \cdots.$$

Proof. We can draw these integral points on the (α, β)-plane as follows:

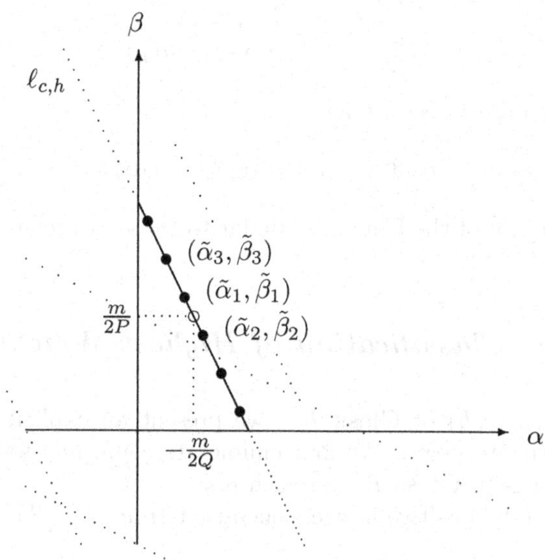

Hence, it is obvious that $\{(\tilde{\alpha}_i, \tilde{\beta}_i)\} = \{(\alpha_i, \beta_i)\}$ as a set. Similarly to the proof of Lemma 5.4, by the assumption (5.16), one can directly check the inequalities. □

Class R^- of **Type** *II*: Let $\{(\alpha_i, \beta_i) | 1 \leq i \leq \sharp\tilde{D}(c,h)\}$ be the set of integral points of (5.8). We assume that

$$\alpha_1\beta_1 = \max_{1 \leq k \leq \sharp\tilde{D}(c,h)} \{\alpha_k\beta_k\}. \tag{5.18}$$

We choose $\ell_{c,h}$ as

$\ell_{c,h} : \alpha + \beta = m$, where $m \in \mathbb{Z}_{>0}$.

Note that there are at most two integral points on $\ell_{c,h}$ nearest to the point $(\frac{m}{2}, \frac{m}{2})$, and (α_1, β_1) is one of the two points. Now, we assume that

$$\alpha_1 \leq \frac{m}{2}.$$

Lemma 5.7. *Suppose that highest weight* (c,h) *belongs to* **Class** R^- *of* ***Type*** *II. Then, we have*

$$(\alpha_i, \beta_i) = \begin{cases} (\alpha_1, \beta_1) + \frac{1}{2}i(1,-1) & i \equiv 0 \mod 2 \\ (\alpha_1, \beta_1) - \frac{1}{2}(i-1)(1,-1) & i \equiv 1 \mod 2 \end{cases}. \tag{5.19}$$

Moreover, we have

1. if m is an odd integer, then

$$\tilde{\alpha}_1\tilde{\beta}_1 = \tilde{\alpha}_2\tilde{\beta}_2 > \tilde{\alpha}_3\tilde{\beta}_3 = \tilde{\alpha}_4\tilde{\beta}_4 > \cdots,$$

2. if m is an even integer, then

$$\tilde{\alpha}_1\tilde{\beta}_1 > \tilde{\alpha}_2\tilde{\beta}_2 = \tilde{\alpha}_3\tilde{\beta}_3 > \tilde{\alpha}_4\tilde{\beta}_4 = \cdots.$$

Proof. The proof of the lemma is similar to those of previous ones. □

5.1.4 Fine Classification of Highest Weights: Class R^+

Here, for each (c, h) of **Class** R^+, we present an explicit form of $D(c, h)$ (**Step** 3 in the strategy). We first enumerate conformal weight h such that $(c_{p,q}, h)$ belongs to **Class** R^+ for each p, q.

Let p and q be positive integers such that $(p, q) = 1$. We set

$$K^+_{p,q} := \left\{ (r,s) \in \mathbb{Z}^2 \,\middle|\, \begin{matrix} 0 \le r < p, \\ 0 \le s \le q, \end{matrix} rq + sp \le pq \right\} \tag{5.20}$$

and

$$(K^+_{p,q})^\circ := \left\{ (r,s) \in \mathbb{Z}^2 \,\middle|\, \begin{matrix} 0 < r < p, \\ 0 < s < q, \end{matrix} rq + sp < pq \right\}. \tag{5.21}$$

For each $(r, s) \in K^+_{p,q}$ and $i \in \mathbb{Z}$, we set

$$h_{p,q:r,s:i} := \begin{cases} h_{-ip+r,s}\left(\dfrac{q}{p}\right) & i \equiv 0 \mod 2 \\ h_{-(i+1)p+r,-s}\left(\dfrac{q}{p}\right) & i \equiv 1 \mod 2 \end{cases}, \tag{5.22}$$

where $h_{\alpha,\beta}(t)$ is defined in (5.4). We sometimes abbreviate $h_{p,q:r,s:i}$ to h_i.

$K^+_{p,q} \times \mathbb{Z}$ parameterises highest weights of **Class** R^+. In fact, we have

Lemma 5.8. *For any highest weight (c, h) of* **Class** R^+*, there exist $p, q \in \mathbb{Z}_{>0}$ $((p, q) = 1)$, unique $(r, s) \in K^+_{p,q}$ and $i \in \mathbb{Z}$ such that*

$$(c, h) = (c_{p,q}, h_{p,q,:r,s:i}).$$

Note that $i \in \mathbb{Z}$ is not uniquely determined (see Lemma 5.10). The next technical lemma is the key step to show Lemma 5.8:

Lemma 5.9. *Let p, $q \in \mathbb{Z}_{>0}$ be as above. Then, we have*

$$\{\pm(rq + sp) \bmod 2pq,\ \pm(rq - sp) \bmod 2pq | (r, s) \in K^+_{p,q}\} = \mathbb{Z}/2pq\mathbb{Z}. \tag{5.23}$$

Proof. We divide the set $K^+_{p,q}$ as follows: $K^+_{p,q} = K_1 \sqcup K_2 \sqcup K_3$ (disjoint), where $K_1 := (K^+_{p,q})^\circ$, $K_2 := \{(r,0) \in \mathbb{Z}^2 | 0 < r < p\} \sqcup \{(0,s) \in \mathbb{Z}^2 | 0 < s < q\}$ and $K_3 := \{(0,0),\ (0,q)\}$. We further introduce N_i $(i = 1,2,3)$ by

$$N_i := \{\pm(rq - sp) \bmod 2pq,\ \pm(rq + sp) \bmod 2pq | (r,s) \in K_i\}, \qquad (5.24)$$

regarding them as sets without multiplicity. Then, we have

$$\begin{aligned}\sharp N_1 &= 4 \times \sharp K_1 = 2(p-1)(q-1),\\ \sharp N_2 &= 2 \times \sharp K_2 = 2(p+q-2),\\ \sharp N_3 &= \sharp K_3 = 2\end{aligned}$$

and thus, $\sharp N_1 + \sharp N_2 + \sharp N_3 = 2pq$. Since N_1, N_2 and N_3 are disjoint, we have $N_1 \sqcup N_2 \sqcup N_3 = \mathbb{Z}/2pq\mathbb{Z}$. Hence, the lemma holds. □

Proof of Lemma 5.8. Recall that if (c,h) belongs to **Class** R^+, then $c = c_{p,q}$ and

$$h = h_{p,q:m} = \frac{m^2 - (p-q)^2}{4pq}$$

for some $p, q \in \mathbb{Z}_{>0}$ such that $(p,q) = 1$ and $m \in \mathbb{Z}$. By the definition of $h_{p,q:r,s:i}$ in (5.22), it is enough to check that

$$\{\pm(2ipq + rq - sp),\ \pm(2ipq + rq + sp) | i \in \mathbb{Z},\ (r,s) \in K^+_{p,q}\} = \mathbb{Z}.$$

This is an immediate consequence of Lemma 5.9. The uniqueness of (r,s) easily follows from the definition of $K^+_{p,q}$. □

From Lemma 5.8, we see that for each central charge $c = c_{p,q}$, the set $K^+_{p,q} \times \mathbb{Z}$ parameterises the set of the conformal weights h such that (c,h) belongs to **Class** R^+. As the next step, we check the degeneration of these conformal weights, i.e., when two conformal weights $h_{p,q:r,s:i}$ and $h_{p,q:r',s':i'}$ coincide.

In order to describe the degeneration, we divide $K^+_{p,q}$ as follows:

Case 1^+: $0 < r < p$ and $0 < s < q$,
Case 2^+: $r = 0$ and $0 < s < q$,
Case 3^+: $0 < r < p$ and $s = 0$,
Case 4^+: $(r,s) = (0,0),\ (0,q)$.

Since by Lemma 5.8, (r,s) is uniquely determined for each highest weight, we define

Definition 5.1 *Suppose that $(c_{p,q}, h)$ is a highest weight of* **Class** R^+. *We say that $(c_{p,q}, h)$ (or h) is in* **Case** $*^+$ $(* \in \{1,2,3,4\})$, *if $(r,s) \in K^+_{p,q}$, such that $h = h_{p,q:r,s:i}$ for some $i \in \mathbb{Z}$, is of* **Case** $*^+$.

Lemma 5.10. *For each case, the degeneration of the conformal weights $\{h_{p,q:r,s:i} | i \in \mathbb{Z}\}$ can be described as follows:*

Case 1^+: *no degeneration,*
Case 2^+: $h_{-i-1} = h_i \ (i \in \mathbb{Z}_{\geq 0})$,
Case 3^+: $h_{2i} = h_{2i-1} \ (i \in \mathbb{Z})$,
Case 4^+: $\begin{matrix} h_{-2i-1} = h_{-2i} = h_{2i-1} = h_{2i} & (r,s) = (0,0) \\ h_{-2i-2} = h_{-2i-1} = h_{2i+1} = h_{2i} & (r,s) = (0,q) \end{matrix} \ (i \in \mathbb{Z}_{\geq 0})$.

Hence, the following list exhausts the conformal weights h such that $(c_{p,q}, h)$ belongs to **Class** R^+:

Case 1^+	$h_i \ (i \in \mathbb{Z})$
Case 2^+	$h_i \ (i \in \mathbb{Z}_{\geq 0})$
Case 3^+	$h_{(-1)^{i-1}i} \ (i \in \mathbb{Z}_{\geq 0})$
Case 4^+	$h_{2i} \ (i \in \mathbb{Z}_{\geq 0})$

Below, we restrict the range of i as in Lemma 5.10.

Remark 5.3 *The relation between two classification;* ***Type*** $*$ *and* **Case** $*'$ *can be described as follows:*

1. **Case** 1^+, **Case** 2^+ *and* **Case** 3^+: *These cases are of* ***Type*** *I.*
2. **Case** 4^+:

	$s \neq 0 \vee i \neq 0$	$s = 0 \wedge i = 0$
$(p,q) \neq (1,1)$	***Type*** *I*	***Type*** *III* $(c \neq 1)$
$(p,q) = (1,1)$	***Type*** *II*	***Type*** *III* $(c = 1)$

Finally, using the list in the above lemma, we describe the set $D(c,h)$ and $a(n)$ $(n \in D(c,h))$ explicitly.

Lemma 5.11. *For each highest weight (c,h) of* **Class** R^+ *such that $c = c_{p,q}$ $(p, q \in \mathbb{Z}_{>0}, (p,q) = 1)$, $D(c,h)$ and $a(n)$ $(n \in D(c,h))$ are given as follows:*

1. **Case** 1^+: *For $i \in \mathbb{Z}$,*
$$D(c,h_i) = \{h_k - h_i | k \in \mathbb{Z}, \ |k| > |i|, \ k - i \equiv 1 \mod 2\},$$
and $a(n) = 1$ for any $n \in D(c,h_i)$.
2. **Case** 2^+: *For $i \in \mathbb{Z}_{\geq 0}$,*
$$D(c,h_i) = \{h_k - h_i | k \in \mathbb{Z}_{>0}, \ k > i\},$$
and $a(n) = 1$ for any $n \in D(c,h_i)$.
3. **Case** 3^+: *For $i \in \mathbb{Z}_{\geq 0}$,*
$$D(c, h_{(-1)^{i-1}i}) = \{h_{(-1)^{k-1}k} - h_{(-1)^{i-1}i} | k \in \mathbb{Z}_{>0}, \ k > i\},$$
and $a(n) = 1$ for any $n \in D(c,h_i)$.

4. **Case** 4^+: *For* $i \in \mathbb{Z}_{\geq 0}$,

$$D(c, h_{2i}) = \{h_{2k} - h_{2i} | k \in \mathbb{Z}_{>0},\ k > i\},$$

and for $n \in D(c, h_{2i})$,

$$a(n) = \begin{cases} 1 & \textit{if } (p,q) = (1,1) \vee (s = 0 \wedge i = 0) \\ 2 & \textit{otherwise} \end{cases}.$$

Proof. This lemma is a consequence of Lemmas 5.4 and 5.5 (and Remark 5.3). For the reader's convenience, we give the list of the lines $\ell_{c,h}$ and the integral points (α_k, β_k) for each case in § 5.A. □

5.1.5 Special Highest Weights

In this subsection, we list the highest weights of some special irreducible highest weight representations, which are important in mathematical and theoretical physics. For physical background, see, e.g., [ID].

1. **BPZ (Belavin–Polyakov–Zamolodchikov) series** [BPZ1], [BPZ2]: (**Case** 1^+)

$$\left\{(c_{p,q}, h_{p,q:r,s:0}) \,\middle|\, p, q \in \mathbb{Z}_{>1}\ (p,q) = 1, (r,s) \in (K^+_{p,q})^\circ \right\},$$

   where the set $(K^+_{p,q})^\circ$ is defined in (5.21).
   a. **Minimal series** [BPZ1], [BPZ2], [FQS1]:
      $(p,q) = (m+1, m+2)$ $(m \in \mathbb{Z}_{\geq 2})$.
      i. **Ising model** [Is], [Len]:
         $c = \frac{1}{2}$, i.e., $(p,q) = (3,4)$.
      ii. **Tri-critical Ising model** [BEG], [NBRS]:
         $c = \frac{7}{10}$, i.e., $(p,q) = (4,5)$.
      iii. **3-state Potts model** [P]:
         $c = \frac{4}{5}$, i.e., $(p,q) = (5,6)$.
   b. **Yang–Lee edge singularity** [YL]:
      $c = -\frac{22}{5}$, i.e, $(p,q) = (2,5)$.
2. **Logarithmic series:** (**Case** 2^+)

$$\{(c_{p,1}, h_{p,1:r,0:0}) | p \in \mathbb{Z}_{>2},\ r \in \mathbb{Z},\ 1 \leq r < p\}.$$

   a. **Free fermionic point:**
      $c = -2$, i.e, $(p,q) = (2,1)$.
3. $\mathbb{Z}_N$ **Parafermionic model** $(N > 1)$ [ZF1], [ZF2]:

$$c = \frac{2(N-1)}{N+2}.$$

4. **Gaussian model**: $(p,q) = (1,1)$,

$$c = 1.$$

Remark 5.4 *There seems to be confusion about 'minimal series representations'. For one case, BPZ series are called* **minimal series**, *and discrete series are called* **minimal unitary series**. *For the other case,* **minimal series** *simply indicates discrete series.*

5.1.6 Fine classification of Highest Weights: Class R^-

In this subsection, we classify highest weights of **Class** R^- in a way similar to **Class** R^+.

Suppose that (c,h) is a highest weight which belongs to **Class** R^-. Then, central charge c can be written as

$$c = c_{p,-q}$$

for some $p, q \in \mathbb{Z}_{>0}$ such that $(p,q) = 1$.

Similarly to **Class** R^+, we introduce a set $K^-_{p,q}$ to parameterise conformal weight h. For each $p, q \in \mathbb{Z}_{>0}$, we set

$$K^-_{p,q} := \left\{ (r,s) \in \mathbb{Z}^2 \,\middle|\, \begin{matrix} 0 \le r < p, \\ 0 \le -s \le q, \end{matrix} rq - sp \le pq \right\}. \tag{5.25}$$

For each $(r,s) \in K^-_{p,q}$ and $i \in \mathbb{Z}$, we put

$$h_{p,q:r,s:i} := \begin{cases} h_{-ip+r,s}\left(-\dfrac{q}{p}\right) & i \equiv 0 \mod 2 \\ h_{-(i+1)p+r,-s}\left(-\dfrac{q}{p}\right) & i \equiv 1 \mod 2 \end{cases}, \tag{5.26}$$

where $h_{\alpha,\beta}(t)$ is defined in (5.4). We often abbreviate $h_{p,q:r,s:i}$ to h_i for simplicity.

Then, we have

Lemma 5.12. *For any highest weight (c,h) of* **Class** R^-, *there exist $p, q \in \mathbb{Z}_{>0}$ such that $(p,q) = 1$, unique $(r,s) \in K^-_{p,q}$ and $i \in \mathbb{Z}$ such that*

$$(c,h) = (c_{p,-q}, h_{p,q:r,s:i}).$$

Proof. We can prove the lemma similarly to the proof of Lemma 5.8. □

Next, we describe degeneration of the conformal weights $h_{p,q:r,s:i}$. Similarly to **Class** R^+, we divide $K^-_{p,q}$ as follows:

Case 1^-: $0 < r < p$ and $0 < -s < q$,
Case 2^-: $r = 0$ and $0 < -s < q$,
Case 3^-: $0 < r < p$ and $s = 0$,
Case 4^-: $(r, s) = (0, 0), (0, -q)$,

Accordingly, we define

Definition 5.2 *Suppose that $(c_{p,-q}, h)$ is a highest weight of* **Class** R^-. *We say that $(c_{p,-q}, h)$ (or h) is in* **Case** $*^-$ $(* \in \{1, 2, 3, 4\})$, *if $(r, s) \in K^-_{p,q}$, such that $h = h_{p,q:r,s:i}$ for some $i \in \mathbb{Z}$, is of* **Case** $*^-$.

One can easily check the lemma below:

Lemma 5.13. *For each case, the degeneration of the conformal weights $\{h_{p,q,:r,s:i} | i \in \mathbb{Z}\}$ can be described as follows:*

Case 1^-: *no degeneration,*
Case 2^-: $h_{-i-1} = h_i \quad (i \in \mathbb{Z}_{\geq 0})$,
Case 3^-: $h_{2i} = h_{2i-1} \quad (i \in \mathbb{Z})$,
Case 4^-: $\begin{array}{ll} h_{-2i-1} = h_{-2i} = h_{2i-1} = h_{2i} & (r,s) = (0,0) \\ h_{-2i-2} = h_{-2i-1} = h_{2i+1} = h_{2i} & (r,s) = (0,-q) \end{array} \quad (i \in \mathbb{Z}_{\geq 0})$.

Hence, the following list exhausts the conformal weights h such that $(c_{p,-q}, h)$ belongs to **Class** R^-:

Case 1^-	$h_i \ (i \in \mathbb{Z} \setminus \{0\})$
Case 2^-	$h_i \ (i \in \mathbb{Z}_{>0})$
Case 3^-	$h_{(-1)^{i-1}i} \ (i \in \mathbb{Z}_{>0})$
Case 4^-	$h_{2i} \ (i \in \mathbb{Z}_{>0})$

Remark 5.5 *Since $M(c_{p,-q}, h_0)$ is irreducible, highest weight $(c_{p,-q}, h_0)$ belongs to* **Class** V.

We may restrict the range of i as in Lemma 5.13.

Remark 5.6 *The relation between two classification;* ***Type*** $*$ *and* **Case** $*'$ *is as follows:*

1. **Case** 1^-, **Case** 2^- *and* **Case** 3^-: *These cases are of* ***Type*** *I*.
2. **Case** 4^-:

$(p, q) \neq (1, 1)$	***Type I***
$(p, q) = (1, 1)$	***Type II***

Finally, we list the $D(c, h)$ for each (c, h).

Lemma 5.14. *For each highest weight (c, h) of* **Class** R^- *such that $c = c_{p,-q}$ $(p, q \in \mathbb{Z}_{>0}, (p, q) = 1)$, $D(c, h)$ and $a(n)$ $(n \in D(c, h))$ are given as follows:*

1. **Case** 1^-: *For* $i \in \mathbb{Z} \setminus \{0\}$,

$$D(c,h_i) = \{h_k - h_i | k \in \mathbb{Z},\ |k| < |i|,\ i-k \equiv 1 \mod 2\},$$

and $a(n) = 1$ *for any* $n \in D(c,h_i)$.
2. **Case** 2^-: *For* $i \in \mathbb{Z}_{>0}$,

$$D(c,h_i) = \{h_k - h_i | k \in \mathbb{Z}_{>0},\ k < i\},$$

and $a(n) = 1$ *for any* $n \in D(c,h_i)$.
3. **Case** 3^-: *For* $i \in \mathbb{Z}_{>0}$,

$$D(c,h_{(-1)^{i-1}i}) = \{h_{(-1)^{k-1}k} - h_{(-1)^{i-1}i} | k \in \mathbb{Z}_{>0},\ k < i\},$$

and $a(n) = 1$ *for any* $n \in D(c,h_i)$.
4. **Case** 4^-: *For* $i \in \mathbb{Z}_{>0}$,

$$D(c,h_{2i}) = \{h_{2k} - h_{2i} | k \in \mathbb{Z}_{>0},\ k < i\},$$

and for $n \in D(c,h_{2i})$

$$a(n) = \begin{cases} 1 & \text{if } (p,q) = (1,1) \vee (s = 0 \wedge n = h_0 - h_{2i}) \\ 2 & \text{otherwise} \end{cases}.$$

Proof. This lemma is a direct consequence of Lemmas 5.6 and 5.7 (and Remark 5.6). Note that if $s = 0$ in **Case** 4^-, then $\tilde{\alpha}_1 = -\frac{m}{2Q}$ in Lemma 5.17. Hence, in this case,

$$a(n) = \begin{cases} 2 & n < h_0 - h_{2i} \\ 1 & n = h_0 - h_{2i} \end{cases}.$$

For the explicit forms of the integral points (α_k, β_k), see § 5.A. □

Remark 5.7 *In [FeFu4], Feigin and Fuchs classified highest weights in a different way. Our classification corresponds to theirs as follows:*

Class V	Class I	Case $1^\pm$	Case $2^\pm$, $3^\pm$	Case $4^\pm$
I, II_0, II_-	II_+	$\mathrm{III}_\mp$	$\mathrm{III}_\mp{}^0$	$\mathrm{III}_\mp{}^{00}$

5.2 Singular Vectors

First, we define a singular vector and a subsingular vector of a $\mathfrak{h}$-semi-simple $\mathfrak{g}$-module M.

Definition 5.3 *1. A weight vector* $v \in M^{\mathfrak{g}^+} \setminus \{0\}$ *is called a* **singular vector**.

2. *A weight vector $v \in M$ is called a* **subsingular vector**, *if there exists a proper submodule N of M such that $v + N \in (M/N)^{\mathfrak{g}^+} \setminus \{0\}$. Here, we allow N to be $\{0\}$.*
In particular, a singular vector is always subsingular.

Here, we say that a singular vector $v \in M(c,h)^{\mathfrak{g}^+}$ is *of level n* if $v \in M(c,h)_{h+n}$.

5.2.1 Uniqueness of Singular Vectors

In order to construct embedding diagrams of Verma modules, we show the following uniqueness of singular vectors of Verma modules.

Proposition 5.1 *For each $n \in \mathbb{Z}_{>0}$, we have*

$$\dim\{M(c,h)_{h+n}\}^{\mathfrak{g}^+} \leq 1.$$

We first introduce some notation. Let $<$ be the total order on $\mathcal{P}_n$ defined as follows: For $\mathbb{I}, \mathbb{J} \in \mathcal{P}_n$ such that $\mathbb{I} = (1^{r_1}2^{r_2}\cdots n^{r_n})$ and $\mathbb{J} = (1^{s_1}2^{s_2}\cdots n^{s_n})$, we define

$$\mathbb{I} < \mathbb{J} \Leftrightarrow \exists m \in \mathbb{Z}_{>0}; (r_k = s_k \ (k < m)) \wedge (r_m < s_m).$$

In the sequel, we denote the maximal element (1^n) of $\mathcal{P}_n$ by $\mathbb{I}_0$. For $\mathbb{I} = (1^{r_1}2^{r_2}\cdots n^{r_p}) \in \mathcal{P}_p$ and $\mathbb{J} = (1^{s_1}2^{s_2}\cdots n^{s_q}) \in \mathcal{P}_q$, we set

$$\mathbb{I} \pm \mathbb{J} := (1^{r_1 \pm s_1}2^{r_2 \pm s_2}\cdots) \in \mathcal{P}_{p+q}.$$

For simplicity, we sometimes denote

$$(1^0 2^0 \cdots (k-1)^0 k^r (k+1)^0 \cdots)$$

by (k^r). In particular, for $\mathbb{I} = (1^{r_1}2^{r_2}\cdots n^{r_n}) \in \mathcal{P}_n$,

$$\mathbb{I} \pm (k^r) = (1^{r_1}2^{r_2}\cdots k^{r_k \pm r}\cdots n^{r_n}) \in \mathcal{P}_{n \pm rk}.$$

Let $e_{\mathbb{I}}$ be the element of $U(\mathfrak{g}^-)$ defined in (4.17). Notice that $\{e_{\mathbb{I}}.v_{c,h} | \mathbb{I} \in \mathcal{P}_n\}$ forms a basis of $M(c,h)_{h+n}$, where $v_{c,h} := 1 \otimes \mathbf{1}_{c,h}$. For a weight vector $w \in M(c,h)_{h+n}$, we express w as follows:

$$w = \sum_{\mathbb{I} \in \mathcal{P}_n} c_{\mathbb{I}}^w e_{\mathbb{I}}.v_{c,h}.$$

To prove the proposition, we show the following 'triangularity':

Lemma 5.15. *Suppose that $n \in \mathbb{Z}_{>0}$ and $w \in M(c,h)_{h+n} \setminus \{0\}$. For $\mathbb{J} = (1^{s_1}2^{s_2}\cdots n^{s_n}) \in \mathcal{P}_n \setminus \{\mathbb{I}_0\}$, let $j > 1$ be the positive integer such that $s_2 =$*

$\cdots = s_{j-1} = 0$ *and* $s_j \neq 0$. *Set* $\mathbb{J}' := \mathbb{J} - (j^1) + (1^1) \in \mathcal{P}_{n-j+1}$ *and* $w' := L_{j-1}.w$. *We express* w' *as*

$$w' = \sum_{\mathbb{I}' \in \mathcal{P}_{n-j+1}} c_{\mathbb{I}'}^{w'} e_{\mathbb{I}'}.v_{c,h}.$$

Then, there exist $\{Q_{\mathbb{I},\mathbb{J}}^{w} | \mathbb{I} > \mathbb{J}\} \subset \mathbb{C}$ *which satisfy*

$$c_{\mathbb{J}'}^{w'} = s_j(2j-1)c_{\mathbb{J}}^{w} + \sum_{\substack{\mathbb{I} \in \mathcal{P}_n \\ \mathbb{I} > \mathbb{J}}} Q_{\mathbb{I},\mathbb{J}}^{w} c_{\mathbb{I}}^{w}.$$

Proof. We verify the following assertion: for any $\mathbb{I} \in \mathcal{P}_n$,

$$\mathbb{I} \leq \mathbb{J} \wedge c_{\mathbb{J}'}^{L_{j-1}e_{\mathbb{I}}.v_{c,h}} \neq 0 \quad \Leftrightarrow \quad \mathbb{I} = \mathbb{J}. \tag{5.27}$$

Let us first prove the 'only if' part of (5.27). Suppose that $\mathbb{I} = (1^{r_1}2^{r_2}\cdots n^{r_n})$ satisfies

$$\mathbb{I} \leq \mathbb{J} \wedge c_{\mathbb{J}'}^{L_{j-1}e_{\mathbb{I}}.v_{c,h}} \neq 0.$$

For the proof, it is convenient to use the following notation: For a weight vector

$$u = \sum_{\mathbb{K} \in \mathcal{P}_n} c_{\mathbb{K}}^{u} e_{\mathbb{K}}.v_{c,h} \in M(c,h)_{h+n},$$

we set

$$\max_{L_{-1}}(u) := \max\{t_1 | c_{\mathbb{K}}^{u} \neq 0,\ \mathbb{K} = (1^{t_1}2^{t_2}\cdots n^{t_n}) \in \mathcal{P}_n\}.$$

We divide the proof of the 'only if' part into three steps.

Step *I* We show that $r_1 = s_1$. Since the condition $\mathbb{I} \leq \mathbb{J}$ implies that $r_1 \leq s_1$, we check that $s_1 \leq r_1$. Since $c_{\mathbb{J}'}^{L_{j-1}e_{\mathbb{I}}.v_{c,h}} \neq 0$, we have

$$s_1 + 1 \leq \max_{L_{-1}}(L_{j-1}e_{\mathbb{I}}.v_{c,h}).$$

On the other hand, by direct calculation, one can show that

$$\max_{L_{-1}}(L_{j-1}e_{\mathbb{I}}.v_{c,h}) \leq r_1 + 1.$$

Thus, $s_1 \leq r_1$ holds.

Step *II* Let i be the positive integer such that $r_2 = r_3 = \cdots = r_{i-1} = 0$ and $r_i \neq 0$. We show that $i = j$. Since $r_1 = s_1$ and $\mathbb{I} \leq \mathbb{J}$, we have $i \geq j$. We assume that $i > j$ and lead to a contradiction.

Notice that the condition $c_{\mathbb{J}'}^{L_{j-1}e_{\mathbb{I}}.v_{c,h}} \neq 0$ implies that

$$\max_{L_{-1}}(L_{j-1}e_{\mathbb{I}}.v_{c,h}) \geq s_1 + 1 = r_1 + 1.$$

On the other hand, we have

$$\max_{L_{-1}}(L_{j-1}e_{\mathbb{I}}.v_{c,h}) = r_1,$$

since

$$L_{j-1}e_{\mathbb{I}}.v_{c,h} = [L_{j-1}, L_{-n}^{r_n} \cdots L_{-i}^{r_i}]L_{-1}^{r_1}.v_{c,h} + L_{-n}^{r_n} \cdots L_{-i}^{r_i} L_{j-1} L_{-1}^{r_1}.v_{c,h},$$

and

$$\begin{aligned}
&\max_{L_{-1}}([L_{j-1}, L_{-n}^{r_n} \cdots L_{-i}^{r_i}]L_{-1}^{r_1}.v_{c,h}) = r_1,\\
&\max_{L_{-1}}(L_{-n}^{r_n} \cdots L_{-i}^{r_i} L_{j-1} L_{-1}^{r_1}.v_{c,h}) < r_1,
\end{aligned}$$

by the assumption $i > j$. This is a contradiction, and thus, $i = j$.

<u>Step *III*</u> We show that $r_m = s_m$ for any $m > 1$. Since $j = i$, we have

$$\begin{aligned}
&L_{j-1}L_{-n}^{r_n} \cdots L_{-j}^{r_j} L_{-1}^{r_1}.v_{c,h}\\
&= r_j(2j-1)L_{-n}^{r_n} \cdots L_{-j-1}^{r_{j+1}} L_{-j}^{r_j-1} L_{-1}^{r_1+1}.v_{c,h} + u
\end{aligned}$$

for some $u \in M(c,h)_{h+n-j+1}$ such that $\max_{L_{-1}}(u) \leq r_1$. Hence, we have

$$(1^{r_1+1} i^{r_i-1} (i+1)^{r_{i+1}} \cdots n^{r_n}) = \mathbb{J}' = \mathbb{J} - (j^1) + (1^1),$$

since $c_{\mathbb{J}'}^{L_{j-1}e_{\mathbb{I}}.v_{c,h}} \neq 0$. Hence, $r_m = s_m$ for any $m > 1$, and the 'only if' part has been proved.

For the 'if' part of (5.27), by direct calculation we get

$$L_{j-1}e_{\mathbb{J}}.v_{c,h} = s_j(2j-1)e_{\mathbb{J}'}.v_{c,h} + u'.$$

for some $u' \in M(c,h)_{h+n-j+1}$ such that $\max_{L_{-1}}(u') \leq s_1$. Since $s_j(2j-1) \neq 0$, the 'if' part follows. Therefore, we have completed the proof of Lemma 5.15. □

Proof of Proposition 5.1. Let w be a singular vector of level n. Since $L_{j-1}.w = 0$ for any $j > 1$, we have

$$c_{\mathbb{J}}^{w} = -\{s_j(2j-1)\}^{-1} \sum_{\substack{\mathbb{I} \in \mathcal{P}_n \\ \mathbb{I} > \mathbb{J}}} Q_{\mathbb{I},\mathbb{J}}^{w} c_{\mathbb{I}}^{w} \tag{5.28}$$

by Lemma 5.15. This means that the coefficient $c_{\mathbb{J}}^{w}$ of the singular vector w is uniquely determined by the coefficients $\{c_{\mathbb{I}}^{w} | \mathbb{I} > \mathbb{J}\}$, i.e., w is uniquely determined by the coefficient $c_{\mathbb{I}_0}^{w}$ ($\mathbb{I}_0 = (1^n)$). Hence, w is unique up to a scalar. □.

Corollary 5.1 *For any $h, h' \in \mathbb{C}$, the following holds:*

$$\dim \mathrm{Hom}_{\mathfrak{g}}(M(c,h), M(c,h')) \leq 1.$$

5.2.2 Existence of Singular Vectors

In this subsection, we show the existence of homomorphisms between Verma modules to construct embedding diagrams of Verma modules by showing the existence of singular vectors. Practically, we prove the existence of an element, called a Shapovalov element, which was first considered by N. Shapovalov [Sh] for a complex semi-simple Lie algebra and which defines a desired homomorphism. They are also used to define Jantzen filtration of certain quotient modules (§ 5.6).

Through this subsection, we fix $\alpha, \beta \in \mathbb{Z}_{>0}$, and set $n := \alpha\beta$. Let $\mathcal{V}_{\alpha,\beta}$ be the curve in $\mathfrak{h}^*$ ($\mathfrak{h} = \mathbb{C}C \oplus \mathbb{C}L_0$) defined by $\Phi_{\alpha,\beta}(c,h) = 0$. Let $\mathbb{C}[\mathcal{V}_{\alpha,\beta}]$($\simeq \mathbb{C}[\mathfrak{h}^*]/(\Phi_{\alpha,\beta}(c,h))$) be the coordinate ring of the curve $\mathcal{V}_{\alpha,\beta}$. For simplicity, we sometimes denote $\Phi_{\alpha,\beta}(c,h)$ by $\Phi(c,h)$ or Φ.

Proposition 5.2 (cf. [RW3]) *There exists $S_{n,\Phi} \in U(\mathfrak{g}^- \oplus \mathfrak{h})_{-n}$, which is called a* ***Shapovalov element****, such that*

1. for any $k \in \mathbb{Z}_{>0}$,

$$L_k S_{n,\Phi} \in U(\mathfrak{g})\Phi(C, L_0) + U(\mathfrak{g})\mathfrak{g}^+, \tag{5.29}$$

2.

$$S_{n,\Phi} = \sum_{\mathbb{I} \in \mathcal{P}_n} e_{\mathbb{I}} H_{\Phi,\mathbb{I}}, \tag{5.30}$$

where $e_{\mathbb{I}} \in U(\mathfrak{g}^-)$ is defined in (4.17), $H_{\Phi,\mathbb{I}} \in U(\mathfrak{h})(\simeq \mathbb{C}[\mathfrak{h}^*])$ and $H_{\Phi,\mathbb{I}_0} = 1$ for $\mathbb{I}_0 = (1^n)$.

Proof. We first show that there exists an element

$$S_n = \sum_{\mathbb{I} \in \mathcal{P}_n} e_{\mathbb{I}} \otimes c_{\mathbb{I}} \in U(\mathfrak{g}^-)_{-n} \otimes_{\mathbb{C}} \mathbb{C}[\mathcal{V}_{\alpha,\beta}]$$

such that $c_{\mathbb{I}_0} = 1$ ($\mathbb{I}_0 := (1^n)$) and

$$S_n(c,h).v_{c,h} \in \{M(c,h)_{h+n}\}^{\mathfrak{g}^+} \quad (\forall (c,h) \in \mathcal{V}_{\alpha,\beta}). \tag{5.31}$$

The condition (5.31) is equivalent to a system of linear equations in $\{c_{\mathbb{I}}\}$ defined over $\mathbb{C}[\mathcal{V}_{\alpha,\beta}]$, and any solution $\{c_{\mathbb{I}}\}$ of this system lies in $\mathbb{C}(\mathcal{V}_{\alpha,\beta})$, the quotient field of $\mathbb{C}[\mathcal{V}_{\alpha,\beta}]$. We may assume that $\{c_{\mathbb{I}}\} \subset \mathbb{C}[\mathcal{V}_{\alpha,\beta}]$ by multiplying a non-zero element of $\mathbb{C}[\mathcal{V}_{\alpha,\beta}]$. Hence, it is enough to prove that the system of linear equations has a non-trivial solution on a Zariski dense subset of $\mathcal{V}_{\alpha,\beta}$. The set

$$D_{\alpha,\beta} := \mathcal{V}_{\alpha,\beta} \cap \bigcup_{\substack{(r,s)\in(\mathbb{Z}_{>0})^2 \\ rs<\alpha\beta}} \mathcal{V}_{r,s}$$

is a finite set, i.e., it is a Zariski closed subset of $\mathcal{V}_{\alpha,\beta}$. For any $(c,h) \in \mathcal{V}_{\alpha,\beta} \setminus D_{\alpha,\beta}$, we have

$$\begin{aligned} &\det(c,h)_{n'} \neq 0 \quad (\forall n' < n), \\ &\det(c,h)_n = 0. \end{aligned}$$

Hence, the system of linear equations equivalent to (5.31) has a solution for $(c,h) \in \mathcal{V}_{\alpha,\beta} \setminus D_{\alpha,\beta}$ and thus, it has a solution for any $(c,h) \in \mathcal{V}_{\alpha,\beta}$. Moreover, by (5.28), a solution of (5.31) is uniquely determined from the coefficient $c_{\mathbb{I}_0}$ ($\mathbb{I}_0 := (1^n)$). Hence, we may assume that $c_{\mathbb{I}_0} = 1$.

Let $\iota : U(\mathfrak{g}^-) \otimes_{\mathbb{C}} \mathbb{C}[\mathfrak{h}^*] \to U(\mathfrak{g}^- \oplus \mathfrak{h})$ be the linear isomorphism which is defined by $x_1 \otimes x_2 \mapsto x_1 x_2$ for $x_1 \in U(\mathfrak{g}^-)$ and $x_2 \in \mathbb{C}[\mathfrak{h}^*] = U(\mathfrak{h})$. Let $\pi_{\alpha,\beta}$ be the canonical projection

$$U(\mathfrak{g}^-) \otimes_{\mathbb{C}} \mathbb{C}[\mathfrak{h}^*] \to U(\mathfrak{g}^-) \otimes_{\mathbb{C}} \mathbb{C}[\mathcal{V}_{\alpha,\beta}]$$

induced from $\mathbb{C}[\mathfrak{h}^*] \twoheadrightarrow \mathbb{C}[\mathcal{V}_{\alpha,\beta}]$. We set $\pi'_{\alpha,\beta} := \pi_{\alpha,\beta} \circ \iota^{-1}$. Notice that

$$\mathrm{Ker}\pi'_{\alpha,\beta} = U(\mathfrak{g}^- \oplus \mathfrak{h})\Phi, \tag{5.32}$$

since $\Phi = \Phi_{\alpha,\beta}(c,h)$ is irreducible. Let

$$\tilde{S}_n = \sum_{\mathbb{I}\in\mathcal{P}_n} \tilde{c}_{\mathbb{I}} e_{\mathbb{I}}$$

be an element of $U(\mathfrak{g}^- \oplus \mathfrak{h})_{-n}$ such that $\pi_{\alpha,\beta}(\tilde{S}_n) = S_n$ and $\tilde{c}_{\mathbb{I}_0} = 1$. Since the condition (5.31) is equivalent to

$$y.S_n(c,h) \in U(\mathfrak{g})\mathfrak{g}^+ \quad (\forall \beta \in Q^+ \setminus \{0\},\ \forall y \in \mathfrak{g}_\beta)$$

for any $(c,h) \in \mathcal{V}_{\alpha,\beta}$, we see that

$$y.\tilde{S}_n \in U(\mathfrak{g})\Phi + U(\mathfrak{g})\mathfrak{g}^+ \quad (\forall \beta \in Q^+ \setminus \{0\},\ \forall y \in \mathfrak{g}_\beta)$$

by (5.32). Therefore, setting $S_{\Phi,n} := \tilde{S}_n$, we have completed the proof. □

Remark that the universal enveloping algebra of a Lie algebra is an integral domain, since the graded algebra associated with the standard filtration (A.6) is isomorphic to the symmetric algebra on the Lie algebra. Hence, any non-trivial homomorphism between Verma modules is injective, since a Verma module is a free $U(\mathfrak{g}^-)$-module of rank 1. Combining this fact with Corollary 5.1, we have

Corollary 5.2 *Suppose that highest weight (c, h) satisfies $c = c(t)$ and $h = h_{\alpha,\beta}(t)$ for some $\alpha, \beta \in \mathbb{Z}_{>0}$ and $t \in \mathbb{C} \setminus \{0\}$. Then,*

$$\dim \mathrm{Hom}_{\mathfrak{g}}(M(c, h+n), M(c,h)) = 1.$$

Moreover, such a homomorphism is a scaler multiple of the embedding which maps a highest weight vector $v_{c,h+n}$ to $S_{n,\Phi}.v_{c,h}$, where $n := \alpha\beta$ and $\Phi := \Phi_{\alpha,\beta}(c,h)$.

5.3 Embedding Diagrams of Verma Modules

5.3.1 Embedding Diagrams

Below, we denote a non-trivial homomorphism

$$\iota_{h,h'} : M(c, h') \to M(c, h),$$

by $[h'] \rightarrowtail [h]$.

<u>Class V</u>: All Verma modules are irreducible, and there is nothing to do.

<u>Class I</u>: The highest weight (c, h) can be written as $c = c(t)$, $h = h_{\alpha,\beta}(t)$ for some $t \in \mathbb{C} \setminus \mathbb{Q}$ and $\alpha, \beta \in \mathbb{Z}_{>0}$. By Lemma 5.2, $\tilde{D}(c, h) = \{(\alpha, \beta)\}$. By Corollary 5.2, the following holds:

Proposition 5.3 *Suppose that highest weight $(c, h) = (c(t), h_{\alpha,\beta}(t))$ ($t \in \mathbb{C} \setminus \mathbb{Q}$ and $\alpha, \beta \in \mathbb{Z}_{>0}$) belongs to Class I. Then, we have*

$$\begin{array}{c} [h] \\ \uparrow \\ [h+\alpha\beta] \end{array} .$$

Note that the submodule $M(c, h + \alpha\beta)$ is irreducible.

<u>Class R^+</u>: The highest weight (c, h) can be written as $c = c_{p,q}$, $h = h_i$ (5.22) for some $p, q \in \mathbb{Z}_{>0}$ such that $(p, q) = 1$, $(r, s) \in K^+_{p,q}$ and $i \in \mathbb{Z}$.

Proposition 5.4 *For* **Class** R^+, *there exist the commutative embedding diagrams of Verma modules given in Figure 5.4.*

This proposition will be proved in the next subsection.

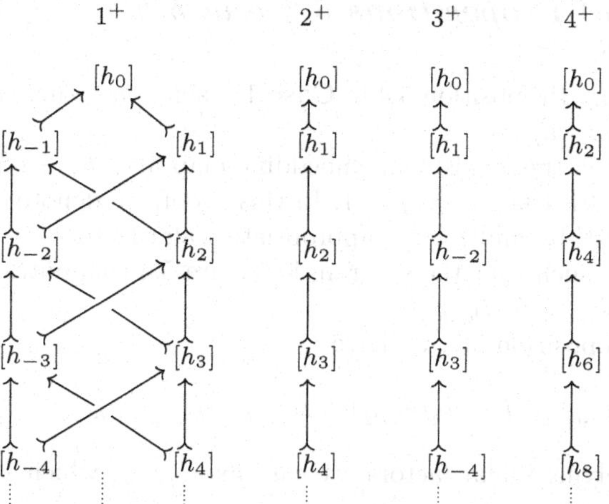

Fig. 5.4 Embedding diagrams for **Class** R^+

<u>**Class** R^-</u>: The highest weight (c, h) can be written as $c = c_{p,-q}$, $h = h_i$ (5.26) for some $p, q \in \mathbb{Z}_{>0}$ such that $(p, q) = 1$, $(r, s) \in K^-_{p,q}$ and $i \in \mathbb{Z} \setminus \{0\}$.

Proposition 5.5 *For* **Class** R^-, *there exist the commutative embedding diagrams of Verma modules given in Figure 5.5.*

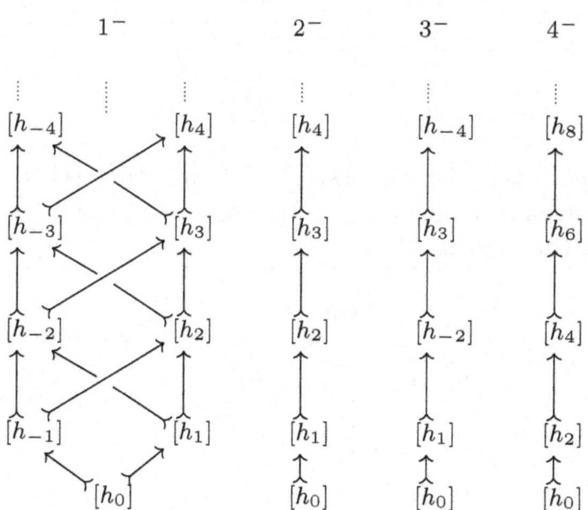

Fig. 5.5 Embedding diagrams for **Class** R^-

<u>**Note.**</u> Although (c, h_0) belongs to **Class** V, to describe embedding diagrams, it is convenient to use the conformal weight h_0.

5.3.2 Proof of Propositions 5.4 and 5.5

Here, we prove only Proposition 5.4 in **Case** 1^+, since the other cases can be proved in a similar way.

By Corollary 5.2, there exists an embedding map $M(c,h_j) \to M(c,h_i)$ for each $(i,j) \in \mathbb{Z}^2$ such that $|i| = |j| - 1$. In this proof, we denote it by $\tilde{\iota}_{h_j.h_i}$. Here, we show that, by multiplying appropriate scalar factors, we can choose embeddings $\iota_{h_j.h_i}$ such that the diagram in Figure 5.4 commutes.

First, we set $\iota_{h_0,h_{\pm 1}} := \tilde{\iota}_{h_0,h_{\pm 1}}$.

Second, by Proposition 5.1, we have

$$\iota_{h_0,h_{-1}} \circ \tilde{\iota}_{h_{-1},h_2}(v_{c,h_2}) \propto \iota_{h_0,h_1} \circ \tilde{\iota}_{h_1,h_2}(v_{c,h_2}).$$

Hence, by multiplying scalar factors, we can take $\tilde{\iota}_{h_{\pm 1},h_2}$ which satisfy

$$\iota_{h_0,h_{-1}} \circ \iota_{h_{-1},h_2} = \iota_{h_0,h_1} \circ \iota_{h_1,h_2}.$$

Similarly, we can choose $\iota_{h_{\pm 1},h_{-2}}$ such that

$$\iota_{h_0,h_{-1}} \circ \iota_{h_{-1},h_{-2}} = \iota_{h_0,h_1} \circ \iota_{h_1,h_{-2}}$$

holds. Hence, we obtain a commutative embedding diagram

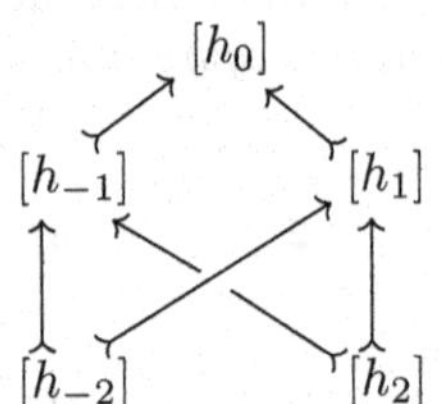

Third, we suppose that for each $(k,l) \in \mathbb{Z}^2$ such that $|k| = |l| - 1$ and $|k| \leq i$, there exist embeddings $\tilde{\iota}_{h_k,h_l} : M(c,h_l) \to M(c,h_k)$ such that the following diagram commutes:

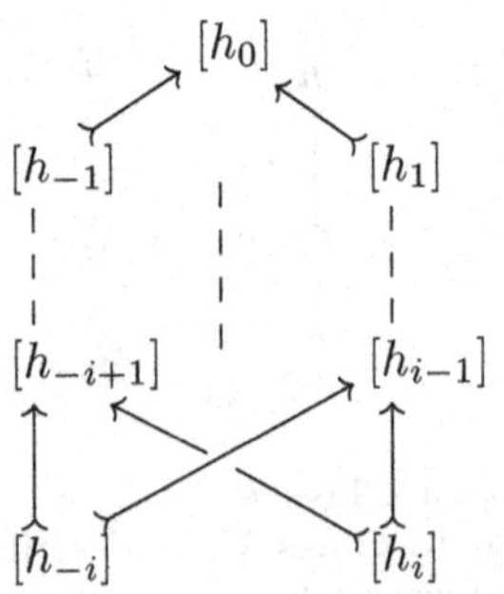

We choose embeddings $\iota_{h_{-i},h_{i+1}}$ and $\iota_{h_i,h_{i+1}}$ such that

$$\iota_{h_{i-1},h_{-i}} \circ \iota_{h_{-i},h_{i+1}} = \iota_{h_{i-1},h_i} \circ \iota_{h_i,h_{i+1}} \tag{5.33}$$

holds, and show the commutativity of the following diagram:

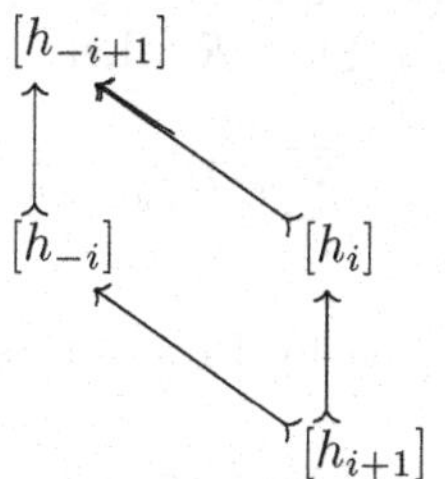

It is enough to see that

$$\iota_{h_{-i+1},h_{-i}} \circ \iota_{h_{-i},h_{i+1}} = \iota_{h_{-i+1},h_i} \circ \iota_{h_i,h_{i+1}}. \tag{5.34}$$

From (5.33), we obtain

$$\iota_{h_{i-2},h_{i-1}} \circ \iota_{h_{i-1},h_{-i}} \circ \iota_{h_{-i},h_{i+1}} = \iota_{h_{i-2},h_{i-1}} \circ \iota_{h_{i-1},h_i} \circ \iota_{h_i,h_{i+1}}.$$

On the other hand, by the inductive assumption we have

$$\iota_{h_{i-2},h_{i-1}} \circ \iota_{h_{i-1},h_{-i}} = \iota_{h_{i-2},h_{-i+1}} \circ \iota_{h_{-i+1},h_{-i}},$$
$$\iota_{h_{i-2},h_{i-1}} \circ \iota_{h_{i-1},h_i} = \iota_{h_{i-2},h_{-i+1}} \circ \iota_{h_{-i+1},h_i}.$$

Hence, we obtain

$$\iota_{h_{i-2},h_{-i+1}} \circ \iota_{h_{-i+1},h_{-i}} \circ \iota_{h_{-i},h_{i+1}} = \iota_{h_{i-2},h_{-i+1}} \circ \iota_{h_{-i+1},h_i} \circ \iota_{h_i,h_{i+1}}.$$

Since $\iota_{h_{i-2},h_{-i+1}}$ is injective, we obtain (5.34). Similarly, there exist embeddings $\iota_{h_i,h_{-i-1}}$ and $\iota_{h_{-i},h_{-i-1}}$ such that

$$\iota_{h_{i-1},h_i} \circ \iota_{h_i,h_{-i-1}} = \iota_{h_{i-1},h_{-i}} \circ \iota_{h_{-i},h_{-i-1}},$$
$$\iota_{h_{-i+1},h_i} \circ \iota_{h_i,h_{-i-1}} = \iota_{h_{-i+1},h_{-i}} \circ \iota_{h_{-i},h_{-i-1}}.$$

Therefore, the existence of the embedding diagram of Verma modules in **Case** 1^+ has been proved.

5.4 Singular Vector Formulae

There are many studies on singular vectors of Verma modules over $\mathfrak{g}$ (e.g., [BS], [Mill]). However, except for some special cases, completely explicit expressions of them are still unknown. In this section, we present two formulae related with singular vectors which play important roles in mathematical and theoretical physics [FeFu5].

Through this section, we fix $\alpha, \beta \in \mathbb{Z}_{>0}$, and set $\Phi := \Phi_{\alpha,\beta}(c,h) \in \mathbb{C}[c,h]$ and $n := \alpha\beta$ for simplicity. Let $\mathcal{R} := \mathbb{C}[\xi, \xi^{-1}]$ be the Laurent polynomial ring. We set $(\tilde{c}, \tilde{h}) := (c(\xi), h_{\alpha,\beta}(\xi)) \in \mathcal{R}^2$. Let

$$S_{n,\Phi} := e_{\mathbb{I}_0} + \sum_{\mathbb{I} < \mathbb{I}_0} e_{\mathbb{I}} H_{\Phi,\mathbb{I}} \quad (\mathbb{I}_0 := (1^n))$$

be the Shapovalov element given by Proposition 5.2, where $H_{\Phi,\mathbb{I}} \in S(\mathfrak{h}) \simeq \mathbb{C}[\mathfrak{h}^*]$. Note that

$$S_{n,\Phi}(\tilde{c}, \tilde{h}) \in U(\mathfrak{g}^-) \otimes_{\mathbb{C}} \mathcal{R}.$$

5.4.1 Formula I

For $a, b \in \mathbb{C}$, let $V_{a,b} = \bigoplus_{u \in \mathbb{Z}} \mathbb{C} v_u$ be the $\mathfrak{g}$-module defined in (1.11), i.e.,

$$\begin{aligned} L_s.v_u &= (as + b - u)v_{u+s}, \\ C.v_u &= 0. \end{aligned}$$

The first singular vector formula describes the action of $S_{n,\Phi}(\tilde{c}, \tilde{h})$ on the module $V_{a,b}$.

Proposition 5.6

$$S_{n,\Phi}(\tilde{c}, \tilde{h}).v_0 = P_{\alpha,\beta}(a, b; \xi) v_{-n},$$

where $P_{\alpha,\beta}(a, b; \xi) \in \mathbb{C}[a, b, \xi, \xi^{-1}]$ *satisfies*

$$\begin{aligned} P_{\alpha,\beta}(a, b; \xi)^2 &= \prod_{k=0}^{\alpha-1} \prod_{l=0}^{\beta-1} Q_{k,l}^{\alpha,\beta}(a, b; \xi), \\ Q_{k,l}^{\alpha,\beta}&(a, b; \xi) \\ &= \Big[(b-a) - (k\xi^{\frac{1}{2}} - l\xi^{-\frac{1}{2}})\{(\alpha-k)\xi^{\frac{1}{2}} - (\beta-l)\xi^{-\frac{1}{2}}\}\Big] \\ &\quad \times \Big[(b-a) - \{(k+1)\xi^{\frac{1}{2}} - (l+1)\xi^{-\frac{1}{2}}\}\{(\alpha-k-1)\xi^{\frac{1}{2}} - (\beta-l-1)\xi^{-\frac{1}{2}}\}\Big] \\ &\quad + \{(\alpha-2k-1)\xi^{\frac{1}{2}} - (\beta-2l-1)\xi^{-\frac{1}{2}}\}^2 a. \end{aligned}$$

We show this proposition after stating the second formula.

Remark 5.8 *By the normalisation of the Shapovalov element,* $P_{\alpha,\beta}(a, b; \xi)$ *is of the form*

$$P_{\alpha,\beta}(a, b; \xi) = b^{\alpha\beta} + (\textit{lower terms}),$$

as a polynomial in b. Hence, $P_{\alpha,\beta}(a,b;\xi)$ is uniquely determined by the above proposition.

5.4.2 Formula II

Let $I_{\leq -3}$ be the two-sided ideal of $U(\mathfrak{g}^-)$ generated by

$$[\mathfrak{g}^-,\mathfrak{g}^-] = \bigoplus_{n\leq -3} \mathbb{C}L_n.$$

Note that

$$\begin{aligned} U(\mathfrak{g}^-)/I_{\leq -3} &\simeq U(\mathfrak{g}^-/[\mathfrak{g}^-,\mathfrak{g}^-]) \\ &\simeq \mathbb{C}[L_{-1},L_{-2}]. \end{aligned}$$

Let $\mathrm{Pr} : U(\mathfrak{g}^-) \to U(\mathfrak{g}^-)/I_{\leq -3}$ be the canonical projection. We denote the map

$$U(\mathfrak{g}^-)\otimes_{\mathbb{C}} \mathcal{R} \longrightarrow \{U(\mathfrak{g}^-)/I_{\leq -3}\}\otimes_{\mathbb{C}} \mathcal{R}$$

induced from the above projection by the same symbol Pr. The second singular vector formula describes the image of the Shapovalov element under this projection.

Proposition 5.7

$$\mathrm{Pr}(S_{n,\Phi}(\tilde{c},\tilde{h})) = R_{\alpha,\beta}(L_{-1},L_{-2};\xi),$$

where $R_{\alpha,\beta}(x,y;\xi)\in\mathbb{C}[x,y,\xi,\xi^{-1}]$ satisfies

$$R_{\alpha,\beta}(x,y;\xi)^2 = \prod_{k=0}^{\alpha-1}\prod_{l=0}^{\beta-1}(x^2-\{(\alpha-2k-1)\xi^{\frac{1}{2}}-(\beta-2l-1)\xi^{-\frac{1}{2}}\}^2 y).$$

5.4.3 Proof of Formula I

For $f\in\mathcal{R}$, let $\xi\text{-}\deg_+ f$ (resp. $\xi\text{-}\deg_- f$) be the highest degree (resp. the lowest degree) of f as a polynomial in ξ. We define $\{c_{\mathbb{I}}\}\subset\mathcal{R}$ by $S_{n,\Phi}.v_{\tilde{c},\tilde{h}}\in M_{\mathcal{R}}(\tilde{c},\tilde{h})$ as follows:

$$S_{n,\Phi}.v_{\tilde{c},\tilde{h}} = \sum_{\mathbb{I}\in\mathcal{P}_n} c_{\mathbb{I}} e_{\mathbb{I}}.v_{\tilde{c},\tilde{h}}.$$

In particular, $c_{\mathbb{I}_0}=1$. To prove the first formula by using embedding diagrams, it is necessary to estimate $\xi\text{-}\deg_+ P_{\alpha,\beta}(a,b;\xi)$ and $\xi\text{-}\deg_- P_{\alpha,\beta}(a,b;\xi)$.

Lemma 5.16. *([AsFu])*

1. $\max_{\mathbb{I}\in\mathcal{P}_n}\{\xi\text{-}\deg_+ c_{\mathbb{I}}\} \le (\alpha-1)\beta$,
2. $\min_{\mathbb{I}\in\mathcal{P}_n}\{\xi\text{-}\deg_- c_{\mathbb{I}}\} \ge -\alpha(\beta-1)$.

We need some preliminaries. For each weight vector $w \in M_{\mathcal{R}}(\tilde{c},\tilde{h})_{\tilde{h}+n}$, we define $\{c_{\mathbb{I}}^w\} \subset \mathcal{R}$ by

$$w = \sum_{\mathbb{I}\in\mathcal{P}_n} c_{\mathbb{I}}^w e_{\mathbb{I}}.v_{\tilde{c},\tilde{h}}.$$

Let $\{F_jU(\mathfrak{g}^+)|j \in \mathbb{Z}_{\ge 0}\}$ be the standard filtration of $U(\mathfrak{g}^+)$ (for the definition, see (A.6)). First, we show a variant of Lemma 5.15.

Lemma 5.17. *For* $\mathbb{J} = (1^{s_1}2^{s_2}\cdots n^{s_n}) \in \mathcal{P}_n \setminus \{\mathbb{I}_0\}$, *let* $j > 1$ *be the positive integer such that* $s_2 = \cdots = s_{j-1} = 0$ *and* $s_j \ne 0$. *Set* $\mathbb{J}' := \mathbb{J} - (j^1) + (1^1) \in \mathcal{P}_{n-j+1}$. *We take*

$$y \in U(\mathfrak{g}^+)_{-j+1} := \{u \in U(\mathfrak{g}^+)|[L_0, u] = (-j+1)u\},$$

and set $w' := y.w$. *We express* w' *as*

$$w' = \sum_{\mathbb{I}'\in\mathcal{P}_{n-j+1}} c_{\mathbb{I}'}^{w'} e_{\mathbb{I}'}.v_{\tilde{c},\tilde{h}}.$$

Then, there exist $\alpha_y \in \mathbb{C}$ *and* $\{Q_{\mathbb{I},\mathbb{J}}^y|\mathbb{I} > \mathbb{J}\} \subset \mathcal{R}$ *satisfying the following :*

1. Let N *be the positive integer such that*

$$y \in F_NU(\mathfrak{g}^+) \setminus F_{N-1}U(\mathfrak{g}^+).$$

Then, $\xi\text{-}\deg_+ Q_{\mathbb{I},\mathbb{J}}^y \le N$ *and* $\xi\text{-}\deg_- Q_{\mathbb{I},\mathbb{J}}^y \ge -N$.

2.

$$c_{\mathbb{J}'}^{w'} = \alpha_y c_{\mathbb{J}}^w + \sum_{\substack{\mathbb{I}\in\mathcal{P}_n \\ \mathbb{I}>\mathbb{J}}} Q_{\mathbb{I},\mathbb{J}}^w c_{\mathbb{I}}^w.$$

Proof. The existence of $\alpha_y \in \mathbb{C}$ and $\{Q_{\mathbb{I},\mathbb{J}}^y|\mathbb{I} > \mathbb{J}\} \subset \mathcal{R}$ with the second condition can be proved in a way similar to Lemma 5.15. From the proof, one can directly check that they also satisfy the first condition. □

Remark 5.9 *If* y *is a 'monomial', i.e.,*

$$y = L_{i_1}L_{i_2}\cdots L_{i_k} \quad (i_1,\cdots,i_k \in \mathbb{Z}_{>0} \wedge i_1 + i_2 + \cdots + i_k = j),$$

then $\alpha_y \ne 0$.

Second, to estimate $\xi\text{-}\deg_+ c_{\mathbb{I}}$ and $\xi\text{-}\deg_- c_{\mathbb{I}}$ for each $\mathbb{I} \in \mathcal{P}_n$, we introduce a function φ_γ $(\gamma \in \mathbb{Z}_{>0})$. For $j \in \mathbb{Z}_{>0}$, we set

$$\varphi_\gamma(j) := j - 1 - \left\lceil \frac{j-1}{\gamma} \right\rceil,$$

where $\lceil x \rceil$ denotes the greatest integer not exceeding x. Further, for $\mathbb{I} = (1^{r_1}2^{r_2}\cdots n^{r_n}) \in \mathcal{P}_n$, we set

$$\varphi_\gamma(\mathbb{I}) := \sum_{k=1}^{n} r_k \varphi_\gamma(k).$$

Then, one can directly check that φ_γ enjoys the following properties.

1. For any $i, j \in \mathbb{Z}_{>0}$,

$$\varphi_\gamma(i+j) \geq \varphi_\gamma(i) + \varphi_\gamma(j) \geq \varphi_\gamma(i+j-1). \tag{5.35}$$

2. For $s \in \mathbb{Z}_{>0}$ and $i_1, i_2, \cdots, i_s \in \mathbb{Z}_{>0}$,

$$\varphi_\gamma(i_1+i_2+\cdots+i_s+1) = \varphi_\gamma(i_1) + \varphi_\gamma(i_2) + \cdots + \varphi_\gamma(i_s), \tag{5.36}$$

   if and only if $s = 1$ and $i_1 \equiv 0 \pmod{\gamma}$.

Note that by (5.35) the inequality

$$\varphi_\gamma(i_1+i_2+\cdots+i_s+1) \geq \varphi_\gamma(i_1) + \varphi_\gamma(i_2) + \cdots + \varphi_\gamma(i_s)$$

always holds.

We show Lemma 5.16 by using the next lemma.

Lemma 5.18. *For any* $\mathbb{I} \in \mathcal{P}_n$,

1. $\xi\text{-}\deg_+ c_{\mathbb{I}} \leq \varphi_\alpha(\mathbb{I})$,
2. $\xi\text{-}\deg_- c_{\mathbb{I}} \geq -\varphi_\beta(\mathbb{I})$.

Proof. We first show 1. of Lemma 5.18 by induction. For $\mathbb{I}_0 := (1^n)$, we have $c_{\mathbb{I}_0} = 1$ and $\varphi_\alpha(\mathbb{I}_0) = n\varphi_\alpha(1) = 0$. Hence, 1. of Lemma 5.18 holds for $\mathbb{I} = \mathbb{I}_0$.

Next, we suppose that $\mathbb{J} < \mathbb{I}_0$ and 1. of Lemma 5.18 holds for any $\mathbb{I}$ such that $\mathbb{J} < \mathbb{I} < \mathbb{I}_0$. Here, for $w := S_{n,\Phi}.v_{\tilde{c},\tilde{h}}$ and $y := L_{j-1}$, we apply Lemma 5.17. Since $w' = 0$, we have

$$c_{\mathbb{J}} = -(\alpha_y)^{-1} \sum_{\mathbb{I}>\mathbb{J}} Q^y_{\mathbb{I},\mathbb{J}} c_{\mathbb{I}},$$

and thus,

$$\xi\text{-}\deg_+ c_{\mathbb{J}} \leq \max_{\mathbb{I}>\mathbb{J}} \{\xi\text{-}\deg_+ c_{\mathbb{I}} + \xi\text{-}\deg_+ Q^y_{\mathbb{I},\mathbb{J}}\}.$$

Moreover, in this case, we have $\xi\text{-}\deg_+ Q^y_{\mathbb{I},\mathbb{J}} \leq 1$. Hence, by the induction hypothesis, it is enough to show that

C1. $\varphi_\alpha(\mathbb{I}) \leq \varphi_\alpha(\mathbb{J}) \quad (\forall \mathbb{I} \in \mathcal{P}_n : \mathbb{I} > \mathbb{J} \wedge Q^y_{\mathbb{I},\mathbb{J}} \neq 0 \wedge \xi\text{-}\deg_+ Q^y_{\mathbb{I},\mathbb{J}} = 0)$,
C2. $\varphi_\alpha(\mathbb{I}) < \varphi_\alpha(\mathbb{J}) \quad (\forall \mathbb{I} \in \mathcal{P}_n : \mathbb{I} > \mathbb{J} \wedge \xi\text{-}\deg_+ Q^y_{\mathbb{I},\mathbb{J}} = 1)$.

Notice that $L_{j-1} e_{\mathbb{I}}.v_{\tilde{c},\tilde{h}} = v_1 + v_2$, where

$$v_1 = [L_{j-1}, L_{-n}^{r_n} \cdots L_{-j}^{r_j}] L_{-j+1}^{r_{j-1}} \cdots L_{-1}^{r_1}.v_{\tilde{c},\tilde{h}},$$
$$v_2 = L_{-n}^{r_n} \cdots L_{-j}^{r_j} [L_{j-1}, L_{-j+1}^{r_{j-1}} \cdots L_{-1}^{r_1}].v_{\tilde{c},\tilde{h}}.$$

Hence, if $Q_{\mathbb{I},\mathbb{J}}^y \neq 0$, i.e., $c_{\mathbb{J}'}^{ye_{\mathbb{I}}.v_{\tilde{c},\tilde{h}}} \neq 0$, then $c_{\mathbb{J}'}^{v_1} \neq 0$ or $c_{\mathbb{J}'}^{v_2} \neq 0$. We divide the proof into the following two cases:

Case $c_{\mathbb{J}'}^{v_1} \neq 0$: In this case, we have $\xi\text{-}\deg_+ Q_{\mathbb{I},\mathbb{J}}^y = 0$ by definition. Since $v_1 = \sum_{k=j}^n v_{1,k}$ where

$$v_{1,k} := L_{-n}^{r_n} \cdots [L_{j-1}, L_{-k}^{r_k}] \cdots L_{-j}^{r_j} L_{-j+1}^{r_{j-1}} \cdots L_{-1}^{r_1}.v_{\tilde{c},\tilde{h}},$$

there exists k $(j \leq k \leq n)$ such that $c_{\mathbb{J}'}^{v_{1,k}} \neq 0$. We first show that, for such integer k,

$$\varphi_\alpha(\mathbb{I} - (k^1) + ((k-j+1)^1)) \leq \varphi_\alpha(\mathbb{J}'). \tag{5.37}$$

Indeed, $c_{\mathbb{J}'}^{v_{1,k}} \neq 0$ means that $e_{\mathbb{J}'}.v_{\tilde{c},\tilde{h}}$ has a non-zero coefficient in the expression of $v_{1,k}$ with respect to the basis $\{e_{\mathbb{K}}.v_{\tilde{c},\tilde{h}} | \mathbb{K} \in \mathcal{P}_{n-j+1}\}$. This implies that $\mathbb{J}'$ is obtained by regrouping the partition $\mathbb{I}' := \mathbb{I} - (k^1) + ((k-j+1)^1)$, namely, if $\mathbb{I}' = (i_1', i_2', \cdots, i_a')$ and $\mathbb{J}' = (j_1', j_2', \cdots, j_b')$ such that

$$i_1' \leq i_2' \leq \cdots \leq i_a' \wedge i_1' + i_2' + \cdots + i_a' = n - j + 1,$$
$$j_1' \leq j_2' \leq \cdots \leq j_b' \wedge j_1' + j_2' + \cdots + j_b' = n - j + 1,$$

then

$$j_l' = \sum_u i_{k_u^{(l)}}' \quad (l = 1, 2, \cdots, b),$$

where

$$\{1, 2, \cdots, a\} = \bigsqcup_{l=1}^b \{k_1^{(l)}, k_2^{(l)}, \cdots\} \text{ (disjoint union)}.$$

Hence, the inequality (5.37) follows from the property (5.35).

Since $\mathbb{J}' = \mathbb{J} - (j^1) + (1^1)$, from (5.37), we obtain

$$\varphi_\alpha(\mathbb{I}) - \varphi_\alpha(k) + \varphi_\alpha(k-j+1) \leq \varphi_\alpha(\mathbb{J}) - \varphi_\alpha(j) + \varphi_\alpha(1).$$

Hence, we have

$$\varphi_\alpha(\mathbb{J}) - \varphi_\alpha(\mathbb{I}) \geq \varphi_\alpha(k-j+1) + \varphi_\alpha(j) - \varphi_\alpha(k) \geq 0$$

by (5.35). Hence, **C1** holds.

Case $c_{\mathbb{J}'}^{v_2} \neq 0$: In this case, we see that if $c_{\mathbb{J}'}^{v_2} \neq 0$, then

$$r_1 \geq s_1 \wedge \sum_{k=1}^{j-1} r_k k = j + s_1 \wedge r_j = s_j - 1 \wedge r_k = s_k \ (j+1 \leq k \leq n),$$

for $\mathbb{J} = (1^{s_1} \cdots n^{s_n})$ and $\mathbb{I} = (1^{r_1} \cdots n^{r_n})$. Hence, by the property (5.35), we have

$$\varphi_\alpha(\mathbb{I}) \leq \varphi_\alpha(\mathbb{J}). \tag{5.38}$$

Further, by (5.36), $\varphi_\alpha(\mathbb{I}) < \varphi_\alpha(\mathbb{J})$ holds for $\mathbb{I}$ such that

$$\mathbb{I} \neq \mathbb{J} - (j^1) + ((j-1)^1) - (1^1) \vee j - 1 \not\equiv 0 \pmod{\alpha}. \tag{5.39}$$

Hence, if $\mathbb{I}$ satisfies (5.39), then **C2** holds.

To complete the proof, it is necessary to show that

$$\xi\text{-}\deg_+ Q^y_{\mathbb{I},\mathbb{J}} = 0 \tag{5.40}$$

holds for $\mathbb{I} \in \mathcal{P}_n$ such that

$$\mathbb{I} = \mathbb{J} - (j^1) + ((j-1)^1) - (1^1) \wedge j - 1 \equiv 0 \pmod{\alpha}.$$

We take $y := (L_\alpha)^{m_1}$ $(m_1 = (j-1)/\alpha)$. Since the inequality (5.38) still holds, if $\xi\text{-}\deg_+ Q^y_{\mathbb{I},\mathbb{J}} = 0$, then **C1** holds. Indeed, from the explicit form of $\mathbb{I}$, we see that $Q^y_{\mathbb{I},\mathbb{J}}$ is a $\mathbb{C}$-linear combination of some products of $f_l \in \mathcal{R}$ $(l \leq n)$, where f_l is given by

$$[L_\alpha, L_{-\alpha}]|_{M_{\mathcal{R}}(\tilde{c},\tilde{h})_{\tilde{h}+l}} = f_l \mathrm{id}_{M_{\mathcal{R}}(\tilde{c},\tilde{h})_{\tilde{h}+l}}.$$

On the other hand, we have $\xi\text{-}\deg_+ f_l = 0$ by direct computation. Hence, $\xi\text{-}\deg_+ Q^y_{\mathbb{I},\mathbb{J}} = 0$ holds. Therefore, we have completed the proof of 1. of Lemma 5.18.

For 2. of Lemma 5.18, since $g_l \in \mathcal{R}$ defined by

$$[L_\beta, L_{-\beta}]|_{M_{\mathcal{R}}(\tilde{c},\tilde{h})_{\tilde{h}+l}} = g_l \mathrm{id}_{M_{\mathcal{R}}(\tilde{c},\tilde{h})_{\tilde{h}+l}}$$

satisfies $\xi\text{-}\deg_- g_l = 0$, one can similarly prove as above. □

Proof of Lemma 5.16. We prove the first statement. For each k $(1 \leq k \leq n)$, there uniquely exist $\gamma_k, \delta_k \in \mathbb{Z}_{\geq 0}$ such that

$$k = \gamma_k \alpha - \delta_k \quad (0 \leq \delta_k < \alpha).$$

Then, we have $\varphi_\alpha(k) = k - \gamma_k$ and $\varphi_\alpha(\mathbb{I}) = \sum_{k=1}^n r_k(k - \gamma_k)$ for $\mathbb{I} = (1^{r_1} 2^{r_2} \cdots n^{r_n}) \in \mathcal{P}_n$. On the other hand, since

$$\alpha\beta = n = \sum_{k=1}^n r_k k = \alpha \sum_{k=1}^n r_k \gamma_k - \sum_{k=1}^n \gamma_k \delta_k,$$

we have $\sum_{k=1}^n r_k \gamma_k \geq \beta$. Hence, $\varphi_\alpha(\mathbb{I}) \leq (\alpha - 1)\beta$ for any $\mathbb{I} \in \mathcal{P}_n$.

For the second statement, expressing each k as

$$k = \gamma'_k \beta - \delta'_k \quad (\gamma'_k, \delta'_k \in \mathbb{Z}_{\geq 0},\ 0 \leq \delta'_k < \beta),$$

one can similarly show the second inequality. Therefore, Lemma 5.16 follows from Lemma 5.18. □

Finally, by using Lemma 5.16, we show the singular vector formulae. Here, we only prove Proposition 5.6, since Proposition 5.7 can be proved similarly.

For the proof, we introduce some notation. For $\xi_0 \in \mathbb{C} \setminus \{0\}$, let $\phi_{\xi;\xi_0} : \mathrm{Mod}_{\mathcal{R}} \to \mathrm{Vect}_{\mathbb{C}}$ be the reduction functor induced from the projection $\mathcal{R} \to \mathcal{R}/(\xi - \xi_0)\mathcal{R} \simeq \mathbb{C}$. For $v \in M$ ($M \in \mathrm{Ob}(\mathrm{Mod}_{\mathcal{R}})$), we set

$$\phi_{\xi;\xi_0} v := v \otimes_{\mathcal{R}} 1 \in \phi_{\xi;\xi_0} M := M \otimes_{\mathcal{R}} (\mathcal{R}/(\xi - \xi_0)\mathcal{R})$$

Proof of Proposition 5.6. Since $S_{n,\Phi}$ is an L_0-weight vector and each weight subspace of the module $V_{a,b}$ is of dimension one, there exists $\tilde{P}_{\alpha,\beta}(a,b;\xi) \in \mathbb{C}[a,b,\xi,\xi^{-1}]$ such that

$$S_{n,\Phi} v_0 = \tilde{P}_{\alpha,\beta}(a,b;\xi) v_{-n}.$$

In the sequel, we show that

$$\tilde{P}_{\alpha,\beta}(a,b;\xi) = P_{\alpha,\beta}(a,b;\xi) \tag{5.41}$$

by induction on $n := \alpha\beta$.

For $n = 1, 2$, one can compute Shapovalov elements explicitly, and thus can directly check Proposition 5.6.

For $n \geq 3$, by the embedding diagrams in Proposition 5.4 and Proposition 5.5, there exist a non-zero rational number ξ_0, $s \in \mathbb{Z}_{\geq 2}$ and $\alpha_k, \beta_k \in \mathbb{Z}_{>0}$ ($k = 1, \cdots, s$) such that

$$\begin{aligned} &\phi_{\xi;\xi_0}(S_{n,\Phi}).v_{c(\xi_0),h_{\alpha,\beta}(\xi_0)} \\ &= \phi_{\xi;\xi_0}(S_{n_s,\Phi_s}) \cdots \phi_{\xi;\xi_0}(S_{n_2,\Phi_2}) \phi_{\xi;\xi_0}(S_{n_1,\Phi_1}).v_{c(\xi_0),h_{\alpha,\beta}(\xi_0)}, \end{aligned} \tag{5.42}$$

where we set $n_k := \alpha_k \beta_k$ and $\Phi_k := \Phi_{\alpha_k,\beta_k}(c,h)$. Hence, by the uniqueness of singular vectors (Proposition 5.1), the induction hypothesis and Remark 5.8, we obtain

$$\tilde{P}_{\alpha,\beta}(a,b;\xi_0) = \prod_{k=1}^{s} P_{\alpha_k,\beta_k}(a,b;\xi_0). \tag{5.43}$$

Moreover, by direct computation, one can check that the right-hand side of (5.43) coincides with $P_{\alpha,\beta}(a,b;\xi_0)$. Hence, (5.41) holds at $\xi = \xi_0$.

On the other hand, as a consequence of Lemma 5.18, we see that

$$\begin{aligned} &\xi\text{-}\deg_+ \tilde{P}_{\alpha,\beta}(a,b;\xi_0) \leq (\alpha - 1)\beta, \\ &\xi\text{-}\deg_- \tilde{P}_{\alpha,\beta}(a,b;\xi_0) \geq -\alpha(\beta - 1). \end{aligned}$$

Hence, to complete the proof, it suffices to show that there exist enough of ξ_0 such that (5.43) holds, i.e.,

$$\sharp\{\xi_0 |\ (5.42) \text{ holds at } \xi=\xi_0\} > 2\alpha\beta-\alpha-\beta. \tag{5.44}$$

To show the inequality (5.44), we first notice that, from the embedding diagrams in **Class** R^+ and **Class** R^-, if (5.43) holds at $\xi=\xi_0\in\mathbb{Q}_{>0}$, then it also holds at $\xi=-\xi_0$.

Next, to estimate the number of $\xi_0\in\mathbb{Q}_{>0}$, we prove the following lemma: Suppose that $p,q\in\mathbb{Z}_{>0}$ such that $(p,q)=1$, and $(c(\frac{q}{p}),h_i)$ belongs to **Class** R^+. Let $\mathcal{D}^{p,q}_{\geq 2}$ be the union of the following sets:

1. **Case** 1^+:
$$\{h_k-h_l|k,l\in\mathbb{Z},\ |k|\geq |l|+2\},$$
2. **Case** 2^+:
$$\{h_k-h_l|k,l\in\mathbb{Z}_{\geq 0},\ k\geq l+2\},$$
3. **Case** 3^+:
$$\{h_{(-1)^{k-1}k}-h_{(-1)^{l-1}l}|k,l\in\mathbb{Z}_{\geq 0},\ k\geq l+2\},$$
4. **Case** 4^+:
$$\{h_{2k}-h_{2l}|k,l\in\mathbb{Z}_{\geq 0},\ k\geq l+2\}.$$

Note that if $n\in\mathcal{D}^{p,q}_{\geq 2}$ $(n:=\alpha\beta)$, then (5.42) holds for $\xi_0=\frac{q}{p}$.

Lemma 5.19.

$$\left\{x,\ ypq\ \middle|\ \begin{matrix} x\in\mathbb{Z}_{>pq}: x\not\equiv 0 \pmod{pq}, \\ y\in\mathbb{Z}_{>0}: \textit{not prime}\end{matrix}\right\}\subset\mathcal{D}^{p,q}_{\geq 2} \tag{5.45}$$

Proof. Let $\tilde{\mathcal{D}}^{p,q}_{\geq 2}$ be the set consists of the following positive integers:

Case 1^+: for $k,l\in\mathbb{Z}_{\geq 0}$, $(k\geq l+1)$:

$$h_{2k}-h_{2l},\ h_{-2k}-h_{-2l},\ h_{2k+1}-h_{2l+1},\ h_{-2k-1}-h_{-2l-1},$$

Case 2^+: for $k,l\in\mathbb{Z}_{\geq 0}$, $(k\geq l+1)$:

$$h_{2k}-h_{2l},\ h_{2k+1}-h_{2l+1},$$

Case 3^+: for $k,l\in\mathbb{Z}_{\geq 0}$, $(k\geq l+1)$:

$$h_{-2k}-h_{-2l},\ h_{2k+1}-h_{2l+1},$$

Case 4^+: for $k,l\in\mathbb{Z}_{\geq 0}$, $(k\geq l+2)$:

$$h_{2k}-h_{2l}.$$

By Lemma 5.9, the left-hand side of (5.45) is a subset of $\tilde{\mathcal{D}}^{p,q}_{\geq 2}$. Since $\tilde{\mathcal{D}}^{p,q}_{\geq 2}\subset\mathcal{D}^{p,q}_{\geq 2}$ by definition, the lemma holds. □

By Lemma 5.19, if $p,q\in\mathbb{Z}_{>0}$ satisfies

$$pq < n \wedge (n \not\equiv 0 (\text{mod } pq) \vee n = ypq \ (\exists y \in \mathbb{Z}_{>0} : \text{not prime})),$$

then (5.42) holds at $\xi = \frac{q}{p}$. One can check that the cardinality of such rational numbers $\frac{q}{p}$ is greater than $\frac{1}{2}(2\alpha\beta - \alpha - \beta)$. Therefore, we have proved Proposition 5.6. □

5.5 Character Sums of Jantzen Filtration of Verma Modules

Using Lemmas 5.11 and 5.14, we compute character sums of Jantzen filtrations of Verma modules.

5.5.1 Notation

Through this section, let $\mathcal{R}$ be the polynomial ring $\mathbb{C}[\xi]$. Let $\phi_\xi : \mathcal{R} \to \mathcal{R}/\xi\mathcal{R} (\simeq \mathbb{C})$ be the canonical projection, and let

$$\begin{aligned} \phi_\xi : \mathrm{Mod}_{\mathcal{R}} &\to \mathrm{Vect}_{\mathbb{C}} \\ M &\mapsto M \otimes_{\mathcal{R}} (\mathcal{R}/\xi\mathcal{R}) \end{aligned}$$

be the reduction functor. For simplicity, we set $\phi_\xi v := v \otimes_{\mathcal{R}} 1$ for $v \in M$ ($M \in \mathrm{Ob}(\mathrm{Mod}_{\mathcal{R}})$). Here, we denote the ξ-adic valuation $\mathcal{R} \to \mathbb{Z}_{\geq 0} \sqcup \{\infty\}$ by ord_ξ.

For $(\tilde{c}, \tilde{h}) \in \mathcal{R}^2$, let

$$M_{\mathcal{R}}(\tilde{c}, \tilde{h}) = \bigoplus_{n \in \mathbb{Z}_{\geq 0}} M_{\mathcal{R}}(\tilde{c}, \tilde{h})_{\tilde{h}+n}$$

be the Verma module over $\mathfrak{g}_{\mathcal{R}}$ with highest weight $(\tilde{c}, \tilde{h})$ (see § 3.2.1). Let

$$\langle \cdot, \cdot \rangle_{\tilde{c},\tilde{h}} : M_{\mathcal{R}}(\tilde{c}, \tilde{h}) \times M_{\mathcal{R}}(\tilde{c}, \tilde{h}) \longrightarrow \mathcal{R} \tag{5.46}$$

be the contravariant bilinear form on $M_{\mathcal{R}}(\tilde{c}, \tilde{h})$, which is normalised as

$$\langle v_{\tilde{c},\tilde{h}}, v_{\tilde{c},\tilde{h}} \rangle_{\tilde{c},\tilde{h}} = 1,$$

where $v_{\tilde{c},\tilde{h}} := 1 \otimes \mathbf{1}_{\tilde{c},\tilde{h}}$. The contravariance of the form $\langle \cdot, \cdot \rangle_{\tilde{c},\tilde{h}}$ implies that

$$\langle M_{\mathcal{R}}(\tilde{c}, \tilde{h})_{\tilde{h}+n}, M_{\mathcal{R}}(\tilde{c}, \tilde{h})_{\tilde{h}+n'} \rangle_{\tilde{c},\tilde{h}} = \{0\} \quad \text{if } n \neq n'.$$

Hence, we consider the discriminant of the form

$$\langle\cdot,\cdot\rangle_{\tilde{c},\tilde{h}}|_{M_{\mathcal{R}}(\tilde{c},\tilde{h})_{\tilde{h}+n}\times M_{\mathcal{R}}(\tilde{c},\tilde{h})_{\tilde{h}+n}}.$$

Remark that the discriminant is determined up to a unit of $\mathcal{R}$. By Theorem 4.2, this discriminant is given by

$$\det(\tilde{c},\tilde{h})_n = \prod_{\substack{\alpha,\beta\in\mathbb{Z}_{>0}\\ \alpha\geq\beta\\ 1\leq\alpha\beta\leq n}} \varPhi_{\alpha,\beta}(\tilde{c},\tilde{h})^{p(n-\alpha\beta)}. \tag{5.47}$$

5.5.2 Character Sum Formula

To deal with the character sum of the Jantzen filtration, we first introduce some notation. Let $\tilde{\mathcal{E}}$ be the $\mathbb{Z}$-algebra associated to $\mathfrak{h}^*_{\mathcal{R}}$ introduced in § 3.2.1. Here, under the identification $\mathfrak{h}^* \ni \lambda \leftrightarrow (c,h) \in \mathbb{C}^2$, where $\lambda(C) = c$ and $\lambda(L_0) = h$, we denote $e(\lambda) \in \tilde{\mathcal{E}}$ by $e(c,h)$.

We set $c := \phi_\xi(\tilde{c})$ and $h := \phi_\xi(\tilde{h})$, and introduce the Jantzen filtration of $M(c,h)(\simeq \phi_\xi M_{\mathcal{R}}(\tilde{c},\tilde{h}))$.

To define the Jantzen filtration on the Verma module $M(c,h)$, the following assumption is important (see § 3.2.2):

Assumption: the contravariant form $\langle\cdot,\cdot\rangle_{\tilde{c},\tilde{h}}$ (on $M_{Q(\mathcal{R})}(\tilde{c},\tilde{h})$) is non-degenerate.

Under this assumption, for $l \in \mathbb{Z}_{>0}$, we set

$$M_{\mathcal{R}}(\tilde{c},\tilde{h})(l) := \{u \in M_{\mathcal{R}}(\tilde{c},\tilde{h}) | \operatorname{ord}_\xi\langle u,v\rangle_{\tilde{c},\tilde{h}} \geq l \ (\forall v \in M_{\mathcal{R}}(\tilde{c},\tilde{h}) \ )\},$$

and

$$M(c,h)(l) := \operatorname{Im}\phi_\xi(\iota_l),$$

where $\iota_l : M_{\mathcal{R}}(\tilde{c},\tilde{h})(l) \hookrightarrow M_{\mathcal{R}}(\tilde{c},\tilde{h})$ is the inclusion. Then, $\{M(c,h)(l)|l \in \mathbb{Z}_{>0}\}$ gives a filtration

$$M(c,h) \supset M(c,h)(1) \supset M(c,h)(2) \supset \cdots \tag{5.48}$$

of $M(c,h)$.

Proposition 5.8

$$\sum_{l\in\mathbb{Z}_{>0}} \operatorname{ch} M(c,h)(l) = \sum_{(\alpha,\beta)\in\tilde{D}(c,h)} \operatorname{ord}_\xi \varPhi_{\alpha,\beta}(\tilde{c},\tilde{h}) \times \operatorname{ch} M(c,h+\alpha\beta),$$

where $\tilde{D}(c,h)$ is defined in (5.1).

Proof. Proposition 3.6 says that

$$\mathrm{ord}_\xi \det(\tilde{c},\tilde{h})_n = \sum_{l\in\mathbb{Z}_{>0}} \dim M(c,h)(l)_{h+n}.$$

Hence, by (5.47), we have

$$\begin{aligned}\sum_{l\in\mathbb{Z}_{>0}} \mathrm{ch}\, M(c,h)(l) &= \sum_{n\in\mathbb{Z}_{\geq 0}} \mathrm{ord}_\xi \det(\tilde{c},\tilde{h})_n\, e(c,h+n)\\ &= \sum_{n\in\mathbb{Z}_{\geq 0}} \mathrm{ord}_\xi \prod_{\substack{\alpha,\beta\in\mathbb{Z}_{>0}\\ \alpha\geq\beta\\ 1\leq\alpha\beta\leq n}} \Phi_{\alpha,\beta}(\tilde{c},\tilde{h})^{p(n-\alpha\beta)}\, e(c,h+n).\end{aligned}$$

Since $p(n-\alpha\beta)=0$ if $\alpha\beta>n$, we obtain

$$\begin{aligned}\sum_{l\in\mathbb{Z}_{>0}} \mathrm{ch}\, M(c,h)(l) &= \sum_{\substack{\alpha,\beta\in\mathbb{Z}_{>0}\\ \alpha\geq\beta}} \mathrm{ord}_\xi\, \Phi_{\alpha,\beta}(\tilde{c},\tilde{h}) \times \sum_{n\in\mathbb{Z}_{\geq 0}} p(n-\alpha\beta)q^{h+n}\\ &= \sum_{\substack{\alpha,\beta\in\mathbb{Z}_{>0}\\ \alpha\geq\beta}} \mathrm{ord}_\xi\, \Phi_{\alpha,\beta}(\tilde{c},\tilde{h}) \times \mathrm{ch}\, M(c,h+\alpha\beta).\end{aligned}$$

Since $\mathrm{ord}_\xi\, \Phi_{\alpha,\beta}(\tilde{c},\tilde{h})=0$ if $\Phi_{\alpha,\beta}(c,h)\neq 0$, the proposition holds. □

In the sequel, for each $(c,h)\in\mathbb{C}^2$, we fix $(\tilde{c},\tilde{h})\in\mathcal{R}^2$ as

$$(\tilde{c},\tilde{h}) := \begin{cases}(c+\xi,h+\xi) & \text{if } c=1,25\\ (c,h+\xi) & \text{if } c\neq 1,25\end{cases}. \tag{5.49}$$

Remark 5.10 *We choose the above 'perturbation' (5.49) of (c,h) so that any curve $\mathcal{V}_{\alpha,\beta}$ (defined in (4.19)), which passes through the point (c,h), transversally intersects with the line $\{(\tilde{c},\tilde{h})|\xi\in\mathbb{C}\}$ at (c,h).*

One can directly check the lemma below.

Lemma 5.20. *For any $\alpha,\beta\in\mathbb{Z}_{>0}$,*

$$\mathrm{ord}_\xi\, \Phi_{\alpha,\beta}(\tilde{c},\tilde{h}) = \begin{cases}1 & \text{if } \Phi_{\alpha,\beta}(c,h)=0\\ 0 & \text{if } \Phi_{\alpha,\beta}(c,h)\neq 0\end{cases}.$$

Hence, the following holds:

Proposition 5.9 *For each highest weight $(c,h)\in\mathbb{C}^2$, we have*

$$\sum_{l\in\mathbb{Z}_{>0}} \mathrm{ch}\, M(c,h)(l) = \sum_{n\in D(c,h)} a(n)\, \mathrm{ch}\, M(c,h+n), \tag{5.50}$$

where $D(c,h)$ and $a(n)$ are defined in (5.2) and (5.3).

5.5.3 Explicit Forms

In this subsection, we list the explicit forms of character sums (5.50).

Class V: In this case, the right-hand side of (5.50) is trivial. Hence, we have
Lemma 5.21. *Suppose that (c,h) belongs to* **Class** V. *Then,*

$$\sum_{l\in\mathbb{Z}_{>0}} \mathrm{ch}\, M(c,h)(l) = 0.$$

Class I. In this case, (c,h) can be written as $c = c(t)$ and $h = h_{\alpha,\beta}(t)$ for some $\alpha, \beta \in \mathbb{Z}_{>0}$ and $t \in \mathbb{C}\backslash\mathbb{Q}$. It was stated in § 5.1.1 that $\tilde{D}(c,h) = \{(\alpha,\beta)\}$.

Lemma 5.22. *Suppose that $(c,h) = (c(t), h_{\alpha,\beta}(t))$ belongs to* **Class** I. *Then,*

$$\sum_{l\in\mathbb{Z}_{>0}} \mathrm{ch}\, M(c,h)(l) = \mathrm{ch}\, M(c, h+\alpha\beta).$$

Class $R^{\pm}$. As consequences of Lemmas 5.11 and 5.14, we obtain the following character sum formulae of Jantzen filtrations of Verma modules.

Lemma 5.23. *Suppose that highest weight $(c,h) = (c_{p,\pm q}, h_i)$ belongs to* **Class** $R^{\pm}$. *For each (c,h), the character sum of Jantzen filtration $\{M(c,h)(l) | l \in \mathbb{Z}_{>0}\}$ is given as follows:*

1. **Class** R^+: $c = c_{p,q}$,

I. **Case** 1^+: $h = h_i$ ($i \in \mathbb{Z}$),

$$\sum_{l>0} \mathrm{ch}\, M(c,h)(l) = \sum_{\substack{|k|>|i| \\ k-i\equiv 1 \mod 2}} \mathrm{ch}\, M(c,h_k),$$

II. **Case** 2^+: $h = h_i$ ($i \in \mathbb{Z}_{\geq 0}$),

$$\sum_{l>0} \mathrm{ch}\, M(c,h)(l) = \sum_{k>i} \mathrm{ch}\, M(c,h_k),$$

III. **Case** 3^+: $h = h_{(-1)^{i-1}i}$ ($i \in \mathbb{Z}_{\geq 0}$),

$$\sum_{l>0} \mathrm{ch}\, M(c,h)(l) = \sum_{k>i} \mathrm{ch}\, M(c,h_{(-1)^{k-1}k}),$$

IV. **Case** 4^+: $h = h_{2i}$ ($i \in \mathbb{Z}_{\geq 0}$),

i. $(p,q) \neq (1,1) \wedge (s = q \vee i \neq 0)$,

$$\sum_{l>0} \mathrm{ch}\, M(c,h)(l) = 2\sum_{k>i} \mathrm{ch}\, M(c,h_{2k}),$$

ii. $(p,q)=(1,1)\vee(s=0\wedge i=0)$,

$$\sum_{l>0}\operatorname{ch}M(c,h)(l)=\sum_{k>i}\operatorname{ch}M(c,h_{2k}),$$

2. **Class** R^-: $c=c_{p,-q}$,

I. **Case** 1^-: $h=h_i\ (\ i\in\mathbb{Z}\setminus\{0\}\ )$,

$$\sum_{l>0}\operatorname{ch}M(c,h)(l)=\sum_{\substack{|k|<|i|\\ k-i\equiv 1 \mod 2}}\operatorname{ch}M(c,h_k),$$

II. **Case** 2^-: $h=h_i\ (\ i\in\mathbb{Z}_{>0}\ )$,

$$\sum_{l>0}\operatorname{ch}M(c,h)(l)=\sum_{0\leq k<i}\operatorname{ch}M(c,h_k),$$

III. **Case** 3^-: $h=h_{(-1)^{i-1}i}\ (\ i\in\mathbb{Z}_{>0}\ )$,

$$\sum_{l>0}\operatorname{ch}M(c,h)(l)=\sum_{0\leq k<i}\operatorname{ch}M(c,h_{(-1)^{k-1}k}),$$

IV. **Case** 4^-: $h=h_{2i}\ (\ i\in\mathbb{Z}_{>0}\ )$,

i. $(p,q)\neq(1,1)\wedge s=0$,

$$\sum_{l>0}\operatorname{ch}M(c,h)(l)=2\sum_{0<k<i}\operatorname{ch}M(c,h_{2k})+\operatorname{ch}M(c,h_0),$$

ii. $(p,q)\neq(1,1)\wedge s=-q$,

$$\sum_{l>0}\operatorname{ch}M(c,h)(l)=2\sum_{0\leq k<i}\operatorname{ch}M(c,h_{2k}),$$

iii. $(p,q)=(1,1)$,

$$\sum_{l>0}\operatorname{ch}M(c,h)(l)=\sum_{0\leq k<i}\operatorname{ch}M(c,h_{2k}).$$

5.6 Character Sums of the Jantzen Filtration of Quotient Modules

In this section, following the idea stated in § 3.5, we compute the character sums of the Jantzen filtration of quotients of Verma modules. The results of this section will be applied only to **Case** $4^\pm$.

5.6.1 Integral Forms of Quotient Modules

Let $\mathcal{A}$ and $\mathcal{R}$ be the following commutative algebra over $\mathbb{C}$:

$$\mathcal{R} := \{f \in \mathbb{C}(\xi) | f \text{ has no pole at } \xi = 0\}, \quad \mathcal{A} := Q(\mathcal{R}) \otimes_{\mathbb{C}} \mathbb{C}[[\zeta]],$$

and let $\phi_\xi : \mathcal{R} \to \mathcal{R}/\xi\mathcal{R}(\simeq \mathbb{C})$ (resp. $\phi_\zeta : \mathcal{A} \to \mathcal{A}/\zeta\mathcal{A}(\simeq Q(\mathcal{R})))$ be the canonical projections. We denote the reduction functors

$$\begin{aligned} \mathrm{Mod}_{\mathcal{R}} &\longrightarrow \mathrm{Vect}_{\mathbb{C}} \quad (M \mapsto M \otimes_{\mathcal{R}} (\mathcal{R}/\xi\mathcal{R})), \\ \mathrm{Mod}_{\mathcal{A}} &\longrightarrow \mathrm{Vect}_{Q(\mathcal{R})} \quad (N \mapsto N \otimes_{\mathcal{A}} (\mathcal{A}/\zeta\mathcal{A})) \end{aligned}$$

by the same notation ϕ_ξ and ϕ_ζ respectively. For an element v of $M \in \mathrm{Ob}(\mathrm{Mod}_{\mathcal{R}})$ (resp. an element w of $N \in \mathrm{Ob}(\mathrm{Mod}_{\mathcal{A}})$), we set $\phi_\xi v := v \otimes_{\mathcal{R}} 1 \in \phi_\xi(M)$ (resp. $\phi_\zeta w := w \otimes_{\mathcal{A}} 1 \in \phi_\zeta(N)$). We denote the ξ-adic valuation on $\mathcal{R}$ (resp. ζ-adic valuation on $\mathcal{A}$) by ord_ξ (resp. ord_ζ).

Until the end of the next subsection, we fix $(c,h) \in \mathbb{C}^2 \simeq \mathfrak{h}^*$ such that $M(c,h)$ is reducible. Then, there exist $\alpha, \beta \in \mathbb{Z}_{>0}$ and $\xi_0 \in \mathbb{C} \setminus \{0\}$ such that $(c,h) = (c(\xi_0), h_{\alpha,\beta}(\xi_0))$. Here, we also fix α, β and ξ_0.

Notice that, by Corollary 5.2, there exists an embedding $M(c, h+\alpha\beta) \hookrightarrow M(c,h)$. In the sequel, we introduce the Jantzen filtration of

$$\overline{M}(c,h) := M(c,h)/M(c,h+\alpha\beta),$$

and describe its character sum. To define the Jantzen filtration, we consider the following perturbation of highest weight (c,h) along the curve $\Phi_{\alpha,\beta}(c,h) = 0$:

$$(\tilde{c}, \tilde{h}) := (c(\xi_0 + \xi), h_{\alpha,\beta}(\xi_0 + \xi)) \in \mathcal{R}^2 \simeq \mathfrak{h}^*_{\mathcal{R}}. \tag{5.51}$$

For simplicity, we set $n := \alpha\beta$ and denote $\Phi_{\alpha,\beta}(c,h)$ by $\Phi(c,h)$ or Φ. Then, Proposition 5.2 implies that there exists the Shapovalov element

$$S_{n,\Phi} = e_{\mathbb{I}_0} + \sum_{\mathbb{I} \in \mathcal{P}_n \setminus \{\mathbb{I}_0\}} e_{\mathbb{I}} H_{\Phi,\mathbb{I}} \quad (H_{\Phi,\mathbb{I}} \in U(\mathfrak{h})),$$

where we set $\mathbb{I}_0 := (1^n)$.

We set $v_{(\tilde{c},\tilde{h}):n} := S_{n,\Phi} v_{\tilde{c},\tilde{h}} \in M_{\mathcal{R}}(\tilde{c},\tilde{h})_{\tilde{h}+n}$ and define $\overline{M}_{\mathcal{R}}(\tilde{c},\tilde{h})$ by

$$\overline{M}_{\mathcal{R}}(\tilde{c},\tilde{h}) := M_{\mathcal{R}}(\tilde{c},\tilde{h}) \Big/ U(\mathfrak{g})_{\mathcal{R}} v_{(\tilde{c},\tilde{h}):n} \,.$$

Moreover, for $k \in \mathbb{Z}_{\geq 0}$, we set

$$\overline{M}_{\mathcal{R}}(\tilde{c},\tilde{h})_{\tilde{h}+k} := (M_{\mathcal{R}}(\tilde{c},\tilde{h})_{\tilde{h}+k} + U(\mathfrak{g})_{\mathcal{R}} v_{(\tilde{c},\tilde{h}):n})/U(\mathfrak{g})_{\mathcal{R}} v_{(\tilde{c},\tilde{h}):n}$$

and

$$\overline{v}_{\tilde{c},\tilde{h}} := v_{\tilde{c},\tilde{h}} + U(\mathfrak{g})_{\mathcal{R}} v_{\tilde{c},\tilde{h}} \in \overline{M}_{\mathcal{R}}(\tilde{c},\tilde{h})_{\tilde{h}}. \tag{5.52}$$

Then, we have

$$\overline{M}_{\mathcal{R}}(\tilde{c},\tilde{h}) = \bigoplus_{k\in\mathbb{Z}_{\geq 0}} \overline{M}_{\mathcal{R}}(\tilde{c},\tilde{h})_{\tilde{h}+k}$$

and $\overline{M}_{\mathcal{R}}(\tilde{c},\tilde{h})_{\tilde{h}} = \mathcal{R}\overline{v}_{\tilde{c},\tilde{h}}$.

We show that each weight subspace $\overline{M}_{\mathcal{R}}(\tilde{c},\tilde{h})_{\tilde{h}+k}$ is an $\mathcal{R}$-free module of finite rank (cf. [Ja1]).

Lemma 5.24. *For each $k \in \mathbb{Z}_{\geq 0}$, set*

$$\overline{\mathcal{P}}_{k:n} := \{\mathbb{I} = (1^{n_1}2^{n_2}\cdots) \in \mathcal{P}_k | n_1 < n\}.$$

Then, $\overline{M}_{\mathcal{R}}(\tilde{c},\tilde{h})_{\tilde{h}+k}$ is an $\mathcal{R}$-free module with an $\mathcal{R}$-free basis $\{e_{\mathbb{I}}.\overline{v}_{\tilde{c},\tilde{h}} | \mathbb{I} \in \overline{\mathcal{P}}_{k:n}\}$.

Proof. By definition, one can show that

$$e_{\mathbb{J}}.v_{(\tilde{c},\tilde{h}):n} \in e_{\mathbb{I}_0+\mathbb{J}} v_{\tilde{c},\tilde{h}} + \sum_{\mathbb{I}<\mathbb{I}_0+\mathbb{J}} \mathcal{R} e_{\mathbb{I}} v_{\tilde{c},\tilde{h}}.$$

This formula implies that if $k > n$ then the determinant of the transition matrix from the $\mathcal{R}$-free basis $\{e_{\mathbb{I}}.v_{\tilde{c},\tilde{h}} | \mathbb{I} \in \mathcal{P}_k\}$ to

$$\{e_{\mathbb{I}'} v_{\tilde{c},\tilde{h}} | \mathbb{I}' \in \overline{\mathcal{P}}_{k:n}\} \sqcup \{e_{\mathbb{I}''} v_{(\tilde{c},\tilde{h}):n} | \mathbb{I}'' \in \mathcal{P}_{k-n}\} \tag{5.53}$$

is equal to 1. Hence, (5.53) forms an $\mathcal{R}$-free basis of $M_{\mathcal{R}}(\tilde{c},\tilde{h})_{\tilde{h}+k}$, and thus, the lemma follows. □

This lemma implies that

$$\phi_\xi \overline{M}_{\mathcal{R}}(\tilde{c},\tilde{h}) \simeq \overline{M}(c,h).$$

5.6.2 Definition

We set

$$(\tilde{\tilde{c}},\tilde{\tilde{h}}) := \begin{cases} (\tilde{c}, \tilde{h}+\zeta) & \text{if } \ c(\xi_0) \neq 1, 25 \\ (\tilde{c}+\zeta, \tilde{h}+\zeta) & \text{if } \ c(\xi_0) = 1, 25 \end{cases}. \tag{5.54}$$

Then, $(\tilde{\tilde{c}},\tilde{\tilde{h}}) \in \mathcal{A}^2 \simeq \mathfrak{h}^*_{\mathcal{A}}$ and $(\phi_\zeta(\tilde{\tilde{c}}), \phi_\zeta(\tilde{\tilde{h}})) = (\tilde{c},\tilde{h})$. Since $\Phi_{\alpha'\beta'}(\tilde{\tilde{c}},\tilde{\tilde{h}}) \neq 0$ for any α', $\beta' \in \mathbb{Z}_{>0}$, the following lemma holds.

Lemma 5.25. *$M_{Q(\mathcal{A})}(\tilde{\tilde{c}},\tilde{\tilde{h}})$ is an irreducible $\mathfrak{g}_{Q(\mathcal{A})}$-module.*

For $l \in \mathbb{Z}_{>0}$, we set

$$M_{\mathcal{A}}(\approx{c},\approx{h})(l) := \left\{ v \in M_{\mathcal{A}}(\approx{c},\approx{h}) \,\middle|\, \mathrm{ord}_{\zeta}\langle v, w\rangle_{\approx{c},\approx{h}} \geq l \ \ (\forall w \in M_{\mathcal{A}}(\approx{c},\approx{h})) \right\}, \tag{5.55}$$

and define the Jantzen filtration of $M_{Q(\mathcal{R})}(\tilde{c},\tilde{h})$ by setting

$$M_{Q(\mathcal{R})}(\tilde{c},\tilde{h})(l) := \mathrm{Im}\phi_{\zeta}(\iota_l),$$

where $\iota_l : M_{\mathcal{A}}(\approx{c},\approx{h})(l) \hookrightarrow M_{\mathcal{A}}(\approx{c},\approx{h})$ is the inclusion. Using this filtration, one can show the following lemma.

Lemma 5.26. *1.* $(U(\mathfrak{g})_{\mathcal{R}} v_{(\tilde{c},\tilde{h}):n}) \otimes_{\mathcal{R}} Q(\mathcal{R})$ *is an irreducible* $\mathfrak{g}_{Q(\mathcal{R})}$*-module.*
2. $\overline{M}_{\mathcal{R}}(\tilde{c},\tilde{h}) \otimes_{\mathcal{R}} Q(\mathcal{R})$ *is an irreducible* $\mathfrak{g}_{Q(\mathcal{R})}$*-module.*

Proof. By direct computation, we have $\Phi_{\alpha',\beta'}(\tilde{c},\tilde{h}+n) \neq 0$ for any α', $\beta' \in \mathbb{Z}_{>0}$. Hence, by Proposition 3.4, $M_{Q(\mathcal{R})}(\tilde{c},\tilde{h}+n)$ is irreducible. Since $(U(\mathfrak{g})_{\mathcal{R}} v_{(\tilde{c},\tilde{h}):n}) \otimes_{\mathcal{R}} Q(\mathcal{R})$ is a highest weight $\mathfrak{g}_{Q(\mathcal{R})}$-module with highest weight $(\tilde{c},\tilde{h}+n)$, we have

$$(U(\mathfrak{g})_{\mathcal{R}} v_{(\tilde{c},\tilde{h}):n}) \otimes_{\mathcal{R}} Q(\mathcal{R}) \simeq M_{Q(\mathcal{R})}(\tilde{c},\tilde{h}+n), \tag{5.56}$$

and thus, the first statement follows.

To show the second statement, we remark the following two facts:

1. $\Phi_{\alpha',\beta'}(\tilde{c},\tilde{h}) \neq 0$ for any $(\alpha',\beta') \neq (\alpha,\beta)$,
2. $\mathrm{ord}_{\zeta}\, \Phi_{\alpha,\beta}(\approx{c},\approx{h}) = 1$.

Hence, similarly to Lemma 5.22, we have

$$\sum_{l=1}^{\infty} \mathrm{ch}\, M_{Q(\mathcal{R})}(\tilde{c},\tilde{h})(l) = \mathrm{ch}\, M_{Q(\mathcal{R})}(\tilde{c},\tilde{h}+n) \in \tilde{\mathcal{E}}. \tag{5.57}$$

This implies that

$$(U(\mathfrak{g})_{\mathcal{R}} v_{(\tilde{c},\tilde{h}):n}) \otimes_{\mathcal{R}} Q(\mathcal{R}) = M_{Q(\mathcal{R})}(\tilde{c},\tilde{h})(1).$$

Indeed, since by Proposition 3.5, $M_{Q(\mathcal{R})}(\tilde{c},\tilde{h})(1)$ is the the maximal proper submodule of $M_{Q(\mathcal{R})}(\tilde{c},\tilde{h})$, we have

$$(U(\mathfrak{g})_{\mathcal{R}} v_{(\tilde{c},\tilde{h}):n}) \otimes_{\mathcal{R}} Q(\mathcal{R}) \subset M_{Q(\mathcal{R})}(\tilde{c},\tilde{h})(1).$$

Hence, by (5.56) and (5.57), we obtain

$$M_{Q(\mathcal{R})}(\tilde{c},\tilde{h})(l) = \begin{cases} M_{Q(\mathcal{R})}(\tilde{c},\tilde{h}+n) & l = 1 \\ 0 & l > 1 \end{cases}. \tag{5.58}$$

Hence, $(U(\mathfrak{g})_{\mathcal{R}} v_{(\tilde{c},\tilde{h}):n}) \otimes_{\mathcal{R}} Q(\mathcal{R})$ is the maximal proper submodule of $M_{Q(\mathcal{R})}(\tilde{c},\tilde{h})$. The second statement follows. □

By the above argument, we have

$$\mathrm{rad}\langle\ ,\ \rangle_{\tilde{c},\tilde{h}} = U(\mathfrak{g})_{Q(\mathcal{R})}.v_{(\tilde{c},\tilde{h}):n},$$

where $\langle\ ,\ \rangle_{\tilde{c},\tilde{h}}$ is the contravariant form on $M_{Q(\mathcal{R})}(\tilde{c},\tilde{h})$. Hence, $\langle\ ,\ \rangle_{\tilde{c},\tilde{h}}$ induces a non-degenerate contravariant bilinear form on $\overline{M}_{\mathcal{R}}(\tilde{c},\tilde{h}) \otimes_{\mathcal{R}} Q(\mathcal{R})$. We denote this contravariant form by the same notation $\langle\ ,\ \rangle_{\tilde{c},\tilde{h}}$. We set

$$\overline{\det}(\tilde{c},\tilde{h})_k := \det\left(\langle e_{\mathbb{I}}\overline{v}_{\tilde{c},\tilde{h}}, e_{\mathbb{J}}\overline{v}_{\tilde{c},\tilde{h}}\rangle_{\tilde{c},\tilde{h}}\right)_{\mathbb{I},\mathbb{J}\in\overline{\mathcal{P}}_{k:n}}.$$

Let us define the Jantzen filtration of $\overline{M}(c,h)$. For $l \in \mathbb{Z}_{\geq 0}$, we set

$$\overline{M}_{\mathcal{R}}(\tilde{c},\tilde{h})(l) := \{v \in \overline{M}_{\mathcal{R}}(\tilde{c},\tilde{h}) | \langle v, w\rangle_{\tilde{c},\tilde{h}} \in \xi^l\mathcal{R}\ (\forall w \in \overline{M}_{\mathcal{R}}(\tilde{c},\tilde{h}))\},$$

and

$$\overline{M}(c,h)(l) := \mathrm{Im}\phi_\xi(\iota_l),$$

where $\iota_l : \overline{M}_{\mathcal{R}}(\tilde{c},\tilde{h})(l) \hookrightarrow \overline{M}_{\mathcal{R}}(\tilde{c},\tilde{h})$ is the inclusion. Then, we obtain the **Jantzen filtration**

$$\overline{M}(c,h) \supset \overline{M}(c,h)(1) \supset \overline{M}(c,h)(2) \supset \cdots \tag{5.59}$$

of $\overline{M}(c,h)$ (associated with the perturbation $(\tilde{c},\tilde{h}) \in \mathfrak{h}^*_{\mathcal{R}}$). Since $\mathcal{R}$ is a principal integral domain, by Proposition 3.5 and Proposition 3.6, we obtain

Proposition 5.10 *The filtration (5.59) satisfies the following:*

1. $\overline{M}(c,h)(1)$ *coincides with the maximal proper submodule of* $\overline{M}(c,h)$.
2. *For each* $k \in \mathbb{Z}_{\geq 0}$, *there exists a non-degenerate contravariant form on* $\overline{M}(c,h)(k)/\overline{M}(c,h)(k+1)$.
3. *For any* $k \in \mathbb{Z}_{\geq 0}$,

$$\mathrm{ord}_\xi(\overline{\det}(\tilde{c},\tilde{h})_k) = \sum_{l=1}^{\infty} \dim \overline{M}(c,h)(l)_{h+k}.$$

5.6.3 Character Sum Formula

In order to compute the character sum of the Jantzen filtration of $\overline{M}(c,h)$, we describe the valuation $\mathrm{ord}_\xi(\overline{\det}(\tilde{c},\tilde{h})_k)$ by using the determinants of the contravariant forms on $M_{\mathcal{R}}(\tilde{c},\tilde{h})$ and $M_{\mathcal{R}}(\tilde{c},\tilde{h}+n)$.

We start with two technical lemmas.

Lemma 5.27. *1.* $S_{n,\Phi}v_{\approx{c},\approx{h}} \in M_{\mathcal{A}}(\approx{c},\approx{h})(1) \setminus M_{\mathcal{A}}(\approx{c},\approx{h})(2)$.

2. For any $x \in \mathfrak{g}_{\mathcal{A}}^{+}$*, we have* $x.S_{n,\Phi}v_{\approx{c},\approx{h}} \in \zeta M_{\mathcal{A}}(\approx{c},\approx{h})$.

Proof. We show the first statement. Since $S_{n,\Phi} \in U(\mathfrak{g})$, we have

$$\phi_\zeta S_{n,\Phi}v_{\approx{c},\approx{h}} = S_{n,\Phi}\phi_\zeta v_{\approx{c},\approx{h}} = S_{n,\Phi}v_{\tilde{c},\tilde{h}} \in M_{Q(\mathcal{R})}(\tilde{c},\tilde{h})^{\mathfrak{g}_{\mathcal{R}}^{+}} \setminus Q(\mathcal{R})v_{\tilde{c},\tilde{h}},$$

Since $\phi_\zeta S_{n,\Phi}v_{\approx{c},\approx{h}}$ is a non-zero singular vector and is not a highest weight vector, we have $S_{n,\Phi}v_{\approx{c},\approx{h}} \in M_{\mathcal{A}}(\approx{c},\approx{h})(1)$. Hence, it suffices to see that

$$S_{n,\Phi}v_{\approx{c},\approx{h}} \notin M_{\mathcal{A}}(\approx{c},\approx{h})(2). \tag{5.60}$$

By (5.57), we have

$$\begin{aligned}\sum_{l=1}^{\infty} \mathrm{ch}\, \mathrm{Im}\phi_\zeta(\iota_l) &= \sum_{l=1}^{\infty} \mathrm{ch}\, M_{Q(\mathcal{R})}(\tilde{c},\tilde{h})(l) \\ &= \mathrm{ch}\, M_{Q(\mathcal{R})}(\tilde{c},\tilde{h}+n).\end{aligned} \tag{5.61}$$

On the other hand, if $S_{n,\Phi}v_{\approx{c},\approx{h}} \in M_{\mathcal{A}}(\approx{c},\approx{h})(2)$ holds, then

$$\sum_{l=1}^{\infty} \dim_{Q(\mathcal{R})} M_{Q(\mathcal{R})}(\tilde{c},\tilde{h})(l)_{\tilde{h}+n} \geq 2,$$

where $M_{Q(\mathcal{R})}(\tilde{c},\tilde{h})(l)_{\tilde{h}+n} := M_{Q(\mathcal{R})}(\tilde{c},\tilde{h})(l) \cap M_{Q(\mathcal{R})}(\tilde{c},\tilde{h})_{\tilde{h}+n}$. This contradicts (5.61). We have proved the first statement.

For $x \in \mathfrak{g}_{\mathcal{A}}^{+}$, we have

$$\phi_\zeta x.S_{n,\Phi}v_{\approx{c},\approx{h}} = (\phi_\zeta x).S_{n,\Phi}v_{\tilde{c},\tilde{h}} = 0,$$

since $\phi_\zeta x \in \mathfrak{g}_{Q(\mathcal{R})}^{+}$ and $S_{n,\Phi}v_{\tilde{c},\tilde{h}}$ is a singular vector. Hence, the second statement holds. □

Lemma 5.28. *1.* $\mathrm{ord}_\zeta(\langle S_{n,\Phi}v_{\approx{c},\approx{h}}, S_{n,\Phi}v_{\approx{c},\approx{h}}\rangle_{\approx{c},\approx{h}}) = 1$.
2. Suppose that $\det(c,h)_k \neq 0$ *for* $k < n$*. Then,*

$$\mathrm{ord}_\xi\left(\phi_\zeta(\zeta^{-1}\langle S_{n,\Phi}v_{\approx{c},\approx{h}}, S_{n,\Phi}v_{\approx{c},\approx{h}}\rangle_{\approx{c},\approx{h}})\right) = \mathrm{ord}_\xi\left(\phi_\zeta\left(\zeta^{-1}\det(\approx{c},\approx{h})_n\right)\right). \tag{5.62}$$

Proof. We show the first statement. By Lemma 5.27, we have

$$\langle S_{n,\Phi} v_{\widetilde{\widetilde{c}},\widetilde{\widetilde{h}}}, S_{n,\Phi} v_{\widetilde{\widetilde{c}},\widetilde{\widetilde{h}}} \rangle_{\widetilde{\widetilde{c}},\widetilde{\widetilde{h}}} \in \zeta \mathcal{A}.$$

Hence, it suffices to show that $\langle S_{n,\Phi} v_{\widetilde{\widetilde{c}},\widetilde{\widetilde{h}}}, S_{n,\Phi} v_{\widetilde{\widetilde{c}},\widetilde{\widetilde{h}}} \rangle_{\widetilde{\widetilde{c}},\widetilde{\widetilde{h}}} \notin \zeta^2 \mathcal{A}$.

Recall that by the proof of Lemma 5.24,

$$\{S_{n,\Phi} v_{\widetilde{\widetilde{c}},\widetilde{\widetilde{h}}}\} \cup \{e_{\mathbb{I}} v_{\widetilde{\widetilde{c}},\widetilde{\widetilde{h}}} | \mathbb{I} \in \mathcal{P}_n \setminus \{\mathbb{I}_0\} \}$$

forms an $\mathcal{A}$-free basis of $M_{\mathcal{A}}(\widetilde{\widetilde{c}}, \widetilde{\widetilde{h}})_{\widetilde{\widetilde{h}}+n}$. Hence, up to $\mathcal{A}^\times$ (the set of the units of $\mathcal{A}$), $\det(\widetilde{\widetilde{c}}, \widetilde{\widetilde{h}})_n$ is expressed as

$$\det(\widetilde{\widetilde{c}}, \widetilde{\widetilde{h}})_n = \det \left[\begin{array}{c|c} \langle S_{n,\Phi} v_{\widetilde{\widetilde{c}},\widetilde{\widetilde{h}}}, S_{n,\Phi} v_{\widetilde{\widetilde{c}},\widetilde{\widetilde{h}}} \rangle_{\widetilde{\widetilde{c}},\widetilde{\widetilde{h}}} & A \\ \hline B & C \end{array} \right], \tag{5.63}$$

where A, B and C are $1 \times (p(n)-1)$, $(p(n)-1) \times 1$ and $(p(n)-1) \times (p(n)-1)$ blocks ($p(n) := \sharp \mathcal{P}_n$ is the partition number of n) given by

$$\begin{aligned} A &:= (\langle S_{n,\Phi} v_{\widetilde{\widetilde{c}},\widetilde{\widetilde{h}}}, e_{\mathbb{J}} v_{\widetilde{\widetilde{c}},\widetilde{\widetilde{h}}} \rangle_{\widetilde{\widetilde{c}},\widetilde{\widetilde{h}}})_{\mathbb{J} \in \mathcal{P}_n \setminus \{\mathbb{I}_0\}}, \\ B &:= (\langle e_{\mathbb{I}} v_{\widetilde{\widetilde{c}},\widetilde{\widetilde{h}}}, S_{n,\Phi} v_{\widetilde{\widetilde{c}},\widetilde{\widetilde{h}}} \rangle_{\widetilde{\widetilde{c}},\widetilde{\widetilde{h}}})_{\mathbb{I} \in \mathcal{P}_n \setminus \{\mathbb{I}_0\}}, \\ C &:= (\langle e_{\mathbb{I}} v_{\widetilde{\widetilde{c}},\widetilde{\widetilde{h}}}, e_{\mathbb{J}} v_{\widetilde{\widetilde{c}},\widetilde{\widetilde{h}}} \rangle_{\widetilde{\widetilde{c}},\widetilde{\widetilde{h}}})_{\mathbb{I},\mathbb{J} \in \mathcal{P}_n \setminus \{\mathbb{I}_0\}}. \end{aligned}$$

From Lemma 5.27. 1, we see that

$$\langle S_{n,\Phi} v_{\widetilde{\widetilde{c}},\widetilde{\widetilde{h}}}, e_{\mathbb{I}} v_{\widetilde{\widetilde{c}},\widetilde{\widetilde{h}}} \rangle_{\widetilde{\widetilde{c}},\widetilde{\widetilde{h}}} \in \zeta \mathcal{A}. \tag{5.64}$$

Hence, if $\langle S_{n,\Phi} v_{\widetilde{\widetilde{c}},\widetilde{\widetilde{h}}}, S_{n,\Phi} v_{\widetilde{\widetilde{c}},\widetilde{\widetilde{h}}} \rangle_{\widetilde{\widetilde{c}},\widetilde{\widetilde{h}}} \in \zeta^2 \mathcal{A}$ holds, then by the definition of determinants we have $\det(\widetilde{\widetilde{c}}, \widetilde{\widetilde{h}})_n \in \zeta^2 \mathcal{A}$.

On the other hand, by direct computation, we have

$$\mathrm{ord}_\zeta(\det(\widetilde{\widetilde{c}}, \widetilde{\widetilde{h}})_n) = 1. \tag{5.65}$$

This is a contradiction. Hence, we have $\langle S_{n,\Phi} v_{\widetilde{\widetilde{c}},\widetilde{\widetilde{h}}}, S_{n,\Phi} v_{\widetilde{\widetilde{c}},\widetilde{\widetilde{h}}} \rangle_{\widetilde{\widetilde{c}},\widetilde{\widetilde{h}}} \notin \zeta^2 \mathcal{A}$ (in particular, $\langle S_{n,\Phi} v_{\widetilde{\widetilde{c}},\widetilde{\widetilde{h}}}, S_{n,\Phi} v_{\widetilde{\widetilde{c}},\widetilde{\widetilde{h}}} \rangle_{\widetilde{\widetilde{c}},\widetilde{\widetilde{h}}} \neq 0$), and thus the first statement holds.

We show the second statement. By (5.63) and (5.64), we have

$$\phi_\zeta \left(\zeta^{-1} \det(\widetilde{\widetilde{c}}, \widetilde{\widetilde{h}})_n \right) = \phi_\zeta \left(\zeta^{-1} \langle S_{n,\Phi} v_{\widetilde{\widetilde{c}},\widetilde{\widetilde{h}}}, S_{n,\Phi} v_{\widetilde{\widetilde{c}},\widetilde{\widetilde{h}}} \rangle_{\widetilde{\widetilde{c}},\widetilde{\widetilde{h}}} \det C \right).$$

On the other hand, we see that

$$\phi_\xi(\phi_\zeta(\det C)) = \det(\langle e_{\mathbb{I}} \overline{v}_{c,h}, e_{\mathbb{J}} \overline{v}_{c,h} \rangle_{c,h})_{\mathbb{I},\mathbb{J} \in \mathcal{P}_n \setminus \{\mathbb{I}_0\}},$$

where $\overline{v}_{c,h}$ is the highest weight vector of $\overline{M}(c,h)$. Here, we notice that under the assumption $\det(c,h)_k \neq 0$ for any $k < n$, the uniqueness of the singular vectors (Proposition 5.1) implies that

$$\mathrm{rad}(\langle\ ,\ \rangle_{c,h}|_{M(c,h)_{h+n}\times M(c,h)_{h+n}}) = \mathbb{C}S_{n,\Phi}v_{c,h}.$$

Hence, we have $\phi_\xi(\phi_\zeta(\det C)) \neq 0$. Now, we have proved the second statement. □

We describe $\mathrm{ord}_\xi(\overline{\det}(\tilde{c},\tilde{h})_k)$ as follows:

Proposition 5.11 *Suppose that* $\det(c,h)_l \neq 0$ *for any* $l < n$.

1. *If* $k \geq n$, *then*

$$\begin{aligned}\mathrm{ord}_\xi(\overline{\det}(\tilde{c},\tilde{h})_k) =& \,\mathrm{ord}_\xi\left(\phi_\zeta(\zeta^{-p(k-n)}\det(\overset{\approx}{c},\overset{\approx}{h})_k)\right)\\ &- mp(k-n) - \mathrm{ord}_\xi(\det(\tilde{c},\tilde{h}+n)_{k-n}),\end{aligned}$$

where we set

$$m := \mathrm{ord}_\xi\left(\phi_\zeta\left(\zeta^{-1}\det(\overset{\approx}{c},\overset{\approx}{h})_n\right)\right).$$

2. *If* $k < n$, *then*

$$\mathrm{ord}_\xi(\overline{\det}(\tilde{c},\tilde{h})_k) = \mathrm{ord}_\xi(\det(\tilde{c},\tilde{h})_k).$$

Proof. In the case where $k < n$, the proposition follows from (5.56) and $M_{Q(\mathcal{R})}(\tilde{c},\tilde{h}+n)_{\tilde{h}+k} = \{0\}$. Hence, we may assume that $k \geq n$.

Note that, up to $\mathcal{A}^\times$, the determinant $\det(\overset{\approx}{c},\overset{\approx}{h})_k$ is expressed as follows:

$$\det(\overset{\approx}{c},\overset{\approx}{h})_k = \det\left[\begin{array}{c|c} A & B \\ \hline C & D \end{array}\right],$$

where the blocks A, B, C and D are given by

$$\begin{aligned}
A &:= (\langle e_{\mathbb{I}}v_{\overset{\approx}{c},\overset{\approx}{h}}, e_{\mathbb{J}}v_{\overset{\approx}{c},\overset{\approx}{h}}\rangle_{\overset{\approx}{c},\overset{\approx}{h}})_{\mathbb{I},\mathbb{J}\in\overline{\mathcal{P}}_{k:n}},\\
B &:= (\langle e_{\mathbb{I}}v_{\overset{\approx}{c},\overset{\approx}{h}}, e_{\mathbb{J}}S_{n,\Phi}v_{\overset{\approx}{c},\overset{\approx}{h}}\rangle_{\overset{\approx}{c},\overset{\approx}{h}})_{\mathbb{I}\in\overline{\mathcal{P}}_{k:n},\mathbb{J}\in\mathcal{P}_{k-n}},\\
C &:= (\langle e_{\mathbb{I}}S_{n,\Phi}v_{\overset{\approx}{c},\overset{\approx}{h}}, e_{\mathbb{J}}v_{\overset{\approx}{c},\overset{\approx}{h}}\rangle_{\overset{\approx}{c},\overset{\approx}{h}})_{\mathbb{I}\in\mathcal{P}_{k-n},\mathbb{J}\in\overline{\mathcal{P}}_{k:n}},\\
D &:= (\langle e_{\mathbb{I}}S_{n,\Phi}v_{\overset{\approx}{c},\overset{\approx}{h}}, e_{\mathbb{J}}S_{n,\Phi}v_{\overset{\approx}{c},\overset{\approx}{h}}\rangle_{\overset{\approx}{c},\overset{\approx}{h}})_{\mathbb{I},\mathbb{J}\in\mathcal{P}_{k-n}},
\end{aligned}$$

and the set $\overline{\mathcal{P}}_{k:n}$ was introduced in Lemma 5.24.

By Lemma 5.27. 1, we see that

$$\langle e_{\mathbb{I}}S_{n,\Phi}v_{\overset{\approx}{c},\overset{\approx}{h}}, e_{\mathbb{J}}v_{\overset{\approx}{c},\overset{\approx}{h}}\rangle_{\overset{\approx}{c},\overset{\approx}{h}} \in \zeta\mathcal{A},$$

and thus, the matrix elements in the blocks in B, C and D are divisible by ζ. Hence, by the definition of determinants, we have

$$\det(\approx{c}, \approx{h})_k = \det A \det D + O(\zeta^{p(k-n)+1}) \tag{5.66}$$

up to $\mathcal{A}^\times$.

From Lemma 5.24, we see that $\phi_\zeta(\det A) = \overline{\det}(\tilde{c}, \tilde{h})_k \neq 0$. Hence, we show that $\det D \in \zeta^{p(k-n)}\mathcal{A}^\times$. Notice that

$$\langle e_{\mathbb{I}} S_{n,\Phi} v_{\approx{c},\approx{h}}, e_{\mathbb{J}} S_{n,\Phi} v_{\approx{c},\approx{h}} \rangle_{\approx{c},\approx{h}} = \langle \sigma(e_{\mathbb{J}}) e_{\mathbb{I}} S_{n,\Phi} v_{\approx{c},\approx{h}}, S_{n,\Phi} v_{\approx{c},\approx{h}} \rangle_{\approx{c},\approx{h}}.$$

Since by the definition of the Shapovalov form F of $\mathfrak{g}$ in § 3.1.1, we have

$$\sigma(e_{\mathbb{J}}) e_{\mathbb{I}} = F(e_{\mathbb{I}}, e_{\mathbb{J}}) + X \quad (\exists X \in \mathfrak{g}^- U(\mathfrak{g}) + U(\mathfrak{g})\mathfrak{g}^+).$$

Moreover, since $\sigma(e_{\mathbb{J}}) e_{\mathbb{I}} \in U(\mathfrak{g})_0$, by the Poincaré–Birkhoff–Witt theorem, we may assume that

$$X \in \mathfrak{g}^- U(\mathfrak{g})\mathfrak{g}^+.$$

Here, we notice that, by Lemma 5.27. 2, for $x_\pm \in \mathfrak{g}^\pm$ and $u \in U(\mathfrak{g})$

$$\langle S_{n,\Phi} v_{\approx{c},\approx{h}}, x_- u x_+ S_{n,\Phi} v_{\approx{c},\approx{h}} \rangle_{\approx{c},\approx{h}} = \langle \sigma(x_-) S_{n,\Phi} v_{\approx{c},\approx{h}}, u x_+ S_{n,\Phi} v_{\approx{c},\approx{h}} \rangle_{\approx{c},\approx{h}} \in \zeta^2 \mathcal{A}$$

holds. Hence, by Lemma 5.28. 1, we have

$$\langle e_{\mathbb{I}} S_{n,\Phi} v_{\approx{c},\approx{h}}, e_{\mathbb{J}} S_{n,\Phi} v_{\approx{c},\approx{h}} \rangle_{\approx{c},\approx{h}} = F(e_{\mathbb{I}}, e_{\mathbb{J}})(\approx{c}, \approx{h}+n) \langle S_{n,\Phi} v_{\approx{c},\approx{h}}, S_{n,\Phi} v_{\approx{c},\approx{h}} \rangle_{\approx{c},\approx{h}} + O(\zeta^2).$$

This formula implies that

$$\det D = (\langle S_{n,\Phi} v_{\approx{c},\approx{h}}, S_{n,\Phi} v_{\approx{c},\approx{h}} \rangle_{\approx{c},\approx{h}})^{p(k-n)} \det(\approx{c}, \approx{h}+n)_{k-n} + O(\zeta^{p(k-n)+1})$$

since $\det(\approx{c}, \approx{h}+n)_{k-n} \in \mathcal{A}^\times$ by Lemma 5.26. Hence, $\det D \in \zeta^{p(k-n)}\mathcal{A}^\times$, and thus,

$$\mathrm{ord}_\zeta(\det A \det D) = p(k-n).$$

Multiplying $\zeta^{-p(k-n)}$ to both sides of (5.66) and applying ϕ_ζ, we have

$$\begin{aligned} &\phi_\zeta\left(\zeta^{-p(k-n)} \det(\approx{c}, \approx{h})_k\right) \\ &= \phi_\zeta(\det A)\phi_\zeta(\det D) \\ &= \overline{\det}(\tilde{c}, \tilde{h})_k \det(\tilde{c}, \tilde{h}+n)_{k-n} \phi_\zeta\left(\zeta^{-1}\langle S_{n,\Phi} v_{\approx{c},\approx{h}}, S_{n,\Phi} v_{\approx{c},\approx{h}} \rangle_{\approx{c},\approx{h}}\right)^{p(n-k)} \end{aligned} \tag{5.67}$$

up to $\mathcal{R}^\times$. By Lemma 5.26, $\overline{\det}(\tilde{c}, \tilde{h})_k \neq 0$ and $\det(\tilde{c}, \tilde{h}+n)_{k-n} \neq 0$. Therefore, from Lemma 5.28. 2, we obtain the conclusion. □

Applying Proposition 5.11 to the Jantzen filtration of $\overline{M}(c,h)$ associated with $(\tilde{c},\tilde{h})$, we obtain the following character sum formula:

Proposition 5.12 *Suppose that* $\det(c,h)_k \neq 0$ *for* $k < n$. *Then,*

$$\begin{aligned}
&\sum_{l=1}^{\infty} \mathrm{ch}\, \overline{M}(c,h)(l) \\
&= \mathrm{ord}_\xi\{\phi_\zeta(\zeta^{-1}\Phi_{\alpha,\beta}(\approx{c},\approx{h}))\} \times \mathrm{ch}\, M(c,h+n) \\
&\quad + \sum_{(\alpha',\beta')\in\tilde{D}(c,h)\setminus\{(\alpha,\beta)\}} \mathrm{ord}_\xi\{\Phi_{\alpha',\beta'}(\tilde{c},\tilde{h})\} \times \mathrm{ch}\, M(c,h+\alpha'\beta') \\
&\quad - m\, \mathrm{ch}\, M(c,h+n) \\
&\quad - \sum_{(\alpha',\beta')\in\tilde{D}(c,h+n)} \mathrm{ord}_\xi\{\Phi_{\alpha',\beta'}(\tilde{c},\tilde{h}+n)\} \times \mathrm{ch}\, M(c,h+n+\alpha'\beta').
\end{aligned}$$

Proof. First, we notice that, since $\det(c,h)_k \neq 0$ for $k < n$,

$$\mathrm{ord}_\xi(\det(\tilde{c},\tilde{h})_k) = 0.$$

Hence, combining Proposition 3.6 with Proposition 5.11, we have

$$\begin{aligned}
\sum_{l=1}^{\infty} \mathrm{ch}\, \overline{M}(c,h)(l) &= \sum_{k\geq n} \mathrm{ord}_\xi \left\{\phi_\zeta(\zeta^{-p(k-n)} \det(\approx{c},\approx{h})_k)\right\} e(c,h+k) \\
&\quad - m \sum_{k\geq n} p(k-n) e(c,h+k) \\
&\quad - \sum_{k\geq n} \mathrm{ord}_\xi \det(\tilde{c},\tilde{h}+n)_{k-n} e(c,h+k).
\end{aligned}$$

Since one can directly check that

$$\mathrm{ord}_\zeta\, \Phi_{\alpha',\beta'}(\approx{c},\approx{h}) = \begin{cases} 1 & \text{if } (\alpha',\beta') = (\alpha,\beta) \\ 0 & \text{otherwise} \end{cases},$$

we see that

$$\begin{aligned}
\mathrm{ord}_\xi \left\{\phi_\zeta(\zeta^{-p(k-n)} \det(\approx{c},\approx{h})_k)\right\} &= \sum_{\substack{\alpha',\beta'\in\mathbb{Z}_{>0} \\ \alpha'\geq\beta' \\ 1\leq\alpha'\beta'\leq k \\ (\alpha',\beta')\neq(\alpha,\beta)}} p(k-\alpha'\beta')\, \mathrm{ord}_\xi\, \Phi_{\alpha',\beta'}(\tilde{c},\tilde{h}) \\
&\quad + p(k-n)\, \mathrm{ord}_\xi\{\phi_\zeta(\zeta^{-1}\Phi_{\alpha,\beta}(\approx{c},\approx{h}))\}.
\end{aligned}$$

Hence, by an argument similar to the proof of Lemma 5.8, we obtain

$$\sum_{k\geq n} \mathrm{ord}_\xi \left\{ \phi_\zeta(\zeta^{-p(k-n)} \det(\tilde{\tilde{c}}, \tilde{\tilde{h}})_k) \right\} e(c, h+k)$$
$$= \sum_{(\alpha',\beta')\in\tilde{D}(c,h)\setminus\{(\alpha,\beta)\}} \mathrm{ord}_\xi\{\Phi_{\alpha',\beta'}(\tilde{c},\tilde{h})\} \times \mathrm{ch}\, M(c, h+\alpha'\beta')$$
$$+ \mathrm{ord}_\xi\{\phi_\zeta(\zeta^{-1}\Phi_{\alpha,\beta}(\tilde{\tilde{c}}, \tilde{\tilde{h}}))\} \times \mathrm{ch}\, M(c, h+n).$$

For the other terms, we have

$$\sum_{k\geq n} p(k-n) e(c, h+k) = \mathrm{ch}\, M(c, h+n),$$
$$\sum_{k\geq n} \mathrm{ord}_\xi \det(\tilde{c}, \tilde{h}+n)_{k-n} e(c, h+k)$$
$$= \sum_{(\alpha',\beta')\in\tilde{D}(c,h+n)} \mathrm{ord}_\xi\{\Phi_{\alpha',\beta'}(\tilde{c}, \tilde{h}+n)\} \times \mathrm{ch}\, M(c, h+n+\alpha'\beta').$$

Therefore, the proposition has been proved. □

5.6.4 Explicit Forms

First, we should notice that the character sum formulae of quotient modules are necessary only for **Case** $4^\pm$ of **Type** *I* (see Remarks 5.3 and 5.6), i.e., $(c,h) = (c_{p,\pm q}, h_{2i})$ with

$$(p,q) \neq (1,1) \wedge (s \neq 0 \vee i \neq 0).$$

Hence, in this subsection, we only deal with **Case** $4^\pm$ of **Type** *I*.

Let $(\tilde{\alpha}, \tilde{\beta})$ be an integral point of $\tilde{D}(c,h)$ such that

$$\tilde{\alpha}\tilde{\beta} = \min_{(\alpha_k,\beta_k)\in\tilde{D}(c,h)} \{\alpha_k\beta_k\}.$$

For example, one can choose $(\tilde{\alpha}, \tilde{\beta})$ as (α_1, β_1) in the data of § 5.A. To study the structures of Jantzen filtrations in **Case** $4^\pm$ of **Type** *I*, it is enough to consider the case where $(\alpha, \beta) = (\tilde{\alpha}, \tilde{\beta})$, i.e.,

$$\overline{M}(c, h_{2i}) \simeq M(c, h_{2i})/M(c, h_{2(i\pm 1)}).$$

Lemma 5.29. *Suppose that the highest weight $(c,h) = (c_{\pm p,q}, h_{2i})$ belongs to* **Case** $4^\pm$ *of* ***Type*** *I, and $(\alpha,\beta) = (\tilde{\alpha}, \tilde{\beta})$. We set $n := \alpha\beta$. Then, we have*

1. $\mathrm{ord}_\xi \left\{ \phi_\zeta(\zeta^{-1}\Phi_{\alpha,\beta}(\tilde{\tilde{c}}, \tilde{\tilde{h}})) \right\} = 0.$
2. $\mathrm{ord}_\xi\, \Phi_{\alpha',\beta'}(\tilde{c}, \tilde{h}) = 1$ *for* $(\alpha',\beta') \in \tilde{D}(c,h) \setminus \{(\alpha,\beta)\}$.

3. $\mathrm{ord}_\xi\left(\phi_\zeta\left(\zeta^{-1}\det(\tilde{\tilde{c}},\tilde{\tilde{h}})_n\right)\right)=1.$
4. $\det(c,h)_k\neq 0$ *for any* $k<n$.
5. $\mathrm{ord}_\xi\,\Phi_{\alpha',\beta'}(\tilde{c},\tilde{h}+n)=1$ *for* $(\alpha',\beta')\in\tilde{D}(c,h+n)$.

Proof. One can directly check the first statement.

Let $\mathcal{V}_{\alpha,\beta}$ be the curve in $\mathfrak{h}^*$ defined by $\Phi_{\alpha,\beta}(c,h)=0$. In **Case** $4^\pm$ of **Type** I, for any $(\alpha',\beta')\in\tilde{D}(c,h)\setminus\{(\alpha,\beta)\}$, $\mathcal{V}_{\alpha',\beta'}$ transversally intersects with $\mathcal{V}_{\alpha,\beta}$ at $(c_{p,\pm q},h_{2i})$. Hence, the second statement holds.

Notice that, by Lemmas 5.11 and 5.14, there uniquely exists $(\alpha',\beta')\in\tilde{D}(c,h)\setminus\{(\alpha,\beta)\}$ such that $(c_{p,\pm q},h_{2i})\in\mathcal{V}_{\alpha',\beta'}$ and $\alpha'\beta'=n$. Hence, the third statement follows from the first two statements and the determinant formula (Theorem 4.2).

The fourth statement immediately follows from Lemmas 5.11 and 5.14. Using these lemmas, one can check the last statement by direct computation. We have proved the lemma. □

As a corollary of Proposition 5.12 and Lemma 5.29, we obtain

Lemma 5.30. *Suppose that the highest weight* $(c,h)=(c_{p,\pm q},h_{2i})$ *belongs to* **Case** $4^\pm$ *of* ***Type*** *I and* $(\alpha,\beta)=(\tilde{\alpha},\tilde{\beta})$. *Then, the following holds:*

$$\sum_{l=1}^{\infty}\mathrm{ch}\,\overline{M}(c,h)(l)=0.$$

5.7 Bibliographical Notes and Comments

In 1983, B. L. Feigin and D. B. Fuchs [FeFu2] announced that they had obtained the complete structure theorem of Verma modules. Their idea was to classify the highest weights of Verma modules, and for each series of highest weights, they constructed embedding diagrams. Its detailed and expanded version [FeFu4] appeared in 1990. Here, we have classified the highest weights of Verma modules following their ideas.

In 1997, A. Astashkevich [As] reviewed the uniqueness of singular vectors due to D. B. Fuchs. The uniqueness and the existence of a Shapovalov element was proved for $c=0$ by A. Rocha-Caridi and N. R. Wallach [RW3] in 1984.

Two concerete formulae of singular vectors given in this chapter were obtained by B. L. Feigin and D. B. Fuchs in [FeFu3]. For the proof of these formulae, we have followed arguments due to A. Astashkevich and D. B. Fuchs [AsFu].

5.A Appendix: Integral Points on $\ell_{c,h}$

For the reader's convenience, we give the list of integral points on $\ell_{c,h}$, which corresponds to $\tilde{D}(c,h)$ under the bijection in Lemma 5.3. In fact, we take the line $\ell_{c,h}$ explicitly for each highest weight (c,h), enumerate the points $\{(\alpha_k,\beta_k)\}$, and calculate $h+\alpha_k\beta_k$.

We first give the data for **Class** R^+.

Case 1$^+$ $\quad (r,s)\in\left\{(\alpha,\beta)\in\mathbb{Z}^2 \,\middle|\, \begin{matrix}0<\alpha<p\\ 0<\beta<q\end{matrix}, q\alpha+p\beta\le pq\right\}$,

1. $h=h_{2i-1}$ $(i\in\mathbb{Z}_{>0})$

   i. Case $rq-sp<0$:

$$\ell_{c,h}: q\alpha-p\beta=-2ipq+rq+sp,$$

$$(\alpha_1,\beta_1)=(r,2iq-s),$$

$$(\alpha_k,\beta_k)=\begin{cases}(\alpha_1,\beta_1)+\frac{1}{2}(k-1)(p,q) & k\equiv 1 \mod 2\\ (\alpha_1,\beta_1)-(\frac{1}{2}k+2i-1)(p,q) & k\equiv 0 \mod 2\end{cases},$$

$$h_{2i-1}+\alpha_k\beta_k=\begin{cases}h_{-2i-k+1} & k\equiv 1 \mod 2\\ h_{2i+k-2} & k\equiv 0 \mod 2\end{cases}.$$

   ii. Case $rq-sp>0$:

$$\ell_{c,h}: q\alpha-p\beta=2ipq-rq-sp,$$

$$(\alpha_1,\beta_1)=(2ip-r,s),$$

$$(\alpha_k,\beta_k)=\begin{cases}(\alpha_1,\beta_1)+\frac{1}{2}(k-1)(p,q) & k\equiv 1 \mod 2\\ (\alpha_1,\beta_1)-(\frac{1}{2}k+2i-1)(p,q) & k\equiv 0 \mod 2\end{cases},$$

$$h_{2i-1}+\alpha_k\beta_k=\begin{cases}h_{2i+k-1} & k\equiv 1 \mod 2\\ h_{-2i-k+2} & k\equiv 0 \mod 2\end{cases}.$$

2. $h=h_{2i}$ $(i\in\mathbb{Z}_{\ge 0})$:

$$\ell_{c,h}: q\alpha-p\beta=-2ipq+rq-sp$$

$$(\alpha_1,\beta_1)=(r,2iq+s),$$

$$(\alpha_k,\beta_k)=\begin{cases}(\alpha_1,\beta_1)+\frac{1}{2}(k-1)(p,q) & k\equiv 1 \mod 2\\ (\alpha_1,\beta_1)-(\frac{1}{2}k+2i)(p,q) & k\equiv 0 \mod 2\end{cases},$$

$$h_{2i}+\alpha_k\beta_k=\begin{cases}h_{-2i-k} & k\equiv 1 \mod 2\\ h_{2i+k-1} & k\equiv 0 \mod 2\end{cases}.$$

3. $h = h_{-2i+1}$ $(i \in \mathbb{Z}_{>0})$
   i. Case $rq - sp < 0$:

$$\begin{aligned}
&\ell_{c,h} : q\alpha - p\beta = 2(i-1)pq + rq + sp,\\
&(\alpha_1, \beta_1) = ((2i-1)p + r, q - s),\\
&(\alpha_k, \beta_k) = \begin{cases} (\alpha_1, \beta_1) + \frac{1}{2}(k-1)(p,q) & k \equiv 1 \mod 2 \\ (\alpha_1, \beta_1) - (\frac{1}{2}k + 2i - 1)(p,q) & k \equiv 0 \mod 2 \end{cases},\\
&h_{-2i+1} + \alpha_k\beta_k = \begin{cases} h_{-2i-k+1} & k \equiv 1 \mod 2 \\ h_{2i+k-2} & k \equiv 0 \mod 2 \end{cases}.
\end{aligned}$$

   ii. Case $rq - sp > 0$:

$$\begin{aligned}
&\ell_{c,h} : q\alpha - p\beta = -2(i-1)pq - rq - sp,\\
&(\alpha_1, \beta_1) = (p - r, (2i-1)q + s),\\
&(\alpha_k, \beta_k) = \begin{cases} (\alpha_1, \beta_1) + \frac{1}{2}(k-1)(p,q) & k \equiv 1 \mod 2 \\ (\alpha_1, \beta_1) - (\frac{1}{2}k + 2i - 1)(p,q) & k \equiv 0 \mod 2 \end{cases},\\
&h_{-2i+1} + \alpha_k\beta_k = \begin{cases} h_{2i+k-1} & k \equiv 1 \mod 2 \\ h_{-2i-k+2} & k \equiv 0 \mod 2 \end{cases}.
\end{aligned}$$

4. h_{-2i} $(i \in \mathbb{Z}_{\geq 0})$:

$$\begin{aligned}
&\ell_{c,h} : q\alpha - p\beta = 2ipq + rq - sp,\\
&(\alpha_1, \beta_1) = (2ip + r, s),\\
&(\alpha_k, \beta_k) = \begin{cases} (\alpha_1, \beta_1) + \frac{1}{2}(k-1)(p,q) & k \equiv 1 \mod 2 \\ (\alpha_1, \beta_1) - (\frac{1}{2}k + 2i)(p,q) & k \equiv 0 \mod 2 \end{cases},\\
&h_{-2i} + \alpha_k\beta_k = \begin{cases} h_{-2i-k} & k \equiv 1 \mod 2 \\ h_{2i+k-1} & k \equiv 0 \mod 2 \end{cases}.
\end{aligned}$$

In particular, for any $i \in \mathbb{Z}$,

$$D(c, h_i) = \{h_k - h_i | k \in \mathbb{Z},\ |k| > |i|,\ k - i \equiv 1 \mod 2\}.$$

Case 2^+ $\quad r = 0 \wedge 0 < s < q$,

1. $h = h_{2i-1}$ $(i \in \mathbb{Z}_{>0})$:

$$\ell_{c,h} : q\alpha - p\beta = 2ipq - sp,$$
$$(\alpha_1, \beta_1) = (2ip, s),$$
$$(\alpha_k, \beta_k) = \begin{cases} (\alpha_1, \beta_1) + \frac{1}{2}(k-1)(p,q) & k \equiv 1 \mod 2 \\ (\alpha_1, \beta_1) - (\frac{1}{2}k + 2i)(p,q) & k \equiv 0 \mod 2 \end{cases},$$
$$h_{2i-1} + \alpha_k\beta_k = h_{2i+k-1}.$$

2. $h = h_{2i}$ $(i \in \mathbb{Z}_{\geq 0})$:

$$\ell_{c,h} : q\alpha - p\beta = 2ipq + sp,$$
$$(\alpha_1, \beta_1) = ((2i+1)p, q - s),$$
$$(\alpha_k, \beta_k) = \begin{cases} (\alpha_1, \beta_1) + \frac{1}{2}(k-1)(p,q) & k \equiv 1 \mod 2 \\ (\alpha_1, \beta_1) - (\frac{1}{2}k + 2i + 1)(p,q) & k \equiv 0 \mod 2 \end{cases},$$
$$h_{2i} + \alpha_k\beta_k = h_{2i+k}.$$

In particular, for any $i \in \mathbb{Z}_{\geq 0}$,

$$D(c, h_i) = \{h_k - h_i | k \in \mathbb{Z}_{>0},\ k > i\}.$$

Case 3^+ $\quad 0 < r < p \wedge s = 0$,

1. $h = h_{2i-1}$ $(i \in \mathbb{Z}_{>0})$:

$$\ell_{c,h} : q\alpha - p\beta = -2ipq + rq,$$
$$(\alpha_1, \beta_1) = (r, 2iq),$$
$$(\alpha_k, \beta_k) = \begin{cases} (\alpha_1, \beta_1) + \frac{1}{2}(k-1)(p,q) & k \equiv 1 \mod 2 \\ (\alpha_1, \beta_1) - (\frac{1}{2}k + 2i)(p,q) & k \equiv 0 \mod 2 \end{cases},$$
$$h_{2i-1} + \alpha_k\beta_k = h_{(-1)^k(2i+k-1)}.$$

2. $h = h_{-2i}$ $(i \in \mathbb{Z}_{\geq 0})$:

$$\ell_{c,h} : q\alpha - p\beta = -2ipq - rq,$$
$$(\alpha_1, \beta_1) = (p - r, (2i+1)q),$$
$$(\alpha_k, \beta_k) = \begin{cases} (\alpha_1, \beta_1) + \frac{1}{2}(k-1)(p,q) & k \equiv 1 \mod 2 \\ (\alpha_1, \beta_1) - (\frac{1}{2}k + 2i + 1)(p,q) & k \equiv 0 \mod 2 \end{cases},$$
$$h_{-2i} + \alpha_k\beta_k = h_{(-1)^{k-1}(2i+k)}.$$

In particular, for any $i \in \mathbb{Z}_{\geq 0}$,

$$D(c, h_{(-1)^{i-1}i}) = \{h_{(-1)^{k-1}k} - h_{(-1)^{i-1}i} | k \in \mathbb{Z}_{>0},\ k > i\}.$$

Case 4$^+$ $\quad r = 0 \wedge s \in \{0, q\}$,

1. $h = h_{2i}$ $(i \in \mathbb{Z}_{\geq 0})$, $(p,q) \neq (1,1) \wedge (i > 0 \vee s = q)$:

$$\ell_{c,h} : q\alpha - p\beta = (2i + \frac{s}{q})pq,$$
$$(\alpha_1, \beta_1) = ((2i + 1 + \frac{s}{q})p, q),$$
$$(\alpha_k, \beta_k) = \begin{cases} (\alpha_1, \beta_1) + \frac{1}{2}(k-1)(p,q) & k \equiv 1 \mod 2 \\ (\alpha_1, \beta_1) - (\frac{1}{2}k + i + 1 + \frac{s}{q})(p,q) & k \equiv 0 \mod 2 \end{cases},$$
$$h_{2i} + \alpha_k \beta_k = h_{2i + 2\lceil \frac{k+1}{2} \rceil},$$

where $\lceil x \rceil$ denotes the greatest integer not exceeding x.

2. $h = h_{2i}$ $(i \in \mathbb{Z}_{\geq 0})$, $(p,q) = (1,1) \wedge (i > 0 \vee s = q)$ (**Type** *II*):

$$\ell_{c,h} : \alpha - \beta = 2i + s,$$
$$(\alpha_k, \beta_k) = (2i + 1 + \frac{s}{q}, 1) + (k-1)(1,1),$$
$$h_{2i} + \alpha_k \beta_k = h_{2i+2k}.$$

3. $h = h_0$, $i = 0 \wedge s = 0$ (**Type** *III*): $\quad \ell_{c,h} : q\alpha - p\beta = 0,$

$$(\alpha_k, \beta_k) = k(p,q),$$
$$h_0 + \alpha_k \beta_k = h_{2k}.$$

In particular, for any $i \in \mathbb{Z}_{\geq 0}$,

$$D(c, h_{2i}) = \{h_{2k} - h_{2i} | k \in \mathbb{Z}_{>0},\ k > i\}.$$

Next, we consider **Class** R^-.

Case 1^- $\quad (r, -s) \in \left\{ (\alpha, \beta) \in \mathbb{Z}^2 \,\middle|\, \begin{matrix} 0 < \alpha < p \\ 0 < \beta < q \end{matrix}, q\alpha + p\beta \le pq \right\},$

1. $h = h_{2i-1}$ $(i \in \mathbb{Z}_{>0})$:

$$\ell_{c,h} : q\alpha + p\beta = 2ipq - rq + sp,$$
$$(\alpha_1, \beta_1) = (ip - r, iq + s),$$

   i. Case $rq + sp < 0$:

$$(\alpha_k, \beta_k) = \begin{cases} (\alpha_1, \beta_1) + \frac{1}{2}(k-1)(p, -q) & k \equiv 1 \mod 2 \\ (\alpha_1, \beta_1) - \frac{1}{2}k(p, -q) & k \equiv 0 \mod 2 \end{cases},$$

$$h_{2i-1} + \alpha_k \beta_k = \begin{cases} h_{k-1} & k \equiv 1 \mod 2 \\ h_{-k} & k \equiv 0 \mod 2 \end{cases},$$

   where $1 \le k \le 2i - 1$.

   ii. Case $rq + sp > 0$:

$$(\alpha_k, \beta_k) = \begin{cases} (\alpha_1, \beta_1) - \frac{1}{2}(k-1)(p, -q) & k \equiv 1 \mod 2 \\ (\alpha_1, \beta_1) + \frac{1}{2}k(p, -q) & k \equiv 0 \mod 2 \end{cases},$$

$$h_{2i-1} + \alpha_k \beta_k = \begin{cases} h_{-k+1} & k \equiv 1 \mod 2 \\ h_k & k \equiv 0 \mod 2 \end{cases},$$

   where $1 \le k \le 2i - 1$.

2. $h = h_{2i}$ $(i \in \mathbb{Z}_{>0})$:

$$\ell_{c,h} : q\alpha + p\beta = 2ipq - rq - sp,$$
$$(\alpha_1, \beta_1) = (ip - r, iq - s),$$
$$(\alpha_k, \beta_k) = \begin{cases} (\alpha_1, \beta_1) - \frac{1}{2}(k-1)(p, -q) & k \equiv 1 \mod 2 \\ (\alpha_1, \beta_1) + \frac{1}{2}k(p, -q) & k \equiv 0 \mod 2 \end{cases},$$
$$h_{2i} + \alpha_k \beta_k = \begin{cases} h_{-k} & k \equiv 1 \mod 2 \\ h_{k-1} & k \equiv 0 \mod 2 \end{cases},$$

   where $1 \le k \le 2i$.

3. $h = h_{-2i+1}$ $(i \in \mathbb{Z}_{>0})$:

$$\ell_{c,h} : q\alpha + p\beta = 2(i-1)pq + rq - sp,$$
$$(\alpha_1, \beta_1) = ((i-1)p + r, (i-1)q - s),$$

i. Case $rq + sp < 0$:

$$(\alpha_k, \beta_k) = \begin{cases} (\alpha_1, \beta_1) - \frac{1}{2}(k-1)(p, -q) & k \equiv 1 \mod 2 \\ (\alpha_1, \beta_1) + \frac{1}{2}k(p, -q) & k \equiv 0 \mod 2 \end{cases},$$

$$h_{-2i+1} + \alpha_k\beta_k = \begin{cases} h_{k-1} & k \equiv 1 \mod 2 \\ h_{-k} & k \equiv 0 \mod 2 \end{cases},$$

where $1 \le k \le 2i - 1$.

ii. Case $rq + sp > 0$:

$$(\alpha_k, \beta_k) = \begin{cases} (\alpha_1, \beta_1) + \frac{1}{2}(k-1)(p, -q) & k \equiv 1 \mod 2 \\ (\alpha_1, \beta_1) - \frac{1}{2}k(p, -q) & k \equiv 0 \mod 2 \end{cases},$$

$$h_{-2i+1} + \alpha_k\beta_k = \begin{cases} h_{-k+1} & k \equiv 1 \mod 2 \\ h_k & k \equiv 0 \mod 2 \end{cases},$$

where $1 \le k \le 2i - 1$,

4. $h = h_{-2i}$ $(i \in \mathbb{Z}_{>0})$:

$$\ell_{c,h} : q\alpha + p\beta = 2ipq + rq + sp,$$
$$(\alpha_1, \beta_1) = (ip + r, iq + s),$$
$$(\alpha_k, \beta_k) = \begin{cases} (\alpha_1, \beta_1) + \frac{1}{2}(k-1)(p, -q) & k \equiv 1 \mod 2 \\ (\alpha_1, \beta_1) - \frac{1}{2}k(p, -q) & k \equiv 0 \mod 2 \end{cases},$$

$$h_{-2i} + \alpha_k\beta_k = \begin{cases} h_{-k} & k \equiv 1 \mod 2 \\ h_{k-1} & k \equiv 0 \mod 2 \end{cases},$$

where $1 \le k \le 2i$.

In particular, for any $i \in \mathbb{Z} \setminus \{0\}$,

$$D(c, h_i) = \{h_k - h_i | k \in \mathbb{Z},\ |k| < |i|,\ i - k \equiv 1 \mod 2\}.$$

Case 2^{-} $r = 0 \wedge 0 < -s < q$,

1. $h = h_{2i-1}$ $(i \in \mathbb{Z}_{>0})$:

$$\begin{aligned}
&\ell_{c,h} : q\alpha + p\beta = 2ipq + sp, \\
&(\alpha_1, \beta_1) = (ip, iq + s), \\
&(\alpha_k, \beta_k) = \begin{cases} (\alpha_1, \beta_1) + \frac{1}{2}(k-1)(p, -q) & k \equiv 1 \mod 2 \\ (\alpha_1, \beta_1) - \frac{1}{2}k(p, -q) & k \equiv 0 \mod 2 \end{cases}, \\
&h_{2i-1} + \alpha_k\beta_k = h_{k-1},
\end{aligned}$$

where $1 \leq k \leq 2i - 1$.

2. $h = h_{2i}$ $(i \in \mathbb{Z}_{>0})$:

$$\begin{aligned}
&\ell_{c,h} : q\alpha + p\beta = 2ipq - sp, \\
&(\alpha_1, \beta_1) = (ip, iq - s), \\
&(\alpha_k, \beta_k) = \begin{cases} (\alpha_1, \beta_1) - \frac{1}{2}(k-1)(p, -q) & k \equiv 1 \mod 2 \\ (\alpha_1, \beta_1) + \frac{1}{2}k(p, -q) & k \equiv 0 \mod 2 \end{cases}, \\
&h_{2i} + \alpha_k\beta_k = h_{k-1},
\end{aligned}$$

where $1 \leq k \leq 2i$.

In particular, for any $i \in \mathbb{Z}_{>0}$,

$$D(c, h_i) = \{h_k - h_i | k \in \mathbb{Z}_{>0},\ k < i\}.$$

Case 3^{-} $0 < r < p \wedge s = 0$,

1. $h = h_{2i-1}$ $(i \in \mathbb{Z}_{>0})$:

$$\begin{aligned}
&\ell_{c,h} : q\alpha + p\beta = 2ipq - rq, \\
&(\alpha_1, \beta_1) = (ip - r, iq), \\
&(\alpha_k, \beta_k) = \begin{cases} (\alpha_1, \beta_1) - \frac{1}{2}(k-1)(p, -q) & k \equiv 1 \mod 2 \\ (\alpha_1, \beta_1) + \frac{1}{2}k(p, -q) & k \equiv 0 \mod 2 \end{cases}, \\
&h_{2i-1} + \alpha_k\beta_k = h_{(-1)^k(k-1)},
\end{aligned}$$

where $1 \leq k \leq 2i - 1$.

2. $h = h_{-2i}$ ($i \in \mathbb{Z}_{>0}$):

$$\ell_{c,h} : q\alpha + p\beta = 2ipq + rq,$$
$$(\alpha_1, \beta_1) = (ip + r, iq),$$
$$(\alpha_k, \beta_k) = \begin{cases} (\alpha_1, \beta_1) + \frac{1}{2}(k-1)(p, -q) & k \equiv 1 \mod 2 \\ (\alpha_1, \beta_1) - \frac{1}{2}k(p, -q) & k \equiv 0 \mod 2 \end{cases},$$
$$h_{-2i} + \alpha_k \beta_k = h_{(-1)^k(k-1)},$$

where $1 \leq k \leq 2i$.

In particular, for any $i \in \mathbb{Z}_{>0}$,

$$D(c, h_{(-1)^{i-1}i}) = \{h_{(-1)^{k-1}k} - h_{(-1)^{i-1}i} | k \in \mathbb{Z}_{>0},\ k < i\}.$$

Case 4^- $\quad r = 0 \wedge s \in \{0, -q\}$,

1. $h = h_{2i}$ ($i \in \mathbb{Z}_{>0}$), $(p, q) \neq (1, 1)$,

$$\ell_{c,h} : q\alpha + p\beta = (2i - \frac{s}{q})pq,$$
$$(\alpha_1, \beta_1) = ((i - \frac{s}{q})p, iq),$$
$$(\alpha_k, \beta_k) = \begin{cases} (\alpha_1, \beta_1) + \frac{1}{2}(k-1)(p, -q) & k \equiv 1 \mod 2 \\ (\alpha_1, \beta_1) - \frac{1}{2}k(p, -q) & k \equiv 0 \mod 2 \end{cases},$$
$$h_{2i} + \alpha_k \beta_k = h_{2\left\lceil \frac{k + \frac{s}{q}}{2} \right\rceil},$$

where $1 \leq k \leq 2i - 1 - \frac{s}{q}$.

2. $h = h_{2i}$ ($i \in \mathbb{Z}_{>0}$), $(p, q) = (1, 1)$ (**Type** II):

$$\ell_{c,h} : \alpha + \beta = 2i - s,$$
$$(\alpha_k, \beta_k) = (i - \frac{s}{q}, i) + (k-1)(1, -1),$$
$$h_{2i} + \alpha_k \beta_k = h_{2k-2},$$

where $1 \leq k \leq i$.

In particular, for any $i \in \mathbb{Z}_{>0}$,

$$D(c, h_{2i}) = \{h_{2k} - h_{2i} | k \in \mathbb{Z}_{>0},\ k < i\}.$$

Chapter 6
Verma Modules II: Structure Theorem

We will completely reveal the structure of Jantzen filtration of Verma modules over the Virasoro algebra by means of the embedding diagrams and the character sums described in the previous chapter.

The structure theorems of Jantzen filtration presented in this chapter give us much information about the structures of Verma modules. For example, here, we will classify singular vectors and submodules of Verma modules. Moreover, we will construct BGG (Bernstein–Gelfand–Gelfand) type resolutions, and will compute the characters of the irreducible highest weight modules.

6.1 Structures of Jantzen Filtration

6.1.1 Class V *and* Class I

First, we consider **Class** V. Since the Verma module $M(c,h)$ is irreducible, the following holds:

Theorem 6.1 *Suppose that* (c,h) *belongs to* **Class** V*. Then, we have*

$$M(c,h)(l)=\{0\} \quad (l\in\mathbb{Z}_{>0}).$$

Next, we consider **Class** I.

Theorem 6.2 *Suppose that* (c,h) *belongs to* **Class** I*, i.e., there exists* $t\in\mathbb{C}\setminus\mathbb{Q}$ *and* $\alpha,\beta\in\mathbb{Z}_{>0}$ *such that* $(c,h)=(c(t),h_{\alpha,\beta}(t))$*. Then, we have*

$$M(c,h)(l)\simeq\begin{cases}M(c,h+\alpha\beta) & \text{if } l=1\\ \{0\} & \text{if } l>1\end{cases}.$$

Proof. By Corollary 3.1 and Proposition 5.3, we have

K. Iohara, Y. Koga, *Representation Theory of the Virasoro Algebra*, Springer Monographs in Mathematics, DOI 10.1007/978-0-85729-160-8_6,

$$M(c,h)(1) \supset M(c,h+\alpha\beta).$$

Hence, by Lemma 5.22, we see that $M(c,h)(1) = M(c,h+\alpha\beta)$ and $M(c,h)(l) = \{0\}$ for $l \geq 2$. □

6.1.2 Class R^+

Until the end of § 6.1.5, we assume that

$$c = c_{p,q} \text{ for } p,q \in \mathbb{Z}_{>0} \text{ such that } (p,q) = 1.$$

For simplicity, we introduce the following notation: We set

$$\xi_i := \begin{cases} h_i & \textbf{Case } 1^+ \\ h_i & \textbf{Case } 2^+ \\ h_{(-1)^{i-1}i} & \textbf{Case } 3^+ \\ h_{2i} & \textbf{Case } 4^+ \end{cases}, \tag{6.1}$$

where $i \in \mathbb{Z}$ in **Case** 1^+ and $i \in \mathbb{Z}_{\geq 0}$ for the other cases. By using the above notation, the embedding diagrams in Figure 5.4 can be described as follows:

1^+ $\quad$ $2^+, 3^+, 4^+$ (6.2)

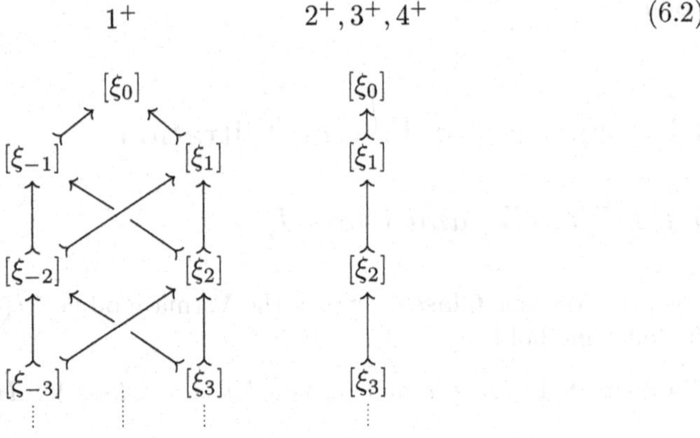

Let $\iota_{\xi_i,\xi_j} : M(c,\xi_j) \hookrightarrow M(c,\xi_i)$ be the embedding map defined by a composition of embeddings in the above diagrams. To state the structure theorem of Jantzen filtration, we introduce an auxiliary filtration

$$M(c,\xi_i) \supset N(c,\xi_i)(1) \supset N(c,\xi_i)(2) \supset \cdots$$

of $M(c,\xi_i)$ by

Case 1^+: for $i \in \mathbb{Z}$ and $l \in \mathbb{Z}_{>0}$,

$$N(c,\xi_i)(l) := \iota_{\xi_i,\xi_{|i|+l}} M(c,\xi_{|i|+l}) + \iota_{\xi_i,\xi_{-|i|-l}} M(c,\xi_{-|i|-l}), \tag{6.3}$$

Case 2^+, 3^+, 4^+: for $i \in \mathbb{Z}_{\geq 0}$ and $l \in \mathbb{Z}_{>0}$,

$$N(c,\xi_i)(l) := \iota_{\xi_i,\xi_{i+l}} M(c,\xi_{i+l}). \tag{6.4}$$

By means of these filtrations, we can describe the structure of Jantzen filtration as follows:

Theorem 6.3 *Suppose that* $(c,h) = (c_{p,q},\xi_i)$ $(i \in \mathbb{Z})$ *belongs to* **Class** R^+. *Let* $\{M(c,\xi_i)(l) | l \in \mathbb{Z}_{>0}\}$ *be the Jantzen filtration of Verma module* $M(c,\xi_i)$ *introduced in § 5.5. Then, we have*

(i) **Case** 1^+: ($i \in \mathbb{Z}$),

$$M(c,\xi_i)(l) = N(c,\xi_i)(l),$$

(ii) **Case** 2^+, 3^+: ($i \in \mathbb{Z}_{\geq 0}$),

$$M(c,\xi_i)(l) = N(c,\xi_i)(l),$$

(iii) **Case** 4^+: ($i \in \mathbb{Z}_{\geq 0}$ *and* $(p,q) \neq (1,1) \wedge (s = q \vee i \neq 0)$),

$$M(c,\xi_i)(l) = N(c,\xi_i)(\lceil \frac{l+1}{2} \rceil),$$

(iv) **Case** 4^+: ($i \in \mathbb{Z}_{\geq 0}$ *and* $(p,q) = (1,1) \vee (s = 0 \wedge i = 0)$),

$$M(c,\xi_i)(l) = N(c,\xi_i)(l),$$

where $\lceil x \rceil$ *denotes the greatest integer not exceeding* x.

We prove this theorem in the following three subsections.

Remark 6.1 *Pictorially, the structures of Jantzen filtration are described as in Figure 6.1.*

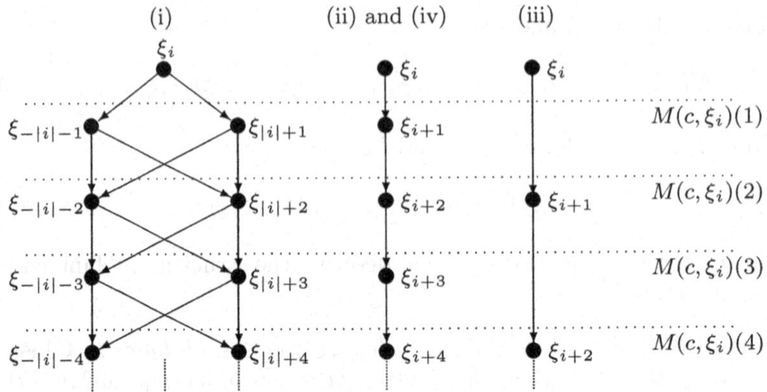

Each $\bullet_\xi$ signifies a singular vector, say u_ξ, of L_0-weight ξ, and $\bullet_\xi \to \bullet_{\xi'}$ means $u_{\xi'} \in U(\mathrm{Vir}).u_\xi$.

Fig. 6.1 Jantzen filtration of Class R^+

6.1.3 Proof of (i) in Theorem 6.3

In the proof of the theorem, for $i, j \in \mathbb{Z}$ such that $|i| < |j|$, we identify $M(c,\xi_j)$ with its image via the embedding map $\iota_{\xi_i,\xi_j} : M(c,\xi_j) \hookrightarrow M(c,\xi_i)$.

In order to prove statement (i) in the theorem, we first show the following lemma.

Lemma 6.1. *For any $i \in \mathbb{Z}$ and $l \in \mathbb{Z}_{>0}$, we have*

$$M(c,\xi_i)(l) \supset N(c,\xi_i)(l).$$

Proof. We prove the lemma by induction on l. Suppose that $l = 1$. Since $M(c,\xi_i)(1)$ is the maximal proper submodule of $M(c,\xi_i)$ (Corollary 3.1), $M(c,\xi_i)(1) \supset N(c,\xi_i)(1)$ follows.

Next, we suppose that the lemma holds for $l < l_0$. Let $v_{c,\xi_{|i|+l_0}}$ and $v_{c,\xi_{-|i|-l_0}}$ be highest weight vectors of $M(c,\xi_{|i|+l_0})$ and $M(c,\xi_{-|i|-l_0})$ respectively. These highest weight vectors satisfy

$$v_{c,\xi_{|i|+l_0}}, v_{c,\xi_{-|i|-l_0}} \in \{M(c,\xi_i)(l_0-1)\}^{\mathrm{Vir}^+},$$

since $v_{c,\xi_{|i|+l_0}}, v_{c,\xi_{-|i|-l_0}} \in \{N(c,\xi_i)(l_0-1)\}^{\mathrm{Vir}^+}$ by definition and $N(c,\xi_i)(l_0-1) \subset M(c,\xi_i)(l_0-1)$ by the induction hypothesis (cf. Lemma 1.12).

On the other hand, Proposition 3.5 says that there exists a non-degenerate contravariant form on $M(c,\xi_i)(l_0-1)/M(c,\xi_i)(l_0)$. This implies that

$$v_{c,\xi_{|i|+l_0}}, v_{c,\xi_{-|i|-l_0}} \in M(c,\xi_i)(l_0).$$

Hence, we have $N(c,\xi_i)(l_0) \subset M(c,\xi_i)(l_0)$ by the definition of $N(c,\xi_i)(l_0)$. □

PROOF OF THEOREM 6.3 (i). We show that the theorem holds on each L_0-weight subspace of weight $\xi_i + n$ ($n \in \mathbb{Z}_{\geq 0}$) by induction on n, i.e., we prove the following statement $P(n)$:

$$P(n) \ : \ M(c,\xi_i)(l)_{\xi_i+n} = N(c,\xi_i)(l)_{\xi_i+n} \ (\forall i \in \mathbb{Z}, \ \forall l \in \mathbb{Z}_{>0}).$$

Here, we use the following auxiliary statement:

$$Q(n) \ : \ \begin{array}{l} M(c,\xi_{|i|+l})_{\xi_i+n} \cap M(c,\xi_{-|i|-l})_{\xi_i+n} \\ = M(c,\xi_{|i|+l+1})_{\xi_i+n} + M(c,\xi_{-|i|-l-1})_{\xi_i+n} \end{array} \ (\forall i \in \mathbb{Z}, \ \forall l \in \mathbb{Z}_{>0}),$$

and show $P(n)$ by the following steps:

STEP I : $P(0)$ and $Q(0)$.
STEP II : $P(m)$ ($\forall m < n$) $\Rightarrow Q(n)$.
STEP III: $Q(n) \Rightarrow P(n)$.

STEP I: Since

$$\begin{aligned} M(c,\xi_i)(l)_{\xi_i} &= \{0\} = N(c,\xi_i)(l)_{\xi_i}, \\ M(c,\xi_{\pm(|i|+l)})_{\xi_i} &= \{0\} \quad (l \in \mathbb{Z}_{>0}), \end{aligned}$$

both $P(0)$ and $Q(0)$ are obvious.
STEP II: Since

$$M(c,\xi_{\sigma(|i|+l)}) \supset M(c,\xi_{\tau(|i|+l+1)}) \quad (\forall \sigma,\tau \in \{\pm 1\}),$$

the inclusion

$$\begin{aligned} &M(c,\xi_{|i|+l})_{\xi_i+n} \cap M(c,\xi_{-|i|-l})_{\xi_i+n} \\ &\supset M(c,\xi_{|i|+l+1})_{\xi_i+n} + M(c,\xi_{-|i|-l-1})_{\xi_i+n} \end{aligned}$$

holds. On the other hand, we see that

$$M(c,\xi_{|i|+l}) \cap M(c,\xi_{-|i|-l}) \subset M(c,\xi_{|i|+l})(1), \tag{6.5}$$

since $M(c,\xi_{|i|+l})(1)$ is the maximal proper submodule of $M(c,\xi_{|i|+l})$. By the induction hypothesis, we have

$$M(c,\xi_{|i|+l})(1)_{\xi_i+n} = N(c,\xi_{|i|+l})(1)_{\xi_i+n}. \tag{6.6}$$

Indeed, since for $l > 0$

$$\xi_i + n = \xi_{|i|+l} + (n + \xi_i - \xi_{|i|+l}) \text{ and } \xi_i - \xi_{|i|+l} + n \lneq n,$$

$P(n + \xi_i - \xi_{|i|+l})$ implies (6.6). Moreover, by definition, we have

$$N(c,\xi_{|i|+l})(1) = M(c,\xi_{|i|+l+1}) + M(c,\xi_{-|i|-l-1}). \tag{6.7}$$

Hence, by (6.5), (6.6) and (6.7), we obtain the opposite inclusion which implies $Q(n)$.
<u>STEP III</u>: Combining the short exact sequence

$$\begin{aligned}0 \to M(c,\xi_{|i|+l}) \cap M(c,\xi_{-|i|-l}) &\to M(c,\xi_{|i|+l}) \oplus M(c,\xi_{-|i|-l}) \\ &\to M(c,\xi_{|i|+l}) + M(c,\xi_{-|i|-l}) \to 0\end{aligned}$$

with $Q(n)$, we obtain the following exact sequence

$$\begin{aligned}0 \to N(c,\xi_i)(l+1)_{\xi_i+n} &\to M(c,\xi_{|i|+l})_{\xi_i+n} \oplus M(c,\xi_{-|i|-l})_{\xi_i+n} \\ &\to N(c,\xi_i)(l)_{\xi_i+n} \to 0\end{aligned} \tag{6.8}$$

by the induction hypothesis.

Hence, we have

$$\begin{aligned}&\dim N(c,\xi_i)(l)_{\xi_i+n} + \dim N(c,\xi_i)(l+1)_{\xi_i+n} \\ &= \dim M(c,\xi_{|i|+l})_{\xi_i+n} + \dim M(c,\xi_{-|i|-l})_{\xi_i+n},\end{aligned}$$

and thus,

$$\begin{aligned}\sum_{l=1}^{\infty} \dim N(c,\xi_i)(l)_{\xi_i+n} = \sum_{k=1}^{\infty} \{ &\dim M(c,\xi_{|i|+2k-1})_{\xi_i+n} \\ &+ \dim M(c,\xi_{-|i|-2k+1})_{\xi_i+n}\}.\end{aligned}$$

From the character sums in Lemma 5.23, we obtain

$$\sum_{l=1}^{\infty} \dim N(c,\xi_i)(l)_{\xi_i+n} = \sum_{l=1}^{\infty} \dim M(c,\xi_i)(l)_{\xi_i+n}.$$

Hence, by Lemma 6.1,

$$N(c,\xi_i)(l)_{\xi_i+n} = M(c,\xi_i)(l)_{\xi_i+n}$$

holds for any $i \in \mathbb{Z}$ and $l \in \mathbb{Z}_{>0}$. Therefore, we have completed the proof. □

6.1.4 Proof of (ii) and (iv) in Theorem 6.3

In these cases, the following lemma holds:

Lemma 6.2.

$$M(c,\xi_i)(l) \supset N(c,\xi_i)(l).$$

Proof. One can show this lemma in a way similar to the proof of Lemma 6.1. □

PROOF OF THEOREM 6.3 (ii) AND (iv). By definition, we see that

$$\operatorname{ch} N(c,\xi_i)(l) = \operatorname{ch} M(c,\xi_{i+l}).$$

Combining this with the character sum formulae in Lemma 5.23, we have

$$\sum_{l=1}^{\infty} \dim N(c,\xi_i)(l) = \sum_{l=1}^{\infty} \dim M(c,\xi_i)(l).$$

Hence, Lemma 6.2 implies that

$$N(c,\xi_i)(l) = M(c,\xi_i)(l). \qquad \square$$

6.1.5 Proof of (iii) in Theorem 6.3

First, we notice that in **Case** 4^+,

$$(p,q) \neq (1,1) \wedge (s = q \vee i \neq 0) \ \Leftrightarrow\ \textbf{Type } I.$$

(See Remark 5.3.) In this case, multiplicity 2 appears in the character sums in Lemma 5.23, and an argument similar to the previous subsections does not work. As a consequence of Lemma 5.30, we have

Lemma 6.3. *Suppose that highest weight (c,ξ_i) $(i \in \mathbb{Z}_{\geq 0})$ belongs to* **Case** 4^+ *of* **Type** *I. Then, $M(c,\xi_{i+1})$ is the maximal proper submodule of $M(c,\xi_i)$.*

PROOF OF THEOREM 6.3 (iii). We prove the assertion (iii) by induction on l. We first show the theorem for $l = 1, 2$. By Lemma 6.3, $N(c,\xi_i)(1)$ is the maximal proper submodule of $M(c,\xi_i)$. Hence, it is enough to show that

$$M(c,\xi_i)(2) \supset N(c,\xi_i)(1). \tag{6.9}$$

By Lemma 5.23, we have

$$\sum_{k=1}^{\infty} \dim M(c,\xi_i)(k)_h = 0 \quad (\forall h < \xi_{i+1}).$$

Hence, for $k \in \mathbb{Z}_{>0}$ we have

$$M(c,\xi_i)(k)_{\xi_{i+1}} \subset \{M(c,\xi_i)_{\xi_{i+1}}\}^{\mathrm{Vir}_+}.$$

Moreover, by Lemma 5.23,

$$\sum_{k=1}^{\infty} \dim M(c,\xi_i)(k)_{\xi_{i+1}} = 2.$$

Since by Proposition 5.1, $\dim\{M(c,\xi_i)_{\xi_{i+1}}\}^{\mathrm{Vir}_+} \leq 1$, we see that

$$\dim M(c,\xi_i)(1)_{\xi_{i+1}} = 1 \quad \text{and} \quad \dim M(c,\xi_i)(2)_{\xi_{i+1}} = 1.$$

Hence, we have

$$M(c,\xi_i)(1)_{\xi_{i+1}} = M(c,\xi_i)(2)_{\xi_{i+1}} = \mathbb{C}v_{c,\xi_{i+1}},$$

where $v_{c,\xi_{i+1}}$ is a highest weight vector of $M(c,\xi_{i+1})$. Thus, (6.9) holds. The theorem for $l = 1, 2$ is proved.

Next, we suppose that the theorem holds for $l < 2l_0 - 1$. By the character sum formulae in Lemma 5.23 and the induction hypothesis, we have

$$\sum_{k=2l_0-1}^{\infty} \mathrm{ch}\, M(c,\xi_i)(k) = 2\sum_{k=l_0}^{\infty} \mathrm{ch}\, M(c,\xi_{i+k}).$$

In particular, for $h \leq \xi_{i+l_0}$

$$\sum_{k=2l_0-1}^{\infty} \dim M(c,\xi_i)(k)_h = \begin{cases} 0 & \text{if } h < \xi_{i+l_0} \\ 2 & \text{if } h = \xi_{i+l_0} \end{cases}. \tag{6.10}$$

Hence, by an argument similar to the proof of (6.9), we have

$$M(c,\xi_i)(2l_0) \supset N(c,\xi_i)(l_0).$$

On the other hand, Lemma 6.3 implies that $N(c,\xi_i)(l_0)(\simeq M(c,\xi_{i+l_0}))$ is the maximal proper submodule of $N(c,\xi_i)(l_0-1)(\simeq M(c,\xi_{i+l_0-1}))$. Therefore, we have

$$M(c,\xi_i)(2l_0-1) = M(c,\xi_i)(2l_0) = N(c,\xi_i)(l_0),$$

since $M(c,\xi_i)(2l_0-1) \subsetneq N(c,\xi_i)(l_0-1)$ by (6.10). Therefore, the theorem holds for $l = 2l_0-1, 2l_0$, and we have completed the proof. □

6.1.6 Class R^-

Through this subsection, we assume that

$$c = c_{p,-q} \text{ for } p, q \in \mathbb{Z}_{>0} \text{ such that } (p,q) = 1.$$

Similarly to **Class** R^+, we introduce L_0-weights $\{\xi_i\}$ by

$$\xi_i := \begin{cases} h_i & \textbf{Case } 1^- \\ h_i & \textbf{Case } 2^- \\ h_{(-1)^{i-1}i} & \textbf{Case } 3^- \\ h_{2i} & \textbf{Case } 4^- \end{cases}, \tag{6.11}$$

where $i \in \mathbb{Z}$ in **Case** 1^- and $i \in \mathbb{Z}_{\geq 0}$ for the other cases. (Although (c, h_0) belongs to **Class** V, we use ξ_0 as in Figure 5.5.) By using this notation, the embedding diagrams of Verma modules in Figure 5.5 can be described as follows:

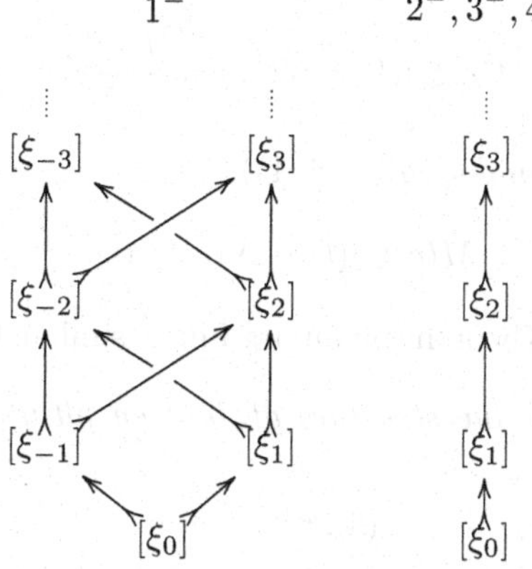

Moreover, we define an auxiliary filtration $\{N(c,\xi_i)(l) | l \in \mathbb{Z}_{>0}\}$ of the Verma module $M(c, \xi_i)$ as follows:

Case 1^- : for $i \in \mathbb{Z} \setminus \{0\}$ and $l \in \mathbb{Z}_{>0}$,

$$N(c,\xi_i)(l) := \begin{cases} \iota_{\xi_i,\xi_{|i|-l}} M(c, \xi_{|i|-l}) + \iota_{\xi_i,\xi_{-|i|+l}} M(c, \xi_{-|i|+l}) & l < |i| \\ \iota_{\xi_i,\xi_0} M(c, \xi_0) & l = |i| \\ \{0\} & l > |i| \end{cases},$$

Case 2^-, 3^- and 4^- : for $i \in \mathbb{Z}_{>0}$ and $l \in \mathbb{Z}_{>0}$,

$$N(c,\xi_i)(l) := \begin{cases} \iota_{\xi_i,\xi_{i-l}} M(c, \xi_{i-l}) & l \leq i \\ \{0\} & l > i \end{cases}.$$

The structure of Jantzen filtration of $M(c, \xi_i)$ is described as follows:

Theorem 6.4 *Suppose that* $(c, h) = (c_{p,-q}, \xi_i)$ *belongs to* **Class** R^-. *Let* $\{M(c,\xi_i)(l) | l \in \mathbb{Z}_{>0}\}$ *be the Jantzen filtration of Verma module* $M(c, \xi_i)$ *introduced in § 5.5. Then, we have*

(i) **Case** 1^-: $(i \in \mathbb{Z} \setminus \{0\})$,

$$M(c,\xi_i)(l) = N(c,\xi_i)(l),$$

(ii) **Case** 2^-, 3^-: $(i \in \mathbb{Z}_{>0})$,

$$M(c,\xi_i)(l) = N(c,\xi_i)(l),$$

(iii) **Case** 4^-: $(i \in \mathbb{Z}_{>0}$ *and* $(p, q) \neq (1, 1) \wedge s = 0)$,

$$M(c,\xi_i)(l) = \begin{cases} N(c,\xi_i)(\lceil\frac{l+1}{2}\rceil) & l \neq 2i \\ \{0\} & l = 2i \end{cases},$$

(iv) **Case** 4^-: ($i \in \mathbb{Z}_{>0}$ *and* $(p,q) \neq (1,1) \wedge s = -q$),

$$M(c,\xi_i)(l) = N(c,\xi_i)(\lceil\frac{l+1}{2}\rceil),$$

(v) **Case** 4^-: ($i \in \mathbb{Z}_{>0}$ *and* $(p,q) = (1,1)$),

$$M(c,\xi_i)(l) = N(c,\xi_i)(l).$$

Proof. We can show the theorem by an argument similar to Theorem 6.3. □

Remark 6.2 *Pictorially, the structure of Jantzen filtration is described as in Figure 6.2.*

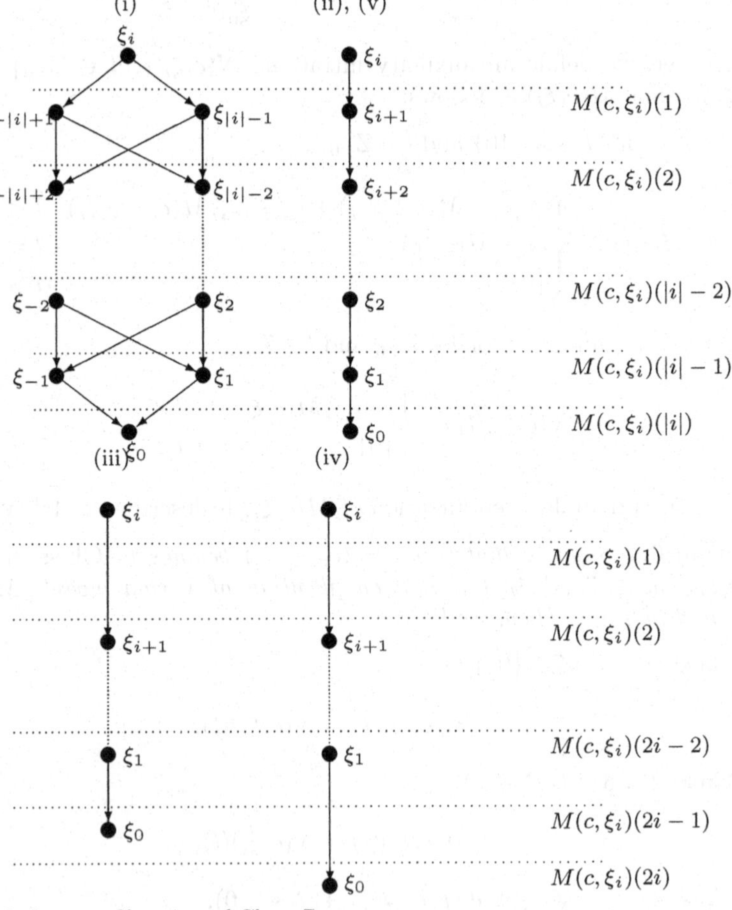

Fig. 6.2 Jantzen filtration of Class R^-

6.2 Structures of Verma Modules

In this section, we prove two theorems as applications of the structure theorems of Jantzen filtration of Verma modules. One is on structures of submodules of Verma modules, and the other is related to the existence of non-trivial homomorphisms between Verma modules.

6.2.1 Main Results

The first theorem we are going to prove in the next subsection is as follows:

Theorem 6.5 *Any non-trivial proper submodule of a Verma module over* Vir *is generated by at most two singular vectors.*

In fact, in the next subsection, we classify all submodules of given Verma modules. This theorem is a consequence of this classification.

The other consequence of the classification is the second theorem we are going to prove.

By Corollary 5.1, the following theorem completely describes necessary and sufficient conditions of the existence of a non-trivial homomorphism between Verma modules.

Theorem 6.6 *For each* $(c,h) \in \mathfrak{h}^*$, *the following list exhausts the conformal weights* h' *such that*

$$\dim \mathrm{Hom}_{\mathrm{Vir}}(M(c,h'), M(c,h)) = 1.$$

Class V: *For* (c,h) *which belongs to* ***Class*** V,

$$h' = h.$$

Class I: *For* $(c,h) = (c(t), h_{\alpha,\beta}(t))$ $(t \in \mathbb{C} \setminus \mathbb{Q}$ *and* $\alpha, \beta \in \mathbb{Z}_{>0})$,

$$h' = h, \quad h + \alpha\beta.$$

Class R^+: *For* $c = c_{p,q}$ $(p, q \in \mathbb{Z}_{>0}$ *such that* $(p,q) = 1)$,

Case 1^+: $h = \xi_i$ $(i \in \mathbb{Z})$,

$$h' = \xi_k \quad (|k| \geq |i| \wedge k \neq -i).$$

Case 2^+, 3^+ *and* 4^+: $h = \xi_i$ $(i \in \mathbb{Z}_{\geq 0})$,

$$h' = \xi_k \quad (k \geq i).$$

Class R^-: *For* $c = c_{p,-q}$ $(p, q \in \mathbb{Z}_{>0}$ *such that* $(p,q) = 1)$,

Case 1^-: $h = \xi_i$ $(i \in \mathbb{Z} \setminus \{0\})$,

$$h' = \xi_k \quad (|k| \leq |i| \wedge k \neq -i).$$

Case 2^-, 3^- *and* 4^-: $h = \xi_i$ $(i \in \mathbb{Z}_{\geq 0})$,

$$h' = \xi_k \quad (0 \leq k \leq i).$$

6.2.2 Proof of Theorems 6.5 and 6.6

The above two theorems are direct consequences of the following classification of submodules of Verma modules.

Proposition 6.1 *Class V*: *Suppose that (c, h) belongs to* **Class** *V. Then, $M(c, h)$ is irreducible.*
Class I: *Suppose that $(c, h) = (c(t), h_{\alpha,\beta}(t))$ for some $t \in \mathbb{C}\backslash\mathbb{Q}$ and $\alpha, \beta \in \mathbb{Z}_{>0}$. Then, any non-trivial proper submodule of $M(c, h)$ is isomorphic to $M(c, h + \alpha\beta)$.*
Class R^+: *Suppose that highest weight $(c, h) = (c, \xi_i)$ ($i \in \mathbb{Z}$) belongs to* **Class** *R^+. Then, any non-trivial proper submodule of $M(c, \xi_i)$ is isomorphic to one of the modules*

Case 1^+:

$$M(c, \xi_{\pm l}) \text{ and } M(c, \xi_l) + M(c, \xi_{-l}),$$

for $l \in \mathbb{Z}_{>0}$ such that $l > |i|$.
Case 2^+, 3^+, 4^+: $(i \geq 0)$

$$M(c, \xi_l)$$

for $l \in \mathbb{Z}_{>0}$ such that $l > i$.

Class R^-: *Suppose that highest weight $(c, h) = (c, \xi_i)$ ($i \in \mathbb{Z} \setminus \{0\}$) belongs to* **Class** *R^-. Then, any non-trivial proper submodule of $M(c, \xi_i)$ is isomorphic to one of the modules*

Case 1^-:

$$M(c, \xi_{\pm l}) \text{ and } M(c, \xi_{l'}) + M(c, \xi_{-l'}),$$

for $l, l' \in \mathbb{Z}_{\geq 0}$ such that $0 \leq l < |i|$ and $0 < l' < |i|$,
Case 2^-, 3^-, 4^-: $(i > 0)$

$$M(c, \xi_l)$$

for $l \in \mathbb{Z}_{\geq 0}$ such that $0 \leq l < i$.

Proof. We show this proposition in **Case** 1^+, since the other cases can be proved similarly.

Let v_{c,ξ_j} be a highest weight vector of $M(c, \xi_j)$. Then, the image of v_{c,ξ_j} under the embedding map $M(c, \xi_j) \hookrightarrow M(c, \xi_i)$ given in Figure 6.1 is a singular vector of $M(c,\xi_i)$. Here and after, we identify $M(c,\xi_j)$ with a submodule of $M(c,\xi_i)$ for $|i|<|j|$ via the embedding diagram in **Case** 1^+.

Let M be a non-trivial proper submodule of $M(c,\xi_i)$. By Theorem 6.3, there exists a positive integer such that $M \subset M(c,\xi_i)(k)$ and $M \not\subset M(c,\xi_i)(k+1)$, since the filtration $\{M(c,\xi_i)(l)\}$ is a decreasing filtration and $\bigcap_{l=1}^{\infty} M(c,\xi_j)(l)=\{0\}$. Theorem 6.3 and Proposition 3.5 imply that

$$M(c,\xi_i)(k)/M(c,\xi_i)(k+1) \simeq L(c,\xi_{|i|+k}) \oplus L(c,\xi_{-|i|-k}).$$

Hence, we have

$$\begin{aligned}&\{M+M(c,\xi_i)(k+1)\}/M(c,\xi_i)(k+1)\\ &\simeq L(c,\xi_{|i|+k}) \oplus L(c,\xi_{-|i|-k}) \quad \text{or} \quad L(c,\xi_{\pm(|i|+k)}).\end{aligned}$$

This implies that

$$M+M(c,\xi_i)(k+1) \simeq M(c,\xi_{|i|+k})+M(c,\xi_{-|i|-k}) \quad \text{or} \quad M(c,\xi_{\pm(|i|+k)}). \tag{6.12}$$

In any case, by Theorem 6.3,

$$v_{c,\xi_{|i|+k}} \in M \quad \text{or } v_{c,\xi_{-|i|-k}} \in M,$$

and hence, $M(c,\xi_i)(k+1) \subset M$. Thus, (6.12) implies the result. □

6.3 Bernstein–Gelfand–Gelfand Type Resolutions

In this section, as an application of Theorems 6.1 – 6.4, we construct BGG type resolutions, i.e., resolutions of irreducible highest weight representations by Verma modules (cf. [BGG2] for a semi-simple Lie algebra). Here, we denote the canonical projection

$$M(c,h) \twoheadrightarrow L(c,h)$$

by π_h.

6.3.1 Class *V and* Class *I*

First, we consider **Class** V. In this case, the Verma module $M(c,h)$ is irreducible. Hence, the following holds:

Theorem 6.7 *Suppose that (c,h) belongs to* **Class** *V. Then, there exists the following exact sequence:*

$$0 \longrightarrow M(c,h) \longrightarrow L(c,h) \longrightarrow 0.$$

Second, we consider **Class** I.

Theorem 6.8 *Suppose that* $(c,h) = (c(t), h_{\alpha,\beta}(t))$, *where* $t \in \mathbb{C} \setminus \mathbb{Q}$ *and* $\alpha, \beta \in \mathbb{Z}_{>0}$, *belongs to* **Class** I. *Then, there exists the following resolution of* $L(c,h)$:

$$0 \longrightarrow M(c, h+\alpha\beta) \xrightarrow{d_1} M(c,h) \xrightarrow{d_0} L(c,h) \longrightarrow 0,$$

where $d_0 = \pi_h$ *and* d_1 *is the embedding map*

$$\iota_{h,h+\alpha\beta} : M(c,h+\alpha\beta) \hookrightarrow M(c,h).$$

Proof. By Theorem 6.2, the maximal proper submodule of $M(c,h)$ is isomorphic to $M(c,h+\alpha\beta)$. □

6.3.2 Class R^+

Suppose that $c = c_{p,q}$ for $p, q \in \mathbb{Z}_{>0}$ such that $(p,q) = 1$. Let $\{\xi_i\}$ be the L_0-weights defined in (6.1). For $i, j \in \mathbb{Z}$ such that $|i| < |j|$, let

$$\iota_{\xi_i,\xi_j} : M(c,\xi_j) \to M(c,\xi_i)$$

be the embedding map given by the embedding diagrams in Figure 5.4.

Theorem 6.9 *Suppose that highest weight* $(c,h) = (c_{p,q}, \xi_i)$ *belongs to* ***Class*** R^+. *Then, there exists the following resolutions of* $L(c,h)$:

1. **Case** 1^+: $i \in \mathbb{Z}$,

$$\cdots \xrightarrow{d_{k+1}} M(c, \xi_{|i|+k}) \oplus M(c, \xi_{-|i|-k}) \xrightarrow{d_k} \cdots$$

$$\cdots \xrightarrow{d_2} M(c, \xi_{|i|+1}) \oplus M(c, \xi_{-|i|-1}) \xrightarrow{d_1} M(c,h) \xrightarrow{d_0} L(c,h) \to 0,$$

where the maps d_k *are given by*

$k = 0 : d_0 = \pi_h$,
$k = 1$: *for* $(x,y) \in M(c,\xi_{|i|+1}) \oplus M(c, \xi_{-|i|-1})$,

$$d_1((x,y)) := \iota_{h,\xi_{|i|+1}}(x) + \iota_{h,\xi_{-|i|-1}}(y),$$

$k > 1$: *for* $(x,y) \in M(c,\xi_{|i|+k}) \oplus M(c,\xi_{-|i|-k})$,

$$d_k((x,y)) := (\ \iota_{\xi_{|i|+k-1},\xi_{|i|+k}}(x) + \iota_{\xi_{|i|+k-1},\xi_{-|i|-k}}(y), \\ - \iota_{\xi_{-|i|-k+1},\xi_{|i|+k}}(x) - \iota_{\xi_{-|i|-k+1},\xi_{-|i|-k}}(y)\ ),$$

2. **Case** 2^+, 3^+, 4^+: $i \in \mathbb{Z}_{\geq 0}$,

$$0 \longrightarrow M(c,\xi_{i+1}) \xrightarrow{d_1} M(c,h) \xrightarrow{d_0} L(c,h) \longrightarrow 0,$$

where the maps d_i are given by

$k=0 : d_0 := \pi_h,$
$k=1 : d_1 := \iota_{h,\xi_{i+1}}.$

Proof. Here, we show this theorem in **Case** 1^+. By Theorem 6.3 and Corollary 3.1, we have the following exact sequence:

$$E(0): \ 0 \to N(c,\xi_i)(1) \to M(c,\xi_i) \to L(c,\xi_i) \to 0.$$

On the other hand, for each $l \in \mathbb{Z}_{>0}$, there exists a short exact sequence

$$E(l): 0 \to N(c,\xi_i)(l+1) \to M(c,\xi_{|i|+l}) \oplus M(c,\xi_{-|i|-l}) \to N(c,\xi_i)(l) \to 0,$$

since (6.8) holds for any $n \in \mathbb{Z}_{\geq 0}$. Taking the Yoneda products of $E(l)$'s (cf. § A.2.2), i.e.,

$$E(0) \circ E(1) \circ E(2) \circ \cdots,$$

we obtain the result. □

6.3.3 Class R^-

Next, we consider the case where (c,h) belongs to **Class** R^-. We define the L_0-weights $\{\xi_i\}$ as in (6.11).

Theorem 6.10 *Suppose that highest weight $(c,h) = (c_{p,-q},\xi_i)$ belongs to* **Class** R^-. *Then, there exists the following resolutions of $L(c,h)$:*

1. **Case** 1^-: $i \in \mathbb{Z}\setminus\{0\}$,

$$0 \to M(c,\xi_0) \xrightarrow{d_{|i|}} M(c,\xi_1) \oplus M(c,\xi_{-1}) \xrightarrow{d_{|i|-1}} \cdots$$

$$\cdots \to M(c,\xi_{|i|-1}) \oplus M(c,\xi_{-|i|+1}) \xrightarrow{d_1} M(c,h) \xrightarrow{d_0} L(c,h) \to 0,$$

where the maps d_k are given by

$k=0 : d_0 = \pi_h,$
$k=1 :$ *for* $(x,y) \in M(c,\xi_{|i|-1}) \oplus M(c,\xi_{-|i|+1}),$

$$d_1((x,y)) := \iota_{h,\xi_{|i|-1}}(x) + \iota_{h,\xi_{-|i|+1}}(y),$$

$1 < k < |i| :$ *for* $(x,y) \in M(c,\xi_{|i|-k}) \oplus M(c,\xi_{-|i|+k}),$

$$d_k((x,y)) := (\ \iota_{\xi_{|i|-k+1},\xi_{|i|-k}}(x) + \iota_{\xi_{|i|-k+1},\xi_{-|i|+k}}(y),$$
$$- \iota_{\xi_{-|i|+k-1},\xi_{|i|-k}}(x) - \iota_{\xi_{-|i|+k-1},\xi_{-|i|+k}}(y)\ ),$$

$k = |i|$: *for* $x \in M(c, \xi_0)$,

$$d_{|i|}(x) := (\ \iota_{\xi_1,\xi_0}(x), \quad -\iota_{\xi_{-1},\xi_0}(x)\ ),$$

2. **Case** 2^-, 3^-, 4^-: $i \in \mathbb{Z}_{>0}$,

$$0 \longrightarrow M(c,\xi_{i-1}) \xrightarrow{d_1} M(c,h) \xrightarrow{d_0} L(c,h) \longrightarrow 0,$$

where the maps d_k *are given by*

$k = 0 : d_0 := \pi_h$,
$k = 1 : d_1 := \iota_{h,\xi_{i-1}}$.

Proof. The proof is similar to the case of **Class** R^+. We omit the detail. □

6.4 Characters of Irreducible Highest Weight Representations

As an application of the BGG type resolutions, we compute the character of $L(c,h)$ for any $(c,h) \in \mathbb{C}^2$.

6.4.1 Normalised Character

First, we introduce the *normalised character* $\chi_M(\tau)$ for $M \in \mathrm{Ob}(\mathcal{O})$ with weight space decomposition

$$M = \bigoplus_{\lambda \in \mathfrak{h}^*} M_\lambda.$$

To discuss modular invariance property of the normalised characters, we use the following convention: Let $\mathbb{H}$ be the Siegel upper half-plane $\{a + b\sqrt{-1} \in \mathbb{C} | a, b \in \mathbb{R}, b > 0\}$. For $\tau \in \mathbb{H}$, we set $q := e^{2\pi\sqrt{-1}\tau}$, and for $(c,h) \in \mathbb{C}^2 \simeq \mathfrak{h}^*$, we specialise $e(c,h) \in \tilde{\mathcal{E}}$ to $q^{h-\frac{1}{24}c}$.

Definition 6.1 *For* $M \in \mathrm{Ob}(\mathcal{O})$, *we define* $\chi_M(\tau)$ *as follows:*

$$\chi_M(\tau) := \sum_{\lambda \in \mathfrak{h}^*} \dim M_\lambda \ q^{\lambda(L_0) - \frac{1}{24}\lambda(C)}. \tag{6.13}$$

In the next subsection, we express $\chi_{L(c,h)}(\tau)$ by using the **Dedekind η-function** $\eta(\tau)$

$$\eta(\tau) := q^{\frac{1}{24}} \prod_{n\in\mathbb{Z}_{>0}} (1-q^n), \tag{6.14}$$

and the **classical theta function** $\Theta_{n,m}(\tau)$ ($m \in \mathbb{Z}_{>0}$, $n \in \mathbb{Z}/2m\mathbb{Z}$)

$$\Theta_{n,m}(\tau) := \sum_{k\in\mathbb{Z}} q^{m\left(k+\frac{n}{2m}\right)^2}. \tag{6.15}$$

By the definition of the normalised character, we have

Lemma 6.4.

$$\chi_{M(c,h)}(\tau) = q^{h-\frac{1}{24}(c-1)}\eta(\tau)^{-1}.$$

6.4.2 Characters of the Irreducible Highest Weight Representations

In this subsection, we compute the characters of the irreducible highest weight representations.

Theorem 6.11 (Class V)

$$\chi_{L(c,h)}(\tau) = \chi_{M(c,h)}(\tau) = q^{h-\frac{1}{24}(c-1)}\eta(\tau)^{-1}. \tag{6.16}$$

Theorem 6.12 (Class I) *$c = c(t)$ and $h = h_{\alpha,\beta}(t)$ for some $t \in \mathbb{C}\setminus\mathbb{Q}$ and $\alpha, \beta \in \mathbb{Z}_{>0}$,*

$$\chi_{L(c,h)}(\tau) = \left[q^{\frac{1}{4}(\alpha t^{\frac{1}{2}}-\beta t^{-\frac{1}{2}})^2} - q^{\frac{1}{4}(\alpha t^{\frac{1}{2}}+\beta t^{-\frac{1}{2}})^2}\right]\eta(\tau)^{-1}. \tag{6.17}$$

Theorem 6.13 (Class R^+) *$c = c_{p,q}$ ($p, q \in \mathbb{Z}_{>0}$ such that $(p,q) = 1$),*

Case 1^+: *$0 < r < p \wedge 0 < s < q$ and $h = h_i$ ($i \in \mathbb{Z}$),*

$$\chi_{L(c,h)}(\tau) = (-1)^i\left[\Theta_{rq-sp,pq}(\tau) - \Theta_{rq+sp,pq}(\tau) - r_i(\tau)\right]\eta(\tau)^{-1}, \tag{6.18}$$

where

$$r_i(\tau) := \sum_{\substack{|k|\le|i|\\ k\neq i}} (-1)^k q^{\frac{1}{4pq}\left\{2\lceil\frac{k+1}{2}\rceil pq-rq+(-1)^k sp\right\}^2},$$

Case 2^+: *$r = 0 \wedge 0 < s < q$ and $h = h_i$ ($i \in \mathbb{Z}_{\ge 0}$),*

$$\chi_{L(c,h)}(\tau) = \Big[q^{\frac{1}{4pq}\{2\lceil\frac{i+1}{2}\rceil pq+(-1)^i sp\}^2} - q^{\frac{1}{4pq}\{2\lceil\frac{i+2}{2}\rceil pq-(-1)^i sp\}^2}\Big]\eta(\tau)^{-1},$$

Case 3^+: $0 < r < p \wedge s = 0$ *and* $h = h_{(-1)^{i-1}i}$ $(i \in \mathbb{Z}_{\geq 0})$,

$$\chi_{L(c,h)}(\tau) = \left[q^{\frac{1}{4pq}\left\{2\lceil\frac{(-1)^{i-1}i+1}{2}\rceil pq-rq\right\}^2} - q^{\frac{1}{4pq}\left\{2\lceil\frac{(-1)^i(i+1)+1}{2}\rceil pq-rq\right\}^2}\right]\eta(\tau)^{-1},$$

Case 4^+*:* $(r,s) = (0,0),(0,q)$ *and* $h = h_{2i}$ $(i \in \mathbb{Z}_{\geq 0})$,

$$\chi_{L(c,h)}(\tau) = \left[q^{\frac{1}{4}pq\left(2i+\frac{s}{q}\right)^2} - q^{\frac{1}{4}pq\left(2i+2+\frac{s}{q}\right)^2}\right]\eta(\tau)^{-1},$$

where $\lceil x\rceil$ *denotes the greatest integer not exceeding* x.

Proof. We only prove **Case** 1^+. By applying the Euler–Poincaré principle to the BGG type resolution of $L(c,h_i)$ in Theorem 6.9, we obtain

$$\chi_{L(c,h_i)}(\tau) = \sum_{\substack{|k|\geq|i| \\ k\neq -i}} (-1)^{i+k}\chi_{M(c,h_k)}(\tau).$$

Indeed, this is possible, since for each weight subspace, the right-hand side is a finite sum. Hence, we have

$$\begin{aligned}
&\chi_{L(c,h_i)}(\tau)\\
&= \left[\sum_{\substack{|k|\geq|i| \\ k\neq -i}} (-1)^{i+k}q^{h_k-\frac{1}{24}(c-1)}\right]\eta(\tau)^{-1}\\
&= (-1)^i\left[\sum_{k\in\mathbb{Z}}(-1)^k q^{h_k-\frac{1}{24}(c-1)} - \sum_{\substack{|k|\leq|i| \\ k\neq i}}(-1)^k q^{h_k-\frac{1}{24}(c-1)}\right]\eta(\tau)^{-1}.
\end{aligned}$$

Since

$$r_i(\tau) = \sum_{\substack{|k|\leq|i| \\ k\neq i}}(-1)^k q^{h_k-\frac{1}{24}(c-1)},$$

we obtain the formulae. □

Similarly, we can prove the following theorem:

Theorem 6.14 (Class R^-) $c = c_{p,-q}$ *($p, q \in \mathbb{Z}_{>0}$ such that $(p,q) = 1$),*

Case 1^-: $0 < r < p \wedge 0 < -s < q$ *and* $h = h_i$ $(i \in \mathbb{Z} \setminus \{0\})$,

$$\chi_{L(c,h)}(\tau) = (-1)^i \left[\sum_{\substack{|k| \le |i| \\ k \ne -i}} (-1)^k q^{-\frac{1}{4pq}\{2\lceil \frac{k+1}{2} \rceil pq - rq - (-1)^k sp\}^2} \right] \eta(\tau)^{-1},$$

Case 2^-: $r = 0 \wedge 0 < -s < q$, *and* $h = h_i$ $(i \in \mathbb{Z}_{>0})$,

$$\chi_{L(c,h)}(\tau) = \Big[q^{-\frac{1}{4pq}\{2\lceil \frac{i+1}{2} \rceil pq - (-1)^i sp\}^2} - q^{-\frac{1}{4pq}\{2\lceil \frac{i}{2} \rceil pq + (-1)^i sp\}^2} \Big] \eta(\tau)^{-1},$$

Case 3^-: $0 < r < p \wedge s = 0$ *and* $h = h_{(-1)^{i-1}i}$ $(i \in \mathbb{Z}_{>0})$,

$$\chi_{L(c,h)}(\tau) = \left[q^{-\frac{1}{4pq}\left\{2\lceil \frac{(-1)^{i-1}i+1}{2} \rceil pq - rq\right\}^2} - q^{-\frac{1}{4pq}\left\{2\lceil \frac{(-1)^{i}(i-1)+1}{2} \rceil pq - rq\right\}^2} \right] \eta(\tau)^{-1},$$

Case 4^-: $(r,s) = (0,0), (0,-q)$ *and* $h = h_{2i}$ $(i \in \mathbb{Z}_{>0})$,

$$\chi_{L(c,h)}(\tau) = \left[q^{-\frac{1}{4}pq(2i - \frac{s}{q})^2} - q^{-\frac{1}{4}pq(2i-2-\frac{s}{q})^2} \right] \eta(\tau)^{-1}.$$

Corollary 6.1 Characters of the BPZ series representations: *Suppose that (c,h) is a highest weight of the BPZ series representations, i.e., $c = c_{p,q}$ for $p, q \in \mathbb{Z}_{>1}$ such that $(p,q) = 1$ and $h = h_{r,s}(\frac{q}{p})$ for $r, s \in \mathbb{Z}_{>0}$ such that $r < p$ and $s < q$. Then, the normalised character of the irreducible representation $L(c,h)$ is given by*

$$\chi_{L(c,h)}(\tau) = [\Theta_{rq-sp,pq}(\tau) - \Theta_{rq+sp,pq}(\tau)]\, \eta(\tau)^{-1}. \tag{6.19}$$

The special case $(p,q) = (2,3)$ and $(r,s) = (1,1)$ of this corollary provides us an interesting formula. In fact, we have $(c,h) = (0,0)$ in this case and $L(0,0)$ is a trivial one-dimensional Vir-module. Hence, $\chi_{L(0,0)} = 1$ by definition and Corollary 6.1 gives us the denominator identity for the Virasoro algebra. Explicitly, it shows $\eta(\tau) = \Theta_{1,6}(\tau) - \Theta_{5,6}(\tau)$ which is equivalent to the formula

$$\prod_{n=1}^{\infty}(1-q^n)=\sum_{m\in\mathbb{Z}}(-1)^m q^{\frac{1}{2}m(3m-1)}$$

by (6.14) and (6.15). This is the formula known as **Euler's pentagonal number theorem**. For its combinatorial proof, see, e.g., [And].

6.4.3 Multiplicity

As a corollary of Theorems 6.9 and 6.10, we compute the multiplicity $[M(c,h):L(c',h')]$ for any $(c,h),\ (c',h')\in\mathbb{C}^2$.

Since $[M(c,h):L(c',h')]=0$ if $c\neq c'$, we list the pairs (h,h') such that $[M(c,h):L(c,h')]\neq 0$ for each central charge c.

Proposition 6.2 **Class** V:

$$[M(c,h):L(c,h')]=\begin{cases}1 & h'=h\\ 0 & otherwise\end{cases}.$$

Class I: *Suppose that* $c=c(t)$ *and* $h=h_{\alpha,\beta}(t)$ *for* $t\in\mathbb{C}\setminus\mathbb{Q}$. *Then*

$$[M(c,h):L(c,h')]=\begin{cases}1 & h'=h,\ h+\alpha\beta\\ 0 & otherwise\end{cases}.$$

Class R^+: *Suppose that* $c=c_{p,q}$ *for* $p,q\in\mathbb{Z}_{>0}$ *such that* $(p,q)=1$.

Case 1^+: $h=h_i\quad(i\in\mathbb{Z})$,

$$[M(c,h):L(c,h')]=\begin{cases}1 & h'=h_k\ (k\in\mathbb{Z},\ |k|\geq|i|,\ k\neq -i)\\ 0 & otherwise\end{cases},$$

Case 2^+: $h=h_i\quad(i\in\mathbb{Z}_{\geq 0})$,

$$[M(c,h):L(c,h')]=\begin{cases}1 & h'=h_k\ (k\in\mathbb{Z}_{\geq 0},\ k\geq i)\\ 0 & otherwise\end{cases},$$

Case 3^+: $h=h_{(-1)^{i-1}i}\quad(i\in\mathbb{Z}_{\geq 0})$,

$$[M(c,h):L(c,h')]=\begin{cases}1 & h'=h_{(-1)^{k-1}k}\ (k\in\mathbb{Z}_{\geq 0},\ k\geq i)\\ 0 & otherwise\end{cases},$$

Case 4^+: $h=h_{2i}\quad(i\in\mathbb{Z}_{\geq 0})$,

$$[M(c,h):L(c,h')] = \begin{cases} 1 & h' = h_{2k} \ (k \in \mathbb{Z}_{\geq 0}, \ k \geq i) \\ 0 & otherwise \end{cases},$$

where L_0*-weights* $\{h_i\}$ *are defined in (5.22).*

Class R^-: *Suppose* $c = c_{p,-q}$ *for* $p, q \in \mathbb{Z}_{>0}$ *such that* $(p,q) = 1$.

Case 1^-: $h = h_i \quad (i \in \mathbb{Z} \setminus \{0\})$,

$$[M(c,h):L(c,h')] = \begin{cases} 1 & h' = h_k \ (k \in \mathbb{Z}, \ |k| \leq |i|, \ k \neq -i) \\ 0 & otherwise \end{cases},$$

Case 2^-: $h = h_i \quad (i \in \mathbb{Z}_{>0})$,

$$[M(c,h):L(c,h')] = \begin{cases} 1 & h' = h_k \ (k \in \mathbb{Z}_{\geq 0}, \ k \leq i) \\ 0 & otherwise \end{cases},$$

Case 3^-: $h = h_{(-1)^{i-1}i} \quad (i \in \mathbb{Z}_{>0})$,

$$[M(c,h):L(c,h')] = \begin{cases} 1 & h' = h_{(-1)^{k-1}k} \ (k \in \mathbb{Z}_{\geq 0}, \ k \leq i) \\ 0 & otherwise \end{cases},$$

Case 4^-: $h = h_{2i} \quad (i \in \mathbb{Z}_{>0})$,

$$[M(c,h):L(c,h')] = \begin{cases} 1 & h' = h_{2k} \ (k \in \mathbb{Z}_{\geq 0}, \ k \leq i) \\ 0 & otherwise \end{cases},$$

where L_0*-weights* $\{h_i\}$ *are defined in (5.26).*

Proof. Here, we show this proposition in **Case** 1^+. By Theorem 6.9 and Proposition 1.8, for $i, j \in \mathbb{Z}$ such that $|j| \geq |i|$ we have

$$\delta_{i,j} = [M(c,\xi_i):L(c,\xi_j)] + \sum_{\substack{|j|\geq|k|>|i| \\ k\neq -j}} (-1)^{k-i}[M(c,\xi_k):L(c,\xi_j)].$$

Hence, by induction on $|j| - |i|$, we obtain the results. For the other cases, one can similarly check this proposition. □

6.4.4 Modular Transformation

We first state some fundamental properties of the modular forms $\eta(\tau)$ and $\Theta_{n,m}(\tau)$. Recall that $SL_2(\mathbb{Z})$ acts on the upper half-plane $\mathbb{H}$ by

$$\begin{pmatrix} a & b \\ c & d \end{pmatrix} \cdot \tau = \frac{a\tau + b}{c\tau + d}.$$

Moreover, if we set

$$S := \begin{pmatrix} 0 & -1 \\ 1 & 0 \end{pmatrix}, \quad T := \begin{pmatrix} 1 & 1 \\ 0 & 1 \end{pmatrix},$$

then S and T generate the group $SL_2(\mathbb{Z})$ and

$$T \cdot \tau = \tau + 1, \quad S \cdot \tau = -\frac{1}{\tau}.$$

From now on, for $\tau \in \mathbb{H}$, we fix the branch of τ as $0 < \arg\tau < \pi$. With respect to this action, $\eta(\tau)$ and $\Theta_{n,m}(\tau)$ transform as follows:

Lemma 6.5. *1.*

$$\eta(\tau+1) = e^{\frac{1}{12}\pi\sqrt{-1}}\eta(\tau), \quad \eta(-\frac{1}{\tau}) = \left(\frac{\tau}{\sqrt{-1}}\right)^{\frac{1}{2}} \eta(\tau).$$

2.

$$\Theta_{n,m}(\tau+1) = e^{\frac{n^2}{2m}\pi\sqrt{-1}}\Theta_{n,m}(\tau),$$

$$\Theta_{n,m}(-\frac{1}{\tau}) = \left(\frac{\tau}{2m\sqrt{-1}}\right)^{\frac{1}{2}} \sum_{n' \in \mathbb{Z}/2m\mathbb{Z}} e^{-\frac{nn'}{m}\pi\sqrt{-1}}\Theta_{n',m}(\tau).$$

Proof. These facts are well known (cf. [Chan]). □

Here, let us state the **modular invariance property** of the normalised characters $\chi_{L(c,h)}(\tau)$ of the BPZ series representations $L(c,h)$. Suppose that $c = c_{p,q}$ for $p, q \in \mathbb{Z}_{>1}$ such that $(p,q) = 1$. Recall that (c,h) is the highest weight of a BPZ series representation if and only if there exists

$$(r,s) \in (K^+_{p,q})^\circ := \left\{ (r,s) \in \mathbb{Z}^2 \,\middle|\, \begin{matrix} 0 < r < p, \\ 0 < s < q \end{matrix} rq + sp \le pq \right\}$$

such that $h = h_{r,s}(\frac{q}{p})$.

For simplicity, we set

$$\chi_{r,s}(\tau) := \chi_{L(c_{p,q},h_{r,s}(\frac{q}{p}))}(\tau).$$

By Corollary 6.1, we have

$$\chi_{r,s}(\tau) = [\Theta_{rq-sp,pq}(\tau) - \Theta_{rq+sp,pq}(\tau)]\,\eta(\tau)^{-1}.$$

Now, the transformation law of the normalised characters $\chi_{r,s}(\tau)$ is described as follows:

Proposition 6.3 *Let us take positive integers p, q, r and s as above. Then, we have*

$$\chi_{r,s}(\tau+1) = e^{\{\frac{(rq-sp)^2}{2pq}-\frac{1}{12}\}\pi\sqrt{-1}}\chi_{r,s}(\tau), \tag{6.20}$$

$$\chi_{r,s}(-\frac{1}{\tau}) = \sum_{(r',s')\in(K_{p,q}^+)^\circ} S_{(r,s),(r',s')}\chi_{r',s'}(\tau), \tag{6.21}$$

where

$$S_{(r,s),(r',s')} := \sqrt{\frac{8}{pq}}(-1)^{(r+s)(r'+s')}\sin\left(\frac{\pi r r'}{p}(p-q)\right)\sin\left(\frac{\pi s s'}{q}(p-q)\right).$$

Proof. The transformation law for $\tau \mapsto \tau+1$ is easy. Indeed, we have

$$\chi_{r,s}(\tau+1) = \Big[e^{\{\frac{(rq-sp)^2}{2pq}-\frac{1}{12}\}\pi\sqrt{-1}}\Theta_{rq-sp,pq}(\tau) \\ - e^{\{\frac{(rq+sp)^2}{2pq}-\frac{1}{12}\}\pi\sqrt{-1}}\Theta_{rq+sp,pq}(\tau)\Big]\,\eta(\tau)^{-1}.$$

Since

$$\frac{(rq-sp)^2}{2pq}-\frac{1}{12} \equiv \frac{(rq+sp)^2}{2pq}-\frac{1}{12} \mod 2,$$

we obtain (6.20).

Next, we show the transformation law for $\tau \mapsto -\frac{1}{\tau}$.

Let N_i be the sets defined as (5.24). Set $N^\circ := N_1$, $\partial N := N_2 \cup N_3$. By the proof of Lemma 5.9, the following hold:

Fact 6.1 *1.* $N^\circ = \{\pm(rq+sp) \pmod{2pq},\ \pm(rq-sp) \pmod{2pq} \mid (r,s)\in (K_{p,q}^+)^\circ\}$ *and* $\sharp N^\circ = 4\times\sharp(K_{p,q}^+)^\circ$,
2. $\mathbb{Z}/2pq\mathbb{Z} = N^\circ \sqcup \partial N$ *(disjoint).*

Lemma 6.5 implies

$$\chi_{r,s}(-\frac{1}{\tau}) = \sum_{n'\in\mathbb{Z}/2pq\mathbb{Z}} \frac{1}{\sqrt{2pq}}\left\{e^{-\frac{(rq-sp)n'}{pq}\pi\sqrt{-1}} - e^{-\frac{(rq+sp)n'}{pq}\pi\sqrt{-1}}\right\}\Theta_{n',pq}(\tau)\eta(\tau)^{-1} \\ = \frac{2\sqrt{-1}}{\sqrt{2pq}}\sum_{n'\in\mathbb{Z}/2pq\mathbb{Z}} e^{-\frac{rn'}{p}\pi\sqrt{-1}}\sin\left(\frac{sn'}{q}\pi\right)\Theta_{n',pq}(\tau)\eta(\tau)^{-1}.$$

Noticing $\Theta_{n',pq}(\tau) = \Theta_{-n',pq}(\tau)$, we have

$$\chi_{r,s}(-\frac{1}{\tau}) = \frac{2\sqrt{-1}}{\sqrt{2pq}}\sum_{n'\in\mathbb{Z}/2pq\mathbb{Z}} \frac{1}{2}\{e^{-\frac{rn'}{p}\pi\sqrt{-1}} - e^{\frac{rn'}{p}\pi\sqrt{-1}}\}\sin\left(\frac{sn'}{q}\pi\right)\Theta_{n',pq}(\tau)\eta(\tau)^{-1} \\ = \sqrt{\frac{2}{pq}}\sum_{n'\in\mathbb{Z}/2pq\mathbb{Z}} \sin\left(\frac{rn'\pi}{p}\right)\sin\left(\frac{sn'\pi}{q}\right)\Theta_{n',pq}(\tau)\eta(\tau)^{-1}.$$

If $n' \in \partial N$, then $\sin\left(\frac{rn'\pi}{p}\right)\sin\left(\frac{sn'\pi}{q}\right) = 0$. Hence, by Fact 6.1, we have

$$\chi_{r,s}(-\frac{1}{\tau}) = \sqrt{\frac{2}{pq}} \sum_{n' \in N^\circ} \sin\left(\frac{rn'\pi}{p}\right)\sin\left(\frac{sn'\pi}{q}\right)\Theta_{n',pq}(\tau)\eta(\tau)^{-1}.$$

Moreover, we have, for $n' \in N^\circ$ such that $n' = \pm(r'q - s'p)$,

$$\begin{aligned}&\sin\left(\frac{rn'\pi}{p}\right)\sin\left(\frac{sn'\pi}{q}\right) = (-1)^{rs'+r's}\sin\left(\frac{rr'q\pi}{p}\right)\sin\left(-\frac{ss'p\pi}{q}\right)\\ &= (-1)^{(r+r')(s+s')}\sin\left(\frac{\pi rr'}{p}(p-q)\right)\sin\left(\frac{\pi ss'}{q}(p-q)\right),\end{aligned}$$

and for $n' \in N^\circ$ such that $n' = \pm(r'q + s'p)$,

$$\begin{aligned}&\sin\left(\frac{rn'\pi}{p}\right)\sin\left(\frac{sn'\pi}{q}\right) = (-1)^{rs'+r's}\sin\left(\frac{rr'q\pi}{p}\right)\sin\left(\frac{ss'p\pi}{q}\right)\\ &= -(-1)^{(r+r')(s+s')}\sin\left(\frac{\pi rr'}{p}(p-q)\right)\sin\left(\frac{\pi ss'}{q}(p-q)\right).\end{aligned}$$

Hence, Fact 6.1 implies

$$\begin{aligned}&\chi_{r,s}(-\frac{1}{\tau})\\ =&2\sqrt{\frac{2}{pq}} \sum_{(r',s') \in (K^+_{p,q})^\circ} (-1)^{(r+r')(s+s')}\sin\left(\frac{\pi rr'}{p}(p-q)\right)\sin\left(\frac{\pi ss'}{q}(p-q)\right)\\ &\qquad \times [\Theta_{r'q-s'p,pq}(\tau) - \Theta_{r'q+s'p,pq}(\tau)]\eta(\tau)^{-1}\\ =& \sum_{(r',s') \in (K^+_{p,q})^\circ} S_{(r,s),(r',s')}\chi_{r',s'}(\tau)\end{aligned}$$

and (6.21) holds. □

6.4.5 Asymptotic Dimension

In this section, we consider the behaviour of the normalised character $\chi_{L(c,h)}(\tau)$ under the limit $\tau \downarrow 0$, where $\tau \downarrow 0$ means the limit $T \to 0$ with $\tau := \sqrt{-1}T$ ($T \in \mathbb{R}_{>0}$). We use the symbol $f(\tau) \underset{\tau\downarrow 0}{\sim} g(\tau)$ which signifies

$$\lim_{T \to 0} \frac{f(\tau)}{g(\tau)} = 1.$$

Following [KW2], let us introduce the asymptotic dimension of $L(c,h)$.

Definition-Lemma 6.1 *For each irreducible highest weight module $L(c,h)$, there exist constants A, B and C such that*

$$\chi_{L(c,h)}(\tau) \underset{\tau\downarrow 0}{\sim} A\left(\frac{\tau}{\sqrt{-1}}\right)^{\frac{B}{2}} e^{\frac{\pi\sqrt{-1}}{12\tau}C}.$$

The triplet (A,B,C) is called the **asymptotic dimension** *of $L(c,h)$.*

Remark that one can compute the asymptotic behaviour of $\dim L(c,h)_n$ $(n\to\infty)$ by means of the asymptotic dimension of $L(c,h)$. For the detail, see Theorem 9.9.

The existence of the triplet (A,B,C) for each $L(c,h)$ is guaranteed by the following theorem:

Theorem 6.15 *The following list gives the asymptotic dimension of $L(c,h)$:*

Class	Case	(c,h)	(A,B,C)
V		(c,h)	$(1,1,1)$
I		$(c(t),h_{\alpha,\beta}(t))$	$(2\pi\alpha\beta,3,1)$
R^+	1^+	$(c_{p,q},h_0)$	$(S_{(r,s),(r',s')},0,1-\frac{6}{pq})$
		$(c_{p,q},h_i)$ $(i\in\mathbb{Z}_{>0})$	$(2\pi irs_{\overline{i+1}},3,1)$
		$(c_{p,q},h_i)$ $(i\in\mathbb{Z}_{<0})$	$(2\pi(-i)(p-r)s_{\bar{i}},3,1)$
	2^+	$(c_{p,q},h_i)$ $(i\in\mathbb{Z}_{\geq 0})$	$(2\pi(i+1)ps_{\overline{i+1}},3,1)$
	3^+	$(c_{p,q},h_{(-1)^{i-1}i})$ $(i\in\mathbb{Z}_{\geq 0})$	$(2\pi(i+1)r_{\overline{i+1}}q,3,1)$
	4^+	$(c_{p,q},h_{2i})$ $(i\in\mathbb{Z}_{\geq 0})$	$(2\pi(2i+1+\frac{s}{q})pq,3,1)$
R^-	1^-	$(c_{p,-q},h_i)$ $(i\in\mathbb{Z}_{>0})$	$(2\pi i(p-r)s'_{\bar{i}},3,1)$
		$(c_{p,-q},h_i)$ $(i\in\mathbb{Z}_{<0})$	$(2\pi(-i)rs'_{\overline{i+1}},3,1)$
	2^-	$(c_{p,-q},h_i)$ $(i\in\mathbb{Z}_{>0})$	$(2\pi ips'_{\bar{i}},3,1)$
	3^-	$(c_{p,-q},h_{(-1)^{i-1}i})$ $(i\in\mathbb{Z}_{>0})$	$(2\pi ir_{\bar{i}}q,3,1)$
	4^-	$(c_{p,-q},h_{2i})$ $(i\in\mathbb{Z}_{>0})$	$(2\pi(2i-1-\frac{s}{q})pq,3,1)$

where (r',s') in **Case** 1^+ *is a unique element of $(K^+_{p,q})^\circ$ such that $r'q-s'p\in\{\pm1\}$ and $\bar{i}:=i+2\mathbb{Z}\in\mathbb{Z}/2\mathbb{Z}$,*

$$r_\sigma := \begin{cases} r & (\sigma=\bar{0}) \\ p-r & (\sigma=\bar{1}) \end{cases}, \quad s_\sigma := \begin{cases} s & (\sigma=\bar{0}) \\ q-s & (\sigma=\bar{1}) \end{cases}, \quad s'_\sigma := \begin{cases} -s & (\sigma=\bar{0}) \\ q+s & (\sigma=\bar{1}) \end{cases}.$$

In the following, we show Theorem 6.15. We first look at the asymptotic behaviour of $\eta(\tau)$ and $\Theta_{n,m}(\tau)$ by using their modular transformations.

For the eta function, we have $\eta(\tau)=\left(\frac{\tau}{\sqrt{-1}}\right)^{-\frac{1}{2}}\eta(-\frac{1}{\tau})$ by Lemma 6.5. Set $x:=e^{-\frac{2\pi\sqrt{-1}}{\tau}}$. Then, $x\to 0$ as $\tau\downarrow 0$. Hence, $\eta(-\frac{1}{\tau})=x^{\frac{1}{24}}\prod_{n=1}^{\infty}(1-x^n)\underset{\tau\downarrow 0}{\sim} x^{\frac{1}{24}}$, and thus,

$$\eta(\tau)\underset{\tau\downarrow 0}{\sim}\left(\frac{\tau}{\sqrt{-1}}\right)^{-\frac{1}{2}} e^{-\frac{\pi\sqrt{-1}}{12\tau}}. \tag{6.22}$$

For the theta function, by Lemma 6.5, we have

$$\Theta_{n,m}(\tau) = \left(\frac{\sqrt{-1}}{2m\tau}\right)^{\frac{1}{2}} \sum_{n' \in \mathbb{Z}/2m\mathbb{Z}} e^{-\frac{nn'}{m}\pi\sqrt{-1}} \Theta_{n',m}(-\frac{1}{\tau}). \tag{6.23}$$

We first show that

$$\Theta_{n',m}(-\frac{1}{\tau}) \underset{\tau\downarrow 0}{\sim} \begin{cases} x^{\frac{(n'')^2}{4m}} & (n' \not\equiv m \bmod 2m) \\ 2x^{\frac{(n'')^2}{4m}} & (n' \equiv m \bmod 2m) \end{cases}. \tag{6.24}$$

Here, $n'' \in \{\tilde{n} | -m < \tilde{n} \le m\}$ is the representatives of $n' \in \mathbb{Z}/2m\mathbb{Z}$. We may assume $0 \le n'' \le m$ since $\Theta_{-n'',m}(-\frac{1}{\tau}) = \Theta_{n'',m}(-\frac{1}{\tau})$. Since $(k + \frac{n''}{2m})^2 \ge |k + \frac{n''}{2m}|$ for $k \neq 0, -1$, we have

$$\Theta_{n',m}(-\frac{1}{\tau}) = \sum_{k\in\mathbb{Z}} x^{m(k+\frac{n''}{2m})^2} \le \sum_{k=0,-1} x^{m(k+\frac{n''}{2m})^2} + \sum_{k\neq 0,-1} x^{m|k+\frac{n''}{2m}|}$$

for $0 < x < 1$. Moreover,

$$\sum_{k\neq 0,-1} x^{m|k+\frac{n''}{2m}|} = (x^{m(2-\frac{n''}{2m})} + x^{m(1+\frac{n''}{2m})})/(1-x^m),$$

$(\frac{n''}{2m})^2 \le (-1 + \frac{n''}{2m})^2 \le 1$ and $(\frac{n''}{2m})^2 < (-1 + \frac{n''}{2m})^2$ for $n'' \neq m$. Hence, we obtain (6.24). Thus, by (6.23), the following formula holds:

$$\Theta_{n,m}(\tau) \underset{\tau\downarrow 0}{\sim} \frac{1}{\sqrt{2m}} \left(\frac{\tau}{\sqrt{-1}}\right)^{-\frac{1}{2}}. \tag{6.25}$$

Next, we show Theorem 6.15.

<u>Class V</u> In this class, $\chi_{L(c,h)}(\tau)$ is given by (6.16). Since $q \to 1$ as $\tau \downarrow 0$, by (6.22) we have

$$\chi_{L(c,h)}(\tau) \underset{\tau\downarrow 0}{\sim} \left(\frac{\tau}{\sqrt{-1}}\right)^{\frac{1}{2}} e^{\frac{\pi\sqrt{-1}}{12\tau}},$$

and hence, Theorem 6.15 for **Class** V holds.

<u>Class I</u> The normalised character of $L(c(t), h_{\alpha,\beta}(t))$ is given by (6.17). For the numerator of $\chi_{L(c(t),h_{\alpha,\beta}(t))}(\tau)$,

$$q^{\frac{1}{4}(\alpha t^{\frac{1}{2}} - \beta t^{-\frac{1}{2}})^2} - q^{\frac{1}{4}(\alpha t^{\frac{1}{2}} + \beta t^{-\frac{1}{2}})^2} = q^{\frac{1}{4}(\alpha t^{\frac{1}{2}} - \beta t^{-\frac{1}{2}})^2}(1 - e^{-2\pi\alpha\beta T}) \underset{\tau\downarrow 0}{\sim} 2\pi\alpha\beta T. \tag{6.26}$$

Hence, (6.22) implies

$$\chi_{L(c(t),h_{\alpha,\beta}(t))}(\tau) \underset{\tau\downarrow 0}{\sim} 2\pi\alpha\beta\left(\frac{\tau}{\sqrt{-1}}\right)^{\frac{3}{2}} e^{\frac{\pi\sqrt{-1}}{12\tau}},$$

and thus, Theorem 6.15 for **Class** I holds.

<u>Class R</u> Here, we demonstrate the proof for **Case** 1^+, since for the other cases, the proof is easier or an argument similar to **Case** 1^+ works. For simplicity, we set $c := c_{p,q}$.

First, we consider the case where $i = 0$. In this case, the normalised character $\chi_{r,s}(\tau)$ is given by (6.19). Noticing the modular transformation (6.21), we look at the asymptotic behaviour of $\chi_{r',s'}(-\frac{1}{\tau})$ for $(r',s') \in (K^+_{p,q})^\circ$.

Recall that $|r'q \pm s'p| < pq$ for $(r',s') \in (K^+_{p,q})^\circ$. Hence, (6.24) implies

$$\Theta_{r'q-s'p,pq}(-\frac{1}{\tau}) \underset{\tau\downarrow 0}{\sim} x^{\frac{(r'q-s'p)^2}{4pq}}, \quad \Theta_{r'q+s'p,pq}(-\frac{1}{\tau}) \underset{\tau\downarrow 0}{\sim} x^{\frac{(r'q+s'p)^2}{4pq}},$$

where $x := e^{-\frac{2\pi\sqrt{-1}}{\tau}}$. Since $(r'q - s'p)^2 < (r'q + s'p)^2$, we have

$$\chi_{r',s'}(-\frac{1}{\tau}) \underset{\tau\downarrow 0}{\sim} x^{\frac{(r'q-s'p)^2}{4pq} - \frac{1}{24}} = e^{\frac{\pi\sqrt{-1}}{12\tau}(1-\frac{6(r'q-s'p)^2}{pq})}.$$

By the proof of Lemma 5.9, there uniquely exists $(r',s') \in (K^+_{p,q})^\circ$ such that $r'q - s'p \in \{\pm 1\}$. By (6.21), we have

$$\chi_{r,s}(\tau) \underset{\tau\downarrow 0}{\sim} S_{(r,s),(r',s')} e^{\frac{\pi\sqrt{-1}}{12\tau}(1-\frac{6}{pq})}. \tag{6.27}$$

Hence, we have shown Theorem 6.15 for **Case** 1^+ $(i = 0)$.

Next, we consider the case where $i \neq 0$. In this case, the normalised character of $L(c, h_i)$ is given by (6.18). Notice that $r_i(\tau)$ in the character formula can be written as

$$r_i(\tau) = \begin{cases} \sum_{k=0}^{i-1}(-1)^k(q^{h_k - \frac{1}{24}(c-1)} - q^{h_{-k-1} - \frac{1}{24}(c-1)}) & (i > 0) \\ \sum_{k=0}^{-i-1}(-1)^k(q^{h_{-k} - \frac{1}{24}(c-1)} - q^{h_{k+1} - \frac{1}{24}(c-1)}) & (i < 0) \end{cases}.$$

By (6.26), we have $r_i(\tau) \underset{\tau\downarrow 0}{\sim} 2\pi T A(i)$, where

$$A(i) := \begin{cases} \sum_{k=0}^{i-1}(-1)^k(h_{-k-1} - h_k) & (i > 0) \\ \sum_{k=0}^{-i-1}(-1)^k(h_{k+1} - h_{-k}) & (i < 0) \end{cases}.$$

Hence, by (6.27) and (6.22), the following holds:

$$\chi_{L(c,h_i)}(\tau) \underset{\tau\downarrow 0}{\sim} (-1)^i S_{(r,s),(r',s')} e^{\frac{\pi\sqrt{-1}}{12\tau}(1-\frac{6}{pq})} - (-1)^i 2\pi A(i)\left(\frac{\tau}{\sqrt{-1}}\right)^{\frac{3}{2}} e^{\frac{\pi\sqrt{-1}}{12\tau}}.$$

Since $1 - \frac{6}{pq} < 1$, we obtain

$$\chi_{L(c,h_i)}(\tau) \underset{\tau \downarrow 0}{\sim} (-1)^{i+1} 2\pi A(i) \left(\frac{\tau}{\sqrt{-1}} \right)^{\frac{3}{2}} e^{\frac{\pi\sqrt{-1}}{12\tau}}.$$

Now, it is sufficient to calculate $A(i)$ explicitly by the definition of h_i (cf. (5.22)).

6.5 Bibliographical Notes and Comments

In 1978, V. Kac [Kac1] announced the character formulae of the irreducible highest weight modules for $c = 1$. In 1983, A. Rocha-Caridi and N. R. Wallach [RW3] constructed the BGG type resolutions of trivial representation and the irreducible highest weight modules in the series. Almost at the sime time, B. L. Feigin and D. B. Fuchs completely analyzed the structure of Verma modules in general, as we explained in § 5.7.

As in [RW3], A. Rocha-Caridi and N. R. Wallach [RW4] constructed the BGG type resolution of the irreducible highest weight modules for $c = 0, 1, 25, 26$ completely in 1984. Here, they did not use the results announced in [FeFu2]. In 1984, B. L. Feigin and D. B. Fuchs [FeFu3] stated the general cases. The modular property of the characters of the minimal series representations were discovered by C. Itzykson and J. B. Zuber [IZ] in 1986.

Here, we have described the structure of Verma modules following an idea due to F. Malikov [Mal], where he dealt with rank 2 Kac–Moody algebras.

In 1988, V. Kac and M. Wakimoto [KW2] introduced the asymptotic dimension for the so-called positive energy representations over infinite dimensional Lie (super)algebras and stated a conjecture on a necessary and sufficient condition for their characters to have the modular invariance property.

Chapter 7
A Duality among Verma Modules

By the structure theorem of Verma modules stated in the previous two chapters, the reader may notice a similarity between **Class** R^+ and **Class** R^-. The purpose of this chapter is to give an explanation of such similarity by the categorical equivalence called the **tilting equivalence** [Ark], [So]. Here, following W. Soergel [So], we construct a categorical equivalence and apply it to the case of the Virasoro algebra.

7.1 Semi-regular Bimodule

Throughout this section, we assume that $\mathbb{K}$ is a field whose characteristic is zero, and $\mathfrak{g} = \bigoplus_{i \in \mathbb{Z}} \mathfrak{g}^i$ is a $\mathbb{Z}$-graded Lie algebra in the sense of § 1.2.1 (with $\Gamma = \mathbb{Z}$). Moreover, we assume that

$$\dim \mathfrak{g}^i < \infty \quad (\forall i \in \mathbb{Z}). \tag{7.1}$$

7.1.1 Preliminaries

We first recall some notation for $\mathbb{Z}$-graded $\mathbb{K}$-vector spaces and $\mathbb{Z}$-graded Lie algebras from Chapter 1.

For $\mathbb{Z}$-graded $\mathbb{K}$-vector spaces $V = \bigoplus_{n \in \mathbb{Z}} V^n$ and $W = \bigoplus_{n \in \mathbb{Z}} W^n$, we set

$$\overline{\mathrm{Hom}_{\mathbb{K}}}(V, W) := \bigoplus_{n \in \mathbb{Z}} \mathrm{Hom}_{\mathbb{K}}(V, W)^n,$$

where

K. Iohara, Y. Koga, *Representation Theory of the Virasoro Algebra*, Springer Monographs in Mathematics, DOI 10.1007/978-0-85729-160-8_7,

$$\begin{aligned}\overline{\mathrm{Hom}}_{\mathbb{K}}(V,W)^j &:= \prod_{n\in\mathbb{Z}} \mathrm{Hom}_{\mathbb{K}}(V^n, W^{n+j})\\ &= \{\varphi \in \mathrm{Hom}_{\mathbb{K}}(V,W) | \varphi(V^n) \subset W^{n+j}\ (n\in\mathbb{Z})\}.\end{aligned}$$

For later use, here, we introduce the formal character $\mathrm{ch}' V$ for a $\mathbb{Z}$-graded $\mathbb{K}$-vector space $V = \bigoplus_{i\in\mathbb{Z}} V^i$ such that $\dim V^i < \infty$. Set

$$\mathrm{ch}' V := \sum_{i\in\mathbb{Z}} (\dim V^i) q^i \in \mathbb{Z}[[q, q^{-1}]]. \tag{7.2}$$

Suppose that $\mathfrak{a}$ and $\mathfrak{b}$ are $\mathbb{Z}$-graded Lie algebras over $\mathbb{K}$, and V (resp. W) is a $\mathbb{Z}$-graded $(\mathfrak{a},\mathfrak{b})$-bimodule (resp. a $\mathbb{Z}$-graded $\mathbb{K}$-vector space). Throughout this chapter, except for the case where we consider the antipode duals, we regard the space $\overline{\mathrm{Hom}}_{\mathbb{K}}(V,W)$ as $(\mathfrak{b},\mathfrak{a})$-bimodule in the following way:

1. The left $\mathfrak{b}$-module structure:

$$(b.\varphi)(v) := \varphi(v.b) \quad (\varphi \in \overline{\mathrm{Hom}}_{\mathbb{K}}(V,W),\ b\in\mathfrak{b},\ v\in V). \tag{7.3}$$

2. The right $\mathfrak{a}$-module structure:

$$(\varphi.a)(v) := \varphi(a.v) \quad (\varphi \in \overline{\mathrm{Hom}}_{\mathbb{K}}(V,W),\ a\in\mathfrak{a},\ v\in V). \tag{7.4}$$

Moreover, let $\mathfrak{s}$ be a $\mathbb{Z}$-graded subalgebra of $\mathfrak{a}$ and let W be a left $\mathfrak{s}$-module.

$$\overline{\mathrm{Hom}}_{\mathfrak{s}}(V,W) := \{\varphi \in \overline{\mathrm{Hom}}_{\mathbb{K}}(V,W) | \varphi(s.v) = s.(\varphi(v))\ (\forall s\in\mathfrak{s},\ v\in V)\}.$$

Remark that $\overline{\mathrm{Hom}}_{\mathfrak{s}}(V,W)$ is a $\mathbb{Z}$-graded left $\mathfrak{b}$-submodule of $\overline{\mathrm{Hom}}_{\mathbb{K}}(V,W)$.
The following two lemmas are useful in this chapter.

Lemma 7.1. *Let $\mathfrak{a}$, $\mathfrak{b}$ and $\mathfrak{c}$ be a $\mathbb{Z}$-graded Lie algebra. Let L be a $\mathbb{Z}$-graded $(\mathfrak{a},\mathfrak{b})$-bimodule, M be a $\mathbb{Z}$-graded $(\mathfrak{b},\mathfrak{c})$-bimodule and N be a $\mathbb{Z}$-graded left $\mathfrak{a}$-module. Then, there exists an isomorphism*

$$\overline{\mathrm{Hom}}_{\mathfrak{a}}(L\otimes_{\mathfrak{b}} M, N) \simeq \overline{\mathrm{Hom}}_{\mathfrak{b}}(M, \overline{\mathrm{Hom}}_{\mathfrak{a}}(L,N)) \tag{7.5}$$

of left $\mathfrak{c}$-modules, where we regard both sides of (7.5) as left $\mathfrak{c}$-module via (7.3).

Proof. By the same argument as the proof of (1.26), one can show that (7.5) is an isomorphism of $\mathbb{Z}$-graded $\mathbb{K}$-vector spaces. The isomorphism is given by

$$\begin{aligned}\Psi : \overline{\mathrm{Hom}}_{\mathfrak{a}}(L\otimes_{\mathfrak{b}} M, N) &\longrightarrow \overline{\mathrm{Hom}}_{\mathfrak{b}}(M, \overline{\mathrm{Hom}}_{\mathfrak{a}}(L,N)),\\ \Psi(\psi)(m)(l) &:= \psi(l\otimes m),\end{aligned}$$

where $\psi \in \overline{\mathrm{Hom}}_{\mathfrak{a}}(L\otimes_{\mathfrak{b}} M, N)$, $m\in M$ and $l\in L$. We show that Ψ is a homomorphism of left $\mathfrak{c}$-modules. We have

$$\Psi(a.\psi)(m)(l) = (a.\psi)(l \otimes m) = \psi((l \otimes m).a) = \psi(l \otimes (m.a))$$
$$= \Psi(\psi)(m.a)(l) = (a.\Psi(\psi))(m)(l).$$

Hence, the lemma holds. □

Lemma 7.2. *Let $\mathfrak{g}$ and $\mathfrak{c}$ be $\mathbb{Z}$-graded Lie algebras over $\mathbb{K}$, and let $\mathfrak{s}$ be a $\mathbb{Z}$-graded Lie subalgebra of $\mathfrak{g}$. Let M be a $\mathbb{Z}$-graded $(\mathfrak{s}, \mathfrak{c})$-bimodule, and let N be a left $\mathfrak{g}$-module. Then, there exists an isomorphism*

$$\overline{\mathrm{Hom}}_{\mathfrak{g}}(\mathrm{Ind}_{\mathfrak{s}}^{\mathfrak{g}} M, N) \simeq \overline{\mathrm{Hom}}_{\mathfrak{s}}(M, \mathrm{Res}_{\mathfrak{s}}^{\mathfrak{g}} N) \tag{7.6}$$

of left $\mathfrak{c}$-modules, where we regard both sides as left $\mathfrak{c}$-modules via (7.3).

Proof. We apply the above lemma to the case where $\mathfrak{a} = \mathfrak{g}$, $\mathfrak{b} = \mathfrak{s}$ and $L = U(\mathfrak{g})$. Then, from (1.27), the lemma follows. □

We recall some notation. For a $\mathbb{Z}$-graded Lie algebra $\mathfrak{g}$ and an integer j, we set

$$\mathfrak{g}^{\geq j} := \bigoplus_{i \geq j} \mathfrak{g}^i, \quad \mathfrak{g}^{\leq j} := \bigoplus_{i \leq j} \mathfrak{g}^i. \tag{7.7}$$

Moreover, we denote $\mathfrak{g}^{\geq 1}$, $\mathfrak{g}^{\leq -1}$, $\mathfrak{g}^{\geq 0}$ and $\mathfrak{g}^{\leq 0}$ by $\mathfrak{g}^+$, $\mathfrak{g}^-$, $\mathfrak{g}^{\geq}$ and $\mathfrak{g}^{\leq}$ respectively. For simplicity, we set

$$\mathfrak{b} := \mathfrak{g}^{\geq}, \quad \mathfrak{n} := \mathfrak{g}^-.$$

Notice that $\mathfrak{g} = \mathfrak{n} \oplus \mathfrak{b}$. For $n \in \mathbb{Z}$, let $\mathbb{K}^{(n)}$ be the $\mathbb{Z}$-graded one-dimensional $\mathbb{K}$-vector space such that

$$\dim(\mathbb{K}^{(n)})^i = \begin{cases} 1 & i = n \\ 0 & i \neq n \end{cases}.$$

7.1.2 Semi-infinite Character

Let $\pi^- : \mathfrak{g} \to \mathfrak{n}$ be the projection with respect to the decomposition $\mathfrak{g} = \mathfrak{n} \oplus \mathfrak{b}$, and let $i^- : \mathfrak{n} \hookrightarrow \mathfrak{g}$ be the inclusion. Let $\pi : \mathfrak{g} \to \mathfrak{g}$ be the map defined by the composition

$$\pi := i^- \circ \pi^- : \mathfrak{g} \xrightarrow{\pi^-} \mathfrak{n} \xrightarrow{i^-} \mathfrak{g}.$$

Definition 7.1 *The **critical cocycle** $\omega \in \mathrm{Hom}_{\mathbb{K}}(\mathfrak{g} \wedge \mathfrak{g}, \mathbb{K})$ of a $\mathbb{Z}$-graded Lie algebra $\mathfrak{g}$ is defined by*

$$\omega(x, y) := \mathrm{tr}_{\mathfrak{g}}([\pi \circ \mathrm{ad}x, \pi \circ \mathrm{ad}y] - \pi \circ [\mathrm{ad}x, \mathrm{ad}y]),$$

where $\mathrm{tr}_{\mathfrak{g}}$ denotes the trace on $\mathfrak{g}$.

First, we show that ω is well-defined.

Lemma 7.3. *For $x \in \mathfrak{g}^{n_1}$ and $y \in \mathfrak{g}^{n_2}$, the following hold:*

1. *If $n_1 + n_2 \neq 0$, then $\omega(x, y) = 0$.*
2. *If $n_1 = n_2 = 0$, then $\omega(x, y) = 0$.*
3. *If $n_1 = n$ and $n_2 = -n$ for $n \in \mathbb{Z}_{>0}$, then*

$$\omega(x, y) = \mathrm{tr}_{\bigoplus_{m=1}^{n} \mathfrak{g}^{-m}}(\mathrm{ad}y \circ \mathrm{ad}x).$$

Proof. By definition, the first statement follows. If $n_1 = n_2 = 0$, then

$$[\pi \circ \mathrm{ad}x, \pi \circ \mathrm{ad}y] = \pi \circ [\mathrm{ad}x, \mathrm{ad}y],$$

and thus, the second statement follows. We show the third statement.

Suppose that $z \in \mathfrak{g}^m$. In the case where $m \geq 0$, we have

$$[\pi \circ \mathrm{ad}x, \pi \circ \mathrm{ad}y](z) = 0 = \pi \circ [\mathrm{ad}x, \mathrm{ad}y](z).$$

In the case where $m < -n$, we have

$$[\pi \circ \mathrm{ad}x, \pi \circ \mathrm{ad}y](z) = \pi \circ [\mathrm{ad}x, \mathrm{ad}y](z).$$

Moreover, in the case where $-n \leq m \leq -1$, noticing that $\mathrm{ad}x(z) \in \mathfrak{b}$, we have

$$[\pi \circ \mathrm{ad}x, \pi \circ \mathrm{ad}y](z) = \pi \circ \mathrm{ad}x(\pi \circ \mathrm{ad}y(z)) = \mathrm{ad}x \circ \mathrm{ad}y(z).$$

Hence, we obtain

$$[\pi \circ \mathrm{ad}x, \pi \circ \mathrm{ad}y](z) - \pi \circ [\mathrm{ad}x, \mathrm{ad}y](z) = \mathrm{ad}y \circ \mathrm{ad}x(z).$$

We complete the proof. □

By the assumption (7.1), $\bigoplus_{m=1}^{n} \mathfrak{g}^{-m}$ is finite dimensional, and thus, $\omega \in \mathrm{Hom}_{\mathbb{K}}(\mathfrak{g} \otimes_{\mathbb{K}} \mathfrak{g}, \mathbb{K}^{(0)})^0$. Moreover, by the following lemma, the critical cocycle ω is well-defined and is, indeed, a 2-cocycle of $\mathfrak{g}$.

Lemma 7.4. *The map ω satisfies the 2-cocycle conditions.*

Proof. For $u, v \in \mathfrak{g}$, $\omega(u, v) = -\omega(v, u)$ by definition. Hence, it suffices to prove that

$$\omega([u, v], w) + \omega([w, u], v) + \omega([v, w], u) = 0 \tag{7.8}$$

holds for any $u \in \mathfrak{g}^l$, $v \in \mathfrak{g}^m$, $w \in \mathfrak{g}^n$ such that $l + m + n = 0$.

It is enough to check the following three cases:

1. $l > 0$, $m > 0$ and $n < 0$.
2. $l > 0$, $m = 0$ and $n < 0$.
3. $l > 0$, $m < 0$ and $n < 0$.

Here, we show the first case, since the second and third cases can be proved similarly.

By Lemma 7.3, we have

$$\begin{aligned}\omega([u,v],w) &= \operatorname{tr}_{\bigoplus_{s=1}^{-n}\mathfrak{g}^{-s}}(\mathrm{ad}w\circ\mathrm{ad}u\circ\mathrm{ad}v - \mathrm{ad}w\circ\mathrm{ad}v\circ\mathrm{ad}u),\\ \omega([w,u],v) &= -\operatorname{tr}_{\bigoplus_{s=1}^{m}\mathfrak{g}^{-s}}(\mathrm{ad}w\circ\mathrm{ad}u\circ\mathrm{ad}v - \mathrm{ad}u\circ\mathrm{ad}w\circ\mathrm{ad}v),\\ \omega([v,w],u) &= -\operatorname{tr}_{\bigoplus_{s=1}^{l}\mathfrak{g}^{-s}}(\mathrm{ad}v\circ\mathrm{ad}w\circ\mathrm{ad}u - \mathrm{ad}w\circ\mathrm{ad}v\circ\mathrm{ad}u).\end{aligned}$$

Thus, we obtain

$$\begin{aligned}&\omega([u,v],w)+\omega([w,u],v)+\omega([v,w],u)\\ &= \operatorname{tr}_{\bigoplus_{s=m+1}^{-n}\mathfrak{g}^{-s}}(\mathrm{ad}w\circ\mathrm{ad}u\circ\mathrm{ad}v) - \operatorname{tr}_{\bigoplus_{s=l+1}^{-n}\mathfrak{g}^{-s}}(\mathrm{ad}w\circ\mathrm{ad}v\circ\mathrm{ad}u)\\ &\quad + \operatorname{tr}_{\bigoplus_{s=1}^{m}\mathfrak{g}^{-s}}(\mathrm{ad}u\circ\mathrm{ad}w\circ\mathrm{ad}v) - \operatorname{tr}_{\bigoplus_{s=1}^{l}\mathfrak{g}^{-s}}(\mathrm{ad}v\circ\mathrm{ad}w\circ\mathrm{ad}u).\end{aligned}$$

By the cyclic property of the trace map, we have

$$\begin{aligned}\operatorname{tr}_{\bigoplus_{s=1}^{m}\mathfrak{g}^{-s}}(\mathrm{ad}u\circ\mathrm{ad}w\circ\mathrm{ad}v) &= \operatorname{tr}_{\bigoplus_{s=l+1}^{-n}\mathfrak{g}^{-s}}(\mathrm{ad}w\circ\mathrm{ad}v\circ\mathrm{ad}u),\\ \operatorname{tr}_{\bigoplus_{s=1}^{l}\mathfrak{g}^{-s}}(\mathrm{ad}v\circ\mathrm{ad}w\circ\mathrm{ad}u) &= \operatorname{tr}_{\bigoplus_{s=m+1}^{-n}\mathfrak{g}^{-s}}(\mathrm{ad}w\circ\mathrm{ad}u\circ\mathrm{ad}v).\end{aligned}$$

Hence, the lemma holds. □

In the sequel, let

$$\{x_k|k\in I^+\},\quad \{h_k|k\in I^0\},\quad \{y_k|k\in I^-\} \tag{7.9}$$

be $\mathbb{K}$-bases of $\mathfrak{g}^+$, $\mathfrak{g}^0$ and $\mathfrak{n}=\mathfrak{g}^-$ consisting of homogeneous vectors. For x, $y\in\mathfrak{g}$, we denote the coefficients of $\{x_k\}$, $\{h_k\}$ and $\{y_k\}$ in $[x,y]$ expanded with respect to these basis vectors by $C_{x,y}^{x_k}$, $C_{x,y}^{h_k}$ and $C_{x,y}^{y_k}$ respectively, i.e.,

$$[x,y]=\sum_{k\in I^+}C_{x,y}^{x_k}x_k+\sum_{k\in I^0}C_{x,y}^{h_k}h_k+\sum_{k\in I^-}C_{x,y}^{y_k}y_k.$$

Lemma 7.5. *Suppose that $x\in\mathfrak{g}^n$, $y\in\mathfrak{g}^{-n}$ for $n\in\mathbb{Z}_{>0}$. Then,*

$$\omega(x,y)=\sum_{k\in I^0}\sum_{l\in I^-}C_{y_l,x}^{h_k}C_{h_k,y}^{y_l}+\sum_{k\in I^+}\sum_{l\in I^-}C_{y_l,x}^{x_k}C_{x_k,y}^{y_l}.$$

Proof. Let $\tilde{I}^-$ be a subset of I^- such that $\{y_k|k\in\tilde{I}^-\}$ forms a basis of $\bigoplus_{1\le j\le n}\mathfrak{g}^{-j}$. Since $[x,y_l]\in\mathfrak{g}^{\geq}$ for $l\in\tilde{I}^-$, by Lemma 7.3, we have

$$\begin{aligned}\omega(x,y)&=\operatorname{tr}_{\bigoplus_{1\le j\le n}\mathfrak{g}^{-j}}(\mathrm{ad}y\circ\mathrm{ad}x)\\ &=\sum_{k\in I^0}\sum_{l\in\tilde{I}^-}C_{y,h_k}^{y_l}C_{x,y_l}^{h_k}+\sum_{k\in I^+}\sum_{l\in\tilde{I}^-}C_{y,x_k}^{y_l}C_{x,y_l}^{x_k}.\end{aligned}$$

Since $[h_k, y]$, $[x_k, y] \in \mathfrak{g}^{\geq -n}$, we have $C^{y_l}_{h_k,y} = 0 = C^{y_l}_{x_k,y}$ for $l \notin \tilde{I}^-$, and thus,

$$\omega(x,y) = \sum_{k\in I^0}\sum_{l\in I^-} C^{y_l}_{y,h_k} C^{h_k}_{x,y_l} + \sum_{k\in I^+}\sum_{l\in I^-} C^{y_l}_{y,x_k} C^{x_k}_{x,y_l}.$$

Using $C^{y_l}_{y,h_k} = -C^{y_l}_{h_k,y}$ etc., we obtain the lemma. □

We give some examples of the critical cocycle.

The Virasoro algebra: Let $\mathrm{Vir} = \bigoplus_{n\in\mathbb{Z}} \mathbb{K}L_n \oplus \mathbb{K}C$ be the Virasoro algebra. The explicit form of the critical cocycle ω of the Virasoro algebra can be described as follows.

Proposition 7.1

$$\omega(L_m, L_n) = \delta_{m+n,0}(-\frac{13}{6}m^3 + \frac{1}{6}m).$$

Proof. By Lemma 7.3, it suffices to compute $\omega(L_m, L_{-m})$ for $m > 0$. We have

$$\omega(L_m, L_{-m}) = \mathrm{tr}_{\bigoplus_{n=1}^m \mathbb{K}L_{-n}}(\mathrm{ad}L_{-m} \circ \mathrm{ad}L_m).$$

Since

$$\mathrm{ad}L_{-m} \circ \mathrm{ad}L_m(L_{-n}) = (m+n)(n-2m)L_{-n},$$

we obtain

$$\omega(L_m, L_{-m}) = \sum_{n=1}^{m}(m+n)(n-2m) = -\frac{13}{6}m^3 + \frac{1}{6}m.$$

Hence, this proposition holds. □

Kac–Moody algebras: Let $\mathfrak{g} = \mathfrak{h} \oplus \bigoplus_{\alpha\in\Delta} \mathfrak{g}^\alpha$ be a Kac–Moody algebra (Chapter 1 of [Kac4]). Let Δ be the root system of $\mathfrak{g}$, and let $\Pi = \{\alpha_i | i \in I\} \subset \Delta$ (resp. $\{\alpha_i^\vee | i \in I\}$) be a set of simple roots (resp. coroots), which are indexed by a finite set I. Let e_i and f_i be Chavalley generators of $\mathfrak{g}$, i.e.,

$$[e_i, f_j] = \delta_{i,j}\alpha_i^\vee, \ [h, e_i] = \langle \alpha_i, h\rangle e_i, \ [h, f_i] = -\langle \alpha_i, h\rangle f_i.$$

Here, we regard $\mathfrak{g}$ as $\mathbb{Z}$-graded Lie algebra via the principal gradation, i.e.,

$$\mathfrak{g} = \bigoplus_{m\in\mathbb{Z}} \mathfrak{g}^m \quad \mathfrak{g}^m := \bigoplus_{\alpha\in\Delta:\mathrm{ht}(\alpha)=m} \mathfrak{g}^\alpha,$$

where $\mathrm{ht}(\alpha) := \sum_{i\in I} k_i$ for $\alpha = \sum_i k_i\alpha_i \in \Delta$. Let $\rho \in \mathfrak{h}^*$ be an element which satisfies

$$\langle \rho, \alpha_i^\vee \rangle = 1 \quad (\forall i \in I).$$

Then, we have

Proposition 7.2 ([Ark]) *Let ω be the critical cocycle of $\mathfrak{g}$. For $x \in \mathfrak{g}^{\alpha}$ and $y \in \mathfrak{g}^{\beta}$, we have*

$$\omega(x,y) = \begin{cases} 2\rho([x,y]) & \alpha+\beta=0 \\ 0 & \alpha+\beta\neq 0 \end{cases}.$$

Proof. We may assume that $x \in \mathfrak{g}^m$, $y \in \mathfrak{g}^{-m}$ for $m \in \mathbb{Z}_{>0}$. Let us show this proposition by induction on m.

For $m=1$, since

$$\mathfrak{g}^1 = \bigoplus_{i\in I} \mathbb{K}e_i, \quad \mathfrak{g}^{-1} = \bigoplus_{i\in I} \mathbb{K}f_i$$

this proposition follows from Lemma 7.3 and the commutation relations

$$[f_j,[e_i,f_k]] = \delta_{i,k}\langle \alpha_j, \alpha_i^\vee\rangle f_j \quad (i,j,k \in I).$$

For $m>1$, we may assume that $x=[e_i,x']$ for some $x' \in \mathfrak{g}^{m-1}$ and $i \in I$. Then, by Lemma 7.4, we have

$$\begin{aligned} \omega(x,y) &= -\omega([x',y],e_i) - \omega([y,e_i],x') \\ &= -2\rho([[x',y],e_i]) - 2\rho([[y,e_i],x']) \\ &= 2\rho([x,y]). \end{aligned}$$

Hence, the proposition holds. □

In order to introduce a semi-regular bimodule of $\mathfrak{g}$, we suppose the following assumption on $\mathfrak{g}$:

Assumption There exists $\eta \in \overline{\mathrm{Hom}_{\mathbb{K}}}(\mathfrak{g}, \mathbb{K}^{(0)})$ such that

$$\omega(x,y) = d\eta(x,y)(:= -\eta([x,y])) \tag{7.10}$$

for any x, $y \in \mathfrak{g}$.

It is needless to say that the cohomology class of ω (in a suitable second cohomology) is trivial.

Under the assumption (7.10), Lemma 7.3 implies that $\eta([\mathfrak{g},\mathfrak{g}]\cap\mathfrak{g}^m)=\{0\}$ for $m\neq 0$. Moreover, by Lemma 7.3, η gives a character of $\mathfrak{g}^0$, i.e, it satisfies $\eta([\mathfrak{g}^0,\mathfrak{g}^0])=\{0\}$.

Let $\pi^0 : \mathfrak{g} \to \mathfrak{g}^0$ be the projection with respect to the triangular decomposition of $\mathfrak{g}$. For $\gamma \in \mathrm{Hom}_{\mathbb{K}}(\mathfrak{g}^0,\mathbb{K}^{(0)})$, set $\tilde{\gamma} := \gamma\circ\pi^0 \in \overline{\mathrm{Hom}_{\mathbb{K}}}(\mathfrak{g},\mathbb{K}^{(0)})$.

Definition 7.2 $\gamma \in \mathrm{Hom}_{\mathbb{K}}(\mathfrak{g}^0,\mathbb{K}^{(0)})$ *is called a* ***semi-infinite character*** *of* $\mathfrak{g}$ *if* $\omega = d\tilde{\gamma}$.

Remark that under the assumptions (7.1) and (7.10), a semi-infinite character of $\mathfrak{g}$ exists, and it is unique if $\mathfrak{g}^0 \subset [\mathfrak{g},\mathfrak{g}]$.

The Virasoro algebra and Kac–Moody algebras satisfy (7.1) and (7.10). Indeed, semi-infinite characters uniquely exist for these Lie algebras, and they are explicitly given as follows.

The **Virasoro algebra**: $\gamma \in \mathrm{Hom}_{\mathbb{K}}(\mathrm{Vir}^0, \mathbb{K}^{(0)})$ is given by

$$\gamma(C) = 26, \quad \gamma(L_0) = 1. \tag{7.11}$$

Kac–Moody algebras: Let $\mathfrak{g}$ be a Kac–Moody algebra. By Proposition 7.2, $\gamma \in \mathrm{Hom}_{\mathbb{K}}(\mathfrak{g}^0, \mathbb{K}^{(0)})$ is given by

$$\gamma = -2\rho.$$

7.1.3 Definition

Let $\mathfrak{g}$ be a $\mathbb{Z}$-graded Lie algebra over $\mathbb{K}$, which satisfies the assumptions (7.1) and (7.10), and let γ be a semi-infinite character of $\mathfrak{g}$. In this subsection, we construct the **semi-regular bimodule** of $\mathfrak{g}$ associated to the semi-infinite character γ.

We first notice that $U(\mathfrak{n})$ is a $\mathbb{Z}_{\leq 0}$-graded associative algebra, and each homogeneous component $U(\mathfrak{n})^n$ $(n \in \mathbb{Z}_{\leq 0})$ is finite dimensional by the assumption (7.1).

The $(\mathfrak{n}, \mathfrak{n})$-bimodule introduced here plays important roles in the construction of the semi-regular bimodule of $\mathfrak{g}$. Set

$$U(\mathfrak{n})^{\circledast} := \overline{\mathrm{Hom}}_{\mathbb{K}}(U(\mathfrak{n}), \mathbb{K}^{(0)}),$$

and regard it as $(\mathfrak{n}, \mathfrak{n})$-bimodule via (7.3) and (7.4), i.e.,

$$(u.\varphi)(u_1) := \varphi(u_1 u), \quad (\varphi.u)(u_1) := \varphi(u u_1),$$

where $\varphi \in U(\mathfrak{n})^{\circledast}$ and $u, u_1 \in U(\mathfrak{n})$. Notice that $U(\mathfrak{n})^{\circledast}$ is $\mathbb{Z}_{\geq 0}$-graded, i.e.,

$$U(\mathfrak{n})^{\circledast} = \bigoplus_{i \in \mathbb{Z}_{\geq 0}} (U(\mathfrak{n})^{\circledast})^i, \quad (U(\mathfrak{n})^{\circledast})^i := \mathrm{Hom}_{\mathbb{K}}(U(\mathfrak{n})^{-i}, \mathbb{K}^{(0)}).$$

We set

$$S_\gamma(\mathfrak{g}) := U(\mathfrak{n})^{\circledast} \otimes_{\mathbb{K}} U(\mathfrak{b}). \tag{7.12}$$

In the following, we define a $(\mathfrak{g}, \mathfrak{g})$-bimodule structure on $S_\gamma(\mathfrak{g})$.

<u>The left $\mathfrak{g}$-module structure on $S_\gamma(\mathfrak{g})$</u> Noticing that $\gamma([\mathfrak{g}^0, \mathfrak{g}^0]) = \{0\}$, we define the one-dimensional $\mathbb{Z}$-graded $\mathfrak{b}$-module $\mathbb{K}^{(n)}_{-\gamma} := \mathbb{K}\mathbf{1}_{-\gamma}$ $(n \in \mathbb{Z})$ as follows:

1. $\mathbb{K}^{(n)}_{-\gamma} \simeq \mathbb{K}^{(n)}$ as $\mathbb{Z}$-graded $\mathbb{K}$-vector space,

2. $h.\mathbf{1}_{-\gamma} = -\gamma(h)\mathbf{1}_{-\gamma}$ ($h \in \mathfrak{g}^0$),
3. $e.\mathbf{1}_{-\gamma} = 0$ ($e \in \mathfrak{g}^+$).

Let $\mathbb{K}^{(0)}_{-\gamma} \otimes U(\mathfrak{b})$ be the tensor product of the left $\mathfrak{b}$-modules $\mathbb{K}^{(0)}_{-\gamma}$ and $U(\mathfrak{b})$.

Lemma 7.6. *The following isomorphisms of $\mathbb{Z}$-graded left $\mathfrak{n}$-modules hold.*

$$i_1 : S_\gamma(\mathfrak{g}) \xrightarrow{\sim} \overline{\mathrm{Hom}}_{\mathbb{K}}(U(\mathfrak{n}), U(\mathfrak{b})), \tag{7.13}$$

$$i_2 : \overline{\mathrm{Hom}}_{\mathfrak{b}}(U(\mathfrak{g}), \mathbb{K}^{(0)}_{-\gamma} \otimes_{\mathbb{K}} U(\mathfrak{b})) \xrightarrow{\sim} \overline{\mathrm{Hom}}_{\mathbb{K}}(U(\mathfrak{n}), U(\mathfrak{b})), \tag{7.14}$$

where $\overline{\mathrm{Hom}}_{\mathfrak{b}}(U(\mathfrak{g}), \mathbb{K}^{(0)}_{-\gamma} \otimes_{\mathbb{K}} U(\mathfrak{b}))$ and $\overline{\mathrm{Hom}}_{\mathbb{K}}(U(\mathfrak{n}), U(\mathfrak{b}))$ are regarded as left $\mathfrak{n}$-module via (7.3), and $S_\gamma(\mathfrak{g})$ is regarded as left $\mathfrak{n}$-module via the left multiplication.

Proof. The isomorphism i_1 is given as follows:

$$\begin{aligned} i_1 : S_\gamma(\mathfrak{g}) = U(\mathfrak{n})^{\circledast} \otimes_{\mathbb{K}} U(\mathfrak{b}) &\longrightarrow \overline{\mathrm{Hom}}_{\mathbb{K}}(U(\mathfrak{n}), U(\mathfrak{b})), \\ \varphi \otimes b &\longmapsto \Psi \end{aligned}$$

where Ψ is defined by $\Psi(f) := \varphi(f)b$ for $f \in U(\mathfrak{n})$.

The isomorphism i_2 is given as follows: Since $U(\mathfrak{g}) \simeq U(\mathfrak{b}) \otimes_{\mathbb{K}} U(\mathfrak{n})$ as $(\mathfrak{b}, \mathfrak{n})$-bimodule, by Lemma 7.2, we have an isomorphism of left $\mathfrak{n}$-modules

$$\overline{\mathrm{Hom}}_{\mathfrak{b}}(U(\mathfrak{g}), \mathbb{K}^{(0)}_{-\gamma} \otimes_{\mathbb{K}} U(\mathfrak{b})) \simeq \overline{\mathrm{Hom}}_{\mathbb{K}}(U(\mathfrak{n}), \mathbb{K}^{(0)}_{-\gamma} \otimes_{\mathbb{K}} U(\mathfrak{b})).$$

Moreover, let i_3 be the following isomorphism of $\mathbb{Z}$-graded $\mathbb{K}$-vector spaces:

$$i_3 : \mathbb{K}^{(0)}_{-\gamma} \otimes U(\mathfrak{b}) \simeq U(\mathfrak{b}) \quad (\mathbf{1}_{-\gamma} \otimes b \longmapsto b).$$

Then,

$$\begin{aligned} i_2 : \overline{\mathrm{Hom}}_{\mathfrak{b}}(U(\mathfrak{g}), \mathbb{K}^{(0)}_{-\gamma} \otimes_{\mathbb{K}} U(\mathfrak{b})) &\longrightarrow \overline{\mathrm{Hom}}_{\mathbb{K}}(U(\mathfrak{n}), U(\mathfrak{b})) \\ \varphi &\longmapsto i_3 \circ \varphi|_{U(\mathfrak{n})} \end{aligned}$$

is an isomorphism of left $\mathfrak{n}$-modules. □

Moreover, we regard the space

$$\overline{\mathrm{Hom}}_{\mathfrak{b}}(U(\mathfrak{g}), \mathbb{K}^{(0)}_{-\gamma} \otimes_{\mathbb{K}} U(\mathfrak{b}))$$

as left $\mathfrak{g}$-module via (7.3), and introduce the left $\mathfrak{g}$-module structure on $S_\gamma(\mathfrak{g}) = U(\mathfrak{n})^{\circledast} \otimes_{\mathbb{K}} U(\mathfrak{b})$ through the isomorphisms of Lemma 7.6.

The right $\mathfrak{g}$-module structure on $S_\gamma(\mathfrak{g})$ Using the following lemma, we introduce right $\mathfrak{g}$-module structure.

Lemma 7.7. *There exists the following isomorphism of $\mathbb{Z}$-graded right $\mathfrak{b}$-modules:*

$$U(\mathfrak{n})^{\circledast} \otimes_{\mathbb{K}} U(\mathfrak{b}) \simeq U(\mathfrak{n})^{\circledast} \otimes_{\mathfrak{n}} U(\mathfrak{g}), \tag{7.15}$$

where we regard both sides of (7.15) as right $\mathfrak{b}$-modules via the right multiplication.

The right-hand side of (7.15) is a right $\mathfrak{g}$-module via right multiplication. Hence, we regard $S_\gamma(\mathfrak{g})$ as a right $\mathfrak{g}$-module by the isomorphism (7.15).

7.1.4 Compatibility of Two Actions on $S_\gamma(\mathfrak{g})$

We show that $S_\gamma(\mathfrak{g})$ is a $(\mathfrak{g}, \mathfrak{g})$-bimodule with respect to the two actions introduced in the previous subsection.

We describe the $\mathfrak{g}$-module structure on $S_\gamma(\mathfrak{g})$ explicitly. Let us introduce some notation. For $f \in \mathfrak{n}$, let L_f and $R_f \in \mathrm{End}_{\mathbb{K}}(U(\mathfrak{n}))$ be the maps defined by

$$L_f(u) := fu, \quad R_f(u) := uf \quad (u \in U(\mathfrak{n})).$$

By definition, $L_f,\ R_f \in \overline{\mathrm{Hom}}_{\mathbb{K}}(U(\mathfrak{n}), U(\mathfrak{n}))$. Recall that the $(\mathfrak{n}, \mathfrak{n})$-bimodule structure on $U(\mathfrak{n})^{\circledast}$ can be described as

$$\varphi.f = \varphi \circ L_f, \quad f.\varphi = \varphi \circ R_f \quad (\varphi \in U(\mathfrak{n})^{\circledast},\ f \in \mathfrak{n}). \tag{7.16}$$

Let $\{x_k | k \in I^+\}$ and $\{h_k | k \in I^0\}$ be $\mathbb{K}$-bases of $\mathfrak{g}^+$ and $\mathfrak{g}^0$ in (7.9). Since $U(\mathfrak{g}) \simeq U(\mathfrak{b}) \otimes_{\mathbb{K}} U(\mathfrak{n})$, for each $u \in U(\mathfrak{n})$ and $e \in \mathfrak{g}^+$, there uniquely exist

$$H_e^k(u),\ X_e^k(u),\ N_e(u) \in U(\mathfrak{n})$$

such that

$$[u, e] = \sum_k h_k H_e^k(u) + \sum_k x_k X_e^k(u) + N_e(u).$$

Here, we denote the maps from $U(\mathfrak{n})$ to $U(\mathfrak{n})$ given by $u \mapsto H_X^k(u)$, $u \mapsto X_e^k(u)$ and $u \mapsto N_e(u)$ by H_e^k, X_e^k and N_e respectively. By definition, we have H_e^k, X_e^k, $N_e \in \overline{\mathrm{Hom}}_{\mathbb{K}}(U(\mathfrak{n}), U(\mathfrak{n}))$.

Then, we have the following lemma:

Lemma 7.8. *The left action of $\mathfrak{g}$ on $S_\gamma(\mathfrak{g})$ can be described as follows:*

1. for $e \in \mathfrak{g}^+$,

$$\begin{aligned} e.(\varphi \otimes b) =& \varphi \otimes eb + \sum_k \varphi \circ H_e^k \otimes (-\gamma(h_k) + h_k)b \\ &+ \sum_k \varphi \circ X_e^k \otimes x_k b + \varphi \circ N_e \otimes b, \end{aligned}$$

2. for $h \in \mathfrak{g}^0$,

$$h.(\varphi \otimes b) = -\varphi \circ \mathrm{ad}h \otimes b + \varphi \otimes (-\gamma(h) + h)b,$$

3. for $f \in \mathfrak{n}$,

$$f.(\varphi \otimes b) = \varphi \circ R_f \otimes b,$$

where $\varphi \in U(\mathfrak{n})^{\circledast}$ *and* $b \in U(\mathfrak{b})$ $(\varphi \otimes b \in S_\gamma(\mathfrak{g}))$.

Notice that by the definitions of H_e^k and X_e^k, the right-hand side of the first formula is a finite sum.

Proof. By the isomorphisms of Lemma 7.6, for $\varphi \in U(\mathfrak{n})^{\circledast}$ and $b \in U(\mathfrak{b})$, we regard $\varphi \otimes b$ as an element of $\overline{\mathrm{Hom}}_{\mathfrak{b}}(U(\mathfrak{g}), \mathbb{K}_{-\gamma}^{(0)} \otimes_{\mathbb{K}} U(\mathfrak{b}))$.

Since $\Psi \in \overline{\mathrm{Hom}}_{\mathfrak{b}}(U(\mathfrak{g}), \mathbb{K}_{-\gamma}^{(0)} \otimes_{\mathbb{K}} U(\mathfrak{b}))$ is determined by $\Psi|_{U(\mathfrak{n})}$, we compute

$$(z.(\varphi \otimes b))(u) \quad (u \in U(\mathfrak{n}))$$

for each $z \in \{e,\ h,\ f\}$.

The case $z = e \in \mathfrak{g}^+$:

$$\begin{aligned}(e.(\varphi \otimes b))(u) &= (\varphi \otimes b)(ue)\\ &= (\varphi \otimes b)(eu + [u,e])\\ &= (\varphi \otimes b)(eu + \sum_k h_k H_e^k(u) + \sum_k x_k X_e^k(u) + N_e(u)).\end{aligned}$$

Since $\varphi \otimes b$ is a homomorphism of $\mathfrak{b}$-modules, we have

$$\begin{aligned}&e.((\varphi \otimes b)(u)) + \sum_k h_k.((\varphi \otimes b)(H_e^k(u)))\\ &\qquad + \sum_k x_k.((\varphi \otimes b)(X_e^k(u))) + (\varphi \otimes b)(N_e(u))\\ &= \varphi(u)eb + \sum_k \varphi(H_e^k(u))(-\gamma(h_k) + h_k)b\\ &\qquad + \sum_k \varphi(X_e^k(u))x_k b + \varphi(N_e(u))b\\ &= (\varphi \otimes eb)(u) + \sum_k \left(\varphi \circ H_e^k \otimes (-\gamma(h_k) + h_k)b\right)(u)\\ &\qquad + \left(\sum_k \varphi \circ X_e^k \otimes x_k b\right)(u) + (\varphi \circ N_e \otimes b)(u).\end{aligned}$$

Hence, the first formula follows.

The case $z = h \in \mathfrak{g}^0$: We have

$$\begin{aligned}(h.(\varphi\otimes b))(u) &= (\varphi\otimes b)(uh)\\ &= (\varphi\otimes b)(hu-\mathrm{ad}h(u))\\ &= h.((\varphi\otimes b)(u)) - \varphi(\mathrm{ad}h(u))\otimes b\\ &= \varphi(u)(-\gamma(h)+h)b - \varphi(\mathrm{ad}h(u))b\\ &= (\varphi\otimes(-\gamma(h)+h).b)(u) - (\varphi\circ\mathrm{ad}h\otimes b)(u),\end{aligned}$$

and thus, the second formula follows.

The case $z = f \in \mathfrak{n}$:

$$(f.(\varphi\otimes b))(u) = (\varphi\otimes b)(uf) = \varphi(uf)b = \varphi\circ R_f(u)b = (\varphi\circ R_f\otimes b)(u).$$

Hence, the third formula follows. □

Next, let us describe the right action of $\mathfrak{g}$ on $S_\gamma(\mathfrak{g})$ explicitly. Let $\{y_k | k \in I^-\}$ be the basis of $\mathfrak{n}$ in (7.9). The decomposition $U(\mathfrak{g}) \simeq U(\mathfrak{n}) \otimes_{\mathbb{K}} U(\mathfrak{b})$ implies that for any $b \in U(\mathfrak{b})$ and $f \in \mathfrak{n}$, there uniquely exist $Y_f^k(b) \in U(\mathfrak{b})$ and $B_f(b) \in U(\mathfrak{b})$ such that

$$[b, f] = \sum_{k\in I^-} y_k Y_f^k(b) + B_f(b).$$

Lemma 7.9. *Suppose that $\varphi \in U(\mathfrak{n})$ and $b \in U(\mathfrak{b})$ and $\varphi\otimes b \in S_\gamma(\mathfrak{g})$.*

1. For $e \in \mathfrak{g}^+$,

$$(\varphi\otimes b).e = \varphi\otimes(be).$$

2. For $h \in \mathfrak{g}^0$,

$$(\varphi\otimes b).h = \varphi\otimes(bh).$$

3. For $f \in \mathfrak{n}$,

$$(\varphi\otimes b).f = \varphi\circ L_f\otimes b + \sum_{k\in I^-}\varphi\circ L_{y_k}\otimes Y_f^k(b) + \varphi\otimes B_f(b).$$

Proof. We can directly show this lemma. □

Notice that by the definition of $Y_f^k(b)$, the right-hand side of the third formula is a finite sum.

The two $\mathfrak{g}$-actions on $S_\gamma(\mathfrak{g})$, indeed, define $(\mathfrak{g},\mathfrak{g})$-bimodule structure on it, i.e., the following holds.

Theorem 7.1 *$S_\gamma(\mathfrak{g})$ is a $(\mathfrak{g},\mathfrak{g})$-bimodule, i.e.,*

$$(z_1.(\varphi\otimes b)).z_2 = z_1.((\varphi\otimes b).z_2) \quad (z_1,\ z_2 \in U(\mathfrak{g})). \tag{7.17}$$

Proof. Combining the formulae given in Lemma 7.8 and 7.9 with $L_z\circ R_{z'} = R_{z'}\circ L_z$, one can directly check that if $z_1 \in \mathfrak{n}$ or $z_2 \in \mathfrak{b}$, then (7.17) holds.

Hence, it suffices to show the case where $(z_1, z_2) = (e, f)$, (h, f) for $e \in \mathfrak{g}^+$, $h \in \mathfrak{g}^0$ and $f \in \mathfrak{n}$.

First, we reduce the proof to the case where $b = 1$. Suppose that $b = b_1 b_2$ for $b_1 \in U(\mathfrak{b})$ and $b_2 \in \mathfrak{b}$, and assume that the following holds:

$$(z_1(\varphi \otimes b_1)).z_2 = z_1((\varphi \otimes b_1).z_2) \quad (\forall z_1,\ z_2 \in U(\mathfrak{g})).$$

We have

$$\begin{aligned}(z_1(\varphi \otimes b)).z_2 &= (z_1(\varphi \otimes b_1 b_2)).z_2 \\ &= (z_1.(\varphi \otimes b_1))b_2 z_2 \\ &= (z_1.(\varphi \otimes b_1))(z_2 b_2 + [b_2, z_2]) \\ &= (z_1.((\varphi \otimes b_1).z_2)).b_2 + (z_1.(\varphi \otimes b_1)).[b_2, z_2].\end{aligned}$$

Since $b_2 \in \mathfrak{b}$, we obtain

$$\begin{aligned}&(z_1.((\varphi \otimes b_1).z_2)).b_2 + (z_1.(\varphi \otimes b_1)).[b_2, z_2] \\ &= z_1.(((\varphi \otimes b_1).z_2).b_2) + z_1.((\varphi \otimes b_1).[b_2, z_2]) \\ &= z_1((\varphi \otimes b_1)(z_2 b_2 + [b_2, z_2])) \\ &= z_1((\varphi \otimes b_1)(b_2 z_2)) \\ &= z_1((\varphi \otimes b)z_2)\end{aligned}$$

and thus, (7.17) holds for $b = b_1 b_2$. Hence, we may assume that $b = 1$.

In the sequel, we show the following two formulae:

$$h.((\varphi \otimes 1).f) = (h.(\varphi \otimes 1)).f, \tag{7.18}$$

$$e.((\varphi \otimes 1).f) = (e.(\varphi \otimes 1)).f. \tag{7.19}$$

For the proof, it is convenient to use the following notation: For any $z \in \mathfrak{g}$, we denote the decomposition of z with respect to the triangular decomposition $\mathfrak{g} = \mathfrak{g}^+ \oplus \mathfrak{g}^0 \oplus \mathfrak{g}^-$ by

$$z = z^+ + z^0 + z^- \quad (z^\pm \in \mathfrak{g}^\pm, z^0 \in \mathfrak{g}^0).$$

To show the above two formulae, we need the following lemma.

Lemma 7.10. *For $e \in \mathfrak{g}^+$, $h \in \mathfrak{g}^0$ and $f \in \mathfrak{n}$, the following formulae hold.*

1. $Y_f^k(-\gamma(h) + h) = C_{h,f}^{y_k}$, $B_f(-\gamma(h) + h) = 0$.
2. $Y_f^k(e) = C_{e,f}^{y_k}$, $B_f(e) = [e, f]^+ + [e, f]^0$.
3. *As elements of $\overline{\mathrm{Hom}}_{\mathbb{K}}(U(\mathfrak{n})^\circledast, U(\mathfrak{n})^\circledast)$, the following equalities hold:*

$$H_e^k \circ L_f = L_f \circ H_e^k + \sum_i C_{f,x_i}^{h_k} X_e^i + C_{f,e}^{h_k} \mathrm{id}_{U(\mathfrak{n})},$$

$$X_e^k \circ L_f = L_f \circ X_e^k + \sum_i C_{f,x_i}^{x_k} X_e^i + C_{f,e}^{x_k} \mathrm{id}_{U(\mathfrak{n})},$$

$$N_e \circ L_f = L_f \circ N_e + \sum_k L_{[f,h_k]} \circ H_e^k + \sum_k L_{[f,x_k]^-} \circ X_e^k + L_{[f,e]^-}.$$

Proof. The first two formulae follow from

$$[-\gamma(h) + h, f] = [h, f] = \sum_k C_{h,f}^{y_k} y_k,$$

and the next two follow from

$$[e, f] = \sum_k C_{e,f}^{y_k} y_k + [e, f]^0 + [e, f]^+.$$

We show the last three formulae. For $u \in U(\mathfrak{n})$, we have

$$\begin{aligned}
[L_f(u), e] &= f[u, e] + [f, e]u \\
&= f\{\sum_k h_k H_e^k(u) + \sum_k x_k X_e^k(u) + N_e(u)\} \\
&\quad + \{\sum_k C_{f,e}^{x_k} x_k + \sum_k C_{f,e}^{h_k} h_k + [f, e]^-\}u \\
&= \sum_k h_k \{f H_e^k(u) + \sum_i C_{f,x_i}^{h_k} X_e^i(u) + C_{f,e}^{h_k} u\} \\
&\quad + \sum_k x_k \{f X_e^k(u) + \sum_i C_{f,x_i}^{x_k} X_e^i(u) + C_{f,e}^{x_k} u\} \\
&\quad + f N_e(u) + \sum_k [f, h_k] H_e^k(u) + \sum_k [f, x_k]^- X_e^k(u) + [f, e]^- u. \quad \square
\end{aligned}$$

Proof of (7.18) By Lemmas 7.8 and 7.9, we have

$$h.((\varphi \otimes 1).f) = \varphi \circ L_f \otimes (-\gamma(h) + h) - \varphi \circ L_f \circ \mathrm{ad} h \otimes 1,$$

and by Lemma 7.10,

$$\begin{aligned}
&(h.(\varphi \otimes 1)).f \\
&= (\varphi \otimes (-\gamma(h) + h) - \varphi \circ \mathrm{ad} h \otimes 1).f \\
&= \varphi \circ L_f \otimes (-\gamma(h) + h) + \sum_k \varphi \circ L_{y_k} \otimes C_{h,f}^{y_k} - \varphi \circ \mathrm{ad} h \circ L_f \otimes 1.
\end{aligned}$$

Since $\sum_k C_{h,f}^{y_k} L_{y_k} = L_{[h,f]}$ and $\mathrm{ad} h \circ L_f - L_f \circ \mathrm{ad} h = L_{[h,f]}$, we have

$$\varphi \circ L_f \otimes (-\gamma(h) + h) + \sum_k \varphi \circ L_{y_k} \otimes C_{h,f}^{y_k} - \varphi \circ \mathrm{ad}h \circ L_f \otimes 1$$
$$= h.((\varphi \otimes 1).f).$$

Thus, Formula (7.18) holds.

Proof of (7.19) Using Lemmas 7.8 and 7.9, we compute both sides of (7.19).

For the left-hand side, we have

$$e.((\varphi \otimes 1).f) = \varphi \circ L_f \otimes e + \sum_k \varphi \circ L_f \circ H_e^k \otimes (-\gamma(h_k) + h_k) + \sum_k \varphi \circ L_f \circ X_e^k \otimes x_k + \varphi \circ L_f \circ N_e \otimes 1,$$

and for the right-hand side,

$$\begin{aligned}
&(e.(\varphi \otimes 1)).f \\
&= \varphi \circ L_f \otimes e + \sum_k \varphi \circ L_{y_k} \otimes Y_f^k(e) + \varphi \otimes B_f(e) \\
&\quad + \sum_k \varphi \circ H_e^k \circ L_f \otimes (-\gamma(h_k) + h_k) + \sum_{k,m} \varphi \circ H_e^k \circ L_{y_m} \otimes Y_f^m(-\gamma(h_k) + h_k) \\
&\quad + \sum_k \varphi \circ H_e^k \otimes B_f(-\gamma(h_k) + h_k) + \sum_k \varphi \circ X_e^k \circ L_f \otimes x_k \\
&\quad + \sum_{k,m} \varphi \circ X_e^k \circ L_{y_m} \otimes Y_f^m(x_k) + \sum_k \varphi \circ X_e^k \otimes B_f(x_k) + \varphi \circ N_e \circ L_f \otimes 1.
\end{aligned}$$

By long but direct computation, we obtain

$$\begin{aligned}
&(e.(\varphi \otimes 1)).f - e.((\varphi \otimes 1).f) \\
&= \sum_i \varphi \circ X_e^i \otimes \left\{ \gamma([x_i, f]^0) + \sum_{k,m} C_{y_m,x_i}^{h_k} C_{h_k,f}^{y_m} + \sum_{k,m} C_{y_m,x_i}^{x_k} C_{x_k,f}^{y_m} \right\} \\
&\quad + \varphi \otimes \left\{ \gamma([e, f]^0) + \sum_{k,m} C_{y_m,e}^{h_k} C_{h_k,f}^{y_m} + \sum_{k,m} C_{y_m,e}^{x_k} C_{x_k,f}^{y_m} \right\}.
\end{aligned}$$

Therefore, by Lemma 7.5, if γ is a semi-infinite character of $\mathfrak{g}$, then

$$e.((\varphi \otimes 1).f) = (e.(\varphi \otimes 1)).f.$$

Now, the formula (7.19) has been proved. □

7.1.5 Isomorphisms

In this subsection, we state two isomorphisms related with the semi-regular bimodule $S_\gamma(\mathfrak{g})$ of $\mathfrak{g}$. Let $\iota : U(\mathfrak{n})^{\circledast} \hookrightarrow S_\gamma(\mathfrak{g})$ be the embedding map defined by

$$U(\mathfrak{n})^{\circledast} \ni \varphi \longmapsto \varphi \otimes 1 \in U(\mathfrak{n})^{\circledast} \otimes_{\mathbb{K}} U(\mathfrak{b}) = S_\gamma(\mathfrak{g}).$$

By Lemmas 7.8 and 7.9, one can directly show that the map ι is a homomorphism of $(\mathfrak{n}, \mathfrak{n})$-bimodule.

Theorem 7.2 *Let γ be a semi-infinite character of $\mathfrak{g}$, and let $S_\gamma(\mathfrak{g})$ be the semi-regular bimodule of $\mathfrak{g}$. Then, we have*

1. $$\iota_L : U(\mathfrak{g}) \otimes_{\mathfrak{n}} U(\mathfrak{n})^{\circledast} \longrightarrow S_\gamma(\mathfrak{g}) \quad (u \otimes \varphi \longmapsto u.\iota(\varphi)) \tag{7.20}$$
   is an isomorphism of $(\mathfrak{g}, \mathfrak{n})$-bimodules, where we regard $U(\mathfrak{g}) \otimes_{\mathfrak{n}} U(\mathfrak{n})^{\circledast}$ as a $(\mathfrak{g}, \mathfrak{n})$-bimodule via
   $$x.(u \otimes \varphi).y := (x.u) \otimes (\varphi.y) \quad (x \in \mathfrak{g},\ y \in \mathfrak{n}).$$
2. $$\iota_R : U(\mathfrak{n})^{\circledast} \otimes_{\mathfrak{n}} U(\mathfrak{g}) \longrightarrow S_\gamma(\mathfrak{g}) \quad (\varphi \otimes u \mapsto \iota(\varphi).u) \tag{7.21}$$
   is an isomorphism of $(\mathfrak{n}, \mathfrak{g})$-modules, where we regard $U(\mathfrak{n})^{\circledast} \otimes_{\mathfrak{n}} U(\mathfrak{g})$ as an $(\mathfrak{n}, \mathfrak{g})$-bimodule via
   $$y.(\varphi \otimes u).x := (y.\varphi) \otimes (u.x) \quad (x \in \mathfrak{g},\ y \in \mathfrak{n}).$$

Proof. By using Lemmas 7.8 and 7.9, we can directly check that ι_L (resp. ι_R) is a well-defined homomorphism of $(\mathfrak{g}, \mathfrak{n})$-bimodules (resp. of $(\mathfrak{n}, \mathfrak{g})$-bimodules), since ι is a homomorphism of $(\mathfrak{n}, \mathfrak{n})$-modules. Moreover, by Lemma 7.7, the map ι_R is a bijection. Hence, we have only to show that ι_L is bijective.

Since for any $n \in \mathbb{Z}_{\geq 0}$, $\dim S_\gamma(\mathfrak{g})^n = \dim(U(\mathfrak{g}) \otimes_{\mathfrak{n}} U(\mathfrak{n})^{\circledast})^n$, it suffices to show that ι_L is surjective. Here, we prove that

$$U(\mathfrak{b}).\iota(U(\mathfrak{n})^{\circledast}) = S_\gamma(\mathfrak{g}). \tag{7.22}$$

Remark that (7.22) is an immediate consequence of the 'triangularity' (7.23), stated below, of the left $\mathfrak{b}$-action on $S_\gamma(\mathfrak{g})$.

Let us first introduce filtrations $\{F_i U(\mathfrak{b}) | i \in \mathbb{Z}_{\geq 0}\}$ and $\{F_i^j U(\mathfrak{b}) | i \in \mathbb{Z}_{>0},\ j \in \mathbb{Z}_{\geq -1}\}$ of $U(\mathfrak{b})$. Let $\{F_i U(\mathfrak{b}) | i \in \mathbb{Z}_{\geq 0}\}$ be the standard filtration of $U(\mathfrak{b})$, i.e.,

$$F_i U(\mathfrak{b}) := \mathfrak{b} F_{i-1} U(\mathfrak{b}) + F_{i-1} U(\mathfrak{b}) \ (i > 0), \quad F_0 U(\mathfrak{b}) := \mathbb{K}1.$$

For each $i \in \mathbb{Z}_{>0}$ and $j \in \mathbb{Z}_{\geq -1}$, set

$$F_i^j U(\mathfrak{b}) := \begin{cases} \mathfrak{b}^{\leq j} F_{i-1} U(\mathfrak{b}) + F_{i-1}(\mathfrak{b}) & (j \geq 0) \\ F_{i-1} U(\mathfrak{b}) & (j = -1) \end{cases}.$$

Then, by definition,

$$F_{i-1}U(\mathfrak{b}) = F_i^{-1}U(\mathfrak{b}) \subset F_i^0 U(\mathfrak{b}) \subset F_i^1 U(\mathfrak{b}) \subset F_i^2 U(\mathfrak{b}) \subset \cdots,$$

and

$$\bigcup_{j=-1}^{\infty} F_i^j U(\mathfrak{b}) = F_i U(\mathfrak{b}).$$

We also introduce a filtration of $S_\gamma(\mathfrak{g}) = U(\mathfrak{n})^{\circledast} \otimes_{\mathbb{K}} U(\mathfrak{b})$ by

$$F_i^j S_\gamma(\mathfrak{g}) := U(\mathfrak{n})^{\circledast} \otimes_{\mathbb{K}} F_i^j U(\mathfrak{b}).$$

Then, by Lemma 7.8, for $\varphi \otimes b \in F_i^{-1} S_\gamma(\mathfrak{g})$ and $b_1 \in \mathfrak{b}^j$ $(j \in \mathbb{Z}_{\geq 0})$, we have

$$b_1.(\varphi \otimes b) \equiv \varphi \otimes b_1 b \pmod{F_i^{j-1} S_\gamma(\mathfrak{g})}. \tag{7.23}$$

This fact implies (7.22), and thus, the theorem holds. □

7.2 Tilting Equivalence

Using the semi-regular bimodule $S_\gamma(\mathfrak{g})$, we construct an equivalence of categories, which explains a similarity between structures of Verma modules over the Virasoro algebra with highest weights

$$(c, h) \quad \text{and} \quad (26 - c, 1 - h). \tag{7.24}$$

7.2.1 Preliminaries

In this subsection, we show two isomorphisms related with the $(\mathfrak{n}, \mathfrak{n})$-bimodule $U(\mathfrak{n})^{\circledast}$.

For $y \in U(\mathfrak{n})$, we define $\ell_y \in \overline{\mathrm{Hom}}_{\mathbb{K}}(U(\mathfrak{n})^{\circledast}, U(\mathfrak{n})^{\circledast})$ as follows:

$$\ell_y(\phi) := \phi \circ L_y \quad (\phi \in U(\mathfrak{n})^{\circledast},\ y \in U(\mathfrak{n})).$$

Notice that

$$\ell_y \in \overline{\mathrm{Hom}}_{\mathfrak{n}}(U(\mathfrak{n})^{\circledast}, U(\mathfrak{n})^{\circledast}), \tag{7.25}$$

since for any y and $y_1 \in U(\mathfrak{n})$, we have

$$\ell_y(y_1.\phi) = \ell_y(\phi \circ R_{y_1}) = \phi \circ R_{y_1} \circ L_y = \phi \circ L_y \circ R_{y_1} = y_1.(\ell_y(\phi)).$$

Lemma 7.11. *There exists the following isomorphism of left* $\mathfrak{n}$*-modules:*

$$\ell : U(\mathfrak{n}) \xrightarrow{\sim} \overline{\mathrm{Hom}}_{\mathfrak{n}}(U(\mathfrak{n})^{\circledast}, U(\mathfrak{n})^{\circledast}) \quad (y \longmapsto \ell_y).$$

Proof. The isomorphism ℓ is obtained as the composition of the following isomorphisms ℓ_1, ℓ_2 and ℓ_3 (the isomorphism ℓ_3 follows from Lemma 7.1):

1. $$\ell_1 : U(\mathfrak{n}) \longrightarrow \overline{\mathrm{Hom}}_{\mathbb{K}}(U(\mathfrak{n})^{\circledast}, \mathbb{K}^{(0)}),$$
   where $\ell_1(x)(\varphi) := \varphi(x)$ for $x \in U(\mathfrak{n})$ and $\varphi \in U(\mathfrak{n})^{\circledast}$,
2. $$\ell_2 : \overline{\mathrm{Hom}}_{\mathbb{K}}(U(\mathfrak{n})^{\circledast}, \mathbb{K}^{(0)}) \longrightarrow \overline{\mathrm{Hom}}_{\mathbb{K}}(U(\mathfrak{n}) \otimes_{\mathfrak{n}} U(\mathfrak{n})^{\circledast}, \mathbb{K}^{(0)}),$$
   where $\ell_2(\alpha)(x \otimes \varphi) := \alpha(x.\varphi)$ for $\alpha \in \overline{\mathrm{Hom}}_{\mathbb{K}}(U(\mathfrak{n})^{\circledast}, \mathbb{K}^{(0)})$,
3. $$\ell_3 : \overline{\mathrm{Hom}}_{\mathbb{K}}(U(\mathfrak{n}) \otimes_{\mathfrak{n}} U(\mathfrak{n})^{\circledast}, \mathbb{K}^{(0)}) \longrightarrow \overline{\mathrm{Hom}}_{\mathbb{K}}(U(\mathfrak{n})^{\circledast}, \overline{\mathrm{Hom}}_{\mathfrak{n}}(U(\mathfrak{n}), \mathbb{K}^{(0)})),$$
   where $\ell_3(\beta)(\varphi)(x) := \beta(x \otimes \varphi)$ for $\beta \in \overline{\mathrm{Hom}}_{\mathbb{K}}(U(\mathfrak{n}) \otimes_{\mathfrak{n}} U(\mathfrak{n})^{\circledast}, \mathbb{K}^{(0)})$.

The composition $\ell = \ell_3 \circ \ell_2 \circ \ell_1$ is explicitly given by $\ell(y) = \ell_y$ for $y \in U(\mathfrak{n})$. Indeed, for any $y \in U(\mathfrak{n})$ and $\psi \in U(\mathfrak{n})^{\circledast}$, we have

$$\ell(y)(\psi)(x) = \ell_2 \circ \ell_1(y)(x \otimes \psi) = \ell_1(y)(x.\varphi) = (x.\varphi)(y) = \varphi(y.x) = (\varphi \circ L_y)(x).$$

Hence, $\ell(y) = \ell_y$. □

Remark that ℓ is an anti-automorphism of $\mathbb{K}$-algebras, i.e., $\ell(y_1 y_2) = \ell(y_2)\ell(y_1)$ for $y_1, y_2 \in U(\mathfrak{n})$.

Next, we show an isomorphism between $U(\mathfrak{n})^{\circledast}$ and the antipode dual $U(\mathfrak{n})^{\sharp a}$ of $U(\mathfrak{n})$ (Definition 1.21).

Lemma 7.12. *There exists the following isomorphism of left* $\mathfrak{n}$*-modules:*

$${}^t a : U(\mathfrak{n})^{\circledast} \simeq U(\mathfrak{n})^{\sharp a},$$

where ${}^t a$ *is the transpose of* $a : U(\mathfrak{n}) \to U(\mathfrak{n})$ *and* $U(\mathfrak{n})^{\circledast}$ *is regarded as left* $\mathfrak{n}$*-module via (7.3).*

Proof. To prove the lemma, it is enough to show that ${}^t a$ is a homomorphism of $\mathfrak{n}$-modules. For $y \in \mathfrak{n}$, $\varphi \in U(\mathfrak{n})^{\circledast}$ and $u \in U(\mathfrak{n})$, we have

$$\begin{aligned} {}^t a(y.\varphi)(u) &= (y.\varphi)(a(u)) = \varphi(a(u).y) \\ &= \varphi(a(a(y).u)) = {}^t a(\varphi)(a(y).u) = y.{}^t a(\varphi)(u). \end{aligned}$$

Hence, ${}^t a(y.\varphi) = y.{}^t a(\varphi)$ holds. □

7.2.2 Some Categories

Recall that $\mathrm{Mod}_{\mathfrak{g}}^{\mathbb{Z}}$ is the category of the left $\mathbb{Z}$-graded $\mathfrak{g}$-modules whose morphisms are given by

$$\mathrm{Hom}_{\mathrm{Mod}_{\mathfrak{g}}^{\mathbb{Z}}}(M,N) := \{\phi \in \mathrm{Hom}_{\mathfrak{g}}(M,N) | \phi(M^i) \subset N^i \ (\forall i \in \mathbb{Z})\}$$

for $M, N \in \mathrm{Ob}(\mathrm{Mod}_{\mathfrak{g}}^{\mathbb{Z}})$ (Definition 1.9).

The main result of this section is a categorical equivalence between the following subcategories $\mathcal{M}$ and $\mathcal{K}$ of $\mathrm{Mod}_{\mathfrak{g}}^{\mathbb{Z}}$.

Definition 7.3 *1. $\mathcal{M}$ is the full subcategory of $\mathrm{Mod}_{\mathfrak{g}}^{\mathbb{Z}}$ whose objects M satisfy that there exists a finite dimensional $\mathbb{Z}$-graded $\mathbb{K}$-vector space E such that*

$$\mathrm{Res}_{\mathfrak{n}}^{\mathfrak{g}} M \simeq U(\mathfrak{n}) \otimes_{\mathbb{K}} E.$$

2. $\mathcal{K}$ is the full subcategory of $\mathrm{Mod}_{\mathfrak{g}}^{\mathbb{Z}}$ whose objects K satisfy that there exists a finite dimensional $\mathbb{Z}$-graded $\mathbb{K}$-vector space E such that

$$\mathrm{Res}_{\mathfrak{n}}^{\mathfrak{g}} K \simeq U(\mathfrak{n})^{\circledast} \otimes_{\mathbb{K}} E.$$

Remark 7.1 *$\mathcal{M}$ and $\mathcal{K}$ are additive categories, but they are not abelian categories in general.*

For later use, here, we define the rank $\mathrm{rk}M$ of $M \in \mathrm{Ob}(\mathcal{M})$ by its rank as a $U(\mathfrak{n})$-free module. Though the rank of a free module over a non-commutative ring is not well-defined in general, in our case, $\mathrm{rk}M$ is well-defined. Indeed, for $M \in \mathrm{Ob}(\mathcal{M})$ such that $\mathrm{Res}_{\mathfrak{n}}^{\mathfrak{g}} M \simeq U(\mathfrak{n}) \otimes_{\mathbb{K}} E$, we have $\mathrm{ch}' M = \mathrm{ch}' U(\mathfrak{n}) \,\mathrm{ch}' E$, where $\mathrm{ch}' M$ is defined in (7.2). Hence, $\mathrm{rk}M = \dim_{\mathbb{K}} E$ is well-defined.

Here, we state a characterisation of the objects of the category $\mathcal{M}$. For a finite dimensional left $\mathfrak{g}^0$-module E, we regard it as $\mathfrak{b}$-module via $\mathfrak{g}^+|_E \equiv 0$, and set

$$\Delta(E) := U(\mathfrak{g}) \otimes_{\mathfrak{b}} E. \tag{7.26}$$

By definition, we have

$$\mathrm{Res}_{\mathfrak{n}}^{\mathfrak{g}} \Delta(E) = \mathrm{Res}_{\mathfrak{n}}^{\mathfrak{g}} U(\mathfrak{g}) \otimes_{\mathfrak{b}} E \simeq U(\mathfrak{n}) \otimes_{\mathbb{K}} E,$$

and thus, $\Delta(E) \in \mathrm{Ob}(\mathcal{M})$.

Definition 7.4 *We say that $M \in \mathrm{Ob}(\mathrm{Mod}_{\mathfrak{g}}^{\mathbb{Z}})$ has a finite **Δ-flag**, if there exist $M_k \in \mathrm{Ob}(\mathrm{Mod}_{\mathfrak{g}}^{\mathbb{Z}})$ $(k = 1, 2, \cdots, n)$ such that*

1. *$\{0\} = M_0 \subset M_1 \subset \cdots \subset M_{n-1} \subset M_n = M$,*
2. *for each $k = 1, 2, \cdots, n$, there exists a finite dimensional irreducible $\mathbb{Z}$-graded $\mathfrak{g}^0$-module E_k such that*

$$M_k/M_{k-1} \simeq \Delta(E_k)$$

as left $\mathfrak{g}$*-module.*

Proposition 7.3 *Let* M *be an object of* $\mathrm{Mod}_{\mathfrak{g}}^{\mathbb{Z}}$*. Then,* $M \in \mathrm{Ob}(\mathcal{M})$ *if and only if* $M = \{0\}$ *or* M *has a finite* Δ*-flag. Moreover, any* Δ*-flag* $\{0\} = M_0 \subset M_1 \subset \cdots \subset M_{n-1} \subset M_n = M$ *of* $M \in \mathrm{Ob}(\mathcal{M})$ *satisfies* $M_k \in \mathrm{Ob}(\mathcal{M})$ *for any* k*.*

Proof. We first show that if $M \in \mathrm{Ob}(\mathrm{Mod}_{\mathfrak{g}}^{\mathbb{Z}})$ has a finite Δ-flag, then $M \in \mathrm{Ob}(\mathcal{M})$. Let $\{0\} = M_0 \subset M_1 \subset \cdots \subset M_n = M$ be a finite Δ-flag of M such that $M_k/M_{k-1} \simeq \Delta(E_k)$ for some $\mathbb{Z}$-graded finite dimensional irreducible $\mathfrak{g}^0$-module E_k. Since $\mathrm{Res}_{\mathfrak{n}}^{\mathfrak{g}}\Delta(E_k)$ is $U(\mathfrak{n})$-free, it is a projective $\mathfrak{n}$-module. Moreover, by an argument similar to the proof of Proposition 1.12, one can show that $\mathrm{Mod}_{\mathfrak{n}}$ has enough injectives and projectives. Hence, Proposition A.3 and Lemma A.4 imply that

$$\overline{\mathrm{Ext}}^1_{\mathrm{Mod}_{\mathfrak{n}}}(\mathrm{Res}_{\mathfrak{n}}^{\mathfrak{g}}\Delta(E_j), \mathrm{Res}_{\mathfrak{n}}^{\mathfrak{g}}\Delta(E_k)) = \{0\}$$

for any j. Hence, by Lemma A.3, we obtain

$$\mathrm{Res}_{\mathfrak{n}}^{\mathfrak{g}} M \simeq \bigoplus_{i=1}^{n} \mathrm{Res}_{\mathfrak{n}}^{\mathfrak{g}}\Delta(E_i) \simeq U(\mathfrak{n}) \otimes_{\mathbb{K}} \left(\bigoplus_{i=1}^{n} E_i \right)$$

as left $\mathfrak{n}$-module, and thus, $M \in \mathrm{Ob}(\mathcal{M})$.

Next, we show that any $M \in \mathrm{Ob}(\mathcal{M})$ ($M \neq \{0\}$) has a finite Δ-flag. We use induction on $\mathrm{rk}M$. Let $\{u_1, \cdots, u_r\}$ ($r := \mathrm{rk}M$) be a $U(\mathfrak{n})$-free basis of $\mathrm{Res}_{\mathfrak{n}}^{\mathfrak{g}} M$, and let $E := \bigoplus_{j=1}^{r} \mathbb{K}u_j$. Set $n_0 := \max\{n \in \mathbb{Z} | M^n \neq \{0\}\}$. Then, we have $M^{n_0} \subset E$. Since M^{n_0} is a finite dimensional $\mathfrak{g}^0$-module, there exists an irreducible $\mathfrak{g}^0$-submodule F of M^{n_0}. Since $\mathfrak{g}^+.M^{n_0} = \{0\}$, if we set $N := U(\mathfrak{g}).F$, then

$$N \simeq \Delta(F).$$

Since $\mathrm{Res}_{\mathfrak{n}}^{\mathfrak{g}} M \simeq U(\mathfrak{n}) \otimes_{\mathbb{K}} E$ and $\mathrm{Res}_{\mathfrak{n}}^{\mathfrak{g}} N \simeq U(\mathfrak{n}) \otimes_{\mathbb{K}} F$, we have

$$\mathrm{Res}_{\mathfrak{n}}^{\mathfrak{g}}(M/N) \simeq \mathrm{Res}_{\mathfrak{n}}^{\mathfrak{g}} M / \mathrm{Res}_{\mathfrak{n}}^{\mathfrak{g}} N \simeq (U(\mathfrak{n}) \otimes_{\mathbb{K}} E)/(U(\mathfrak{n}) \otimes_{\mathbb{K}} F)$$

as $\mathbb{Z}$-graded $\mathfrak{n}$-module. Applying the exact functor $U(\mathfrak{n}) \otimes_{\mathbb{K}} (\cdot)$ to the exact sequence $F \hookrightarrow E \twoheadrightarrow E/F$, we have an exact sequence

$$0 \longrightarrow U(\mathfrak{n}) \otimes_{\mathbb{K}} F \longrightarrow U(\mathfrak{n}) \otimes_{\mathbb{K}} E \longrightarrow U(\mathfrak{n}) \otimes_{\mathbb{K}} (E/F) \longrightarrow 0$$

of $\mathbb{Z}$-graded left $\mathfrak{n}$-modules. Hence,

$$\mathrm{Res}_{\mathfrak{n}}^{\mathfrak{g}}(M/N) \simeq U(\mathfrak{n}) \otimes_{\mathbb{K}} (E/F),$$

and thus, $M/N \in \mathrm{Ob}(\mathcal{M})$. Since $\mathrm{rk}(M/N) < \mathrm{rk}M$, by induction hypothesis, M/N has a finite Δ-flag, and so does M. □

Here, we state a lemma, which is an immediate consequence of Lemma 7.12.

Lemma 7.13. *For any* $M \in \mathrm{Ob}(\mathcal{M})$ *(resp.* $K \in \mathrm{Ob}(\mathcal{K})$*), we have* $M^{\sharp a} \in \mathrm{Ob}(\mathcal{K})$ *(resp.* $K^{\sharp a} \in \mathrm{Ob}(\mathcal{M})$*).*

7.2.3 Some Functors

In this subsection, we introduce two functors which give a categorical equivalence between $\mathcal{M}$ and $\mathcal{K}$. The proof of the equivalence will be given in the following subsection.

To introduce the functors, we first show the following lemma.

Lemma 7.14. *For* $M \in \mathrm{Ob}(\mathcal{M})$ *and* $K \in \mathrm{Ob}(\mathcal{K})$*, we have*

$$\begin{aligned} &S_\gamma(\mathfrak{g}) \otimes_{\mathfrak{g}} M \in \mathrm{Ob}(\mathcal{K}), \\ &\overline{\mathrm{Hom}}_{\mathfrak{g}}(S_\gamma(\mathfrak{g}), K) \in \mathrm{Ob}(\mathcal{M}). \end{aligned}$$

Proof. Suppose that $\mathrm{Res}^{\mathfrak{g}}_{\mathfrak{n}} M \simeq U(\mathfrak{n}) \otimes_{\mathbb{K}} E$, where E is a finite dimensional $\mathbb{Z}$-graded $\mathbb{K}$-vector space. By the isomorphism (7.21), we have

$$\begin{aligned} \mathrm{Res}^{\mathfrak{g}}_{\mathfrak{n}}(S_\gamma(\mathfrak{g}) \otimes_{\mathfrak{g}} M) &\simeq U(\mathfrak{n})^{\circledast} \otimes_{\mathfrak{n}} \mathrm{Res}^{\mathfrak{g}}_{\mathfrak{n}} M \\ &\simeq U(\mathfrak{n})^{\circledast} \otimes_{\mathfrak{n}} U(\mathfrak{n}) \otimes_{\mathbb{K}} E \\ &\simeq U(\mathfrak{n})^{\circledast} \otimes_{\mathbb{K}} E \end{aligned}$$

as left $\mathfrak{n}$-module. Hence, $S_\gamma(\mathfrak{g}) \otimes_{\mathfrak{g}} M \in \mathrm{Ob}(\mathcal{K})$.

Next, we suppose that $\mathrm{Res}^{\mathfrak{g}}_{\mathfrak{n}} K \simeq U(\mathfrak{n})^{\circledast} \otimes_{\mathbb{K}} E$. Combining the isomorphism (7.20) with Lemma 7.2, we have

$$\begin{aligned} \mathrm{Res}^{\mathfrak{g}}_{\mathfrak{n}} \overline{\mathrm{Hom}}_{\mathfrak{g}}(S_\gamma(\mathfrak{g}), K) &\simeq \overline{\mathrm{Hom}}_{\mathfrak{n}}(U(\mathfrak{n})^{\circledast}, \mathrm{Res}^{\mathfrak{g}}_{\mathfrak{n}} K) \\ &\simeq \overline{\mathrm{Hom}}_{\mathfrak{n}}(U(\mathfrak{n})^{\circledast}, U(\mathfrak{n})^{\circledast} \otimes_{\mathbb{K}} E) \\ &\simeq \overline{\mathrm{Hom}}_{\mathfrak{n}}(U(\mathfrak{n})^{\circledast}, U(\mathfrak{n})^{\circledast}) \otimes_{\mathbb{K}} E. \end{aligned}$$

Hence, by Lemma 7.11, we obtain

$$\mathrm{Res}^{\mathfrak{g}}_{\mathfrak{n}} \overline{\mathrm{Hom}}_{\mathfrak{g}}(S_\gamma(\mathfrak{g}), K) \simeq U(\mathfrak{n}) \otimes_{\mathbb{K}} E.$$

□

Definition 7.5 *1. Let* $T : \mathcal{M} \to \mathcal{K}$ *be the functor defined by*

$$T(M) := S_\gamma(\mathfrak{g}) \otimes_{\mathfrak{g}} M \quad (M \in \mathrm{Ob}(\mathcal{M})).$$

2. Let $H : \mathcal{K} \to \mathcal{M}$ *be the functor defined by*

$$H(K) := \overline{\mathrm{Hom}}_{\mathfrak{g}}(S_\gamma(\mathfrak{g}), K).$$

Remark that $\mathcal{M}$ and $\mathcal{K}$ are not abelian categories in general, the kernel and the image of a morphism do not necessarily exist (cf. § A.1.3 and A.1.4). Here, we define a short exact sequence in $\mathcal{M}$ or $\mathcal{K}$ as follows:

Definition 7.6 *We call a sequence* $0 \to M_1 \to M_2 \to M_3 \to 0$ *in* $\mathcal{M}$ **a short exact sequence in** $\mathcal{M}$ *if it is an exact sequence in* $\mathrm{Mod}_{\mathfrak{g}}$. *We define a short exact sequence in* $\mathcal{K}$ *similarly.*

Then, we have

Proposition 7.4 *The functors* $T : \mathcal{M} \to \mathcal{K}$ *(resp.* $H : \mathcal{K} \to \mathcal{M}$*) send a short exact sequence in* $\mathcal{M}$ *(resp. in* $\mathcal{K}$*) to a short exact sequence in* $\mathcal{K}$ *(resp. in* $\mathcal{M}$*).*

Proof. Let $0 \longrightarrow M_1 \longrightarrow M_2 \longrightarrow M_3 \longrightarrow 0$ be a short exact sequence in $\mathcal{M}$. Applying the functor T, we have a sequence of left $\mathfrak{g}$-modules

$$0 \longrightarrow T(M_1) \longrightarrow T(M_2) \longrightarrow T(M_3) \longrightarrow 0. \tag{7.27}$$

We show that (7.27) is a short exact sequence in $\mathcal{K}$. It is enough to prove that

$$0 \longrightarrow \mathrm{Res}^{\mathfrak{g}}_{\mathfrak{n}} T(M_1) \longrightarrow \mathrm{Res}^{\mathfrak{g}}_{\mathfrak{n}} T(M_2) \longrightarrow \mathrm{Res}^{\mathfrak{g}}_{\mathfrak{n}} T(M_3) \longrightarrow 0 \tag{7.28}$$

is an exact sequence of left $\mathfrak{n}$-modules. Notice that, by the isomorphism (7.21), for $M \in \mathrm{Ob}(\mathcal{M})$,

$$\mathrm{Res}^{\mathfrak{g}}_{\mathfrak{n}} T(M) \simeq U(\mathfrak{n})^{\circledast} \otimes_{\mathfrak{n}} \mathrm{Res}^{\mathfrak{g}}_{\mathfrak{n}} M.$$

It is clear that

$$0 \longrightarrow \mathrm{Res}^{\mathfrak{g}}_{\mathfrak{n}} M_1 \longrightarrow \mathrm{Res}^{\mathfrak{g}}_{\mathfrak{n}} M_2 \longrightarrow \mathrm{Res}^{\mathfrak{g}}_{\mathfrak{n}} M_3 \longrightarrow 0 \tag{7.29}$$

is an exact sequence of left $\mathfrak{n}$-modules. Since $\mathrm{Res}^{\mathfrak{g}}_{\mathfrak{n}} M_3$ is a $U(\mathfrak{n})$-free module, by Proposition A.3 and Lemma A.4, the sequence (7.29) splits in $\mathrm{Mod}_{\mathfrak{n}}$. Hence,

$$0 \longrightarrow U(\mathfrak{n})^{\circledast} \otimes_{\mathfrak{n}} \mathrm{Res}^{\mathfrak{g}}_{\mathfrak{n}} M_1 \longrightarrow U(\mathfrak{n})^{\circledast} \otimes_{\mathfrak{n}} \mathrm{Res}^{\mathfrak{g}}_{\mathfrak{n}} M_2 \longrightarrow U(\mathfrak{n})^{\circledast} \otimes_{\mathfrak{n}} \mathrm{Res}^{\mathfrak{g}}_{\mathfrak{n}} M_3 \longrightarrow 0$$

also splits, i.e., (7.28) is exact. Hence, T sends a short exact sequence in $\mathcal{M}$ to a short exact sequence in $\mathcal{K}$.

Next, we show the assertion for the functor $H : \mathcal{K} \to \mathcal{M}$. For a short exact sequence $0 \longrightarrow K_1 \longrightarrow K_2 \longrightarrow K_3 \longrightarrow 0$ in $\mathcal{K}$, we prove that

$$0 \longrightarrow H(K_1) \longrightarrow H(K_2) \longrightarrow H(K_3) \longrightarrow 0 \tag{7.30}$$

is a short exact sequence in $\mathcal{M}$. It suffices to show that

$$0 \longrightarrow \mathrm{Res}^{\mathfrak{g}}_{\mathfrak{n}} H(K_1) \longrightarrow \mathrm{Res}^{\mathfrak{g}}_{\mathfrak{n}} H(K_2) \longrightarrow \mathrm{Res}^{\mathfrak{g}}_{\mathfrak{n}} H(K_3) \longrightarrow 0 \tag{7.31}$$

is an exact sequence of left $\mathfrak{n}$-modules. By Lemma 7.13,

$$0 \longrightarrow K_3^{\sharp a} \longrightarrow K_2^{\sharp a} \longrightarrow K_1^{\sharp a} \longrightarrow 0$$

is a short exact sequence in $\mathcal{M}$. Hence, as was seen above, the sequence

$$0 \longrightarrow \mathrm{Res}^{\mathfrak{g}}_{\mathfrak{n}}(K_3^{\sharp a}) \longrightarrow \mathrm{Res}^{\mathfrak{g}}_{\mathfrak{n}}(K_2^{\sharp a}) \longrightarrow \mathrm{Res}^{\mathfrak{g}}_{\mathfrak{n}}(K_1^{\sharp a}) \longrightarrow 0$$

splits in $\mathrm{Mod}_{\mathfrak{n}}$. Noticing that $(\mathrm{Res}^{\mathfrak{g}}_{\mathfrak{n}}(K^{\sharp a}))^{\sharp a'} \simeq \mathrm{Res}^{\mathfrak{g}}_{\mathfrak{n}} K$ holds for any $K \in \mathrm{Ob}(\mathcal{K})$, where a' is the antipode of $U(\mathfrak{n})$ defined by $a' := a|_{U(\mathfrak{n})}$, the sequence

$$0 \longrightarrow \mathrm{Res}^{\mathfrak{g}}_{\mathfrak{n}} K_1 \longrightarrow \mathrm{Res}^{\mathfrak{g}}_{\mathfrak{n}} K_2 \longrightarrow \mathrm{Res}^{\mathfrak{g}}_{\mathfrak{n}} K_3 \longrightarrow 0$$

splits. Thus, the following also splits:

$$\begin{aligned} 0 \longrightarrow \overline{\mathrm{Hom}}_{\mathfrak{n}}(U(\mathfrak{n})^{\circledast}, \mathrm{Res}^{\mathfrak{g}}_{\mathfrak{n}} K_3) &\longrightarrow \overline{\mathrm{Hom}}_{\mathfrak{n}}(U(\mathfrak{n})^{\circledast}, \mathrm{Res}^{\mathfrak{g}}_{\mathfrak{n}} K_2) \\ &\longrightarrow \overline{\mathrm{Hom}}_{\mathfrak{n}}(U(\mathfrak{n})^{\circledast}, \mathrm{Res}^{\mathfrak{g}}_{\mathfrak{n}} K_1) \longrightarrow 0. \end{aligned}$$

This means that (7.31) is a short exact sequence of $\mathfrak{n}$-modules. □

Proposition 7.4 does not ensure that T and H send an injection (resp. a surjection) to an injection (resp. a surjection), since the kernel or the image of a morphism do not necessarily exist in $\mathcal{M}$ and $\mathcal{K}$. The following simple example of a functor on additive categories adequately explains such a situation.

Example 7.1 *Let $\mathcal{C}$ be the category of free abelian groups of finite rank, and let $\mathcal{C}'$ be the category of finite abelian groups. Notice that $\mathcal{C}$ is an additive category, but it is not an abelian category. Let F be a functor defined by*

$$F : \mathcal{C} \to \mathcal{C}' \quad (F(L) := L \otimes_{\mathbb{Z}} (\mathbb{Z}/2\mathbb{Z})).$$

By definition, F sends a short exact sequence to a short exact sequence in a sense similar to Definition 7.6. On the other hand, let $f : \mathbb{Z} \to \mathbb{Z}$ be an injection defined by $f(n) := 2n$ ($n \in \mathbb{Z}$). Then, $F(f) : \mathbb{Z}/2\mathbb{Z} \to \mathbb{Z}/2\mathbb{Z}$ is the zero map, and it is not injection.

Notice that, by Lemma 7.13, contravariant functors

$$\begin{aligned} (\cdot)^{\sharp a} &: \mathcal{M} \to \mathcal{K}, \\ (\cdot)^{\sharp a} &: \mathcal{K} \to \mathcal{M} \end{aligned}$$

are well-defined. Since $(\cdot)^{\sharp a} : \mathrm{Mod}^{\mathbb{Z}}_{\mathfrak{g}} \to \mathrm{Mod}^{\mathbb{Z}}_{\mathfrak{g}}$ is exact, these functors send a short exact sequence in $\mathcal{M}$ (resp. $\mathcal{K}$) to a short exact sequence in $\mathcal{K}$ (resp. $\mathcal{M}$). Hence, we obtain

Lemma 7.15. *The covariant functor*

$$(\cdot)^{\sharp a} : \mathcal{K} \to \mathcal{M}^{\mathrm{opp}} \tag{7.32}$$

sends a short exact sequence in $\mathcal{K}$ to a short exact sequence in $\mathcal{M}^{\mathrm{opp}}$.

7.2.4 Equivalence between $\mathcal{M}$ and $\mathcal{K}$

We first show the following theorem.

Theorem 7.3 *The functor T defines an equivalence of the categories $\mathcal{M}$ and $\mathcal{K}$.*

Proof. We prove that T and H are quasi-inverse functors to each other. Lemma 7.1 implies an isomorphism

$$\overline{\mathrm{Hom}}_{\mathfrak{g}}(T(M), K) \simeq \overline{\mathrm{Hom}}_{\mathfrak{g}}(M, H(K)),$$

of $\mathbb{Z}$-graded $\mathbb{K}$-vector spaces, and by restricting this isomorphism to the homogeneous subspace with degree 0, we have

$$\mathrm{Hom}_{\mathcal{K}}(T(M), K) \simeq \mathrm{Hom}_{\mathcal{M}}(M, H(K)).$$

Hence, (T, H) is an **adjoint pair**, and there exist natural transformations $\eta : \mathrm{id}_{\mathcal{M}} \Rightarrow H \circ T$ and $\epsilon : T \circ H \Rightarrow \mathrm{id}_{\mathcal{K}}$ (cf. § A.1.2). In the sequel, we show that they are natural isomorphisms.

The natural transformation η is given as follows. For each $M \in \mathrm{Ob}(\mathcal{M})$,

$$\begin{gathered}\mathrm{Hom}_{\mathcal{K}}(T(M), T(M)) \simeq \mathrm{Hom}_{\mathcal{M}}(M, H \circ T(M)).\\ \mathrm{id}_{T(M)} \mapsto \eta_M\end{gathered}$$

Since $\mathrm{Res}^{\mathfrak{g}}_{\mathfrak{n}} M \simeq U(\mathfrak{n}) \otimes_{\mathbb{K}} E$ for some finite dimensional $\mathbb{Z}$-graded $\mathbb{K}$-vector space E, by Theorem 7.2 and Lemmas 7.2 and 7.11, we have

$$\begin{aligned}\mathrm{Res}^{\mathfrak{g}}_{\mathfrak{n}} H \circ T(M) &= \mathrm{Res}^{\mathfrak{g}}_{\mathfrak{n}} \overline{\mathrm{Hom}}_{\mathfrak{g}}(S_\gamma(\mathfrak{g}), S_\gamma(\mathfrak{g}) \otimes_{\mathfrak{g}} M)\\ &\simeq \overline{\mathrm{Hom}}_{\mathfrak{g}}(U(\mathfrak{g}) \otimes_{\mathfrak{n}} U(\mathfrak{n})^{\circledast}, S_\gamma(\mathfrak{g}) \otimes_{\mathfrak{g}} M)\\ &\simeq \overline{\mathrm{Hom}}_{\mathfrak{n}}(U(\mathfrak{n})^{\circledast}, \mathrm{Res}^{\mathfrak{g}}_{\mathfrak{n}}(S_\gamma(\mathfrak{g}) \otimes_{\mathfrak{g}} M))\\ &\simeq \overline{\mathrm{Hom}}_{\mathfrak{n}}(U(\mathfrak{n})^{\circledast}, U(\mathfrak{n})^{\circledast} \otimes_{\mathfrak{n}} U(\mathfrak{g}) \otimes_{\mathfrak{g}} M)\\ &\simeq \overline{\mathrm{Hom}}_{\mathfrak{n}}(U(\mathfrak{n})^{\circledast}, U(\mathfrak{n})^{\circledast} \otimes_{\mathfrak{n}} \mathrm{Res}^{\mathfrak{g}}_{\mathfrak{n}} M)\\ &\simeq \overline{\mathrm{Hom}}_{\mathfrak{n}}(U(\mathfrak{n})^{\circledast}, U(\mathfrak{n})^{\circledast} \otimes_{\mathfrak{n}} U(\mathfrak{n}) \otimes_{\mathbb{K}} E)\\ &\simeq \overline{\mathrm{Hom}}_{\mathfrak{n}}(U(\mathfrak{n})^{\circledast}, U(\mathfrak{n})^{\circledast} \otimes_{\mathbb{K}} E)\\ &\simeq \overline{\mathrm{Hom}}_{\mathfrak{n}}(U(\mathfrak{n})^{\circledast}, U(\mathfrak{n})^{\circledast}) \otimes_{\mathbb{K}} E\\ &\simeq U(\mathfrak{n}) \otimes_{\mathbb{K}} E \simeq \mathrm{Res}^{\mathfrak{g}}_{\mathfrak{n}} M.\end{aligned}$$

Hence, η_M is an isomorphism of left $\mathfrak{n}$-modules, and thus, it is an isomorphism of left $\mathfrak{g}$-modules. Hence, η is a natural isomorphism.

The natural transformation ϵ is given as follows: For $K \in \mathrm{Ob}(\mathcal{K})$,

$$\begin{gathered}\mathrm{Hom}_{\mathcal{K}}(T \circ H(K), K) \simeq \mathrm{Hom}_{\mathcal{M}}(H(K), H(K)).\\ \epsilon_K \mapsto \mathrm{id}_{H(K)}\end{gathered}$$

For each $K \in \mathrm{Ob}(\mathcal{K})$, there exists a finite dimensional $\mathbb{Z}$-graded $\mathbb{K}$-vector space E such that $\mathrm{Res}^{\mathfrak{g}}_{\mathfrak{n}} K \simeq U(\mathfrak{n})^{\circledast} \otimes_{\mathbb{K}} E$. Then, by Theorem 7.2 and Lemmas 7.2 and 7.11, there exists an isomorphism

$$\begin{aligned}
\mathrm{Res}^{\mathfrak{g}}_{\mathfrak{n}} T \circ H(K) &= \mathrm{Res}^{\mathfrak{g}}_{\mathfrak{n}}(S_\gamma(\mathfrak{g}) \otimes_{\mathfrak{g}} \overline{\mathrm{Hom}}_{\mathfrak{g}}(S_\gamma(\mathfrak{g}), K)) \\
&\simeq U(\mathfrak{n})^{\circledast} \otimes_{\mathfrak{n}} U(\mathfrak{g}) \otimes_{\mathfrak{g}} \overline{\mathrm{Hom}}_{\mathfrak{g}}(S_\gamma(\mathfrak{g}), K) \\
&\simeq U(\mathfrak{n})^{\circledast} \otimes_{\mathfrak{n}} \mathrm{Res}^{\mathfrak{g}}_{\mathfrak{n}} \overline{\mathrm{Hom}}_{\mathfrak{g}}(S_\gamma(\mathfrak{g}), K) \\
&\simeq U(\mathfrak{n})^{\circledast} \otimes_{\mathfrak{n}} \overline{\mathrm{Hom}}_{\mathfrak{g}}(U(\mathfrak{g}) \otimes_{\mathfrak{n}} U(\mathfrak{n})^{\circledast}, K) \\
&\simeq U(\mathfrak{n})^{\circledast} \otimes_{\mathfrak{n}} \overline{\mathrm{Hom}}_{\mathfrak{n}}(U(\mathfrak{n})^{\circledast}, \mathrm{Res}^{\mathfrak{g}}_{\mathfrak{n}} K) \\
&\simeq U(\mathfrak{n})^{\circledast} \otimes_{\mathfrak{n}} \overline{\mathrm{Hom}}_{\mathfrak{n}}(U(\mathfrak{n})^{\circledast}, U(\mathfrak{n})^{\circledast} \otimes_{\mathbb{K}} E) \\
&\simeq U(\mathfrak{n})^{\circledast} \otimes_{\mathfrak{n}} \overline{\mathrm{Hom}}_{\mathfrak{n}}(U(\mathfrak{n})^{\circledast}, U(\mathfrak{n})^{\circledast}) \otimes_{\mathbb{K}} E \\
&\simeq U(\mathfrak{n})^{\circledast} \otimes_{\mathfrak{n}} U(\mathfrak{n}) \otimes_{\mathbb{K}} E \\
&\simeq U(\mathfrak{n})^{\circledast} \otimes_{\mathbb{K}} E \simeq \mathrm{Res}^{\mathfrak{g}}_{\mathfrak{n}} K.
\end{aligned}$$

Hence, ϵ_K is also an isormorphism of left $\mathfrak{n}$-modules, and thus, it is an isomorphism of left $\mathfrak{g}$-modules. Therefore, ϵ is a natural isomorphism. We have completed the proof. □

Moreover, the following categorical equivalence holds.

Theorem 7.4 *The functor* $(\cdot)^{\sharp a} : \mathcal{K} \to \mathcal{M}^{\mathrm{opp}}$ *in (7.32) defines an equivalence of categories.*

Proof. For an object M of $\mathcal{M}$ or $\mathcal{K}$, we have $(M^{\sharp a})^{\sharp a} \simeq M$. Hence,

$$((\cdot)^{\sharp a})^{\sharp a} \simeq \mathrm{id}_{\mathcal{K}}, \quad ((\cdot)^{\sharp a})^{\sharp a} \simeq \mathrm{id}_{\mathcal{M}^{\mathrm{opp}}}.$$

These facts mean that $(\cdot)^{\sharp a} : \mathcal{K} \to \mathcal{M}^{\mathrm{opp}}$ and $(\cdot)^{\sharp a} : \mathcal{M}^{\mathrm{opp}} \to \mathcal{K}$ are quasi-inverse functors to each other. Hence, $(\cdot)^{\sharp a} : \mathcal{K} \to \mathcal{M}^{\mathrm{opp}}$ defines a categorical equivalence. □

As a consequence of Theorems 7.3 and 7.4, Proposition 7.4 and Lemma 7.15, we obtain what is called the **tilting equivalence**

Theorem 7.5 *Let* Φ *be the functor from* $\mathcal{M}$ *to* $\mathcal{M}^{\mathrm{opp}}$ *defined by*

$$\Phi(M) := T(M)^{\sharp a}.$$

Then, Φ *defines a categorical equivalence, and it maps a short exact sequence in* $\mathcal{M}$ *to a short exact sequence in* $\mathcal{M}^{\mathrm{opp}}$.

Finally, we describe $\Phi(\Delta(E))$.

Proposition 7.5 *Let* E *be an irreducible finite dimensional* $\mathbb{Z}$*-graded left* $\mathfrak{g}^0$*-module. Then, we have*

$$\Phi(\Delta(E)) = \Delta(\mathbb{K}^{(0)}_\gamma \otimes_{\mathbb{K}} E^{\sharp a}) \quad (M \in \mathrm{Ob}(\mathcal{M})).$$

Proof. Since E is irreducible, there exists $n \in \mathbb{Z}$ such that $E^n = E$ and $E^i = \{0\}$ for $i \neq n$. By Theorem 7.2, we have an isomorphism

$$\begin{aligned} T(\Delta(E)) &\simeq S_\gamma(\mathfrak{g}) \otimes_{\mathfrak{g}} U(\mathfrak{g}) \otimes_{\mathfrak{b}} E \\ &\simeq \overline{\mathrm{Hom}_{\mathfrak{b}}}(U(\mathfrak{g}), \mathbb{K}_{-\gamma}^{(0)} \otimes_{\mathbb{K}} U(\mathfrak{b})) \otimes_{\mathfrak{b}} E \end{aligned}$$

of left $\mathfrak{g}$-modules. Hence, we have an isomorphism of left $\mathfrak{g}^0$-modules

$$T(\Delta(E))^n \simeq \mathbb{K}_{-\gamma}^{(0)} \otimes_{\mathbb{K}} E,$$

and $T(\Delta(E))^j = \{0\}$ for $j < n$. These facts imply that

$$\Phi(\Delta(E))^{-n} \simeq \mathbb{K}_{\gamma}^{(0)} \otimes_{\mathbb{K}} E^{\sharp a}$$

as left $\mathfrak{g}^0$-modules, and $\Phi(\Delta(E))^j = \{0\}$ for $j > -n$. Hence,

$$\Phi(\Delta(E))^{\geq -n} \simeq \mathbb{K}_{\gamma}^{(0)} \otimes E^{\sharp a}$$

as $\mathbb{Z}$-graded left $\mathfrak{b}$-modules, and thus, there exists a homomorphism of $\mathbb{Z}$-graded left $\mathfrak{g}$-modules

$$\Delta(\mathbb{K}_{\gamma}^{(0)} \otimes E^{\sharp a}) \longrightarrow \Phi(\Delta(E)). \tag{7.33}$$

On the other hand, by Lemmas 7.7 and 7.12, we have

$$\mathrm{Res}^{\mathfrak{g}}_{\mathfrak{n}} \Phi(\Delta(E)) \simeq U(\mathfrak{n}) \otimes_{\mathbb{K}} (\mathbb{K}_{\gamma}^{(0)} \otimes E^{\sharp a}).$$

Hence, (7.33) is an isomorphism of left $\mathfrak{g}$-modules. Now, we have proved the proposition. □

7.2.5 The Virasoro Case

Similarly to the previous chapters, we identify $(\mathrm{Vir}^0)^*$ with $\mathbb{K}^2$ via $\lambda \mapsto (\lambda(C), \lambda(L_0))$. By (7.11), under the identification, the semi-infinite character of the Virasoro algebra is $(26, 1)$. By Proposition 7.5, we have

Corollary 7.1 *Let $\Phi : \mathcal{M} \to \mathcal{M}^{\mathrm{opp}}$ be the functor in Theorem 7.5, we have the following isomorphism of* Vir*-modules*

$$\Phi(M(c,h)) \simeq M(26 - c, 1 - h).$$

7.3 Bibliographical Notes and Comments

In [Fe], B. Feigin introduced the semi-infinite cohomology, and showed that the (semi-infinite) torsion of Verma modules over the Virasoro algebra does not vanish only if highest weights satisfy the condition (7.24). This result gave a mathematical meaning to the value '26' called the critical dimension in the bosonic string theory [GSW], [Pol].

Motivated by the result, S. Arkhipov [Ark] established the Feigin–Arkhipov –Soergel duality stated in this chapter for $\mathbb{Z}$-graded associative algebras based on the theory of the semi-infinite homological algebra (cf. [Vor1]). Later, W. Soergel [So] simplified the proof of the duality without using the semi-infinite homological algebra in the case where the $\mathbb{Z}$-graded Lie algebra $\mathfrak{g}$ is generated by its partial part $\mathrm{Par}^1_{-1}\mathfrak{g}$ (§ 2.2). Note that in this chapter, we extended his argument to more general $\mathbb{Z}$-graded Lie algebras including the Virasoro algebra.

We make some remarks on the critical cocycle. The critical cocycle coincides with the so-called Japanese cocycle which was discovered through the study of soliton equations [DJKM]. In [IK6], the authors have shown the tilting equivalence for a certain class of $\mathbb{Z}$-graded Lie superalgebra which contains so-called physical conformal superalgebras, classified by V. G. Kac [Kac6] and G. Yamamoto [Y]. It was also shown in [IK6] that the critical cocycle is related to the condition that the square of the BRST charge vanishes.

The Feigin–Arkhipov–Soergel duality is also called the tilting equivalence. In fact, W. Soergel applied the duality to compute characters of tilting modules over symmetrisable Kac–Moody algebras. For more about tilting modules and their related topoics, see e.g., [HHK].

Chapter 8
Fock Modules

The main subject of this chapter is the Virasoro module structure of Fock modules $\mathcal{F}_\lambda^\eta$ studied by B. Feigin and D. Fuchs in [FeFu4]. Similarly to Chapters 5 and 6, the Jantzen filtration plays important roles. In fact, the Jantzen filtration à la Feigin and Fuchs given in Chapter 3 reveals the structure of Fock modules.

We also show that singular vectors of a Fock module $\mathcal{F}_\lambda^\mu$ can be expressed in terms of the Jack symmetric polynomials.

Remark that the Fock modules $\mathcal{F}_\lambda^\eta$ we study in this chapter are the so-called *bosonic* Fock modules. On the other hand, the Fock modules which B. Feigin and D. Fuchs dealt with in [FeFu4] are the Virasoro modules defined on the spaces of semi-infinite forms. At the end of this chapter, we explicitly establish isomorphisms between these Virasoro modules.

8.1 Classification of Weights (λ, η)

Here, we classify the pairs $(\lambda, \eta) \in \mathbb{C}^2$ which parameterise Fock modules.

8.1.1 Coarse Classification

As stated in § 4.5.1, there exist Vir-module homomorphisms

$$\Gamma_{\lambda,\eta} : M(c_\lambda, h_\lambda^\eta) \longrightarrow \mathcal{F}_\lambda^\eta, \quad \mathrm{L}^{\lambda,\eta} : \mathcal{F}_\lambda^\eta \longrightarrow M(c_\lambda, h_\lambda^\eta)^c,$$

where

$$c_\lambda := 1 - 12\lambda^2, \quad h_\lambda^\eta := \frac{1}{2}\eta(\eta - 2\lambda).$$

K. Iohara, Y. Koga, *Representation Theory of the Virasoro Algebra*, Springer Monographs in Mathematics, DOI 10.1007/978-0-85729-160-8_8,

Taking these homomorphisms into account, we classify the weights (λ, η) as follows:

Definition 8.1 *We say that* $(\lambda, \eta) \in \mathbb{C}^2$ *belongs to* **Class** V, **Class** I *and* **Class** $R^{\pm}$, *if* $(c_\lambda, h_\lambda^\eta)$ *belongs to* **Class** V, **Class** I *and* **Class** $R^{\pm}$ *respectively, where the classes for* (c, h) *are defined in Chapter 5.*

To investigate the Vir-module structure of Fock modules, we list the zeros of the determinants $\det(\Gamma_{\lambda,\eta})_n$ and $\det(\mathrm{L}^{\lambda,\eta})_n$ given in Theorem 4.3. For $T \in \mathbb{C} \setminus \{0\}$, set

$$\lambda(T) := \frac{1}{\sqrt{2}}(T - T^{-1}).$$

Notice that, if $T^2 = \frac{P}{Q}$ or $T^2 = \frac{Q}{P}$, then

$$c_{\lambda(T)} = c_{P,Q}, \quad \text{and} \quad h_{\lambda(T)}^{\lambda(T) \pm \frac{m}{\sqrt{2PQ}}} = h_{P,Q;m},$$

where $c_{P,Q}$ and $h_{P,Q;m}$ are defined in (5.5). Hence, if (λ, η) with $\lambda = \lambda(T)$ belongs to **Class** I, **Class** R^+ and **Class** R^-, then $T^2 \notin \mathbb{Q} \setminus \{0\}$, $T^2 \in \mathbb{Q}_{>0}$ and $T^2 \in \mathbb{Q}_{<0}$ respectively.

For each λ, μ, α, $\beta \in \mathbb{C}$, we set

$$\Psi_{\alpha,\beta}^{\pm}(\lambda, \eta) := (\eta - \lambda) \pm (\frac{1}{2}\lambda_+ \alpha + \frac{1}{2}\lambda_- \beta) \quad \left(\lambda_\pm := \lambda \pm \sqrt{\lambda^2 + 2}\right),$$

where we choose the branch of $\sqrt{\lambda^2+2}$ as $-\frac{1}{2}\pi < \arg\sqrt{\lambda^2+2} \leq \frac{1}{2}\pi$ for $\lambda \neq \pm\sqrt{-2}$. Remark that for positive integers α and β, $\Psi_{\alpha,\beta}^{\sigma}(\lambda, \eta)$ $(\sigma = \pm)$ is nothing but the factor of the determinants $\det(\Gamma_{\lambda,\eta})_n$ and $\det(\mathrm{L}^{\lambda,\eta})_n$ given in Theorem 4.3. Similarly to the Verma module case in § 5.1, we introduce the line $\ell_{\lambda,\eta}^{\sigma}$ in the (α, β)-plane by

$$\ell_{\lambda,\eta}^{\sigma} := \{(\alpha, \beta) \in \mathbb{C}^2 | \Psi_{\alpha,\beta}^{\sigma}(\lambda, \eta) = 0\}.$$

For **Class** V, I and R^-, the following lemma holds:

Lemma 8.1. *Suppose that* (λ, η) *belongs to* **Class** V, I *or* R^-. *Then, there exists* $\sigma \in \{\pm\}$ *such that*

$$\ell_{\lambda,\eta}^{\sigma} \cap (\mathbb{Z}_{>0})^2 = \emptyset.$$

Proof. We first recall the factorisation

$$\Phi_{\alpha,\beta}(c_\lambda, h_\lambda^\eta) = \Psi_{\alpha,\beta}^{+}(\lambda, \eta)\Psi_{\alpha,\beta}^{-}(\lambda, \eta)\Psi_{\beta,\alpha}^{+}(\lambda, \eta)\Psi_{\beta,\alpha}^{-}(\lambda, \eta)$$

of the factor $\Phi_{\alpha,\beta}(c, h)$ of the determinant $\det(c, h)_n$ (see the formulae (4.27)). Hence, for $\lambda = \lambda(T)$ with $T^2 = \frac{Q}{P}$ or $\frac{P}{Q}$, the line $\ell_{\lambda,\eta}^{\sigma}$ is the one of the four lines $Q\alpha - P\beta = \pm m$ and $P\alpha - Q\beta = \pm m$ that appeared in § 5.1.1. Moreover, we notice that $\ell_{\lambda,\eta}^{+}$ and $\ell_{\lambda,\eta}^{-}$ are symmetric with respect to the origin. Hence, the lemma holds. □

By this lemma, for **Class** V, I and R^-, at least one of $\Gamma_{\lambda,\eta}$ and $\mathrm{L}^{\lambda,\eta}$ is an isomorphism. Thus, we have

Proposition 8.1 *For* **Class** V, **Class** I *and* **Class** R^-, *at least one of the following holds:*

$$\mathcal{F}_\lambda^\eta \simeq M(c_\lambda, h_\lambda^\eta), \quad \mathcal{F}_\lambda^\eta \simeq M(c_\lambda, h_\lambda^\eta)^c.$$

Hence, in the sequel, we concentrate on **Class** R^+.

8.1.2 Fine Classification: Class R^+

Until the end of this section, suppose that (λ, η) belongs to **Class** R^+, unless otherwise stated. Hence, there exist p and $q \in \mathbb{Z}_{>0}$ such that $(p, q) = 1$ and $\lambda = \lambda(\sqrt{\frac{q}{p}})$. For simplicity, set

$$\lambda_{p,q} := \lambda(\sqrt{\frac{q}{p}}).$$

Let us describe the set of η such that $(\lambda_{p,q}, \eta)$ belongs to **Class** R^+.

We first introduce some notation. For r and $s \in \mathbb{Z}_{\geq 0}$ such that $r < p$, $s \leq q$ and $\sigma = \pm$, we introduce $\eta^\sigma_{r,s:i}$ as follows:

$$\eta^\sigma_{r,s:i} = \lambda_{p,q} + \sigma \mathbf{sgn}(i, sp - rq) \times \begin{cases} \eta_{-ip+r,s}(\sqrt{\frac{q}{p}}) & (i \equiv 0 \bmod 2) \\ \eta_{-(i+1)p+r,-s}(\sqrt{\frac{q}{p}}) & (i \equiv 1 \bmod 2) \end{cases} \quad (8.1)$$

where $\eta_{\alpha,\beta}(T) := (\alpha T - \beta T^{-1})/\sqrt{2}$ and

$$\mathbf{sgn}(i) := \begin{cases} 1 & (i \geq 0) \\ -1 & (i < 0) \end{cases}, \quad \mathbf{sgn}(i, j) := \begin{cases} \mathbf{sgn}(i) & (i \neq 0) \\ \mathbf{sgn}(j) & (i = 0) \end{cases}. \quad (8.2)$$

Noticing the relation

$$\eta^\sigma_{p-r,q-s:-i} = \eta^\sigma_{r,s:i},$$

we may suppose that $(r, s) \in K^+_{p,q}$ defined in (5.20). We often denote $\eta^\sigma_{r,s:i}$ by η^σ_i. Here, it should be mentioned that

$$h^{\eta^\sigma_i}_{\lambda_{p,q}} = h_i$$

holds for any $\sigma = \pm$, where h_i is defined in (5.22).

Lemma 8.2.

$$\{\eta \in \mathbb{C} | (\lambda_{p,q}, \eta) \text{ belongs to } \textbf{Class } R^+\} = \{\eta^\sigma_{r,s:i} | (r, s) \in K^+_{p,q},\ i \in \mathbb{Z},\ \sigma = \pm\}.$$

Proof. Since $(c_{\lambda_{p,q}}, h_{\lambda_{p,q}}^{\eta^{\sigma}_{r,s:i}}) = (c_{p,q}, h_{p,q:r,s:i})$, the inclusion $\supset$ follows from Lemma 5.8. Hence, we show the opposite inclusion. We define the map ϕ by

$$\phi : \mathbb{C}^2 \longrightarrow \mathfrak{h}^* \quad (\lambda, \eta) \mapsto (c_\lambda, h^\eta_\lambda). \tag{8.3}$$

Then, $\phi^{-1}(\phi(\lambda,\eta)) = \{(\lambda,\eta), (-\lambda,-\eta), (\lambda, 2\lambda-\eta), (-\lambda, -2\lambda+\eta)\}$. Since $2\lambda_{p,q} - \eta^{\sigma}_{r,s:i} = \eta^{-\sigma}_{r,s:i}$, we have

$$\phi^{-1}((c_{p,q}, h_{p,q:r,s:i})) = \{\pm(\lambda_{p,q}, \eta^{\sigma}_{r,s:i}) | \sigma = \pm, i \in \mathbb{Z}\}.$$

Hence, the opposite inclusion holds. □

Remark 8.1 *The map ϕ gives a 4-fold cover of $\mathfrak{h}^*$, and it relates to the following isomorphisms of Fock modules in Propositions 4.2 and 4.3:*

1. $\mathcal{F}^\eta_\lambda \simeq \mathcal{F}^{-\eta}_{-\lambda}$,
2. $(\mathcal{F}^\eta_\lambda)^c \simeq \mathcal{F}^{2\lambda-\eta}_\lambda$.

In particular, if $\mathcal{F}^\eta_\lambda$ is isomorphic to its contragredient dual, then (λ,η) lies in the ramification locus of ϕ.

As in § 5.1.4, we divided **Class** R^+ into the following four cases:

1. **Case** 1^+: $0 < r < p$ and $0 < s < q$,
2. **Case** 2^+: $r = 0 \wedge 0 < s < q$,
3. **Case** 3^+: $0 < r < p \wedge s = 0$,
4. **Case** 4^+: $(r,s) = (0,0)$, $(0,q)$.

Lemma 8.3. *For each $(r,s) \in K^+_{p,q}$, the degeneration of $\{\eta^{\sigma}_{r,s:i} | i \in \mathbb{Z}\}$ can be described as follows:*

1. **Case** 1^+: *no degeneration.*
2. **Case** 2^+: $\eta^{\sigma}_{-i-1} = \eta^{\sigma}_{i}$ $(i \in \mathbb{Z}_{\geq 0})$,
3. **Case** 3^+: $\eta^{\sigma}_{2i} = \eta^{\sigma}_{2i-1}$ $(i \in \mathbb{Z})$,
4. **Case** 4^+: $\begin{matrix} \eta^{\sigma}_{-2i-1} = \eta^{\sigma}_{-2i} = \eta^{\sigma}_{2i} = \eta^{\sigma}_{2i-1} & (r,s) = (0,0) \\ \eta^{\sigma}_{-2i-2} = \eta^{\sigma}_{-2i-1} = \eta^{\sigma}_{2i+1} = \eta^{\sigma}_{2i} & (r,s) = (0,q) \end{matrix}$ $(i \in \mathbb{Z}_{\geq 0})$,

Thus, the following list exhausts the set of η such that $(\lambda_{p,q}, \eta)$ belongs to **Class** R^+*:*

Case 1^+	$\eta^{\pm}_{i}$ $(i \in \mathbb{Z})$
Case 2^+	$\eta^{\pm}_{i}$ $(i \in \mathbb{Z}_{\geq 0})$
Case 3^+	$\eta^{\pm}_{(-1)^{i-1}i}$ $(i \in \mathbb{Z}_{\geq 0})$
Case 4^+	$\eta^{\pm}_{2i}$ $(i \in \mathbb{Z}_{\geq 0})$

Remark 8.2 *In* **Case** 4^+, *for $(r,s) = (0,0)$, $i = 0$, we have $\eta^+_0 = \eta^-_0$. Besides this case, the η's in the above table are all distinct.*

8.1.3 Zeros of $\det(\Gamma_{\lambda,\eta})_n$

For the study of the Jantzen filtration defined by the homomorphism $\Gamma_{\lambda,\eta}$, we list the zeros of the determinants $\det(\Gamma_{\lambda,\eta})_n$ $(n \in \mathbb{Z}_{>0})$. We set

$$\tilde{D}_\Gamma(\lambda, \eta) := \{(\alpha, \beta) \in (\mathbb{Z}_{>0})^2 | \Psi^+_{\alpha,\beta}(\lambda, \eta) = 0\}, \tag{8.4}$$

where $\Psi^+_{\alpha,\beta}(\lambda, \eta)$ is defined in Theorem 4.3, and

$$D_\Gamma(\lambda, \eta) := \{\alpha\beta | (\alpha, \beta) \in \tilde{D}_\Gamma(\lambda, \eta)\}. \tag{8.5}$$

Moreover, for $n \in D_\Gamma(\lambda, \eta)$, we set

$$a_\Gamma(n) := \sharp\{(\alpha, \beta) \in \tilde{D}_\Gamma(\lambda, \eta) | \alpha\beta = n\}. \tag{8.6}$$

To describe $D_\Gamma(\lambda, \eta)$, we introduce and recall some notation. In **Case** 1^+, to parameterise the elements of the set $\mathbb{Z}_{<-|i|} \cup \{i\} \cup \mathbb{Z}_{>|i|}$, it is convenient to use the following notation:

$$\ell(i, \sigma; n) := \begin{cases} \sigma\{\mathbf{sgn}(n)i + \mathbf{sgn}(i, sp - rq)n\} & (n \neq 0) \\ i & (n = 0) \end{cases}. \tag{8.7}$$

Indeed, $\ell(i, \sigma; n)$ is one of $\pm(|i| + |n|)$ and

$$\mathbb{Z}_{<-|i|} \cup \{i\} \cup \mathbb{Z}_{>|i|} = \{\ell(i, \sigma; n) | n \in \mathbb{Z}\}.$$

Then, the set $D_\Gamma(\lambda, \eta)$ and $a_\Gamma(n)$ are described as follows:

Lemma 8.4. *1.* **Case** 1^+: $\eta = \eta_i^\sigma$ $(i \in \mathbb{Z})$,

$$D_\Gamma(\lambda, \eta) = \{h_{\ell(i,\sigma;2k-1)} - h_i | k \in \mathbb{Z}_{>0}\}.$$

2. **Case** 2^+: $\eta = \eta_i^\sigma$ $(i \in \mathbb{Z}_{\geq 0})$,

$$D_\Gamma(\lambda, \eta) = \left\{h_{i+2k-\delta_{\sigma,+}} - h_i | k \in \mathbb{Z}_{>0}\right\}.$$

3. **Case** 3^+: $\eta = \eta^\sigma_{(-1)^{i-1}i}$ $(i \in \mathbb{Z}_{\geq 0})$,

$$D_\Gamma(\lambda, \eta) = \left\{h_{(-1)^{i+2k-\delta_{\sigma,-}}(i+2k-\delta_{\sigma,-})} - h_{(-1)^{i-1}i} | k \in \mathbb{Z}_{>0}\right\}.$$

4. **Case** 4^+: $\eta = \eta_{2i}^\sigma$ $(i \in \mathbb{Z}_{\geq 0})$,

$$D_\Gamma(\lambda, \eta) = \{h_{i+2k} - h_{2i} | k \in \mathbb{Z}_{>0}\}.$$

For any $n \in D_\Gamma(\lambda, \eta)$, we have $a_\Gamma(n) = 1$.

8.2 The Jantzen (Co)filtrations of Fock Modules

In this section, we fix some notations on the Jantzen (co)filtrations defined for the maps $\Gamma_{\lambda,\eta}$ and $\mathrm{L}^{\lambda,\eta}$. As an application of the structure theorem of Verma modules in Chapter 6, we determine the structure of the Jantzen (co)filtrations of Fock modules. Throughout this chapter, as in § 5.5.1, let $\mathcal{R}$ be the polynomial ring $\mathbb{C}[t]$ and $\mathcal{Q}$ its quotient field $\mathbb{C}(t)$. We denote the t-adic valuation $\mathcal{Q} \to \mathbb{Z}\cup\{\infty\}$ by ord_t and the canonical projection $\mathcal{R} \to \mathcal{R}/t\mathcal{R} \simeq \mathbb{C}$ by ϕ_t. The functor $\mathrm{Mod}_{\mathcal{R}} \to \mathrm{Vect}_{\mathbb{C}}$ induced from ϕ_t is denoted by the same symbol.

8.2.1 Contragredient Dual of $\mathfrak{g}_{\mathcal{R}}$-Modules

Let us first recall some notations from § 1.2.2 and 3.2.1. Let $(\mathfrak{g}, \mathfrak{h})$ be a Q-graded Lie algebra with a Q-graded anti-involution σ and $\iota : \mathrm{Im}\pi_Q \to Q$ the map defined in (1.6). Recall that $G := Q/\mathrm{Im}\iota$, $\mathfrak{g}_{\mathcal{R}} := \mathfrak{g} \otimes_{\mathbb{K}} \mathcal{R}$, $\mathfrak{h}_{\mathcal{R}} := \mathfrak{h} \otimes_{\mathbb{K}} \mathcal{R}$, $\mathfrak{h}^*_{\mathcal{R}} := (\mathfrak{h}^*) \otimes_{\mathbb{K}} \mathcal{R}$. We denote by $\sigma_{\mathcal{R}}$ the anti-involution on $\mathfrak{g}_{\mathcal{R}}$ induced from σ.

Here, for a $G \times \mathfrak{h}^*_{\mathcal{R}}$-graded $\mathfrak{g}_{\mathcal{R}}$-module $\tilde{V}$, we define the 'contragredient dual' $\tilde{V}^c$ as follows:

Definition 8.2 *Suppose that* $\tilde{V} = \bigoplus_{(\alpha,\tilde{\lambda})\in G\times\mathfrak{h}^*_{\mathcal{R}}} \tilde{V}^{\alpha}_{\tilde{\lambda}}$. *We set*

$$\tilde{V}^c := \bigoplus_{(\alpha,\tilde{\lambda})\in G\times\mathfrak{h}^*_{\mathcal{R}}} \mathrm{Hom}_{\mathcal{R}}(\tilde{V}^{\alpha}_{\tilde{\lambda}}, \mathcal{R})$$

and regard it as $\mathfrak{g}_{\mathcal{R}}$*-module via*

$$(x.f)(v) = f(\sigma_{\mathcal{R}}(x).v) \quad (x \in \mathfrak{g}_{\mathcal{R}},\ f \in \tilde{V}^c,\ v \in \tilde{V}).$$

8.2.2 Fock Modules over $\mathfrak{g}_{\mathcal{R}}$

Let $\mathcal{H} := \bigoplus_{n\in\mathbb{Z}} \mathbb{C}a_n \oplus \mathbb{C}K_{\mathcal{H}}$ be the Heisenberg Lie algebra (see § 1.2.3). For $\tilde{\eta} \in \mathcal{R}$, we define the Fock module $\mathcal{F}^{\tilde{\eta}}_{\mathcal{R}}$ over $\mathcal{H}_{\mathcal{R}} := \mathcal{H} \otimes_{\mathbb{C}} \mathcal{R}$ as follows: set $\mathcal{H}^{\geq}_{\mathcal{R}} := \mathcal{H}^{\geq} \otimes_{\mathbb{C}} \mathcal{R}$, and introduce an $\mathcal{H}^{\geq}_{\mathcal{R}}$ module $\mathcal{R}_{\tilde{\eta}} := \mathcal{R}\mathbf{1}_{\tilde{\eta}}$ of rank one by

1. Q-gradation: $(\mathcal{R}_{\tilde{\eta}})^{\beta} = \begin{cases} \mathcal{R}_{\tilde{\eta}} & (\beta = 0) \\ \{0\} & (\beta \neq 0) \end{cases}$,
2. $\mathcal{H}^{\geq}_{\mathcal{R}}$-action: $a_0.\mathbf{1}_{\tilde{\eta}} = \tilde{\eta}\mathbf{1}_{\tilde{\eta}}$, $K_{\mathcal{H}}.\mathbf{1}_{\tilde{\eta}} = \mathbf{1}_{\tilde{\eta}}$, $\mathcal{H}^{+}_{\mathcal{R}}.\mathbf{1}_{\tilde{\eta}} = \{0\}$.

We set

$$\mathcal{F}_{\mathcal{R}}^{\tilde{\eta}} := U(\mathcal{H}_{\mathcal{R}}) \otimes_{U(\mathcal{H}_{\mathcal{R}}^{\geq})} \mathcal{R}_{\tilde{\eta}}.$$

Next, we introduce the action of the Virasoro algebra $\mathfrak{g}_{\mathcal{R}} := \mathfrak{g} \otimes_{\mathbb{C}} \mathcal{R}$ on $\mathcal{F}_{\mathcal{R}}^{\tilde{\eta}}$. For $\tilde{\lambda} \in \mathcal{R}$, we let $\mathfrak{g}_{\mathcal{R}}$ act on $\mathcal{F}_{\mathcal{R}}^{\tilde{\eta}}$ by

$$\sum_{n\in\mathbb{Z}} L_n z^{-n-2} \longmapsto \frac{1}{2} {}^{\circ}_{\circ} a(z)^2 {}^{\circ}_{\circ} + \tilde{\lambda}\partial a(z), \quad C \longmapsto c_{\tilde{\lambda}} \mathrm{id}_{\mathcal{F}_{\mathcal{R}}^{\tilde{\eta}}},$$

and denote this $\mathfrak{g}_{\mathcal{R}}$-module by $\mathcal{F}_{\tilde{\lambda}\,\mathcal{R}}^{\tilde{\eta}}$.

Let $(\mathcal{F}_{\tilde{\lambda}\,\mathcal{R}}^{\tilde{\eta}})^c$ be the contragredient dual of $\mathcal{F}_{\tilde{\lambda}\,\mathcal{R}}^{\tilde{\eta}}$ (Definition 8.2). By the same argument as in the proof of Corollary 4.1, one can show

Lemma 8.5. *There exists an isomorphism* $(\mathcal{F}_{\tilde{\lambda}\,\mathcal{R}}^{\tilde{\eta}})^c \simeq \mathcal{F}_{\tilde{\lambda}\,\mathcal{R}}^{2\tilde{\lambda}-\tilde{\eta}}$ *of* $\mathfrak{g}_{\mathcal{R}}$*-modules.*

8.2.3 The Jantzen (Co)filtrations defined by $\Gamma_{\lambda,\eta}$ and $\mathrm{L}^{\lambda,\eta}$

Here, for $(\tilde{\lambda}, \tilde{\mu}) \in \mathcal{R}^2$, put $(\tilde{c}, \tilde{h}) := (c_{\tilde{\lambda}}, h_{\tilde{\lambda}}^{\tilde{\mu}})$ and $(c, h) := (\phi_t(\tilde{c}), \phi_t(\tilde{h}))$. Notice that $(\tilde{c}, \tilde{h}) \in \mathcal{R}^2$.

First, we recall some notations for Verma modules over $\mathfrak{g}_{\mathcal{R}}$ from § 5.5.1. Let $M_{\mathcal{R}}(\tilde{c}, \tilde{h})$ be the Verma module over $\mathfrak{g}_{\mathcal{R}}$ with highest weight $(\tilde{c}, \tilde{h})$ and $M_{\mathcal{R}}(\tilde{c}, \tilde{h})^c$ the contragredient dual of $M_{\mathcal{R}}(\tilde{c}, \tilde{h})$ (Definition 8.2). Let $v_{\tilde{c},\tilde{h}}$ be a highest weight vector of $M_{\mathcal{R}}(\tilde{c}, \tilde{h})$ and $\langle \cdot, \cdot \rangle_{\tilde{c},\tilde{h}}$ the contravariant form on $M_{\mathcal{R}}(\tilde{c}, \tilde{h})$ which satisfies $\langle v_{\tilde{c},\tilde{h}}, \cdot v_{\tilde{c},\tilde{h}} \rangle_{\tilde{c},\tilde{h}} = 1$. As in § 3.2.2, we assume that

$\langle \cdot, \cdot \rangle_{\tilde{c},\tilde{h}}$ is non-degenerate, i.e., $\det(\tilde{c}, \tilde{h})_n \neq 0$ ($\forall n \in \mathbb{Z}_{\geq 0}$).

Notice that, for any $(\lambda, \eta) \in \mathbb{C}^2$, if $(\tilde{\lambda}, \tilde{\eta}) \in \mathcal{R}^2$ is so chosen as

$$(\tilde{\lambda}, \tilde{\eta}) := (\lambda + t, \eta + t), \tag{8.8}$$

then, one can directly check that $\langle \cdot, \cdot \rangle_{\tilde{c},\tilde{h}}$ is non-degenerate.

Let us introduce the Jantzen (co)filtrations defined from the homomorphisms $\Gamma_{\lambda,\mu}$ and $\mathrm{L}^{\lambda,\mu}$ under the perturbation (8.8). The universality of $M_{\mathcal{R}}(\tilde{c}, \tilde{h})$ implies that there exists the homomorphism of $\mathfrak{g}_{\mathcal{R}}$-modules

$$\tilde{\Gamma}_{\tilde{\lambda},\tilde{\eta}} : M_{\mathcal{R}}(\tilde{c}, \tilde{h}) \longrightarrow \mathcal{F}_{\tilde{\lambda}\,\mathcal{R}}^{\tilde{\eta}}; \qquad v_{\tilde{c},\tilde{h}} \mapsto 1 \otimes \mathbf{1}_{\tilde{\eta}}.$$

We consider the following transpose of $\tilde{\Gamma}_{\tilde{\lambda},2\tilde{\lambda}-\tilde{\eta}}$:

$$\tilde{\mathrm{L}}^{\tilde{\lambda},\tilde{\eta}} : \mathcal{F}_{\tilde{\lambda}\,\mathcal{R}}^{\tilde{\eta}} \simeq (\mathcal{F}_{\tilde{\lambda}\,\mathcal{R}}^{2\tilde{\lambda}-\tilde{\eta}})^c \overset{{}^t\tilde{\Gamma}_{\tilde{\lambda},2\tilde{\lambda}-\tilde{\eta}}}{\longrightarrow} M_{\mathcal{R}}(\tilde{c}, h_{\tilde{\eta}}^{2\tilde{\lambda}-\tilde{\eta}})^c = M_{\mathcal{R}}(\tilde{c}, \tilde{h})^c.$$

Since $\langle\cdot,\cdot\rangle_{\tilde{c},\tilde{h}}$ is non-degenerate, the $\mathfrak{g}_{\mathcal{Q}}$-modules $M_{\mathcal{Q}}(\tilde{c},\tilde{h}) := M_{\mathcal{R}}(\tilde{c},\tilde{h}) \otimes_{\mathcal{R}} \mathcal{Q}$, $\mathcal{F}^{\tilde{\eta}}_{\tilde{\lambda}\,\mathcal{Q}} := \mathcal{F}^{\tilde{\eta}}_{\tilde{\lambda}\,\mathcal{R}} \otimes_{\mathcal{R}} \mathcal{Q}$, $M_{\mathcal{Q}}(\tilde{c},\tilde{h})^c := M_{\mathcal{Q}}(\tilde{c},\tilde{h}) \otimes_{\mathcal{R}} \mathcal{Q}$ are irreducible, and $\Gamma_{\tilde{\lambda},\tilde{\eta}} \otimes_{\mathcal{R}} \mathcal{Q} : M_{\mathcal{Q}}(\tilde{c},\tilde{h}) \to \mathcal{F}^{\tilde{\eta}}_{\tilde{\lambda}\,\mathcal{Q}}$ and $\mathrm{L}^{\tilde{\lambda},\tilde{\eta}} \otimes_{\mathcal{R}} \mathcal{Q} : \mathcal{F}^{\tilde{\eta}}_{\tilde{\lambda}\,\mathcal{Q}} \to M_{\mathcal{Q}}(\tilde{c},\tilde{h})^2$ are isomorphisms. Hence, we can consider the Jantzen (co)filtrations defined by these homomorphisms as in § 3.3. Here, we collect some notation for those Jantzen (co)filtrations:

1. The Jantzen (co)filtration defined by $\tilde{\Gamma}_{\tilde{\lambda},\tilde{\eta}}$:

$$M(c,h) \supset M(c,h)(1) \supset M(c,h)(2) \supset \cdots, \quad \mathcal{F}^{\eta}_{\lambda} \twoheadrightarrow \mathcal{F}^{\eta}_{\lambda}(1) \twoheadrightarrow \mathcal{F}^{\eta}_{\lambda}(2) \twoheadrightarrow \cdots,$$

   $\Gamma^{(k)}_{\lambda,\eta} : M(c,h)(k] \to \mathcal{F}^{\eta}_{\lambda}(k]$ is the kth derivative, $\pi_{(k]} : \mathcal{F}^{\eta}_{\lambda} \twoheadrightarrow \mathcal{F}^{\eta}_{\lambda}(k]$, $IK(k-1] \subset \mathcal{F}^{\eta}_{\lambda}$ such that $IK(k-1] = \mathrm{Ker}\pi_{(k]}$.
2. The Jantzen (co)filtration defined by $\mathrm{L}^{\tilde{\lambda},\tilde{\eta}}$:

$$\mathcal{F}^{\eta}_{\lambda} \supset \mathcal{F}^{\eta}_{\lambda}[1) \supset \mathcal{F}^{\eta}_{\lambda}[2) \supset \cdots, \quad M(c,h)^c \twoheadrightarrow M(c,h)^c[1) \twoheadrightarrow M(c,h)^c[2) \twoheadrightarrow \cdots,$$

   $\mathrm{L}^{\lambda,\eta}_{[k)} : \mathcal{F}^{\eta}_{\lambda}[k) \to M(c,h)^c[k)$ is the kth derivative, $\pi_{[k)} : M(c,h)^c \twoheadrightarrow M(c,h)^c[k)$ and $IK[k-1) \subset M(c,h)^c$ such that $IK[k-1) = \mathrm{Ker}\pi_{[k)}$.

Here, $k \in \mathbb{Z}$ is assumed to be positive.

8.2.4 Character Sum Formula

Here, we choose $(\tilde{\lambda},\tilde{\eta})$ as (8.8), and set $(c,h) := (\phi_t(c_{\tilde{\lambda}}), \phi_t(h^{\tilde{\eta}}_{\tilde{\lambda}}))$. By Proposition 3.4, one can show the following proposition in a way similar to the proof of Proposition 5.8:

Proposition 8.2

$$\sum_{l=1}^{\infty} \mathrm{ch}\, M(c,h)(l] = \sum_{(\alpha,\beta)\in \tilde{D}_{\Gamma}(\lambda,\eta)} \mathrm{ord}_t\, \Psi^{+}_{\alpha,\beta}(\tilde{\lambda},\tilde{\eta}) \times \mathrm{ch}\, M(c, h+\alpha\beta),$$

where $\tilde{D}_{\Gamma}(\lambda,\eta)$ is define in (8.4).

The value of $\mathrm{ord}_t\, \Psi^{+}_{\alpha,\beta}(\tilde{\lambda},\tilde{\eta})$ in the right-hand side of the above formula is given by

$$\mathrm{ord}_t\, \Psi^{+}_{\alpha,\beta}(\tilde{\lambda},\tilde{\eta}) = \begin{cases} 1 & (\Psi^{+}_{\alpha,\beta}(\lambda,\eta) = 0) \\ 0 & (\Psi^{+}_{\alpha,\beta}(\lambda,\eta) \neq 0) \end{cases}.$$

From now on, suppose that (λ,η) belongs to **Class** R^+, namely, $\lambda = \lambda_{p,q}$ for some $p,q \in \mathbb{Z}_{>0}$ such that $(p,q) = 1$ and

$$\eta = \begin{cases} \eta_i^\sigma & (i \in \mathbb{Z}) & \textbf{Case } 1^+ \\ \eta_i^\sigma & (i \in \mathbb{Z}_{\geq 0}) & \textbf{Case } 2^+ \\ \eta_{(-1)^{i-1}i}^\sigma & (i \in \mathbb{Z}_{\geq 0}) & \textbf{Case } 3^+ \\ \eta_{2i}^\sigma & (i \in \mathbb{Z}_{\geq 0}) & \textbf{Case } 4^+ \end{cases}. \tag{8.9}$$

We recall the following notation introduced in (6.1):

$$\xi_i := \begin{cases} h_i & \textbf{Case } 1^+,\ 2^+ \\ h_{(-1)^{i-1}i} & \textbf{Case } 3^+ \\ h_{2i} & \textbf{Case } 4^+ \end{cases}. \tag{8.10}$$

Hence, we have $h_\lambda^\eta = \xi_i$. By Lemma 8.4, we have

Lemma 8.6. *Suppose that η is chosen as (8.9).*

1. **Case 1^+**:
$$\sum_{l=1}^\infty \mathrm{ch}\, M(c,\xi_i)(l] = \sum_{k=1}^\infty \mathrm{ch}\, M\left(c, \xi_{\ell(i,\sigma;2k-1)}\right).$$

2. **Case 2^+**:
$$\sum_{l=1}^\infty \mathrm{ch}\, M(c,\xi_i)(l] = \sum_{k=1}^\infty \mathrm{ch}\, M\left(c, \xi_{i+2k-\delta_{\sigma,+}}\right).$$

3. **Case 3^+**:
$$\sum_{l=1}^\infty \mathrm{ch}\, M(c,\xi_i)(l] = \sum_{k=1}^\infty \mathrm{ch}\, M\left(c, \xi_{i+2k-\delta_{\sigma,-}}\right).$$

4. **Case 4^+**:
$$\sum_{l=1}^\infty \mathrm{ch}\, M(c,\xi_i)(l] = \sum_{k=1}^\infty \mathrm{ch}\, M\left(c, \xi_{i+k}\right).$$

We omit writing the character sum $\sum_{l=1}^\infty \mathrm{ch}\, M(c,h)^c[l)$ for each case explicitly, since this follows immediately from this lemma and the duality principle explained in Proposition 3.9.

8.2.5 Structures of the Jantzen Filtrations defined by $\Gamma_{\tilde{\lambda},\tilde{\eta}}$ and $L^{\tilde{\lambda},\tilde{\eta}}$

Here, we determine the structures of the Jantzen filtration $\{M(c,\xi_i)(n]\}$ and cofiltration $\{M(c,\xi_i)^c[n)\}$ with the aid of the classification of submodules of $M(c,\xi_i)$ (Proposition 6.1).

Recall that (λ,η) belongs to **Class** R^+ and η_i^σ and ξ_i are given as (8.10) and (8.1). As in § 6.1.3, for $i,j \in \mathbb{Z}$ (in **Case** 1^+), $i,j \in \mathbb{Z}_{\geq 0}$ (in **Case** 2^+,

3^+, 4^+) such that ($|i| < |j|$), we identify $M(c, \xi_j)$ with its image under the embedding $M(c, \xi_j) \hookrightarrow M(c, \xi_i)$.

Proposition 8.3 *Suppose that η is chosen as in (8.9). For any $n \in \mathbb{Z}_{>0}$,*

1. **Case** 1^+: $M(c, \xi_i)(n] = M\left(c, \xi_{\ell(i,\sigma;2n-1)}\right)$.
2. **Case** 2^+: $M(c, \xi_i)(n] = M\left(c, \xi_{i+2n-\delta_{\sigma,+}}\right)$.
3. **Case** 3^+: $M(c, \xi_i)(n] = M\left(c, \xi_{i+2n-\delta_{\sigma,-}}\right)$.
4. **Case** 4^+: $M(c, \xi_i)(n] = M\left(c, \xi_{i+n}\right)$.

Proof. Here, we show the proposition in **Case** 1^+, since the other cases can be proved similarly. By Lemma 8.6, we have

$$\sum_{l=1}^{\infty} \operatorname{ch} M(c, \xi_i)(l] = \sum_{k=1}^{\infty} \operatorname{ch} M\left(c, \xi_{\ell(i,\sigma;2k-1)}\right). \tag{8.11}$$

We prove the statement by induction on n. It follows from (8.11) that

$$\{M(c, \xi_i)(1]_{\xi_{\ell(i,\sigma;1)}}\}^{\mathfrak{g}^+} \neq \{0\}, \quad \left[M(c, \xi_i)(1] : L(c, \xi_{\ell(i,\sigma;-1)})\right] = 0. \tag{8.12}$$

Hence, Proposition 6.1 implies $M(c, \xi_i)(1] \simeq M(c, \xi_{\ell(i,\sigma;1)})$, and the statement holds for $n = 1$. Next, we assume that the statement holds up to $n - 1$. From (8.11) and the inductive hypothesis, we have

$$\sum_{l=n}^{\infty} \operatorname{ch} M(c, \xi_i)(l] = \sum_{k=n}^{\infty} \operatorname{ch} M\left(c, \xi_{\ell(i,\sigma;2k-1)}\right).$$

Hence, by a similar argument to the case of $n = 1$, we obtain $M(c, \xi_i)(n] \simeq M(c, \xi_{\ell(i,\sigma;2n-1)})$. □

Combining Proposition 8.3 with Proposition 3.9, we obtain

Proposition 8.4 *Suppose that η is chosen as in (8.9). For any $n \in \mathbb{Z}_{>0}$,*

1. **Case** 1^+: $M(c, \xi_i)^c[n) \simeq M\left(c, \xi_{\ell(i,\sigma;-2n+1)}\right)^c$.
2. **Case** 2^+: $M(c, \xi_i)^c[n) \simeq M(c, \xi_{i+2n-\delta_{\sigma,-}})^c$.
3. **Case** 3^+: $M(c, \xi_i)^c[n) \simeq M(c, \xi_{i+2n-\delta_{\sigma,+}})^c$.
4. **Case** 4^+: $M(c, \xi_i)^c[n) \simeq M(c, \xi_{i+n})^c$.

8.2.6 Singular Vectors and $M(c, \xi_i)(n]$

Until the end of this subsection, we fix $i \in \mathbb{Z}$ and $\sigma \in \{\pm\}$ (in **Case** 1^+) and $i \in \mathbb{Z}_{\geq 0}$ (in **Case** 2^+, 3^+, 4^+). Here, we give some diagrams describing the structure of $M(c, \xi_i)(n]$. To this end, let us relabel the singular vectors of $M(c, \xi_i)$ given in Proposition 6.1 as follows:

1. **Case 1$^+$**: For $n \in \mathbb{Z}$, $w_n \in \{M(c,\xi_i)_{\xi_{\ell(i,\sigma;n)}}\}^{\mathfrak{g}^+} \setminus \{0\}$.
2. **Case 2$^+$, 3$^+$, 4$^+$**: For $n \in \mathbb{Z}_{\geq 0}$, $w_n \in \{M(c,\xi_i)_{\xi_{i+n}}\}^{\mathfrak{g}^+} \setminus \{0\}$.

Recall the diagrams below that describe the structure of $M(c,\xi_i)$ (see § 6.1.2):

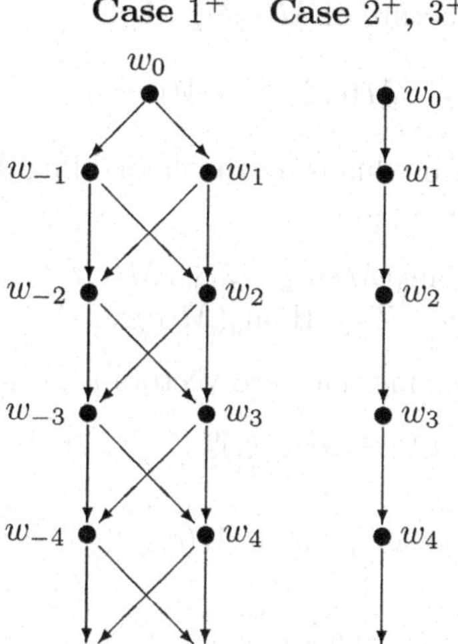

Here, $\bullet w$ denotes a singular vector w, and the arrow $w\bullet \longrightarrow \bullet w'$ indicates that $w' \in U(\mathfrak{g})w$. By using these diagrams, the structure of the Jantzen filtration stated in Proposition 8.3 can be described as follows:

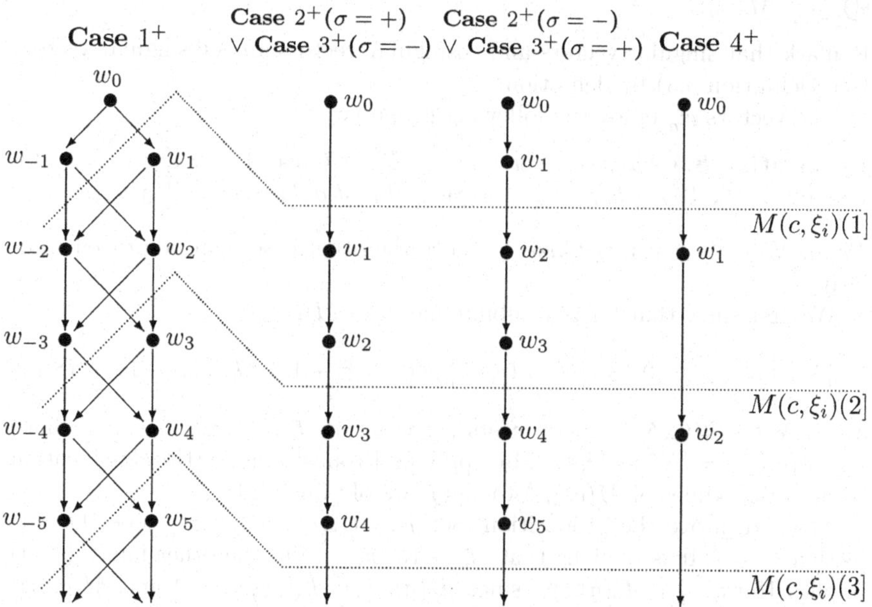

8.2.7 Cosingular Vectors and $M(c, \xi_i)^c[n]$

Using diagrams similar to those in the previous subsection, we describe the Jantzen cofiltration $\{M(c, \xi_i)^c[n]\}$.

We first explain the structure of $M(c, \xi_i)^c$. Let

$$\pi_{k,l} : M(c, \xi_k)^c \twoheadrightarrow M(c, \xi_l)^c \tag{8.13}$$

be the surjection obtained by dualising the embedding $M(c, \xi_l) \hookrightarrow M(c, \xi_k)$. By Theorem 6.6, we have

1. **Case** 1^+: For $n \in \mathbb{Z}$, $\mathrm{Hom}_{\mathfrak{g}}(M(c, \xi_{\ell(i,\sigma;n)}), M(c, \xi_i)) \neq \{0\}$.
2. **Case** 2^+, 3^+, 4^+: For $n \in \mathbb{Z}_{\geq 0}$, $\mathrm{Hom}_{\mathfrak{g}}(M(c, \xi_{i+n}), M(c, \xi_i)) \neq \{0\}$.

Hence, there exist the following non-zero vectors $w_n^c \in M(c, \xi_i)^c$:

1. **Case** 1^+: For $n \in \mathbb{Z}$, $w_n^c \in M(c, \xi_i)^c_{\xi_{\ell(i,\sigma;n)}}$ satisfying $\pi_{i,\ell(i,\sigma;n)}(w_n^c) = v^c_{\ell(i,\sigma;n)}$.
2. **Case** 2^+, 3^+, 4^+: For $n \in \mathbb{Z}_{\geq 0}$, $w_n^c \in M(c, \xi_i)^c_{\xi_{i+n}}$ satisfying $\pi_{i,i+n}(w_n^c) = v^c_{i+n}$.

Here v_k^c is a non-zero vector in $M(c, \xi_k)^c_{\xi_k}$.

These vectors are cosingular vectors of $M(c, \xi_k)^c$:

Definition 8.3 *Suppose that $M = \bigoplus_{h \in \mathbb{C}} M_h$ is a $(\mathfrak{g}, \mathfrak{h})$-module and $u \in M_h$. If $u \notin U(\mathfrak{g}).M_{h-\mathbb{Z}_{>0}}$, then u is called a* **cosingular vector**, *where $M_{h-\mathbb{Z}_{>0}} := \bigoplus_{n \in \mathbb{Z}_{>0}} M_{h-n}$.*

Remark that singular vectors and cosingular vectors are subsingular vectors (see Definition 5.3) by definition.

The vectors w_n^c enjoy the following property:

Proposition 8.5 *Suppose that $m, n \in \mathbb{Z}$ (in* **Case** 1^+*), $m, n \in \mathbb{Z}_{\geq 0}$ (in* **Case** 2^+, 3^+, 4^+*). Then, $w_m^c \in U(\mathfrak{g})w_n^c$ if and only if $w_n \in U(\mathfrak{g})w_m$.*

Proof. We only consider **Case** 1^+ since the other cases can be proved similarly.

We first show that for any submodule N of $M(c, \xi_i)^c$,

$$[N : L(c, \xi_k)] \neq 0 \Rightarrow [N : L(c, \xi_l)] \neq 0 \; (-|k| + 1 \leq \forall l \leq |k| - 1). \tag{8.14}$$

Since $N \hookrightarrow M(c, \xi_i)^c$, there exists a surjection $f : M(c, \xi_i) \twoheadrightarrow N^c$. Since $[N : L(c, \xi_k)] = [N^c : L(c, \xi_k)]$, by applying Proposition 6.1 (the classification of the submodules of $M(c, \xi_i)$) to $\mathrm{Ker} f$, we obtain (8.14).

Now, we prove the proposition. Set $j := \ell(i, \sigma; n)$. Let $\pi_{i,-j}$ be the surjection (8.13). It is obvious that $w_n^c \in \mathrm{Ker}\pi_{i,-j}$. On the other hand, (8.14) implies $\mathrm{Ker}\pi_{i,-j} \subset U(\mathfrak{g}).w_n^c$, since $[U(\mathfrak{g}).w_n^c : L(c, \xi_j)] \neq 0$ and $M(c, \xi_i)^c$ is multiplicity free. Hence, $\mathrm{Ker}\pi_{i,-j} = U(\mathfrak{g}).w_n^c$. Since $w_m^c \in \mathrm{Ker}\pi_{i,-j} \Leftrightarrow |m| < |n| \vee m = n$, the proposition follows. □

The structure of $M(c, \xi_i)^c$ stated in Proposition 8.5 can be described by the following diagrams:

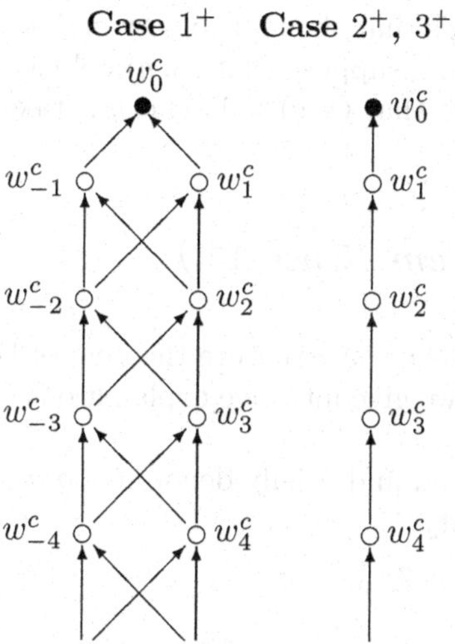

Here, $\bigcirc w$ denotes the cosingular vector w, and $w\bigcirc \longrightarrow \bigcirc w'$ means $w' \in U(\mathfrak{g})w$.

Next, we describe the structure of the Jantzen cofiltration given in Proposition 8.4 by means of the above diagram. The following diagrams indicate the structure of the increasing sequence $IK[0] \subset IK[1] \subset IK[2] \subset \cdots$ of the submodules which satisfy $M(c, \xi_i)^c[k] = M(c, \xi_i)^c / IK[k-1)$.

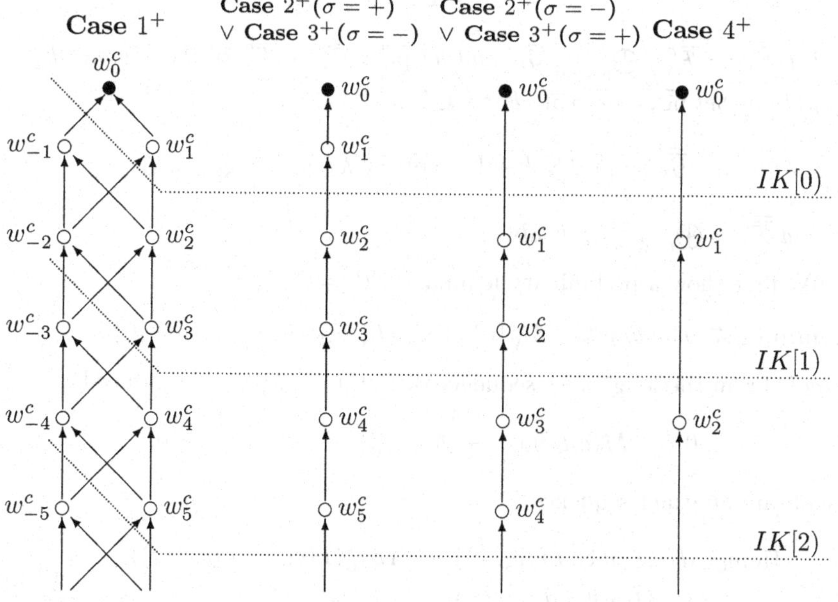

8.3 Structure of Fock Modules (Class R^+)

In this section, we reveal the structure of Fock modules. One of the key facts in the proof given below is that the multiplicity of $L(c,h)$ in $\mathcal{F}^\eta_\lambda$ is at most 1.

Throughout this section, suppose that (λ,η) belongs to **Class** R^+. Hence, $\lambda=\lambda_{p,q}$ (p, $q\in\mathbb{Z}_{>0}$ such that $(p,q)=1$) and η is taken as in (8.9).

8.3.1 Main Theorem (Case 1^+)

In this subsection, we prove the structure theorem of Fock modules in **Case** 1^+. After stating them, we give intuitive explanations of these statements by using diagrams.

We fix $i\in\mathbb{Z}$ and $\sigma=\pm$, and briefly denote $\ell(i,\sigma;n)$ by $\ell(n)$. Here, we set $\mathcal{F}^\eta_\lambda[0]:=\mathcal{F}^\mu_\lambda$ for simplicity.

Theorem 8.1 *1. For $k\in\mathbb{Z}_{>0}$,*

$$\mathcal{G}_k:=IK(k-1]\cap\mathcal{F}^\eta_\lambda[k)\simeq L(c,\xi_{\ell(-2k+1)}).$$

2. Let $\overline{\mathcal{F}}^\eta_\lambda:=\mathcal{F}^\eta_\lambda\Big/\bigoplus_{k\in\mathbb{Z}_{>0}}\mathcal{G}_k$ and let $\overline{\mathrm{pr}}:\mathcal{F}^\eta_\lambda\twoheadrightarrow\overline{\mathcal{F}}^\eta_\lambda$ be the canonical projection. For $k\in\mathbb{Z}_{\geq 0}$,

$$\overline{\mathcal{G}}_k:=\overline{\mathrm{pr}}(IK(k])\cap\overline{\mathrm{pr}}(\mathcal{F}^\eta_\lambda[k))\simeq\begin{cases}L(c,\xi_{\ell(-2k)})\oplus L(c,\xi_{\ell(2k)}) & (k>0)\\ L(c,\xi_{\ell(0)}) & (k=0)\end{cases}.$$

3. Let $\overline{\overline{\mathcal{F}}}{}^\eta_\lambda:=\overline{\mathcal{F}}^\eta_\lambda\Big/\bigoplus_{k\in\mathbb{Z}_{\geq 0}}\overline{\mathcal{G}}_k$ and let $\overline{\mathrm{pr}}':\overline{\mathcal{F}}^\eta_\lambda\twoheadrightarrow\overline{\overline{\mathcal{F}}}{}^\eta_\lambda$ be the canonical projection. Set $\overline{\overline{\mathrm{pr}}}:=\overline{\mathrm{pr}}'\circ\overline{\mathrm{pr}}$. For $k\in\mathbb{Z}_{\geq 0}$,

$$\overline{\overline{\mathcal{G}}}_k:=\overline{\overline{\mathrm{pr}}}(IK(k+1])\cap\overline{\overline{\mathrm{pr}}}(\mathcal{F}^\eta_\lambda[k))\simeq L(c,\xi_{\ell(2k+1)}),$$

and $\overline{\overline{\mathcal{F}}}{}^\eta_\lambda=\bigoplus_{k\in\mathbb{Z}_{\geq 0}}\overline{\overline{\mathcal{G}}}_k$ holds.

We first show a preliminary lemma: set $\mathcal{C}:=\mathcal{C}^\iota_{(\mathfrak{g},\mathfrak{h})}$.

Lemma 8.7. *For any $k\in\mathbb{Z}\setminus\{0\}$, $\mathrm{Ext}^1_{\mathcal{C}}(L(c,\xi_k),L(c,\xi_{-k}))=\{0\}$.*

Proof. From the long exact sequence of $\mathrm{Ext}^\bullet_{\mathcal{C}}(\,\cdot\,,L(c,\xi_{-k}))$ induced by

$$0\longrightarrow M(c,\xi_i)(1)\longrightarrow M(c,\xi_i)\longrightarrow L(c,\xi_i)\longrightarrow 0,$$

we obtain an exact sequence

$$\begin{aligned}&\mathrm{Hom}_{\mathcal{C}}(M(c,\xi_k)(1),L(c,\xi_{-k}))\longrightarrow\mathrm{Ext}^1_{\mathcal{C}}(L(c,\xi_k),L(c,\xi_{-k}))\\&\longrightarrow\mathrm{Ext}^1_{\mathcal{C}}(M(c,\xi_k),L(c,\xi_{-k})).\end{aligned}\tag{8.15}$$

Since $\mathrm{Hom}_{\mathcal{C}}(M(c,\xi_k)(1), L(c,\xi_{-k})) = \{0\}$ by Theorem 6.3, it is enough to show that $\mathrm{Ext}^1_{\mathcal{C}}(M(c,\xi_k), L(c,\xi_{-k})) = \{0\}$. By Proposition 1.13, we have

$$\mathrm{Ext}^1_{\mathcal{C}}(M(c,\xi_k), L(c,\xi_{-k})) \simeq \mathrm{Hom}_{\mathfrak{h}}(\mathbb{C}_{c,\xi_k}, H^1(\mathfrak{g}^+, L(c,\xi_{-k}))),$$

and $H^1(\mathfrak{g}^+, L(c,\xi_k)) \simeq H_1(\mathfrak{g}^-, L(c,\xi_k))$ as $\mathfrak{h}$-module by Proposition 1.14. This can be calculated by using the Bernstein–Gelfand–Gelfand type resolution for $L(c,\xi_k)$ (Theorem 6.9), since it is a $\mathfrak{g}^-$-free resolution and the result looks as follows, which is called the Kostant homology:

$$H_1(\mathfrak{g}^-, L(c,\xi_{-k})) \simeq \mathbb{C}_{c,\xi_{|k|+1}} \oplus \mathbb{C}_{c,\xi_{-|k|-1}}. \tag{8.16}$$

This formula implies

$$\mathrm{Ext}^1_{\mathcal{C}}(M(c,\xi_k), L(c,\xi_{-k})) \simeq \mathrm{Hom}_{\mathfrak{h}}(\mathbb{C}_{c,\xi_k}, \mathbb{C}_{c,\xi_{|k|+1}} \oplus \mathbb{C}_{c,\xi_{-|k|-1}}) = \{0\},$$

hence the lemma is proved. □

Proof of Theorem 8.1. The proof is based on the fact that $\mathcal{F}^\eta_\lambda$ is multiplicity free and the following formulae:

$$\mathrm{ch}\, IK(k'] = \sum_{n=-2k'-1}^{2k'} \mathrm{ch}\, L(c,\xi_{\ell(n)}), \quad \mathrm{ch}\, \mathcal{F}^\eta_\lambda[k'') = \sum_{\substack{|n| \geq 2k''-1 \\ n \neq 2k''-1}} \mathrm{ch}\, L(c,\xi_{\ell(n)}), \tag{8.17}$$

which follow from Propositions 8.3 and 8.4.

The first statement is clear since (8.17) implies

$$[IK(k-1] : L(c,h)] = 1 = [\mathcal{F}^\eta_\lambda[k) : L(c,h)] \iff h = \xi_{\ell(-2k+1)}.$$

We show the second statement. By (8.17), we have

$$\begin{aligned} &[IK(k] : L(c,h)] = 1 = [\mathcal{F}^\eta_\lambda[k) : L(c,h)] \\ &\iff h \in \begin{cases} \{\xi_{\ell(-2k+1)}, \xi_{\ell(-2k)}, \xi_{\ell(-2k-1)}, \xi_{\ell(2k)}\} & (k>0) \\ \{\xi_{\ell(-1)}, \xi_{\ell(0)}\} & (k=0) \end{cases}, \end{aligned}$$

which implies

$$\mathrm{ch}\, \overline{\mathcal{G}}_k = \begin{cases} \mathrm{ch}\, L(c,\xi_{\ell(-2k)}) + \mathrm{ch}\, L(c,\xi_{\ell(2k)}) & (k>0) \\ \mathrm{ch}\, L(c,\xi_{\ell(0)}) & (k=0) \end{cases}$$

by the definition of $\overline{\mathrm{pr}}$. Combining this with Lemma 8.7, we obtain the second statement.

We show the last statement. It follows from (8.17) that

$$[IK(k+1] : L(c,h)] = 1 = [\mathcal{F}^\eta_\lambda[k) : L(c,h)] \iff h \in H_k,$$

where

$$H_k := \begin{cases} \{\xi_{\ell(-2k-j_1)}, \xi_{\ell(2k+j_2)} | j_1 = -1, 0, 1, 2, 3,\ j_2 = 0, 1, 2\} & (k > 0) \\ \{\xi_{\ell(j)} | j = -3, \pm 2, \pm 1, 0\} & (k = 0) \end{cases}.$$

Hence, we see that $\mathrm{ch}\,\overline{\overline{\mathcal{G}}}_k = \mathrm{ch}\,L(c, \xi_{\ell(2k+1)})$, and thus, $\overline{\overline{\mathcal{G}}}_k \simeq L(c, \xi_{\ell(2k+1)})$. By comparing both sides at the level of characters, we have $\overline{\overline{\mathcal{F}}}{}^{\eta}_{\lambda} = \bigoplus_{k \in \mathbb{Z}_{\geq 0}} \overline{\overline{\mathcal{G}}}_k$. Now, we have completed the proof. □

Finally, we make a comment on subsingular vectors of $\mathcal{F}^{\eta}_{\lambda}$.

The following vectors in $w^f_n \in \mathcal{F}^{\eta}_{\lambda}$ ($n \in \mathbb{Z}$) are subsingular vectors:

1. For $k \in \mathbb{Z}_{>0}$, $w^f_{-2k+1} \in (\mathcal{G}^{\mathfrak{g}^+}_k)_{\xi_{\ell(-2k+1)}} \setminus \{0\}$.
2. For $k \in \mathbb{Z}$, $w^f_{\pm 2k} \in \overline{\mathrm{pr}}^{-1}(\overline{\mathcal{G}}^{\mathfrak{g}^+}_k)_{\xi_{\ell(\pm 2k)}} \setminus \{0\}$.
3. For $k \in \mathbb{Z}_{\geq 0}$, $w^f_{2k+1} \in \overline{\overline{\mathrm{pr}}}^{-1}(\overline{\overline{\mathcal{G}}}^{\mathfrak{g}^+}_k)_{\xi_{\ell(2k+1)}} \setminus \{0\}$.

By definition, w^f_{-2k-1} ($k \in \mathbb{Z}_{\geq 0}$) is a singular vector. In particular, we have $\mathrm{Hom}_{\mathfrak{g}}(M(c, \xi_{\ell(-2k-1)}), \mathcal{F}^{\eta}_{\lambda}) \cong \mathbb{C}$. By dualising it for $\eta \mapsto 2\lambda - \eta$, we obtain $\mathrm{Hom}_{\mathfrak{g}}(\mathcal{F}^{\eta}_{\lambda}, M(c, \xi_{\ell(2k+1)})^c) \cong \mathbb{C}$. Hence, one may choose w^f_{2k+1} ($k \in \mathbb{Z}_{\geq 0}$) to be a cosingular vector, as a preimage of a cosingular vector.

In [FeFu4], B. Feigin and D. Fuchs gave an intuitive description on the structure of Fock modules in **Case** 1^+ by making use of the following diagram. Here, the symbol ● (resp. ○ and ◎) signifies a singular (resp. cosingular and subsingular) vector.

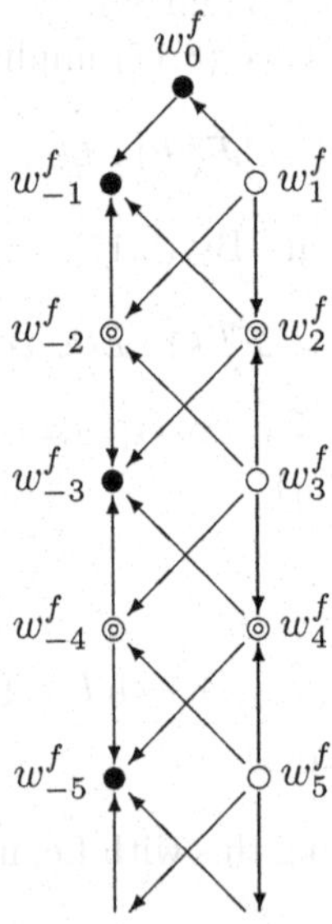

Roughly speaking, the arrows in the above diagram can be drawn by patching the structure of the Jantzen (co)filtration $M(c, \xi_i)(k)$ and $M(c, \xi_i)^c[k]$. By Proposition 3.7, the kth derivative $\Gamma^{(k)}_{\lambda,\eta}$ induces an isomorphism

$$IK(k]/IK(k-1] \simeq M(c,\xi_i)(k]/M(c,\xi_i)(k+1].$$

The structure of $M(c,\xi_i)(k]/M(c,\xi_i)(k+1]$ can be described as

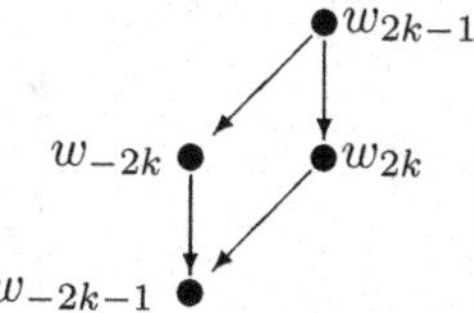

Similarly, the kth derivative $\mathrm{L}^{\lambda,\eta}_{[k)}$ induces an isomorphism

$$\mathcal{F}^{\eta}_{\lambda}[k)/\mathcal{F}^{\eta}_{\lambda}[k+1) \simeq IK[k)/IK[k-1).$$

The structure of $IK[k)/IK[k-1)$ can be described as

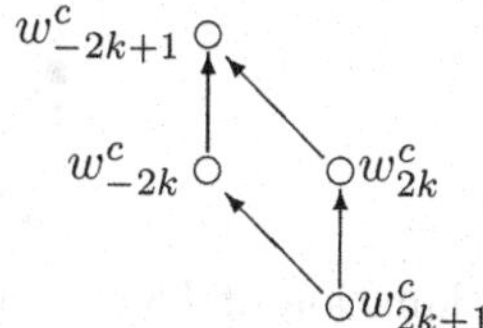

By patching these subdiagrams, one can draw the arrows in the above diagrams.

Next, we give an intuitive explanation on Theorem 8.1 by using the above diagram. The symbol $\times$ means that, under the map $\overline{\mathrm{pr}}$ or $\overline{\overline{\mathrm{pr}}}$, the image of the corresponding vector vanishes.

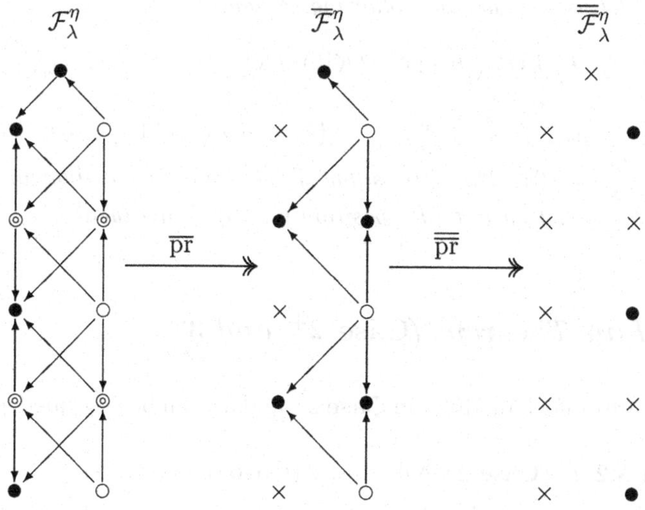

Pictorial explanation of Theorem 8.1

Finally, we describe the structure of $IK(k]$ and $\mathcal{F}_\lambda^\eta[k)$.

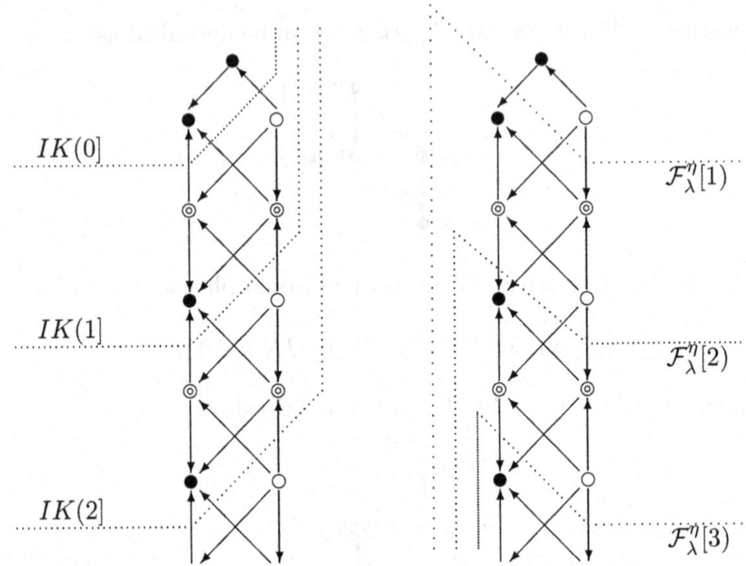

Pictorial explanation of $IK(k]$ and $\mathcal{F}_\lambda^\eta[k)$

Remark 8.3 *Here is an alternative proof of Theorem 8.1. For example, let us show that there exists a submodule $\mathcal{G}'_k \subset \mathcal{F}_\lambda^\eta$ which is isomorphic to $L(c, \xi_{\ell(-2k+1)})$ as follows:*

By Proposition 8.4, we have $\mathcal{F}_\lambda^\eta[k)^{\mathfrak{g}^+}_{\xi_{\ell(-2k+1)}} \neq \{0\}$. Hence, $\mathcal{G}'_k := U(\mathfrak{g}).u_k$ ($u_k \in \mathcal{F}_\lambda^\eta[k)^{\mathfrak{g}^+}_{\xi_{\ell(-2k+1)}} \setminus \{0\}$) is a highest weight module. To show that $\mathcal{G}'_k$ is irreducible, let us consider the following sequence:

$$\mathcal{G}'_k \hookrightarrow \mathcal{F}_\lambda^\eta[k) \twoheadrightarrow \mathcal{F}_\lambda^\eta[k)/\mathcal{F}_\lambda^\eta[k+1) \simeq IK[k)/IK[k-1) \simeq \mathrm{Ker}\pi_{\ell(-2k+1),\ell(-2k-1)}.$$

Since $(\mathrm{Ker}\pi_{\ell(-2k+1),\ell(-2k-1)})^{\mathfrak{g}^+}_{\xi_{\ell(\pm 2k)}} = \{0\}$ and $\mathcal{F}_\lambda^\eta[k+1)_{\xi_{\ell(\pm 2k)}} = \{0\}$, we see that $(\mathcal{G}'_k)^{\mathfrak{g}^+}_{\xi_{\ell(\pm 2k)}} = \{0\}$. This means that $\mathcal{G}'_k$ does not have a singular vector of L_0-weight $\xi_{\ell(\pm 2k)}$. Hence, by Proposition 6.1, $\mathcal{G}'_k$ is irreducible.

8.3.2 Main Theorem (Case 2^+ *and* 3^+)

The structure of Fock modules in **Case** 2^+ and 3^+ can be described as follows:

Theorem 8.2 *1.* (**Case** $2^+ \wedge \sigma = +) \vee ($**Case** $3^+ \wedge \sigma = -)$*:*

a. For $k \in \mathbb{Z}_{\geq 0}$,

$$\mathcal{G}_k := IK(k] \cap \mathcal{F}_\lambda^\eta[k) \simeq L(c, \xi_{i+2k}).$$

b. Let $\overline{\mathcal{F}}^{\eta}_{\lambda} := \mathcal{F}^{\eta}_{\lambda} \Big/ \bigoplus_{k \in \mathbb{Z}_{\geq 0}} \mathcal{G}_k$ *and let* $\overline{\mathrm{pr}} : \mathcal{F}^{\eta}_{\lambda} \twoheadrightarrow \overline{\mathcal{F}}^{\eta}_{\lambda}$ *be the canonical projection. For* $k \in \mathbb{Z}_{\geq 0}$*,*

$$\overline{\mathcal{G}}_k := \overline{\mathrm{pr}}\,(IK(k+1]) \cap \overline{\mathrm{pr}}\,(\mathcal{F}^{\eta}_{\lambda}[k)) \simeq L(c, \xi_{i+2k+1}),$$

and $\overline{\mathcal{F}}^{\eta}_{\lambda} = \bigoplus_{k \in \mathbb{Z}_{\geq 0}} \overline{\mathcal{G}}_k$ *holds.*

2. (**Case** $2^+ \wedge \sigma = -$) $\vee$ (**Case** $3^+ \wedge \sigma = +$)*:*

a. For $k \in \mathbb{Z}_{>0}$*,*

$$\mathcal{G}_k = IK(k-1] \cap \mathcal{F}^{\eta}_{\lambda}[k) \simeq L(c, \xi_{i+2k-1}).$$

b. Let $\overline{\mathcal{F}}^{\eta}_{\lambda} := \mathcal{F}^{\eta}_{\lambda} \Big/ \bigoplus_{k \in \mathbb{Z}_{>0}} \mathcal{G}_k$ *and let* $\overline{\mathrm{pr}} : \mathcal{F}^{\eta}_{\lambda} \twoheadrightarrow \overline{\mathcal{F}}^{\eta}_{\lambda}$ *be the canonical projection. For* $k \in \mathbb{Z}_{\geq 0}$*,*

$$\overline{\mathcal{G}}_k = \overline{\mathrm{pr}}\,(IK(k]) \cap \overline{\mathrm{pr}}\,(\mathcal{F}^{\eta}_{\lambda}[k)) \simeq L(c, \xi_{i+2k}),$$

and $\overline{\mathcal{F}}^{\eta}_{\lambda} = \bigoplus_{k \in \mathbb{Z}_{\geq 0}} \overline{\mathcal{G}}_k$ *holds.*

Proof. In **Case** 2^+ and 3^+, the Fock module $\mathcal{F}^{\eta}_{\lambda}$ is multiplicity free. Hence, arguments similar to the proof of Theorem 8.2 work. We omit the details. □

Similarly to **Case** 1^+, we describe Theorem 8.2 by using diagrams. The structure of Fock modules in **Case** 2^+ and **Case** 3^+ can be described as follows:

(**Case** $2^+ \wedge \sigma = +$) $\vee$(**Case** $3^+ \wedge \sigma = -$)	(**Case** $2^+ \wedge \sigma = -$) $\vee$(**Case** $3^+ \wedge \sigma = +$)

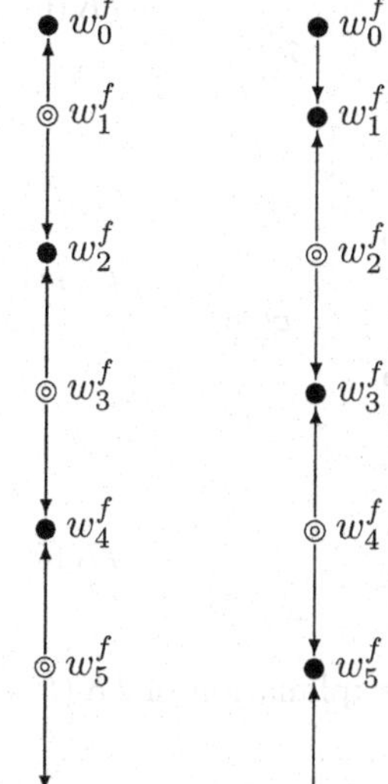

Similarly to **Case** 1^+, the following diagrams explain the statements of Theorem 8.2 and the structure of $IK(k]$ and $\mathcal{F}^{\eta}_{\lambda}[k)$.

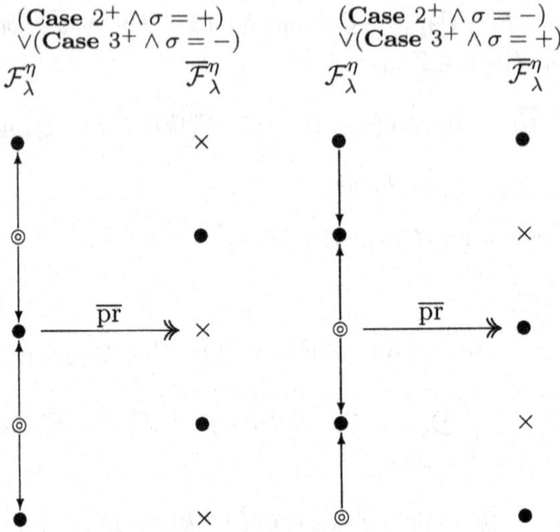

Pictorial explanation of Theorem 8.2

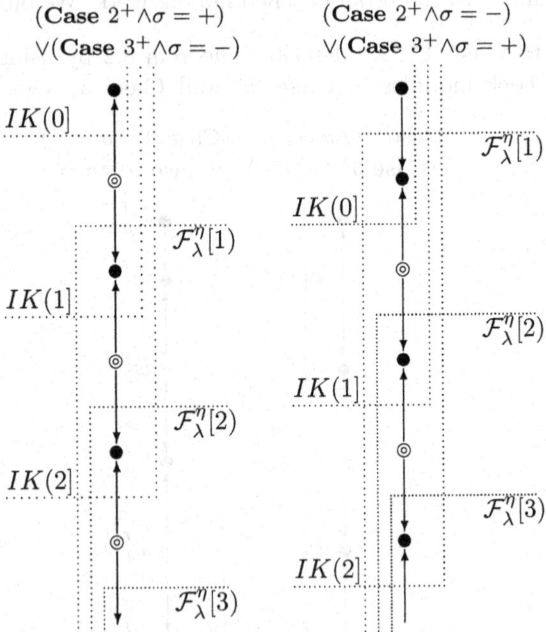

Pictorial explanation of $IK(k]$ and $\mathcal{F}^{\eta}_{\lambda}[k)$.

8.3.3 Main Theorem (Case 4^+)

The structure of Fock modules in **Case** 4^+ can be described as follows:

Theorem 8.3 *For $k \in \mathbb{Z}_{\geq 0}$,*

$$\mathcal{G}_k := IK(k] \cap \mathcal{F}^\eta_\lambda[k) \simeq L(c, \xi_{i+k}),$$

and $\mathcal{F}^\eta_\lambda = \bigoplus_{k \in \mathbb{Z}_{\geq 0}} \mathcal{G}_k$ holds.

Proof. This theorem can be shown in a way similar to Theorem 8.1. □

Similarly to **Case** 1^+, 2^+ and 3^+, one can draw the structure of Fock modules in **Case** 4^+ and those of $IK(k]$ and $\mathcal{F}^\eta_\lambda[k)$ as follows:

$\bullet\ w^f_0$

$\bullet\ w^f_1$

$\bullet\ w^f_2$

$\bullet\ w^f_3$

Pictorial explanation of Theorem 8.3

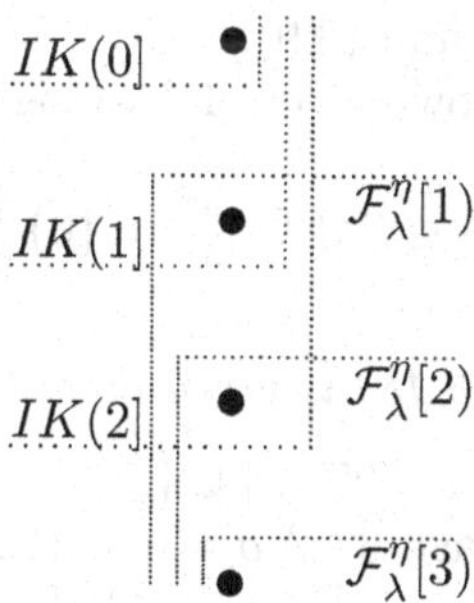

Pictorial explanation of $IK(k]$ and $\mathcal{F}^\eta_\lambda[k)$.

Remark 8.4 *In the case where $p = q = 1$, i.e., $c = 1$ and $\lambda = 0$, Theorem 8.3 immediately follows from the unitarisability of $\mathcal{F}^\eta_0$ and the character formula for $L(1, \xi_i)$ (Theorem 6.13). Indeed, $\mathcal{F}^\eta_0$, as a module over the Heisenberg Lie algebra $\mathcal{H}$, admits a contravariant form $\langle \cdot, \cdot \rangle$ with respect to the anti-linear anti-involution defined by $\omega(a_n) := a_{-n}$, $\omega(K_\mathcal{H}) := K_\mathcal{H}$, i.e., $\langle x.u, v \rangle = \langle u, \omega(x).v \rangle$ for $x \in U(\mathcal{H})$, $u, v \in \mathcal{F}^\eta_0$. It follows from the explicit formula (4.3) with $\lambda = 0$ that $\langle L_n.u, v \rangle = \langle u, L_{-n}.v \rangle$ for any $n \in \mathbb{Z}$. Hence, $\mathcal{F}^\eta_0$ for $\eta \in \mathbb{R}$*

is a unitarisable $\mathfrak{g}$*-module with respect to the anti-linear anti-involution* $\widetilde{\omega}$ *defined by* $\widetilde{\omega}(L_n) = L_{-n}$ *(cf. Chapter 11), and thus, it is completely reducible.*

For example, in the case $\eta_0^\sigma = 0$ *and* $\pm\frac{1}{\sqrt{2}}$*, the corresponding* L_0*-weights are* $h = 0$ *and* $\frac{1}{4}$*, respectively. By comparing the formal characters in both sides, we have*

$$\mathcal{F}_0^0 \simeq \bigoplus_{n\in\mathbb{Z}_{\geq 0}} L(1, n^2), \quad \mathcal{F}_0^{\pm\frac{1}{\sqrt{2}}} \simeq \bigoplus_{n\in\mathbb{Z}_{\geq 0}} L(1, \frac{1}{4}(2n+1)^2).$$

These are special cases of the above theorem. Irreducible decomposition of Fock modules (belonging to **Case** 4^+*) by using their unitarisability can be found in [KR].*

8.3.4 Classification of Singular Vectors

In this subsection, we classify the singular vectors of Fock modules. We first show the uniqueness of singular vectors in Fock modules.

Lemma 8.8. *For any* $h \in \mathbb{C}$, $\dim (\mathcal{F}_\lambda^\eta)_h^{\mathfrak{g}^+} \leq 1$.

Proof. Since $\dim (\mathcal{F}_\lambda^\eta)_h^{\mathfrak{g}^+} \leq [\mathcal{F}_\lambda^\eta : L(c,h)]$ by Lemma 1.9, the lemma follows from the fact that $\mathcal{F}_\lambda^\eta$ is multiplicity free. □

Since for **Class** V, I and R^-, Fock modules are isomorphic either to Verma modules or their contragredient duals, we deal only with **Class** R^+ so that we choose (λ, η) as $\lambda = \lambda_{p,q}$ and (8.9).

Here, we describe the following set of L_0-weights of singular vectors:

$$\left\{ h \in \mathbb{C} \,\middle|\, (\mathcal{F}_\lambda^\eta)_h^{\mathfrak{g}^+} \neq \{0\} \right\}. \tag{8.18}$$

Proposition 8.6 *The set (8.18) is given as follows:*

1. **Case** 1^+: $\{\xi_{\ell(i,\sigma;n)} | n \in -1 - 2\mathbb{Z}_{\geq 0}\} \cup \{\xi_i\}$.
2. (**Case** $2^+ \wedge \sigma = +) \vee ($**Case** $3^+ \wedge \sigma = -)$: $\{\xi_{i+n} | n \in 2\mathbb{Z}_{\geq 0}\}$.
3. (**Case** $2^+ \wedge \sigma = -) \vee ($**Case** $3^+ \wedge \sigma = +)$: $\{\xi_{i+n} | n \in 1 + 2\mathbb{Z}_{\geq 0}\} \cup \{\xi_i\}$.
4. **Case** 4^+: $\{\xi_{i+n} | n \in \mathbb{Z}_{\geq 0}\}$.

Proof. First, we show the proposition in **Case** 1^+. By the first statement of Theorem 8.1, we have

$$\{\xi_{\ell(i,\sigma;n)} | n \in -1 - 2\mathbb{Z}_{\geq 0}\} \cup \{\xi_i\} \subset \{h \in \mathbb{C} | (\mathcal{F}_\lambda^\eta)_h^{\mathfrak{g}^+} \neq \{0\}\}.$$

To show the opposite inclusion, it is enough to prove that

$$\mathcal{F}_\lambda^\eta[k]_h^{\mathfrak{g}^+} = \{0\} \quad (h \in \{\xi_{\ell(-2k)}, \xi_{\ell(2k)}, \xi_{\ell(2k+1)}\}), \tag{8.19}$$

since $[\mathcal{F}_\lambda^\eta : L(c,h)] = 1 = [\mathcal{F}_\lambda^\eta[k] : L(c,h)]$ for the above h. Let us consider

$$\mathcal{F}_\lambda^\eta[k) \twoheadrightarrow \mathcal{F}_\lambda^\eta[k)/\mathcal{F}_\lambda^\eta[k+1) \simeq IK[k)/IK[k-1).$$

Since $(IK[k)/IK[k-1))_h^{\mathfrak{g}^+} = \{0\}$ for any $h \in \{\xi_{\ell(-2k)}, \xi_{\ell(2k)}, \xi_{\ell(2k+1)}\}$, (8.19) holds. Thus, the proposition holds in **Case** 1^+.

In **Case** 2^+ and **Case** 3^+, by Theorem 8.2, one can similarly show the proposition. In **Case** 4^+, the proposition is an immediate consequence of Theorem 8.3. □

Remark 8.5 *The singular vectors of Fock modules in* **Class** R^+ *can be expressed in terms of the Jack symmetric polynomials. This follows from Theorem 8.7 in the next subsection and the following observation: by Proposition 8.6, one can check that* $\dim(\mathcal{F}_\lambda^\eta)_{\xi_i+n} = 1$ $(n \in \mathbb{Z}_{>0})$ *if and only if there exists a pair* $(r,s) \in \mathbb{Z}_{>0}^2$ *such that* $\Psi_{r,s}^-(\lambda,\eta) = 0$ *and* $rs = n$, *where* $\Psi_{r,s}^-(\lambda,\eta)$ *is a factor of the determinant of* $L^{\lambda,\eta}$ *(cf. (4.26)).*

8.4 Jack Symmetric Polynomials and Singular Vectors

In this section, we show that homomorphisms between Fock modules are provided by the so-called screening operators. Here, we recall the definition and make comments on a sufficient condition for which screening operators are non-trivial. As an application, we show that singular vectors of Fock modules can be expressed in terms of the Jack symmetric polynomials.

8.4.1 Completion of Fock Modules and Operators

We introduce screening operators by using the operator $V_\mu(z)$, which was defined in § 4.3 for μ such that $\mu^2 \in \mathbb{Z}_{>0}$. In order to construct screening operators appearing in the following subsections, the operators $V_\mu(z)$ with $\mu \in \mathbb{C}$ and their compositions are necessary. To treat them in a rigorous manner, following [TK2], we introduce completions of Fock spaces and the notion of operators on them.

First, we introduce a completion of $\mathcal{F}^\eta$. We set

$$\hat{\mathcal{F}}^\eta := \prod_{n \in \mathbb{Z}_{\geq 0}} (\mathcal{F}^\eta)^{-n\alpha_{\mathcal{H}}}$$

and regard it as a topological space with the product topology. Note that $\hat{\mathcal{F}}^\eta$ is a complete topological space, and $\mathcal{F}^\eta$ is a dense subset of $\hat{\mathcal{F}}^\eta$. Hence, the actions of $\mathcal{H}$ and Vir on $\mathcal{F}^\eta$ can be extended to those on $\hat{\mathcal{F}}^\eta$ continuously. When we regard $\hat{\mathcal{F}}^\eta$ as a Vir-module, we denote it by $\hat{\mathcal{F}}_\lambda^\eta$. Further, we can uniquely extend the pairing $\langle\, ,\, \rangle : (\mathcal{F}^\eta)^c \times \mathcal{F}^\eta \longrightarrow \mathbb{C}$ to a continuous bilinear form

$$\langle\, ,\, \rangle : (\mathcal{F}^\eta)^c \times \hat{\mathcal{F}}^\eta \longrightarrow \mathbb{C}.$$

We introduce operators on $\hat{\mathcal{F}}^\eta$. We call a linear map $O : \mathcal{F}^{\eta_1} \to \hat{\mathcal{F}}^{\eta_2}$ an **operator**. An operator $O(\xi_1, \cdots, \xi_n)$ from $\mathcal{F}^{\eta_1}$ to $\hat{\mathcal{F}}^{\eta_2}$ which depends on complex variables $\xi_1, \cdots, \xi_n$ is said to be holomorphic if $\langle v^\dagger, O(\xi_1, \cdots, \xi_n)u\rangle$ is a holomorphic function for any $v^\dagger \in (\mathcal{F}^{\eta_2})^c$ and $u \in \mathcal{F}^{\eta_1}$.

Next, we define the composition of operators. It should be noted that the composition of operators $O_1 : \mathcal{F}^{\eta_1} \to \hat{\mathcal{F}}^{\eta_2}$ and $O_2 : \mathcal{F}^{\eta_2} \to \hat{\mathcal{F}}^{\eta_3}$ does not always exist.

In order to define the composition of operators, we recall that

Lemma 8.9. *There exists a one-to-one correspondence between the set of the operators from $\mathcal{F}^{\eta_1}$ to $\hat{\mathcal{F}}^{\eta_2}$ and the set of the bilinear maps from $(\mathcal{F}^{\eta_2})^c \times \mathcal{F}^{\eta_1}$ to $\mathbb{C}$. In fact, an operator O corresponds to the bilinear form O' given by*

$$O'(v^\dagger, u) := \langle v^\dagger, Ou\rangle,$$

where $v^\dagger \in (\mathcal{F}^{\eta_2})^c$ and $u \in \mathcal{F}^{\eta_1}$.

Using this correspondence, we define the composition of operators as follows: Let $\{v_i\}_{i\in\mathcal{I}_n}$ be a basis of the weight subspace $(\mathcal{F}^{\eta_2})^{-n\alpha_\mathcal{H}}$, and let $\{v_i^\dagger\}_{i\in\mathcal{I}_n}$ be the dual basis. We say that operators $O_1 : \mathcal{F}^{\eta_1} \to \hat{\mathcal{F}}^{\eta_2}$ and $O_2 : \mathcal{F}^{\eta_2} \to \hat{\mathcal{F}}^{\eta_3}$ are **composable** if

$$\sum_{n=0}^{\infty} \left| \sum_{i\in\mathcal{I}_n} \langle w^\dagger, O_2 v_i\rangle \langle v_i^\dagger, O_1 u\rangle \right| < \infty$$

hold for any $w^\dagger \in (\mathcal{F}^{\eta_3})^c$ and $u \in \mathcal{F}^{\eta_1}$. For composable operators O_1 and O_2, the composition $O_3 := O_2 O_1 : \mathcal{F}^{\eta_1} \to \hat{\mathcal{F}}^{\eta_3}$ is defined as follows: Let $O_3' : (\mathcal{F}^{\eta_3})^c \times \mathcal{F}^{\eta_1} \to \mathbb{C}$ be the bilinear map given by

$$O_3'(w^\dagger, u) := \sum_{n=0}^{\infty} \sum_{n\in\mathcal{P}_n} \langle w^\dagger, O_2 v_\mathbb{I}\rangle \langle v_\mathbb{I}^\dagger, O_1 u\rangle.$$

We define the composition O_3 as the operator corresponding to O_3' by Lemma 8.9.

8.4.2 Screening Operators

To define screening operators, recall some notation introduced in § 4.3. For η, $\mu \in \mathbb{C}$ and $z \in \mathbb{C}^*$, $e^{\mu q} \in \mathrm{Hom}_{U(\mathcal{H}^-)}(\mathcal{F}^\eta, \mathcal{F}^{\eta+\mu})$ and $z^{\mu a_0} \in \mathrm{End}_{U(\mathcal{H})}(\mathcal{F}^\eta)$ are the linear maps satisfying

$$e^{\mu q}.(1 \otimes \mathbf{1}_\eta) = 1 \otimes \mathbf{1}_{\eta+\mu}, \quad z^{\mu a_0}.(1 \otimes \mathbf{1}_\eta) = z^{\mu\eta}(1 \otimes \mathbf{1}_\eta).$$

We introduce the **vertex operator** $V_\mu(z)$ with $\mu \in \mathbb{C}$, which is an operator in the sense of the previous subsection.

Definition 8.4 *For $\mu \in \mathbb{C}$, we define an operator $V_\mu(z) : \mathcal{F}^\eta \longrightarrow \hat{\mathcal{F}}^{\eta+\mu}$ by*

$$V_\mu(z) := e^{\mu q} z^{\mu a_0} \exp\left(\mu \sum_{k=1}^{\infty} \frac{a_{-k}}{k} z^k\right) \exp\left(-\mu \sum_{k=1}^{\infty} \frac{a_k}{k} z^{-k}\right).$$

Here, it can be verified that vertex operators $V_\mu(z_{a+1-i}) : \mathcal{F}^{\eta+(i-1)\mu} \to \hat{\mathcal{F}}^{\eta+i\mu}$ $(i = 1, \cdots, a)$ are composable (see e.g. [TK2]). In a way similar to the proof of Lemma 4.6, the next formula is valid:

$$\begin{aligned} &V_\mu(z_1) \cdots V_\mu(z_a) \\ &= \prod_{1 \le i < j \le a} (z_i - z_j)^{\mu^2} \prod_{i=1}^{a} \exp\left(\mu \sum_{k=1}^{\infty} \frac{a_{-k}}{k} z_i^k\right) e^{a\mu q} \\ &\quad \times \prod_{i=1}^{a} \exp\left(-\mu \sum_{k=1}^{\infty} \frac{a_k}{k} z_i^{-k}\right) \prod_{i=1}^{a} z_i^{\mu a_0} \end{aligned} \tag{8.20}$$

for $z_1, \cdots, z_a$ satisfying $|z_1| > \cdots > |z_a|$. Taking this formula into account, we set

$$K_\mu(z_1, \cdots, z_a) := V_\mu(z_1) \cdots V_\mu(z_a) \prod_{i=1}^{a} z_i^{-\mu a_0 - \frac{1}{2}(a-1)\mu^2},$$
$$M_a := \{(z_1, \cdots, z_a) \in (\mathbb{C}^*)^a | z_i \neq z_j \ (1 \le i < j \le a)\},$$

and regard K_μ as an operator on M_a via the analytic continuation. We denote the multi-valued part of K_μ by Ψ_μ, i.e.,

$$\Psi_\mu = \Psi_\mu(z_1, \cdots, z_a) := \prod_{1 \le i < j \le a} (z_i - z_j)^{\mu^2} \prod_{i=1}^{a} z_i^{-\frac{1}{2}(a-1)\mu^2}. \tag{8.21}$$

Let $\mathcal{S}_\mu$ and $\mathcal{S}_\mu^\vee$ be the local system on M_a whose local section is given by Ψ_μ^{-1} (resp. Ψ_μ) up to a constant multiple. For $p \in \mathbb{Z}_{\ge 0}$, the pth chain group $C_p(M_a, \mathcal{S}_\mu^\vee)$ with coefficients in $\mathcal{S}_\mu^\vee$ is defined as the complex vector space spanned by $\sigma \otimes (\Psi_\mu)_\sigma$ where σ is a singular p-simplex and $(\Psi_\mu)_\sigma$ is a fixed branch of Ψ_μ on the image of σ. The boundary map $\partial : C_p(M_a, \mathcal{S}_\mu^\vee) \longrightarrow C_{p-1}(M_a, \mathcal{S}_\mu^\vee)$ is defined, where we set $C_{-1}(M_a, \mathcal{S}_\mu^\vee) = \{0\}$, in a natural way, and one can define its (pth) homology which is denoted by $H_p(M_a, \mathcal{S}_\mu^\vee)$ and is called the pth **twisted homology group** (for detail, see, e.g., [AK]).

Lemma 8.10. *Suppose that λ, η and μ satisfy*

$$\lambda = \frac{1}{2}\mu - \mu^{-1}, \quad \eta = \lambda - \frac{1}{2}a\mu - b\mu^{-1} \tag{8.22}$$

for some $a \in \mathbb{Z}_{>0}$, $b \in \mathbb{Z}$. *Then, by setting*

$$S_\Gamma(\mu; a, b) := \int_\Gamma K_\mu(z_1, \cdots, z_a) \prod_{i=1}^{a} z_i^{-b-1} dz_i : \mathcal{F}_\lambda^\eta \longrightarrow \mathcal{F}_\lambda^{\eta+a\mu}$$

the map $S_\Gamma(\mu; a, b) : \mathcal{F}_\lambda^\eta \to \mathcal{F}_\lambda^{\eta+a\mu}$ *is a homomorphism of* $\mathfrak{g}$*-modules, where* $\Gamma \in H_a(M_a, \mathcal{S}_\mu^\vee)$.

Proof. The commutation relation

$$[L_n, V_\mu(z)] = z^n \left\{ z\frac{\partial}{\partial z} + h_\lambda^\mu (n+1) \right\} V_\mu(z)$$

stated in Lemma 4.4 holds for any $\mu \in \mathbb{C}$, which implies

$$\begin{aligned}
&[L_n, K_\mu(z_1, \cdots, z_a)] \\
&= \sum_{i=1}^{a} z_i^n \{ z_i \frac{\partial}{\partial z_i} + h_\lambda^\mu (n+1) + \mu a_0 - \frac{1}{2}(a+1)\mu^2 \} K_\mu(z_1, \cdots, z_a)
\end{aligned}$$

and the lemma holds. □

Corollary 8.1 *Suppose that* λ, η *and* μ *satisfy (8.22) and* $S_\Gamma(\mu; a, b)$ *is non-trivial. Then,*

1. *if* $b > 0$, *then* $S_\Gamma(\mu; a, b)(1 \otimes \mathbf{1}_\eta) \in \mathcal{F}_\lambda^{\eta+a\mu}$ *is a singular vector of level* ab,
2. *if* $b < 0$, *then there exists a cosingular vector* $u \in \mathcal{F}_\lambda^\eta$ *of level* ab *such that* $S_\Gamma(\mu; a, b)(u) = 1 \otimes \mathbf{1}_{\eta+a\mu}$.

The homomorphism $S_\Gamma(\mu; a, b)$ is called a **screening operator** (associated with a twisted cycle Γ).

8.4.3 Non-Triviality of Screening Operators

In this subsection, we state a sufficient condition for the non-triviality of the screening operator $S_\Gamma(\mu; a, b)$.

We will see later in Lemma 8.11 that if the integral $\int_\Gamma \Psi_\mu \prod_{i=1}^{a} z_i^{-1} dz_i$ does not vanish, then $S_\Gamma(\mu; a, b)$ is non-trivial. In fact, the following theorem on this integral holds: Set

$$\Omega_a := \{ x \in \mathbb{C} | d(d+1)x \notin \mathbb{Z} \wedge d(a-d)x \notin \mathbb{Z}\ (1 \leq \forall d \leq a-1) \}. \tag{8.23}$$

Theorem 8.4 ([TK2]) *There exists a twisted cycle* $\Gamma \in H_a(M_a, \mathcal{S}_\mu^\vee)$ *satisfying the following conditions:*

1. For $m_1, \cdots, m_a \in \mathbb{Z}$*, if* $m_1 + \cdots + m_a \neq 0$*, then*

$$\int_\Gamma \Psi_\mu \prod_{i=1}^{a} z_i^{m_i} \frac{dz_i}{z_i} = 0. \tag{8.24}$$

2. Suppose that $\frac{1}{2}\mu^2 \in \Omega_a$*. Then,*

$$\int_\Gamma \Psi_\mu \prod_{i=1}^{a} \frac{dz_i}{z_i} = \frac{1}{\Gamma(a)} \prod_{i=1}^{a-1} \frac{\Gamma(\frac{1}{2}(i-a)\mu^2)\Gamma(\frac{1}{2}(i+1)\mu^2+1)}{\Gamma(\frac{1}{2}\mu^2+1)}. \tag{8.25}$$

Proof. We first prove 1. Since M_a admits a $\mathbb{C}^*$-action and the integrand is homogeneous, the change of variables

$$z_1 = x, \qquad z_i = xy_{i-1} \quad (1 < i \leq a),$$

implies

$$(z_1, \cdots, z_a) \in M_a \iff (x, y_1, \cdots, y_{a-1}) \in \mathbb{C}^* \times Y_{a-1}$$

and

$$\Psi_\mu \prod_{i=1}^{a} z_i^{m_i-1} dz_i = \left\{ x^{m_1+\cdots+m_a-1} \Psi'_\mu \prod_{i=1}^{a-1} y_i^{m_i-1} \right\} dx \prod_{i=1}^{a-1} dy_i,$$

where we set

$$Y_{a-1} = \{(y_1, \cdots, y_{a-1}) \in (\mathbb{C} \setminus \{0,1\})^{a-1} | y_i \neq y_j (1 \leq i < j \leq a-1)\}$$

and

$$\Psi'_\mu = \Psi'_\mu(y_1, \cdots, y_{a-1}) := \prod_{1 \leq i < j \leq a-1} (y_i - y_j)^{\mu^2} \prod_{i=1}^{a-1} (1-y_i)^{\mu^2} y_i^{-\frac{1}{2}(a-1)\mu^2}.$$

Let $\mathcal{L}_\mu^\vee$ be the local system on Y_{a-1} whose local section is given by Ψ'_μ up to a constant multiple. By the Künneth type formula, there is an embedding $H_1(\mathbb{C}^*, \mathbb{C}) \otimes H_{a-1}(Y_{a-1}, \mathcal{L}_\mu^\vee) \hookrightarrow H_a(M_a, \mathcal{S}_\mu^\vee)$ and we denote the image of (Γ_1, Γ_2) via this map by Γ. We have

$$\int_\Gamma \Psi_\mu \prod_{i=1}^{a} z_i^{m_i-1} dz_i = \int_{\Gamma_1} x^{m_1+\cdots+m_a-1} dx \int_{\Gamma_2} \Psi'_\mu \prod_{i=1}^{a-1} y_i^{m_i-1} dy_i.$$

Hence, if $\sum_i m_i \neq 0$, then the integral $\int_{\Gamma_1} x^{m_1+\cdots+m_a-1} dx$ becomes 0 for any Γ_1 which implies 1.

Let us make a brief remark on the proof of 2. The case $a = 2$ is reduced to the case of the **Euler** B**-function**, and for $a > 2$, this is reduced to the so-called **Selberg integral** due to A. Selberg [Sel] given as follows: For

$a, b, c \in \mathbb{C}$ such that $\mathrm{Re}a, \mathrm{Re}b > 0$ and $\mathrm{Re}c > -\min\left\{\frac{1}{n}, \frac{\mathrm{Re}a}{n-1}, \frac{\mathrm{Re}b}{n-1}\right\}$,

$$\frac{1}{n!}\int_{[0,1]^n}\prod_{i=1}^{n} x_i^{a-1}(1-x_i)^{b-1}\prod_{1\leq i<j\leq n}|x_i-x_j|^{2c}dx_1\cdots dx_n$$
$$=\prod_{j=1}^{n}\frac{\Gamma(a+(j-1)c)\Gamma(b+(j-1)c)\Gamma(jc+1)}{\Gamma(a+b+(n+j-2)c)\Gamma(c+1)}.$$

For some proofs of this formula, see, e.g., [AAR]. This implies 2. for some $\mu \in \mathbb{C}$. An analytic continuation of this formula is given by A. Tsuchiya and Y. Kanie [TK2] by showing that a non-trivial twisted cycle exists if $\frac{1}{2}\mu^2 \in \Omega_a$. For detail, the reader may consult, e.g., [TK2], [AK] and [OT]. □

8.4.4 Jack Symmetric Polynomials

Let $\Lambda_n := \mathbb{C}[x_1, \cdots, x_n]^{\mathfrak{S}_n}$ be the $\mathbb{C}$-algebra of the symmetric polynomials in independent variables $x_1, \cdots, x_n$.

The ring Λ_n admits various basis indexed by partitions. A partition $\mathbb{I}$ is a decreasing sequence $\mathbb{I} = (i_1, i_2, \cdots, i_n)$ of non-negative integers; $i_1 \geq i_2 \geq \cdots \geq i_n \geq 0$. The sum of i_k's is the weight of $\mathbb{I}$ denoted by $|\mathbb{I}|$, i.e., $|\mathbb{I}| := \sum_{k=1}^{n} i_k$. Given a partition $\mathbb{I} = (i_1, i_2, \cdots, i_n)$, its conjugate partition $\mathbb{I}'$ is defined by $\mathbb{I}' = (i'_1, i'_2, \cdots, i'_n)$, where $i'_k := \sharp\{j | i_j \geq k\}$. The set of the partitions will be denoted by $\mathcal{P}$.

The non-zero i_k are called parts of $\mathbb{I}$ and the number of parts is the length $l(\mathbb{I})$ of $\mathbb{I}$. If $\mathbb{I}$ has m_r parts equal to r, we may also write $\mathbb{I} = (1^{m_1}2^{m_2}\cdots)$. The dominance ordering in $\mathcal{P}$ is a partial ordering defined as follows:

$$\mathbb{I} \preccurlyeq \mathbb{J} \iff |\mathbb{I}| = |\mathbb{J}| \text{ and } i_1 + \cdots + i_k \leq j_1 + \cdots + j_k \ (\forall k).$$

Notice that for $\mathbb{I}, \mathbb{J} \in \mathcal{P}$, it can be shown that

$$\mathbb{I} \preccurlyeq \mathbb{J} \iff \mathbb{J}' \preccurlyeq \mathbb{I}'.$$

Recall the following basic elements of Λ_n: the monomial symmetric function, the elementary symmetric function and the power sum. For a partition $\mathbb{I}$ with $l(\mathbb{I}) \leq n$, the monomial symmetric function $m_{\mathbb{I}}(x_1, \cdots, x_n)$ is defined by

$$m_{\mathbb{I}}(x_1, \cdots, x_n) := \sum_{(\gamma_1, \cdots, \gamma_n)} x_1^{\gamma_1} \cdots x_n^{\gamma_n},$$

where $(\gamma_1, \cdots, \gamma_n)$ in the right-hand side runs over all of the distinct permutations of $\mathbb{I} = (i_1, i_2, \cdots, i_\ell)$. For $r \in \mathbb{Z}_{>0}$, the elementary symmetric function $e_r(x_1, \cdots, x_n)$ and the power sum $p_r(x_1, \cdots, x_n)$ are defined by

$e_r(x_1, \cdots, x_n) := m_{(1^r)}(x_1, \cdots, x_n)$ and $p_r(x_1, \cdots, x_n) := m_{(r^1)}(x_1, \cdots, x_n)$, namely,

$$e_r(x_1, \cdots, x_n) = \sum_{k_1 < \cdots < k_r} x_{k_1} \cdots x_{k_r}, \quad p_r(x_1, \cdots, x_n) = \sum_{k=1}^{n} x_k^r.$$

For a partition $\mathbb{I} \in \mathcal{P}$, we set $e_{\mathbb{I}} := e_{i_1} e_{i_2} \cdots e_{i_n}$. Notice that

$$e_{\mathbb{I}'} = m_{\mathbb{I}} + \sum_{\mathbb{J} \prec \mathbb{I}} a_{\mathbb{I},\mathbb{J}} m_{\mathbb{J}}$$

for suitable coeficients $a_{\mathbb{I},\mathbb{J}}$.

The Jack symmetric polynomials are characterised as eigenfunctions of the following differential operator (the **Laplace–Beltrami operator**): for $\alpha \in \mathbb{C}$, set

$$D^{(\alpha)} := \frac{\alpha}{2} \sum_{k=1}^{n} x_k^2 \frac{\partial^2}{\partial x_k^2} + \sum_{1 \leq k \neq l \leq n} \frac{x_k^2}{x_k - x_l} \frac{\partial}{\partial x_k}. \tag{8.26}$$

Theorem 8.5 *For each partition $\mathbb{I}$ with $l(\mathbb{I}) \leq n$, there uniquely exists a symmetric polynomial $J_{\mathbb{I}}(x_1, \cdots, x_n; \alpha) \in \Lambda_n$ with parameter α such that*

1. *$J_{\mathbb{I}}(x_1, \cdots, x_n; \alpha)$ is an eigenfunctions of $D^{(\alpha)}$, i.e,*

$$D^{(\alpha)} J_{\mathbb{I}}(x_1, \cdots, x_n; \alpha) = e_{\mathbb{I}}(\alpha) J_{\mathbb{I}}(x_1, \cdots, x_n; \alpha). \tag{8.27}$$

Here, the eigenvalue $e_{\mathbb{I}}(\alpha)$ is given by

$$e_{\mathbb{I}}(\alpha) = \frac{\alpha}{2} \sum_{k=1}^{\ell} i_k(i_k - 1) + \sum_{k=1}^{\ell} (n-k) i_k. \tag{8.28}$$

2.

$$J_{\mathbb{I}}(x_1, \cdots, x_n; \alpha) = \sum_{\mathbb{J} \preccurlyeq \mathbb{I}} a_{\mathbb{I},\mathbb{J}} m_{\mathbb{J}}(x_1, \cdots, x_n), \tag{8.29}$$

where $a_{\mathbb{I},\mathbb{J}} \in \mathbb{C}$ (depending on α) and, in particular, $a_{\mathbb{I},\mathbb{I}} = 1$.

The polynomial $J_{\mathbb{I}}(x_1, \cdots, x_n; \alpha)$ is called the **Jack symmetric polynomial** *indexed by $\mathbb{I}$ with a parameter $\alpha \in \mathbb{C}$.*

For an outline of the proof of this theorem, see [Mac1]. The explicit form of the eigenvalue $e_{\mathbb{I}}(\alpha)$ follows from [Mac2] (Example 3 in Section 3 and Example 2 in Section 4 of Chapter VI).

Remark 8.6 *The Jack symmetric polynomial $J_{\mathbb{I}}$ specialises to some well-known classical polynomial for particular α:*

1. *for $\alpha = 0$, $J_{\mathbb{I}}(x_1, \cdots, x_n; 0) = e_{\mathbb{I}'}(x_1, \cdots, x_n)$ where $\mathbb{I}'$ is the conjugate partition to $\mathbb{I}$,*
2. *for $\alpha = \frac{1}{2}$, they are the zonal spherical functions on $GL_n(\mathbb{H})/U(n, \mathbb{H})$,*

3. *for* $\alpha = 1$, *this is the Schur polynomial* $s_{\mathbb{I}}(x_1, \cdots, x_n)$,
4. *for* $\alpha = 2$, *they are the zonal spherical functions on* $GL_n(\mathbb{R})/O(n)$, *and*
5. *for* $\alpha \mapsto \infty$, *they reduce to the monomial symmetric polynomial* $m_{\mathbb{I}}(x_1, \cdots, x_n)$.

8.4.5 Singular Vectors of Fock Modules

Throughout this subsection, suppose that $a, b \in \mathbb{Z}_{>0}$ and $t \in \mathbb{C}^*$. We further suppose that λ, η and μ satisfy (8.22) and $\mu = \sqrt{2}t^{\frac{1}{2}}$.

Let $P_{a,b}(a_{-1}, a_{-2}, \cdots, a_{-n}; t) \in \mathbb{C}[a_{-1}, \cdots, a_{-n}]$ be the polynomial defined by

$$S_\Gamma(\sqrt{2}t^{\frac{1}{2}}; a, b)(1 \otimes \mathbf{1}_\eta) = P_{a,b}(a_{-1}, a_{-2}, \cdots, a_{-n}; t) \otimes \mathbf{1}_{\eta + a\mu},$$

where $S_\Gamma(\sqrt{2}t^{\frac{1}{2}}; a, b)$ is the screening operator. In this subsection, we show that $P_{a,b}(a_{-1}, a_{-2}, \cdots, a_{-n}; t)$ can be expressed in terms of the Jack symmetric polynomials.

By the formula (8.20), we have

$$\begin{aligned}
&K_\mu(z_1, \cdots, z_a)(1 \otimes \mathbf{1}_\eta) \\
&= \prod_{1 \le i < j \le a} (z_i - z_j)^{\mu^2} \prod_{i=1}^{a} z_i^{-\frac{1}{2}(a-1)\mu^2} \prod_{i=1}^{a} \exp\left(\mu \sum_{k=1}^{\infty} \frac{a_{-k}}{k} z_i^k\right) \mathbf{1}_{\eta + a\mu},
\end{aligned}$$

which implies

$$\begin{aligned}
&P_{a,b}(a_{-1}, \cdots, a_{-n}; t) \\
&= \int_\Gamma \prod_{1 \le i < j \le a} (z_i - z_j)^{2t} \prod_{i=1}^{a} z_i^{-(a-1)t-b-1} \exp\left(\sqrt{2}t^{\frac{1}{2}} \sum_{k=1}^{\infty} \frac{a_{-k}}{k} z_i^k\right) dz_i.
\end{aligned}$$

In order to relate $P_{a,b}(a_{-1}, \cdots, a_{-n}; t)$ with symmetric polynomials, we consider the isomorphism of $\mathbb{C}$-algebras

$$\iota_n : \Lambda_n \to \mathbb{C}[a_{-1}, \cdots, a_{-n}]; \quad p_r(x_1, \cdots, x_n) \mapsto (-1)^{r-1}\sqrt{2}t^{\frac{1}{2}} a_{-r} \; (1 \le r \le n).$$

Recall that Ω_a is the set defined in (8.23).

Theorem 8.6 *Suppose that* a *and* b *satisfy* $ab \le n$. *If* $t \in \Omega_a$, *then the following equality holds up to a scalar multiple:*

$$P_{a,b}(a_{-1}, \cdots, a_{-n}; t) = \iota_n(J_{(a^b)}(x_1, \cdots, x_n; t)).$$

Below, we set $\mathbb{I} := (a^b)$ for simplicity. Setting

$$E(z) := \sum_{r=0}^{n} e_r(x_1, \cdots, x_n) z^r = \prod_{k=1}^{n} (1 + x_k z),$$

we have

$$\iota_n(E(z)) = \exp\left(\sqrt{2} t^{\frac{1}{2}} \sum_{k=1}^{\infty} \frac{b_k}{k} z^k \right),$$

where $b_k \in \mathbb{C}[a_{-1}, \cdots, a_{-n}]$ are defined by $\iota_n(p_k(x_1, \cdots, x_n)) = (-1)^{k-1} \sqrt{2} t^{\frac{1}{2}} b_k$. Notice that $b_k = a_{-k}$ for $0 < k \leq n$. Since the degree of $P_{a,b}$ is at most n by hypothesis, we see that

$$\begin{aligned}
&\int_\Gamma \prod_{1 \leq i < j \leq a} (z_i - z_j)^{2t} \prod_{i=1}^{a} z_i^{-(a-1)t-b-1} \exp\left(\sqrt{2} t^{\frac{1}{2}} \sum_{k=1}^{\infty} \frac{a_{-k}}{k} z_i^k \right) dz_i \\
= &\int_\Gamma \prod_{1 \leq i < j \leq a} (z_i - z_j)^{2t} \prod_{i=1}^{a} z_i^{-(a-1)t-b-1} \exp\left(\sqrt{2} t^{\frac{1}{2}} \sum_{k=1}^{\infty} \frac{b_k}{k} z_i^k \right) dz_i.
\end{aligned}$$

Hence, to prove Theorem 8.6, it is enough to show the following proposition:

Proposition 8.7 *Suppose that a satisfies $a \leq n$. If $t \in \Omega_a$, then*

$$J_{\mathbb{I}}(x_1, \cdots, x_n; t) = C_a(t)^{-1} \int_\Gamma \prod_{1 \leq i < j \leq a} (z_i - z_j)^{2t} \prod_{i=1}^{a} z_i^{-(a-1)t-b-1} E(z_i) dz_i, \tag{8.30}$$

where Γ is the twisted cycle given by Theorem 8.4, and $C_a(t)$ is the constant that appeared in the right-hand side of (8.25), namely,

$$C_a(t) := \frac{1}{\Gamma(a)} \prod_{i=1}^{a-1} \frac{\Gamma((i-a)t)\Gamma((i+1)t+1)}{\Gamma(t+1)}.$$

For simplicity, we denote the integrand in the right-hand side of (8.30) by $\Phi_t = \Phi_t(z_1, \cdots, z_a, x_1, \cdots, x_n)$, i.e.,

$$\Phi_t := \prod_{1 \leq i < j \leq a} (z_i - z_j)^{2t} \prod_{i=1}^{a} z_i^{-(a-1)t-b-1} E(z_i)$$

and denote $dz_1 \cdots dz_n$ by dz. We first show that integral $\int_\Gamma \Phi_t dz$ satisfies the condition (8.29).

Lemma 8.11.

$$\int_\Gamma \Phi_t dz = \sum_{\mathbb{J} \preccurlyeq \mathbb{I}} c_{\mathbb{J}} m_{\mathbb{J}}(x_1, \cdots, x_k), \tag{8.31}$$

where $c_{\mathbb{J}} \in \mathbb{C}$ (depends on t) and $c_{\mathbb{I}} = C_a(t)$.

Proof. By Theorem 8.4.1., we have

$$\int_{\Gamma} \prod_{1\le i<j\le a} (z_i - z_j)^{2t} \prod_{i=1}^{a} z_i^{-(a-1)t-b-1} E(z_i) dz_i = \sum_{\mathbb{J}} c'_{\mathbb{J}} e_{\mathbb{J}}(x_1, \cdots, x_n) \quad (8.32)$$

where $\mathbb{J}$ runs over the set $\{\mathbb{K} \in \mathcal{P}_{ab} | |\mathbb{K}| \le a\}$ and $\mathcal{P}_{ab}$ is the set of the partitions of ab. Since $\mathbb{I} = (a^b)$, by definition, $\mathbb{I}' \preccurlyeq \mathbb{J}'$ holds. Moreover, $e_{\mathbb{J}}(x_1, \cdots, x_k)$ can be expressed as

$$e_{\mathbb{J}}(x_1, \cdots, x_k) = \sum_{\mathbb{K} \preccurlyeq \mathbb{J}'} c_{\mathbb{J},\mathbb{K}} m_{\mathbb{K}}(x_1, \cdots, x_n)$$

where $c_{\mathbb{J},\mathbb{J}'} = 1$. Hence, (8.31) holds. By Theorem 8.4.2., we have $c_{\mathbb{I}} = C_a(t)$. □

Thus, by Theorem 8.5 and Lemma 8.11, it suffices to show that $\int_{\Gamma} \Phi_t dz$ is an eigenfunction of the differential operator $D^{(t)}$ with eigenvalue $e_{\mathbb{I}}(t) = \frac{1}{2}ab\{(a-1)t + 2n - b - 1\}$.

For polynomials f and g in variables $z_1, \cdots, z_a$, $x_1, \cdots, x_n$, we set

$$f \equiv g \Leftrightarrow \int_{\Gamma} f \Phi_t dz = \int_{\Gamma} g \Phi_t dz.$$

Let us denote the total derivative with respect to $z_1, \cdots, z_a$ by d_z.

We show three technical lemmas. The first lemma is

Lemma 8.12.

$$\sum_{i=1}^{a} \sum_{k=1}^{n} \frac{1}{1 + x_k z_i} \equiv a(n-b).$$

Proof. We compute $d_z((-1)^{i-1} z_i \Phi_t dz_1 \cdots \widehat{dz_i} \cdots dz_a)$, where $\widehat{dz_i}$ means dz_i should be omitted. Indeed,

$$\begin{aligned} & d_z((-1)^{i-1} z_i \Phi_t dz_1 \cdots \widehat{dz_i} \cdots dz_a) \\ &= \left\{ -2t \sum_{j(j<i)} \frac{z_i}{z_j - z_i} + 2t \sum_{j(j>i)} \frac{z_i}{z_i - z_j} - (a-1)t - b + \sum_k \frac{x_k z_i}{1 + x_k z_i} \right\} \Phi_t dz \\ &= \left\{ 2t \sum_{j(j \ne i)} \frac{z_i}{z_i - z_j} - (a-1)t - b + n - \sum_k \frac{1}{1 + x_k z_i} \right\} \Phi_t dz. \end{aligned}$$

Here and after, $\sum_j$ (resp. $\sum_k$) for the subscript j (resp. k) of $z_1, \cdots, z_a$ (resp. $x_1, \cdots, x_n$) means the sum over $1 \le j \le a$ (resp. $1 \le k \le n$). By summing up for i and using the formula

$$\sum_{i,j(i\neq j)} \frac{z_i}{z_i - z_j} = \sum_{i,j(i<j)} \left(\frac{z_i}{z_i - z_j} + \frac{z_j}{z_j - z_i} \right) = \frac{a(a-1)}{2},$$

we have

$$d_z \left\{ \sum_i (-1)^{i-1} z_i \Phi_t dz_1 \cdots \widehat{dz_i} \cdots dz_a \right\} = \left\{ a(-b+n) - \sum_{i,k} \frac{1}{1 + x_k z_i} \right\} \Phi_t dz,$$

and thus, the lemma holds. □

Let us show the second lemma.

Lemma 8.13.

$$\sum_{1\leq i<j\leq a} \sum_{k=1}^{n} \frac{2t}{(1+x_k z_i)(1+x_k z_j)} - \sum_{i=1}^{a} \sum_{1\leq k<l\leq n} \frac{2}{(1+x_k z_i)(1+x_l z_i)}$$
$$\equiv \{(a-1)t - n + b + 1\} \sum_{i=1}^{a} \sum_{k=1}^{n} \frac{1}{1+x_k z_i}.$$

Proof. We compute $d_z((-1)^{i-1} z_i (1+x_k z_i)^{-1} \Phi_t dz_1 \cdots \widehat{dz_i} \cdots dz_a)$. By direct computation,

$$d_z \left\{ \frac{(-1)^{i-1} z_i}{1 + x_k z_i} \Phi_t dz_1 \cdots \widehat{dz_i} \cdots dz_a \right\}$$
$$= \frac{1}{1+x_k z_i} d_z((-1)^{i-1} z_i \Phi_t dz_1 \cdots \widehat{dz_i} \cdots dz_a) - \frac{x_k z_i}{(1+x_k z_i)^2} \Phi_t dz$$
$$= \frac{1}{1+x_k z_i} \left\{ 2t \sum_{j(j\neq i)} \frac{z_i}{z_i - z_j} - (a-1)t - b + n - 1 - \sum_{l(l\neq k)} \frac{1}{1+x_l z_i} \right\} \Phi_t dz.$$

By summing up for i and using the formula

$$\sum_{i,j(i\neq j)} \frac{z_i}{(1+x_k z_i)(z_i - z_j)} = \sum_{i,j(i<j)} \left\{ \frac{z_i}{(1+x_k z_i)(z_i - z_j)} + \frac{z_j}{(1+x_k z_j)(z_j - z_i)} \right\}$$
$$= \sum_{i,j(i<j)} \frac{1}{(1+x_k z_i)(1+x_k z_j)},$$

we have

$$d_z \left\{ \sum_i \frac{(-1)^{i-1} z_i}{1 + x_k z_i} \Phi_t dz_1 \cdots \widehat{dz_i} \cdots dz_a \right\}$$

$$= \left\{ \sum_{i,j(i<j)} \frac{2t}{(1 + x_k z_i)(1 + x_k z_j)} + \{n - (a-1)t - b - 1\} \sum_i \frac{1}{1 + x_k z_i} \right.$$

$$\left. - \sum_i \sum_{l(l \neq k)} \frac{1}{(1 + x_k z_i)(1 + x_l z_i)} \right\} \Phi_t dz.$$

By summing up for k, we obtain the lemma. □

We show the third lemma.

Lemma 8.14.

$$D^{(t)} \Phi_t = \left[t \left\{ \frac{a(a-1)n}{2} - (a-1) \sum_{i=1}^{a} \sum_{k=1}^{n} \frac{1}{1 + x_k z_i} \right. \right.$$

$$\left. + \sum_{1 \leq i < j \leq a} \sum_{k=1}^{n} \frac{1}{(1 + x_k z_i)(1 + x_k z_j)} \right\} \tag{8.33}$$

$$\left. + \frac{n(n-1)a}{2} - \sum_{i=1}^{a} \sum_{1 \leq k < l \leq n} \frac{1}{(1 + x_k z_i)(1 + x_l z_i)} \right] \Phi_t,$$

where $D^{(t)}$ is the differential operator defined by (8.26).

Proof. Notice that

$$\frac{\partial}{\partial x_k} \Phi_t = \left(\sum_i \frac{z_i}{1 + x_k z_i} \right) \Phi_t, \quad \frac{\partial^2}{\partial x_k^2} \Phi_t = \left(\sum_{i,j(i \neq j)} \frac{z_i z_j}{(1 + x_k z_i)(1 + x_k z_j)} \right) \Phi_t.$$

Hence, by setting

$$A_k := \sum_{i,j(i \neq j)} \frac{x_k^2 z_i z_j}{(1 + x_k z_i)(1 + x_k z_j)}, \quad B_i := \sum_{k,l(k \neq l)} \frac{x_k^2 z_i}{(x_k - x_l)(1 + x_k z_i)},$$

we have $D^{(t)} \Phi_t = \{\frac{t}{2} \sum_k A_k + \sum_i B_i\} \Phi_t$. Thus, the lemma follows from the following computation:

$$A_k = \sum_{i,j(i\neq j)} \left\{1 - \frac{1}{1+x_kz_i} - \frac{1}{1+x_kz_j} + \frac{1}{(1+x_kz_i)(1+x_kz_j)}\right\}$$

$$= a(a-1) - (a-1)\sum_i \frac{2}{1+x_kz_i} + \sum_{i,j(i<j)} \frac{2}{(1+x_kz_i)(1+x_kz_j)},$$

$$B_i = \sum_{k,l(k<l)} \left\{\frac{x_k^2 z_i}{(x_k - x_l)(1+x_kz_i)} + \frac{x_l^2 z_i}{(x_l - x_k)(1+x_lz_i)}\right\}$$

$$= \sum_{k,l(k<l)} \left\{1 - \frac{1}{(1+x_kz_i)(1+x_lz_i)}\right\}$$

$$= \frac{n(n-1)}{2} - \sum_{k,l(k<l)} \frac{1}{(1+x_kz_i)(1+x_lz_i)}.$$

□

Proof of Proposition 8.7. Finally, we show that $\int_\Gamma \Phi_t dz$ is an eigenfunction of $D^{(t)}$. Lemma 8.13 implies that the factor

$$t\left\{\frac{a(a-1)n}{2} - (a-1)\sum_{i=1}^{a}\sum_{k=1}^{n}\frac{1}{1+x_kz_i} + \sum_{1\le i<j\le a}\sum_{k=1}^{n}\frac{1}{(1+x_kz_i)(1+x_kz_j)}\right\}$$

$$+\frac{n(n-1)a}{2} - \sum_{i=1}^{a}\sum_{1\le k<l\le n}\frac{1}{(1+x_kz_i)(1+x_lz_i)}$$

in the right-hand side of (8.33) is equivalent to

$$\frac{ta(a-1)n}{2} + \frac{n(n-1)a}{2} - \frac{1}{2}\{(a-1)t + n - b - 1\}\sum_{i=1}^{a}\sum_{k=1}^{n}\frac{1}{1+x_kz_i}.$$

By Lemma 8.12, this is equivalent to $\frac{1}{2}ab\{(a-1)t + 2n - b - 1\}$, which is equal to $e_{\mathbb{I}}(t)$ with $\mathbb{I} = (a^b)$. Hence, we obtain

$$D^{(t)}\int_\Gamma \Phi dz = e_{\mathbb{I}}(t)\int_\Gamma \Phi dz$$

and thus, we complete the proof of Proposition 8.7. □

The next theorem is the main result of this subsection:

Theorem 8.7 *Suppose that a and b satisfy $ab \le n$. Then, for any $t \in \mathbb{C}^*$,*

$$\iota_n(J_{(a^b)}(x_1, \cdots, x_n; t)) \otimes \mathbf{1}_{\eta+a\mu} \in \mathcal{F}_\lambda^{\eta+a\mu}$$

is a singular vector of level ab.

Proof. By Theorem 8.6 and Corollary 8.1, the statements holds for $t \in \Omega_a$. We show that the theorem holds for $t \in \mathbb{C}^* \setminus \Omega_a$.

Let $\{a_{\mathbb{I}} \otimes \mathbf{1}_{\eta+a\mu} | \mathbb{I} \in \mathcal{P}_{ab}\}$ be a basis of $(\mathcal{F}_\lambda^{\eta+a\mu})_{h_\lambda^{\eta+a\mu}+ab}$, where $a_{\mathbb{I}}$ is given by (4.5). By the definition of Vir-action on Fock modules, the condition that $\sum_{\mathbb{I}\in\mathcal{P}_{ab}} c_{\mathbb{I}} a_{\mathbb{I}} \otimes \mathbf{1}_{\eta+a\mu}$ is a singular vector is a linear equation with respect to $\{c_{\mathbb{I}} | \mathbb{I} \in \mathcal{P}_{ab}\}$ with coefficients in $\mathbb{C}[t^{\frac{1}{2}}, t^{-\frac{1}{2}}]$. Hence, we may assume that $c_{\mathbb{I}} \in \mathbb{C}[t^{\frac{1}{2}}, t^{-\frac{1}{2}}]$. On the other hand, for $t \in \Omega_a$, $c_{\mathbb{I}}$ is given by the coefficient of $a_{\mathbb{I}}$ in $\iota_n(J_{(a^b)}(x_1, \cdots, x_n; t))$. Since Ω_a is Zariski dense in $\mathbb{C}^*$, we may conclude that the theorem holds for any $t \in \mathbb{C}^*$. □

8.5 Spaces of Semi-infinite Forms and Fock Modules

Let $V_{a,b}$ be the $\mathfrak{g}$-module defined in (1.11). Here and after, we simply denote by $V_{a,b}^\sharp$ its restricted antipode dual (Definition 1.21). Then, one can define the 'natural action' of the Virasoro algebra with non-trivial central charge on the space of semi-infinite forms $\bigwedge^{\frac{\infty}{2}+\bullet} V_{a,b}^\sharp$, whose structure was investigated by B. Feigin and D. Fuchs in [FeFu4]. In fact, certain submodules of $\bigwedge^{\frac{\infty}{2}+\bullet} V_{a,b}^\sharp$ are isomorphic to Fock modules studed in this chapter. In this section, we will establish the isomorphisms between these modules via Fock modules over the Clifford Lie superalgebra.

8.5.1 Space of Semi-infinite Forms

Here, we introduce the space of semi-infinite forms $\bigwedge^{\frac{\infty}{2}+\bullet} V_{a,b}^\sharp$ and discuss its natural $\mathfrak{g}$-module structure. We first recall that $V_{a,b} := \bigoplus_{n\in\mathbb{Z}} \mathbb{C} v_n$ is the $\mathfrak{g}$-module defined by

$$L_s.v_n = (as + b - n)v_{n+s} \ (s \in \mathbb{Z}), \quad C.v_n = 0.$$

Then, $\mathfrak{g}$ acts on the restricted antipode dual $(V_{a,b})^{\sharp a}$ as follows: let $\{v'_n\}_{n\in\mathbb{Z}}$ be the dual basis of $\{v_n\}_{n\in\mathbb{Z}}$, and

$$L_s v'_n = \{-(a+1)s - b + n\} v'_{n-s}, \quad C.v'_n = 0. \tag{8.34}$$

The **space of semi-infinite forms** $\bigwedge^{\frac{\infty}{2}+\bullet} V_{a,b}^\sharp$ is defined as the space spanned by formal semi-infinite exterior products

$$v'_{i_1} \wedge v'_{i_2} \wedge v'_{i_3} \wedge \cdots \quad (i_1, i_2, \cdots \in \mathbb{Z})$$

with the boundary condition $i_{n+1} = i_n - 1$ for $n \gg 0$. More precisely,

$$\bigwedge^{\frac{\infty}{2}+\bullet} V^{\sharp}_{a,b} = \bigoplus_{\substack{i_1>i_2>\cdots \\ i_{n+1}=i_n-1(n\gg 0)}} \mathbb{C}v'_{i_1} \wedge v'_{i_2} \wedge v'_{i_3} \wedge \cdots .$$

For each $i \in \mathbb{Z}$, we set

$$\bigwedge^{\frac{\infty}{2}+i} V^{\sharp}_{a,b} := \bigoplus_{\substack{i_1>i_2>i_3>\cdots \\ i_n=i-n(n\gg 0)}} \mathbb{C}v'_{i_1} \wedge v'_{i_2} \wedge v'_{i_3} \wedge \cdots ..$$

By definition, $\bigwedge^{\frac{\infty}{2}+\bullet} V^{\sharp}_{a,b} = \bigoplus_{i\in\mathbb{Z}} \bigwedge^{\frac{\infty}{2}+i} V^{\sharp}_{a,b}$, and we define $u_i \in \bigwedge^{\frac{\infty}{2}+i} V^{\sharp}_{a,b}$ as follows:

$$u_i := v'_{i-1} \wedge v'_{i-2} \wedge v'_{i-3} \wedge \cdots . \tag{8.35}$$

As we will see in the next subsection, the space of semi-infinite forms $\bigwedge^{\frac{\infty}{2}+\bullet} V^{\sharp}_{a,b}$ equips with the next 'natural' $\mathfrak{g}$-module structure:

$$L_s.(v'_{i_1} \wedge v'_{i_2} \wedge \cdots) := \sum_{k=1}^{\infty} v'_{i_1} \wedge v'_{i_2} \wedge \cdots \wedge L_s.v'_{i_k} \wedge \cdots \quad (\text{for } s \neq 0). \tag{8.36}$$

However, for L_0, the above 'natural' action is not a finite sum, thus, it is not obvious whether there exists a $\mathfrak{g}$-module structure which satisfies (8.36) or not.

Later, we will show the existence of such a $\mathfrak{g}$-action by means of Fock modules over a Clifford algebra. Here, assuming its existence, we compute the actions of L_0 and C by using commutation relations of $\mathfrak{g}$. By $[L_1, L_{-1}] = 2L_0$ and $[L_2, L_{-2}] = 4L_0 + \frac{1}{2}C$, we have

$$L_0 u_i = \frac{1}{2} L_1 L_{-1} u_i = -\frac{1}{2}(a-b+i)(a+b-i+1)u_i \tag{8.37}$$

and $C.u_i = -(12a^2 + 12a + 2)u_i$.

8.5.2 Clifford Algebra and Fermionic Fock Modules

Definition 8.5 *Let $\mathcal{C}$ be the vector superspace*

$$\bigoplus_{m\in\mathbb{Z}} \mathbb{C}\varphi_m \oplus \bigoplus_{m\in\mathbb{Z}} \mathbb{C}\varphi^{\dagger}_m \oplus \mathbb{C}K_{\mathcal{C}}$$

with the parity $|\varphi_m| = |\varphi^{\dagger}_m| = \bar{1}$, $|K| = \bar{0}$, endowed with the Lie superbracket $[\cdot,\cdot]$ (see § C.4.1) defined by

$$[\varphi_m, \varphi_n] = 0 = [\varphi^{\dagger}_m, \varphi^{\dagger}_n], \quad [\varphi_m, \varphi^{\dagger}_n] = \delta_{m+n,0} K_{\mathcal{C}}, \quad [\mathcal{C}, K_{\mathcal{C}}] = \{0\}.$$

The quotient algebra $U(\mathcal{C})/\langle K_\mathcal{C} - 1\rangle$ is called a ***Clifford algebra,*** *where $\langle K_\mathcal{C} - 1\rangle$ signifies the two-sided ideal of $U(\mathcal{C})$ generated by $K_\mathcal{C} - 1$.*

Let us introduce the module $\mathcal{F}^\mathcal{C}$ over $\mathcal{C}$ called a **fermionic Fock module.** We set

$$\mathcal{C}^{\geq} := \bigoplus_{m\in\mathbb{Z}_{>0}} \mathbb{C}\varphi_m \oplus \bigoplus_{m\in\mathbb{Z}_{\geq 0}} \mathbb{C}\varphi_m^\dagger \oplus \mathbb{C}K_\mathcal{C},$$

and introduce a one-dimensional $\mathcal{C}^{\geq}$-module $\mathbb{C}\mathbf{1}_\mathcal{C}$ as follows:

$$|\mathbf{1}_\mathcal{C}| := \bar{0}, \quad \varphi_m.\mathbf{1}_\mathcal{C} := 0 \ (m > 0), \quad \varphi_m^\dagger.\mathbf{1}_\mathcal{C} := 0 \ (m \geq 0), \quad K_\mathcal{C}.\mathbf{1}_\mathcal{C} := \mathbf{1}_\mathcal{C}.$$

The fermionic Fock module $\mathcal{F}^\mathcal{C}$ is the following induced representation:

$$\mathcal{F}^\mathcal{C} := \mathrm{Ind}_{\mathcal{C}^{\geq}}^{\mathcal{C}} \mathbb{C}\mathbf{1}_\mathcal{C}.$$

Next, let us introduce a **vertex superalgebra** (VSA) structure on $\mathcal{F}^\mathcal{C}$. We simply denote $1 \otimes \mathbf{1}_\mathcal{C} \in \mathcal{F}^\mathcal{C}$ by $|0\rangle_\mathcal{C}$.

Definition 8.6 **(Vacuum vector)** $|0\rangle_\mathcal{C}$.
(Translation operator) $T.|0\rangle_\mathcal{C} = 0$,

$$[T, \varphi_m] := -(m-1)\varphi_{m-1}, \quad [T, \varphi_m^\dagger] := -m\varphi_{m-1}^\dagger.$$

(Vertex operators) $Y(|0\rangle_\mathcal{C}, z) := \mathrm{id}_{\mathcal{F}^\mathcal{C}}$,

$$Y(\varphi_0.|0\rangle_\mathcal{C}, z) := \varphi(z), \quad Y(\varphi_{-1}^\dagger.|0\rangle_\mathcal{C}, z) := \varphi^\dagger(z),$$

where

$$\varphi(z) := \sum_{m\in\mathbb{Z}} \varphi_m z^{-m}, \quad \varphi^\dagger(z) := \sum_{m\in\mathbb{Z}} \varphi_m^\dagger z^{-m-1}.$$

A super version of Theorem C.1 implies that the above data uniquely determine a VSA structure on $\mathcal{F}^\mathcal{C}$.

The fermionic Fock module equips with the $\mathfrak{g}$-module structure given as follows: For a, $b \in \mathbb{C}$, we set

$$\begin{aligned} T_{a,b}^\mathcal{C}(z) := &(1+a)\mathop{\circ}\limits_{\circ}\partial\varphi(z)\varphi^\dagger(z)\mathop{\circ}\limits_{\circ} + a\mathop{\circ}\limits_{\circ}\varphi(z)\partial\varphi^\dagger(z)\mathop{\circ}\limits_{\circ} \\ &+ (a-b)z^{-1}\mathop{\circ}\limits_{\circ}\varphi(z)\varphi^\dagger(z)\mathop{\circ}\limits_{\circ} - \frac{1}{2}(a-b)(a+b+1)z^{-2}\mathrm{id}_{\mathcal{F}^\mathcal{C}}, \end{aligned} \tag{8.38}$$

where the normal order $\mathop{\circ}\limits_{\circ} \cdot \mathop{\circ}\limits_{\circ}$ is defined as in Definition C.4. By direct computation, one can check

$$T_{a,b}^\mathcal{C}(z)T_{a,b}^\mathcal{C}(w) \sim \frac{-(6a^2+6a+1)\mathrm{id}_{\mathcal{F}^\mathcal{C}}}{(z-w)^4} + \frac{2T_{a,b}^\mathcal{C}(w)}{(z-w)^2} + \frac{\partial T_{a,b}^\mathcal{C}(w)}{z-w}.$$

This operator product expansion implies

Lemma 8.15. *The following map defines $\mathfrak{g}$-module structure on $\mathcal{F}^{\mathcal{C}}$:*

$$\sum_{s\in\mathbb{Z}} L_s z^{-s-2} \longmapsto T^{\mathcal{C}}_{a,b}(z), \quad C \longmapsto -(12a^2+12a+2)\mathrm{id}_{\mathcal{F}^{\mathcal{C}}}. \tag{8.39}$$

When we regard the $\mathcal{C}$-module $\mathcal{F}^{\mathcal{C}}$ as $\mathfrak{g}$-module via the above action, we denote it by $\mathcal{F}^{\mathcal{C}}_{a,b}$. The actions of L_s $(s\neq 0)$, L_0 and C on $\mathcal{F}^{\mathcal{C}}_{a,b}$ can be written as follows:

$$\begin{aligned} L_s &\mapsto -\sum_{m\in\mathbb{Z}}\{(a+1)s+b-m\}\varphi_{s-m}\varphi^\dagger_m \\ L_0 &\mapsto -\sum_{m\in\mathbb{Z}_{\geq 0}}(b-m)\varphi_{-m}\varphi^\dagger_m + \sum_{m\in\mathbb{Z}_{>0}}(b-m)\varphi^\dagger_m\varphi_{-m} - \frac{(a-b)(a+b+1)}{2}\mathrm{id}_{\mathcal{F}^{\mathcal{C}}} \\ C &\mapsto -(12a^2+12a+2)\mathrm{id}_{\mathcal{F}^{\mathcal{C}}}. \end{aligned} \tag{8.40}$$

Remark 8.7 *In the case where $a=1$ and $b=0$, the $\mathfrak{g}$-module $V_{1,0}$ is isomorphic to its adjoint representation on $\mathfrak{g}/\mathbb{C}C$. In particular, the pair $\left(-\frac{1}{2}(a-b)(a+b+1), -(12a^2+12a+2)\right)$ which appears in the correction term of the above formula specialises to $(-26,-1)=(-\gamma(C),-\gamma(L_0))$, where γ is the semi-infinite character of $\mathfrak{g}$ given in (7.11). For a relation between a semi-infinite character and fermionic Fock modules, see [IK8].*

For later use, let us introduce some notation. Notice that

$$\mathcal{F}^{\mathcal{C}} = \bigoplus_{\substack{s,t\in\mathbb{Z}_{\geq 0}\\ 0\leq m_1<\cdots<m_s\\ 0<n_1<\cdots<n_t}} \mathbb{C}\varphi_{-m_s}\cdots\varphi_{-m_1}\varphi^\dagger_{-n_t}\cdots\varphi^\dagger_{-n_1}.|0\rangle_{\mathcal{C}}$$

by the PBW theorem. For each $i\in\mathbb{Z}$, we set

$$\mathcal{F}^{\mathcal{C}}(i) := \bigoplus_{\substack{s,t\in\mathbb{Z}_{\geq 0}, s-t=i\\ 0\leq m_1<\cdots<m_s\\ 0<n_1<\cdots<n_t}} \mathbb{C}\varphi_{-m_s}\cdots\varphi_{-m_1}\varphi^\dagger_{-n_t}\cdots\varphi^\dagger_{-n_1}.|0\rangle_{\mathcal{C}}. \tag{8.41}$$

Moreover, we define a vector $|i\rangle_{\mathcal{C}}\in\mathcal{F}^{\mathcal{C}}(i)$ as follows:

$$|i\rangle_{\mathcal{C}} := \begin{cases} \varphi_{-i+1}\cdots\varphi_0.|0\rangle_{\mathcal{C}} & (i>0) \\ |0\rangle_{\mathcal{C}} & (i=0) \\ \varphi^\dagger_i\cdots\varphi^\dagger_{-1}.|0\rangle_{\mathcal{C}} & (i<0) \end{cases}. \tag{8.42}$$

By definition, $\mathcal{F}^{\mathcal{C}}(i)$ is a $\mathfrak{g}$-submodule of $\mathcal{F}^{\mathcal{C}}$. We denote it by $\mathcal{F}^{\mathcal{C}}_{a,b}(i)$.

8.5.3 Isomorphism between $\bigwedge^{\frac{\infty}{2}+\bullet}V^{\sharp}_{a,b}$ and $\mathcal{F}^{\mathcal{C}}_{a,b}$

In this subsection, we will show that $\bigwedge^{\frac{\infty}{2}+\bullet}V^{\sharp}_{a,b}$ and $\mathcal{F}^{\mathcal{C}}_{a,b}$ are isomorphic as $\mathcal{C}$-module. Such an isomorphism induces a $\mathfrak{g}$-module structure on $\bigwedge^{\frac{\infty}{2}+\bullet}V^{\sharp}_{a,b}$ from the action on $\mathcal{F}^{\mathcal{C}}_{a,b}$ defined by (8.39). It turns out that this $\mathfrak{g}$-action on $\bigwedge^{\frac{\infty}{2}+\bullet}V^{\sharp}_{a,b}$ is nothing but the 'natural action' (8.36).

We first introduce some notation in order to regard the space $\bigwedge^{\frac{\infty}{2}+\bullet}V^{\sharp}_{a,b}$ as $\mathcal{C}$-module. For $x \in V^{\sharp}_{a,b}$ and $y \in V_{a,b}$, we define $\varepsilon(x)$ and $\iota(y) \in \mathrm{End}(\bigwedge^{\frac{\infty}{2}+\bullet}V^{\sharp}_{a,b})$ as follows:

$$\begin{aligned}\varepsilon(x).(v'_{i_1}\wedge v'_{i_2}\wedge\cdots) &:= x\wedge v'_{i_1}\wedge v'_{i_2}\wedge\cdots,\\ \iota(y).(v'_{i_1}\wedge v'_{i_2}\wedge\cdots) &:= \sum_{k=1}^{\infty}(-1)^{k-1}\langle y, v'_{i_k}\rangle v'_{i_1}\wedge v'_{i_2}\wedge\cdots\wedge\widehat{v'_{i_k}}\wedge\cdots,\end{aligned} \tag{8.43}$$

where $\langle\cdot,\cdot\rangle : V_{a,b}\times V^{\sharp}_{a,b}\to\mathbb{C}$ signifies the dual pairing, and $\widehat{v'_{i_k}}$ means that v'_{i_k} should be omitted.

Lemma 8.16. *The space $\bigwedge^{\frac{\infty}{2}+\bullet}V^{\sharp}_{a,b}$ becomes a $\mathcal{C}$-module via the following map:*

$$\varphi_k\mapsto\varepsilon(v'_{-k}),\quad \varphi^{\dagger}_l\mapsto\iota(v_l),\quad K_{\mathcal{C}}\mapsto \mathrm{id}_{\bigwedge^{\frac{\infty}{2}+\bullet}V^{\sharp}_{a,b}}. \tag{8.44}$$

Proof. By direct computation, one can show

$$\begin{aligned}\varepsilon(v'_k)\varepsilon(v'_l)+\varepsilon(v'_l)\varepsilon(v'_k) &= 0 = \iota(v_k)\iota(v_l)+\iota(v_l)\iota(v_k),\\ \varepsilon(v'_k)\iota(v_l)+\iota(v_l)\varepsilon(v'_k) &= \delta_{k,l}\mathrm{id}_{\bigwedge^{\frac{\infty}{2}+\bullet}V^{\sharp}_{a,b}}.\end{aligned}$$

□

Proposition 8.8 *There exists the following isomorphism of $\mathcal{C}$-modules:*

$$\mathcal{F}^{\mathcal{C}}\simeq\bigwedge^{\frac{\infty}{2}+\bullet}V^{\sharp}_{a,b}\quad (|0\rangle_{\mathcal{C}}\longmapsto u_0) \tag{8.45}$$

where the vector u_0 is defined as in (8.35).

Proof. We have $\varphi_m.u_0 := 0$ $(m>0)$, $\varphi^{\dagger}_m.u_0 := 0$ $(m\geq 0)$ and $K_{\mathcal{C}}.u_0 := u_0$. Hence, the universality of $\mathcal{F}^{\mathcal{C}}$ implies that the surjective homomorphism $\mathcal{F}^{\mathcal{C}}\to\bigwedge^{\frac{\infty}{2}+\bullet}V^{\sharp}_{a,b}$ $(|0\rangle_{\mathcal{C}}\longmapsto u_0)$ of $\mathcal{C}$-modules exists. Since $\mathcal{F}^{\mathcal{C}}$ is irreducible, this homomorphism is injective. □

The $\mathfrak{g}$-action (8.39) on $\mathcal{F}^{\mathcal{C}}=\mathcal{F}^{\mathcal{C}}_{a,b}$ induces a $\mathfrak{g}$-action on $\bigwedge^{\frac{\infty}{2}+\bullet}V^{\sharp}_{a,b}$ via the isomorphism (8.45).

Finally, we show that the above $\mathfrak{g}$-action on $\bigwedge^{\frac{\infty}{2}+\bullet}V^{\sharp}_{a,b}$ coincides with the 'natural' $\mathfrak{g}$-action (8.36). Indeed, for $s\neq 0$, by (8.40), we have

$$
\begin{aligned}
L_s.(v'_{i_1} \wedge v'_{i_2} \wedge \cdots) &= \sum_{k=1}^{\infty} \{-(a+1)s - b + i_k\} v'_{i_1} \wedge v'_{i_2} \wedge \cdots \wedge v'_{i_k - s} \wedge \cdots \\
&= \sum_{k=1}^{\infty} v'_{i_1} \wedge v'_{i_2} \wedge \cdots \wedge L_s.v'_{i_k} \wedge \cdots
\end{aligned} \tag{8.46}
$$

where $L_s.v'_{i_k}$ is given by (8.34). Thus, the action of $\mathfrak{g}$ on $\bigwedge^{\frac{\infty}{2}+\bullet} V^{\sharp}_{a,b}$ coincides with (8.36).

By definition, the subspace $\mathcal{F}^C_{a,b}(i)$ of $\mathcal{F}^C_{a,b}$ is a $\mathfrak{g}$-submodule. Hence, for any a, $b \in \mathbb{C}$ and $i \in \mathbb{Z}$,

$$
\mathcal{F}^C_{a,b}(i) \simeq \bigwedge^{\frac{\infty}{2}+i} V^{\sharp}_{a,b} \tag{8.47}
$$

as $\mathfrak{g}$-module.

The following theorem is the main result of this section:

Theorem 8.8 *For any λ, $\eta \in \mathbb{C}$ and $i \in \mathbb{Z}$, there exists an isomorphism of $\mathfrak{g}$-modules*

$$
\bigwedge^{\frac{\infty}{2}+i} V^{\sharp}_{\lambda-\frac{1}{2},\lambda-\eta-\frac{1}{2}} \simeq \mathcal{F}^{\eta+i}_{\lambda}.
$$

Remark 8.8 *1. By Remark 1.7 and the above theorem, the central charge of the* Vir*-action on $\bigwedge^{\frac{\infty}{2}+\bullet}(\mathbb{C}[t,t^{-1}](dt)^a)^{\sharp}$ is given by $-(12a^2 - 12a + 2) = -12B_2(a)$, where $B_n(x) \in \mathbb{C}[x]$ $(n \in \mathbb{Z}_{\geq 0})$ is the nth* **Bernoulli polynomial** *(cf. [AAR]) defined by*

$$
\frac{te^{xt}}{e^t - 1} = \sum_{n \geq 0} B_n(x) \frac{t^n}{n!}.
$$

2. Let C be a genus g (≥ 2) compact Riemann surface and λ_n $(n \in \mathbb{Z}_{>0})$ the determinant line bundle of the vector bundle on (a compactification of) the moduli space of stable curves of genus g. By Theorem 5.10 of [Mum] due to D. Mumford, one concludes that $c_1(\lambda_n) = (6n^2 - 6n + 1)c_1(\lambda_1)$. Two appearances of the 2nd Bernoulli polynomial B_2 are not a coincidence. In fact, some relations between these two formulae were explained in [ADKP], [BMS], [BS] and [Kon].

8.5.4 Boson–Fermion Correspondence

In this subsection, we prove Theorem 8.8 via the boson–fermion correspondence. By the isomorphism (8.47), it suffices to show

Proposition 8.9 *For any λ, $\eta \in \mathbb{C}$ $i \in \mathbb{Z}$, there exists an isomorphism of $\mathfrak{g}$-modules:*

$$
\mathcal{F}^C_{\lambda-\frac{1}{2},\lambda-\eta-\frac{1}{2}}(i) \simeq \mathcal{F}^{\eta+i}_{\lambda}.
$$

As a preliminary step, we show the existence of an isomorphism of $\mathcal{H}$-modules between the above spaces. To do this, let us introduce an $\mathcal{H}$-module structure on $\mathcal{F}^{C}(i)$. Set

$$a_{\eta}^{C}(z) := {}^{\circ}_{\circ}\varphi(z)\varphi^{\dagger}(z)^{\circ}_{\circ} + \eta z^{-1}\mathrm{id}_{\mathcal{F}^{C}}. \tag{8.48}$$

Then, the operator product expansion

$$a_{\eta}^{C}(z)a_{\eta}^{C}(w) \sim \frac{1}{(z-w)^2}$$

holds. Hence, we have

Lemma 8.17. *For any $\eta \in \mathbb{C}$ and $i \in \mathbb{Z}$, $\mathcal{F}^{C}(i)$ becomes an $\mathcal{H}$-module via the following map:*

$$\sum_{s\in\mathbb{Z}} a_s z^{-s-1} \longmapsto a_{\eta}^{C}(z), \quad K_{\mathcal{H}} \mapsto \mathrm{id}_{\mathcal{F}^{C}(i)}. \tag{8.49}$$

Proposition 8.10 *For any $\eta \in \mathbb{C}$ and $i \in \mathbb{Z}$, there exists an isomorphism of $\mathcal{H}$-modules:*

$$\mathcal{F}^{C}(i) \simeq \mathcal{F}^{\eta+i}.$$

Proof. Since $a_0 = \sum_{m\leq 0}\varphi_m\varphi^{\dagger}_{-m} - \sum_{m>0}\varphi^{\dagger}_{-m}\varphi_m + \eta\mathrm{id}_{\mathcal{F}^{C}}$, we have $a_0.|i\rangle_C = (\eta+i)|i\rangle_C$. Similarly, one can show $\mathcal{H}^{+}.|i\rangle_C = \{0\}$ and $K_{\mathcal{H}}.|i\rangle_C = |i\rangle_C$, thus, by the universality of $\mathcal{F}^{\eta+i}$, we obtain the homomorphism

$$\rho_{\eta,i} : \mathcal{F}^{\eta+i} \longrightarrow \mathcal{F}^{C}(i) \quad (|\eta+i\rangle \mapsto |i\rangle_C)$$

of $\mathcal{H}$-modules. Since $\mathcal{F}^{\eta+i}$ is an irreducible $\mathcal{H}$-module, the map $\rho_{\eta,i}$ is injective. We show that it is surjective.

Let us introduce $\mathbb{Z}_{\geq 0}$-gradations of $\mathcal{F}^{\eta+i}$ and $\mathcal{F}^{C}(i)$ as follows:

1. $\deg(a_{-m_s}\cdots a_{-m_1}|\eta+i\rangle) := \frac{1}{2}i(i-1) + \sum_{k=1}^{s} m_k$,
2. $\deg(\varphi_{-m_s}\cdots\varphi_{-m_1}\varphi^{\dagger}_{-n_t}\cdots\varphi^{\dagger}_{-n_1}|0\rangle_C) := \sum_{i=1}^{s} m_k + \sum_{k=1}^{t} n_k$.

Notice that they coincide with the L_0-eigenvalue when $(\lambda,\eta) = (\frac{1}{2},0)$ and $(a,b) = (0,0)$. We denote the graded subspaces of $\mathcal{F}^{\eta+i}$ and $\mathcal{F}^{C}(i)$ of degree k by $(\mathcal{F}^{\eta+i})^k$ and $\mathcal{F}^{C}(i)^k$ respectively.

Lemma 8.18. *For any $i \in \mathbb{Z}$ and $k \in \mathbb{Z}_{\geq 0}$, $\dim(\mathcal{F}^{\eta+i})^k = \dim\mathcal{F}^{C}(i)^k$ holds.*

Proof. Let us consider generating functions

$$\sum_{i\in\mathbb{Z}}\sum_{k\in\mathbb{Z}_{\geq 0}} \dim(\mathcal{F}^{\eta})^k z^i q^k = \frac{\sum_{i\in\mathbb{Z}} q^{\frac{1}{2}i(i-1)}z^i}{\prod_{n\in\mathbb{Z}_{>0}}(1-q^n)},$$

$$\sum_{i\in\mathbb{Z}}\sum_{k\in\mathbb{Z}_{\geq 0}} \dim\mathcal{F}^{C}(i)^k z^i q^k = \prod_{n\in\mathbb{Z}_{>0}} (1+zq^{n-1})(1+z^{-1}q^n).$$

By the Jacobi triple product identity (cf. (10.4)), these generating functions coincide. Hence, the lemma holds. □

Notice that $\deg(|i\rangle_{\mathcal{C}}) = \frac{1}{2}i(i-1)$, so we have

$$\rho_{\eta,i}((\mathcal{F}^{\eta+i})^k) \subset \mathcal{F}^{\mathcal{C}}(i)^k.$$

Hence, Lemma 8.18 and the injectivety $\rho_{\eta,i}$ imply Proposition 8.10. □

PROOF OF PROPOSITION 8.9. By Proposition 8.10, it suffices to compare the $\mathfrak{g}$-actions on $\mathcal{F}^{\eta+i}$ and $\mathcal{F}^{\mathcal{C}}(i)$, i.e., it is enough to show

$$\frac{1}{2}{}^{\circ}_{\circ} a^{\mathcal{C}}_{\eta}(z)^2 {}^{\circ}_{\circ} + \lambda \partial_z a^{\mathcal{C}}_{\eta}(z) = T^{\mathcal{C}}_{\lambda-\frac{1}{2},\lambda-\eta-\frac{1}{2}}(z). \tag{8.50}$$

Here, we indicate briefly a proof of (8.50). Since $a^{\mathcal{C}}_{\eta}(z) = a^{\mathcal{C}}_{0}(z) + \eta z^{-1}\mathrm{id}_{\mathcal{F}^{\mathcal{C}}(i)}$ and

$$T^{\mathcal{C}}_{\lambda-\frac{1}{2},\lambda-\eta-\frac{1}{2}}(z) = T^{\mathcal{C}}_{\lambda-\frac{1}{2},\lambda-\frac{1}{2}}(z) + \eta z^{-1} a^{\mathcal{C}}_{0}(z) + h^{\eta}_{\lambda} z^{-2}\mathrm{id}_{\mathcal{F}^{\mathcal{C}}(i)}$$

hold, it is sufficient to show the case $\eta = 0$. The case $\eta = 0$ (8.50) can be proved by using the formula

$${}^{\circ\circ}_{\circ\circ}\varphi(z)\varphi^{\dagger}(z){}^{\circ}_{\circ}{}^{\circ}_{\circ}\varphi(z)\varphi^{\dagger}(z){}^{\circ\circ}_{\circ\circ} = {}^{\circ}_{\circ}\partial\varphi(z)\varphi^{\dagger}(z){}^{\circ}_{\circ} - {}^{\circ}_{\circ}\varphi(z)\partial\varphi^{\dagger}(z){}^{\circ}_{\circ}$$

and so on. Hence, Proposition 8.9 holds, and thus, Theorem 8.8 does. □

Remark 8.9 *Here is a remark on an alternative proof of Proposition 8.9 via vertex superalgebras. Let $V_{\mathbb{Z}}$ be the lattice vertex superalgebra introduced in § 4.1.3. It is known that there exists an isomorphism $\rho : \mathcal{F}^{\mathcal{C}} \simeq V_{\mathbb{Z}}$ of vertex superalgebras (see Section 5.3.2 in [FB]). By setting*

$$\omega^{\mathcal{C}}_{\lambda} := \{(\lambda + \frac{1}{2})\varphi_{-1}\varphi^{\dagger}_{-1} + (\lambda - \frac{1}{2})\varphi_{0}\varphi^{\dagger}_{-2}\}.|0\rangle_{\mathcal{C}},$$

one can directly check that $T^{\mathcal{C}}_{\lambda-\frac{1}{2},\lambda-\frac{1}{2}}(z) = Y(\omega^{\mathcal{C}}_{\lambda}, z)$ and $\rho(\omega^{\mathcal{C}}_{\lambda}) = \omega_{\lambda}$, where the vector $\omega_{\lambda} \in V_{\mathbb{Z}}$ is given in (4.1). Hence, ρ is an isomorphism of $\mathfrak{g}$-modules (cf. Section 5.3.4 in [FB]). Recalling $V_{\mathbb{Z}} = \bigoplus_{i\in\mathbb{Z}} \mathcal{F}^i$, we have

$$\rho : \mathcal{F}^{\mathcal{C}}_{\lambda-\frac{1}{2},\lambda-\frac{1}{2}} \simeq \bigoplus_{i\in\mathbb{Z}} \mathcal{F}^i_{\lambda},$$

i.e., ρ is obtained as the direct sum of the isomorphisms in Proposition 8.9 with $\eta = 0$. Moreover, combining the isomorphism ρ with those of bosonic Fock modules given in Proposition 4.2, Proposition 8.9 has been proved.

8.6 Bibliographical Notes and Comments

In this chapter, we described the structure theorem of Fock modules over the Virasoro algebra due to B. Feigin and D. Fucks [FeFu4]. To be precise, B. Feigin and D. Fucks revealed the Virasoro module structure on the spaces of semi-infinite forms, and we applied their arguments to the bosonic Fock modules.

In § 3.3 and 3.4, motivated by [FeFu4], we reformulated the Jantzen filtration of Fock modules, and here, using the Jantzen filtration, we simplified arguments of B. Feigin and D. Fucks. In particular, thanks to Proposition 3.9, we reduced the proofs of the structure theorem related with contragredient Verma modules to those related with Verma modules. The reader who reads the original proof in [FeFu4] should notice some incorrect statements on the structures of Fock modules in Case III^+ (**Class** R^- in this chapter).

In 1986, M. Wakimoto and H. Yamada [WY] found that the singular vector of a Fock module over Vir for $c = 1$ can be expressed in terms of the Schur polynomials. Since then, it was quite natural to expect that the singular vector of a Fock module over Vir for general c can be expressed in terms of a deformation of the Schur polynomials. In 1995, K. Mimachi and Y. Yamada [MY] proved that this is indeed the case, i.e., they proved that the singular vectors can be expressed in terms of the Jack symmetric polynomials, which was explained in § 8.4.

We also proved that bosonic Fock modules and the spaces of semi-infinite forms are isomorphic as modules of the Virasoro algebra. Here, we first showed that fermionic Fock modules and spaces of semi-infinite forms are isomorphic. As we have mentioned in Remark 8.8, this correspondence leads us to find an interesting relation between the Virasoro algebra and (a compactification of) the moduli space of stable curves of genus g (≥ 2). As M. Konstevich [Kon] mentioned, it is an interesting question to study this relation from the viewpoint of the representation theory.

Next, we constructed isomorphisms between bosonic Fock modules and fermionic Fock modules via the so-called boson–fermion correspondence. Combining these isomorphisms, we obtain an isomorphism of Virasoro modules between the spaces of semi-infinite forms and the bosonic Fock modules. As far as the authors know, a reference on such isomorphisms does not exist, though the isomorphisms seem to be well-known to experts.

8.A Appendix: Another Proof of Theorem 8.8

In this section, we give a direct proof of the $\mathfrak{g}$-isomorphism

$$\bigwedge^{\frac{\infty}{2}+0} V^{\sharp}_{\lambda-\frac{1}{2},\lambda-\eta-\frac{1}{2}} \simeq \mathcal{F}^{\eta}_{\lambda}$$

in Theorem 8.8 by combinatorial arguments. To parameterise basis vectors of $\bigwedge^{\frac{\infty}{2}+0} V^{\sharp}_{a,b}$, we introduce

$$\mathcal{I}_0 := \{(i_1, i_2, \cdots) | i_1 > i_2 > \cdots,\ i_n = -n\ (n \gg 0)\},$$

and set $v'_I := v'_{i_1} \wedge v'_{i_2} \wedge v'_{i_3} \wedge \cdots$ for $I := (i_1, i_2, \cdots) \in \mathcal{I}_0$. Moreover, we associate a diagram, called a **Maya diagram**, to each v'_I. It is a sequence of boxes indexed by $\mathbb{Z}$, some of which boxes are filled with a particle. For $I = (i_1, i_2, \cdots) \in \mathcal{I}_0$, we associate the following diagram to v'_I:

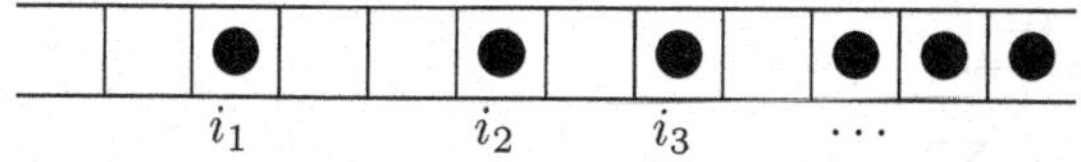

For example, v'_I with $I = (4, 1, -3, -4, \cdots)$ is represented by using

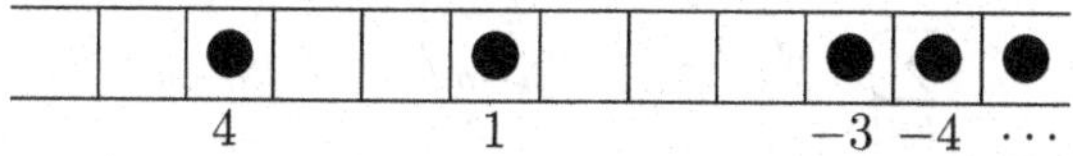

We first introduce an $\mathcal{H}$-module structure on the space $\bigwedge^{\frac{\infty}{2}+0} V^{\sharp}_{a,b}$.

Lemma 8.19. *For $\eta \in \mathbb{C}$, $\mathcal{H}$ acts on the space $\bigwedge^{\frac{\infty}{2}+0} V^{\sharp}_{a,b}$ via*

$$a_n.(v'_{i_1} \wedge v'_{i_2} \wedge \cdots) = \sum_{k=1}^{\infty} v'_{i_1} \wedge v'_{i_2} \wedge \cdots \wedge \overset{k}{v'_{i_k - n}} \wedge \cdots \quad (n \neq 0),$$

$$a_0 \mapsto \eta \mathrm{id}_{\bigwedge^{\frac{\infty}{2}+0} V^{\sharp}_{a,b}}, \quad K_{\mathcal{H}} \mapsto \mathrm{id}_{\bigwedge^{\frac{\infty}{2}+0} V^{\sharp}_{a,b}}.$$

Proof. It is convenient to use the following function: for $I = (i_1, i_2, i_3, \cdots) \in \mathcal{I}_0$ and $x \in \mathbb{Z}$, we set

$$P(I; x) := \begin{cases} 0 & \text{if } i_l = x \text{ for some } l \in \mathbb{Z}_{>0} \\ 1 & \text{otherwise} \end{cases}.$$

Notice that, to show Lemma 8.19, it is enough to check that

$$a_m a_n - a_n a_m = m \delta_{m+n,0} K_{\mathcal{H}} \tag{8.51}$$

holds for $m, n \neq 0$, since the other commutation relations $[a_m, a_0] = 0$ and $[a_m, K_{\mathcal{H}}] = 0$ obviously hold. Hence, we show (8.51) for $m, n \neq 0$.

Suppose that $m, n \neq 0$ and $I = (i_1, i_2, \cdots) \in \mathcal{I}_0$. By definition,

$$a_m a_n . v'_I = \sum_{\substack{k \neq l \\ i_k - n \neq i_l}} v'_{i_1} \wedge \cdots \wedge \overset{k}{v'_{i_k - n}} \wedge \cdots \wedge \overset{l}{v'_{i_l - m}} \wedge \cdots \\ + \sum_k P(I; i_k - n) v'_{i_1} \wedge \cdots \wedge \overset{k}{v'_{i_k - n - m}} \wedge \cdots . \tag{8.52}$$

We divide the first term of the right-hand side as follows:

$$\begin{aligned} &\sum_{\substack{k \neq l \\ i_k - n \neq i_l}} v'_{i_1} \wedge \cdots \wedge \overset{k}{v'_{i_k - n}} \wedge \cdots \wedge \overset{l}{v'_{i_l - m}} \wedge \cdots \\ &= \sum_{\substack{k \neq l \\ i_k - n \neq i_l \\ i_l - m \neq i_k}} v'_{i_1} \wedge \cdots \wedge \overset{k}{v'_{i_k - n}} \wedge \cdots \wedge \overset{l}{v'_{i_l - m}} \wedge \cdots \\ &\quad + \sum_{\substack{k \neq l \\ i_k - n \neq i_l \\ i_l - m = i_k}} v'_{i_1} \wedge \cdots \wedge \overset{k}{v'_{i_k - n}} \wedge \cdots \wedge \overset{l}{v'_{i_l - m}} \wedge \cdots . \end{aligned} \tag{8.53}$$

Here, we consider the following cases:

Case $m + n \neq 0$ $(m, n \neq 0)$. In this case, if $i_l - m = i_k$, then $i_k - n \neq i_l$. Hence, the second term of the right-hand side of (8.53) is written as

$$\begin{aligned} &\sum_{\substack{k \neq l \\ i_k - n \neq i_l \\ i_l - m = i_k}} v'_{i_1} \wedge \cdots \wedge \overset{k}{v'_{i_k - n}} \wedge \cdots \wedge \overset{l}{v'_{i_l - m}} \wedge \cdots \\ &= \sum_{\substack{k \neq l \\ i_l - m = i_k}} v'_{i_1} \wedge \cdots \wedge \overset{k}{v'_{i_k - n}} \wedge \cdots \wedge \overset{l}{v'_{i_l - m}} \wedge \cdots \\ &= - \sum_{\substack{k \neq l \\ i_l - m = i_k}} v'_{i_1} \wedge \cdots \wedge \overset{k}{v'_{i_l - m}} \wedge \cdots \wedge \overset{l}{v'_{i_k - n}} \wedge \cdots \\ &= - \sum_l (1 - P(I; i_l - m)) v'_{i_1} \wedge \cdots \wedge \overset{l}{v'_{i_l - n - m}} \wedge \cdots . \end{aligned}$$

Hence, we obtain

$$
\begin{aligned}
a_m a_n.v'_I = & \sum_{\substack{k\neq l\\ i_k-n\neq i_l\\ i_l-m\neq i_k}} v'_{i_1}\wedge\cdots\wedge \overset{k}{v'_{i_k-n}}\wedge\cdots\wedge \overset{l}{v'_{i_l-m}}\wedge\cdots\\
& -\sum_l v'_{i_1}\wedge\cdots\wedge \overset{l}{v'_{i_l-n-m}}\wedge\cdots\\
& +\sum_l P(I;i_l-m)v'_{i_1}\wedge\cdots\wedge \overset{l}{v'_{i_l-n-m}}\wedge\cdots\\
& +\sum_k P(I;i_k-n)v'_{i_1}\wedge\cdots\wedge \overset{k}{v'_{i_k-n-m}}\wedge\cdots,
\end{aligned}
\tag{8.54}
$$

and thus, (8.51) holds in this case.

Case $m+n=0$ $(m,n\neq 0)$. In this case, $i_k-n=i_l$ if and only if $i_l-m=i_k$. Hence, we have

$$
\begin{aligned}
a_m a_{-m}.v'_I = & \sum_{\substack{k\neq l\\ i_k+m\neq i_l}} v'_{i_1}\wedge\cdots\wedge \overset{k}{v'_{i_k+m}}\wedge\cdots\wedge \overset{l}{v'_{i_l-m}}\wedge\cdots\\
& +\sum_k P(I;i_k+m)v'_{i_1}\wedge v'_{i_2}\wedge v'_{i_3}\wedge\cdots,
\end{aligned}
$$

and thus,

$$
[a_m,a_{-m}].v'_I=\sum_k\{P(I;i_k+m)-P(I;i_k-m)\}v'_I.
$$

Hence, we have only to show

Lemma 8.20. *For $m\in\mathbb{Z}\setminus\{0\}$, the following holds:*

$$
\sum_k\{P(I;i_k+m)-P(I;i_k-m)\}=m. \tag{8.55}
$$

Proof. It is enough to prove for $m>0$. We first consider the case $m=1$.

Let $I=(i_1,i_2,\cdots)\in\mathcal{I}_0$ be a sequence such that $i_1>i_2>\cdots$. There exist $s\in\mathbb{Z}_{>0}$ and $k_1,k_2,\cdots,k_s\in\mathbb{Z}_{>0}$ such that

1. $1=k_1<k_2<k_3<\cdots<k_s$,
2. if $k_i<k<k_{i+1}$, then $i_k=i_{k_i}-(k-k_i)$,
3. $i_{k_i-1}-1>i_{k_i}$,
4. if $k\geq k_s$, then $i_k=i_{k_s}-(k-k_s)=-k$.

In this case, v'_I corresponds to the following Maya diagram

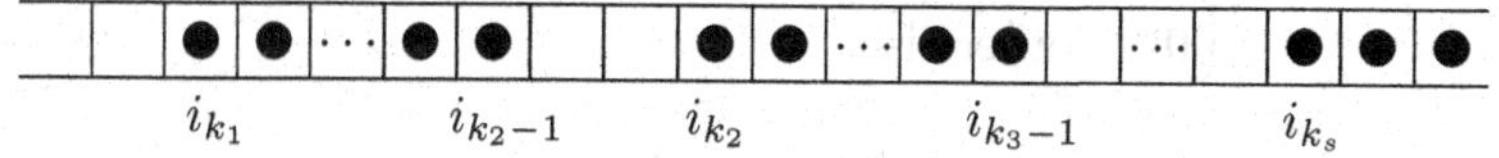

Since by definition

$$
\begin{aligned}
&P(I; i_k + 1) - P(I; i_k - 1) \\
&= \begin{cases} 1 & i_{k-1} - 1 \neq i_k \wedge i_k - 1 = i_{k+1}, \text{ i.e., } \square\underset{i_k}{\bullet}\bullet \\ -1 & i_{k-1} - 1 = i_k \wedge i_k - 1 \neq i_{k+1}, \text{ i.e., } \bullet\underset{i_k}{\bullet}\square \\ 0 & \text{otherwise} \end{cases},
\end{aligned}
$$

we have

$$\sum_{k=k_l}^{k_{l+1}-1} \{P(I; i_k + 1) - P(I; i_k - 1)\} = 0$$

for $l = 1, 2, \cdots, s-1$. Hence, (8.55) for $m = 1$ is given as follows:

$$
\begin{aligned}
\sum_{k \in \mathbb{Z}_{>0}} \{P(I; i_k + 1) - P(I; i_k - 1)\} &= \sum_{k \geq k_s} \{P(I; i_k + 1) - P(I; i_k - 1)\} \\
&= P(I; i_{k_s} + 1) - P(I; i_{k_s} - 1) \\
&= 1,
\end{aligned}
$$

and the lemma for $m = 1$ has been proved.

In the case $m > 1$, for each r such that $0 \leq r < m$, we set

$$I_r := (i_{n_1^{(r)}}, i_{n_2^{(r)}}, i_{n_3^{(r)}}, \cdots)$$

where $n_1^{(r)}, n_2^{(r)}, \cdots \in \mathbb{Z}$ such that $n_1^{(r)} < n_2^{(r)} < \cdots$, $i_{n_l^{(r)}} \equiv r \pmod{m}$ for $l \in \mathbb{Z}_{>0}$, and

$$I = \bigsqcup_{r=0}^{m-1} I_r$$

holds. Here and after, I is regarded not only as a sequence of integers but also as a subset of $\mathbb{Z}$. Then, the left-hand side of (8.55) can be written as follows:

$$\sum_{r=0}^{m-1} \sum_{l \in \mathbb{Z}_{>0}} \{P(I_r; i_{n_l^{(r)}} + m) - P(I_r; i_{n_l^{(r)}} - m)\}.$$

Similarly to the case $m = 1$, for each r, we have

$$
\begin{aligned}
&P(I_r; i_{n_l^{(r)}} + m) - P(I_r; i_{n_l^{(r)}} - m) \\
&= \begin{cases} 1 & i_{n_{l-1}^{(r)}} - m \neq i_{n_l^{(r)}} \wedge i_{n_l^{(r)}} - m = i_{n_{l+1}^{(r)}} \\ -1 & i_{n_{l-1}^{(r)}} - m = i_{n_l^{(r)}} \wedge i_{n_l^{(r)}} - m \neq i_{n_{l+1}^{(r)}} \\ 0 & \text{otherwise} \end{cases},
\end{aligned}
$$

and thus,

$$\sum_{l\in\mathbb{Z}_{>0}} \{P(I_r; i_{n_l^{(r)}} + m) - P(I_r; i_{n_l^{(r)}} - m)\} = 1.$$

Hence, Lemma 8.20 follows. □

We have completed the proof of Lemma 8.19. □

Next, we show the following isomorphism:

Proposition 8.11 *For any $\eta \in \mathbb{C}$, as $\mathcal{H}$-module,*

$$\bigwedge^{\frac{\infty}{2}+0} V_{a,b}^{\sharp} \simeq \mathcal{F}^{\eta}. \tag{8.56}$$

Proof. Let $u_0 \in \bigwedge^{\frac{\infty}{2}+0} V_{a,b}^{\sharp}$ be the vector introduced in (8.35). One can immediately show $a_n.u_0 = 0$ $(n > 0)$, $a_0.u_0 = \eta u_0$, $K_{\mathcal{H}}.u_0 = u_0$. Hence, there exists the $\mathcal{H}$-homomorphism

$$\rho_\eta : \mathcal{F}^{\eta} \longrightarrow \bigwedge^{\frac{\infty}{2}+0} V_{a,b}^{\sharp} \quad (|\eta\rangle \longmapsto u_0),$$

which is injective since $\mathcal{F}^{\eta}$ is irreducible. Hence, we show the surjectivity.

Let $(\mathcal{F}^{\eta})^k$ be the graded subspace of degree k with respect to the $\mathbb{Z}_{\geq 0}$-gradation on $\mathcal{F}^{\eta}$ introduced in the proof of Proposition 8.10. Notice that $\dim(\mathcal{F}^{\eta})^k = p(k)$, where $p(k)$ is the partition number of $k \in \mathbb{Z}_{\geq 0}$. We further introduce a $\mathbb{Z}_{\geq 0}$-gradation on $\bigwedge^{\frac{\infty}{2}+0} V_{a,b}^{\sharp}$ as follows:

$$\deg v'_{i_1} \wedge v'_{i_2} \wedge v'_{i_3} \wedge \cdots := \sum_s (i_s + s),$$

and denote the graded subspace of degree k by $(\bigwedge^{\frac{\infty}{2}+0} V_{a,b}^{\sharp})^k$. By definition, we have $\rho_\eta((\mathcal{F}^{\eta})^k) \subset (\bigwedge^{\frac{\infty}{2}+0} V_{a,b}^{\sharp})^k$.

Here, we remark that there exists a bijection

$$\mathcal{I}_0 \simeq \bigcup_{k\in\mathbb{Z}_{\geq 0}} \mathcal{P}_k \quad ((i_1, i_2, \cdots) \mapsto \{i_s + s | i_s \neq -s\})$$

where $\mathcal{P}_k$ signifies the set of the partitions of k. This bijection implies that $\dim\left(\bigwedge^{\frac{\infty}{2}+0} V_{a,b}^{\sharp}\right)^k = p(k)$. Since ρ_η preserves the above $\mathbb{Z}_{\geq 0}$-gradations, we conclude that ρ_η is surjective. □

Finally, we show that the $\mathcal{H}$-isomorphism (8.56) between $\bigwedge^{\frac{\infty}{2}+0} V_{a,b}^{\sharp}$ and $\mathcal{F}_\lambda^{\eta}$ becomes a $\mathfrak{g}$-homomorphism when $(a, b) = (\lambda - \frac{1}{2}, \lambda - \eta - \frac{1}{2})$. As the $\mathfrak{g}$-module structure $(L_n \longmapsto L_n^{\lambda})$ on $\mathcal{F}_\lambda^{\eta}$ is given by (4.3), to show that (8.56) is a homomorphism of $\mathfrak{g}$-modules, it suffices to prove the following lemma:

Lemma 8.21. *For any $n \in \mathbb{Z} \setminus \{0\}$, the action of L_n^{λ} on $(\bigwedge^{\frac{\infty}{2}+0} V_{a,b}^{\sharp})^k$ coincides with the action (8.46).*

Proof. For $n \neq 0$,

$$L_n^\lambda = \frac{1}{2} \sum_{\substack{m \in \mathbb{Z} \\ m \neq 0, n}} a_{n-m} a_m + a_n a_0 - (n+1)\lambda a_n. \tag{8.57}$$

By (8.54), for $I = (i_1, i_2, \cdots) \in \mathcal{I}_0$,

$$a_{n-m} a_m . v'_I = \sum_{\substack{k \neq l \\ i_k - m \neq i_l \\ i_l - (n-m) \neq i_k}} v'_{i_1} \wedge \cdots \wedge \overset{k}{v'_{i_k - m}} \wedge \cdots \wedge \overset{l}{v'_{i_l - n + m}} \wedge \cdots$$
$$+ \sum_{k \in \mathbb{Z}_{>0}} \{P(I; i_k - m) + P(I; i_k - n + m) - 1\} v'_{i_1} \wedge \cdots \wedge \overset{k}{v'_{i_k - n}} \wedge \cdots.$$

For the first term on the right-hand side, by setting $p := i_k - i_l - m + n$, we have $i_k - m \neq i_l \Leftrightarrow p \neq n$, $i_l - (n-m) \neq i_k \Leftrightarrow p \neq 0$, $m \neq 0 \Leftrightarrow i_l - (n-p) \neq i_k$ and $m \neq n \Leftrightarrow i_k - p \neq i_l$. Hence, we have

$$\sum_{\substack{m \in \mathbb{Z} \\ m \neq 0, n}} \sum_{\substack{k \neq l \\ i_k - m \neq i_l \\ i_l - (n-m) \neq i_k}} v'_{i_1} \wedge \cdots \wedge \overset{k}{v'_{i_k - m}} \wedge \cdots \wedge \overset{l}{v'_{i_l - n + m}} \wedge \cdots$$
$$= \sum_{\substack{p \in \mathbb{Z} \\ p \neq 0, n}} \sum_{\substack{k \neq l \\ i_k - p \neq i_l \\ i_l - (n-p) \neq i_k}} v'_{i_1} \wedge \cdots \wedge \overset{k}{v'_{i_l - n + p}} \wedge \cdots \wedge \overset{l}{v'_{i_k - p}} \wedge \cdots,$$

and thus,

$$\sum_{\substack{m \in \mathbb{Z} \\ m \neq 0, n}} \sum_{\substack{k \neq l \\ i_k - m \neq i_l \\ i_l - (n-m) \neq i_k}} v'_{i_1} \wedge \cdots \wedge \overset{k}{v'_{i_k - m}} \wedge \cdots \wedge \overset{l}{v'_{i_l - n + m}} \wedge \cdots = 0.$$

Hence, by (8.57), for $n \neq 0$, we obtain

$$L_n^\lambda . v'_I = \sum_{k \in \mathbb{Z}_{>0}} \left\{ \frac{1}{2} \sum_{\substack{m \in \mathbb{Z} \\ m \neq 0, n}} \{P(I; i_k - m) + P(I; i_k - n + m) - 1\} + \eta - (n+1)\lambda \right\}$$
$$\times v'_{i_1} \wedge \cdots \wedge \overset{k}{v'_{i_k - n}} \wedge \cdots.$$

Here, we notice that

$$v'_{i_1} \wedge \cdots \wedge \overset{k}{v'_{i_k - n}} \wedge \cdots \neq 0 \ \Leftrightarrow \ P(I; i_k - n) = 1.$$

Hence, to show Lemma 8.21, it suffices to see that

Lemma 8.22. *For any $k \in \mathbb{Z}_{>0}$ and $n \in \mathbb{Z} \setminus \{0\}$ such that $P(I; i_k - n) = 1$, we have*

$$\sum_{\substack{m \in \mathbb{Z} \\ m \neq 0,n}} \{P(I; i_k - m) + P(I; i_k - n + m) - 1\} = 2i_k - n + 1. \tag{8.58}$$

Proof. First, we notice that

$$\begin{aligned} &P(I; i_k - m) + P(I; i_k - n + m) - 1 \\ &= \begin{cases} 1 & \text{if } \underset{j \qquad\quad j'}{\begin{array}{|c|c|c|}\hline & \cdots & \\ \hline \end{array}} \\ 0 & \text{if } \underset{j \qquad\quad j'}{\begin{array}{|c|c|c|}\hline \bullet & \cdots & \\ \hline \end{array}} \text{ or } \underset{j \qquad\quad j'}{\begin{array}{|c|c|c|}\hline & \cdots & \bullet \\ \hline \end{array}}, \\ -1 & \text{if } \underset{j \qquad\quad j'}{\begin{array}{|c|c|c|}\hline \bullet & \cdots & \bullet \\ \hline \end{array}} \end{cases} \end{aligned} \tag{8.59}$$

where $j := i_k - m$ and $j' := i_k - n + m$. (Notice that j is not necessarily greater than j'.)

For I, suppose that there exist $i \in \mathbb{Z}$ and $s \in \mathbb{Z}_{>0}$ such that

1. $i_s \neq i_k, i_k - n$,
2. $i < i_s$ and $i \notin I \cup \{i_k - n\}$.

For such i and s, let $\tilde{I} = (\tilde{i}_1, \tilde{i}_2, \cdots)$ be a sequence of integers such that $\tilde{i}_1 > \tilde{i}_2 > \cdots$, $\tilde{i}_l = -l$ for $l \gg 0$ and

$$\tilde{I} = (I \setminus \{i_s\}) \cup \{i\} \tag{8.60}$$

as set. Notice that, since the right-hand side of (8.60) is multiplicity free, $\tilde{I}$ is uniquely determined from i and I. Remark that, by using Maya diagrams, the procedure $I \mapsto \tilde{I}$ is expressed as

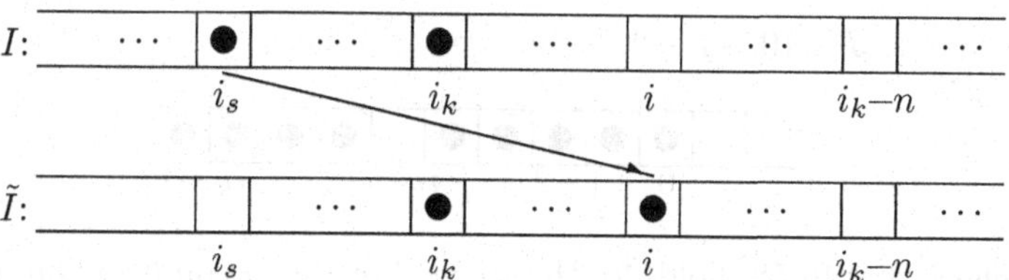

For this procedure $I \mapsto \tilde{I}$, the left-hand side of (8.58) is invariant, i.e.,

$$\begin{aligned} &\sum_{\substack{m \in \mathbb{Z} \\ m \neq 0,n}} \{P(\tilde{I}; i_k - m) + P(\tilde{I}; i_k - n + m) - 1\} \\ &= \sum_{\substack{m \in \mathbb{Z} \\ m \neq 0,n}} \{P(I; i_k - m) + P(I; i_k - n + m) - 1\}. \end{aligned} \tag{8.61}$$

Indeed, this is a direct consequence of formula (8.59).

Let $J = (j_1, j_2, \cdots)$ be the sequence of integers obtained from I by iterating the above procedure as far as possible. By virtue of formula (8.61), it is enough to check (8.22) for such J. The list of possible J's is as follows:

1. For $n > 0$,

   a. $i_k \geq n-1$: $J = (i_k, -2, -3, \cdots)$, i.e.,

   $i_k \quad i_k - n \quad -2\,-3\,-4\cdots$

   b. $0 < i_k < n-1$: $J = (i_k, -1, -2, \cdots, \widehat{i_k - n}, \cdots)$, i.e.,

   $i_k \quad -1\,-2\,-3\cdots \quad i_k - n \quad \cdots$

   c. $i_k \leq 0$: $J = (0, -1, -2, \cdots, i_k, \cdots, \widehat{i_k - n}, \cdots)$, i.e.,

   $0\;-1\,-2\cdots i_k \cdots \quad i_k - n \quad \cdots$

2. For $n < 0$,

   a. $i_k \geq 0$: $J = (i_k, -2, -3, -4, \cdots)$, i.e.,

   $i_k - n \quad i_k \quad -2\,-3\,-4\cdots$

   b. $n-1 < i_k < 0$: $J = (-1, -2, -3, \cdots, i_k, \cdots)$, i.e.,

   $i_k - n \quad -1\,-2\,-3\cdots i_k \cdots$

   c. $i_k \leq n-1$: $J = (0, -1, -2, \cdots, \widehat{i_k - n}, \cdots)$, i.e.,

   $0\;-1\,-2\cdots \quad i_k - n \quad \cdots \; i_k \cdots$

One can check that (8.58) holds for the above J by direct computation. Hence, we have proved Lemma 8.22. □

As stated above, Lemma 8.21 follows from Lemma 8.22. Hence, the isomorphism (8.56) of $\mathcal{H}$-modules turns out to be an isomorphism of Vir-modules. We have completed the proof of Theorem 8.8. □

8.B Appendix: List of the Integral Points on $\ell^{\pm}_{\lambda,\eta}$

Here, we explicitly describe the set $\ell^{\pm}_{\lambda,\eta} \cap (\mathbb{Z}_{>0})^2$ with $(\lambda, \eta) = (\lambda_{p,q}, \eta^{\sigma}_{r,s}(i))$ belonging to **Class** R^+. Since $\ell^{-}_{\lambda,\eta} \cap (\mathbb{Z}_{>0})^2$ is obtained from $\ell^{+}_{\lambda,\eta} \cap (\mathbb{Z}_{>0})^2$ by replacing σ to $-\sigma$, we only consider $\ell^{+}_{\lambda,\eta} \cap (\mathbb{Z}_{>0})^2$.

Let $\{(\alpha_k, \beta_k) | k \in \mathbb{Z}_{>0}\}$ be the set $\ell^{\sigma}_{\lambda,\eta} \cap (\mathbb{Z}_{>0})^2$, where we arrange the points (α_k, β_k) as

$$\alpha_1\beta_1 < \alpha_2\beta_2 < \alpha_3\beta_3 < \cdots .$$

The integral points (α_k, β_k) satisfy $(\alpha_k, \beta_k) = (\alpha_1, \beta_1) + (k-1)(p, q)$, and here we list (α_1, β_1). We further compute $h_i + \alpha_k\beta_k$.

1. **Case** 1^+: $\eta = \eta^{\sigma}_i$ $(i \in \mathbb{Z})$,

   a. $\sigma = +$:

$$(\alpha_1, \beta_1) = \begin{cases} (ip + p - r, q - s) & (i \equiv 0 \bmod 2) \wedge \mathbf{sgn}(i, sp - qr) = 1 \\ (-ip + r, s) & (i \equiv 0 \bmod 2) \wedge \mathbf{sgn}(i, sp - qr) = -1 \\ (ip + p - r, s) & (i \equiv 1 \bmod 2) \wedge i > 0 \\ (-ip + r, q - s) & (i \equiv 1 \bmod 2) \wedge i < 0 \end{cases}.$$

   b. $\sigma = -$:

$$(\alpha_1, \beta_1) = \begin{cases} (r, iq + s) & (i \equiv 0 \bmod 2) \wedge \mathbf{sgn}(i, sp - qr) = 1 \\ (p - r, -iq + q - s) & (i \equiv 0 \bmod 2) \wedge \mathbf{sgn}(i, sp - qr) = -1 \\ (r, iq + q - s) & (i \equiv 1 \bmod 2) \wedge i > 0 \\ (p - r, -iq + s) & (i \equiv 1 \bmod 2) \wedge i < 0 \end{cases}.$$

   In any case, the following holds:

$$h_i + \alpha_k\beta_k = h_{\sigma\{i + \mathbf{sgn}(i, sp - qr)(2k-1)\}}.$$

2. **Case** 2^+: $\eta = \eta^{\sigma}_i$ $(i \in \mathbb{Z}_{\geq 0})$,

   a. $\sigma = +$:

$$(\alpha_1, \beta_1) = \begin{cases} (ip + p, q - s) & i \equiv 0 \bmod 2 \\ (ip + p, s) & i \equiv 1 \bmod 2 \end{cases},$$

$$h_i + \alpha_k\beta_k = h_{i+2k-1}.$$

   b. $\sigma = -$:

$$(\alpha_1, \beta_1) = \begin{cases} (p, (i+1)q + s) & i \equiv 0 \bmod 2 \\ (p, (i+2)q - s) & i \equiv 1 \bmod 2 \end{cases},$$

$$h_i + \alpha_k\beta_k = h_{i+2k}.$$

3. **Case 3^+**: $\eta = \eta^{\sigma}_{(-1)^{i-1}i}$ $(i \in \mathbb{Z}_{\geq 0})$,

   a. $\sigma = +$:

$$(\alpha_1, \beta_1) = \begin{cases} ((i+1)p + r, q) & i \equiv 0 \bmod 2 \\ ((i+2)p - r, q) & i \equiv 1 \bmod 2 \end{cases},$$

$$h_{(-1)^{i-1}i} + \alpha_k \beta_k = h_{(-1)^{i+2k-1}(i+2k)}.$$

   b. $\sigma = -$:

$$(\alpha_1, \beta_1) = \begin{cases} (p - r, (i+1)q) & i \equiv 0 \bmod 2 \\ (r, (i+1)q) & i \equiv 1 \bmod 2 \end{cases},$$

$$h_{(-1)^{i-1}i} + \alpha_k \beta_k = h_{(-1)^{i+2k}(i+2k-1)}.$$

4. **Case 4^+**: $\eta = \eta^{\sigma}_{2i}$ $(i \in \mathbb{Z}_{\geq 0})$,

   a. $(r, s) = (0, 0)$ and $\sigma = +$: $(\alpha_1, \beta_1) = ((2i+1)p, q)$.
   b. $(r, s) = (0, 0)$ and $\sigma = -$: $(\alpha_1, \beta_1) = (p, (2i+1)q)$.
   c. $(r, s) = (0, q)$ and $\sigma = +$: $(\alpha_1, \beta_1) = ((2i+2)p, q)$.
   d. $(r, s) = (0, q)$ and $\sigma = -$: $(\alpha_1, \beta_1) = (p, (2i+2)q)$.

   In any case, the following holds:

$$h_{2i} + \alpha_k \beta_k = h_{2(i+k)}.$$

Chapter 9
Rational Vertex Operator Algebras

In this chapter, we will consider a vertex algebra structure on a highest weight Vir-module V_c with highest weight $(c, 0) \in \mathbb{C}^2$. We will prove that the irreducible quotient $L(c, 0)$ of V_c is rational if and only if it is a BPZ series. (Notice that one should exclude the case $c = 0$, since $L(c, 0) = \mathbb{C}$ in this case and it does not contain a conformal vector.) We will also compute the Fusion rule for the rational cases. It will be shown that $L(c, 0)$ is C_2-cofinite for a BPZ series and a geometric meaning of the C_2-cofiniteness will also be explained.

In Section 9.1, we will introduce a vertex algebra structure on V_c. In Section 9.2, we will compute the Zhu algebra of V_c. As a consequence, it will be shown that V_c cannot be rational. In Section 9.3, we will prove that $L(c, 0)$ for BPZ series is rational. We will also compute the Fusion rule in this case. Section 9.4 is devoted to a geometric characterisation of the BPZ series, namely, we will show that the irreducible highest weight Vir-module $L(c, h)$ with highest weight $(c, h) \in \mathbb{C}^2$ is a BPZ series if and only if its associated variety is a point. For the reader's convenience, we will recall the definition and some basic facts about associated varieties in the appendix to this chapter. A Tauberian theorem will be also recalled with its simple but beautiful application.

9.1 Vertex Operator Algebra Structure

In this section, we introduce a vertex algebra structure on a highest weight Vir-module V_c with highest weight $(c, 0) \in \mathbb{C}^2$. We also summarise some properties of V_c for later use.

For $c \in \mathbb{C}$, let $\mathbb{C}_c = \mathbb{C}\mathbf{1}_c$ be the $\mathrm{Vir}^{\geq -1} := \bigoplus_{n \geq -1} \mathbb{C}L_n \oplus \mathbb{C}C$-module defined by

$$L_n.\mathbf{1}_c := 0, \qquad C.\mathbf{1}_c := c\mathbf{1}_c.$$

The Vir-module

K. Iohara, Y. Koga, *Representation Theory of the Virasoro Algebra*, Springer Monographs in Mathematics, DOI 10.1007/978-0-85729-160-8_9,

$$V_c := \mathrm{Ind}_{\mathrm{Vir}^{\geq -1}}^{\mathrm{Vir}} \mathbb{C}_c$$

is a highest weight module with highest weight $(c, 0)$. Hence, there is a surjective Vir-module map $M(c,0) \twoheadrightarrow V_c$. By definition, it follows that the vector $L_{-1}v_{c,0} \in M(c,0)$, where $v_{c,0}$ is a highest weight vector of $M(c,0)$, lies in the kernel of this map. By the universality of V_c, this implies

Lemma 9.1. *One has the isomorphism*

$$V_c \cong M(c,0)/U(\mathrm{Vir}^-)L_{-1}v_{c,0}.$$

We call V_c the **vacuum representation with central charge** c. Let $\pi : V_c \twoheadrightarrow L(c,0)$ be the canonical projection. By Chapter 6, one has the following structure theorem:

Theorem 9.1 *1.* $\mathrm{Ker}\pi \neq \{0\}$ *if and only if there exist coprime integers* $p, q \in \mathbb{Z}_{>1}$ *such that* $c = c_{p,q} := 1 - 6\dfrac{(p-q)^2}{pq}$.
2. In the case 1., $\mathrm{Ker}\pi$ *is generated by a singular vector* $v_{p-1,q-1}$ *of* V_c *which is unique up to a scalar. It follows that* $L_0.v_{p-1,q-1} = (p-1)(q-1)v_{p-1,q-1}$.

Now, we introduce a $\mathbb{Z}$-graded vertex algebra structure on V_c. For simplicity, we set $v_c := 1 \otimes \mathbf{1}_c$.

1. (**Gradation**) For $j_1 \leq j_2 \leq \cdots \leq j_k < -1$, set

$$\deg(L_{j_1}L_{j_2}\cdots L_{j_k}.v_c) := -\sum_{i=1}^{k} j_i.$$

2. (**Vacuum vector**) Set $|0\rangle := v_c$.
3. (**Translation operator**) Set $T := L_{-1}$.
4. (**Vertex operators**) Set $Y(|0\rangle, z) := \mathrm{id}$ and

$$Y(L_{-2}.v_c, z) := T(z) := \sum_{n\in\mathbb{Z}} L_n z^{-n-2},$$

and in general, set

$$Y(L_{j_1}L_{j_2}\cdots L_{j_k}.v_c, z) := {}^{\circ}_{\circ}\partial_z^{(-j_1-2)}T(z)\cdots\partial_z^{(-j_k-2)}T(z){}^{\circ}_{\circ}.$$

Here, we set $\partial_z^{(n)} := \dfrac{1}{n!}\partial_z^n$ for $n \in \mathbb{Z}_{\geq 0}$.

By the strong reconstruction theorem (cf. Theorem C.1), these data define a vertex algebra structure on V_c. Moreover, setting $\omega := L_{-2}.v_c$, one has

Proposition 9.1 (V_c, ω) *is a vertex operator algebra.*

By Theorem 9.1, one has the following corollary:

Corollary 9.1 *If $c \neq 0$, then $L(c,0)$ has a vertex operator algebra structure which is induced from that of (V_c, ω).*

9.2 The Zhu Algebra of V_c

In this section, we will compute the Zhu algebra of V_c explicitly. As a consequence, we will see that V_c cannot be a rational vertex operator algebra.

9.2.1 Preliminary

Here, we recall three formulae which will be used to determine the Zhu algebra $A(V_c)$. Notice that these formulae are also used in the proof of Theorem C.6. We assume that (V, ω) is a vertex operator algebra. For notation, see Appendix C.

Lemma 9.2. *For any $a \in V$, one has $L_{-1}a + L_0 a \in O(V)$.*

Proof. We may assume that $a \in V$ is homogeneous. By definition, one has

$$L_0 a = (\deg a)a, \qquad a_{(-2)}|0\rangle = [T, a_{(-1)}]|0\rangle = Ta,$$

which implies

$$L_{-1}a + L_0 a = \operatorname{Res}_z \left(Y(a,z) \frac{(1+z)^{\deg a}}{z^2} |0\rangle \right) \in O(V). \qquad \square$$

Lemma 9.3. *For any homogeneous $a \in V$ and integers $m \geq n \geq 0$, one has*

$$\operatorname{Res}_z \left(Y(a,z) \frac{(1+z)^{\deg a + n}}{z^{2+m}} b \right) \in O(V).$$

Proof. By the equality

$$\frac{(1+z)^{\deg a + n}}{z^{2+m}} = \sum_{i=0}^{n} \binom{n}{i} \frac{(1+z)^{\deg a}}{z^{2+m-i}},$$

it suffices to prove this lemma in the cases $n = 0$ and $m \geq 0$. For $m = 0$, this is just the definition of $O(V)$. Suppose that this lemma holds up to $m - 1$. By the induction hypothesis, one has

$$\operatorname{Res}_z \left(Y(L_{-1}a, z) \frac{(1+z)^{\deg a + 1}}{z^{m+1}} b \right) \in O(V).$$

On the other hand, one has

$$\begin{aligned}
&\mathrm{Res}_z\left(Y(L_{-1}a,z)\frac{(1+z)^{\deg a+1}}{z^{m+1}}b\right)\\
=&\mathrm{Res}_z\left(\frac{d}{dz}Y(a,z)\frac{(1+z)^{\deg a+1}}{z^{m+1}}b\right)=-\mathrm{Res}_z\left(Y(a,z)\frac{d}{dz}\frac{(1+z)^{\deg a+1}}{z^{m+1}}b\right)\\
=&-(\deg a-m)\mathrm{Res}_z\left(Y(a,z)\frac{(1+z)^{\deg a}}{z^{m+1}}b\right)+(m+1)\mathrm{Res}_z\left(Y(a,z)\frac{(1+z)^{\deg a}}{z^{m+2}}b\right).
\end{aligned}$$

The first term in the right-hand side is an element of $O(V)$ by the induction hypothesis, hence, so is the second term. □

Lemma 9.4. *For homogeneous $a, b \in V$, one has the following:*

$$a*b \equiv \mathrm{Res}_z\left(Y(b,z)\frac{(1+z)^{\deg b-1}}{z}a\right) \mod O(V), \tag{9.1}$$

$$a*b-b*a \equiv \mathrm{Res}_z(Y(a,z)(1+z)^{\deg a-1}b) \mod O(V). \tag{9.2}$$

Proof. We first show the equality

$$Y(a,z)b=(1+z)^{-\deg a-\deg b}Y\left(b,\frac{-z}{1+z}\right)a \mod O(V). \tag{9.3}$$

By the skew-symmetry (Proposition C.1) and Lemma C.1. i), one has

$$Y(a,z)b=e^{zT}\sum_{i\in\mathbb{Z}}b_{(i)}a(-z)^{-i-1}=\sum_{i\in\mathbb{Z}}(-z)^{-i-1}\sum_{n\geq 0}(b_{(i)}a)_{(-n-1)}|0\rangle z^n.$$

By Lemma 9.2 and Lemma C.1. i), it follows that

$$\begin{aligned}
(b_{(i)}a)_{(-n-1)}|0\rangle =&\frac{1}{n}T(b_{(i)}a)_{(-n)}|0\rangle\\
\equiv&-\frac{1}{n}(\deg a+\deg b+n-i-2)(b_{(i)}a)_{(-n)}|0\rangle\\
\equiv&\binom{-\deg a-\deg b+i+1}{n}b_{(i)}a \qquad \mod O(V).
\end{aligned}$$

Hence, one has

$$\begin{aligned}
Y(a,z)b\equiv&\sum_{i\in\mathbb{Z}}\left\{\sum_{n\geq 0}\binom{-\deg a-\deg b+i+1}{n}z^n\right\}b_{(i)}a(-z)^{-i-1}\\
=&\sum_{i\in\mathbb{Z}}b_{(i)}a(1+z)^{-\deg a-\deg b+i+1}(-z)^{-i-1}\\
=&(1+z)^{-\deg a-\deg b}Y\left(b,\frac{-z}{1+z}\right)a,
\end{aligned}$$

and (9.3) is proved. Now, by (9.3), it follows that

$$\begin{aligned} a * b =& \mathrm{Res}_z \left(Y(a,z) \frac{(1+z)^{\deg a}}{z} b \right) \\ \equiv& \mathrm{Res}_z \left((1+z)^{-\deg a - \deg b} Y\left(b, \frac{-z}{1+z}\right) a \frac{(1+z)^{\deg a}}{z} \right) \\ =& \mathrm{Res}_w \left(Y(b,w) \frac{(1+w)^{\deg b - 1}}{w} a \right). \end{aligned}$$

Here, in the last line, we have made the change of variable $w := -\dfrac{z}{1+z}$. Thus, (9.1) is proved. Let us show (9.2). By definition and (9.1), one has

$$\begin{aligned} a * b - b * a \equiv& \mathrm{Res}_z \left(Y(a,z) \frac{(1+z)^{\deg a}}{z} b \right) - \mathrm{Res}_z \left(Y(a,z) \frac{(1+z)^{\deg a - 1}}{z} b \right) \\ =& \mathrm{Res}_z \left(Y(a,z)(1+z)^{\deg a - 1} b \right), \end{aligned}$$

which implies the result. □

9.2.2 $A(V_c)$

The main result in this subsection is

Proposition 9.2 ([FZ]) *There exists an isomorphism of associative algebras:*

$$A(V_c) \cong \mathbb{C}[x]; \qquad [\omega]^n \longmapsto x^n \quad (n \in \mathbb{Z}_{\geq 0}).$$

Proof. By Lemma 9.3, it follows that, for $n \in \mathbb{Z}_{\geq 0}$, one has

$$(L_{-n-3} + 2L_{-n-2} + L_{-n-1})b = \mathrm{Res}_z \left(Y(\omega, z) \frac{(1+z)^2}{z^{2+n}} b \right) \in O(V_c). \quad (9.4)$$

Hence, for $n \geq 2$, one has

$$(L_{-n}+L_{-n+1})b \equiv -(L_{-n+1}+L_{-n+2})b \equiv \cdots \equiv (-1)^{n-2}(L_{-2}+L_{-1})b \mod O(V_c),$$

which implies

$$\begin{aligned} & L_{-n}b \\ =& (L_{-n} + L_{-n+1})b - (L_{-n+1} + L_{-n+2})b + \cdots + (-1)^{n-2}(L_{-2} + L_{-1})b \\ & + (-1)^{n-1} L_{-1} b \\ \equiv& (-1)^n (n-1)(L_{-2} + L_{-1})b + (-1)^{n-1}(L_{-1} + L_0)b + (-1)^n L_0 b \mod O(V_c) \\ \equiv& (-1)^n \{(n-1)(L_{-2} + L_{-1}) + L_0\} b \mod O(V_c). \end{aligned}$$

By (9.1) in Lemma 9.4, one has

$$[b] * [\omega] = \left[\mathrm{Res}_z\left(Y(\omega, z)\frac{1+z}{z}b\right)\right] = [(L_{-2} + L_{-1})b] \in A(V_c).$$

Taking the following formula into account

$$L_0 L_{-j_1} L_{-j_2} \cdots L_{-j_k}.v_c = (\sum_{i=1}^{k} j_i) L_{-j_1} L_{-j_2} \cdots L_{-j_k}.v_c,$$

one obtains

$$[L_{-n}b] = (-1)^n[\{(n-1)(L_{-2}+L_{-1})+L_0\}b] = (-1)^n\{(n-1)[b]*[\omega]+(\deg b)[b]\}$$

for $n \in \mathbb{Z}_{>1}$ and homogeneous $b \in V$. Thus, by induction, one can show that there exists a polynomial $P \in \mathbb{C}[x]$ satisfying

$$[L_{-j_1} L_{-j_2} \cdots L_{-j_k}.v_c] = P([\omega]).$$

Since V_c is spanned by the elements of the form

$$L_{-j_1} L_{-j_2} \cdots L_{-j_k}.v_c \qquad j_1 \geq j_2 \geq \cdots \geq j_k > 1,$$

it follows that the map of associative algebras

$$F : \mathbb{C}[x] \longrightarrow A(V_c); \qquad x \longmapsto [\omega]$$

is surjective.

To prove that the map F is injective, it suffices to prove that $O(V_c)$ is spanned by the elements of the form (9.4). Indeed, this implies there is no non-trivial relation for $[\omega] \in A(V_c)$.

Let $V_c = \bigoplus_{N \in \mathbb{Z}_{\geq 0}} (V_c)_N$ be the $\mathbb{Z}_{\geq 0}$-graded decomposition. We show

$$\mathrm{Res}_w \left[Y(a', w)\frac{(1+w)^N}{w^k}b\right] \in O'(V_c) \qquad \forall\, k \geq 2 \tag{9.5}$$

for $a' \in (V_c)_N$ by induction on N. Here, we set

$$O'(V_c) := \{(L_{-n-3} + 2L_{-n-2} + L_{-n-1})b |\ n \geq 0,\ b \in V_c\}.$$

For $N = 0, 1$, the left-hand side of (9.5) is 0 and there is nothing to prove. Assume that (9.5) holds up to $N-1$. For N, it is sufficient to prove (9.5) in the case $a' = L_{-n}a$ where $a \in (V_c)_{N-n}$ and $n \geq 2$ since V_c is spanned by the vectors of the form $L_{-j_1} L_{-j_2} \cdots L_{-j_k}.v_c$ $(j_1 \geq j_2 \geq \cdots \geq j_k > 1)$.

Hence, we consider the elements

$$\begin{aligned}
T :=&\mathrm{Res}_w \left[Y(L_{-n}a, w) \frac{(1+w)^{\deg a+n}}{w^k} b \right] \\
=&\mathrm{Res}_{z-w}\mathrm{Res}_w \left[Y(Y(\omega, z-w)a, w)(z-w)^{-n+1} \frac{(1+w)^{\deg a+n}}{w^k} b \right] \\
=&T_1 - T_2, \\
T_1 :=&\mathrm{Res}_z\mathrm{Res}_w \left[Y(\omega, z)Y(a, w)\iota_{z,w}(z-w)^{-n+1} \frac{(1+w)^{\deg a+n}}{w^k} b \right], \\
T_2 :=&\mathrm{Res}_z\mathrm{Res}_w \left[Y(a, w)Y(\omega, z)\iota_{w,z}(z-w)^{-n+1} \frac{(1+w)^{\deg a+n}}{w^k} b \right].
\end{aligned}$$

(Here, we have used the Jacobi identity (cf. Theorem C.4).) Since one has

$$\begin{aligned}
\iota_{w,z}(z-w)^{-n+1} =& \sum_{i\geq 0} (-1)^{-n+1-i} \binom{1-n}{i} z^i w^{-n+1-i} \\
=& \sum_{i>0} (-1)^{-n+1-i} \binom{1-n}{i} z^i w^{-n+1-i} + (-1)^{-n+1} w^{-n+1}
\end{aligned}$$

and $(1+w)^n w^{-n+1-i} \in \mathbb{C}[w^{-1}]$ for $i > 0$, it follows that

$$\mathrm{Res}_z\mathrm{Res}_w \left[Y(a, w)Y(\omega, z) \sum_{i>0} (-1)^{-n+1-i} \binom{1-n}{i} z^i \frac{w^{-n+1-i}(1+w)^{\deg a+n}}{w^k} b \right]$$

is an element of $O'(V_c)$ by the induction hypothesis. Hence, we obtain

$$\begin{aligned}
T_2 \equiv&\mathrm{Res}_z\mathrm{Res}_w \left[Y(a, w)Y(\omega, z)(-1)^{-n+1} \frac{(1+w)^{\deg a+n}}{w^{n-1+k}} b \right] \mod O'(V_c) \\
=&\mathrm{Res}_w \left[Y(a, w)(-1)^{-n+1} \frac{(1+w)^{\deg a+n}}{w^{n-1+k}} L_{-1}b \right].
\end{aligned}$$

By the identity

$$\frac{(1+w)^n}{w^{n-1}} \cdot \frac{(1+w)^{\deg a}}{w^k} = \frac{w(1+w)^{\deg a}}{w^k} + \sum_{i=1}^{n} \binom{n}{i} \frac{(1+w)^{\deg a}}{w^{k+i-1}},$$

the induction hypothesis and Lemma C.1. ii), we obtain

$$\begin{aligned}
T_2 \equiv & \mathrm{Res}_w \left[Y(a,w)(-1)^{-n+1} \frac{w(1+w)^{\deg a}}{w^k} L_{-1} b \right] \\
\equiv & \mathrm{Res}_w \left[Y(a,w)(-1)^{-n+1} \frac{(1+w)^{\deg a+1}}{w^k} L_{-1} b \right] \mod O'(V_c) \\
= & T_{21} - T_{22}, \\
T_{21} := & L_{-1}.\mathrm{Res}_w \left[Y(a,w)(-1)^{-n+1} \frac{(1+w)^{\deg a+1}}{w^k} b \right], \\
T_{22} := & \mathrm{Res}_w \left[Y(L_{-1}a,w)(-1)^{-n+1} \frac{(1+w)^{\deg a+1}}{w^k} b \right] \in O'(V_c).
\end{aligned}$$

Now, by definition, one has

$$\begin{aligned}
T_1 = & \sum_{i \geq 0} (-1)^i \binom{1-n}{i} \mathrm{Res}_z \mathrm{Res}_w \left[Y(\omega, z) Y(a,w) z^{-n+1-i} \frac{w^i (1+w)^{\deg a+n}}{w^k} b \right] \\
= & \sum_{i \geq 0} (-1)^i \binom{1-n}{i} \mathrm{Res}_w \left[L_{-n-i} Y(a,w) \frac{w^i (1+w)^{\deg a+n}}{w^k} b \right].
\end{aligned}$$

Since we have $L_{-M} b' \equiv (-1)^M \{(M-1)(L_{-2} + L_{-1}) - L_{-1}\} b' \mod O'(V_c)$ for $M \geq 2$ as we have seen before, we obtain

$$\begin{aligned}
T_1 \equiv & (-1)^n \sum_{i \geq 0} \binom{1-n}{i} \\
& \times \mathrm{Res}_w \left[\{(n+i-1)(L_{-2}+L_{-1}) - L_{-1}\} Y(a,w) \frac{w^i (1+w)^{\deg a+n}}{w^k} b \right] \\
= & T_{11} - T_{12}, \\
T_{11} := & (-1)^n (n-1)(L_{-2} + L_{-1}) \mathrm{Res}_w \left[Y(a,w) \frac{(1+w)^{\deg a}}{w^k} b \right], \\
T_{12} := & (-1)^n L_{-1} \mathrm{Res}_w \left[Y(a,w) \frac{(1+w)^{\deg a+1}}{w^k} b \right].
\end{aligned}$$

Here, we have used the identity

$$\sum_{i \geq 0} \binom{1-n}{i} (n+i-1) w^i = (n-1)(1+w)^{-n}.$$

Thus, we see that $T \equiv T_{11} - T_{12} - T_{21} = T_{11} \mod O'(V_c)$. By the induction hypothesis, one has

$$\mathrm{Res}_w \left[Y(a,w) \frac{(1+w)^{\deg a}}{w^k} b \right] \in O'(V_c).$$

Hence by definition, it is enough to prove that $(L_{-2}+L_{-1})v \in O'(V_c)$ for $v=(L_{-N-3}+2L_{-N-2}+L_{-N-1})b' \in O'(V_c)$ $(N \geq 0)$. By direct calculation, it follows that

$$\begin{aligned}&(L_{-2}+L_{-1})v\\ =&\{(N+1)(L_{-N-5}+2L_{-N-4}+L_{-N-3})+N(L_{-N-4}+2L_{-N-3}+L_{-N-2})\\ &+(L_{-N-3}+2L_{-N-2}+L_{-N-1})(L_{-2}+L_{-1})\}b' \in O'(V_c),\end{aligned}$$

which implies $T_{11} \equiv 0$, namely, $T \equiv 0 \mod O'(V_c)$. □

Remark 9.1 *From the above proof, it follows that there exists an isomorphism*

$$A(V_c) \cong H_0(\mathcal{L}, V_c) := V_c/\mathcal{L}.V_c,$$

where we set $\mathcal{L} := \bigoplus_{n \geq 0} \mathbb{C}(L_{-n-1}+2L_{-n-2}+L_{-n-3}) \subset \mathrm{Vir}^-$.

By Proposition 9.2, Theorem C.7 and Theorem 9.1, one has the following corollary:

Corollary 9.2 *If c is not a BPZ series, then the vertex operator algebra $L(c,0) \cong V_c$ is not rational.*

9.3 Rationality and the Fusion Algebra of BPZ Series

In this section, we will show that the vertex operator algebra $L(c,0)$ is rational for a BPZ series. Moreover, we will compute the fusion algebra and will show that it is isomorphic to the so-called Verlinde algebra. Hence, in this section, we assume that c is a BPZ series and $c \neq 0$, namely, there exists $p,q \in \mathbb{Z}_{>1}$ such that

$$c = c_{p,q} = 1 - 6\frac{(p-q)^2}{pq}$$

and $(p,q) \neq (2,3),(3,2)$.

9.3.1 Coinvariants I

In this subsection, we will compute some spaces of coinvariants which are related to the Zhu algebra $A(L(c,0))$ and their bimodules.

Recall that an $L(c,0)$-module is a Vir-module and any subquotient of an $L(c,0)$-module as Vir-module is again an $L(c,0)$-module. In particular, it shows that any irreducible $L(c,0)$-module is of the form $L(c,h)$ for some $h \in \mathbb{C}$. Indeed, it will be shown in the next subsection that $L(c,h)$ is a simple $L(c,0)$-module if and only if there exist integers $0<r<p, 0<s<q$ such that

$$h = h_{r,s} := \frac{(rq - sp)^2 - (p-q)^2}{4pq}.$$

Here, we identify the $A(L(c,0))$-bimodule $A(L(c,h_{r,s}))$ with a space of coinvariants (cf. Theorem C.9):

Lemma 9.5. *There is an isomorphism*

$$A(L(c,h_{r,s})) \cong H_0(\mathcal{L}, L(c,h_{r,s})).$$

Proof. The proof of this lemma can be carried out in a way similar to the proof of injectivity of the map F in the proof of Proposition 9.2. The only necessary modification is that one has to show

$$\mathrm{Res}_w \left[Y(a',w)\frac{(1+w)^{\deg a'}}{w^k} b \right] \in \mathcal{L}.L(c,h) \qquad \forall\, k \geq 2,$$

also for $a' = L_{-1}a$. By the induction hypothesis, it follows that

$$\begin{aligned}
&\mathrm{Res}_w \left[Y(L_{-1}a,w)\frac{(1+w)^{\deg a+1}}{w^k} b \right] = \mathrm{Res}_w \left[\frac{d}{dw} Y(a,w)\frac{(1+w)^{\deg a+1}}{w^k} b \right] \\
&= -\,\mathrm{Res}_w \left[Y(a,w)\frac{d}{dw}\frac{(1+w)^{\deg a+1}}{w^k} b \right] \\
&=(k - \deg a - 1)\mathrm{Res}_w \left[Y(a,w)\frac{(1+w)^{\deg a}}{w^k} b \right] + k\mathrm{Res}_w \left[Y(a,w)\frac{(1+w)^{\deg a}}{w^{k+1}} b \right] \\
&\in \mathcal{L}.L(c,h).
\end{aligned}$$

□

Hence, it suffices to compute $H_0(\mathcal{L}, L(c,h_{r,s}))^* \cong H^0(\mathcal{L}, L(c,h_{r,s})^{*a})$. (In general, for a module V over a Lie algebra $\mathfrak{g}$, we define a $\mathfrak{g}$-module structure on the full-dual $V^* = \mathrm{Hom}_{\mathbb{C}}(V,\mathbb{C})$ by $(X.f)(v) := -f(X.v)$ for $X \in \mathfrak{g}$, $v \in V$ and $f \in V^*$ and denote this $\mathfrak{g}$-module by V^{*a}, cf. § 1.2.7.)

The dual of the surjection $M(c,h_{r,s}) \twoheadrightarrow L(c,h_{r,s})$ induces an injection

$$H^0(\mathcal{L}, L(c,h_{r,s})^{*a}) \hookrightarrow H^0(\mathcal{L}, M(c,h_{r,s})^{*a}), \tag{9.6}$$

and it follows from Theorem 6.3 that its image can be described by

$$\{F \in H^0(\mathcal{L}, M(c,h_{r,s})^{*a}) | F(v_{r,s}) = 0, \quad F(v_{p-r,q-s}) = 0\}, \tag{9.7}$$

where $v_{r,s} \in M(c,h_{r,s})^{\mathrm{Vir}^+}_{h_{r,s}+rs} \setminus \{0\}$ and $v_{p-r,q-s} \in M(c,h_{r,s})^{\mathrm{Vir}^+}_{h_{r,s}+(p-r)(q-s)} \setminus \{0\}$ are non-zero vectors. Denoting $v_{r,s} = S_{r,s}.v_{c,h_{r,s}}$ and $v_{p-r,q-s} = S_{p-r,q-s}.v_{c,h_{r,s}}$, where we set $v_{c,h_{r,s}} := 1 \otimes \mathbf{1}_{c,h_{r,s}}$, we fix the normalisation of $S_{r,s}$ and $S_{p-r,q-s}$ as in § 5.4.

We will study the eigenvalues of the action of $[\omega]$ on $A(L(c,h_{r,s}))$ in detail. By definition, one has

$$\omega * v = (L_{-2} + 2L_{-1} + L_0)v, \qquad v * \omega = (L_{-2} + L_{-1})v,$$

for $v \in L(c, h_{r,s})$, hence we set

$$L_0' := L_{-2} + 2L_{-1} + L_0, \qquad L_0'' := L_{-2} + L_{-1}.$$

Let us compute the action of L_0' and L_0'' on $H^0(\mathcal{L}, M(c, h_{r,s})^{*a})$. For $\alpha, \beta \in \mathbb{C}$, we denote the simultaneous eigenspace of the action of L_0' and L_0'' with the eigenvalues α (resp. β) by $H^0(\mathcal{L}, M(c, h_{r,s})^{*a})^{(\alpha,\beta)}$. For a non-zero $F \in H^0(\mathcal{L}, M(c, h_{r,s})^{*a})^{(\alpha,\beta)}$, we set $F_j := F|_{M(c,h_{r,s})_{h_{r,s}+j}}$ where $M(c, h_{r,s}) = \bigoplus_{j \in \mathbb{Z}_{\geq 0}} M(c, h_{r,s})_{h_{r,s}+j}$ is the eigenspace decomposition with respect to the action of L_0. By definition, one has

$$L_0'.F = \alpha F, \qquad L_0''.F = \beta F, \qquad (L_{-n-1} + 2L_{-n-2} + L_{-n-3}).F = 0 \quad n \geq 0.$$

By restricting these equations to the weight subspaces of $M(c, h_{r,s})$ and solving the recurrence relations, one obtains

$$L_{-n}.F_j = (-1)^n \{-j - \alpha - h_{r,s} + n(\beta + 1)\} F_{j-n} \qquad n > 0, \ j \geq 0,$$

where we set $F_j := 0$ for $j \in \mathbb{Z}_{<0}$. Hence, one has a projection $V_{-\beta-1,-\alpha-h_{r,s}+N} \twoheadrightarrow \bigoplus_{j \geq 0} \mathbb{C}F_j$ of Vir^--modules for any $N \in \mathbb{Z}$. (See § 1.2.6 for the definition of the module in the left-hand side.) In particular, one sees that $\dim H^0(\mathcal{L}, M(c, h_{r,s})^{*a})^{(\alpha,\beta)} = 1$.

It follows from (9.7) that $F \in H^0(\mathcal{L}, M(c, h_{r,s})^{*a})^{(\alpha,\beta)}$ lies in the image of the map (9.6) if and only if F satisfies the equations

$$a(S_{r,s}).F_{rs} = 0, \qquad a(S_{p-r,q-s}).F_{(p-r)(q-s)} = 0. \tag{9.8}$$

Hence, we will consider a variant of Proposition 5.6 below.

For $t \in \mathbb{C}^*$ and $\alpha, \beta \in \mathbb{Z}_{>0}$, it follows from Corollary 5.2 that there is a non-zero vector $S_{\alpha,\beta}(t).v_{c(t),h_{\alpha,\beta}(t)} \in M(c(t), h_{\alpha,\beta}(t))^{\mathrm{Vir}^+}_{h_{\alpha,\beta}(t)+\alpha\beta}$. We normalise $S_{\alpha,\beta}(t) \in U(\mathrm{Vir}^-)$ as in § 5.4.

Proposition 9.3 *For $a, b \in \mathbb{C}$, we let $V_{a,b} = \bigoplus_{n \in \mathbb{Z}} \mathbb{C}v_n$ be the* Vir*-module defined in § 1.2.6 . One has*

$$a(S_{\alpha,\beta}(t)).v_0 = P_{\alpha,\beta}(-(a+1), -b - \alpha\beta; t).v_{-\alpha\beta},$$

where $P_{\alpha,\beta}(-(a+1), -b - \alpha\beta; t) \in \mathbb{C}[t^{\pm 1}]$ is a Laurent polynomial defined in Proposition 5.6.

Proof. For $i_1, i_2, \cdots, i_k \in \mathbb{Z}_{<0}$ such that $\sum_{j=1}^k i_j = -\alpha\beta$, we set

$$R(a, b)v_{-\alpha\beta} := L_{i_1} L_{i_2} \cdots L_{i_k}.v_0.$$

By definition, one has

$$R(a,b) = (ai_k + b)(ai_{k-1} + b - i_k) \cdots (ai_1 + b - (i_2 + \cdots + i_k)).$$

On the other hand, one also has

$$\begin{aligned} a(L_{i_1} L_{i_2} \cdots L_{i_k}).v_0 &= (-1)^k L_{i_k} \cdots L_{i_2} L_{i_1}.v_0 \\ &= (-1)^k (ai_1 + b)(ai_2 + b - i_1) \cdots (ai_k + b - (i_1 + \cdots + i_{k-1}))v_{-\alpha\beta}, \end{aligned}$$

and it can be checked that

$$(-1)^k (ai_1+b)(ai_2+b-i_1) \cdots (ai_k+b-(i_1+\cdots+i_{k-1})) = R(-(a+1), -b-\alpha\beta).$$

Hence, the result follows from Proposition 5.6. □

9.3.2 Rationality of $L(c,0)$

In this subsection, we will determine the Zhu algebra $A(L(c,0))$ explicitly and will show that $L(c,0)$ is rational. Moreover, we will show that $L(c,0)$ is C_2-cofinite.

By Theorem 9.1, there is a singular vector $v_{p-1,q-1} \in V_c$ such that $L_0.v_{p-1,q-1} = (p-1)(q-1)v_{p-1,q-1}$ and $v_{p-1,q-1}$ generates the maximal proper submodule of V_c. By definition, V_c is spanned by the elements of the form $L_{i_1}^{m_1} L_{i_2}^{m_2} \cdots L_{i_k}^{m_k}.v_c$, where $i_1 < i_2 < \cdots < i_k \leq -2$ and $m_1, m_2, \cdots, m_k \in \mathbb{Z}_{>0}$. Writing $v_{p-1,q-1}$ as a linear combination of these elements, we have

Lemma 9.6. *The coefficient of* $L_{-2}^{\frac{1}{2}(p-1)(q-1)}$ *of* $v_{p-1,q-1} \in V_c$ *is not zero.*

Proof. It follows from Chapter 6 that there is a singular vector $S_{p-1,q-1}.v_{c,0} \in M(c,0)$ such that $L_0.S_{p-1,q-1}.v_{c,0} = (p-1)(q-1)v_{p-1,q-1}$. We normalise $S_{p-1,q-1} \in U(\mathrm{Vir}^-)$ as in § 5.4. Let

$$\pi : U(\mathrm{Vir}^-) \twoheadrightarrow U(\mathrm{Vir}^-/[\mathrm{Vir}^-, \mathrm{Vir}^-])$$

be the canonical projection. By Proposition 5.7, we see that

$$\pi(S_{p-1,q-1})^2 = \prod_{k=0}^{p-2} \prod_{l=0}^{q-2} \left[L_{-1}^2 - \frac{4}{pq}\{(k+1)q - (l+1)p\}^2 L_{-2} \right].$$

Since p and q are coprime integers, the coefficient of $L_{-2}^{\frac{1}{2}(p-1)(q-1)}$ is

$$\left(-\frac{4}{pq}\right)^{\frac{(p-1)(q-1)}{2}} \prod_{k=0}^{p-2} \prod_{l=0}^{q-2} ((k+1)q - (l+1)p) \neq 0.$$

Since $v_{p-1,q-1} \in V_c$ is a scalar multiple of the image of $S_{p-1,q-1}v_{c,0}$ under the canonical projection $M(c,0) \twoheadrightarrow V_c$ (cf. Lemma 9.1), we obtained the result. □

Now, we fix the normalisation of $v_{p-1,q-1}$ so that the coefficient of $L_{-2}^{\frac{1}{2}(p-1)(q-1)}$ is 1. By Proposition 9.2, there exists $G_{p,q}(x) \in \mathbb{C}[x]$ satisfying $v_{p-1,q-1} \equiv G_{p,q}([\omega]) \mod O(V_c)$. Hence, by Proposition C.3, the following statement is a corollary to Lemma 9.6:

Lemma 9.7. *One has* $\deg G_{p,q} = \frac{1}{2}(p-1)(q-1)$ *and* $A(L(c,0)) \cong \mathbb{C}[x]/(G_{p,q}(x))$.

Proof. For each monomial of V_c of the type

$$L_{i_1}^{m_1} L_{i_2}^{m_2} \cdots L_{i_k}^{m_k}.v_c \qquad i_1 < i_2 < \cdots < i_k \leq -2,$$

we call $\sum_{j=1}^{k} m_j$ the length of the monomial. By the proof of Proposition 9.2, one sees that a monomial of the length n of V_c corresponds to a polynomial of degree n via the map $F^{-1} : A(V_c) \longrightarrow \mathbb{C}[x]$.

Since V_c is isomorphic to $U(\bigoplus_{n\leq -2} \mathbb{C}L_n)$ as graded vector space, it follows that the monomial in $v_{p-1,q-1}$ that gives the maximal length is $L_{-2}^{\frac{1}{2}(p-1)(q-1)}$ by Lemma 9.6. Hence, we see that $\deg G_{p,q}(x) = \frac{1}{2}(p-1)(q-1)$. By the proof of Proposition 9.2, it can be seen that for any $g \in U(\mathrm{Vir}^-)$, there exists $P \in \mathbb{C}[x]$ such that $g.v_{p-1,q-1} \equiv P([\omega])G_{p,q}([\omega]) \mod O(V_c)$, hence the result follows by Proposition C.3. □

Now, we can describe the Zhu algebra $A(L(c,0))$ explicitly, and the result is given as follows:

Proposition 9.4 $A(L(c,0)) \cong \mathbb{C}[x]/(G_{p,q}(x))$, $G_{p,q}(x)^2 = \prod_{k=0}^{p-1}\prod_{l=0}^{q-1}(x - h_{k,l})$.

Proof. By Lemma 9.5, we have $A(L(c,0)) \cong H_0(\mathcal{L}, L(c,0))$, where the Lie algebra $\mathcal{L}$ is defined in Remark 9.1. As in § 9.3.1, we will compute the action of L_0' and L_0'' on the dual $H_0(\mathcal{L}, L(c,0))^* \cong H^0(\mathcal{L}, L(c,0)^{*a})$.

The dual of the surjection $V_c \twoheadrightarrow L(c,0)$ induces an injection

$$H^0(\mathcal{L}, L(c,0)^{*a}) \hookrightarrow H^0(\mathcal{L}, (V_c)^{*a})$$

and its image can be describe by

$$\{F \in H^0(\mathcal{L}, (V_c)^{*a}) | F(v_{p-1,q-1}) = 0\}.$$

Since $A(L(c,0))$ is commutative, it can be seen that the eigenvalues of the action of L_0' and L_0'' on $H^0(\mathcal{L}, L(c,0)^{*a})$ coincide. Hence, for $\alpha \in \mathbb{C}$, we denote the eigenspace of the action of L_0' and L_0'' on $H^0(\mathcal{L}, (V_c)^{*a})$ with the eigenvalue α by $H^0(\mathcal{L}, (V_c)^{*a})^{\alpha}$. For a non-zero $F \in H^0(\mathcal{L}, (V_c)^{*a})^{\alpha}$, we set $F_j := F|_{(V_c)_j}$ where $V_c = \bigoplus_{j\geq 0}(V_c)_j$ is the $\mathbb{Z}_{\geq 0}$-graded decomposition. As in (9.8), we have only to consider the equation

$$a(S_{p-1,q-1}).F_{(p-1)(q-1)} = 0.$$

By Proposition 9.3 (cf. set $a \mapsto -\alpha - 1$, $b \mapsto -\alpha - (p-1)(q-1)$, $\alpha \mapsto p-1$, $\beta \mapsto q-1$ and $t \mapsto \frac{q}{p}$ in the proposition), we obtain

$$a(S_{p-1,q-1}).F_{(p-1)(q-1)} = P_{p-1,q-1}\left(\alpha, \alpha, \frac{q}{p}\right) F_0.$$

Explicitly, one has

$$P_{p-1,q-1}\left(\alpha, \alpha, \frac{q}{p}\right)^2 = \left(\frac{4}{pq}\right)^{(p-1)(q-1)} \prod_{k=1}^{p-1} \prod_{l=1}^{q-1} (\alpha + h_{k,l})(kq - lp)^2.$$

Thus, we obtain

$$H^0(\mathcal{L}, L(c,0)^{*a}) = \bigoplus_{\substack{0<k<p, 0<l<q \\ kq+lp<pq}} H^0(\mathcal{L}, L(c,0)^{*a})^{-h_{k,l}},$$

which implies the result. □

The following theorem is one of the main theorems of this chapter:

Theorem 9.2 ([Wan]) *Let $c = c_{p,q}$ be a BPZ series such that $c \neq 0$.*

1. *$L(c,0)$ is a rational vertex operator algebra.*
2. *Any simple module of $L(c,0)$-module is of the form $L(c_{p,q}, h_{r,s})$ $(0 < r < p,\ 0 < s < q)$.*

Proof. Recall that any $L(c,0)$-module is a Vir-module and any subquotient of an $L(c,0)$-module as Vir-module is again an $L(c,0)$-module. Hence, any simple $L(c,0)$-module is of the form $L(c,h)$ for some $h \in \mathbb{C}$. Thus, by Theorem C.7, the second statement is proved.

Hence, to prove the first statement, we have only to show that any finitely generated $L(c,0)$-module is semi-simple.

Let V be a highest weight Vir-module with highest weight $(c, h_{r,s})$. If V is not irreducible, then V cannot be an $L(c,0)$-module because it contains a submodule which is not an $L(c,0)$-module by what we have proved in Chapter 6. Notice that one can show the vanishing of the extension $\mathrm{Ext}^1_{\mathcal{C}}(L(c_{p,q}, h_{r,s}), L(c_{p,q}, h_{r',s'}))$ $(0 < r, r' < p,\ 0 < s, s' < q)$ in exactly the same way as in Lemma 8.7. Hence, by Lemma A.3 and Proposition A.3, any finitely generated $L(c,0)$-module is a finite extension of simple modules from BPZ series, which implies its semi-simplicity. □

In the rest of this subsection, we will see that $L(c,0)$ is C_2-cofinite. (See, Appendix C for definition.)

Let us show that $C_2(L(c,0)) = [\mathrm{Vir}^-, \mathrm{Vir}^-].L(c,0)$. Indeed, if we set $A := T^n\omega$ for $n \in \mathbb{Z}_{\geq 0}$, we have $A_{(-2)} = (n+1)!L_{-n-3}$ by definition which implies

$[\mathrm{Vir}^-, \mathrm{Vir}^-].L(c,0) \subset C_2(L(c,0))$. Conversely, the opposite inclusion follows from the definition of the vertex operators and the normal ordered product of $L(c,0)$.

Hence, we obtain the following lemma:

Lemma 9.8. *One has an isomorphism:*

$$L(c,0)/C_2(L(c,0)) \cong H_0([\mathrm{Vir}^-, \mathrm{Vir}^-], L(c,0)).$$

The dimension of the right-hand side of this lemma can be easily computed as an application of Proposition 5.7 and the result is given by

$$\dim H_0([\mathrm{Vir}^-, \mathrm{Vir}^-], L(c,0)) = \frac{1}{2}(p-1)(q-1).$$

(See § 9.4.1 for the detail of its proof.) Thus, we obtain

Proposition 9.5 *Let $c = c_{p,q}$ be a BPZ series such that $c \neq 0$. The rational vertex operator algebra $L(c,0)$ is C_2-cofinite.*

Therefore, Thoerem C.8 implies that the $\mathbb{C}$-span of the characters $\{\chi_{r,s}(\tau) | (r,s) \in (K_{p,q}^+)^\circ\}$, where we set

$$\chi_{r,s}(\tau) := \mathrm{tr}_{L(c,h_{r,s})} q^{L_0 - \frac{1}{24}C} \qquad q = e^{2\pi\sqrt{-1}\tau},$$

is stable under the action of $SL(2,\mathbb{Z})$. This fact is compatible with what we have obtained in § 6.4.4 by direct computation.

9.3.3 Fusion Algebra

In this subsection, we will compute the fusion algebra of $L(c,0)$ for a BPZ series such that $c \neq 0$. Moreover, we will define the so-called Verlinde algebra of $L(c,0)$ and show that it is isomorphic to the fusion algebra of $L(c,0)$.

For $(r_i, s_i) \in (K_{p,q}^+)^\circ$ $(i = 1,2,3)$, (see (5.21) for the definition of $(K_{p,q}^+)^\circ$), we set

$$N_{(r_1,s_1),(r_2,s_2)}^{(r_3,s_3)} := \dim I \begin{pmatrix} & L(c,h_{r_3,s_3}) & \\ L(c,h_{r_1,s_1}) & & L(c,h_{r_2,s_2}) \end{pmatrix}.$$

This is the so-called **fusion rule** (cf. § C.3.)

Definition 9.1 *The* ***fusion algebra*** *of $L(c,0)$ is the free $\mathbb{Z}$-module $\mathcal{A}_{p,q} = \bigoplus_{(r,s)\in(K_{p,q}^+)^\circ} \mathbb{Z}[L(c,h_{r,s})]$ with the multiplication $*$ defined by*

$$[L(c,h_{r_1,s_1})] * [L(c,h_{r_2,s_2})] := \sum_{(r_3,s_3)\in(K_{p,q}^+)^\circ} N_{(r_1,s_1),(r_2,s_2)}^{(r_3,s_3)} [L(c,h_{r_3,s_3})].$$

We will compute the fusion rule $N_{(r_1,s_1),(r_2,s_2)}^{(r_3,s_3)}$ ($(r_i, s_i) \in (K_{p,q}^+)^0$) explicitly.

By Theorem C.10 and Lemma 9.5, we have an isomorphism

$$I\begin{pmatrix} & L(c,h_{r_3,s_3}) & \\ L(c,h_{r_1,s_1}) & & L(c,h_{r_2,s_2}) \end{pmatrix}^* \cong H^0(\mathcal{L}, L(c,h_{r_1,s_1})^{*a})^{(-h_{r_3,s_3},-h_{r_2,s_2})}, \tag{9.9}$$

where the right-hand side is the simultaneous eigenspace of the action of L_0' and L_0'' on $H^0(\mathcal{L}, L(c,h_{r_1,s_1})^{*a})$ with eigenvalues α (resp. β).

As was discussed in § 9.3.1, we will solve the equations (9.8), regarded as a system of equations on r_3 and s_3, on $H^0(\mathcal{L}, M(c,h_{r_1,s_1})^{*a})^{(-h_{r_3,s_3},-h_{r_2,s_2})}$.

By Proposition 9.3, we obtain the equations

$$\begin{aligned} &P_{r_1,s_1}\left(-h_{r_2,s_2}, -h_{r_3,s_3}+h_{r_1,s_1}, \frac{q}{p}\right) = 0 \\ =&P_{p-r_1-1,q-s_1-1}\left(-h_{r_2,s_2}, -h_{r_3,s_3}+h_{r_1,s_1}, \frac{q}{p}\right), \end{aligned}$$

i.e.,

$$\begin{aligned} \prod_{k=0}^{r_1-1}\prod_{l=0}^{s_1-1} &[(r_1-r_2+r_3-2k-1)q+(s_1+s_2-s_3-2l-1)p] \\ &\times[(r_1+r_2+r_3-2k-1)q+(s_1-s_2-s_3-2l-1)p] \\ &\times[(r_1-r_2-r_3-2k-1)q+(s_1+s_2+s_3-2l-1)p] \\ &\times[(r_1+r_2-r_3-2k-1)q+(s_1-s_2+s_3-2l-1)p]=0, \end{aligned}$$

$$\begin{aligned} \prod_{k'=0}^{p-r_1-1}\prod_{l'=0}^{q-s_1-1} &[(r_1+r_2-r_3-p+2k'+1)q+(s_1-s_2+s_3-q+2l'+1)p] \\ &\times[(r_1-r_2-r_3-p+2k'+1)q+(s_1+s_2+s_3-q+2l'+1)p] \\ &\times[(r_1+r_2+r_3-p+2k'+1)q+(s_1-s_2-s_3-q+2l'+1)p] \\ &\times[(r_1-r_2+r_3-p+2k'+1)q+(s_1+s_2-s_3-q+2l'+1)p]=0. \end{aligned}$$

It can be checked that these equations are equivalent to the equations

$$\prod_{k=0}^{r_1-1}\prod_{l=0}^{s_1-1}(h_{r_3,s_3}-h_{-r_1+r_2+2k+1,s_1+s_2-2l-1})(h_{r_3,s_3}-h_{r_1+r_2-2k-1,-s_1+s_2+2l+1})=0,$$

$$\prod_{k'=0}^{p-r_1-1}\prod_{l'=0}^{q-s_1-1}(h_{r_3,s_3}-h_{r_1+r_2-p+2k'+1,-s_1+s_2+q-2l'-1})$$

$$\times(h_{r_3,s_3}-h_{-r_1+r_2+p-2k'-1,s_1+s_2-q+2l'+1})=0.$$

By the definition of $h_{r,s}$'s, it can be seen that they are equivalent to the system of equations:

$$\prod_{k=0}^{r_1-1}\prod_{l=0}^{s_1-1}(h_{r_3,s_3}-h_{r_1+r_2-2k-1,s_1+s_2-2l-1})=0,$$

$$\prod_{k'=0}^{p-r_1-1}\prod_{l'=0}^{q-s_1-1}(h_{r_3,s_3}-h_{r_1-r_2+2k'+1,s_1-s_2+2l'+1})=0.$$

Hence, the solutions (r_3, s_3) are given by

$$\begin{aligned}&|r_1-r_2|+1\le r_3\le \max\{r_1+r_2, 2p-(r_1+r_2)\}-1,\\ &|s_1-s_2|+1\le s_3\le \max\{s_1+s_2, 2q-(s_1+s_2)\}-1,\\ &r_3-1\equiv r_1+r_2,\quad s_3-1\equiv s_1+s_2 \mod 2.\end{aligned}$$

Thus, we obtained the following theorem:

Theorem 9.3 *Let $c=c_{p,q}$ be a BPZ series such that $c\neq 0$. The multiplication $*$ of $\mathcal{A}_{p,q}$ is given by*

$$\begin{aligned}&[L(c,h_{r_1,s_1})]*[L(c,h_{r_2,s_2})]\\ &=\sum_{\substack{r_3=|r_1-r_2|+1\\ r_3-1\equiv r_1+r_2 \mod 2}}^{\min\{r_1+r_2,2p-(r_1+r_2)\}-1}\ \sum_{\substack{s_3=|s_1-s_2|+1\\ s_3-1\equiv s_1+s_2 \mod 2}}^{\min\{s_1+s_2,2q-(s_1+s_2)\}-1}[L(c,h_{r_3,s_3})].\end{aligned}$$

In particular, $\mathcal{A}_{p,q}$ is a commutative associative $\mathbb{Z}$-algebra.

We remark that the above associativity is a corollary of Theorem 9.4.

This fusion algebra can be described in terms of the fusion algebra of $SU(2)$ WZNW (Wess–Zumino–Novikov–Witten) models as follows.

Recall that the fusion algebra of level $k\in\mathbb{Z}_{\geq 0}$ $SU(2)$ WZNW model is the $\mathbb{Z}$-algebra $\mathcal{A}_k=\bigoplus_{i=0}^k\mathbb{Z}\phi_i^k$ whose multiplication $\dot{\otimes}_k$ is given by

$$\phi_i^k\dot{\otimes}_k\phi_j^k=\sum_{\substack{l=|i-j|\\ l\equiv i+j \mod 2}}^{\min\{i+j,2k-i-j\}}\phi_l^k.$$

By Theorem 9.3, we see that there exists a surjection

$$\pi : \mathcal{A}_{p-2} \otimes \mathcal{A}_{q-2} \twoheadrightarrow \mathcal{A}_{p,q}; \quad \phi_i^k \otimes \phi_j^l \longmapsto [L(c, h_{i+1,j+1})].$$

It follows from the symmetry of the Kac table that the kernel of π is given by

$$\mathrm{Ker}\pi = \bigoplus_{(r,s)\in(K_{p,q}^+)^\circ} \mathbb{Z}(\phi_{r-1}^{p-2} \otimes \phi_{s-1}^{q-2} - \phi_{p-r-1}^{p-2} \otimes \phi_{q-s-1}^{q-2}).$$

We denote the image of $\phi_i^{p-2} \otimes \phi_j^{q-2}$ in $\mathcal{A}_{p-2} \otimes \mathcal{A}_{q-2}/\mathrm{Ker}\pi$ by $[\phi_i^{p-2}, \phi_j^{q-2}]$. Identifying $[\phi_{r-1}^{p-2}, \phi_{s-1}^{q-2}]$ with $[L(c, h_{r,s})]$ via the isomorphism $\mathcal{A}_{p-2}\otimes\mathcal{A}_{q-2}/\mathrm{Ker}\pi \cong \mathcal{A}_{p,q}$ induced from π, we obtain the following description:

Theorem 9.4 ([FeMa]) *The multiplication rule of $\mathcal{A}_{p,q}$ can be described as follows:*

$$[\phi_{i_1}^{p-2}, \phi_{j_1}^{q-2}] * [\phi_{i_2}^{p-2}, \phi_{j_2}^{q-2}] = [\phi_{i_1}^{p-2} \dot{\otimes}_{p-2} \phi_{i_2}^{p-2}, \phi_{j_1}^{q-2} \dot{\otimes}_{q-2} \phi_{j_2}^{q-2}].$$

We will compare the fusion algebra $\mathcal{A}_{p,q}$ with the so-called Verlinde algebra.

For $(r_i, s_i) \in (K_{p,q}^+)^\circ$ $(i = 1, 2, 3)$, we set

$$\overline{N}_{(r_1,s_1),(r_2,s_2)}^{(r_3,s_3)} := \sum_{(r,s)\in(K_{p,q}^+)^\circ} \frac{S_{(r_1,s_1),(r,s)} S_{(r_2,s_2),(r,s)} S_{(r_3,s_3),(r,s)}}{S_{(1,1),(r,s)}},$$

where the matrix $S = (S_{(r,s),(r's')})$ is given in Proposition 6.3.

Definition 9.2 *The* ***Verlinde algebra*** *is the free $\mathbb{Z}$-module $\mathcal{V}_{p,q} = \bigoplus_{(r,s)\in(K_{p,q}^+)^\circ} \mathbb{Z}\phi_{r,s}$ with the multiplication $\circ$ defined by*

$$\phi_{r_1,s_1} \circ \phi_{r_2,s_2} = \sum_{(r_3,s_3)\in(K_{p,q}^+)^\circ} \overline{N}_{(r_1,s_1),(r_2,s_2)}^{(r_3,s_3)} \phi_{r_3,s_3}.$$

Remark 9.2 *One has to modify the definition of the Verlinde algebra for other models such as WNZW models. See, e.g., [Wak] for details.*

We will compute the structure constant $\overline{N}_{(r_1,s_1),(r_2,s_2)}^{(r_3,s_3)}$ explicitly.

As a preliminary step, one has

Lemma 9.9. *Let $k \in \mathbb{Z}_{>1}$ and $l \in \mathbb{Z}$ be relatively prime integers and $a_1, a_2, b \in \mathbb{Z}_{>0}$ be integers such that $a_1, a_2, b < k$. The following identity holds:*

$$\frac{\sin\left(\frac{\pi a_1 b}{k} l\right) \sin\left(\frac{\pi a_2 b}{k} l\right)}{\sin\left(\frac{\pi b}{k} l\right)} = \sum_{\substack{|a_1-a_2|<a<\min\{a_1+a_2, 2k-(a_1+a_2)\} \\ a \not\equiv |a_1-a_2| \mod 2}} \sin\left(\frac{\pi ab}{k} l\right).$$

Proof. One has

$$\frac{\sin\left(\frac{\pi a_1 b}{k}l\right)\sin\left(\frac{\pi a_2 b}{k}l\right)}{\sin\left(\frac{\pi b}{k}l\right)}$$
$$=\frac{1}{2\sqrt{-1}}\cdot\frac{e^{\frac{\pi(a_1+a_2)b}{k}l}-e^{\frac{\pi(a_1-a_2)b}{k}l}-e^{\frac{\pi(-a_1+a_2)b}{k}l}+e^{\frac{-\pi(a_1+a_2)b}{k}l}}{e^{\frac{\pi b}{k}l}-e^{\frac{-\pi b}{k}l}}.$$

Hence, taking the following formulae into account, one obtains the result:

$$e^{\frac{\pi(a_1-a_2)b}{k}l}+e^{\frac{\pi(-a_1+a_2)b}{k}l}=e^{\frac{\pi|a_1-a_2|b}{k}l}+e^{-\frac{\pi|a_1-a_2|b}{k}l},$$
$$e^{\frac{\pi(a_1+a_2)b}{k}l}+e^{-\frac{\pi(a_1+a_2)b}{k}l}$$
$$=e^{\frac{\pi\min\{a_1+a_2,2k-(a_1+a_2)\}b}{k}l}+e^{-\frac{\pi\min\{a_1+a_2,2k-(a_1+a_2)\}b}{k}l}.$$ □

Hence, setting

$$I_{(r_1,s_1),(r_2,s_2)}:=\left\{(r,s)\in\mathbb{Z}^2\;\middle|\;\begin{array}{l}|r_1-r_2|<r<\min\{r_1+r_2,2p-(r_1+r_2)\},\\ |s_1-s_2|<s<\min\{s_1+s_2,2q-(s_1+s_2)\},\\ r-1\equiv|r_1-r_2|,\quad s-1\equiv|s_1-s_2|\mod 2\end{array}\right\}$$

for $(r_i,s_i)\in\mathbb{Z}^2$ $(i=1,2)$ such that $0<r_i<p$, $0<s_i<q$, one has the following lemma which is a corollary to Proposition 6.3 and Lemma 9.9:

Lemma 9.10. *Let $(r_i,s_i),(r,s)\in\mathbb{Z}^2$ $(i=1,2)$ be pairs of integers such that $0<r_i,r<p$, $0<s_i,s<q$. One has*

$$\frac{S_{(r_1,s_1),(r,s)}S_{(r_2,s_2),(r,s)}}{S_{(1,1),(r,s)}}=\sum_{(r',s')\in I_{(r_1,s_1),(r_2,s_2)}}S_{(r',s'),(r,s)}.$$

On the other hand, by the definition of the matrix $S=(S_{(r,s),(r',s')})$, it satisfies $S^2=\mathrm{id}$. Moreover, by Proposition 6.3, S is a real symmetric matrix which implies that S is a unitary matrix, that is, one has the formula:

$$\sum_{(r',s')\in(K^+_{p,q})^\circ}S_{(r,s),(r',s')}S_{(r'',s''),(r',s')}=\delta_{(r,s),(r'',s'')}. \tag{9.10}$$

Hence, by Lemma 9.10 and (9.10), one obtains

$$\begin{aligned}\overline{N}^{(r_3,s_3)}_{(r_1,s_1),(r_2,s_2)}&=\sum_{(r,s)\in(K^+_{p,q})^\circ}\frac{S_{(r_1,s_1),(r,s)}S_{(r_2,s_2),(r,s)}S_{(r_3,s_3),(r,s)}}{S_{(1,1),(r,s)}}\\&=\sum_{(r',s')\in I_{(r_1,s_1),(r_2,s_2)}}\sum_{(r,s)\in(K^+_{p,q})^\circ}S_{(r',s'),(r,s)}S_{(r_3,s_3),(r,s)}\\&=\sum_{(r',s')\in I_{(r_1,s_1),(r_2,s_2)}}\delta_{(r',s'),(r_3,s_3)}.\end{aligned}$$

Thus, combining this result with Theorem 9.3, one obtains the following theorem:

Theorem 9.5 *There is an isomorphism of $\mathbb{Z}$-algebras:*

$$\mathcal{V}_{p,q} \cong \mathcal{A}_{p,q}; \qquad \phi_{r,s} \longmapsto [L(c, h_{r,s})].$$

Remark 9.3 *In general, for a certain class of vertex operator algebras, Y.-Z. Huang [Hu1], [Hu2] has shown that the Verlinde algebra and the fusion algebra are isomorphic, i.e., the structure constant of the fusion algebra is given by the Verlinde formula. Here, we proved this fact directly for the rational vertex operator algebra $L(c_{p,q}, 0)$.*

9.4 Characterisations of BPZ Series

In this section, we will provide some characterisations of the BPZ series due to [FeFu5] and [BFM]. Here, we will consider a completion of Vir:

$$\mathfrak{g} := \bigoplus_{n \leq 0} \mathbb{C}L_n \oplus \prod_{n>0} \mathbb{C}L_n \oplus \mathbb{C}C.$$

$\mathfrak{g}$ admits a triangular decomposition $\mathfrak{g} = \mathfrak{g}^+ \oplus \mathfrak{g}^0 \oplus \mathfrak{g}^-$, where we set

$$\mathfrak{g}^+ := \prod_{n>0} \mathbb{C}L_n, \qquad \mathfrak{g}^0 := \mathbb{C}L_0 \oplus \mathbb{C}C, \qquad \mathfrak{g}^- := \bigoplus_{n<0} \mathbb{C}L_n.$$

It should be noticed that any object of the category $\mathcal{O}$ of Vir can be extended to a $\mathfrak{g}$-module and the structure of the Verma modules over $\mathfrak{g}$ are the same as those over Vir.

9.4.1 Coinvariants II

Here, we will explain a characterisation of BPZ series in terms of a certain space of coinvariants that are related to the C_2-cofiniteness.

The main result of this subsection is

Theorem 9.6 *For $(c, h) \in \mathbb{C}^2$, one has*

$$\dim H_0([\mathrm{Vir}^-, \mathrm{Vir}^-], L(c, h)) < \infty$$

if and only if (c, h) is a BPZ series (cf. § 5.1.5).

Proof. Since Verma modules are $U([\mathrm{Vir}^-, \mathrm{Vir}^-])$-free, one can use the BGG type resolution proved in § 6.3 to compute the homology

$H_n([[\mathrm{Vir}^-,\mathrm{Vir}^-],L(c,h))$ for $n\in\mathbb{Z}_{\geq 0}$. It follows that $\dim H_0([\mathrm{Vir}^-,\mathrm{Vir}^-],L(c,h))<\infty$ only if (c,h) belongs to Case $1^\pm$ of Class $R^\pm$ since $H_0([\mathrm{Vir}^-,\mathrm{Vir}^-],M(c,h))\cong\mathbb{C}[L_{-1},L_{-2}]$. In this case, the space $H_0([\mathrm{Vir}^-,\mathrm{Vir}^-],L(c,h))$ can be identified with the cokernel of the map

$$H_0([\mathrm{Vir}^-,\mathrm{Vir}^-],M(c,\xi_{|i|\pm 1})\oplus M(c,\xi_{-(|i|\pm 1)}))\longrightarrow H_0([\mathrm{Vir}^-,\mathrm{Vir}^-],M(c,\xi_i)),$$

where we assume that $h=\xi_i$ for some $i\in\mathbb{Z}$. The images of the highest weight vectors $v_{c,\xi_{\pm(|i|\pm 1)}}\in M(c,\xi_{\pm(|i|\pm 1)})$ can be described by Proposition 5.7 and it can be checked that if $i\neq 0$, then the images of the highest weight vectors $v_{c,\xi_{(|i|\pm 1)}},v_{c,\xi_{-(|i|\pm 1)}}$ contain at least one common factor which implies that $\dim H_0([\mathrm{Vir}^-,\mathrm{Vir}^-],L(c,h))=\infty$. Hence, the only possible cases are Case 1^+ of Class R^+ and $h=\xi_0$, and in this case, it can be checked that the images of the highest weight vectors $v_{c,\pm\xi_1}$ contain no common factor and

$$\dim H_0([\mathrm{Vir}^-,\mathrm{Vir}^-],L(c_{p,q},h_{r,s}))=\frac{1}{2}rs(p-r)(q-s)$$

as an application of Proposition 5.7. □

9.4.2 Lisse Modules

In this subsection, we briefly recall the definition of lisse modules and some properties of them for a general Lie algebra. For notation, see § 9.A.

Definition 9.3 *Let M be a finitely generated $\mathfrak{g}$-module.*

1. *If $\mathcal{V}_{\mathfrak{g}}M=\{0\}$ holds, M is said to be* ***lisse****.*
2. *If $\mathcal{V}_{\mathfrak{g}}M\cap\mathfrak{l}^\perp=\{0\}$ for some subspace $\mathfrak{l}\subset\mathfrak{g}$, M is said to be* ***lisse along*** *$\mathfrak{l}$.*

It follows from (9.12) that if a finitely generated $\mathfrak{g}$-module is lisse, then its subquotient is also lisse. In particular, if $\dim M<\infty$, then M is lisse.

Let $\mathfrak{k}\subset\mathfrak{g}$ be a subalgebra and K be its algebraic group.

Lemma 9.11. *Let M be a finitely generated $(\mathfrak{g},\mathfrak{k})$-module. Suppose that a subspace $\mathfrak{n}\subset\mathfrak{g}$ satisfies*

1. *$\dim\mathfrak{g}/\mathfrak{n}+\mathfrak{k}<\infty$ and*
2. *M is lisse along $\mathfrak{n}$.*

Then, one has $\dim M/\mathfrak{n}.M<\infty$.

Proof. Let F be a $\mathfrak{k}$-invariant good filtration on M. We consider the induced filtration on $M/\mathfrak{n}.M$. It suffices to show that $\dim\mathrm{gr}^F(M/\mathfrak{n}.M)<\infty$.

By Lemma 9.15 2., it follows that

$$\mathrm{gr}^F_k(M/\mathfrak{n}.M)=M_k+\mathfrak{n}.M/M_{k-1}+\mathfrak{n}.M\cong M_k/M_{k-1}+M_k\cap\mathfrak{n}.M,$$

and

$$(\mathrm{gr}^F M/\mathfrak{n}.\mathrm{gr}^F M)_k = M_k/M_{k-1} + \mathfrak{n}.M_{k-1}.$$

Hence, $\mathrm{gr}^F(M/\mathfrak{n}.M)$ is a quotient of $\mathrm{gr}^F M/\mathfrak{n}.\mathrm{gr}^F M$.

$S(\mathfrak{g}/\mathfrak{n}+\mathfrak{k})$ is finitely generated by the assumption 1. and $\mathrm{gr}^F M/\mathfrak{n}.\mathrm{gr}^F M$ is finitely generated $S(\mathfrak{g}/\mathfrak{n}+\mathfrak{k})$-module since $\mathrm{gr}^F M$ is a finitely generated $S(\mathfrak{g}/\mathfrak{k})$-module.

On the other hand, it follows from the assumption 2. and (9.12) that $V(\mathrm{Ann}(\mathrm{gr}^F M/\mathfrak{n}.\mathrm{gr}^F M)) \cap \mathfrak{n}^\perp = \{0\}$. In addition, one also has $V(\mathrm{Ann}(\mathrm{gr}^F M/\mathfrak{n}.\mathrm{gr}^F M)) \subset (\mathfrak{n}+\mathfrak{k})^\perp$ since $\mathrm{gr}^F M/\mathfrak{n}.\mathrm{gr}^F M$ is an $S(\mathfrak{g}/\mathfrak{n}+\mathfrak{k})$-module. Hence, it follows that $V(\mathrm{Ann}(\mathrm{gr}^F M/\mathfrak{n}.\mathrm{gr}^F M)) = \{0\}$.

Thus, one obtains $\dim \mathrm{gr}^F M/\mathfrak{n}.\mathrm{gr}^F M < \infty$. □

Now, we state a simple application of the involutivity theorem (cf. Theorem 9.8):

Lemma 9.12. *Let M be a finitely generated $(\mathfrak{g},\mathfrak{k})$-module such that $\mathcal{V}_\mathfrak{g} M$ is finite codimensional in $\mathfrak{k}^\perp$. Then, $J(M)$ is involutive.*

The next corollary follows from Lemma 9.17:

Corollary 9.3 *Suppose that the pair $(\mathfrak{g},\mathfrak{k})$ satisfies the condition: Any Ad^*K-invariant Zariski closed subset of $\mathfrak{k}^\perp$ is either $\{0\}$ or finite codimensional. Then, for any finitely generated $(\mathfrak{g},\mathfrak{k})$-module M, $J(M)$ is involutive.*

9.4.3 Finiteness Condition

In this subsection, we will characterise the BPZ series in several ways. Hence, we assume that $\mathfrak{g}$ is the completion of Vir as we discussed before § 9.4.1.

We set $\mathfrak{g}^{\geq} = \mathfrak{g}^{\geq 0} := \mathfrak{g}^+ \oplus \mathfrak{g}^0$ and $\mathfrak{g}^{\geq n} := \mathfrak{g}^{\geq} \oplus \bigoplus_{i=n}^{-1} \mathbb{C}L_i$ for $n \in \mathbb{Z}_{<0}$.

Lemma 9.13. *Let M be a finitely generated $(\mathfrak{g},\mathfrak{g}^{\geq})$-module. Then, $\mathcal{V}_\mathfrak{g} M$ is given by one of the three: $\{0\}$, $(\mathfrak{g}^{\geq})^\perp$ or $(\mathfrak{g}^{\geq -1})^\perp$.*

Proof. Let K be a pro-unipotent pro-algebraic group of $\mathfrak{g}^{\geq}$. By Lemma 9.17, $\mathcal{V}_\mathfrak{g} M \subset (\mathfrak{g}^{\geq})^\perp$ is a K-invariant. It is clear that a K-invariant Zariski closed subset of $(\mathfrak{g}^{\geq})^\perp$ is given by either $\{0\}$ or $(\mathfrak{g}^{\geq n})^\perp$ for some $n \in \mathbb{Z}_{<0}$. It follows from Corollary 9.3 that $(\mathfrak{g}^{\geq n})^\perp$ can be $\mathcal{V}_\mathfrak{g} M$ if $\mathfrak{g}^{\geq n}$ is a Lie subalgebra and this happens only for $n = 0, -1$. □

Corollary 9.4 *Let (c,h) be a BPZ series. Then, $\mathcal{V}_\mathfrak{g} L(c,h)$ is either $\{0\}$ or $(\mathfrak{g}^{\geq -1})^\perp$.*

Proof. By definition, one has $\mathcal{V}_\mathfrak{g} M(c,h) = (\mathfrak{g}^{\geq})^\perp$. Since $L(c,h)$ is a non-trivial quotient of $M(c,h)$ by assumption, there is $\varphi \in U(\mathfrak{g}^-)$ such that $\varphi.v_{c,h} = 0$ where $v_{c,h} \in L(c,h)$ is a highest weight vector. Hence, its symbol $\sigma(\varphi)$ vanishes on $\mathcal{V}_\mathfrak{g} L(c,h)$ which implies that $\mathcal{V}_\mathfrak{g} L(c,h) \neq (\mathfrak{g}^{\geq})^\perp$. Thus, the result follows from Lemma 9.13. □

Now, we can state the main theorem of this section as follows:

Theorem 9.7 ([BFM]) *We let $(c,h) \in \mathbb{C}^2$ and $v_{c,h} \in L(c,h)$ be a highest weight vector. The following conditions are equivalent:*

1. *$L(c,h)$ is lisse.*
2. *$L(c,h)$ is a BPZ series.*
3. $\dim H_0([\mathrm{Vir}^-, \mathrm{Vir}^-], L(c,h)) < \infty$.
4. $\dim H^0([\mathrm{Vir}^-, \mathrm{Vir}^-], L(c,h)) < \infty$.
5. *There exists $\varphi \in U([\mathrm{Vir}^-, \mathrm{Vir}^-])$ such that $\varphi.v_{c,h} = 0$.*

To prove this theorem, we prepare a technical lemma. For $(c,h) \in \mathbb{C}^2$, we denote the eigenspace decomposition of $L(c,h)$ and $T := U([\mathrm{Vir}^-, \mathrm{Vir}^-]).\tilde{v}_{c,h} \subset M(c,h)$ with respect to the action of L_0 by $L(c,h) = \bigoplus_{n\geq 0} L(c,h)_{h+n}$ (resp. $T = \bigoplus_{n\geq 0} T_{h+n}$.) Here, we set $\tilde{v}_{c,h} := 1 \otimes \mathbf{1}_{c,h} \in M(c,h)$.

Lemma 9.14. *Let $c = c_{p,q}$ be a BPZ series and $h = h_{r,s}$ for some $0 < r < p$ and $0 < s < q$. Then, we have*

$$\dim L(c, h_{r,s})_{h_{r,s}+n} \underset{n\to\infty}{\sim} \frac{1}{\sqrt{2}} S_{(r,s),(r_0,s_0)} \left(\frac{1-\frac{6}{pq}}{24} \right)^{\frac{1}{4}} n^{-\frac{3}{4}} e^{\pi\sqrt{\frac{2}{3}(1-\frac{6}{pq})n}},$$

$$\dim T_{h_{r,s}+n} \underset{n\to\infty}{\sim} \frac{\pi^2}{12\sqrt{3}n^2} e^{\pi\sqrt{\frac{2}{3}n}},$$

where $S = (S_{(r,s),(r',s')})$ is given in Proposition 6.3 and $(r_0, s_0) \in (K^+_{p,q})^\circ$ such that $|r_0 q - s_0 p| = 1$.

Proof. By Theorem 9.9, it is enough to compute the asymptotic $\tau \downarrow 0$ of $\chi_{r,s}(\tau)$ and $\mathrm{tr}_T q^{L_0 - \frac{1}{24}C}$.

The first formula follows from Theorem 6.15. For the second formula, we see that

$$\mathrm{tr}_T q^{L_0-\frac{1}{24}C} = q^{\frac{(rq-sp)^2}{4pq}} (1-q)(1-q^2)\eta(\tau)^{-1},$$

which implies

$$\mathrm{tr}_T q^{L_0-\frac{1}{24}C} \underset{\tau\downarrow 0}{\sim} 8\pi^2 \left(\frac{\tau}{\sqrt{-1}} \right)^{\frac{5}{2}} e^{\frac{\pi\sqrt{-1}}{12\tau}}$$

by (9.13). □

Now, we return to the proof of Theorem 9.7.

1. ⇒ 3. follows from Lemma 9.11. 3. ⇔ 4. follows from Proposition 1.14 and 2. ⇔ 3. is Theorem 9.6. 5. ⇒ 1. follows from Corollary 9.4 since the symbol $\sigma(\varphi)$ vanishes on $\mathcal{V}_{\mathfrak{g}} L(c,h)$. Hence, it is sufficient to prove 2. ⇒ 5. We have only to show that the composition of the maps $T \hookrightarrow M(c,h) \twoheadrightarrow L(c,h)$ cannot be surjective which is a corollary of Lemma 9.14. □

9.5 Bibliographical Notes and Comments

In 1988, B. Feigin and D. Fuchs [FeFu5] determined the fusion rule of BPZ series by purely representation theoretical methods. Later, in 1993, a part of their results were re-interpreted in the framework of vertex operator algebras by W. Wang [Wan]. In particular, it was shown that the vertex operator algebra $L(c,0)$ is rational if and only if c is a BPZ series (and $c \neq 0$). Moreover, W. Wang re-computed the fusion rule for BPZ series. But a nice expression of the fusion rule was discovered by B. Feigin and F. Malikov [FeMa] in 1997. These results are explained in § 10.1 – 10.3. There, we have also shown that the fusion algebra of BPZ series is isomorphic to the so-called Verlinde algebra [Ver]. § 10.4 is one of the beautiful results explained in A. Beilinson, B. Feigin and B. Mazur [BFM] which was partially announced by [BeSch] and [FeFu5].

9.A Appendix: Associated Variety

In this section, we will briefly recall the definition of the associated variety etc. In particular, following [Gab], the Gabber involutivity theorem will be stated in a less familiar form. Here, we assume that a ring always contains the identity.

9.A.1 Filtration

Let A be a **filtered ring**, i.e., $A = \bigcup_{k\in\mathbb{Z}} A_k$ where the A_k's are additive subgroups of A such that

$$A_k \subset A_{k+1} \qquad \forall\, k \in \mathbb{Z},$$
$$A_k \cdot A_l \subset A_{k+l} \qquad \forall\, k, l \in \mathbb{Z},$$

and $1 \in A_0$. We let $\mathrm{gr}_k A := A_k/A_{k-1}$ and $\sigma_k : A_k \twoheadrightarrow \mathrm{gr}_k A$ be the canonical projection. For $x \in A_k \setminus A_{k-1}$, we set $\sigma(x) := \sigma_k(x)$ and call it the **symbol of** x. We set

$$\mathrm{gr}(A) := \bigoplus_{k\in\mathbb{Z}} \mathrm{gr}_k A.$$

Then, $\mathrm{gr}(A)$ inherits a graded ring structure, called the **associated graded ring**, induced from the filtered ring structure of A. In this section, we assume that

$$\mathrm{gr}(A) \textbf{ is commutative.}$$

$\operatorname{gr}(A)$ has a Lie algebra structure defined as follows: Let $f \in \operatorname{gr}_k A$ and $g \in \operatorname{gr}_l A$. Choose any $x \in A_k$ with $\sigma_k(x) = f$ and $y \in A_l$ with $\sigma_l(y) = g$. It follows from the commutativity of $\operatorname{gr}(A)$ that $xy - yx \in A_{k+l-1}$. Set

$$\{f, g\} := \sigma_{k+l-1}(xy - yx). \tag{9.11}$$

This bracket is well-defined and is called the **Poisson bracket**. Indeed, it satisfies

$$\begin{aligned} &\{f, f\} = 0, \\ &\{f, \{g_1 g_2\}\} = \{\{f, g_1\}, g_2\} + \{g_1, \{f, g_2\}\}, \\ &\{f, g_1 g_2\} = \{f, g_1\} g_2 + g_1 \{f, g_2\}, \end{aligned}$$

for any $f, g_1, g_2 \in \operatorname{gr}(A)$.

Let M be an A-module. A **filtration** F **of** M is, by definition, a family $\{F_n M\}_{n\in\mathbb{Z}}$ of subgroups of M such that

$$\begin{aligned} &F_n M \subset F_{n+1} M \qquad \forall\, n \in \mathbb{Z}, \\ &M = \bigcup_{n\in\mathbb{Z}} F_n M, \\ &A_k . F_n M \subset F_{n+k} M \qquad \forall\, k, n \in \mathbb{Z}. \end{aligned}$$

Two filtrations F and F' are said to be **equivalent** if there exists $c_1, c_2 \in \mathbb{Z}$ such that

$$F_{n+c_1} M \subset F'_n M \subset F_{n+c_2} M \qquad \forall\, n \in \mathbb{Z}.$$

Let F be a filtration on an A-module M. Define

$$\operatorname{gr}_n^F M := F_n M / F_{n-1} M, \qquad \operatorname{gr}^F M := \bigoplus_{n\in\mathbb{Z}} \operatorname{gr}_n^F M.$$

One defines a $\operatorname{gr}(A)$-module structure on $\operatorname{gr}^F M$.

Definition 9.4 *A filtration F on an A-module M is called* **good** *if the corresponding $\operatorname{gr}(A)$-module $\operatorname{gr}^F M$ is finitely generated.*

For example, suppose that an A-module M is finitely generated and $m_1, \cdots, m_r$ are its generators. Pick integers $k_1, \cdots, k_r$ and define a filtration with this choice of generators and integers by setting

$$F_n M := \sum_{i=1}^{r} A_{n-k_i} . m_i .$$

By definition, this is a good filtration. In general, one has the following lemma:

Lemma 9.15. *Let M be an A-module which admits a good filtration.*

1. Any two good filtrations of M are equivalent.

2. *Let M' be an A submodule of M. If F is a good filtration on M, then $F_nM' := F_nM \cap M'$ is a good filtration on M' and $F_n(M/M') := (F_nM + M')/M'$ is a good filtration on M/M'.*

For the second case in this lemma, one obtains a short exact sequence

$$0 \longrightarrow \mathrm{gr}^F M' \longrightarrow \mathrm{gr}^F M \longrightarrow \mathrm{gr}^F(M/M') \longrightarrow 0.$$

9.A.2 Associated Variety

Let A be a filtered ring and M an A-module as in the previous subsection. For a filtration F on M, we set

$$I_F := \mathrm{Ann}_{\mathrm{gr}(A)}(\mathrm{gr}^F M).$$

I_F is known to be a graded ideal of $\mathrm{gr}(A)$. Hence, in particular, it implies that $\sqrt{I_F}$ is also a graded ideal of $\mathrm{gr}(A)$.

Lemma 9.16. *For any filtration F' on M equivalent to F, one has*

$$\sqrt{I_{F'}} = \sqrt{I_F}.$$

Thanks to this lemma, taking a good filtration F of M, we may set

$$J(M) := \sqrt{I_F} = \bigcap_{\mathfrak{p} \in \mathbb{V}(I_F)} \mathfrak{p},$$

where we set $\mathbb{V}(\mathfrak{a}) := \{\mathfrak{p} \in \mathrm{Spec}(\mathrm{gr}(A)) | \mathfrak{a} \subset \mathfrak{p}\}$ for an ideal $\mathfrak{a}$ of $\mathrm{gr}(A)$. One can check that the **support of** $\mathrm{gr}^F M$

$$\mathrm{Supp}(\mathrm{gr}^F M) := \{\mathfrak{p} \in \mathrm{Spec}(\mathrm{gr}(A)) | (\mathrm{gr}^F M)_{\mathfrak{p}} \neq \{0\}\}$$

coincides with $\mathbb{V}(\mathrm{Ann}_{\mathrm{gr}(A)}(\mathrm{gr}^F M))$ for a good filtration F on M. For later use, we set

$$\mathrm{Supp}_0(\mathrm{gr}^F M) := \{\mathfrak{p} : \text{ minimal in } \mathrm{Supp}(\mathrm{gr}^F M)\},$$

where the order is taken with respect to the inclusion. $\mathrm{Supp}_0(\mathrm{gr}^F M)$ is sometimes called the **singular support of** $\mathrm{gr}^F M$.

Now, let $\mathfrak{g}$ be a Lie algebra over $\mathbb{C}$ and $A := U(\mathfrak{g})$ be its enveloping algebra. By the standard filtration on A (cf. (A.6)), the PBW theorem implies that $\mathrm{gr}(A) = S(\mathfrak{g}) = \bigoplus_{n \geq 0} S^n(\mathfrak{g})$. For a subset $J \subset A$, we set

$$V(J) := \{\varphi \in \mathfrak{g}^* | f(\varphi) = 0 \ (\forall\ f \in J)\}.$$

Definition 9.5 *Let M be a $\mathfrak{g}$-module with a good filtration. The set*

$$\mathcal{V}_{\mathfrak{g}} M := V(J(M))$$

*is called the **associated variety of** M.*

Since $J(M)$ is a graded ideal, $\mathcal{V}_{\mathfrak{g}} M$ is, in general, a cone in $\mathfrak{g}^*$. If M' is a submodule of M, it follows that

$$\mathcal{V}_{\mathfrak{g}}(M) = \mathcal{V}_{\mathfrak{g}}(M') \cup \mathcal{V}_{\mathfrak{g}}(M/M'). \tag{9.12}$$

Let $\mathfrak{k} \subset \mathfrak{g}$ be a Lie subalgebra and M be $\mathfrak{g}$-module.

Definition 9.6 *M is called a $(\mathfrak{g}, \mathfrak{k})$-module iff $\mathfrak{k}$ acts on M locally finitely, i.e., $\dim U(\mathfrak{k}).m < \infty$ for all $m \in M$.*

We assume that M is a $(\mathfrak{g}, \mathfrak{k})$-module which is finitely generated as $\mathfrak{g}$-module. There is a $\mathfrak{k}$-invariant good filtration on M, that is, a filtration F on M such that $\mathfrak{k}.F_nM \subset F_nM$ for $n \in \mathbb{Z}$. Indeed, this filtration can be defined as follows: Let $m_1, \cdots, m_r \in M$ be generators and set $F_0M := \sum_{i=1}^r U(\mathfrak{k}).m_i$. We set $F_nM := U_n(\mathfrak{g}).F_0M$ for $n \in \mathbb{Z}_{>0}$. With this filtration, we see that $\mathrm{gr}^F M$ is $\mathfrak{k}$-invariant, i.e., $\mathfrak{k}.\mathrm{gr}^F M = \{0\}$ which implies that $\mathrm{gr}^F M$ can be regarded as an $S(\mathfrak{g}/\mathfrak{k})$-module. Moreover, $J(M)$ is $\mathrm{Ad}K$-invariant, where K is an algebraic group of $\mathfrak{k}$. Hence, the next lemma follows:

Lemma 9.17. *Let M be a $(\mathfrak{g}, \mathfrak{k})$-module which admits a good filtration. Then, $\mathcal{V}_{\mathfrak{g}} M$ is an Ad^*K-invariant Zariski closed subset of $\mathfrak{k}^{\perp} := \{\varphi \in \mathfrak{g}^* | \varphi(\mathfrak{k}) = \{0\}\}$.*

For more information about associated varieties, see, e.g., [Ja2].

9.A.3 Involutivity

Let A be a filtered ring and M an A-module with a good filtration as in § 9.A.1. A $\mathrm{gr}(A)$-module $\mathrm{gr}^F M$ is called **weakly Noetherian** if it satisfies

1. $\mathrm{gr}^F M$ is finitely generated, and
2. for every $\mathfrak{p} \in \mathrm{Supp}_0(\mathrm{gr}^F M)$, $(\mathrm{gr}^F M)_{\mathfrak{p}}$ is a $\mathrm{gr}(A)_{\mathfrak{p}}$-module of finite length.

We remark that a Noetherian module over a commutative ring is weakly Noetherian.

Now, a theorem about the integrability of characteristics due to O. Gabber [Gab] can be formulated as follows:

Theorem 9.8 *Let A be a filtered ring such that $\mathrm{gr}(A)$ is a commutative $\mathbb{Q}$-algebra and M be a filtered A-module such that $\mathrm{gr}^F M$ is a weakly Noetherian $\mathrm{gr}(A)$-module. Then, $J(M) = \sqrt{\mathrm{Ann}_{\mathrm{gr}(A)}(\mathrm{gr}^F M)}$ is **involutive**, i.e., it is closed under the Poisson brackets in $\mathrm{gr}(A)$. Here, $\mathrm{gr}^F M$ is constructed via any good filtration F on M.*

9.B Appendix: Tauberian Theorem

In this section, we briefly recall the so-called Tauberian theorem à la Ingham [In]. A simple but important example of an application of this theorem is also given.

Let us define some symbols:

1. $\tau \downarrow 0$ means the limit $\tau \to 0$ along the half-line $\sqrt{-1}\mathbb{R}_{>0}$.
2. $f(\tau) \underset{\tau\downarrow 0}{\sim} g(\tau)$ means $\dfrac{f(\tau)}{g(\tau)} \to 1$ as $\tau \downarrow 0$.
3. $a_n \underset{n\to\infty}{\sim} b_n$ signifies $\displaystyle\lim_{n\longrightarrow\infty} \frac{b_n}{a_n} = 1$.

The following theorem is proved by E. Ingham [In] and is called a Tauberian theorem:

Theorem 9.9 *Suppose that*

$$f(\tau) = q^{\lambda} \sum_{n=0}^{\infty} a_n q^n \quad (q = e^{2\pi\sqrt{-1}\tau})$$

is a holomorphic function on $\mathbb{H} = \{\tau \in \mathbb{C} | \mathrm{Im}\tau > 0\}$, *and satisfies the following two conditions:*

i. $a_n \in \mathbb{R}$ *and* $a_n \leq a_{n+1}$ *for all* n.
ii. There exist $A \in \mathbb{C}$, $B \in \mathbb{R}$ *and* $C \in \mathbb{R}_{>0}$ *satisfying*

$$f(\tau) \underset{\tau\downarrow 0}{\sim} A \cdot \left(\frac{\tau}{\sqrt{-1}}\right)^{\frac{1}{2}B} e^{\frac{\pi\sqrt{-1}}{12\tau}C}.$$

Then

$$a_n \underset{n\to\infty}{\sim} \frac{A}{\sqrt{2}} \left(\frac{C}{24}\right)^{\frac{1}{4}(B+1)} n^{-\frac{1}{4}(B+3)} e^{\pi\sqrt{\frac{2}{3}Cn}}.$$

We will show a typical application of this theorem which was also treated by E. Ingham.

By the definition of the Dedekind eta function (6.14) and Lemma 6.5, it follows that

$$\eta(\tau)^{-1} = q^{-\frac{1}{24}} \sum_{n\geq 0} p(n) q^n = \left(\frac{\tau}{\sqrt{-1}}\right)^{\frac{1}{2}} \eta\left(-\frac{1}{\tau}\right)^{-1},$$

where $p(n)$ is the partition number of $n \in \mathbb{Z}_{\geq 0}$. Hence, it can be checked that

$$\eta(\tau)^{-1} \underset{\tau\downarrow 0}{\sim} \left(\frac{\tau}{\sqrt{-1}}\right)^{\frac{1}{2}} e^{\frac{\pi\sqrt{-1}}{12\tau}}. \tag{9.13}$$

Thus, by Theorem 9.9, one obtains the following well-known asymptotic behaviour:

$$p(n) \sim \frac{1}{4\sqrt{3}n} e^{\pi\sqrt{\frac{2}{3}n}} \qquad (n \to \infty).$$

Chapter 10
Coset Constructions for $\hat{\mathfrak{sl}}_2$

In this chapter, we will construct some irreducible highest weight Vir-modules out of highest weight $\hat{\mathfrak{sl}}_2$-modules, which is called the coset construction. In particular, we will construct all of the BPZ series representations of Vir (cf. Section 5.1) in terms of $\hat{\mathfrak{sl}}_2$-modules.

In Section 10.1, we will recall some basic facts about the affine Lie algebra $\hat{\mathfrak{sl}}_2$ such as the character formulae of the admissible representations and in Section 10.2, the so-called Segal–Sugawara construction will be recalled in terms of vertex algebra. In Section 10.3, we will consider the irreducible decomposition of the following two cases:

1. (Level 1 integrable module) $\otimes$ (admissible representation) as $\hat{\mathfrak{sl}}_2{}' \oplus$ Vir-module,
2. (Level 1 integrable module) as $\mathfrak{sl}_2 \oplus$ Vir-module.

In Section 10.4, we will show that the discrete series representations of Vir (cf. Section 5.1) are unitarisable.

10.1 Admissible Representations

In this section, after a brief review on $\hat{\mathfrak{sl}}_2$, we recall some basic facts about its representation theory. The proofs of most of the statements will not be given here, and the reader should consult, e.g., [MP], for a detailed account.

10.1.1 Affine Lie Algebra $\hat{\mathfrak{sl}}_2$

Affine Lie algebras are defined as Kac–Moody Lie algebras with generalised Cartan matrices of affine type. But we do not go into the general theory of Kac–Moody Lie algebras here, and recall some concepts for later use.

K. Iohara, Y. Koga, *Representation Theory of the Virasoro Algebra*, Springer Monographs in Mathematics, DOI 10.1007/978-0-85729-160-8_10,

Let $\bar{\mathfrak{g}}$ be the simple Lie algebra $\mathfrak{sl}_2$ over $\mathbb{C}$, and let $\{h, e, f\}$ be a basis of $\bar{\mathfrak{g}}$ which satisfies

$$[e, f] = h, \quad [h, e] = 2e, \quad [h, f] = -2f.$$

We set

$$\bar{\mathfrak{h}} := \mathbb{C}h, \quad \bar{\mathfrak{g}}^+ := \mathbb{C}e, \quad \bar{\mathfrak{g}}^- := \mathbb{C}f.$$

Let $(\cdot, \cdot)$ be the non-degenerate symmetric invariant bilinear form on $\bar{\mathfrak{g}}$ normalised by $(h, h) = 2$.

The affine Lie algebra $\widehat{\mathfrak{sl}}_2$ is the following Lie algebra:

$$\hat{\mathfrak{g}} := \bar{\mathfrak{g}} \otimes_{\mathbb{C}} \mathbb{C}[t, t^{-1}] \oplus \mathbb{C}K \oplus \mathbb{C}d$$

with the commutation relations

$$\begin{aligned} &[x \otimes t^k, y \otimes t^l] = [x, y] \otimes t^{k+l} + k\delta_{k+l,0}(x, y)K, \\ &[\hat{\mathfrak{g}}, K] = \{0\}, \quad [d, x \otimes t^k] = kx \otimes t^k, \end{aligned}$$

where $x, y \in \bar{\mathfrak{g}}$. Let $\hat{\mathfrak{g}}'$ be the derived subalgebra of $\hat{\mathfrak{g}}$, i.e.,

$$\hat{\mathfrak{g}}' := [\hat{\mathfrak{g}}, \hat{\mathfrak{g}}] = \bar{\mathfrak{g}} \otimes \mathbb{C}[t, t^{-1}] \oplus \mathbb{C}K.$$

We set

$$\hat{\mathfrak{h}} := \mathbb{C}h \otimes 1 \oplus \mathbb{C}K \oplus \mathbb{C}d,$$

and call it a Cartan subalgebra of $\hat{\mathfrak{g}}$. From now on, we denote $x \otimes t^k$ by $x(k)$ for simplicity, and regard $\bar{\mathfrak{g}} \subset \hat{\mathfrak{g}}$ via the map $x \mapsto x(0)$ $(x \in \bar{\mathfrak{g}})$.

We recall some notation of affine Lie algebras. Set

$$\begin{aligned} &h_0 := K - h, \; e_0 := f(1), \; f_0 := e(-1), \\ &h_1 := h, \qquad e_1 := e, \qquad f_1 := f. \end{aligned}$$

Then, it is easy to see that $\hat{\mathfrak{g}}'$ is generated by these elements. They are called the Chevalley generators. Let $\{\Lambda_0, \alpha_0, \alpha_1\}$ be the basis of $\hat{\mathfrak{h}}^*$ such that

$$\begin{pmatrix} \langle \Lambda_0, d\rangle & \langle \Lambda_0, h_0\rangle & \langle \Lambda_0, h_1\rangle \\ \langle \alpha_0, d\rangle & \langle \alpha_0, h_0\rangle & \langle \alpha_0, h_1\rangle \\ \langle \alpha_1, d\rangle & \langle \alpha_1, h_0\rangle & \langle \alpha_1, h_1\rangle \end{pmatrix} = \begin{pmatrix} 0 & 1 & 0 \\ 1 & 2 & -2 \\ 0 & -2 & 2 \end{pmatrix},$$

where $\langle \cdot, \cdot \rangle$ denotes the dual pairing between $\hat{\mathfrak{h}}^*$ and $\hat{\mathfrak{h}}$. Further, we set $\Lambda_1 := \Lambda_0 + \frac{1}{2}\alpha_1$. $\{h_0, h_1\}$, $\{\alpha_0, \alpha_1\}(=: \Pi)$ and $\{\Lambda_0, \Lambda_1\}$ are called the sets of *simple coroots*, *simple roots* and *fundamental weights* of $\widehat{\mathfrak{sl}}_2$, respectively. For later use, we set $\delta := \alpha_0 + \alpha_1$ and $\rho := \Lambda_0 + \Lambda_1$. Recall that the number $\langle \rho, K \rangle = 2 =: h^\vee$ is called the dual Coxeter number of $\widehat{\mathfrak{sl}}_2$.

Next we recall symmetric bilinear forms on $\hat{\mathfrak{h}}$. Let $(\cdot, \cdot)$ be the symmetric bilinear form on $\hat{\mathfrak{h}}$ defined by

$$\begin{pmatrix} (K,K) & (K,d) & (K,h) \\ (d,K) & (d,d) & (d,h) \\ (h,K) & (h,d) & (h,h) \end{pmatrix} = \begin{pmatrix} 0 & 1 & 0 \\ 1 & 0 & 0 \\ 0 & 0 & 2 \end{pmatrix}.$$

Note that $(\cdot,\cdot)$ is non-degenerate. Hence, we identify $\hat{\mathfrak{h}}$ and $\hat{\mathfrak{h}}^*$ by this bilinear form, and denote this identification by $\nu : \hat{\mathfrak{g}} \to \hat{\mathfrak{g}}^*$, i.e.,

$$\nu(h_i) = \alpha_i \ (i = 0, 1), \quad \nu(K) = \delta, \quad \nu(d) = \Lambda_0.$$

Further, we define $(\cdot,\cdot)$ on $\hat{\mathfrak{h}}^*$ by $(\nu(h), \nu(h')) := (h, h')$ for $h, h' \in \hat{\mathfrak{h}}$.

The Weyl group W of $\hat{\mathfrak{g}}$ is defined as follows: For $i = 0, 1$, we define $r_i \in \mathrm{GL}(\hat{\mathfrak{h}}^*)$ by

$$r_i(\lambda) := \lambda - \langle \lambda, h_i \rangle \alpha_i.$$

Let W be the subgroup of $\mathrm{GL}(\hat{\mathfrak{h}}^*)$ generated by $\{r_0, r_1\}$. We call it the Weyl group. Recall that the following affine action of W on $\hat{\mathfrak{h}}^*$ is called the 'shift action':

$$w \circ \lambda := w(\lambda + \rho) - \rho \quad (w \in W).$$

$\hat{\mathfrak{g}}$ has the following Q-graded Lie algebra structure. Set $Q := \mathbb{Z}\alpha_0 \oplus \mathbb{Z}\alpha_1$ and call it the *root lattice* of $\hat{\mathfrak{g}}$. We also set $Q^{\pm} := \pm(\mathbb{Z}_{\geq 0}\alpha_0 \oplus \mathbb{Z}_{\geq 0}\alpha_1)$. Then, $\hat{\mathfrak{g}}$ decomposes as $\hat{\mathfrak{g}} = \bigoplus_{\alpha \in Q} \hat{\mathfrak{g}}_\alpha$. Further, we set

$$\begin{aligned} \Delta :=& \{\alpha \in Q \setminus \{0\} | \hat{\mathfrak{g}}_\alpha \neq \{0\}\} \\ =& \{m_1\delta, \ \pm\alpha_1 + m_2\delta | m_1, m_2 \in \mathbb{Z}, \ m_1 \neq 0\}, \end{aligned}$$

$\Delta^{\pm} := \Delta \cap Q^{\pm}$ and

$$\Delta_{\mathrm{re}} := W\Pi = \{\pm\alpha_1 + n\delta | n \in \mathbb{Z}\}.$$

They are called the sets of *roots*, *positive roots*, *negative roots* and *real roots* respectively. We also set $\Delta^{\pm}_{\mathrm{re}} := \Delta_{\mathrm{re}} \cap \Delta^{\pm}$. For $\beta = w(\alpha_i) \in \Delta_{\mathrm{re}}$ ($w \in W$ and $i = 0, 1$), we use the following notation: $r_\beta := w r_i w^{-1} \in W$.

We define an anti-involution $\hat{\sigma}$ of $\hat{\mathfrak{g}}$. Let $\bar{\sigma} : U(\bar{\mathfrak{g}}) \to U(\bar{\mathfrak{g}})$ be the anti-involution defined by $\bar{\sigma}(e) = f$, $\bar{\sigma}(h) = h$ and $\bar{\sigma}(f) = e$. We further define $\hat{\sigma} : U(\hat{\mathfrak{g}}) \to U(\hat{\mathfrak{g}})$ by

$$\hat{\sigma}(x(k)) := (\bar{\sigma}(x))(-k) \ (x \in \bar{\mathfrak{g}}), \quad \hat{\sigma}(K) = K, \quad \hat{\sigma}(d) = d.$$

Then, the quadruple $(\hat{\mathfrak{g}}, \hat{\mathfrak{h}}, Q, \hat{\sigma})$ becomes a Q-graded Lie algebra. From now on, we fix a triangular decomposition

$$\hat{\mathfrak{g}} = \hat{\mathfrak{g}}^- \oplus \hat{\mathfrak{h}} \oplus \hat{\mathfrak{g}}^+$$

of $\hat{\mathfrak{g}}$ as in § 1.2.2.

Let $M(\Lambda)$ be the Verma module over $\hat{\mathfrak{g}}$ with highest weight $\Lambda \in \hat{\mathfrak{h}}^*$, and let $L(\Lambda)$ be its irreducible quotient (see § 1.2.5). Then they have the following weight space decompositions:

$$M(\Lambda) = \bigoplus_{\alpha \in Q^+} M(\Lambda)_{\Lambda-\alpha}, \quad L(\Lambda) = \bigoplus_{\alpha \in Q^+} L(\Lambda)_{\Lambda-\alpha}.$$

10.1.2 Admissible Representations

In this subsection, we recall the definition of admissible representation, and classify the admissible weights of $\hat{\mathfrak{sl}}_2$. In particular, the character formulae of the admissible representations are also given.

For $\Lambda \in \hat{\mathfrak{h}}^*$, we set

$$\Delta_\Lambda^+ := \left\{ \alpha \in \Delta_{\mathrm{re}}^+ \,\middle|\, \frac{2(\Lambda+\rho,\alpha)}{(\alpha,\alpha)} \in \mathbb{Z} \right\}, \qquad \Pi_\Lambda := \Delta_\Lambda^+ \setminus (\Delta_\Lambda^+ + \Delta_\Lambda^+).$$

Definition 10.1 (**[KW2]**) *$\Lambda \in \hat{\mathfrak{h}}^*$ is called an* **admissible weight**, *if*

1. $\frac{2(\Lambda+\rho,\alpha)}{(\alpha,\alpha)} \in \mathbb{Z}_{>0}$ *for any* $\alpha \in \Delta_\Lambda^+$,
2. $\mathbb{Q}\Pi_\Lambda = \mathbb{Q}\Pi$.

The irreducible highest weight representation $L(\Lambda)$ with admissible highest weight $\Lambda \in \hat{\mathfrak{h}}^*$ is called an **admissible representation**. *Moreover, $k \in \mathbb{C}$ is called an* **admissible level**, *if there exists an admissible weight Λ such that $k = \langle \Lambda, K \rangle$.*

Suppose that k is an admissible level. We denote the set of the admissible weights of level k by P_k^{adm}. Then, we have

Lemma 10.1. *1. $k \in \mathbb{C}$ is an admissible level if and only if there exists $t \in \mathbb{Z}_{\geq 2}$ and $s \in \mathbb{Z}_{>0}$ such that $(s,t) = 1$ (coprime) and $k+2 = \frac{t}{s}$.*
2. Let $k = -2 + \frac{t}{s}$ be admissible level. Then,

$$P_k^{\mathrm{adm}} = \{\Lambda(k;m,n) + x\delta | m,n \in \mathbb{Z},\ 0 < m < t,\ 0 \leq n < s,\ x \in \mathbb{C}\}, \quad (10.1)$$

where

$$\Lambda(k;m,n) := (k - \lambda_{m,n}^k)\Lambda_0 + \lambda_{m,n}^k \Lambda_1$$
$$\lambda_{m,n}^k := (m-1) - (k+2)n.$$

3. Let $\Lambda = \Lambda(k;m,n)$ be an admissible weight. Then, we have

$$\Pi_\Lambda = \{\alpha_1 + n\delta, -\alpha_1 + (s-n)\delta\}.$$

Proof. For $k, \lambda, x \in \mathbb{C}$, let

$$\Lambda = (k-\lambda)\Lambda_0 + \lambda\Lambda_1 + x\delta$$

be an admissible weight of level k, and denote $\Pi_\Lambda = \{\beta_1, \beta_2\}$. Since $\Pi_\Lambda \subset \Delta^+_{\mathrm{re}}$ and $(\beta_1, \beta_2) < 0$, we may assume that

$$\beta_1 = \alpha_1 + n_1\delta, \quad \beta_2 = -\alpha_1 + n_2\delta$$

for some $n_1 \in \mathbb{Z}_{\geq 0}$ and $n_2 \in \mathbb{Z}_{>0}$.

We set $m_i := (\Lambda + \rho, \beta_i)$ for $i = 1, 2$. By definition, $m_i \in \mathbb{Z}_{>0}$. Note that

$$m_1 + m_2 = (\Lambda + \rho, (n_1 + n_2)\delta) = (k+2)(n_1 + n_2).$$

Hence, there exist $t, s, r \in \mathbb{Z}_{>0}$ such that $n_1 + n_2 = rs$, $m_1 + m_2 = rt$ and $k + 2 = \frac{t}{s}$.

Here, we assume that $r > 1$ and lead to a contradiction. If $r > 1$, then either $n_1 \geq s$ or $n_2 > s$ holds, and thus

$$(\alpha_1 + (n_1 - s)\delta \in \Delta^+_\Lambda) \vee (-\alpha_1 + (n_2 - s)\delta \in \Delta^+_\Lambda).$$

This contradicts the definition of Π_Λ. Hence, we have $r = 1$, $n_1 + n_2 = s$, $m_1 + m_2 = t$, and thus $s \geq 1$ and $t \geq 2$. Moreover, $0 \leq n_1 < s$ and $0 < m_1 < t$ and

$$m_1 = (\Lambda + \rho, \beta_1) = (k+2)n_1 + \lambda + 1.$$

Hence, $\lambda = m_1 - 1 - n_1(k+2)$. Now, we have proved the lemma. □

For $w \in W$, we set $\epsilon(w) := \det(w)$ and for $\Lambda \in \hat{\mathfrak{h}}^*$, we set

$$W_\Lambda := \{r_\alpha | \alpha \in \Delta^+_\Lambda\}.$$

We remark that $(W_\Lambda, \{r_\alpha | \alpha \in \Pi_\Lambda\})$ is a Coxeter system.

Theorem 10.1 ([KW2]) *Suppose that $\Lambda \in \hat{\mathfrak{h}}^*$ is an admissible weight. Then, we have*

$$\mathrm{ch}\, L(\Lambda) = \frac{\sum_{w \in W_\Lambda} \epsilon(w) e^{w \circ \Lambda}}{\sum_{w \in W} \epsilon(w) e^{w \circ 0}}.$$

Remark 10.1 *This character formula can be shown in a way similar to the case of the Virasoro algebra, at least for the rank 2 affine Lia algebras, as an application of the uniqueness of the singular vectors proved in [KT].*

For later convenience, we introduce the normalised character. For $\Lambda \in \hat{\mathfrak{h}}^*$ such that $k = \langle \Lambda, K \rangle \neq -h^\vee$ (recall $h^\vee = 2$), we set

$$m_\Lambda := \frac{|\Lambda + \rho|^2}{2(k + h^\vee)} - \frac{|\rho|^2}{2h^\vee}.$$

The number m_Λ is called the **modular anomaly** of Λ. Note that $m_{\Lambda + x\delta} = m_\Lambda + x$.

Definition 10.2 *For $\Lambda \in \hat{\mathfrak{h}}^*$, we set*

$$\chi_\Lambda := e^{-m_\Lambda \delta} \operatorname{ch} L(\Lambda),$$

and call it the normalised character of $L(\Lambda)$.

Normalised characters of admissible representations can be expressed by using the classical theta functions defined as follows: Let

$$\bar{\cdot} : \hat{\mathfrak{h}}^* \longrightarrow \bar{\mathfrak{h}}^*$$

be the projection with respect to $\hat{\mathfrak{h}}^* = \bar{\mathfrak{h}}^* \oplus \mathbb{C}\delta \oplus \mathbb{C}\Lambda_0$.

Definition 10.3 *For $\Lambda \in \hat{\mathfrak{h}}^*$ which satisfies $k := \langle \Lambda, K \rangle = \frac{t}{s}$ for $s, t \in \mathbb{Z}_{>0}$ such that $(s,t) = 1$, we set*

$$\Theta_\Lambda := e^{-\frac{|\Lambda|^2}{2k}\delta} \sum_{\alpha \in M} e^{t_\alpha(\Lambda)},$$

where $M := s\mathbb{Z}\alpha_1$, and $t_\alpha : \hat{\mathfrak{h}}^* \to \hat{\mathfrak{h}}^*$ for $\alpha \in \bar{\mathfrak{h}}$ is defined by

$$t_\alpha(\Lambda) = \Lambda + (\Lambda, \delta)\alpha - \{(\Lambda, \alpha) + \frac{1}{2}|\alpha|^2(\Lambda, \delta)\}\delta.$$

Θ_Λ is called the **classical theta function of degree k with characteristic $\bar{\Lambda}$.**

By definition, one can check

$$\Theta_\Lambda = \sum_{\alpha \in M} e^{\Lambda + k\alpha - \frac{1}{2k}|\Lambda + k\alpha|^2\delta} = \sum_{n \in \mathbb{Z}} e^{\Lambda + tn\alpha_1 - \frac{1}{2k}|\Lambda + tn\alpha_1|^2\delta}. \tag{10.2}$$

In particular, for $k \in \mathbb{Z}_{>0}$ and $j \in \mathbb{Z}$, the classical theta function $\Theta_{j,k} := \Theta_{(k-j)\Lambda_0 + j\Lambda_1}$ has the following expression (cf. (6.15)):

$$\Theta_{j,k} = e^{k\Lambda_0} \sum_{n \in \mathbb{Z} + \frac{j}{2k}} e^{kn\alpha_1 - kn^2\delta}. \tag{10.3}$$

Now, we have

Lemma 10.2. *Suppose that $k = -2 + \frac{t}{s}$ for $t \in \mathbb{Z}_{\geq 2}$ and $s \in \mathbb{Z}_{>0}$. Let $\Lambda = \Lambda(k; m, n)$ be an admissible weight, i.e., $0 < m < t$ and $0 \leq n < s$. Then, we have*

$$\chi_\Lambda = \frac{\Theta_{\Lambda(k;m,n)+\rho} - \Theta_{\Lambda(k;-m,n)+\rho}}{\Theta_{\Lambda(0;1,0)+\rho} - \Theta_{\Lambda(0;-1,0)+\rho}}.$$

Proof. We set

$$A := e^{-\frac{|\Lambda+\rho|^2}{2(k+2)}\delta} \sum_{w \in W_\Lambda} e(w) e^{w \circ \Lambda},$$

$$B := e^{-\frac{|\rho|^2}{4}\delta} \sum_{w \in W} \epsilon(w) e^{w \circ 0}.$$

Then by Theorem 10.1, we have $\chi_\Lambda = \frac{A}{B}$. We compute A and B explicitly.

By Lemma 10.1, we have

$$\Pi_\Lambda = \{\alpha_1 + n\delta, \ -\alpha_1 + (s-n)\delta\}.$$

We set $\beta_1 := \alpha_1 + n\delta$, $\beta_2 := -\alpha_1 + (s-n)\delta$ and $T := r_{\beta_2} r_{\beta_1}$. Note that for $l \in \mathbb{Z}_{\geq 0}$,

$$T^l(\lambda) = \lambda + (\lambda, \delta) l s \alpha_1 - (\lambda, l s \alpha_1 + (ls)^2 \delta)\delta = t_{ls\alpha_1}(\lambda).$$

Hence by direct computation, we have

$$A = (\Theta_{\Lambda(k;m,n)+\rho} - \Theta_{\Lambda(k;-m,n)+\rho}) e^{-\rho},$$

$$B = (\Theta_{\Lambda(0;1,0)+\rho} - \Theta_{\Lambda(0;-1,0)+\rho}) e^{-\rho}.$$

Thus, we have obtained the lemma. □

10.2 Sugawara Construction

In this section, we briefly recall a vertex algebra structure on the $\hat{\mathfrak{sl}}_2$-module $L(k\Lambda_0)$. In particular, we will show that $L(k\Lambda_0)$ is a conformal vertex algebra if $k \neq 0, -2$.

10.2.1 Vertex Algebra Structure of Vacuum $\hat{\mathfrak{sl}}_2$-modules

Here, we first define a vertex algebra structure on the so-called vacuum representation $V_{k\Lambda_0}$ of level $k \in \mathbb{C}$.

For $k \in \mathbb{C}$, let $\mathbb{C}_k := \mathbb{C}\mathbf{1}_{k\Lambda_0}$ be the $\hat{\mathfrak{g}}^{\geq} := \overline{\mathfrak{g}} \otimes \mathbb{C}[t] \oplus \mathbb{C}K \oplus \mathbb{C}d$-module defined by

$$\overline{\mathfrak{g}} \otimes \mathbb{C}[t].\mathbf{1}_{k\Lambda_0} := \{0\}, \qquad K.\mathbf{1}_{k\Lambda_0} := k\mathbf{1}_{k\Lambda_0}, \qquad d.\mathbf{1}_{k\Lambda_0} := 0.$$

The induced $\hat{\mathfrak{sl}}_2$-module

$$V_{k\Lambda_0} := \mathrm{Ind}_{\hat{\mathfrak{g}}^{\geq}}^{\hat{\mathfrak{g}}} \mathbb{C}_k$$

is a highest weight $\widehat{\mathfrak{sl}}_2$-module with highest weight $k\Lambda_0$. We introduce a $\mathbb{Z}$-graded vertex algebra structure on $V_{k\Lambda_0}$ as follows.

For simplicity, we set $v_k := 1 \otimes \mathbf{1}_{k\Lambda_0}$.

1. (**Gradation**) For $n_1 \leq n_2 \leq \cdots \leq n_m < 0$ and $x_1, x_2, \cdots, x_m \in \bar{\mathfrak{g}}$, set
$$\deg(x_1(n_1)x_2(n_2)\cdots x_m(n_m).v_k) := -\sum_{i=1}^{m} n_i.$$
2. (**Vacuum vector**) v_k.
3. (**Translation operator**) Set $Tv_k := 0$ and $[T, x(n)] := -nx(n-1)$ for $x \in \bar{\mathfrak{g}}$.
4. (**Vertex operators**) Set $Y(v_k, z) := \mathrm{id}$ and
$$Y(x(-1).v_k, z) := x(z) := \sum_{n \in \mathbb{Z}} x(n) z^{-n-1},$$
for $x \in \bar{\mathfrak{g}}$.

By the strong reconstruction theorem (cf. Theorem C.1), these data define a $\mathbb{Z}$-graded vertex algebra structure on $V_{k\Lambda_0}$. Since, the maximal proper submodule of $V_{k\Lambda_0}$ is a $\mathbb{Z}$-graded vertex ideal, a vertex algebra structure on $V_{k\Lambda_0}$ induces a $\mathbb{Z}$-graded vertex algebra structure on $L(k\Lambda_0)$. In the next subsection, we will construct a conformal vector on $V_{k\Lambda_0}$ and hence on $L(k\Lambda_0)$.

10.2.2 Segal–Sugawara Operator

Here, we briefly recall the so-called Segal–Sugawara operator and some of its consequences.

Let $\{u_1, u_2, u_3\}$ be a basis of $\bar{\mathfrak{g}}$ and $\{u^1, u^2, u^3\}$ be the dual basis with respect to the form $(\cdot,\cdot)$ on $\bar{\mathfrak{g}}$.

Definition 10.4 *Set*
$$\begin{aligned} T(z) :=& Y(\sum_{i=1}^{3} u_i(-1)u^i(-1).v_k, z) \\ =& \sum_{i=1}^{3} {}^\circ_\circ u_i(z)u^i(z) {}^\circ_\circ =: \sum_{n \in \mathbb{Z}} T_n z^{-n-2} \in \mathrm{End}(V_{k\Lambda_0})[[z, z^{-1}]]. \end{aligned}$$

It is called the **Segal–Sugawara operator**.

By abuse of the language the vertex operator on $L(k\Lambda_0)$ induced by $T(z)$ on $V_{k\Lambda_0}$ is also called the Segal–Sugawara operator. The following lemma is a simple application of Theorem C.5:

Lemma 10.3. *Suppose that $k \neq -2$. Set $L_n := \frac{1}{2(k+2)} T_n$. We have*

$$[L_m, L_n] = (m-n)L_{m+n} + \delta_{m+n,0}\frac{m^3-m}{12}c_k,$$
$$[L_m, x(n)] = -nx(n+m),$$

where $c_k := \frac{3k}{k+2}$.

In particular, this lemma implies that, for $k \in \mathbb{C}$ such that $k \neq 2$, the vertex algebra $V_{k\Lambda_0}$ is a conformal vertex algebra with the conformal vector

$$\omega_k := \frac{1}{2(k+2)}\sum_{i=1}^{3} u_i(-1)u^i(-1).v_k \in V_{k\Lambda_0}.$$

The image of ω_k under the canonical projection $V_k \twoheadrightarrow L(k\Lambda_0)$ is a non-zero element if and only if $k \neq 0$. Hence, the next corollary follows:

Corollary 10.1 *Let $k \in \mathbb{C}$ such that $k \neq 0, -2$. Then, $(L(k\Lambda_0), \omega_k)$ is a conformal vertex algebra.*

In [FZ], I. Frenkel and Y. Zhu showed that the conformal vertex algebra $(L(k\Lambda_0), \omega)$ becomes a rational vertex operator algebra for $k \in \mathbb{Z}_{>0}$. But for an admissible k which is not an integer, this is not the case. However, we should mention the following:

Remark 10.2 *Let $k \in \mathbb{C}$ be an admissible level such that $k \notin \mathbb{Z}_{\geq 0}$.*

1. *In [AdM1], D. Adamović and A. Milas showed that the conformal vertex algebra $(L(k\Lambda_0), \omega_k)$ is rational in the category $\mathcal{O}$, that is, if we allow only $L(k\Lambda_0)$-modules that are objects of $\mathcal{O}$, then it is rational.*
2. *In [DLM1], C. Dong, H. Li and G. Mason showed that the vertex algebra $L(k\Lambda_0)$ with the conformal vector*

$$\omega_k(a) := \omega_k - \frac{1}{2}ah(-2).v_k$$

for $a \in \mathbb{Q}_{>0}$ such that $a < 1$ is a rational $\mathbb{Q}$-graded vertex algebra, i.e., any $\mathbb{Q}_{\geq 0}$-graded weak $L(k\Lambda_0)$-module is completely reducible, where the central charge of the Virasoro algebra is given by $c_k - 6ka^2$.

In both cases, any irreducible module (with the obvious restriction for 1.) is of the form $L(\Lambda)$ $(\Lambda \in P_k^{\mathrm{adm}})$.

10.3 Coset Constructions

In this section, after rewriting the fundamental characters of $\hat{\mathfrak{sl}}_2$, we consider the so-called coset constructions of two different kinds.

10.3.1 Fundamental Characters of $\hat{\mathfrak{sl}}_2$

First, we recall two classical formulae which are equivalent to the denominator identities of $A_1^{(1)}$ and $A_2^{(2)}$, respectively (see, e.g., [Chan] for their analytic proof):

$$\prod_{m=1}^{\infty}(1-q^m)(1-q^{m-1}w)(1-q^m w^{-1}) = \sum_{n\in\mathbb{Z}}(-1)^n q^{\frac{1}{2}n(n-1)}w^n, \tag{10.4}$$

which is called the **Jacobi triple product identity** and

$$\begin{aligned}&\prod_{m=1}^{\infty}(1-q^m)(1-q^{m-1}w)(1-q^m w^{-1})(1-q^{2m-1}w^2)(1-q^{2m-1}w^{-2})\\ &=\sum_{n\in\mathbb{Z}}\{q^{\frac{1}{2}n(3n-1)}w^{3n}-q^{\frac{1}{2}n(3n+1)}w^{3n+1}\}=\sum_{n\in\mathbb{Z}}q^{\frac{1}{2}n(3n-1)}(w^{3n}-w^{-3n+1}),\end{aligned} \tag{10.5}$$

which is called the **Watson quintuple identity**. Based on these identities, we rewrite the fundamental characters χ_{Λ_0} and χ_{Λ_1} as follows.

We recall that the Weyl–Kac character formula for the integrable module is written in terms of the classical theta function (10.3) as follows: for $j,k\in\mathbb{Z}_{\geq 0}$ such that $0\leq j\leq k$,

$$\chi_{(k-j)\Lambda_0+j\Lambda_1}=\frac{\Theta_{j+1,k+2}-\Theta_{-(j+1),k+2}}{\Theta_{1,2}-\Theta_{-1,2}}. \tag{10.6}$$

Clearly, the denominator of the above formula can be factorised as

$$\Theta_{1,2}-\Theta_{-1,2}=e^{\rho-\frac{1}{8}\delta}\prod_{n>0}(1-e^{-n\delta})(1-e^{-(n-1)\delta-\alpha_1})(1-e^{-n\delta+\alpha_1}), \tag{10.7}$$

which is equivalent to (10.4). Hence, we calculate the numerator of (10.6) below. For $j=0,1$, we have

$$\begin{aligned}&\Theta_{j+1,3}-\Theta_{-(j+1),3}\\ &=e^{3\Lambda_0}\sum_{n\in\mathbb{Z}}\left\{e^{3\left(n+\frac{1}{6}(j+1)\right)\alpha_1-3\left(n+\frac{1}{6}(j+1)\right)^2\delta}-e^{3\left(n-\frac{1}{6}(j+1)\right)\alpha_1-3\left(n-\frac{1}{6}(j+1)\right)^2\delta}\right\}\\ &=e^{3\Lambda_0+\frac{1}{2}(j+1)\alpha_1-\frac{1}{12}(j+1)^2\delta}\sum_{n\in\mathbb{Z}}e^{-n(3n-(j+1))\delta}\left\{e^{-3n\alpha_1}-e^{(3n-(j+1))\alpha_1}\right\}.\end{aligned}$$

Case-by-case calculations show that

$$\sum_{n\in\mathbb{Z}} e^{-n(3n-(j+1))\delta}\left\{e^{-3n\alpha_1} - e^{(3n-(j+1))\alpha_1}\right\}$$
$$=\sum_{n\in\mathbb{Z}} e^{-n(3n-1)\delta}\left\{e^{-3n\alpha_{1-j}} - e^{(3n-1)\alpha_{1-j}}\right\},$$

hence, by (10.5), it follows that

$$\begin{aligned}&\Theta_{j+1,3} - \Theta_{-(j+1),3}\\ =&e^{3\Lambda_0+\frac{1}{2}(j+1)\alpha_1-\frac{1}{12}(j+1)^2\delta}\sum_{n\in\mathbb{Z}} e^{-n(3n-1)\delta}\left\{e^{-3n\alpha_{1-j}} - e^{(3n-1)\alpha_{1-j}}\right\}\\ =&e^{3\Lambda_0+\frac{1}{2}(j+1)\alpha_1-\frac{1}{12}(j+1)^2\delta}\prod_{m>0}(1-e^{-2m\delta})(1-e^{-2(m-1)\delta-\alpha_{1-j}})(1-e^{-2m\delta+\alpha_{1-j}})\\ &\qquad(1-e^{-2(2m-1)\delta-2\alpha_{1-j}})(1-e^{-2(2m-1)\delta+2\alpha_{1-j}}).\end{aligned}$$

Simple calculations show

$$\begin{aligned}&\prod_{m>0}(1-e^{-2(m-1)\delta-\alpha_{1-j}})(1-e^{-2(2m-1)\delta-2\alpha_{1-j}})\\ =&\prod_{m>0}(1-e^{-(m-1)\delta-\alpha_{1-j}})(1+e^{-(2m-1)\delta-\alpha_{1-j}})\\ =&\prod_{m>0}(1-e^{-m\delta+\alpha_j})(1+e^{-2m\delta+\alpha_j}),\\ &\prod_{m>0}(1-e^{-2m\delta+\alpha_{1-j}})(1-e^{-2(2m-1)\delta+2\alpha_{1-j}})\\ =&\prod_{m>0}(1-e^{-m\delta+\alpha_{1-j}})(1+e^{-(2m-1)\delta+\alpha_{1-j}})\\ =&\prod_{m>0}(1-e^{-(m-1)\delta-\alpha_j})(1+e^{-2(m-1)\delta-\alpha_j}),\end{aligned}$$

which implies

$$\begin{aligned}&\Theta_{j+1,3} - \Theta_{-(j+1),3}\\ =&e^{3\Lambda_0+\frac{1}{2}(j+1)\alpha_1-\frac{1}{12}(j+1)^2\delta}\prod_{m>0}(1-e^{-2m\delta})(1-e^{-(m-1)\delta-\alpha_1})(1-e^{-m\delta+\alpha_1})\\ &\qquad(1+e^{-2(m-1)\delta-\alpha_j})(1+e^{-2m\delta+\alpha_j}).\end{aligned}$$

Hence, it follows from (10.6) and (10.7) that

$$\chi_{\Lambda_j} = \frac{e^{\Lambda_0+\frac{1}{2}j\alpha_1-\frac{1}{12}(j^2+2j)\delta}\prod_{m>0}(1-e^{-2m\delta})(1+e^{-2(m-1)\delta-\alpha_j})(1+e^{-2m\delta+\alpha_j})}{\eta(e^{-\delta})},$$

where $\eta(q) := q^{\frac{1}{24}} \prod_{m>0}(1-q^m)$ is the Dedekind eta-function (cf. (6.14)). It follows from (10.4) and the case-by-case computations that

$$\prod_{m>0}(1-e^{-2m\delta})(1+e^{-2(m-1)\delta-\alpha_j})(1+e^{-2m\delta+\alpha_j})$$
$$=\sum_{n\in\mathbb{Z}} e^{-(n^2-n)\delta-n\alpha_j} = \sum_{n\in\mathbb{Z}} e^{-n(n+j)\delta+n\alpha_1},$$

from which we conclude

$$\chi_{\Lambda_j} = \frac{e^{\Lambda_0+\frac{1}{6}j(j-1)\delta}\sum_{n\in\mathbb{Z}+\frac{1}{2}j} e^{-n^2\delta+n\alpha_1}}{\eta(e^{-\delta})}.$$

Thus, by (10.3), we have shown

Lemma 10.4. *For $j=0,1$, we have*

$$\chi_{\Lambda_j} = \frac{\Theta_{\Lambda_j}}{\eta(e^{-\delta})}.$$

10.3.2 Coset $(\hat{\mathfrak{sl}}_2)_1 \times (\hat{\mathfrak{sl}}_2)_k/(\hat{\mathfrak{sl}}_2)_{k+1}$

In the sequel, we assume that $k \neq -2, -3$, unless otherwise stated.

We first recall that, for vertex algebras $(A, \deg_A, |0\rangle_A, T_A, Y_A)$ and $(B, \deg_B, |0\rangle_B, T_B, Y_B)$, the pair $(A\otimes B, \deg_A + \deg_B, |0\rangle_A \otimes |0\rangle_B, T_A \otimes \mathrm{id} + \mathrm{id}\otimes T_B, Y_{A\otimes B})$, where we set $Y_{A\otimes B}(u\otimes v, z) := Y_A(u,z)\otimes Y_B(v,z)$, becomes a vertex algebra.

Let $V^{(1)}$ (resp. $V^{(k)}$) be an $L(\Lambda_0)$-module (resp. $L(k\Lambda_0)$-module) such that $V^{(1)}, V^{(k)} \in \mathrm{Ob}(\mathcal{O})$.

Definition 10.5 *We define $L_n \in \mathrm{End}(V^{(1)}\otimes V^{(k)})$ by*

$$\sum_{n\in\mathbb{Z}} L_n z^{-n-2} = Y_{V^{(1)}}(\omega_1, z)\otimes \mathrm{id} + \mathrm{id}\otimes Y_{V^{(k)}}(\omega_k, z)$$
$$- \frac{1}{2(k+3)}\sum_{i=1}^{3} Y_{V^{(1)}\otimes V^{(k)}}(\Delta(u_i(-1)u^i(-1)).v_1\otimes v_k, z),$$

where Δ is the coproduct of $U(\hat{\mathfrak{g}})$.

Below, we omit indicating the modules on vertex operators, for simplicity. We remark that

$$\frac{1}{2(k+3)}\sum_{i=1}^{3}Y(\Delta(u_i(-1)u^i(-1)).v_1\otimes v_k,z)$$
$$=\frac{3}{k+3}Y(\omega_1,z)\otimes\mathrm{id}+\frac{k+2}{k+3}\mathrm{id}\otimes Y(\omega_k,z)+\frac{1}{k+3}\sum_{i=1}^{3}u_i(z)\otimes u^i(z).$$

Hence, the operators $\{L_n\}$ satisfy

$$\sum_{n\in\mathbb{Z}}L_nz^{-n-2}=\frac{1}{k+3}\left\{kY(\omega_1,z)\otimes\mathrm{id}+\mathrm{id}\otimes Y(\omega_k,z)-\sum_{i=1}^{3}u_i(z)\otimes u^i(z)\right\}.$$

We have

Proposition 10.1 *Suppose that $k\neq -2,-3$. Let $L_n\in\mathrm{End}(V^{(1)}\otimes V^{(k)})$ as above.*

1. For any $x\in\bar{\mathfrak{g}}$ and $m,n\in\mathbb{Z}$,

$$[L_m,\Delta(x(n))]=0.$$

2.

$$[L_m,L_n]=(m-n)L_{m+n}+\delta_{m+n,0}\frac{1}{12}(m^3-m)c\,\mathrm{id}_{V^{(1)}\otimes V^{(k)}},$$

where $c=1-\frac{6}{(k+2)(k+3)}$.

Hence, Vir *and $\hat{\mathfrak{g}}'$ simultaneously act on the space $V^{(1)}\otimes V^{(k)}$, and their actions commute.*

Proof. Here, we denote $Y(\omega_1,z)=\sum_{n\in\mathbb{Z}}L_n^{(1)}z^{-n-2}$ and $Y(\omega_k,z)=\sum_{n\in\mathbb{Z}}L_n^{(k)}z^{-n-2}$, respectively.

First, we notice that, by Lemma 10.3, the Fourier modes of the vertex operator

$$\frac{1}{2(k+3)}\sum_{i=1}^{3}Y(\Delta(u_i(-1)u^i(-1)).v_1\otimes v_k,z)=:\sum_{n\in\mathbb{Z}}L_n^{(k+1)}z^{-n-2}$$

satisfy the commutation relation of the Virasoro algebra with central charge c_{k+1}. By definition, we have

$$\begin{aligned}[L_m,\Delta(x(n))]=&[L_m^{(1)},x(n)]\otimes\mathrm{id}+\mathrm{id}\otimes[L_m^{(k)},x(n)]-[L_m^{(k+1)},\Delta(x(n))]\\=&-nx(m+n)\otimes\mathrm{id}-n\mathrm{id}\otimes x(m+n)+n\Delta(x(m+n))=0.\end{aligned}$$

Next, we show the second statement. The first statement implies that $[L_m,L_n^{(k+1)}]=0$. Hence, we get

$$
\begin{aligned}
[L_m, L_n] &= [L_m, L_n^{(1)} \otimes 1 + 1 \otimes L_n^{(k)}] \\
&= [L_m^{(1)}, L_n^{(1)}] \otimes 1 + 1 \otimes [L_m^{(k)}, L_n^{(k)}] - [L_m^{(k+1)}, L_n^{(1)} \otimes 1 + 1 \otimes L_n^{(k)}].
\end{aligned}
$$

Further, we have

$$[L_m^{(k+1)}, L_n^{(1)} \otimes 1 + 1 \otimes L_n^{(k)}] = [L_m^{(k+1)}, L_n + L_n^{(k+1)}] = [L_m^{(k+1)}, L_n^{(k+1)}].$$

Hence the following commutation relation holds:

$$[L_m, L_n] = (m-n)L_{m+n} + \delta_{m+n,0}\frac{1}{12}(m^3 - m)(c_1 + c_k - c_{k+1})\mathrm{id}_{V^{(1)}\otimes V^{(k)}}.$$

Since $c_1 + c_k - c_{k+1} = 1 - \frac{6}{(k+2)(k+3)}$, we have completed the proof. □

From now on, we take an admissible representation $L(\Lambda)$ as $V^{(k)}$ and $L(\Lambda_0) \oplus L(\Lambda_1)$ as $V^{(1)}$. Now we can state the main result of this section.

Theorem 10.2 *Suppose that $k \in \mathbb{Q}_{>-2}$ is an admissible level. Let p, q, m, n be integers such that $q - p > 0$, $p \geq 2$, $(p, q) = 1$ (coprime), $k + 2 = \frac{p}{q-p}$, $0 < m < p$ and $0 \leq n < q - p$. Then we have*

$$
\begin{aligned}
&[L(\Lambda_0) \oplus L(\lambda_1)] \otimes L(\Lambda(k; m, n)) \\
&\simeq \bigoplus_{0<m'<q} L(\Lambda(k+1; m', n)) \boxtimes L(c_{p,q}, h_{m,m'}(\tfrac{q}{p}))
\end{aligned}
$$

as $\hat{\mathfrak{g}}' \oplus \mathrm{Vir}$-module.

For the proof, we first verify the above isomorphism at the level of their normalised characters as follows.

For this purpose, we fix a coordinate of $\hat{\mathfrak{h}}$ as follows: For

$$v = 2\pi\sqrt{-1}\left(-\tau d + \frac{1}{2}zh + uK\right) \in \hat{\mathfrak{h}}, \tag{10.8}$$

we denote v by (τ, z, u). For $(\tau, z, u) \in \mathbb{H} \times \mathbb{C}^2$, we set

$$\Theta_\Lambda(\tau, z, u) = e^{-\frac{|\Lambda|^2}{2k}\langle\delta, v\rangle} \sum_{\alpha \in M} e^{\langle t_\alpha(\Lambda), v\rangle},$$

and in general, for a normalised character $\chi_\Lambda = \sum_\lambda c_\lambda e^\lambda$, we set

$$\chi_\Lambda(\tau, z, u) := \sum_\lambda c_\lambda e^{\langle\lambda, v\rangle}.$$

Recall that $\chi_{L(c,h)}(\tau)$ denotes the normalised character of the Virasoro algebra defined in § 6.4.

The first step to the proof of Theorem 10.2 is

Lemma 10.5 ([KW2]).

$$\{\chi_{\Lambda_0}(\tau,z,u)+\chi_{\Lambda_1}(\tau,z,u)\}\chi_{\Lambda(k;m,n)}(\tau,z,u)$$
$$=\sum_{0<m'<q}\chi_{\Lambda(k+1;m',n)}(\tau,z,u)\chi_{L(c_{p,q},h_{m,m'}(\frac{q}{p}))}(\tau).$$

The proof of this lemma will be carried out in the next subsection, after recalling some facts about the classical theta functions.

Proof of Theorem 10.2. In this proof, for simplicity, we set $\Lambda'_m := \Lambda(k+1;m',s)$, $c=c_{p,q}$ and $h_{m'} := h_{m,m'}(\frac{q}{p})$.

By the preliminary lemma, the irreducible representations $L(\Lambda_{m'})\boxtimes L(c,h_{m'})$ appear in $[L(\Lambda_0)\oplus L(\Lambda_1)]\otimes L(\Lambda(k;m,s))$ with multiplicity one, and they exhaust its irreducible factors. Hence, if we show that

$$\mathrm{Ext}^1_{\hat{\mathfrak{g}}'\oplus\mathrm{Vir}}(L(\Lambda_{m'_1})\boxtimes L(c,h_{m'_1}),L(\Lambda_{m'_2})\boxtimes L(c,h_{m'_2}))=\{0\},$$

then we obtain the theorem as a consequence of Lemma A.3. Indeed, we have

$$\begin{aligned}&\mathrm{Ext}^1_{\hat{\mathfrak{g}}'\oplus\mathrm{Vir}}(L(\Lambda_{m'_1})\boxtimes L(c,h_{m'_1}),L(\Lambda_{m'_2})\boxtimes L(c,h_{m'_2}))\\ &\simeq \mathrm{Hom}_{\hat{\mathfrak{g}}'}((L(\Lambda_{m'_1}),L(\Lambda_{m'_2}))\otimes\mathrm{Ext}^1_{\mathrm{Vir}}(L(c,h_{m'_1}),L(c,h_{m'_2}))\\ &\quad\oplus\mathrm{Ext}^1_{\hat{\mathfrak{g}}'}((L(\Lambda_{m'_1}),L(\Lambda_{m'_2}))\otimes\mathrm{Hom}_{\mathrm{Vir}}(L(c,h_{m'_1}),L(c,h_{m'_2}))\\ &\simeq\{0\}.\end{aligned}$$

Hence, we have arrived at the conclusion. □

10.3.3 Properties of the Theta Function and Proof of Lemma 10.5

First, we recall some properties of the classical theta function Θ_Λ.

Let $\Lambda\in\hat{\mathfrak{h}}^*$ be such that $\langle\Lambda,K\rangle=k$, where $k=\frac{t}{s}$ for some $t,s\in\mathbb{Z}_{>0}$ such that $(s,t)=1$.

Lemma 10.6. *If $\Lambda'\in\hat{\mathfrak{h}}^*$ satisfies $\Lambda-\Lambda'\in t\mathbb{Z}\alpha_1\oplus\mathbb{C}\delta$, then $\Theta_\Lambda=\Theta_{\Lambda'}$.*

Proof. This follows from (10.2) and the remark that for $\lambda\in\hat{\mathfrak{h}}^*$ such that $\langle\lambda,K\rangle=k\neq 0$, $\lambda':=\lambda-\frac{(\lambda,\lambda)}{2k}\delta$ is normalised so as to satisfy $(\lambda',\lambda')=0$. □

Lemma 10.7. *Suppose that $\Lambda'\in\hat{\mathfrak{h}}^*$ satisfies $\langle\Lambda',K\rangle=k'\in\mathbb{Z}_{>0}$. Then, we have*

$$\Theta_\Lambda\Theta_{\Lambda'}=\sum_{i=0}^{t+k's-1}\Theta_{\Lambda+\Lambda'+k'i\alpha_1}\psi_i^{(k,k')}(\Lambda,\Lambda'),$$

where

$$\psi_i^{(k,k')}(\Lambda,\Lambda') = \sum_{\substack{j\in\mathbb{Z} \\ j\equiv i \mod (t+k's)}} e^{-\frac{|k'\bar{\Lambda}-k\bar{\Lambda}'-jkk'\alpha_1|^2}{2kk'(k+k')}\delta}.$$

Proof. By Lemma 10.6 and its proof, we may assume that $|\Lambda|^2 = |\Lambda'|^2 = 0$. Hence, by definition, we have

$$\Theta_\Lambda \Theta_{\Lambda'} = \sum_{\beta\in s\mathbb{Z}\alpha_1} \sum_{\gamma\in\mathbb{Z}\alpha_1} e^{t_\beta(\Lambda)+t_\gamma(\Lambda')}.$$

By the additivity $t_{\beta+\gamma} = t_\beta \circ t_\gamma$, shifting γ to $\gamma+\beta$, we obtain

$$\begin{aligned}\Theta_\Lambda \Theta_{\Lambda'} &= \sum_{\beta\in s\mathbb{Z}\alpha_1} \sum_{\gamma\in\mathbb{Z}\alpha_1} e^{t_\beta(\Lambda+t_\gamma(\Lambda'))} \\ &= \sum_{\gamma\in\mathbb{Z}\alpha_1} e^{\frac{1}{2(k+k')}|\Lambda+t_\gamma(\Lambda')|^2\delta}\Theta_{\Lambda+t_\gamma(\Lambda')}.\end{aligned}$$

Since $t_{\gamma_1}(\Lambda') - t_{\gamma_2}(\Lambda') \equiv k'(\gamma_1-\gamma_2) \mod \mathbb{C}\delta$, it follows from Lemma 10.6 that

$$\Theta_\Lambda \Theta_{\Lambda'} = \sum_{\gamma\in\mathbb{Z}\alpha_1 \mod (t+k's)\mathbb{Z}\alpha_1} \Theta_{\Lambda+t_\gamma(\Lambda')} \sum_{\gamma'\in(t+k's)\mathbb{Z}\alpha_1} e^{\frac{1}{2(k+k')}|\Lambda+t_{\gamma+\gamma'}(\Lambda')|^2\delta}.$$

By $|\Lambda|^2 = |\Lambda'|^2 = 0$, we have

$$|\Lambda + t_{\gamma+\gamma'}(\Lambda')|^2 = -\frac{1}{kk'}|k'\Lambda - kt_{\gamma+\gamma'}(\Lambda')|^2.$$

Since

$$k'\Lambda - kt_{\gamma+\gamma'}(\Lambda') \equiv k'\bar{\Lambda} - k\bar{\Lambda}' - kk'(\gamma+\gamma') \mod \mathbb{C}\delta,$$

we see that

$$|k'\Lambda - kt_{\gamma+\gamma'}(\Lambda')|^2 = |k'\bar{\Lambda} - k\bar{\Lambda}' - kk'(\gamma+\gamma')|^2.$$

Since $\Lambda + t_\gamma(\Lambda') \equiv \Lambda + \Lambda' + k'\gamma \mod \mathbb{C}\delta$, putting $\gamma = i\alpha_1$ $(0 \leq i < t+k's)$ and $\gamma+\gamma' = j\alpha_1$, we obtain the conclusion. □

PROOF OF LEMMA 10.5. Recall that by Corollary 6.1, one has

$$\chi_{L(c_{p,q},h_{m,m'}(\frac{q}{p}))}(\tau) = \eta(\tau)^{-1}\left[\Theta_{qm-pm',pq}(\tau) - \Theta_{qm+pm',pq}(\tau)\right].$$

Hence, by Lemmas 10.2 and 10.4, it is enough to show that

$$\left[\Theta_{\Lambda_0}(\tau,z,u)+\Theta_{\Lambda_1}(\tau,z,u)\right]\left[\Theta_{\Lambda(k;m,n)+\rho}(\tau,z,u)-\Theta_{\Lambda(k;-m,n)+\rho}(\tau,z,u)\right]$$
$$=\sum_{0<m'<q}\left[\Theta_{\Lambda(k+1;m',n)+\rho}(\tau,z,u)-\Theta_{\Lambda(k+1;-m',n)+\rho}(\tau,z,u)\right]$$
$$\times\left[\Theta_{qm-pm',pq}(\tau)-\Theta_{qm+pm',pq}(\tau)\right],$$

For $\sigma\in\{\pm 1\}$ and $l\in\{0,1\}$, by Lemma 10.7, we have

$$\Theta_{\Lambda(k;\sigma m,n)+\rho}\Theta_{\Lambda_l}=\sum_{i=0}^{q-1}\Theta_{\Lambda(k;\sigma m,n)+\Lambda_l+\rho+i\alpha_1}\psi_i^{(k+2,1)}(\Lambda(k;\sigma m,n)+\rho,\Lambda_l)$$

where

$$\psi_i^{(k+2,1)}(\Lambda(k;\sigma m,n)+\rho,\Lambda_l)=\sum_{\substack{j\in\mathbb{Z}\\ j\equiv i \mod q}}e^{-\frac{(p-q)^2}{4pq}[\sigma m-\frac{p}{q-p}(n+l+2j)]^2\delta},$$

since $k+2=\frac{p}{q-p}$ and $\Lambda_l=\Lambda_0+\frac{1}{2}l\alpha_1$. Hence, if we set

$$\psi_{i,\pm}:=\sum_{\substack{j\in\mathbb{Z}\\ j\equiv i \mod 2q}}e^{-\frac{(p-q)^2}{4pq}[\pm m-\frac{p}{q-p}(n+j)]^2\delta},$$

for $0\leq i<2q$, then we get

$$\left[\Theta_{\Lambda_0}+\Theta_{\Lambda_1}\right]\Theta_{\Lambda(k;\sigma m,n)+\rho}=\sum_{i=0}^{2q-1}\Theta_{\Lambda_0+\Lambda(k;\sigma m,n)+\rho+\frac{1}{2}i\alpha_1}\psi_{i;\sigma}.$$

Here, we notice that

$$\Lambda(k;\sigma m,n)+\Lambda_0+\frac{1}{2}i\alpha_1=\Lambda(k+1;\sigma m+i+n,n)$$

by definition. Further, since $k+3=\frac{q}{q-p}$, Lemma 10.6 implies

$$\Theta_{\Lambda(k+1,m'+2q,n)+\rho}=\Theta_{\Lambda(k+1,m',n)+\rho}. \tag{10.9}$$

Hence, we see that

$$\begin{aligned}\left[\Theta_{\Lambda_0}+\Theta_{\Lambda_1}\right]\Theta_{\Lambda(k;\pm m,n)+\rho}&=\sum_{i=0}^{2q-1}\Theta_{\Lambda(k+1;\pm m+i+n,n)+\rho}\psi_{i,\pm}\\&=\sum_{i=0}^{2q-1}\Theta_{\Lambda(k+1;\pm(m+i),n)+\rho}\psi_{\pm i-n,\pm}.\end{aligned}$$

Since

$$\widetilde{\psi}_i := \psi_{i-n,+} = \psi_{-i-n,-} = \sum_{\substack{j\in\mathbb{Z} \\ j\equiv i \mod 2q}} e^{-\frac{(p-q)^2}{4pq}[m-\frac{pj}{q-p}]^2\delta},$$

we obtain

$$\begin{aligned}
&[\Theta_{\Lambda_0} + \Theta_{\Lambda_1}]\left[\Theta_{\Lambda(k;m,n)+\rho} - \Theta_{\Lambda(k;-m,n)+\rho}\right] \\
&= \sum_{i=0}^{2q-1} \left[\Theta_{\Lambda(k+1;m+i,n)+\rho} - \Theta_{\Lambda(k+1;-m-i,n)+\rho}\right]\widetilde{\psi}_i \\
&= \left\{ \sum_{m\le m'<q} + \sum_{m'=q,2q} + \sum_{q<m'<2q} + \sum_{2q<m'<2q+m} \right\} \\
&\quad \times \left[\Theta_{\Lambda(k+1;m',n)+\rho} - \Theta_{\Lambda(k+1;-m',n)+\rho}\right]\widetilde{\psi}_{m'-m}.
\end{aligned}$$

For the first and the fourth term, by (10.9) and $\widetilde{\psi}_{i+2q} = \widetilde{\psi}_i$,

$$\begin{aligned}
&\sum_{2q<m'<2q+m} \left[\Theta_{\Lambda(k+1;m',n)+\rho} - \Theta_{\Lambda(k+1;-m',n)+\rho}\right]\widetilde{\psi}_{m'-m} \\
&= \sum_{0<m'<m} \left[\Theta_{\Lambda(k+1;m',n)+\rho} - \Theta_{\Lambda(k+1;-m',n)+\rho}\right]\widetilde{\psi}_{m'-m},
\end{aligned}$$

and thus,

$$\begin{aligned}
&\left\{ \sum_{m\le m'<q} + \sum_{2q<m'<2q+m} \right\}\left[\Theta_{\Lambda(k+1;m',n)+\rho} - \Theta_{\Lambda(k+1;-m',n)+\rho}\right]\widetilde{\psi}_{m'-m} \\
&= \sum_{0<m'<q} \left[\Theta_{\Lambda(k+1;m',n)+\rho} - \Theta_{\Lambda(k+1;-m',n)+\rho}\right]\widetilde{\psi}_{m'-m}.
\end{aligned}$$

For the second term, by (10.9),

$$\sum_{m'=q,2q} \left[\Theta_{\Lambda(k+1;m',n)+\rho} - \Theta_{\Lambda(k+1;-m',n)+\rho}\right]\widetilde{\psi}_{m'-m} = 0.$$

For the third term,

$$\begin{aligned}
&\sum_{q<m'<2q} \left[\Theta_{\Lambda(k+1;m',n)+\rho} - \Theta_{\Lambda(k+1;-m',n)+\rho}\right]\widetilde{\psi}_{m'-m} \\
&= \sum_{-q<m'<0} \left[\Theta_{\Lambda(k+1;m',n)+\rho} - \Theta_{\Lambda(k+1;-m',n)+\rho}\right]\widetilde{\psi}_{m'-m}.
\end{aligned}$$

By putting $m'' := -m'$, we have

$$\sum_{-q<m'<0} \left[\Theta_{\Lambda(k+1;m',n)+\rho} - \Theta_{\Lambda(k+1;-m',n)+\rho}\right] \widetilde{\psi}_{m'-m}$$
$$= - \sum_{0<m''<q} \left[\Theta_{\Lambda(k+1;m'',n)+\rho} - \Theta_{\Lambda(k+1;-m'',n)+\rho}\right] \widetilde{\psi}_{-m''-m}.$$

Therefore, we obtain

$$\begin{aligned}
&[\Theta_{\Lambda_0} + \Theta_{\Lambda_1}] \left[\Theta_{\Lambda(k;m,n)+\rho} - \Theta_{\Lambda(k;-m,n)+\rho}\right] \\
&= \sum_{0<m'<q} \left[\Theta_{\Lambda(k+1;m',n)+\rho} - \Theta_{\Lambda(k+1;-m',n)+\rho}\right] [\widetilde{\psi}_{m'-m} - \widetilde{\psi}_{-m'-m}].
\end{aligned}$$

By evaluating $\widetilde{\psi}_{\pm m'-m}$ at $v \in \hat{\mathfrak{h}}$ as in (10.8), we see that

$$\widetilde{\psi}_{\pm m'-m}(v) = \Theta_{qm\mp pm',pq}(\tau),$$

from which we have completed the proof. □

Remark 10.3 *One may prove, in a way similar to the above proof, that for* $l = 0, 1$, *one has the isomorphism*

$$L(\Lambda_l) \otimes L(\Lambda(k;m,n)) \cong \bigoplus_{\substack{0<m'<q \\ m' \equiv l+m+n \mod 2}} L(\Lambda(k+1;m',n)) \boxtimes L(c_{p,q}, h_{m,m'}(\frac{q}{p}))$$

as $\hat{\mathfrak{g}}' \oplus \mathrm{Vir}$*-module.*

10.3.4 Level 1 $\hat{\mathfrak{sl}}_2$-modules as Vir $\oplus\, \mathfrak{sl}_2$-module

Here, we decompose the level 1 integrable $\hat{\mathfrak{sl}}_2$-modules as Vir $\oplus\, \mathfrak{sl}_2$-module.

By Lemma 10.3, the Virasoro algebra Vir and $\mathfrak{sl}_2$ act on $L(\Lambda_l)$ $(l = 0, 1)$ simultaneously and their actions commute. Hence, we may regard $L(\Lambda_l)$ as Vir $\oplus\, \mathfrak{sl}_2$-module.

For $m \in \mathbb{Z}_{\geq 0}$, let T_m be the irreducible $(m+1)$-dimensional $\mathfrak{sl}_2$-module.

Theorem 10.3 *For* $l = 0, 1$, *we have the following decomposition as* Vir $\oplus$ $\mathfrak{sl}_2$*-module*

$$L(\Lambda_l) \cong \bigoplus_{n \in \mathbb{Z}_{\geq 0} + \frac{1}{2}l} L(1, n^2) \boxtimes T_{2n}.$$

In particular, one has

$$L(\Lambda_0) \oplus L(\Lambda_1) \cong \bigoplus_{n \in \mathbb{Z}_{\geq 0}} L(1, \frac{1}{4}n^2) \boxtimes T_n.$$

Proof. Once, the relations between their characters are verified, the rest of the argument is given in exactly the same way as in the previous section. Hence, we verify the relation at the level of the character and leave the rest of the argument to the reader.

By Theorem 6.13, one has

$$\chi_{L(1,\frac{1}{4}m^2)}(\tau) = (q^{\frac{1}{4}m^2} - q^{\frac{1}{4}(m+2)^2})\eta(\tau)^{-1}.$$

Now, by Lemma 10.4, it suffices to indicate the following formula:

$$\sum_{n\in\mathbb{Z}+\frac{1}{2}l} e^{n\alpha_1-n^2\delta} = \sum_{n\in\mathbb{Z}_{\geq 0}+\frac{1}{2}l} (e^{-n^2\delta}-e^{-(n+1)^2\delta})\sum_{k=0}^{2n} e^{(n-k)\alpha_1}. \qquad \square$$

10.4 Unitarisable Vir-modules

Here, as an application of Theorem 10.2, we will show that the discrete series representations of Vir is unitarisable. (For the definition of unitarisability, see Definition 11.1.)

10.4.1 Unitarisable Representations of $\hat{\mathfrak{sl}}_2$

Let $\bar{\omega}$ be an anti-linear anti-involution of $\bar{\mathfrak{g}}$ defined by

$$\bar{\omega}(e) = f, \quad \bar{\omega}(f) = e, \quad \bar{\omega}(h) = h.$$

Then an anti-linear map $\hat{\omega} : \hat{\mathfrak{g}} \to \hat{\mathfrak{g}}$ defined by

$$\hat{\omega}(x(n)) := \bar{\omega}(x)(-n) \quad (x \in \bar{\mathfrak{g}},\ n \in \mathbb{Z}), \quad \hat{\omega}(K) := K, \quad \hat{\omega}(d) := d \tag{10.10}$$

gives an anti-involution of $\hat{\mathfrak{g}}$. For our purpose, it is enough to consider this anti-involution only. Hence, in the sequel, we fix the anti-linear anti-involution of $\hat{\mathfrak{g}}$.

First, we recall the existence of contravariant Hermitian form on highest weight modules.

Proposition 10.2 *Suppose that* $\Lambda \in \hat{\mathfrak{h}}^*_{\mathbb{R}}$*, where* $\hat{\mathfrak{h}}^*_{\mathbb{R}} := \mathbb{R}\Lambda_0 \oplus \mathbb{R}d \oplus \mathbb{R}\delta$*. Then, there exists a unique Hermitian form*

$$(\cdot|\cdot)_\Lambda : M(\Lambda) \times M(\Lambda) \longrightarrow \mathbb{C}.$$

such that

1. $(v_\Lambda|v_\Lambda)_\Lambda = 1$*, where* $v_\Lambda := 1 \otimes \mathbf{1}_\Lambda$ *is the highest weight vector of* $M(\Lambda)$*.*

2. $(\cdot|\cdot)_\Lambda$ *is contravariant with respect to* $\hat{\omega}$.

Moreover, this Hermitian form satisfies

$$\mathrm{rad}(\cdot|\cdot)_\Lambda = J(\Lambda),$$

where $J(\Lambda)$ *is the maximal proper submodule of* $M(\Lambda)$.

Proof. The proof is similar to § 3.1.1. □

We denote the Hermitian form on $L(\Lambda)$ induced from $(\cdot|\cdot)_\Lambda$ by the same notation.

From the following proposition, we see which $L(\Lambda)$ is unitary with respect to $\hat{\omega}$.

Proposition 10.3 *Suppose that* $\Lambda \in \hat{\mathfrak{h}}^*$. *Then,* $L(\Lambda)$ *is unitarisable with respect to* $\hat{\omega}$, *if and only if* Λ *is a dominant integral weight.*

Proof. See Chapter 11 of [Kac4]. □

10.4.2 Unitarisable Representations of Vir

For $\Lambda, \Lambda' \in \hat{\mathfrak{h}}^*$. one can define a Hermitian form $(\cdot|\cdot)$ on $L(\Lambda) \otimes L(\Lambda')$ by

$$(u \otimes v|u' \otimes v) := (u|u')_\Lambda (v|v')_{\Lambda'} \qquad u, u' \in L(\Lambda),\ v, v' \in L(\Lambda').$$

Hence, we consider the Hermitian form defined on

$$[L(\Lambda_0) \oplus L(\Lambda_1)] \otimes L(\Lambda(k; m, n))$$

where k is an admissible level such that $k+2 = \frac{p}{q-p}$ with $p, q \in \mathbb{Z},\ (p, q) = 1$ and $0 < m < p$, $0 \le n < q-p$. It can be checked by direct computation that, with respect to the Virasoro action on this space defined in Proposition 10.1, the form $(\cdot|\cdot)$ is Hermitian with respect to the anti-involution θ defined by

$$\theta(L_n) = L_{-n}, \qquad \theta(C) = C.$$

Thus, in particular, Theorem 10.2 together with Proposition 10.3 implies the following theorem:

Theorem 10.4 *Let* $k \in \mathbb{Z}_{>0}$ *be a positive integer and* $r, s \in \mathbb{Z}$ *be integers such that* $0 < r < k+2$ *and* $0 < s < k+3$. *Then, the* Vir-*module* $L(c_{k+2,k+3}, h_{r,s}(\frac{k+3}{k+2}))$ *is unitarisable.*

This theorem implies that the second condition of Theorem 11.1 is sufficient.

10.5 Bibliographical Notes and Comments

The main theorem of this chapter, i.e., Theorem 10.2, was proved by P. Goddard, A. Kent and D. Olive [GKO1], [GKO2] in 1985 for $k \in \mathbb{Z}_{>0}$ to show that the discrete series representations of the Virasoro algebra is unitarisable (cf. Theorem 10.4). In 1988, V. G. Kac and M. Wakimoto [KW2] obtained an identity stated in Lemma 10.5, but Theorem 10.2 itself is not treated there. This theorem seems to be known among the experts although the authors could not find any appropriate literature.

Theorem 10.3 was stated by V. G. Kac [Kac2] in 1979.

Theorem 10.4 was proved independently by P. Goddard, A. Kent and D. Olive [GKO2], V. G. Kac and M. Wakimoto [KW1] and A. Tsuchiya and Y. Kanie [TK1].

Chapter 11
Unitarisable Harish-Chandra Modules

In this chapter, we will classify the unitarisable Harish-Chandra modules over the Virasoro algebra which is given by [CP2]. As it follows from the classification theorem of Harish-Chandra Vir-modules (cf. Theorem 2.1), it is sufficient to consider three classes of Vir-modules, highest weight modules, lowest weight modules, and intermediate series.

Here, we analyze intermediate series by rather direct computations and highest weight modules following mainly R. P. Langlands' work [La]. The case of lowest weight modules follows from the result for highest weight modules by a simple observation which will also be explained.

11.1 Definition of Unitarisable Representations

In this section, we briefly recall the definition and some basic properties of unitarisable representations over a complex Lie algebra $\mathfrak{a}$.

Let ω be an anti-linear anti-involution of $\mathfrak{a}$, namely,

1. $\omega(\lambda a) = \bar{\lambda}\omega(a)$ for $\lambda \in \mathbb{C}$ and $a \in \mathfrak{a}$,
2. $\omega([a, b]) = [\omega(b), \omega(a)]$,
3. $\omega^2 = \mathrm{id}_{\mathfrak{a}}$.

Let V be an $\mathfrak{a}$-module with a Hermitian form

$$(\cdot|\cdot) : V \times V \longrightarrow \mathbb{C}.$$

In this book, a Hermitian form is anti-linear with respect to the first component and linear with respect to the second component. We say that $(\cdot|\cdot)$ on V is **contravariant** with respect to ω, if it satisfies

$$(a.u|v) = (u|\omega(a).v)$$

for any $a \in \mathfrak{a}$ and $u, v \in V$.

K. Iohara, Y. Koga, *Representation Theory of the Virasoro Algebra*, Springer Monographs in Mathematics, DOI 10.1007/978-0-85729-160-8_11,

Definition 11.1 *An $\mathfrak{a}$-module V is called a* ***unitarisable representation****, if there exists an anti-involution ω of $\mathfrak{a}$ and a positive definite Hermitian form which is contravariant with respect to ω.*

For later use, we remark some fundamental properties of unitarisable representations.

Lemma 11.1. *1. Let V_1 and V_2 be unitarisable representations of $\mathfrak{a}$ with respect to an anti-linear anti-involution ω. Then $V_1 \otimes V_2$ is unitarisable with respect to ω.*
2. Let V be a unitarisable representation of $\mathfrak{a}$. Suppose that V has a composition series of finite length. Then V is a semi-simple $\mathfrak{a}$-module.

Proof. The first statement is clear. Hence, we only show the second one. We denote a positive definite Hermitian form on V by $(\cdot|\cdot)$. We may assume that V is not irreducible. Let W be a proper submodule of V. If we set

$$W^{\perp} := \{v \in V | (v|w) = 0 \ (\forall w \in W)\},$$

then it is easy to see that $W^{\perp}$ is an $\mathfrak{a}$-module and $V = W \oplus W^{\perp}$. Hence the second statement has been proved. □

The next result is also simple but useful:

Lemma 11.2. *Let V be a Harish-Chandra module equipped with a Hermitian form $(\cdot|\cdot)$.*

1. *There exists a Hermitian form $(\cdot|\cdot)^*$ on V^{*a}.*
2. *$(V^{*a}, (\cdot|\cdot)^*)$ is unitarisable if and only if $(V, (\cdot|\cdot))$ is unitarisable. In particular, if $(V^{*a}, (\cdot|\cdot)^*)$ is unitarisable, then its restriction to the restricted-dual $(V^{\sharp a}, (\cdot|\cdot)^*|_{V^{\sharp a} \times V^{\sharp a}})$ is also unitarisable.*

Recall that an $\mathfrak{a}$-module structure on the full-dual V^* was defined in § 9.3.1.

Proof. Since the Hermitian form $(\cdot|\cdot)$ is non-degenerate by assumption, we may identify V^{*a} with V as graded vector space, i.e, for each $f \in V^*$, there uniquely exists $u_f \in V$ such that $f(v) = (u_f|v)$ holds for any $v \in V$. Now, we define the Hermitian form $(\cdot|\cdot)^*$ by

$$(f|g)^* := (u_f|u_g) \qquad \forall\, f, g \in V^{*a}.$$

It is easy to see that this sesqui-linear form is well-defined and satisfies the contravariance. Hence, the first statement is proved. The second statement follows by the definition of $(\cdot|\cdot)^*$. □

11.2 Anti-linear Anti-involutions of Vir

In this section, we classify the anti-linear anti-involutions of Vir. We also classify the anti-linear anti-involutions with which simple Harish-Chandra modules can be unitarisable.

11.2.1 The Classification

In this subsection, we classify the anti-linear anti-involutions of Vir.

The first technical lemma we show is

Lemma 11.3. *Let $\tilde{\mathfrak{h}}$ be a maximal abelian subalgebra which semi-simply acts on* Vir *via the adjoint action. Then $\tilde{\mathfrak{h}}$ coincides with the Cartan subalgebra of* Vir, *i.e.,* $\tilde{\mathfrak{h}} = \mathfrak{h}(:= \mathbb{C}L_0 \oplus \mathbb{C}C)$.

Proof. Suppose that there exists

$$x = \alpha C + \sum_{n=n_0}^{n_1} \alpha_n L_n \in \tilde{\mathfrak{h}} \setminus \{0\},$$

where $n_0 \leq n_1$ and $\alpha_{n_0} \neq 0$, $\alpha_{n_1} \neq 0$. It is sufficient to see that $n_0 = n_1 = 0$ since the maximality of $\tilde{\mathfrak{h}}$ implies the conclusion.

Let us suppose that $n_0 \neq 0$ or $n_1 \neq 0$. We first assume that $n_1 > 0$. Since $\tilde{\mathfrak{h}}$ acts on Vir semi-simply, there exists

$$y = \beta C + \sum_{n=m_0}^{m_1} \beta_n L_n \in \mathrm{Vir}$$

such that $m_0 \leq m_1$, $\beta_{m_0} \neq 0$, $\beta_{m_1} \neq 0$, $m_1 \neq n_1$ and

$$[x, y] = \gamma y \quad (\exists \gamma \in \mathbb{C}). \tag{11.1}$$

Since the left-hand side of (11.1) is of the form

$$\alpha_{n_1}\beta_{m_1}(n_1 - m_1)L_{m_1+n_1} + \sum_{n<n_1+m_1} \delta_n L_n + \delta C,$$

this is a contradiction. Hence $n_1 \leq 0$. One can similarly show that $n_0 \geq 0$, and thus $n_0 = n_1 = 0$. □

It is easy to see that the Virasoro algebra possesses the following anti-linear anti-involutions $\theta_\alpha^\pm$:

1. For $\alpha \in \mathbb{R} \setminus \{0\}$, let $\theta_\alpha^+ : \mathrm{Vir} \to \mathrm{Vir}$ be the anti-linear map defined by

$$\theta_\alpha^+(L_n) = \alpha^n L_{-n}, \quad \theta_\alpha^+(C) = C.$$

2. For $\alpha \in \mathbb{C}$ ($|\alpha| = 1$), let $\theta_\alpha^- : \mathrm{Vir} \to \mathrm{Vir}$ be the anti-linear map defined by

$$\theta_\alpha^-(L_n) = -\alpha^n L_n, \quad \theta_\alpha^-(C) = -C.$$

Conversely, the next proposition holds:

Proposition 11.1 ([CP2]) *Any anti-linear anti-involution of* Vir *is one of the form* θ_α^+ *or* θ_α^-.

Proof. Let θ be the anti-linear anti-involution of Vir. By Lemma 11.3 and the fact that $\mathbb{C}C$ is the centre of the Virasoro algebra, we see that $\theta(\mathfrak{h}) = \mathfrak{h}$ and $\theta(\mathbb{C}C) = \mathbb{C}C$. Further, since $\theta^2 = \mathrm{id}_{\mathrm{Vir}}$, we see that

$$\theta(C) = \delta C, \quad \theta(L_0) = \beta_0 L_0 + \alpha_0 C,$$

where δ and β_0 satisfy $\delta\bar{\delta} = 1$ and $\beta_0\overline{\beta_0} = 1$.

Here we set

$$\theta(L_n) = \sum_{m \in \mathbb{Z}} \beta_{n,m} L_m + \alpha_n C$$

and look at the conditions that $\beta_{n,m}$ and α_n are satisfied. By applying θ to $[L_0, L_n] = -nL_n$, we have

$$\beta_0 \left(\sum_{m \in \mathbb{Z}} m\beta_{n,m} L_m \right) = -n \sum_{m \in \mathbb{Z}} \beta_{n,m} L_m - n\alpha_n C.$$

This implies that $\alpha_n = 0$ for $n \neq 0$ and $\beta_0 = \pm 1$.

In the case $\beta_0 = 1$, we have $\beta_{n,m} = 0$ if $m \neq -n$. We set $\beta_n := \beta_{-n,n}$. Then $\theta(L_n) = \beta_n L_{-n}$ for $n \neq 0$. Applying θ to both sides of

$$[L_n, L_{-n}] = 2nL_0 + \frac{1}{12}(n^3 - n)C, \tag{11.2}$$

we get $\beta_n \beta_{-n} = 1$ ($n \neq 0$), $\delta = 1$ and $\alpha_0 = 0$. Moreover, from

$$\theta^2(L_n) = L_n, \tag{11.3}$$

we obtain $\overline{\beta_n}\beta_{-n} = 1$, and from the commutation relation

$$[L_m, L_n] = (m - n)L_{m+n}, \tag{11.4}$$

we obtain $\beta_m \beta_n = \beta_{m+n}$ for $m \neq -n$. Hence we have $\beta_m \in \mathbb{R} \setminus \{0\}$ and $\beta_m = (\beta_1)^m$ for $m \neq 0$, i.e., $\theta = \theta_{\beta_1}^+$.

In the case $\beta_0 = -1$, we have $\beta_{n,m} = 0$ if $m \neq n$. Putting $\beta_n := \beta_{n,n}$, we have $\theta(L_n) = \beta_n L_n$ for $n \neq 0$. Similarly to the above, by (11.2), we have $\beta_n \beta_{-n} = 1$ ($n \neq 0$), $\delta = -1$ and $\alpha_0 = 0$. Further by (11.3), we have $\overline{\beta_n}\beta_n = 1$, and by (11.4), we have $\beta_m \beta_n = -\beta_{m+n}$. Hence we see that $|\beta_m| = 1$ and $\beta_m = -(-\beta_1)^m$ for $m \neq 0$, i.e., $\theta = \theta_{\beta_1}^-$. Now we have completed the proof. □

11.2.2 Anti-linear Anti-involutions admitting Unitarisable Vir*-Modules*

In this subsection, we see which anti-involution, indeed, corresponds to unitarisable representations over Vir.

The result is as follows:

Proposition 11.2 *Let V be a non-trivial Harish-Chandra module of* Vir.

1. *If V is unitarisable with respect to an anti-linear anti-involution θ, then $\theta = \theta^+_\alpha$ for some $\alpha \in \mathbb{R}_{>0}$.*
2. *If V is unitarisable with respect to θ^+_α, then V is unitarisable with respect to $\theta = \theta^+_1$.*

Proof. Suppose that $C|_V = c\,\mathrm{id}_V$ and $V = \bigoplus_{\mu \in \mathbb{C}} V_\mu$, where $V_\mu := \{v \in V | L_0.v = \mu v\}$.

First we assume that V is unitarisable with respect to θ^-_α and lead to a contradiction. Since $\theta^-_\alpha(L_0) = -L_0$, if $V_\mu \neq \{0\}$, then $\mu \in \sqrt{-1}\mathbb{R}$. On the other hand, since V is a Harish-Chandra module, we have $V = \bigoplus_{n\in\mathbb{Z}} V_{\mu+n}$. Hence $V_{\mu+n} = \{0\}$ for $n \neq 0$. This is a contradiction, since V is not trivial.

Next we assume that V is a unitarisable representation with respect to θ^+_α ($\alpha \in \mathbb{R} \setminus \{0\}$). Let us denote a positive definite contravariant Hermitian form on V by $\langle\cdot,\cdot\rangle^{(\alpha)}$. Using this contravariant form, we define a new bilinear form $\langle\cdot,\cdot\rangle$ on V by

$$\langle v, w\rangle = |\alpha|^n \langle v, w\rangle^{(\alpha)} \quad (v, w \in V_{\mu+n}).$$

Then it is easy to see that $\langle\cdot,\cdot\rangle$ defines a positive definite contravariant Hermitian form with respect to $\theta^+_{\alpha/|\alpha|}$.

Finally we assume that V is unitarisable with respect to θ^+_{-1} and lead to a contradiction. We set $\mathfrak{a} := \mathbb{C}L_{-1} \oplus \mathbb{C}L_0 \oplus \mathbb{C}L_1$ and $\mathfrak{a}_\theta := \{a \in \mathfrak{a} | \theta^+_{-1}(a) = -a\}$. Note that $\mathfrak{a} \simeq \mathfrak{sl}(2,\mathbb{C})$, $\mathfrak{a}_\theta \simeq \mathfrak{su}(2)$ and any unitarisable representation of $\mathfrak{su}(2)$ is a direct sum of finite dimensional representations. Since intermediate series are not direct sum of finite dimensional $\mathfrak{su}(2)$-modules, V is the highest or lowest weight module. First we consider the case where V is the highest weight module $L(c,h)$. Here, we notice that it is enough to consider the case $(c,h) \in \mathbb{R}^2$ since a Hermitian form which is contravariant with respect to θ^+_{-1} exists only if $(c,h) \in \mathbb{R}^2$. (See, e.g., the next section.) Let v be a highest weight vector of V. Then

$$\langle L_{-n}.v, L_{-n}.v\rangle = (-1)^n\{2nh + \frac{1}{12}(n^3 - n)c\}\langle v, v\rangle > 0$$

for any $n \in \mathbb{Z}_{>0}$. This implies that $c = h = 0$, i.e., V is a trivial representation. In the case where V is a lowest weight module, one can similarly show that V is trivial. Therefore, we have proved the proposition. □

Therefore, without loss of generality, we will consider the only Vir-modules that admit Hermitian forms contravariant with respect to θ^+_1. From now on, for simplicity, we set $\omega := \theta_1^+$. We may naturally extend ω to $U(\mathrm{Vir})$ which we denote by the same symbol ω, for simplicity.

11.3 Hermitian Form on Harish-Chandra Modules

In this section, we explicitly construct Hermitian forms on Harish-Chandra modules contravariant with respect to ω. We also study some of their basic properties.

11.3.1 Intermediate Series

Recall that, for $a, b \in \mathbb{C}$, the Vir-module

$$V_{a,b} := \bigoplus_{n\in\mathbb{Z}} \mathbb{C}v_n$$

is defined by

$$\begin{aligned} L_s.v_n &:= (as + b - n)v_{n+s}, \\ C.v_n &:= 0. \end{aligned}$$

(See, e.g., (1.11).) Depending on the irreducibility of $V_{a,b}$, we defined the intermediate series $V'_{a,b}$ as follows (cf. Definition 1.20):

$$V'_{a,b} := \begin{cases} V_{a,b} & a \neq 0, -1 \ \vee \ b \notin \mathbb{Z}, \\ V_{a,b}/\mathbb{C}v_b & a = 0 \ \wedge \ b \in \mathbb{Z}, \\ \bigoplus_{n\neq b} \mathbb{C}v_n & a = -1 \ \wedge \ b \in \mathbb{Z}. \end{cases}$$

In this subsection, we study the case when $V'_{a,b}$ admits a Hermitian form and construct it explicitly.

Suppose that there exists a non-trivial Hermitian form on $V'_{a,b}$

$$(\cdot|\cdot) : V'_{a,b} \times V'_{a,b} \longrightarrow \mathbb{C}$$

which is contravariant with respect to ω. By the contravariance, we have $(L_{m-n}.v_n|v_m) = (v_n|L_{n-m}.v_m)$, i.e.,

$$(\overline{a}m - (\overline{a}+1)n + \overline{b})(v_m|v_m) = (-(a+1)m + an + b)(v_n|v_n) \tag{11.5}$$

for any m, n.

The case $a \neq 0, -1 \ \vee \ b \notin \mathbb{Z}$ Setting $m = 0$ in (11.5), we obtain

$$(an+b)(v_n|v_n) = (-(\overline{a}+1)n+\overline{b})(v_0|v_0). \tag{11.6}$$

By assumption, $an+b$ can be zero at most for one $n \in \mathbb{Z}$. Hence, there are two possibilities:

1. $an+b \notin \mathbb{Z}$ for any $n \in \mathbb{Z}$, i.e., $b \notin a\mathbb{Z}$.
2. $\exists 1\ k \in \mathbb{Z}$ such that $ak+b=0$, i.e., $b \in a\mathbb{Z}$.

The case $b \notin a\mathbb{Z}$. In this case, we have $(v_0|v_0) \neq 0$ by the non-triviality of $(\cdot|\cdot)$. Hence, setting $n=0$ in (11.6), we obtain $b \in \mathbb{R}$. By (11.5) and (11.6), we obtain

$$(a+\overline{a}+1)a(\overline{a}+1)mn(m-n) + \{(a+\overline{a}+1)+(a^2-(\overline{a}+1)^2)\}bmn \\ - \{a(a+1)-\overline{a}(\overline{a}+1)\}bm^2 = 0$$

for any $m, n \in \mathbb{Z}$, which is equivalent to the system of equations:

$$\begin{cases} (a+\overline{a}+1)a(\overline{a}+1) = 0, \\ (a+\overline{a}+1)+(a^2-(\overline{a}+1)^2) = 0, \\ a(a+1)-\overline{a}(\overline{a}+1) = 0. \end{cases}$$

The solutions of this system are given by

$$a \in -\frac{1}{2}+\sqrt{-1}\mathbb{R} \quad \vee \quad a=0 \quad \vee \quad a=-1.$$

In the case $a \in -\frac{1}{2}+\sqrt{-1}\mathbb{R}$, the equation (11.5) is valid for all $m, n \in \mathbb{Z}$. By (11.6), we obtain

$$(v_n|v_n) = (v_0|v_0) \qquad \forall n \in \mathbb{Z}.$$

In the case $a=0$, the equation (11.5) is consistent for all $m, n \in \mathbb{Z}$, and we obtain

$$(v_n|v_n) = \frac{b-n}{b}(v_0|v_0) \qquad \forall n \in \mathbb{Z}.$$

In the case $a=-1$, the equation (11.5) is consistent for all $m, n \in \mathbb{Z}$, and we obtain

$$(v_n|v_n) = \frac{b}{b-n}(v_0|v_0) \qquad \forall n \in \mathbb{Z}.$$

The case $b \in a\mathbb{Z}$. In this case, by assumption, we have $a \neq 0, -1$. In the case $b=0$, (11.5) and (11.6) imply

$$a \in -\frac{1}{2}+\sqrt{-1}\mathbb{R},$$

and in this case, we have

$$(v_n|v_n) = (v_0|v_0) \qquad \forall n \in \mathbb{Z}.$$

In the case $b = -ak$ for some $k \in \mathbb{Z} \setminus \{0\}$, setting $n = k$ in (11.6), we obtain $(v_0|v_0) = 0$, which implies $(v_n|v_n) = 0$ for any $n \in \mathbb{Z} \setminus \{k\}$. Setting $n = k$ in (11.5), we see that this implies $(v_n|v_n) = 0$ for any $n \in \mathbb{Z}$ which contradicts to the non-triviality of $(\cdot|\cdot)$.

The case $a = 0 \ \wedge \ b \in \mathbb{Z}$ Similarly to the above argument, we obtain a consistent Hermitian form satisfying

$$(v_n|v_n) = \frac{b-n}{b}(v_0|v_0) \qquad \forall n \in \mathbb{Z} \setminus \{b\}.$$

The case $a = -1 \ \wedge \ b \in \mathbb{Z}$ Similarly to the above argument, we obtain a consistent Hermitian form satisfying

$$(v_n|v_n) = \frac{b}{b-n}(v_0|v_0) \qquad \forall n \in \mathbb{Z} \setminus \{b\}.$$

Summarising the above discussion, we obtain the following proposition:

Proposition 11.3 *For $a, b \in \mathbb{C}$, let $V'_{a,b}$ be the intermediate series. $V'_{a,b}$ admits a Hermitian form if and only if $b \in \mathbb{R}$ and one of the following conditions are satisfied:*

$$1.\quad a \in -\frac{1}{2} + \sqrt{-1}\mathbb{R}, \qquad 2.\quad a = 0, \qquad \textit{and} \qquad 3.\quad a = -1.$$

In each case, a Hermitian form on $V'_{a,b}$ satisfies

$$\begin{aligned}
&1. \quad (v_n|v_n) = (v_0|v_0) && \forall\, n \in \mathbb{Z},\\
&2. \quad (v_n|v_n) = \frac{b-n}{b}(v_0|v_0) && \forall\, n \in \mathbb{Z} \setminus \{b\},\\
&3. \quad (v_n|v_n) = \frac{b}{b-n}(v_0|v_0) && \forall\, n \in \mathbb{Z} \setminus \{b\}.
\end{aligned}$$

Notice that the above description is consistent with Lemma 11.2.

11.3.2 Verma Modules

In this subsection, we construct a Hermitian form which is contravariant with respect to ω on Verma modules.

Recall that the Poincaré–Birkhoff–Witt theorem implies the decomposition

$$U(\mathrm{Vir}) = S(\mathrm{Vir}^0) \oplus \{\mathrm{Vir}^- U(\mathrm{Vir}) + U(\mathrm{Vir})\mathrm{Vir}^+\},$$

and we denote the canonical projection $U(\mathrm{Vir}) \twoheadrightarrow S(\mathrm{Vir}^0) \cong \mathbb{C}[(\mathrm{Vir}^0)^*]$ with respect to the above decomposition by π. For $(c, h) \in \mathbb{C}^2 \cong (\mathrm{Vir}^0)^*$, we denote the canonical projection $\mathbb{C}[(\mathrm{Vir}^0)^*] \twoheadrightarrow \mathbb{C}[(\mathrm{Vir}^0)^*]/(C - c, L_0 - h) \cong \mathbb{C}$

by $\mathrm{ev}_{c,h}$, where $(C-c, L_0-h) \subset \mathbb{C}[(\mathrm{Vir}^0)^*]$ is the maximal ideal generated by $C-c$ and L_0-h.

Now, on a Verma module $M(c,h)$, we define the Hermitian form $(\cdot|\cdot)_{c,h}$ by

$$(x.v_{c,h}|y.v_{c,h})_{c,h} := \mathrm{ev}_{c,h} \circ \pi(\omega(x)y) \qquad x, y \in U(\mathrm{Vir}), \tag{11.7}$$

where $v_{c,h} := 1 \otimes \mathbf{1}_{c,h} \in M(c,h)$ is a highest weight vector.

As in § 3.1.2, one can show that this sesqui-linear form is well-defined if and only if the ideal $(C-c, L_0-h) \subset S(\mathrm{Vir}^0)$ is ω-stable, i.e., $(c,h) \in \mathbb{R}^2$. Thus, in the rest of this chapter, we may assume that $(c,h) \in \mathbb{R}^2$. Finally, we state some properties of this form whose proof can be given as in § 3.1.2.

Proposition 11.4 *For $(c,h) \in \mathbb{R}^2$, let $M(c,h)$ be the Verma module with highest weight (c,h) and $(\cdot|\cdot)_{c,h}$ be the Hermitian form on $M(c,h)$ defined by (11.7). This Hermitian form enjoys the following properties:*

1. *Normalisation:* $(v_{c,h}|v_{c,h})_{c,h} = 1.$
2. *Contravariance:* $(x.u|v)_{c,h} = (u|\omega(x).v)_{c,h}$ *for any $x \in U(\mathrm{Vir})$ and $u, v \in M(c,h)$.*
3. *Anti-symmetry:* $(v|u)_{c,h} = \overline{(u|v)_{c,h}}$ *for any $u, v \in M(c,h)$.*

As a corollary, we have

Corollary 11.1 *For $(c,h) \in \mathbb{R}^2$, the radical $\mathrm{rad}(\cdot|\cdot)_{c,h}$ is the maximal proper submodule of $M(c,h)$, i.e., we have*

$$L(c,h) \cong M(c,h)/\mathrm{rad}(\cdot|\cdot)_{c,h}.$$

11.4 Main Results

In Chapter 2, we have shown that the simple Harish-Chandra Vir-modules are classified into three classes: the irreducible highest weight modules, the irreducible lowest weight modules and the intermediate series. In this section, we state the main result of this chapter, that is, the classification of the unitarisable Vir-modules among them.

The first main theorem is the classification of the unitarisable irreducible highest weight modules:

Theorem 11.1 *The highest weight representation $L(c,h)$ is a unitarisable representation if and only if one of the following conditions holds:*

(i) $c \geq 1$ *and* $h \geq 0$,
(ii) there exists $m \in \mathbb{Z}_{\geq 2}$ and $r, s \in \mathbb{Z}$, $1 \leq s \leq r < m$ such that

$$c = c_{m,m+1} = 1 - \frac{6}{m(m+1)}, \quad h = h_{m,m+1:r,s:0} = \frac{((m+1)r - ms)^2 - 1}{4m(m+1)}.$$

Remark 11.1 *The second case in Theorem 11.1 for $m \geq 3$ exhausts all of the discrete series (cf. § 5.1.5). Indeed, for such an (r,s), either (r,s) or $(m-r, m+1-s)$ belongs to $(K^+_{m,m+1})^\circ$, which is a consequence of the symmetry*

$$h_{r,s}\left(\frac{m+1}{m}\right) = h_{m-r,m+1-s}\left(\frac{m+1}{m}\right).$$

For the definition of $h_{\alpha,\beta}(t)$, see (5.4).

For $(c,h) \in \mathbb{C}^2 \cong (\mathrm{Vir}^0)^*$, let $L^-(c,h)$ be the irreducible lowest weight Vir-module with lowest weight (c,h). Notice that $L^-(c,h)$ is the irreducible quotient of $M^-(c,h) := \mathrm{Ind}_{\mathrm{Vir}^- \oplus \mathrm{Vir}^0}^{\mathrm{Vir}} \mathbb{C}^-_{c,h}$ where $\mathbb{C}^-_{c,h} := \mathbb{C}\mathbf{1}^-_{c,h}$ is the $\mathrm{Vir}^- \oplus \mathrm{Vir}^0$-module defined by

$$\mathrm{Vir}^-.\mathbf{1}^-_{c,h} := \{0\}, \qquad L_0.\mathbf{1}^-_{c,h} := h\mathbf{1}^-_{c,h}, \quad C.\mathbf{1}^-_{c,h} := c\mathbf{1}^-_{c,h}.$$

The second main theorem is the classification of the unitarisable irreducible lowest weight modules:

Theorem 11.2 *The lowest weight representation $L^-(c,h)$ is a unitarisable representation if and only if one of the following conditions holds:*

(i) $c \leq -1$ and $h \leq 0$,
(ii) there exists $m \in \mathbb{Z}_{\geq 2}$ and $r, s \in \mathbb{Z}$, $1 \leq s \leq r < m$ such that

$$c = -1 + \frac{6}{m(m+1)}, \qquad h = -\frac{((m+1)r - ms)^2 - 1}{4m(m+1)}.$$

The third main theorem is the classification of the unitarisable intermediate series:

Theorem 11.3 *The intermediate series $V'_{a,b}$ is a unitarisable representation if and only if*

$$a \in -\frac{1}{2} + \sqrt{-1}\mathbb{R} \qquad \text{and} \qquad b \in \mathbb{R}.$$

11.5 Proof of Main Results

In this section, we prove three main results stated in the previous section. Theorem 11.3 is a direct consequence of Proposition 11.3, and Lemma 11.2 implies that Theorem 11.1 and Theorem 11.2 are equivalent since we have $L(c,h)^{\sharp a} \cong L^-(-c,-h)$ for any $(c,h) \in \mathbb{C}^2$. Hence, it is sufficient to prove Theorem 11.1 and this section is devoted to its proof.

11.5.1 Determinant Formulae

In this subsection, we compute the determinant of the form $(\cdot|\cdot)_{c,h}$ restricted to each weight subspace $M(c,h)_{h+n}$ $(n \in \mathbb{Z}_{>0})$ of the Verma module $M(c,h)$. We also compute the determinant of the form induced on a quotient of $M(c,h)$.

Fix a positive integer $n \in \mathbb{Z}_{>0}$. For $\mathbb{I} = (1^{r_1}2^{r_2}\cdots n^{r_n}) \in \mathcal{P}_n$, we set

$$e_{\mathbb{I}} := L_{-n}^{r_n}\cdots L_{-2}^{r_2}L_{-1}^{r_1}.$$

The elements $\{e_{\mathbb{I}}.v_{c,h}|\mathbb{I} \in \mathcal{P}_n\}$ form a basis of $M(c,h)_{h+n}$. Let us compute the determinant of the matrix

$$H(c,h)_n := ((e_{\mathbb{I}}.v_{c,h}|e_{\mathbb{J}}.v_{c,h})_{c,h})_{\mathbb{I},\mathbb{J}\in\mathcal{P}_n}. \tag{11.8}$$

We have

Theorem 11.4 *For $n \in \mathbb{Z}_{>0}$, there exists $A_n \in \mathbb{R}_{>0}$ such that*

$$\det H(c,h)_n = A_n \prod_{\substack{r,s\in\mathbb{Z}_{>0}\\ s\le r\\ 1\le rs\le n}} \Phi_{r,s}(c,h)^{p(n-rs)}.$$

Except for the proof of positivity of A_n, this theorem has already been established by Theorem 4.2, since, for $(c,h) \in \mathbb{R}^2$, $\det(c,h)_n$ is proportional to $\det(H(c,h)_n)$. The following lemma guarantees the positivity.

Lemma 11.4. *For $\mathbb{I}$, $\mathbb{J} \in \mathcal{P}_n$,*

1. *h-$\deg(e_{\mathbb{I}}.v_{c,h}|e_{\mathbb{J}}.v_{c,h})_{c,h} \le \min\{|\mathbb{I}|,|\mathbb{J}|\}$,*
2. *h-$\deg(e_{\mathbb{I}}.v_{c,h}|e_{\mathbb{J}}.v_{c,h})_{c,h} \le |\mathbb{I}|-1$, if $|\mathbb{I}| = |\mathbb{J}|$ and $\mathbb{I} \neq \mathbb{J}$,*
3. *$(e_{\mathbb{I}}.v_{c,h}|e_{\mathbb{I}}.v_{c,h})_{c,h} = A_{\mathbb{I}}h^{|\mathbb{I}|} + f_{\mathbb{I}}(c,h)$, for some $A_{\mathbb{I}} > 0$ and $f_{\mathbb{I}}(c,h) \in \mathbb{R}[z,h]$ such that h-$\deg f_{\mathbb{I}}(z,h) < |\mathbb{I}|$.*

Proof. The proof is similar to that of Lemma 4.9. □

We fix $\alpha_0, \beta_0 \in \mathbb{Z}_{>0}$ and $\xi_0 \in \mathbb{R}\setminus\{0\}$, and set $n_0 := \alpha_0\beta_0$,

$$(c,h) := (c(\xi_0), h_{\alpha_0,\beta_0}(\xi_0)),$$

where $c(\xi)$ and $h_{\alpha,\beta}(\xi)$ is defined by (5.4).

Let us compute the determinant of the form induced on the quotient

$$\overline{M}(c,h) := M(c,h)/M(c,h+n_0),$$

which we again denote by the same symbol $(\cdot|\cdot)_{c,h}$.

By Lemma 5.24, setting $\bar{v}_{c,h} := v_{c,h} + M(c,h+n_0) \in \overline{M}(c,h)$ and

$$\bar{\mathcal{P}}_m := \{\mathbb{I} = (1^{r_1}2^{r_2}\cdots) \in \mathcal{P}_m | r_1 < n_0\},$$

$\{e_{\mathbb{I}}.\bar{v}_{c,h}|\mathbb{I} \in \bar{\mathcal{P}}_m\}$ forms a basis of $\overline{M}(c,h)_{h+m}$ for $m \in \mathbb{Z}_{\geq 0}$. Set

$$\overline{H}(c,h)_m := ((e_{\mathbb{I}}.\bar{v}_{c,h}|e_{\mathbb{J}}.\bar{v}_{c,h})_{c,h})_{\mathbb{I},\mathbb{J}\in\bar{\mathcal{P}}_m}.$$

Then by using Lemma 11.4, one can show the following proposition similarly to (5.67) in the proof of Proposition 5.11.

Proposition 11.5 *Suppose that $n \geq n_0$. Then, we have*

$$\det \overline{H}(c,h)_n \times \det H(c,h+n_0)_{n-n_0} = A_{n,\alpha_0,\beta_0} \prod_{\substack{r,s\in\mathbb{Z}_{>0} \\ s\leq r \\ 1\leq rs\leq n \\ (r,s)\neq(\alpha_0,\beta_0)}} \Phi_{r,s}(c,h)^{p(n-rs)}$$

for some positive constant A_{n,α_0,β_0}.

11.5.2 Proof of Theorem 11.1

We divide the proof of Theorem 11.1 into several steps.

11.5.2.1 Step *I*

Lemma 11.5. *If $(\cdot|\cdot)_{c,h}$ is positive semi-definite, then $c \geq 0$ and $h \geq 0$.*

Proof. For any $n \in \mathbb{Z}_{>0}$,

$$(L_{-n}v_{c,h}|L_{-n}v_{c,h})_{c,h} = 2nh + \frac{1}{12}n(n^2-1)c.$$

If $(\cdot|\cdot)_{c,h}$ is positive semi-definite, then these are non-negative for any n. Putting $n := 1$ we have $h \geq 0$, and taking n large enough we have $c \geq 0$. □

11.5.2.2 Step *II*

Lemma 11.6. *If $c \geq 1$ and $h \geq 0$, then $(\cdot|\cdot)_{c,h}$ is positive semi-definite.*

Proof. Let $\{e_i(c,h)|i = 1,\cdots,p(n)\}$ be the set of eigenvalues of $H(c,h)_n$. Since $H(c,h)_n$ is symmetric for $c,h \in \mathbb{R}$, we see that $e_i(c,h) \in \mathbb{R}$. Here we show that $e_i(c,h) \geq 0$ $(i = 1,\cdots,p(n))$ for $c \geq 1$ and $h \geq 0$.

It follows from Lemma 11.4 that, for each $c > 1$, if $h \gg 0$ (very large), then $e_i(c,h) > 0$ $(i = 1,\cdots,p(n))$. One the other hand, one can easily show that $\det H(c,h)_n \neq 0$ for any $c > 1$ and $h > 0$. Combining these facts, we see that $e_i(c,h) > 0$ for $c > 1$ and $h > 0$, and thus $e_i(c,h) \geq 0$ for $c \geq 1$ and $h \geq 0$. □

As a consequence of Lemmas 11.5 and 11.6, it suffices to analyze the case $0 \le c < 1$ and $h \ge 0$. By Theorem 10.4, we have only to prove that if $(\cdot|\cdot)_{c,h}$ is positive semi-definite for $0 \le c < 1$ and $h \ge 0$, then (c,h) satisfies (ii) of Theorem 11.1. Hence, in the following, we suppose that $0 \le c < 1$ and $h \ge 0$, and exclude the set of weights (c,h) such that $L(c,h)$ cannot be unitarisable.

11.5.2.3 Step *III*

If $\det H(c,h)_n < 0$ for some $n \in \mathbb{Z}_{>0}$, then $(\cdot|\cdot)_{c,h}$ is not positive semi-definite. By assumption, there uniquely exist $m \in \mathbb{R}_{\ge 2}$ and $M \in \mathbb{R}_{\ge 1}$ such that $c = c_{m,m+1}$ and $h = h_{m,m+1:M}$. With such correspondence, we set

$$E(c,h) := \left\{ (\alpha,\beta) \in \mathbb{R}^2 \,\middle|\, \begin{array}{c} -M \le m\alpha - (m+1)\beta \le M, \\ (m+1)\alpha - m\beta \ge M, \\ \alpha \ge \beta > 0 \end{array} \right\} \tag{11.9}$$

and look at the integral points of $E(c,h)$. By the factorisation (5.6), we see that for $(\alpha,\beta) \in (\mathbb{Z}_{>0})^2$ such that $\alpha > \beta$, $(\alpha,\beta) \in E(c,h)$ if and only if $\Phi_{\alpha,\beta}(c,h) \le 0$. Similarly, for $\alpha \in \mathbb{Z}_{>0}$, $\Phi_{\alpha,\alpha}(c,h) \le 0$ if and only if $\alpha \ge M$. $E(c,h)$ is given as the domain indicated in Figure 11.1:

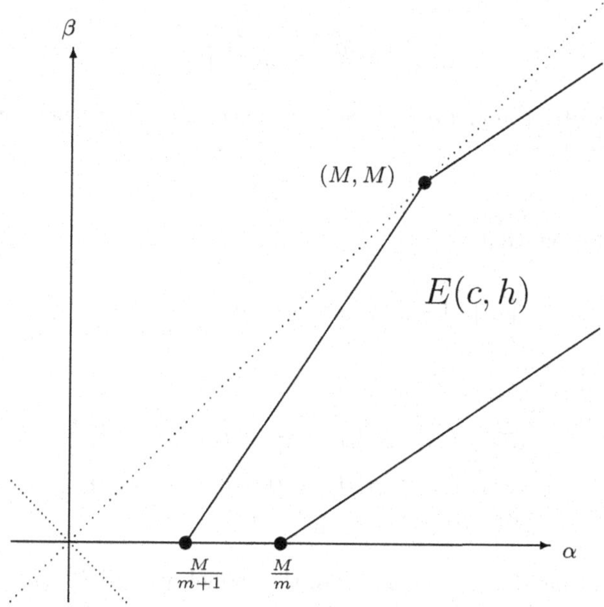

Fig. 11.1 $E(c,h)$

Let us show that for any highest weight (c, h) ($0 \leq c \leq 1$ and $h \geq 0$), the region $E(c, h)$ contains at least one integral point.

Lemma 11.7. *$E(c, h) \cap \mathbb{Z}^2 \neq \emptyset$.*

Proof. Let m and M be positive real numbers such that $c = c_{m,m+1}$ and $h = h_{m,m+1:M}$ where $c_{P,Q}$ and $h_{P,Q:M}$ is defined as (5.5). Notice that $m \geq 2$ and $M \geq 1$, since $0 \leq c \leq 1$ and $h \geq 0$.

For $\alpha \in \mathbb{Z}$ such that $\alpha > M$,

$$(\alpha, \beta) \in E(c, h) \Leftrightarrow a_- \leq \beta \leq a_+,$$

where we set $a_\pm := \frac{m\alpha \pm M}{m+1}$. In particular, for $k \in \mathbb{Z}_{\geq 0}$, we have

$$(\alpha + k, \beta) \in E(c, h) \Leftrightarrow a_- + \frac{km}{m+1} \leq \beta \leq a_+ + \frac{km}{m+1}.$$

Hence, it is enough to see that

$$\left[a_- + \frac{km}{m+1}, a_+ + \frac{km}{m+1}\right] \cap \mathbb{Z} \neq \emptyset,$$

for some $k \in \mathbb{Z}_{\geq 0}$.

By taking $\alpha \in \mathbb{Z}$ appropriately, we may assume that

$$a_- - [a_-] < \frac{1}{m+1},$$

where $[x]$ denotes the greatest integer not exceeding x. Further, since $M \geq 1$, we have

$$[a_-] + 1 \leq a_- + 1 < a_+ + \frac{m}{m+1},$$

and hence, we see that

$$[a_-]+1 \in \left[a_- + \frac{m}{m+1}, a_+ + \frac{m}{m+1}\right].$$

□

We set

$$\alpha(c, h) := \min\{\alpha | (\alpha, \beta) \in E(c, h) \cap \mathbb{Z}^2\},$$
$$\beta(c, h) := \min\{\beta | (\alpha, \beta) \in E(c, h) \cap \mathbb{Z}^2\},$$
$$P(z, h) := (\alpha(c, h), \beta(c, h)).$$

It is obvious that $P(c, h) \in E(c, h)$. In the sequel, we call $P(c, h)$ the *minimal* integral point of $E(c, h)$.

11.5.2.4 Step *IV*

Here we show that if $(\cdot|\cdot)_{c,h}$ is positive semi-definite, then $P(c,h)$ is a point on $\partial E(c,h)$ (the boundary of $E(c,h)$).

Lemma 11.8. *If $P(c,h) \in \mathring{E}(c,h)$ (the interior of $E(c,h)$), then $(\cdot|\cdot)_{c,h}$ is not positive semi-definite.*

Proof. We show that the assumption of the lemma implies $\det H(c,h)_n < 0$ for some n. Set $(\alpha_0, \beta_0) := P(c,h)$ and $n_0 := \alpha_0\beta_0$. Since $(\alpha_0, \beta_0) \in \mathring{E}(c,h)$, we have

$$\Phi_{\alpha_0,\beta_0}(c,h) < 0. \tag{11.10}$$

Moreover, for $(\alpha, \beta) \in (\mathbb{Z}_{>0})^2$ such that $\alpha\beta \le n_0$, $\beta < \alpha$ and $(\alpha,\beta) \ne (\alpha_0,\beta_0)$, we see that $(\alpha,\beta) \notin E(c,h)$ since (α_0,β_0) is the minimal integral point of $E(c,h)$. Hence, we have

$$\Phi_{\alpha,\beta}(c,h) > 0. \tag{11.11}$$

Now, in order to determine the signature of $\det H(c,h)_n$, we study the signature of $\Phi_{\alpha,\alpha}(c,h)$. Set

$$E_1 := \{\alpha \in \mathbb{Z}_{>0} | \alpha^2 \le n_0 \wedge \Phi_{\alpha,\alpha}(c,h) < 0\}.$$

If $E_1 = \emptyset$, then by (11.10) and (11.11), we have $\det H(c,h)_{n_0} < 0$.

We suppose that $E_1 \ne \emptyset$. Let us put $\alpha_1 := \min E_1$, $n_1 := \alpha_1^2$, and show that $\det H(c,h)_{n_1} < 0$. By the assumption on α_1, we have $\Phi_{\alpha,\alpha}(c,h) \ge 0$ for any $\alpha < \alpha_1$. But, if $\Phi_{\alpha,\alpha}(c,h) = 0$, then $(\alpha,\alpha) = (M,M) \in E(c,h)$ which contradicts the hypothesis on $(\alpha_0,\beta_0) := P(c,h)$. Hence, we have $\Phi_{\alpha,\alpha}(c,h) > 0$ for any $\alpha < \alpha_1$, and thus $\det H(c,h)_{n_1} < 0$ by (11.11). □

By Lemma 11.8, it suffices to consider the case where $P(c,h) \in \partial E(c,h)$.

11.5.2.5 Step *V*

Here we introduce three subcases of the case where $P(c,h) \in \partial E(c,h)$.

We denote the half lines or segment of $\partial E(c,h)$ in Figure 11.2 by $\ell_A(m)$, $\ell_B(m)$ and $\ell_C(m)$:

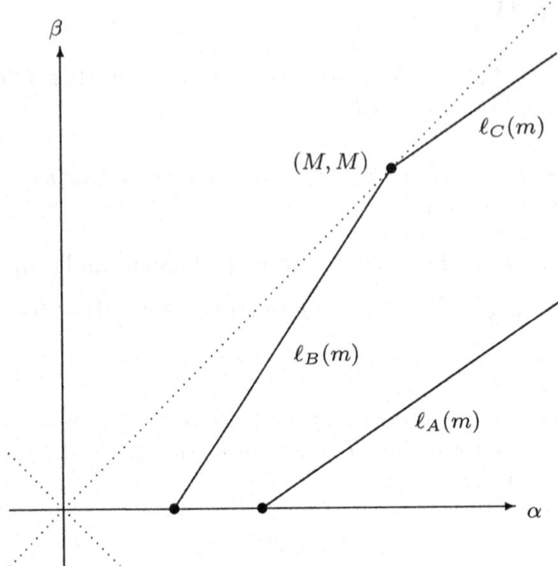

Fig. 11.2 $\partial E(c, h)$

We consider $(M, M) \in \ell_B(m)$ and $(M, M) \notin \ell_C(m)$, namely,

$$\begin{aligned}
&\ell_A(m) : m\alpha - (m+1)\beta = M \quad \alpha > \tfrac{M}{m}, \\
&\ell_B(m) : (m+1)\alpha - m\beta = M \quad \tfrac{M}{m+1} < \alpha \leq M, \\
&\ell_C(m) : m\alpha - (m+1)\beta = -M \quad \alpha > M.
\end{aligned}$$

We sometimes abbreviate $\ell_X(m)$ $(X = A, B, C)$ to ℓ_X for simplicity.

In the sequel, in each case where $P(c, h) \in \ell_X(m)$ $(X = A, B, C)$, we describe the highest weight (c, h). First, we prove

Lemma 11.9. *$P(c, h) \notin \ell_C(m)$.*

Proof. Set $(\alpha_0, \beta_0) := P(c, h)$. Here, we assume that $(\alpha_0, \beta_0) \in \ell_C(m)$, and lead to a contradiction. By assumption, we have $M < \beta_0 < \alpha_0$.

Since $M \geq 1$, we see that $\beta_0 > 1$. But then, $(\alpha_0 - 1, \beta_0 - 1) \in E(c, h)$ holds. Indeed, combining $M = -m\alpha_0 + (m+1)\beta_0$, $\alpha_0 > \beta_0$ and $m \geq 2$, we have

$$\begin{aligned}
M > -m(\alpha_0 - 1) + (m+1)(\beta_0 - 1) > -M, \\
(m+1)(\alpha_0 - 1) - m(\beta_0 - 1) > M.
\end{aligned}$$

This contradicts the definition of (α_0, β_0). □

Hence, it is enough to consider the following cases

Case A : $P(c,h) \in \ell_A(m)$, or
Case B : $P(c,h) \in \ell_B(m)$.

11.5.2.6 Step VI

In this subsubsection, we obtain the condition for m that $P(c,h) \in \ell_A(m)$ or $P(c,h) \in \ell_B(m)$.

In the sequel, we fix $(\alpha_0, \beta_0) \in (\mathbb{Z}_{>0})^2$ such that $\beta_0 \le \alpha_0$, and suppose that the weight (c,h) satisfies

$$(\alpha_0, \beta_0) = P(c,h).$$

We notice that if $(c,h) = (c_{m,m+1}, h_{m,m+1:M})$ $(m \ge 2,\ M \ge 1)$ belongs to **Case** A or **Case** B, then

Case A : $h = h_{m,m+1:\beta_0,\alpha_0:0}$,
Case B : $h = h_{m,m+1:\alpha_0,\beta_0:0}$.

For simplicity, we sometimes use the following notation:

$$c[m] := c_{m,m+1}, \quad h_{\alpha,\beta}[m] := h_{m,m+1:\alpha,\beta:0}. \tag{11.12}$$

Then, in the sequel, we assume that

$$(c,h) = \begin{cases} (c[m], h_{\beta_0,\alpha_0}[m]) & \textbf{Case } A \\ (c[m], h_{\alpha_0,\beta_0}[m]) & \textbf{Case } B \end{cases}$$

and $(\alpha_0, \beta_0) = P(c,h)$.

Lemma 11.10. *1.* **Case** A*: Suppose that* $(c,h) = (c[m], h_{\beta_0,\alpha_0}[m])$*. Then*

$$(\alpha_0, \beta_0) \in \ell_A(m) \iff m > \alpha_0 + \beta_0 - 1.$$

2. **Case** B*: Suppose that* $(c,h) = (c[m], h_{\alpha_0,\beta_0}[m])$*. Then*

$$(\alpha_0, \beta_0) \in \ell_B(m) \iff \begin{cases} m > \alpha_0 + \beta_0 - 1 & \text{if } (\alpha_0, \beta_0) \ne (1,1) \\ m \ge 2 & \text{if } (\alpha_0, \beta_0) = (1,1) \end{cases}.$$

Proof. First, we prove the lemma for **Case** A.
Since $(c,h) = (c[m], h_{\beta_0,\alpha_0}[m])$ satisfies $(\alpha_0, \beta_0) = P(c,h)$ and $(\alpha_0, \beta_0) \in \ell_A(m)$, the lines $\ell_A(m)$ and $\ell_B(m)$ are given as in Figure 11.3:

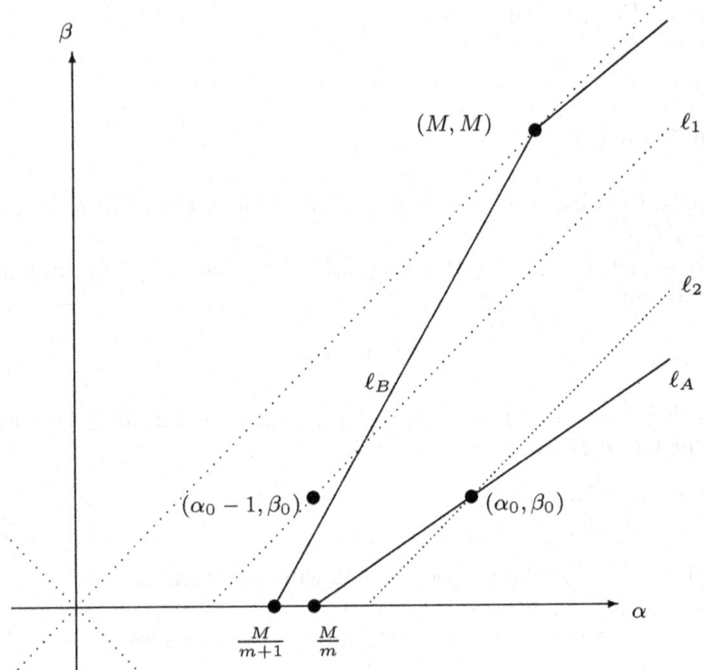

Here, ℓ_1 and ℓ_2 are the lines whose slopes are equal to 1, and they pass through the points $(\alpha_0 - 1, \beta_0)$ and (α_0, β_0) respectively.

Fig. 11.3 Case *A*

We show that

$$\begin{aligned}&P(c[m], h_{\beta_0,\alpha_0}[m]) \in \ell_A(m)\\ \Longrightarrow &P(c[m'], h_{\beta_0,\alpha_0}[m']) = P(c[m], h_{\beta_0,\alpha_0}[m]) \in \ell_A(m') \quad \forall\, m' \geq m.\end{aligned}$$

To show this fact, we notice the following:

1. For any $m' > m$,

$$(\alpha_0, \beta_0) \in \partial E(c[m'], h_{\beta_0,\alpha_0}[m']),$$

   since we have $\Phi_{\alpha_0,\beta_0}(c[m'], h_{\beta_0,\alpha_0}[m']) = 0$.
2. The slopes of the lines $\ell_A(m)$ and $\ell_B(m)$ tend to 1 as m tends to ∞.
3. The interior of the domain bounded by the lines ℓ_1 and ℓ_2 has no integral point.

Therefore, from the above figure, we see that this fact follows from

$$M, \ \frac{M}{m}, \ \frac{M}{m+1} \ \text{ are increasing functions of } m. \tag{11.13}$$

Indeed, these follow from $M = m(\alpha_0 - \beta_0) - \beta_0$, $\frac{M}{m} = \alpha_0 - \beta_0 - \frac{\beta_0}{m}$ and $\frac{M}{m+1} = \alpha_0 - \beta_0 - \frac{\alpha_0}{m+1}$.

Furthermore, from Figure 11.3 and (11.13), we obtain

$$\begin{aligned} &(\alpha_0, \beta_0) = P(c[m], h_{\beta_0,\alpha_0}[m]) \wedge (\alpha_0, \beta_0) \in \ell_A(m) \\ \Leftrightarrow\ &(\alpha_0 - 1, \beta_0) \notin E(c[m], h_{\beta_0,\alpha_0}[m]) \\ \Leftrightarrow\ &(m+1)(\alpha_0 - 1) - m\beta_0 < M \\ \Leftrightarrow\ &m > \alpha_0 + \beta_0 - 1. \end{aligned}$$

Here, we remark that, in this case,

$$\begin{aligned} M &= m(\alpha_0 - \beta_0) - \beta_0 \\ &> (\alpha_0 + \beta_0 - 1)(\alpha_0 - \beta_0) - \beta_0 \\ &= (\alpha_0 - 1)^2 - \beta_0^2 + \alpha_0 - 1 \geq 1. \end{aligned}$$

Thus, we have proved the first statement.

Second, we prove the lemma for **Case** B. If $(c, h) = (c[m], h_{\alpha_0,\beta_0}[m])$ satisfies $(\alpha_0, \beta_0) = P(c, h)$ and $(\alpha_0, \beta_0) \in \ell_B(m)$, then the lines $\ell_A(m)$ and $\ell_B(m)$ are as in Figure 11.4.

Similarly to the previous case, we prove that

$$\begin{aligned} &P(c[m], h_{\alpha_0,\beta_0}[m]) \in \ell_B(m) \\ \Longrightarrow &P(c[m'], h_{\alpha_0,\beta_0}[m']) = P(c[m], h_{\alpha_0,\beta_0}[m]) \in \ell_B(m') \quad \forall\, m' \geq m. \end{aligned}$$

Indeed, since $(\alpha_0, \beta_0) \in \partial E(c[m'], h_{\alpha_0,\beta_0}[m'])$, this fact follows from

$$M \text{ is increasing, and } \frac{M}{m}, \ \frac{M}{m+1} \ \text{ are decreasing functions of } m. \tag{11.14}$$

Furthermore, by Figure 11.4 and (11.14), we have

$$\begin{aligned} &(\alpha_0, \beta_0) = P(c[m], h_{\alpha_0,\beta_0}[m]) \wedge (\alpha_0, \beta_0) \in \ell_B(m) \\ \Leftrightarrow\ &(\alpha_0, \beta_0 - 1) \notin E(c[m], h_{\alpha_0,\beta_0}[m]) \\ \Leftrightarrow\ &m\alpha_0 - (m+1)(\beta_0 - 1) > M \ \Leftrightarrow\ m > \alpha_0 + \beta_0 - 1. \end{aligned}$$

Since $m \geq 2$, we obtain the conclusion. □

Hence, until the end of this section, we assume that

$$\begin{cases} m > \alpha_0 + \beta_0 - 1 & \textbf{Case } A \vee (\textbf{Case } B \wedge (\alpha_0, \beta_0) \neq (1,1)) \\ m \geq 2 & \textbf{Case } B \wedge (\alpha_0, \beta_0) = (1,1) \end{cases}$$

unless otherwise specified, and determine which m may correspond to a unitary representation.

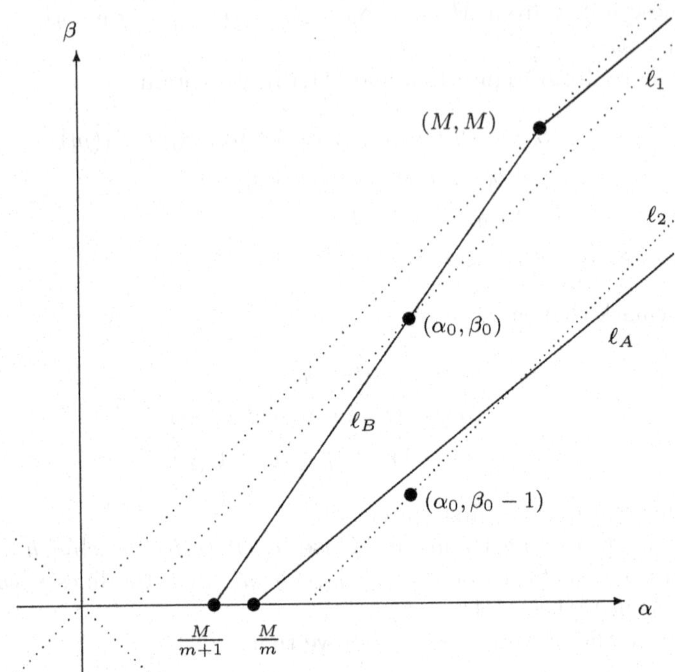

Here, ℓ_1 and ℓ_2 are the lines whose slopes are equal to 1, and they pass through the points (α_0, β_0) and $(\alpha_0, \beta_0 - 1)$ respectively.

Fig. 11.4 Case B

11.5.2.7 Step VII

In this subsubsection, we see that if m is an integer, which satisfies the previous lemma, then the corresponding weight is given in Theorem 11.1 (ii).

For each m, we introduce points $(r(m), s(m))$ and (α_0', β_0') as follows:

Case A. Let $(r(m), s(m))$ be the solution of

$$(m+1)\alpha - m\beta = M,$$
$$\alpha - \beta = \alpha_0 - \beta_0 - 1,$$

i.e., the intersection of $\ell_B(m)$ and ℓ_1 in Figure 11.3. Let (α_0', β_0') be positive integers such that

$$\alpha_0' - 1 < r(m) \leq \alpha_0',$$
$$\alpha_0' - \beta_0' = \alpha_0 - \beta_0 - 1.$$

Case B. Let $(r(m), s(m))$ be the solution of

$$m\alpha - (m+1)\beta = M,$$
$$\alpha - \beta = \alpha_0 - \beta_0 + 1,$$

i.e., the intersection of $\ell_A(m)$ and ℓ_2 in Figure 11.4. Let (α_0', β_0') be positive integers such that

$$\alpha_0' - 1 < r(m) \leq \alpha_0',$$
$$\alpha_0' - \beta_0' = \alpha_0 - \beta_0 + 1.$$

Further, we set

$$n_0 := \alpha_0\beta_0, \quad n_0' := \alpha_0'\beta_0'.$$

For simplicity, we sometimes use the following notation.

$$h[m] :\doteq \begin{cases} h_{\beta_0,\alpha_0}[m] & \textbf{Case } A \\ h_{\alpha_0,\beta_0}[m] & \textbf{Case } B \end{cases}.$$

We check that if $r(m)$ is an integer, then the corresponding weight is one of the highest weights in Theorem 11.1 (ii).

Lemma 11.11. *If $r(m) \in \mathbb{Z}$, i.e., $r(m) = \alpha_0'$, then (c, h) is one of the highest weights given in Theorem 11.1 (ii).*

Proof. In **Case** A, if $r(m) = \alpha_0'$, then $s(m) = \alpha_0' - \alpha_0 + \beta_0 + 1$. On the other hand, since $(r(m), s(m)) \in \ell_B(m)$ and $M = m\alpha_0 - (m+1)\beta_0$, we have $m = \alpha_0' + \beta_0$, and thus $m \in \mathbb{Z}_{\geq 2}$. Since $(\alpha_0', \beta_0') = (r(m), s(m))$, we have $h_{\beta_0,\alpha_0}[m] = h_{\alpha_0',\beta_0'}[m]$. Hence $(c, h) = (c[m], h_{\alpha_0',\beta_0'}[m])$ for some $m \in \mathbb{Z}_{\geq 2}$ and $\alpha_0', \beta_0' \in \mathbb{Z}_{>0}$ such that $\beta_0' \leq \alpha_0' < m$.

In **Case** B, if $r(m) = \alpha_0'$, then $s(m) = \alpha_0' - \alpha_0 + \beta_0 - 1$ and $m = \alpha_0 + \beta_0'$. Hence $m \in \mathbb{Z}_{\geq 2}$, and $(c, h) = (c[m], h_{\alpha_0,\beta_0}[m])$ for some $m \in \mathbb{Z}_{\geq 2}$ and $\alpha_0, \beta_0 \in \mathbb{Z}_{>0}$ such that $\beta_0 \leq \alpha_0 < m$. □

11.5.2.8 Step *VIII*

Consequently, to complete the proof of Theorem 11.1 it is enough to show

Proposition 11.6 *If $r(m) \notin \mathbb{Z}$, then $(\cdot|\cdot)_{c,h}$ is not positive semi-definite.*

Here and after, we assume that $m \in \mathbb{R}$ satisfies $\alpha_0' - 1 < r(m) < \alpha_0'$.

Lemma 11.12. *If $(\alpha, \beta) \in \partial E(c, h) \cap (\mathbb{Z}_{>0})^2$ and $\alpha\beta \leq n_0'$, then*

$$(\alpha, \beta) = (\alpha_0, \beta_0).$$

Proof. We first consider **Case** A. In this case, the lines $\ell_A(m)$ and $\ell_B(m)$ in the (α, β)-plane are as in Figure 11.5. By Figure 11.5, it is enough to show that

$$s - t < 2 \text{ and } t' - r < 2. \tag{11.15}$$

In fact, if (11.15) hold, then in the region $\beta_0 < \beta \leq \beta_0' - 1$, $\ell_A(m)$ lies in between ℓ_2 and ℓ_3, and $\ell_B(m)$ lies in between ℓ_1 and ℓ_2, since the interiors of the regions bounded by ℓ_2 and ℓ_3 or ℓ_1 and ℓ_2 do not contain integral points, and (α_0, β_0) is the minimal integral point of $E(c, h)$. Hence, we see that (11.15) implies the first statement of the lemma in **Case** A.

Since $(r, t) \in \ell_A(m)$ and $(r, s) \in \ell_B(m)$, we have $s - t = \frac{r+t}{m}$. By the definitions of (r, s) and (α_0, β_0), we also have $m = r + \beta_0$. Hence, we have

$$s - t = 1 + \frac{t - \beta_0}{r + \beta_0} = 1 + \frac{r - \alpha_0}{r + \beta_0}\frac{t - \beta_0}{r - \alpha_0} < 2,$$

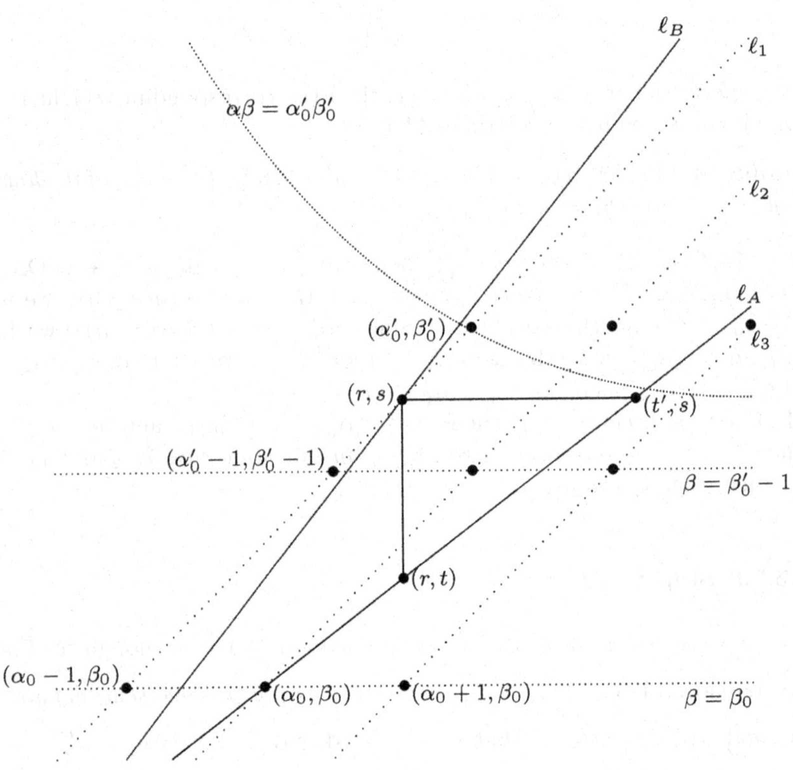

Here, ℓ_i $(i = 1, 2, 3)$ are the lines defined by $\ell_i : \ \alpha - \beta = \alpha_0 - \beta_0 - (2 - i)$. We set $(r, s) := (r(m), s(m))$. (r, t) (resp. (t', s)) is the intersection point of $\ell_A(m)$ and $\alpha = r$ (resp. $\beta = s$).

Fig. 11.5 **Case** A

since $\frac{t-\beta_0}{r-\alpha_0} < 1$ is the slope of $\ell_A(m)$. Similarly, $r = s + \alpha_0 - \beta_0 - 1$ implies

$$t' - r = \frac{r+s}{m} = 1 + \frac{s - \beta_0}{s + \alpha_0 - 1} < 2.$$

Next, we show the lemma in **Case** B. Here, we have the next figure:

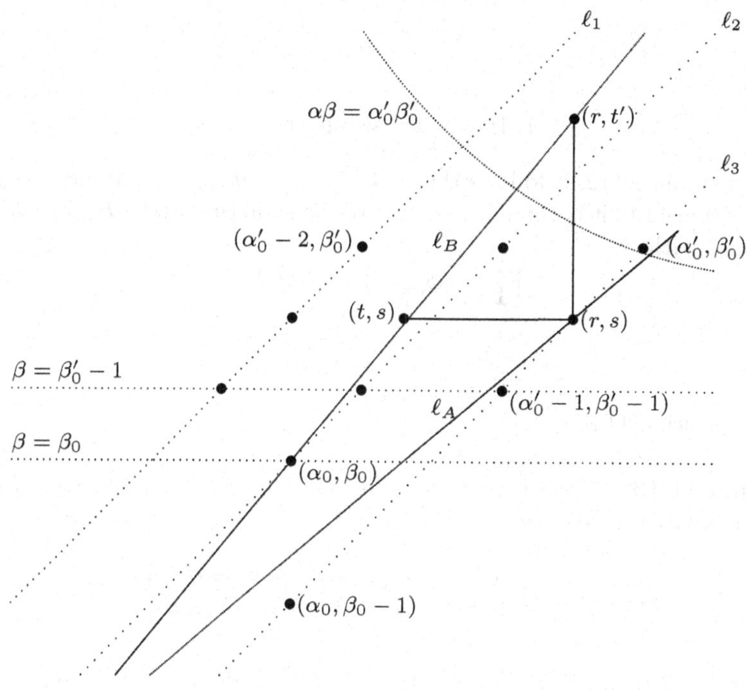

Here, ℓ_i $(i = 1, 2, 3)$ are the lines defined by $\ell_i : \alpha - \beta = \alpha_0 - \beta_0 - (2 - i)$. We set $(r, s) := (r(m), s(m))$. (r, t') (resp. (t, s)) is the intersection point of $\ell_B(m)$ and $\alpha = r$ (resp. $\beta = s$).

Fig. 11.6 Case B

Similarly to **Case** A, it suffices to see that

$$r - t < 2 \text{ and } t' - s < 2. \tag{11.16}$$

Since $(t, s) \in \ell_B(m)$, $(r, s) \in \ell_A(m)$ and $m = s + \alpha_0$, we have

$$r - t = \frac{s+t}{m} = 1 + \frac{t - \alpha_0}{s + \alpha_0} = 1 + \frac{s - \beta_0}{s + \alpha_0} \frac{t - \alpha_0}{s - \beta_0}.$$

Since $\frac{s-\beta_0}{t-\alpha_0}$ is equal to the slope of the line $\ell_B(m)$, we have $r - t < 2$. Similarly,

$$t' - s = \frac{r+s}{m} = 1 + \frac{s - \beta_0 + 1}{s + \alpha_0},$$

since $r = s + \alpha_0 - \beta_0 + 1$. Hence, we also have $t' - s < 2$. □

By Lemma 11.12, it follows that $\det \overline{H}(c[m], h[m])_{n_0'} \neq 0$. Hence, to prove Proposition 11.6, it is enough to compare the signature of $\det H(c[m], h[m] + n_0)_{n_0' - n_0}$ with

$$\prod_{\substack{r,s \in \mathbb{Z}_{>0} \\ s \leq r \\ 1 \leq rs \leq n_0' \\ (r,s) \neq (\alpha_0, \beta_0)}} \Phi_{r,s}(c[m], h[m])^{p(n_0' - rs)}$$

by Proposition 11.5.

Lemma 11.13. *1. The set* $\overset{\circ}{E}(c[m], h[m]) \cap \{(\alpha, \beta) \in (\mathbb{Z}_{>0})^2 | \alpha\beta \leq n_0'\}$ *can be described as follows:*

$$\left\{ (\alpha_0, \beta_0) + k(1,1) \middle| \begin{matrix} 0 < k \leq \alpha_0' - \alpha_0 & \textit{for Case } A, \\ 0 < k \leq \beta_0' - \beta_0 & \textit{for Case } B \end{matrix} \right\} \cup \{(\alpha_0', \beta_0')\}.$$

2. The set $\overset{\circ}{E}(c[m], h[m] + n_0) \cap \{(\alpha, \beta) \in (\mathbb{Z}_{>0})^2 | \alpha\beta \leq n_0' - n_0\}$ *can be described as follows:*

$$\left\{ (\alpha_0 + \beta_0, 0) + k(1,1) \middle| \begin{matrix} 0 < k \leq \alpha_0' - \alpha_0 & \textit{for Case } A, \\ 0 < k \leq \beta_0' - \beta_0 & \textit{for Case } B \end{matrix} \right\}.$$

Proof. The first statement of this lemma has been essentially proved in the proof of Lemma 11.12. Hence, we will only prove the second statement.

We first consider **Case** A. In this case, we have Figure 11.7.

From this figure, we see that the following inclusion holds:

$$\begin{aligned} &\overset{\circ}{E}(c[m], h_{\beta_0, \alpha_0}[m] + \alpha_0\beta_0) \cap \{(\alpha, \beta) \in (\mathbb{Z}_{>0})^2 | \alpha\beta \leq \alpha_0'\beta_0' - \alpha_0\beta_0\} \\ \supset &\{(\alpha_0 + \beta_0, 0) + k(1,1) | 0 < k \leq \alpha_0' - \alpha_0\}. \end{aligned}$$

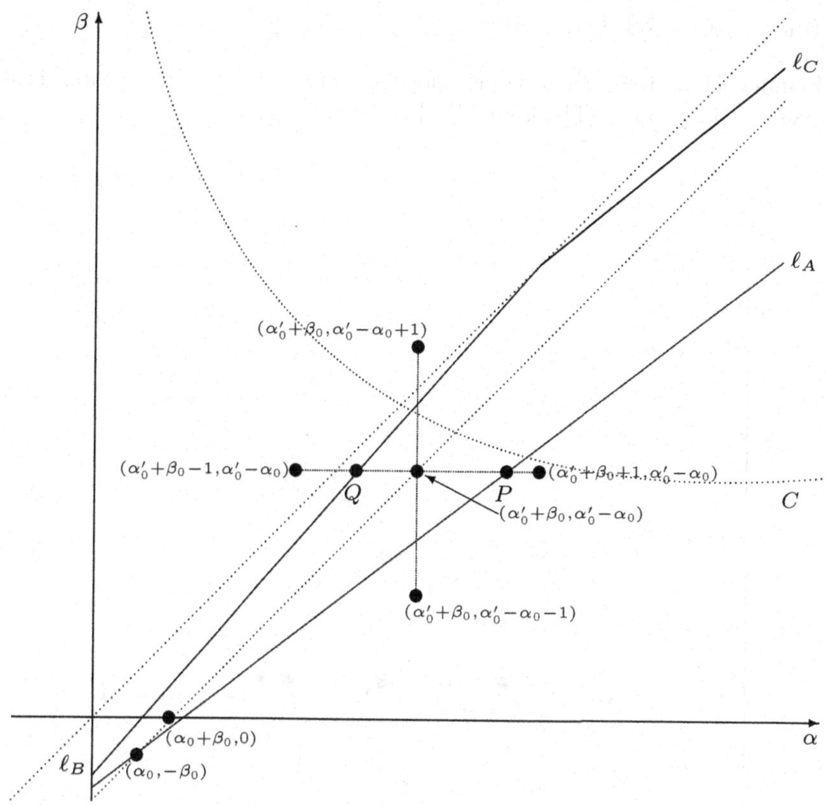

where $M := m\alpha_0 + (m+1)\beta_0$,

$$\ell_A : m\alpha - (m+1)\beta = M, \qquad \ell_B : (m+1)\alpha - m\beta = M,$$
$$\ell_C : m\alpha - (m+1)\beta = -M,$$
$$C : \alpha\beta = \alpha_0'\beta_0' - \alpha_0\beta_0 = (\alpha_0' + \beta_0)(\alpha_0' - \alpha_0) + \alpha_0',$$

$P(\alpha_0' + \beta_0 + \frac{\beta_0'-1}{m}, \alpha_0' - \alpha_0)$ and $Q(\alpha_0' + \beta_0 - \frac{\alpha_0'}{m+1}, \alpha_0' - \alpha_0)$.

Fig. 11.7 Case A

By the assumption $\alpha_0' - 1 < r(m) = m - \beta_0$, it follows that the above inclusion is, in fact, the equality.

Similarly, for **Case** B, Figure 11.8 and the assumption $\alpha_0' - 1 < r(m) = m + 1 - \beta_0$ imply the equality:

$$\overset{\circ}{E}(c[m], h_{\alpha_0,\beta_0}[m] + \alpha_0\beta_0) \cap \{(\alpha, \beta) \in (\mathbb{Z}_{>0})^2 | \alpha\beta \le \alpha_0'\beta_0' - \alpha_0\beta_0\}$$
$$= \{(\alpha_0 + \beta_0, 0) + k(1,1) | 0 < k \le \beta_0' - \beta_0\}. \qquad \square$$

Lemma 11.13 together with Proposition 11.5 implies Proposition 11.6. Therefore, the proof of Theorem 11.1 has been completed.

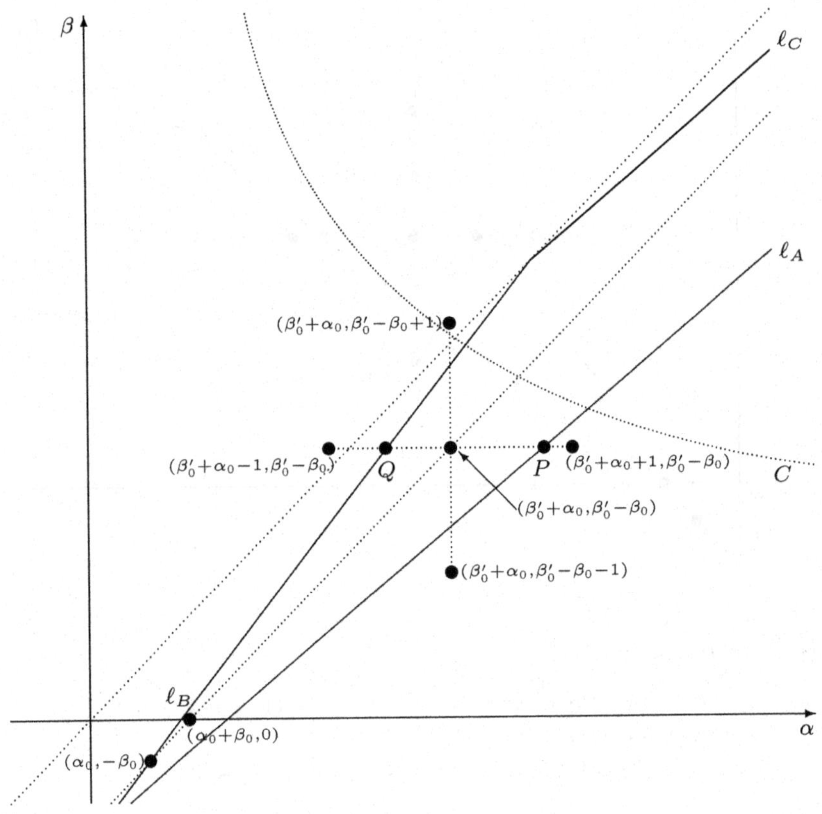

where $M := (m+1)\alpha_0 + m\beta_0$,

$$\ell_A : m\alpha - (m+1)\beta = M, \qquad \ell_B : (m+1)\alpha - m\beta = M,$$
$$\ell_C : m\alpha - (m+1)\beta = -M,$$
$$C : \alpha\beta = \alpha_0'\beta_0' - \alpha_0\beta_0 = (\alpha_0 + \beta_0')(\beta_0' - \beta_0) + \beta_0',$$

$P(\beta_0' + \alpha_0 + \frac{\alpha_0'-1}{m}, \beta_0' - \beta_0)$ and $Q(\beta_0' + \alpha_0 - \frac{\beta_0'}{m+1}, \beta_0' - \beta_0)$.

Fig. 11.8 Case B

11.6 Bibliographical Notes and Comments

In 1984, D. Friedan, Z. Qiu and S. Shenker [FQS1] has announced a necessary condition of the irreducible highest weight modules to be unitarisable, and their detailed proof appeared in [FQS2] in 1986. A different proof was given by R. Langlands [La] in 1988; the proof given in this chapter basically follows his arguments except for the proof of Proposition 11.6 in Step VIII which is our original. R. Langlands' proof relies on the analysis of the contravariant form as $m \to \infty$, whose limit is a Vir-module with central charge 1, with the aid of its structure of the Jantzen filtration we have discussed in Chapter 6.

As we have explained in Chapter 10, the sufficiency of the above-mentioned condition was proved by P. Goddard, A. Kent and D. Olive [GKO1], [GKO2] in 1985.

One of the important applications of Theorem 11.1, which was proved by R. Goodman and N. R. Wallach [GW1], [GW2] (see also [To]), says that some completeions of descrete series representations of the Virasoro algebra can be lifted to projective representations of the group $\mathrm{Diff}_+(S^1)$ of orientation preserving diffeomorphisms of the circle S^1.

Appendix A
Homological Algebras

In this appendix, for the reader's convenience we briefly recall basic facts about homological algebras used in this book. In particular, some statements concerning Lie algebras are explained with proofs.

In Section $A.1$, we will recall the definition and some fundamental properties of categories and functors. In Section $A.2$, we will recall derived functors. In Section $A.3$, we will explain Lie algebra homology and cohomology.

A.1 Categories and Functors

In this section, we recall some basic notions such as categories, functors etc.. In particular, we will explain some subtleness when one treats additive categories. For detail, the reader may consult [HiSt], [Wei].

A.1.1 Categories

Here, we recall the definition of the category.

Definition A.1 *A* ***category*** $\mathcal{C}$ *is a collection of three pieces of data:*

1. *a family* $\mathrm{Ob}(\mathcal{C})$*, whose members are called* ***objects****,*
2. *for all pairs* (X, Y) *of* $\mathrm{Ob}(\mathcal{C})$*, a set* $\mathrm{Hom}_{\mathcal{C}}(X, Y)$*, whose elements are called* ***morphisms*** *from* X *to* Y*,*
3. *for any triple* (X, Y, Z) *of* $\mathrm{Ob}(\mathcal{C})$*, a map*

$$\mathrm{Hom}_{\mathcal{C}}(X, Y) \times \mathrm{Hom}_{\mathcal{C}}(Y, Z) \longrightarrow \mathrm{Hom}_{\mathcal{C}}(X, Z); \qquad (f, g) \longmapsto f \circ g$$

called the ***composition map****.*

These data satisfy

K. Iohara, Y. Koga, *Representation Theory of the Virasoro Algebra*,
Springer Monographs in Mathematics, DOI 10.1007/978-0-85729-160-8,

1. *the composition of morphisms is associative,*
2. *for any* $X \in \mathrm{Ob}(\mathcal{C})$, *there exists* $\mathrm{id}_X \in \mathrm{Hom}_{\mathcal{C}}(X, X)$ *such that* $f \circ \mathrm{id}_X = f$ *and* $\mathrm{id}_X \circ g = g$ *for any* $f \in \mathrm{Hom}_{\mathcal{C}}(X, Y)$ *and any* $g \in \mathrm{Hom}_{\mathcal{C}}(Z, X)$.

Notice that the morphism id_X is uniquely determined.

A morphism $f \in \mathrm{Hom}_{\mathcal{C}}(X, Y)$ is called a **monomorphism** (resp. an **epimorphism**) if the equality $f \circ g_1 = f \circ g_2 \in \mathrm{Hom}_{\mathcal{C}}(W, Y)$ ($g_1 \circ f = g_2 \circ f \in \mathrm{Hom}_{\mathcal{C}}(X, Z)$) implies $g_1 = g_2 \in \mathrm{Hom}_{\mathcal{C}}(W, X)$ (resp. $g_1 = g_2 \in \mathrm{Hom}_{\mathcal{C}}(Y, Z)$), and is called an **isomorphism** if there exists $g \in \mathrm{Hom}_{\mathcal{C}}(Y, X)$ such that $f \circ g = \mathrm{id}_Y$ and $g \circ f = \mathrm{id}_X$.

A **subcategory** $\mathcal{C}'$ of $\mathcal{C}$ is a category $\mathcal{C}'$ such that $\mathrm{Ob}(\mathcal{C}') \subset \mathrm{Ob}(\mathcal{C})$ and for any pair (X, Y) of $\mathrm{Ob}(\mathcal{C}')$, $\mathrm{Hom}_{\mathcal{C}'}(X, Y) \subset \mathrm{Hom}_{\mathcal{C}}(X, Y)$ with the induced composition law, and $\mathrm{id}_X \in \mathrm{Hom}_{\mathcal{C}'}(X, X)$.

If in addition $\mathrm{Hom}_{\mathcal{C}'}(X, Y) = \mathrm{Hom}_{\mathcal{C}}(X, Y)$ always holds, then $\mathcal{C}'$ is called a **full subcategory** of $\mathcal{C}$.

For any category $\mathcal{C}$, the **opposite category** $\mathcal{C}^{\mathrm{opp}}$ is the category defined by $\mathrm{Ob}(\mathcal{C}^{\mathrm{opp}}) = \mathrm{Ob}(\mathcal{C})$ and $\mathrm{Hom}_{\mathcal{C}^{\mathrm{opp}}}(X, Y) := \mathrm{Hom}_{\mathcal{C}}(Y, X)$ with obvious composition laws.

A.1.2 Functors

Here, we recall the definition of functors.

Definition A.2 *Let* $\mathcal{C}$ *and* $\mathcal{C}'$ *be two categories. A* ***functor*** F *from* $\mathcal{C}$ *to* $\mathcal{C}'$ *consists of the following data and rules:*

1. *a map* $F : \mathrm{Ob}(\mathcal{C}) \longrightarrow \mathrm{Ob}(\mathcal{C}')$;
2. *for any pair* (X, Y) *of* $\mathrm{Ob}(\mathcal{C})$, *a map* $F : \mathrm{Hom}_{\mathcal{C}}(X, Y) \longrightarrow \mathrm{Hom}_{\mathcal{C}'}(F(X), F(Y))$,

satisfying

$$F(\mathrm{id}_X) = \mathrm{id}_{F(X)},$$
$$F(f \circ g) = F(f) \circ F(g).$$

In particular, one says that F *is a* ***covariant functor****, and a functor from* $\mathcal{C}^{\mathrm{opp}}$ *to* $\mathcal{C}'$ *is called a* ***contravariant functor*** *from* $\mathcal{C}$ *to* $\mathcal{C}'$.

For any two categories $\mathcal{C}$ and $\mathcal{C}'$, the family of functors from $\mathcal{C}$ to $\mathcal{C}'$ forms a category if one defines the morphisms of functors in an appropriate way. This is done as follows:

Definition A.3 *Let* F *and* G *be two functors from* $\mathcal{C}$ *to* $\mathcal{C}'$. *A* ***morphism of functors*** *(or a* ***natural transformation****)* t *from* F *to* G *consists of following data:*

For any $X \in \mathrm{Ob}(\mathcal{C})$, *an element* $t(X) \in \mathrm{Hom}_{\mathcal{C}'}(F(X), G(X))$, *such that the following diagram commutes for any $f \in \mathrm{Hom}_{\mathcal{C}}(X, Y)$:*

$$\begin{array}{ccc} F(X) & \xrightarrow{t(X)} & G(X) \\ \downarrow{\scriptstyle F(f)} & & \downarrow{\scriptstyle G(f)} \\ F(Y) & \xrightarrow[t(Y)]{} & G(Y). \end{array}$$

A functor F from $\mathcal{C}$ to $\mathcal{C}'$ is said to be a **categorical equivalence** iff there exists a functor G from $\mathcal{C}'$ to $\mathcal{C}$ such that $F \circ G \cong \mathrm{id}_{\mathcal{C}'}$ and $G \circ F \cong \mathrm{id}_{\mathcal{C}}$. In this case, G is called a **quasi-inverse** of F.

In such case, it quite often happens that (F, G) is an **adjoint pair**, namely, there exists an isomorphism of bifunctors $\mathcal{C}^{\mathrm{opp}} \times \mathcal{C}'$ to the category of sets $\mathrm{Hom}_{\mathcal{C}'}(F(X), Y)) \xrightarrow{\sim} \mathrm{Hom}_{\mathcal{C}}(X, G(Y))$. We say that G is a **right adjoint** to F and that F is a **left adjoint** to G. However, this is equivalent only to the following: there exist natural transformations

$$\alpha : F \circ G \longrightarrow \mathrm{id}_{\mathcal{C}'}, \qquad \beta : \mathrm{id}_{\mathcal{C}} \longrightarrow G \circ F,$$

such that the composition $G(Y) \overset{\beta(G(Y))}{\longrightarrow} G \circ F \circ G(Y) \overset{G(\alpha(Y))}{\longrightarrow} G(Y)$ is equal to $\mathrm{id}_{G(Y)}$ for any $Y \in \mathrm{Ob}(\mathcal{C}')$
and the composition $F(X) \overset{F(\beta(X))}{\longrightarrow} F \circ G \circ F(X) \overset{\alpha(F(X))}{\longrightarrow} F(X)$ is equal to $\mathrm{id}_{F(X)}$ for any $X \in \mathrm{Ob}(\mathcal{C})$.

A.1.3 Additive Categories

Here, we briefly recall the definition of the additive category and the additive functor.

Definition A.4 *A category $\mathcal{C}$ is called an* **additive category** *if it satisfies the following:*

1. *for any pair (X, Y) of $\mathrm{Ob}(\mathcal{C})$, $\mathrm{Hom}_{\mathcal{C}}(X, Y)$ has a structure of abelian group, and the composition law is bilinear,*
2. *there exists an object 0 such that $\mathrm{Hom}_{\mathcal{C}}(0, 0) = 0$,*
3. *for any pair (X, Y) of $\mathrm{Ob}(\mathcal{C})$, their product, called the* **direct sum** *of X and Y and denoted by $X \oplus Y$, exists.*

Here, for a family $\{X_i\}_{i \in I}$ of the objects of the category $\mathcal{C}$ indexed by I, a **product** $(X; \pi_i)$ of the objects X_i is an object X, together with morphisms $\pi_i : X \longrightarrow X_i$ called **projections**, with the universal property: for any $Y \in \mathrm{Ob}(\mathcal{C})$ and morphisms $f_i : Y \longrightarrow X_i$, there exists a unique morphism $f : Y \longrightarrow X$ with $f_i = \pi_i \circ f$.

Definition A.5 *Let* $f \in \mathrm{Hom}_{\mathcal{C}}(X,Y)$.

1. *A pair* (K,i) *is called the* **kernel** *of* f *if it satisfies*
   i. $f \circ i = 0$ *and*
   ii. *for any morphism such that* $f \circ j = 0$ *there exists a unique morphism* k *such that* $j = i \circ k$.

   In such case, the object K *is often denoted by* $\mathrm{Ker} f$.
2. *A pair* (C,p) *is called the* **cokernel** *of* f *if it satisfies*
   i. $p \circ f = 0$ *and*
   ii. *for any morphism such that* $q \circ f = 0$ *there exists a unique morphism* r *such that* $q = r \circ p$.

   In such case, the object C *is often denoted by* $\mathrm{Coker} f$.

Notice that the kernel and the cokernel of a morphism may not exist, but if they exist they are unique up to isomorphism. We also remark that in an additive category, a monomorphism (resp. an epimorphism) is a morphism having zero kernel (resp. zero cokernel).

The cokernel of the kernel of a morphism f, if it exists, is called the **coimage** of f and is denoted by $\mathrm{Coim} f$. Similarly, the kernel of the cokernel of f, if it exists, is called the **image** of f and is denoted by $\mathrm{Im} f$.

Let $\mathcal{C}$ and $\mathcal{C}'$ be additive categories. A functor between $\mathcal{C}$ and $\mathcal{C}'$ often satisfies an additional property:

Definition A.6 *A functor* F *from* $\mathcal{C}$ *to* $\mathcal{C}'$ *is called* **additive** *if for any pair* (X,Y) *of* $\mathrm{Ob}(\mathcal{C})$, *the map* $F : \mathrm{Hom}_{\mathcal{C}}(X,Y) \longrightarrow \mathrm{Hom}_{\mathcal{C}'}(F(X),F(Y))$ *is a homomorphism of groups.*

Notice that this condition is equivalent to the condition that F preserves the direct sum.

A.1.4 Abelian Categories

Here, we briefly recall the definition of the abelian category and the exactness of functors.

Definition A.7 *An additive category* $\mathcal{C}$ *is called an* **abelian category** *if it satisfies the following conditions.*

1. *For any morphism* $f : X \longrightarrow Y$, $\mathrm{Ker} f$ *and* $\mathrm{Coker} f$ *exist.*
2. *The canonical morphism* $\mathrm{Coim} f \longrightarrow \mathrm{Im} f$ *is an isomorphism.*

An abelian category is a generalisation of the category of abelian groups. In fact,

Remark A.1 *It has been proved by B. Mitchell [Mit] that every small abelian category is equivalent to a full subcategory of the category of A-modules for some ring A. Here, a category is said to be small if its class of objects forms a set.*

We also remark that a morphism in an abelian category is an isomorphism if and only if it is a monomorphism and an epimorphism.

Definition A.8 *A sequence of morphisms*

$$X \xrightarrow{f} Y \xrightarrow{g} Z$$

in $\mathcal{C}$ is called an ***exact sequence*** *if it satisfies*

1. $g \circ f = 0$ *and*
2. *the natural morphism* $\mathrm{Im} f \to \mathrm{Ker} g$ *is an isomorphism.*

More generally, a sequence of morphisms is called ***exact*** *if any successive pair of arrows is exact.*

We remark that a sequence $0 \to X \xrightarrow{f} Y$ (resp. $X \xrightarrow{f} Y \to 0$) is exact if and only if f is a monomorphism (resp. an epimorphism).

Let $\mathcal{C}$ and $\mathcal{C}'$ be abelian categories.

Definition A.9 *Let F be an additive functor from $\mathcal{C}$ to $\mathcal{C}'$. F is said to be* ***left exact*** *(resp.* ***right exact****) if*

$$0 \to F(X) \to F(X') \to F(X"),$$
$$(\textit{resp.}\ \ F(X) \to F(X') \to F(X") \to 0)$$

is exact in $\mathcal{C}'$ for any exact sequence $0 \to X \to X' \to X"$ (resp. $X \to X' \to X" \to 0$) in $\mathcal{C}$. In particular, F is said to be ***exact*** *if it is both left exact and right exact.*

If an additive functor between two abelian categories is one-sided exact, i.e., either left exact or right exact, then one may measure how far it is from the functor being exact. This will be the subject of the next section.

A.2 Derived Functors

In this section, we briefly recall the definition of derived functors and their properties. Throughout this section, we assume that all the categories we consider are abelian unless otherwise stated.

A.2.1 Definition

After recalling some basic notion such as projective objects and injective objects, we define the derived functors of either a left exact or a right exact additive functor between abelian categories.

Let $\mathcal{C}$ be an abelian category.

Definition A.10 *1. Let $X \in \mathrm{Ob}(\mathcal{C})$. X is called* ***projective*** *(resp.* ***injective****), if the functor $\mathrm{Hom}_{\mathcal{C}}(X, \cdot)$ (resp. $\mathrm{Hom}_{\mathcal{C}}(\cdot, X)$) is exact.*

2. We say that $\mathcal{C}$ ***has enough projectives****, if, for any object $X \in \mathrm{Ob}(\mathcal{C})$, there is a projective object $P \in \mathrm{Ob}(\mathcal{C})$ and an epimorphism $P \twoheadrightarrow X$. Dually, we say that $\mathcal{C}$* ***has enough injectives****, if $\mathcal{C}^{\mathrm{opp}}$ has enough projectives.*

Remark A.2 *Let $X \in \mathrm{Ob}(\mathcal{C})$. In general, the functor $\mathrm{Hom}_{\mathcal{C}}(X, \cdot)$ is only left exact. Hence, the condition that $P \in \mathrm{Ob}(\mathcal{C})$ is projective is equivalent to the following: if for any X, $Y \in \mathrm{Ob}(\mathcal{C})$, $f \in \mathrm{Hom}_{\mathcal{C}}(P, Y)$ and an epimorphism $g \in \mathrm{Hom}_{\mathcal{C}}(X, Y)$, there exists $h \in \mathrm{Hom}_{\mathcal{C}}(P, X)$ such that the following diagram commutes:*

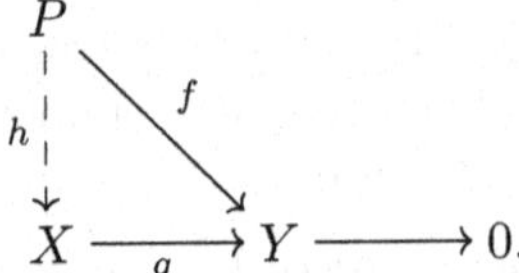

The dual statement to the above also holds.

Definition A.11 *Let $X \in \mathrm{Ob}(\mathcal{C})$.*

1. An exact sequence

$$\to P_n \xrightarrow{\partial_n} P_{n-1} \to \cdots \to P_2 \xrightarrow{\partial_2} P_1 \xrightarrow{\partial_1} P_0 \xrightarrow{\epsilon} X \to 0 \qquad \text{(A.1)}$$

consisting of projective objects P_n, is called a ***projective resolution*** *of X.*

2. An exact sequence

$$0 \to X \xrightarrow{\epsilon} I^0 \xrightarrow{d^0} I^1 \xrightarrow{d^1} I^2 \to \cdots \to I^n \xrightarrow{d^n} I^{n+1} \to \cdots, \qquad \text{(A.2)}$$

consisting of injective objects I_n, is called an ***injective resolution*** *of X.*

By definition, we have

Lemma A.1. *Suppose that $\mathcal{C}$ has enough projectives (resp. injectives). Then, any $X \in \mathrm{Ob}(\mathcal{C})$ admits a projective (resp. an injective) resolution of M.*

Let $\mathcal{C}$ and $\mathcal{C}'$ be abelian categories and $F : \mathcal{C} \to \mathcal{C}'$ be a covariant additive functor. We define functors $R^n F$ and $L_n F$ as follows:

1. Suppose that F is left exact and $\mathcal{C}$ has enough injectives. For $X \in \mathrm{Ob}(\mathcal{C})$, take an injective resolution of X:

$$0 \to X \xrightarrow{\epsilon} I^0 \xrightarrow{d^0} I^1 \xrightarrow{d^1} I^2 \to \cdots \to I^n \xrightarrow{d^n} I^{n+1} \to \cdots .$$

Consider the cocomplex

$$0 \to F(I^0) \to F(I^1) \to F(I^2) \to \cdots \to F(I^n) \xrightarrow{F(d^n)} F(I^{n+1}) \to \cdots ,$$

and define $R^nF(X) := \mathrm{Ker}F(d^n)/\mathrm{Im}F(d^{n-1})$.

2. Suppose that F is right exact and $\mathcal{C}$ has enough projectives. For $X \in \mathrm{Ob}(\mathcal{C})$, take a projective resolution of X:

$$\to P_n \xrightarrow{\partial_n} P_{n-1} \to \cdots \to P_2 \xrightarrow{\partial_2} P_1 \xrightarrow{\partial_1} P_0 \xrightarrow{\epsilon} X \to 0.$$

Consider the complex

$$\to F(P_n) \xrightarrow{F(\partial_n)} F(P_{n-1}) \to \cdots \to F(P_2) \to F(P_1) \to F(P_0) \to 0,$$

and define $L_nF(X) := \mathrm{Ker}F(\partial_n)/\mathrm{Im}F(\partial_{n+1})$.

For a contravariant additive functor $F : \mathcal{C} \to \mathcal{C}'$, one may similarly define functors R^nF and L_nF. (Regard F as a covariant functor from $\mathcal{C}^{\mathrm{opp}}$ to $\mathcal{C}'$ and apply the above definition.)

Notice that since any two projective (resp. injective) resolutions of an object are homotopic, the above definition does not depend on the choice of a projective (resp. an injective) resolution.

A particular case of the derived functor is given by the following lemma:

Lemma A.2. *1. If F is left exact, then $R^0F(X) \simeq F(X)$ for any $X \in \mathrm{Ob}(\mathcal{C})$.*
2. If F is right exact, then $L_0F(X) \simeq F(X)$ for any $X \in \mathrm{Ob}(\mathcal{C})$.

Definition A.12 *1. For a left exact additive functor F, the functors R^nF are called the nth* **right derived functors** *of F.*
2. For a right exact additive functor F, the functors L_nF is called the nth **left derived functors** *of F.*

Let us summarise the definition of the derived functors as follows:

derived functor	functor		resolution
R^nF	covariant	left exact	injective
L_nF	covariant	right exact	projective
R^nF	contravariant	left exact	projective
L_nF	contravariant	right exact	injective

The following propositions is a practical tool to calculate the nth derived functors.

Proposition A.1 *Let F be a covariant one-side exact functor from $\mathcal{C}$ to $\mathcal{C}'$. Suppose that $0 \to X \to Y \to Z \to 0$ is an exact sequence in $\mathcal{C}$. Then, we have the following long exact sequence in $\mathcal{C}'$:*

1. *F: left exact,*

$$\begin{aligned} 0 \to & F(X) \to F(Y) \to F(Z) \to \\ \to & R^1F(X) \to R^1F(Y) \to R^1F(Z) \to \cdots \\ \cdots \to & R^iF(X) \to R^iF(Y) \to R^iF(Z) \to \cdots . \end{aligned}$$

2. *F: right exact,*

$$\begin{aligned} \cdots \to & L_iF(X) \to L_iF(Y) \to L_iF(Z) \to \cdots \\ \cdots \to & L_1F(X) \to L_1F(Y) \to L_1F(Z) \to \\ \to & L_0F(X) \to F(Y) \to F(Z) \to 0. \end{aligned}$$

Here is a simple corollary of the Grothendieck spectral sequence [Gro]:

Proposition A.2 *1. Let $G : \mathcal{C} \to \mathcal{C}'$ be a covariant one-side exact functor, and let $F : \mathcal{C}' \to \mathcal{C}''$ be an exact functor.*

(i) Suppose that F is covariant. If G is left exact, then $R^n(F \circ G) = F \circ R^nG$.

(ii) Suppose that F is contravariant. If G is right exact, then $R^n(F \circ G) = F \circ L_nG$.

2. Let $G : \mathcal{C}' \to \mathcal{C}''$ be a covariant one-side exact functor, and let $F : \mathcal{C} \to \mathcal{C}'$ be an exact functor.

(i) Suppose that F is a covariant functor, which maps injectives to injectives. If G is left exact, then $G \circ F$ is left exact and $R^n(G \circ F) = R^nG \circ F$.

(ii) Suppose that F is a contravariant functor, which maps projectives to injectives. If G is right exact, then $G \circ F$ is left exact and $R^n(G \circ F) = L_nG \circ F$.

A.2.2 Extension of Modules

One of the most typical examples relevant to this book is the extension of modules we are going to review here.

Let $\mathcal{C}$ be an abelian category. For $A, B \in \mathrm{Ob}(\mathcal{C})$, an exact sequence

$$0 \to B \to E_{n-1} \to \cdots \to E_0 \to A \to 0$$

is called an **nth extension of A by B**.

We first introduce an equivalence relation on the set of nth extensions of A by B. It should be noticed that, at this moment, $\mathcal{C}$ does not necessarily have enough projectives and injectives.

For a pair (A, B) of $\mathrm{Ob}(\mathcal{C})$, let

$$E : 0 \to B \to E_{n-1} \to \cdots \to E_0 \to A \to 0,$$
$$E' : 0 \to B \to E'_{n-1} \to \cdots \to E'_0 \to A \to 0$$

be nth extensions of A by B. Let us denote $E \rightarrowtail E'$, if there exists a commutative diagram

$$\begin{array}{lccccccccccc} E: & 0 \longrightarrow & B & \longrightarrow & E_{n-1} & \longrightarrow \cdots \longrightarrow & E_0 & \longrightarrow & A & \longrightarrow 0\,. \\ & & \| & & \downarrow & & \downarrow & & \| & \\ E': & 0 \longrightarrow & B & \longrightarrow & E'_{n-1} & \longrightarrow \cdots \longrightarrow & E'_0 & \longrightarrow & A & \longrightarrow 0 \end{array}$$

Note that for $n \geq 2$ the relation $E \rightarrowtail E'$ is not symmetric. Let $\sim$ be an equivalence relation on the set of n-extensions of A by B generated by the relation $\rightarrowtail$, i.e.,

> $E \sim E'$ if and only if there exists a sequence $E(0), \cdots, E(m)$ such that $E = E(0)$, $E' = E(m)$ and $E(i) \rightarrowtail E(i+1)$ or $E(i+1) \rightarrowtail E(i)$ hold for any $i = 1, \cdots, m-1$.

We denote the equivalence classes of E by $[E]$, and set

$$\overline{\mathrm{Ext}}^n_{\mathcal{C}}(A, B) := \{n\text{th extensions of } A \text{ by } B\}/\sim.$$

It is known that for any $[E], [E'] \in \overline{\mathrm{Ext}}^n_{\mathcal{C}}(A, B)$, one can define $[E] + [E']$ in terms of extensions, and $\overline{\mathrm{Ext}}^n_{\mathcal{C}}(A, B)$ becomes an abelian group. Moreover, $\overline{\mathrm{Ext}}^n_{\mathcal{C}}(\cdot, \cdot)$ is an additive bi-functor from $\mathcal{C}$ to the category of abelian groups.

For two exact sequences

$$E : 0 \to B \xrightarrow{d} E_1 \to \cdots E_r \to A \to 0,$$
$$F : 0 \to C \to F_1 \to \cdots \to F_s \xrightarrow{d'} B \to 0,$$

in $\mathcal{C}$, we obtain a new exact sequence $E \circ F$, called the **Yoneda product** of E and F:

$$0 \to C \to F_1 \to \cdots \to F_s \xrightarrow{d' \circ d} E_1 \to \cdots \to E_r \to A \to 0.$$

Note that Yoneda product defines

$$\overline{\mathrm{Ext}}^r_{\mathcal{C}}(A, B) \times \overline{\mathrm{Ext}}^s_{\mathcal{C}}(B, C) \longrightarrow \overline{\mathrm{Ext}}^{r+s}_{\mathcal{C}}(A, C).$$

For the details, see, e.g., [HiSt].

Using $\overline{\mathrm{Ext}}^n_{\mathcal{C}}(\cdot,\cdot)$, we see the splitting of modules whose composition factors do not have any non-trivial first extensions.

Lemma A.3 ([RW1]). *Let $M = M_0 \supset M_1 \supset \cdots \supset M_d \supset M_{d+1} = (0)$ be a filtration in $\mathcal{C}$. Put $W_i := M_i/M_{i+1}$ for $i = 0, \cdots, d$. Suppose that*

$$\overline{\mathrm{Ext}}^1_{\mathcal{C}}(W_i, W_j) = 0$$

for any $1 \leq i < j \leq d$. Then M is isomorphic to the direct sum of $\{W_i\}$, i.e.,

$$M \simeq \bigoplus_{i=0}^{d} W_i.$$

Proof. In the case where $d = 1$, we have $M = M_0 \supset M_1 \supset M_2 = (0)$ and $\overline{\mathrm{Ext}}^1_{\mathcal{C}}(W_0, W_1) = 0$. Hence the short exact sequence

$$0 \to W_1 \to M \to W_0 \to 0$$

splits, and thus $M \simeq W_0 \oplus W_1$. By induction on d, the lemma is proved. □

In the case when $\mathcal{C}$ has enough injectives and projectives, $\overline{\mathrm{Ext}}^n_{\mathcal{C}}(A, B)$ can be interpreted as the Ext bi-functor defined as derived functors. Noting that $F^A_{\mathcal{C}} := \mathrm{Hom}_{\mathcal{C}}(A, \cdot)$ (resp. $F'^B_{\mathcal{C}} := \mathrm{Hom}_{\mathcal{C}}(\cdot, B)$) is a covariant (resp. a contravariant) left exact functor, we define the **Ext bi-functor** $\mathrm{Ext}^n_{\mathcal{C}}(\cdot,\cdot)$ as follows:

Definition A.13 *1.* $\mathrm{Ext}^n_{\mathcal{C}}(A, B) := (R^n F^A_{\mathcal{C}})(B)$ *if $\mathcal{C}$ has enough injectives, and*

2. $\mathrm{Ext}^n_{\mathcal{C}}(A, B) := (R^n F'^B_{\mathcal{C}})(A)$ *if $\mathcal{C}$ has enough projectives.*

By definition, the next lemma follows:

Lemma A.4. *Let $X \in \mathrm{Ob}(\mathcal{C})$.*

1. If $P \in \mathrm{Ob}(\mathcal{C})$ is projective, then $\mathrm{Ext}^n_{\mathcal{C}}(P, X) = 0$ for any $n > 0$.
2. If $I \in \mathrm{Ob}(\mathcal{C})$ is injective, then $\mathrm{Ext}^n_{\mathcal{C}}(X, I) = 0$ for any $n > 0$.

In the case where $\mathcal{C}$ has enough injectives and projectives, these definitions coincide:

Proposition A.3 *Suppose that $\mathcal{C}$ has enough injectives and projectives. Then for any $A, B \in \mathrm{Ob}(\mathcal{C})$, we have*

$$\overline{\mathrm{Ext}}^n_{\mathcal{C}}(A, B) \simeq (R^n F^A_{\mathcal{C}})(B) \simeq (R^n F'^B_{\mathcal{C}})(A).$$

A.3 Lie Algebra Homology and Cohomology

In § 1.3.3, we introduced Lie algebra (co)homology groups as derived functors. On the other hand, they can be defined by means of the Chevalley–Eilenberg (co)complex. Here, we recall the definition and see that two different definitions lead to the same result.

Through this section, let $\mathfrak{a}$ be a (not necessarily Q-graded) Lie algebra over a field $\mathbb{K}$ of characteristic 0.

A.3.1 Chevalley–Eilenberg (Co)complex and Lie Algebra (Co)homology

First, we recall the Chevalley–Eilenberg complex of $\mathfrak{a}$. (See, [CE] and [Kosz1] for details.)

Let V be an $\mathfrak{a}$-module. For $n \in \mathbb{Z}_{>0}$, we let $\Lambda^n \mathfrak{a}$ be the pth exterior product of $\mathfrak{a}$. In particular, we set $\Lambda^0 \mathfrak{a} := \mathbb{K}$ and $\Lambda^{-1}\mathfrak{a} := \{0\}$. The natural left $\mathfrak{a}$-module structure on $\Lambda^\bullet \mathfrak{a} \otimes V$ is denoted by θ, i.e., for $x \in \mathfrak{a}$, we set

$$\begin{aligned}\theta(x) :&\Lambda^n \mathfrak{a} \otimes V \longrightarrow \Lambda^n \mathfrak{a} \otimes V;\\ &x_1 \wedge \cdots \wedge x_n \otimes v \mapsto \sum_{i=1}^{n} x_1 \wedge \cdots \wedge [x, x_i] \wedge \cdots \wedge x_n \otimes v\\ &\qquad\qquad + x_1 \wedge \cdots \wedge x_n \otimes x.v.\end{aligned}$$

Indeed, it satisfies

$$\theta(x)\theta(y) - \theta(y)\theta(x) = \theta([x, y]) \qquad x, y \in \mathfrak{a}.$$

The exterior product $\varepsilon(x)$ of $x \in \mathfrak{a}$ is defined as follows:

$$\begin{aligned}\varepsilon(x) :&\Lambda^n \mathfrak{a} \otimes V \longrightarrow \Lambda^{n+1}\mathfrak{a} \otimes V;\\ &x_1 \wedge \cdots \wedge x_n \otimes v \mapsto x \wedge x_1 \wedge \cdots \wedge x_n \otimes v.\end{aligned}$$

One can check

$$\theta(x)\varepsilon(y) - \varepsilon(y)\theta(x) = \varepsilon([x, y]) \qquad x, y \in \mathfrak{a}.$$

Now, for $n \in \mathbb{Z}_{\geq 0}$, we define

$$\partial_n : \Lambda^n \mathfrak{a} \otimes V \longrightarrow \Lambda^{n-1}\mathfrak{a} \otimes V,$$

inductively by

$$\partial_n \circ \varepsilon(x) + \varepsilon(x) \circ \partial_{n-1} = -\theta(x) \qquad x \in \mathfrak{a}.$$

By direct calculation, one can verify

Lemma A.5. $\{\partial_n\}_{n \in \mathbb{Z}_{\geq 0}}$ *satisfy the following:*

1. $\partial_n \circ \theta(x) = \theta(x) \circ \partial_n$ *holds for any* $n \in \mathbb{Z}_{\geq 0}$ *and* $x \in \mathfrak{a}$.

2. $\partial_{n-1} \circ \partial_n = 0$ *holds for any* $n \in \mathbb{Z}_{>0}$.

Notice that the second formula can be proved only by using the relations given above. Hence, we define

Definition A.14 *Let* V *be an* $\mathfrak{a}$*-module. The complex of* $\mathfrak{a}$*-modules*

$$\cdots \to \Lambda^n \mathfrak{a} \otimes V \xrightarrow{\partial_n} \Lambda^{n-1}\mathfrak{a} \otimes V \to \cdots \to \Lambda^0 \mathfrak{a} \otimes V \to 0, \tag{A.3}$$

is called the ***Chevalley–Eilenberg complex*** *of* V*. The boundary operator* ∂_n *is explicitly described as follows:*

$$\begin{aligned}
&\partial_n(x_1 \wedge \cdots \wedge x_n \otimes v)\\
&= \sum_{i=1}^{n} (-1)^i x_1 \wedge \cdots \widehat{x_i} \cdots \wedge x_n \otimes x_i.v\\
&\quad + \sum_{1 \le i<j \le n} (-1)^{i+j} [x_i, x_j] \wedge x_1 \wedge \cdots \widehat{x_i} \cdots \widehat{x_j} \cdots \wedge x_n \otimes v,
\end{aligned}$$

where $x_1 \wedge \cdots \wedge x_n \in \Lambda^n \mathfrak{a}$*,* $v \in V$*, and the symbol* $\widehat{x_i}$ *indicates* x_i *to be omitted. The* ***Lie algebra homology*** *of* V *is defined by*

$$H_n(\mathfrak{a}, V) := \mathrm{Ker}\partial_n / \mathrm{Im}\partial_{n+1} \qquad n \in \mathbb{Z}_{\ge 0}.$$

Next, we introduce the dual version of the above construction.

The natural left $\mathfrak{a}$-module structure on $\mathrm{Hom}(\Lambda^\bullet \mathfrak{a}, V)$ is denoted by L, i.e., for $x \in \mathfrak{a}$, we set

$$\begin{aligned}
L_x :&\mathrm{Hom}(\Lambda^n \mathfrak{a}, V) \longrightarrow \mathrm{Hom}(\Lambda^n \mathfrak{a}, V);\\
&(L_x.f)(x_1, \cdots, x_n) := x.(f(x_1, \cdots, x_n)) - \sum_{i=1}^{n} f(x_1, \cdots, [x, x_i], \cdots, x_n)).
\end{aligned}$$

Indeed, it satisfies

$$L_x L_y - L_y L_x = L_{[x,y]} \qquad x, y \in \mathfrak{a}.$$

The interior product $\iota(x)$ of $x \in \mathfrak{a}$ is defined by

$$\begin{aligned}
\iota(x) :&\mathrm{Hom}(\Lambda^n \mathfrak{a}, V) \longrightarrow \mathrm{Hom}(\Lambda^{n-1}\mathfrak{a}, V);\\
&(\iota(x)f)(x_1, \cdots, x_{n-1}) := f(x, x_1, \cdots, x_{n-1}).
\end{aligned}$$

One can check

$$L_x \iota(y) - \iota(y) L_x = \iota([x, y]) \qquad x, y \in \mathfrak{a}.$$

Now, for $n \in \mathbb{Z}_{\ge 0}$, we define

$$\partial^n : \mathrm{Hom}(\Lambda^n \mathfrak{a}, V) \longrightarrow \mathrm{Hom}(\Lambda^{n+1}\mathfrak{a}, V),$$

inductively by

$$\iota(x) \circ \partial^n + \partial^{n-1} \circ \iota(x) = L_x \qquad x \in \mathfrak{a}.$$

By direct calculation, one can verify

Lemma A.6. $\{\partial^n\}_{n\in\mathbb{Z}_{\geq 0}}$ *satisfy the following:*

1. $\partial^n \circ L_x = L_x \circ \partial^n$ *holds for any* $n \in \mathbb{Z}_{\geq 0}$ *and* $x \in \mathfrak{a}$.
2. $\partial^{n+1} \circ \partial^n = 0$ *holds for any* $n \in \mathbb{Z}_{\geq 0}$.

Notice that the second formula can be proved only by using the relations given above. Hence, we define

Definition A.15 *Let* V *an* $\mathfrak{a}$*-module. The cocomplex of* $\mathfrak{a}$*-modules*

$$\cdots \leftarrow \mathrm{Hom}_{\mathbb{K}}(\Lambda^{n+1}\mathfrak{a}, V) \overset{\partial^n}{\leftarrow} \mathrm{Hom}_{\mathbb{K}}(\Lambda^n \mathfrak{a}, V) \leftarrow \cdots \leftarrow \mathrm{Hom}_{\mathbb{K}}(\Lambda^0 \mathfrak{a}, V) \leftarrow 0 \tag{A.4}$$

is called the ***Chevalley–Eilenberg cocomplex*** *of* V*. The coboundary operator* ∂^n *is explicitly described as follows:*

$$\begin{aligned}
&(\partial^n f)(x_1 \wedge \cdots \wedge x_{n+1}) \\
&= \sum_{i=1}^{n+1} (-1)^{i+1} x_i . f(x_1 \wedge \cdots \widehat{x_i} \cdots \wedge x_{n+1}) \\
&\quad + \sum_{1\leq i<j\leq n+1} (-1)^{i+j} f([x_i, x_j] \wedge x_1 \wedge \cdots \widehat{x_i} \cdots \widehat{x_j} \cdots \wedge x_{n+1}),
\end{aligned}$$

where $x_1 \wedge \cdots \wedge x_n \in \Lambda^n \mathfrak{a}$ *and the symbol* $\widehat{x_i}$ *indicates* x_i *to be omitted. The* ***Lie algebra cohomology*** *of* V *is defined by*

$$H^n(\mathfrak{a}, V) := \mathrm{Ker}\partial^n / \mathrm{Im}\partial^{n-1}.$$

A.3.2 Koszul Complex

Here we recall the Koszul complex, intorduced in [Kosz2] for an abelian case, and show that the two definitions of Lie algebra (co)homology coincide.

First, we recall the Koszul complex of a Lie algebra. We denote the natural right module structure on $U(\mathfrak{a}) \otimes_{\mathbb{K}} \Lambda^{\bullet}\mathfrak{a}$ by θ^r, i.e., for $x \in \mathfrak{a}$, we set

$$\begin{aligned}
\theta^r(x) : &U(\mathfrak{a}) \otimes_{\mathbb{K}} \Lambda^n \mathfrak{a} \longrightarrow U(\mathfrak{a}) \otimes_{\mathbb{K}} \Lambda^n \mathfrak{a}; \\
&p \otimes x_1 \wedge \cdots \wedge x_n \mapsto px \otimes x_1 \wedge \cdots \wedge x_n \\
&\qquad\qquad - \sum_{i=1}^{n} p \otimes x_1 \wedge \cdots \wedge [x, x_i] \wedge \cdots \wedge x_n.
\end{aligned}$$

Indeed, it satisfies

$$\theta^r(x)\theta^r(y) - \theta^r(y)\theta^r(x) = -\theta^r([x, y]) \qquad x, y \in \mathfrak{a}.$$

Notice that the module $U(\mathfrak{a}) \otimes_{\mathbb{K}} \Lambda^\bullet \mathfrak{a}$ has a compatible left $\mathfrak{a}$-module structure which is described as follows: Regard $U(\mathfrak{a})$ as a left module via the left multiplication and $\Lambda^\bullet \mathfrak{a}$ as trivial modules.

The exterior product $\varepsilon(x)$ of $x \in \mathfrak{a}$ is defined as follows:

$$\begin{aligned} \varepsilon(x) :& U(\mathfrak{a}) \otimes_{\mathbb{K}} \Lambda^n \mathfrak{a} \longrightarrow U(\mathfrak{a}) \otimes_{\mathbb{K}} \Lambda^{n+1} \mathfrak{a}; \\ & p \otimes x_1 \wedge \cdots \wedge x_n \mapsto p \otimes x \wedge x_1 \wedge \cdots \wedge x_n. \end{aligned}$$

One can check

$$\theta^r(x)\varepsilon(y) - \varepsilon(y)\theta^r(x) = -\varepsilon([x, y]) \qquad x, y \in \mathfrak{a}.$$

Now, for $n \in \mathbb{Z}_{>0}$, we define

$$d_n : U(\mathfrak{a}) \otimes_{\mathbb{K}} \Lambda^n \mathfrak{a} \longrightarrow U(\mathfrak{a}) \otimes_{\mathbb{K}} \Lambda^{n-1} \mathfrak{a},$$

inductively by

$$d_n \circ \varepsilon(x) + \varepsilon(x) \circ d_{n-1} = \theta^r(x) \qquad x \in \mathfrak{a}.$$

By direct calculation, one can verify

Lemma A.7. *$\{d_n\}_{n \in \mathbb{Z}_{>0}}$ satisfy the following:*

1. *$d_n \circ \theta^r(x) = \theta^r(x) \circ d_n$ holds for any $n \in \mathbb{Z}_{>0}$ and $x \in \mathfrak{a}$.*
2. *$d_n \circ d_{n+1} = 0$ holds for any $n \in \mathbb{Z}_{>0}$.*

Hence, we define

Definition A.16 *For $n \in \mathbb{Z}_{\geq 0}$, we set*

$$D_n := U(\mathfrak{a}) \otimes_{\mathbb{K}} \Lambda^n \mathfrak{a}.$$

The complex of right $\mathfrak{a}$-modules

$$D : \quad \cdots \to D_n \xrightarrow{d_n} D_{n-1} \to \cdots \to D_1 \to D_0 \to \mathbb{K} \to 0, \tag{A.5}$$

*is called the **Koszul complex** of $\mathfrak{a}$. Here, the boundary operator d_n is explicitly described as follows:*

$$\begin{aligned} & d_n(p \otimes x_1 \wedge \cdots \wedge x_n) \\ & = \sum_{i=1}^{n} (-1)^{i+1} p x_i \otimes x_1 \wedge \cdots \widehat{x_i} \cdots \wedge x_n \\ & \quad + \sum_{1 \leq i < j \leq n} (-1)^{i+j} p \otimes [x_i, x_j] \wedge x_1 \wedge \cdots \widehat{x_i} \cdots \widehat{x_j} \cdots \wedge x_n, \end{aligned}$$

where $n \in \mathbb{Z}_{>0}$, $p \in U(\mathfrak{a})$ *and* $x_i \in \mathfrak{a}$. *For* $n = 0$,

$$d_0 : D_0 \longrightarrow \mathbb{K}$$

is defined as the ***augmentation*** *of* $U(\mathfrak{a})$, *i.e., the algebra homomorphism defined by* $d_0(1) := 1$ *and* $d_0(\mathfrak{a}) = \{0\}$.

From the explicit formula of $\{d_n\}$, it follows that this complex is also a complex of left $\mathfrak{a}$-module with the module structure explained above. Hence, D can be regarded as a complex of $(\mathfrak{a}, \mathfrak{a})$-bimodules.

Proposition A.4 *1.* D_n *are projectives of* $\mathrm{Mod}_{\mathfrak{a}}$.
2. The sequence (A.5) is exact.

Hence, the sequence D *gives a projective resolution of the trivial* $\mathfrak{a}$*-module* $\mathbb{K}$.

The proof of this proposition can be reduced to the case when $\mathfrak{a}$ is abelian because of the standard filtration on $U(\mathfrak{a})$ defined as follows.

For $N \in \mathbb{Z}_{\geq 0}$, we set

$$T_N(\mathfrak{a}) := \begin{cases} \mathbb{K} & N = 0, \\ \mathbb{K} \oplus \bigoplus_{n=1}^{N} \mathfrak{a}^{\otimes N} & N > 0. \end{cases}$$

$\{T_N(\mathfrak{a})\}_{N \in \mathbb{Z}_{\geq 0}}$ defines an increasing filtration on the tensor algebra $T(\mathfrak{a})$ of $\mathfrak{a}$ and it induces a filtration on $U(\mathfrak{a})$, called the **standard filtration**,

$$F_N U(\mathfrak{a}) := \pi(T_N(\mathfrak{a})), \tag{A.6}$$

where $\pi : T(\mathfrak{a}) \twoheadrightarrow U(\mathfrak{a})$ is the canonical projection. For detail, see, e.g., [HiSt].

Next, we discuss the relation between the two definitions of Lie algebra homology. Below, we regard D as a complex of left $\mathfrak{a}$-modules unless otherwise stated. Using the Koszul complex, we can construct a projective resolution of $V \in \mathrm{Ob}(\mathrm{Mod}_{\mathfrak{a}})$ as follows:

$$\to D_n \otimes_{\mathbb{K}} V \overset{d_n \otimes \mathrm{id}_V}{\longrightarrow} D_{n-1} \otimes_{\mathbb{K}} V \to \cdots \to D_0 \otimes_{\mathbb{K}} V \to V \to 0. \tag{A.7}$$

Since each D_n is a free $U(\mathfrak{a})$-module, the tensor identity in Corollary A.1 implies that $D_n \otimes V$ is also $U(\mathfrak{a})$-free. Hence, the complex (A.7) gives a projective resolution of V.

Applying the coinvariant functor to the above complex, we obtain

$$\begin{aligned} \to \mathbb{K} \otimes_{U(\mathfrak{a})} (D_n \otimes_{\mathbb{K}} V) \overset{\tilde{d}_n}{\longrightarrow} \mathbb{K} \otimes_{U(\mathfrak{a})} (D_{n-1} \otimes_{\mathbb{K}} V) \to \cdots \\ \cdots \to \mathbb{K} \otimes_{U(\mathfrak{a})} (D_0 \otimes_{\mathbb{K}} V) \to 0, \end{aligned}$$

where $\tilde{d}_n := \mathrm{id}_{\mathbb{K}} \otimes d_n \otimes \mathrm{id}_V$, and we regard $\mathbb{K}$ as a trivial right $\mathfrak{a}$-module. By definition, we have isomorphisms of vector spaces

$$\Phi_n : \mathbb{K} \otimes_{U(\mathfrak{a})} (D_n \otimes V) \longrightarrow \Lambda^n \mathfrak{a} \otimes_{\mathbb{K}} V;$$
$$1 \otimes \{[p \otimes (x_1 \wedge \cdots \wedge x_n)] \otimes v\} \longmapsto x_1 \wedge \cdots \wedge x_n \otimes a(p)v,$$

where $a : U(\mathfrak{a}) \longrightarrow U(\mathfrak{a})$ is the antipode of the standard Hopf algebra structure on $U(\mathfrak{a})$. Moreover, by direct computation, we have

Lemma A.8.
$$\partial_n \circ \Phi_n = \Phi_{n-1} \circ \tilde{d}_n \quad (n \in \mathbb{Z}_{>0}).$$

Therefore, the two definitions of Lie algebra homology coincide.

Next, we consider the case of Lie algebra cohomology. Applying the functor $\mathrm{Hom}_{\mathbb{K}}(\cdot, V)$ to the Koszul complex, we have

$$\leftarrow \mathrm{Hom}_{\mathbb{K}}(D_{n+1}, V) \overset{\tilde{d}^n}{\leftarrow} \mathrm{Hom}_{\mathbb{K}}(D_n, V) \leftarrow \cdots \leftarrow \mathrm{Hom}_{\mathbb{K}}(D_0, V) \leftarrow V \leftarrow 0, \tag{A.8}$$

where we define $\tilde{d}^n : \mathrm{Hom}_{\mathbb{K}}(D_n, V) \to \mathrm{Hom}_{\mathbb{K}}(D_{n+1}, V)$ by

$$(\tilde{d}^n f)(p \otimes x_1 \wedge \cdots \wedge x_{n+1}) := f(d_{n+1}(p \otimes x_1 \wedge \cdots \wedge x_{n+1}))$$

for $n \in \mathbb{Z}_{\geq 0}$ and define $\tilde{d}^{-1} : V \to \mathrm{Hom}_{\mathbb{K}}(D_0, V)$ by $\tilde{d}^{-1}(v)(p) := p.v$.

Since the functor $\mathrm{Hom}_{\mathbb{K}}(\cdot, V)$ is contravariant and exact, the sequence (A.8) gives an injective resolution of V. It can be checked that this complex restricts to its $\mathfrak{a}$-invariant. Hence, we obtain

$$\cdots \leftarrow \mathrm{Hom}_{\mathfrak{a}}(D_{n+1}, V) \overset{d^n}{\leftarrow} \mathrm{Hom}_{\mathfrak{a}}(D_n, V) \leftarrow \cdots \leftarrow \mathrm{Hom}_{\mathfrak{a}}(D_0, V) \leftarrow 0, \tag{A.9}$$

where d^n is the restriction of $\tilde{d}^n$ to the invariants. One can easily show that

Lemma A.9. *The isomorphisms of $\mathbb{K}$-vector spaces*

$$\Psi_n : \mathrm{Hom}_{\mathfrak{a}}(D_n, V) \longrightarrow \mathrm{Hom}_{\mathbb{K}}(\Lambda^n \mathfrak{a}, V)$$

defined by $\Psi_n(f)(x_1 \wedge \cdots \wedge x_n) := f(1 \otimes x_1 \wedge \cdots \wedge x_n)$ *satisfy* $\Psi_{n+1} \circ d^n = \partial^n \circ \Psi_n$.

Hence, the two definitions of Lie algebra cohomology also coincide.

A.3.3 Tensor Identity

Here, we explain a general isomorphism which is valid for any Hopf algebra, called the tensor identity.

To state the tensor identity, we use the following notation of Hopf algebras. Let $\mathcal{A}$ be a Hopf algebra over the field $\mathbb{K}$ with the coproduct $\Delta : \mathcal{A} \to \mathcal{A} \otimes \mathcal{A}$, the antipode $a : \mathcal{A} \to \mathcal{A}$, the counit $\epsilon : \mathcal{A} \to \mathbb{K}$ and the unit $u : \mathbb{K} \to \mathcal{A}$. For simplicity we use the convention

$$\Delta(x) = \sum_i x_i \otimes x_i' = x_{(1)} \otimes x_{(2)}.$$

By the co-associativity of Δ, we can write

$$(\Delta \otimes 1)\Delta(x) = (1 \otimes \Delta)\Delta(x) = x_{(1)} \otimes x_{(2)} \otimes x_{(3)}.$$

Proposition A.5 *Let $\mathcal{A}$ be a Hopf algebra over $\mathbb{K}$ and let $\mathcal{B}$ be a Hopf subalgebra of $\mathcal{A}$ over $\mathbb{K}$. Let M be a $\mathcal{B}$-module and N be an $\mathcal{A}$-module. Then we have*

$$(\mathcal{A} \otimes_{\mathcal{B}} M) \otimes_{\mathbb{K}} N \simeq \mathcal{A} \otimes_{\mathcal{B}} (M \otimes_{\mathbb{K}} N)$$

as an $\mathcal{A}$-module, where the module M in the right-hand side is regarded as a $\mathcal{B}$-module via the restriction.

Proof. Define linear maps ϕ and ψ as follows:

$$\begin{aligned}
\phi :&\mathcal{A} \otimes_{\mathcal{B}} (M \otimes_{\mathbb{K}} N) \longrightarrow (\mathcal{A} \otimes_{\mathcal{B}} M) \otimes_{\mathbb{K}} N,\\
&x \otimes (m \otimes n) \longmapsto (x_{(1)} \otimes m) \otimes x_{(2)}n,\\
\psi :&(\mathcal{A} \otimes_{\mathcal{B}} M) \otimes_{\mathbb{K}} N \longrightarrow \mathcal{A} \otimes_{\mathcal{B}} (M \otimes_{\mathbb{K}} N),\\
&(x \otimes m) \otimes n \longmapsto x_{(1)} \otimes (m \otimes a(x_{(2)})n),
\end{aligned}$$

where $x \in \mathcal{A}$, $m \in M$ and $n \in N$. First, one can check that these maps are well-defined, i.e., for $y \in \mathcal{B}$,

1. $\phi(x \otimes y(m \otimes n)) = \phi(xy \otimes (m \otimes n))$,
2. $\psi((xy \otimes m) \otimes n) = \psi((x \otimes ym) \otimes n)$.

Hence, it is enough to show that

1. ϕ and ψ are $\mathcal{A}$-homomorphisms,
2. $\psi \circ \phi = \mathrm{id}_{\mathcal{A} \otimes_{\mathcal{B}} (M \otimes_{\mathbb{K}} N)}$ and $\phi \circ \psi = \mathrm{id}_{(\mathcal{A} \otimes_{\mathcal{B}} M) \otimes_{\mathbb{K}} N}$.

This follows by direct verification. □

Corollary A.1 *Let M and N be $\mathcal{A}$-modules. If M is $\mathcal{A}$-free, then $M \otimes_k N$ becomes an $\mathcal{A}$-free module.*

Proof. Since M is $\mathcal{A}$-free, there exists a $\mathbb{K}$-vector space V such that $M \simeq \mathcal{A} \otimes_{\mathbb{K}} V$ as an $\mathcal{A}$-module. By Proposition A.5, we have

$$M \otimes_{\mathbb{K}} N \simeq (\mathcal{A} \otimes_{\mathbb{K}} V) \otimes_{\mathbb{K}} N \simeq \mathcal{A} \otimes_{\mathbb{K}} (V \otimes_{\mathbb{K}} N).$$

Since $\mathcal{A} \otimes_{\mathbb{K}} (V \otimes_{\mathbb{K}} N)$ is $\mathcal{A}$-free, we obtain the conclusion. □

Appendix B
Lie p-algebras

In this appendix, we will collect some definitions and some basic concepts in the representation theory of the Lie p-algebra. Further, we will state a few properties of irreducible representations over a completely solvable Lie p-algebra, which have used in Chapter 2.

B.1 Basic Objects

In this section, we first define the Lie p-algebra, and introduce its *restricted enveloping algebra.* Further, we introduce the notion of the central character and the induced representation twisted by a central character. For these basics in the representation theory of Lie p-algebras, see e.g. [Jac] and [SF]. Throughout this section, let $\mathbb{K}$ be a field whose characteristic is $p > 0$.

B.1.1 Definition of a Lie p-algebra

A Lie p-algebra $\mathfrak{g}$ is defined as a Lie algebra over the field $\mathbb{K}$ with a mapping $(\cdot)^{[p]} : \mathfrak{g} \to \mathfrak{g}$ called a *pth power operation.* This operation has properties similar to the Frobenius map $x \mapsto x^p$. Here, we first recall some properties of the map defined by $x \mapsto x^p$ on an associative algebra over $\mathbb{K}$.

Lemma B.1. *Let $\mathcal{R}$ be an associative algebra. For $n \in \mathbb{Z}_{>0}$ and $x, y \in \mathcal{R}$, we have*

$$(\mathrm{ad}x)^n(y) = \sum_{i=0}^{n} (-1)^{n-i} \binom{n}{i} x^i y x^{n-i}, \tag{B.1}$$

where we set

$$\binom{n}{i} := \frac{n!}{i!(n-i)!}.$$

K. Iohara, Y. Koga, *Representation Theory of the Virasoro Algebra*,
Springer Monographs in Mathematics, DOI 10.1007/978-0-85729-160-8,

Proof. Let L_x (resp. R_x) be the left (resp. right) multiplication by x. Then, $\mathrm{ad}x = L_x - R_x$. Since L_x and R_x commute, we have

$$(L_x - R_x)^n = \sum_{i=0}^{n} (-1)^{n-i} \binom{n}{i} L_x^i R_x^{n-i}.$$

Hence, this lemma follows. □

From now on, let R be an associative algebra over $\mathbb{K}$. We have

Lemma B.2. *For $x, y \in \mathcal{R}$, the following hold:*

1. $(\mathrm{ad}x)^p(y) = (\mathrm{ad}x^p)(y)$,
2. $(\mathrm{ad}x)^{p-1}(y) = \sum_{i=0}^{p-1} x^i y x^{p-1-i}$.

Proof. Since $\binom{p}{i} = 0$ $(0 < i < p)$, by setting $n := p$ in (B.1), we obtain the first assertion. The second assertion also follows from (B.1), since $\binom{p-1}{i} = (-1)^i$ and $(-1)^{p-1} = 1$ in $\mathbb{K}$. □

To define the Lie p-algebra, we introduce the following notation: Let λ be an indeterminate. For $i = 1, 2, \cdots, p-1$ and $x, y \in \mathcal{R}$, we define $s_i(x, y) \in \mathcal{R}$ by

$$(x\lambda + y)^p = x^p \lambda^p + y^p + \sum_{i=1}^{p-1} s_i(x, y) \lambda^i. \tag{B.2}$$

Remark B.1 *By Lemma B.3, we see that $s_i(x, y)$ can be expressed by using commutators of x and y.*

Indeed, for $p = 2$ and $p = 3$, $s_i(x, y)$ are written as follows:

1. For $p = 2$;
$$s_1(x, y) = xy + yx = yx - xy = [y, x],$$
2. For $p = 3$;
$$\begin{aligned} s_1(x, y) &= xy^2 + yxy + y^2x = xy^2 - 2yxy + y^2x = [y, [y, x]], \\ 2s_2(x, y) &= 2(x^2y + xyx + yx^2) = -x^2y + 2xyx - yx^2 = [x, [y, x]]. \end{aligned}$$

In general, we have

Lemma B.3.

$$\{\mathrm{ad}(x\lambda + y)\}^{p-1}(x) = \sum_{i=1}^{p-1} i s_i(x, y) \lambda^{i-1}.$$

Proof. By differentiating (B.2) with respect to λ, we have

$$\sum_{i=0}^{p-1}(x\lambda+y)^i x(x\lambda+y)^{p-1-i}=\sum_{i=1}^{p-1} is_i(x,y)\lambda^{i-1}.$$

On the other hand, by (B.1),

$$\{\mathrm{ad}(x\lambda+y)\}^{p-1}(x)=\sum_{i=1}^{p-1}(-1)^{p-1-i}\binom{p-1}{i} is_i(x,y)\lambda^{i-1}.$$

Since $(-1)^{p-1-i}\binom{p-1}{i}=1$ in $\mathbb{K}$, we obtain the lemma. □

Remark that, by setting $\lambda := 1$ in (B.2), we obtain

$$(x+y)^p = x^p + y^p + \sum_{i=1}^{p-1} s_i(x,y).$$

On the other hand, for $c \in \mathbb{K}$ we have

$$(cx)^p = c^p x^p.$$

Now, these formulae and $(\mathrm{ad}x)^p(y) = (\mathrm{ad}x^p)(y)$ in Lemma B.2 lead us to the following definition of the Lie p-algebra.

Definition B.1 *A Lie algebra $\mathfrak{g}$ over $\mathbb{K}$ is called a* **Lie p-algebra** *if it equips with a map $(\cdot)^{[p]} : \mathfrak{g} \to \mathfrak{g}$ which satisfies*

$$(\mathrm{ad}x^{[p]})(y) = (\mathrm{ad}x)^p(y), \tag{B.3}$$

$$(cx)^{[p]} = c^p x^{[p]}, \tag{B.4}$$

$$(x+y)^{[p]} = x^{[p]} + y^{[p]} + \sum_{i=1}^{p-1} s_i(x,y). \tag{B.5}$$

The map $(\cdot)^{[p]} : \mathfrak{g} \to \mathfrak{g}$ is called a **pth power operation** of $\mathfrak{g}$.

We give some examples of Lie p-algebras.

Example B.1 *We set $\mathfrak{g} := \mathfrak{gl}_n(\mathbb{K})$ and regard it as a Lie algebra via the usual commutator $[x,y] := xy - yx$. Let E_{ij} be a matrix unit of $\mathfrak{gl}_n(\mathbb{K})$. Notice that $(\mathrm{ad}E_{ii})^p = \mathrm{ad}E_{ii}$ and $(\mathrm{ad}E_{ij})^p = 0$ for $i \neq j$. Hence, by Proposition B.2,*

$$E_{ij}^{[p]} := \begin{cases} E_{ii} & i = j \\ 0 & i \neq j \end{cases},$$

gives a Lie p-algebra structure of $\mathfrak{g}$.

Example B.2 *Let $\mathcal{R}$ be an associative algebra over $\mathbb{K}$. We regard it as a Lie algebra via the commutator $[x, y] := xy - yx$. $\mathcal{R}$ admits a pth power operation given by*

$$x^{[p]} := x^p \qquad (x \in \mathcal{R}), \tag{B.6}$$

which defines a Lie p-algebra structure on $\mathfrak{g}$.

In the following, we regard an associative $\mathbb{K}$-algebra as a Lie p-algebra via (B.6).

Next, we recall the definitions of a homomorphism, a subalgebra and an ideal of Lie p-algebras. Let $\mathfrak{g}$ and $\mathfrak{g}'$ be Lie p-algebras. A linear map $f : \mathfrak{g} \to \mathfrak{g}'$ is called a **homomorphism** of Lie p-algebras if it satisfies

1. f is a homomorphism of Lie algebras, and
2. $f(x^{[p]}) = f(x)^{[p]}$ for any $x \in \mathfrak{g}$.

Let $\mathfrak{g}$ be a Lie p-algebra and let $\mathfrak{a}$ be a subalgebra (resp. an ideal) of a Lie algebra $\mathfrak{g}$. We call $\mathfrak{a}$ a p**-subalgebra** (resp. a p**-ideal**) of $\mathfrak{g}$ if $x^{[p]} \in \mathfrak{a}$ for any $x \in \mathfrak{a}$.

B.1.2 Restricted Enveloping Algebra

Let $\mathfrak{g}$ be a Lie p-algebra over $\mathbb{K}$ with pth power operation $(\cdot)^{[p]}$. A restricted enveloping algebra of $\mathfrak{g}$ is defined as follows:

Definition B.2 *Let $\bar{U}$ be a unital associative algebra over $\mathbb{K}$, and let $i : \mathfrak{g} \to \bar{U}$ be a homomorphism of Lie p-algebras. We call $(\bar{U}, i)$ a* **restricted enveloping algebra** *of $\mathfrak{g}$ if*

> *for any unital associative algebra $\mathcal{R}$ over $\mathbb{K}$ and any homomorphism $f : \mathfrak{g} \to \mathcal{R}$ of Lie p-algebras, there exists a homomorphism $\tilde{f} : \bar{U} \to \mathcal{R}$ of associative $\mathbb{K}$-algebras such that the following diagram commutes:*

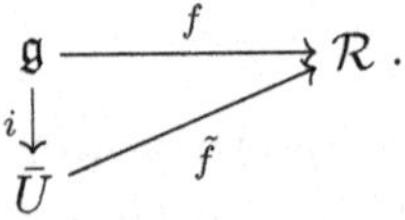

Let $\mathfrak{g}$ and $\mathfrak{g}'$ be Lie p-algebras and $\phi : \mathfrak{g} \to \mathfrak{g}'$ be a homomorphism of Lie algebras, which is not necessarily a homomorphism of Lie p-algebras. For $x \in \mathfrak{g}$, we set

$$\xi_\phi(x) := \phi(x)^{[p]} - \phi(x^{[p]}). \tag{B.7}$$

In order to construct a restricted enveloping algebra of a Lie p-algebra $\mathfrak{g}$, here, we show the following lemma.

Lemma B.4. *1. ξ_ϕ is* ***semi-linear****, i.e., for any $x, y \in \mathfrak{g}$ and $c \in \mathbb{K}$,*

$$\xi_\phi(x+y) = \xi_\phi(x) + \xi_\phi(y), \qquad \xi_\phi(cx) = c^p \xi_\phi(x).$$

2. For any $x, y \in \mathfrak{g}$,

$$[\xi_\phi(x), \phi(y)] = 0.$$

Proof. By definition, one can check this lemma. □

In the case where $\mathfrak{g}'$ is the universal enveloping algebra of $\mathfrak{g}$ and i is the canonical inclusion map $\mathfrak{g} \to U(\mathfrak{g})$, we denote

$$\xi_i : \mathfrak{g} \longrightarrow U(\mathfrak{g})$$

by ξ for simplicity. By the second statement in Lemma B.4, we see that

$$\mathrm{Im}\xi \subset Z(\mathfrak{g})$$

where $Z(\mathfrak{g})$ denotes the centre of $U(\mathfrak{g})$. We set

$$J(\mathfrak{g}) := U(\mathfrak{g}).\mathrm{Im}\xi. \tag{B.8}$$

It is a two-sided ideal of $U(\mathfrak{g})$. We define $(\bar{U}(\mathfrak{g}), \bar{i})$ by

$$\begin{aligned} &\bar{U}(\mathfrak{g}) := U(\mathfrak{g})/J(\mathfrak{g}), \\ &\bar{i} : \mathfrak{g} \overset{i}{\hookrightarrow} U(\mathfrak{g}) \overset{\pi}{\twoheadrightarrow} \bar{U}(\mathfrak{g}), \end{aligned} \tag{B.9}$$

where π denotes the canonical projection. Then, the following proposition holds.

Proposition B.1 *$(\bar{U}(\mathfrak{g}), \bar{i})$ is a restricted enveloping algebra of $\mathfrak{g}$.*

Proof. Let $\mathcal{R}$ be a unital associative algebra and let $f : \mathfrak{g} \to \mathcal{R}$ be a homomorphism of Lie p-algebras. By the universality of $U(\mathfrak{g})$, there exists a homomorphism $\tilde{f} : U(\mathfrak{g}) \to \mathcal{R}$ such that $f = \tilde{f} \circ i$. Since the pth power operation on $U(\mathfrak{g})$ is the pth power operation as an associative algebra, $\tilde{f}$ is a homomorphism of associative algebras and f is a homomorphism of Lie p-algebras,

$$\begin{aligned} \tilde{f}(\xi(x)) &= \tilde{f}(i(x)^{[p]}) - \tilde{f}(i(x^{[p]})) = \tilde{f}(i(x)^p) - f(x^{[p]}) \\ &= \tilde{f}(i(x))^p - f(x^{[p]}) = f(x)^p - f(x^{[p]}) \\ &= f(x)^{[p]} - f(x^{[p]}) = 0. \end{aligned}$$

Hence, $\tilde{f}(\mathrm{Im}\xi) = \{0\}$, and thus there exists $\bar{f} : \bar{U}(\mathfrak{g}) \to \mathcal{R}$ such that the following diagram commutes:

$$\begin{array}{ccccc} \mathfrak{g} & \xrightarrow{i} & U(\mathfrak{g}) & \xrightarrow{\pi} & \bar{U}(\mathfrak{g}) \\ & {\scriptstyle f}\searrow & \downarrow{\scriptstyle \tilde{f}} & \swarrow{\scriptstyle \bar{f}} & \\ & & \mathcal{R} & & \end{array}$$

□

Remark that, by the universal property, if a restricted enveloping algebra of $\mathfrak{g}$ exists, then it is unique up to an isomorphism of associative $\mathbb{K}$-algebras.

Next, we describe PBW type basis of $\bar{U}(\mathfrak{g})$. Let $\{x_i | i \in I\}$ be a basis of $\mathfrak{g}$, where I is a totally ordered set. By the PBW theorem, the following holds:

Lemma B.5. *The following set gives basis of* $U(\mathfrak{g})$:

$$\left\{ x_{i_1}^{m_1} \cdots x_{i_r}^{m_r} \xi(x_{j_1})^{n_1} \cdots \xi(x_{j_s})^{n_s} \,\middle|\, \begin{array}{l} i_1 < i_2 < \cdots < i_r \;\; (r \geq 0) \\ j_1 < j_2 < \cdots < j_s \;\; (s \geq 0) \\ 1 \leq m_k < p, \quad n_l \geq 1 \end{array} \right\}.$$

Since $\bar{U}(\mathfrak{g}) := U(\mathfrak{g})/U(\mathfrak{g}).\mathrm{Im}\xi$, this lemma implies the following theorem:

Theorem B.1 *The following set gives a basis of* $\bar{U}(\mathfrak{g})$:

$$\left\{ \pi(x_{i_1})^{m_1} \cdots \pi(x_{i_r})^{m_r} \,\middle|\, \begin{array}{l} i_1 < i_2 < \cdots < i_r \\ 1 \leq m_k < p \end{array} \right\}.$$

Hence, we obtain the following:

Corollary B.1 *The map* $\bar{\imath} : \mathfrak{g} \to \bar{U}(\mathfrak{g})$ *is injective.*

From the following proposition, we see when a Lie algebra becomes a Lie p-algebra.

Proposition B.2 *Let* $\mathfrak{g}$ *be a Lie algebra over the field* $\mathbb{K}$. *For a given basis* $\{x_i | i \in I\}$ *of* $\mathfrak{g}$ *indexed by a totally ordered set* I, *there exist* $\{y_i | i \in I\}$ *such that*

$$(\mathrm{ad}x_i)^p = \mathrm{ad}y_i \qquad (\forall i \in I).$$

Then, $\mathfrak{g}$ *has a unique Lie* p*-algebra structure such that* $x_i^{[p]} = y_i$.

Proof. We first notice that $x_i^p - y_i \in Z(\mathfrak{g})$, by Lemma B.2. Let J' be the two-sided ideal of $U(\mathfrak{g})$ generated by $\{x_i^p - y_i | i \in I\}$. We set $\bar{U}' := U(\mathfrak{g})/J'$, $\bar{\imath}' := \pi' \circ i : \mathfrak{g} \to \bar{U}'$, where $\pi' : U(\mathfrak{g}) \to \bar{U}'$ is the canonical projection. Similarly to the above corollary, we see that $\bar{\imath}'$ is injective. Hence, there exists a homomorphism $\mathfrak{g} \to \mathrm{Im}\bar{\imath}'$ of Lie algebras. Here, we notice that $\mathrm{Im}\bar{\imath}'$ is a Lie p-subalgebra of $\bar{U}'$, since

$$\bar{\imath}'(x_i)^{[p]} = \bar{\imath}'(x_i)^p = (x_i + J')^p = x_i^p + J' = y_i + J' = \bar{\imath}'(y_i).$$

Hence, if we define a Lie p-algebra structure on $\mathfrak{g}$ by the induced one from the isomorphism $\mathfrak{g} \to \mathrm{Im}\bar{\imath}'$, then it satisfies $x_i^{[p]} = y_i$ for any $i \in I$. The uniqueness follows from (B.4) and (B.5). □

B.1.3 Central Character

From now on, we assume that the field $\mathbb{K}$ is algebraically closed.

Let M be an irreducible $\mathfrak{g}$-module. Since $x^p - x^{[p]} \in Z(\mathfrak{g})$ for any $x \in \mathfrak{g}$, Schur's lemma implies that for each $x \in \mathfrak{g}$, there exists $\lambda_x \in \mathbb{K}$ such that

$$(x^p - x^{[p]}).v = \lambda_x v \quad (\forall v \in M).$$

Here, we recall that for $a \in \mathbb{K}$, a pth root of a is unique. We set $\chi(x) := \lambda_x^{\frac{1}{p}}$. The following lemma follows from Lemma B.4.

Lemma B.6. $\chi \in \mathfrak{g}^*$.

Motivated by this lemma, we define

Definition B.3 *Let M be a (not necessarily irreducible) $\mathfrak{g}$-module. We say that M has* **central character** $\chi \in \mathfrak{g}^*$ *if*

$$(x^p - x^{[p]}).v = \chi(x)^p v \quad (\forall x \in \mathfrak{g},\ \forall v \in M).$$

B.1.4 Induced Representations

To introduce induced representations, we first define a χ-reduced enveloping algebra of $\mathfrak{g}$.

Definition B.4 *Let $\bar{U}_\chi$ be a unital associative algebra over $\mathbb{K}$, and let $\bar{i}_\chi : \mathfrak{g} \to \bar{U}_\chi$ be a homomorphism of Lie algebras such that*

$$\bar{i}_\chi(x)^p - \bar{i}_\chi(x^{[p]}) = \chi(x)^p.1 \quad (\forall x \in \mathfrak{g}).$$

$(\bar{U}_\chi, \bar{i}_\chi)$ is called a **χ-reduced enveloping algebra** *of $\mathfrak{g}$ if*

for any unital associative algebra $\mathcal{R}$ over $\mathbb{K}$ and any homomorphism $f : \mathfrak{g} \to \mathcal{R}$ of Lie algebras such that

$$f(x)^p - f(x^{[p]}) = \chi(x)^p.1 \quad (\forall x \in \mathfrak{g}),$$

there exists a homomorphism $\tilde{f} : \bar{U}_\chi \to \mathcal{R}$ of associative $\mathbb{K}$-algebras such that $f = \tilde{f} \circ \bar{i}_\chi$.

Let $J_\chi(\mathfrak{g})$ be the two-sided ideal of $U(\mathfrak{g})$ generated by $\{\xi(x) - \chi(x)^p | x \in \mathfrak{g}\}$, where $\xi(x) := x^p - x^{[p]}$. We set

$$\bar{U}_\chi(\mathfrak{g}) := U(\mathfrak{g})/J_\chi(\mathfrak{g}),$$

and define $\bar{i}_\chi$ by the composition

$$\bar{i}_\chi : \mathfrak{g} \overset{i}{\hookrightarrow} U(\mathfrak{g}) \overset{\pi_\chi}{\twoheadrightarrow} \bar{U}_\chi(\mathfrak{g}),$$

where i is the canonical inclusion and π_χ is the canonical projection.

One can show the following proposition similarly to the proof of Proposition B.1.

Proposition B.3 *$(\bar{U}_\chi(\mathfrak{g}), \bar{i}_\chi)$ is a χ-reduced enveloping algebra of $\mathfrak{g}$.*

A χ-reduced enveloping algebra is unique up to isomorphisms of associative algebras. Remark that $(\bar{U}_0(\mathfrak{g}), \bar{i}_0)$ is just the restricted enveloping algebra of $\mathfrak{g}$.

Moreover, the following theorem holds:

Theorem B.2 *The following set gives a basis of $\bar{U}_\chi(\mathfrak{g})$:*

$$\left\{ \pi_\chi(x_{i_1})^{m_1} \cdots \pi_\chi(x_{i_r})^{m_r} \,\middle|\, \begin{array}{l} i_1 < i_2 < \cdots < i_r \quad (r \geq 0) \\ 1 \leq m_k < p \end{array} \right\},$$

where $\{x_i | i \in I\}$ is a basis of $\mathfrak{g}$ indexed by a totally ordered set I.

Corollary B.2 *The map $\bar{i}_\chi : \mathfrak{g} \to \bar{U}_\chi(\mathfrak{g})$ is injective.*

Next, we define induced representations twisted by a central character. Let $\mathfrak{a}$ be a p-subalgebra of a Lie p-algebra $\mathfrak{g}$. In the sequel, for $\chi \in \mathfrak{g}^*$, we denote the restriction of χ to $\mathfrak{a}$ by $\chi|_\mathfrak{a}$. Let M be an $\mathfrak{a}$-module with central character $\chi|_\mathfrak{a}$.

Definition B.5 *We set*

$$\mathrm{Ind}_\mathfrak{a}^\mathfrak{g}(M;\chi) := \bar{U}_\chi(\mathfrak{g}) \otimes_{U_{\chi|_\mathfrak{a}}} M \quad (U_{\chi|_\mathfrak{a}} := \bar{U}_{\chi|_\mathfrak{a}}(\mathfrak{a})),$$

and regard it as a left $\mathfrak{g}$-module via

$$x.(u \otimes m) := (\bar{i}_\chi(x)u) \otimes m \qquad (x \in \mathfrak{g},\ u \in \bar{U}_\chi(\mathfrak{g}),\ m \in M).$$

The following lemma follows from the PBW theorem.

Lemma B.7. *Let $\{y_i | j \in I\} \sqcup \{y'_i | i \in I'\}$ be a basis of $\mathfrak{g}$ such that $\{y_i | i \in I\}$ form a basis of $\mathfrak{a}$, where I and I' are totally ordered sets. Then, $\bar{U}_\chi(\mathfrak{g})$ is a free $\bar{U}_{\chi|_\mathfrak{a}}(\mathfrak{a})$-module with a basis*

$$\left\{ \pi_\chi(y'_{i_1})^{m_1} \pi_\chi(y'_{i_2})^{m_2} \cdots \pi_\chi(y'_{i_r})^{m_r} \,\middle|\, \begin{array}{l} i_l \in I',\ i_1 < i_2 < \cdots < i_r \quad (r \geq 0), \\ 0 \leq m_k < p \end{array} \right\}.$$

By the above lemma and the definition, one can directly show the following proposition.

Proposition B.4 *Let $\mathfrak{g}$ be a Lie p-algebra, and let $\mathfrak{a}$ be a p-subalgebra of $\mathfrak{g}$. Let M be an $\mathfrak{a}$-module with central character $\chi|_\mathfrak{a} \in \mathfrak{a}^*$, where $\chi \in \mathfrak{g}$.*

1. *Assume that* $\dim \mathfrak{g} < \infty$ *and* $\dim M < \infty$. *Then,*

$$\dim \mathrm{Ind}_\mathfrak{a}^\mathfrak{g}(M;\chi) = p^{\dim \mathfrak{g} - \dim \mathfrak{a}} \dim M.$$

2. *Let* $\mathfrak{b}$ *be a p-subalgebra of* $\mathfrak{g}$ *such that* $\mathfrak{b} \subset \mathfrak{a}$*, and let* N *be a* $\mathfrak{b}$*-module. Then,*
$$\mathrm{Ind}_{\mathfrak{a}}^{\mathfrak{g}}(\mathrm{Ind}_{\mathfrak{b}}^{\mathfrak{a}}(N;\chi|_{\mathfrak{a}});\chi) \simeq \mathrm{Ind}_{\mathfrak{b}}^{\mathfrak{g}}(N;\chi).$$
3. *Let* V *be a* $\mathfrak{g}$*-module with central character* $\chi \in \mathfrak{g}^*$*, and let* $\psi : M \to V$ *be a non-trivial homomorphism of* $\mathfrak{a}$*-modules. Then, the following map is a* $\mathfrak{g}$*-homomorphism:*
$$\Psi : \mathrm{Ind}_{\mathfrak{a}}^{\mathfrak{g}}(M;\chi) \longrightarrow V, \quad \Psi(u \otimes m) := u.\psi(m),$$
where $u \in \bar{U}_\chi(\mathfrak{g})$ *and* $m \in M$*. Hence, if* V *is irreducible and* ψ *is non-trivial, then* V *is a quotient of* $\mathrm{Ind}_{\mathfrak{a}}^{\mathfrak{g}}(M;\chi)$.

B.2 Completely Solvable Lie Algebras

In this section, we define completely solvable Lie algebras and prove some general facts about them. For these topics, the reader may consult [BGR] and [Di]. In the first two subsections, the base field $\mathbb{K}$ does not necessarily have positive characteristic unless otherwise stated.

B.2.1 Definition

Let $\mathfrak{g}$ be a finite dimensional Lie algebra over $\mathbb{K}$.

Definition B.6 *A finite dimensional Lie algebra* $\mathfrak{g}$ *is said to be* ***completely solvable*** *if there exists a chain*
$$\{0\} = \mathfrak{g}_0 \subset \mathfrak{g}_1 \subset \cdots \subset \mathfrak{g}_d = \mathfrak{g} \quad (d := \dim \mathfrak{g}), \tag{B.10}$$
of ideals of $\mathfrak{g}$ *such that* $\dim \mathfrak{g}_i = i$ *for any* $1 \le i \le d$.

By definition, we have

Proposition B.5 *Suppose that* $\mathbb{K}$ *is an algebraically closed field of characteristic* 0*. Then,*
$$\mathfrak{g} : \textit{solvable} \Leftrightarrow \mathfrak{g} : \textit{completely solvable}.$$

B.2.2 Polarisation

First, we recall the definition of a polarisation of a Lie algebra $\mathfrak{g}$. For $f \in \mathfrak{g}^*$, we define $df : \mathfrak{g} \times \mathfrak{g} \to \mathbb{K}$ by $df(x, y) := f([x, y])$, and set

$$\mathfrak{c}_{\mathfrak{g}}(f) := \mathrm{Ker} df = \{x \in \mathfrak{g} | df(x, y) = 0 \ (\forall y \in \mathfrak{g})\}.$$

Definition B.7 *A subalgebra* $\mathfrak{p}$ *of* $\mathfrak{g}$ *is called a* ***polarisation*** *of* $\mathfrak{g}$ *at* f *if* $\mathfrak{p}$ *is a maximal totally isotropic subspace for* df*, i.e.,*

$$df(\mathfrak{p}, \mathfrak{p}) = \{0\}, \quad \dim \mathfrak{p} = \frac{1}{2}(\dim \mathfrak{g} + \dim \mathfrak{c}_{\mathfrak{g}}(f)).$$

We denote the set of polarisations of $\mathfrak{g}$ at f by $\mathcal{P}_{\mathfrak{g}}(f)$.

Next, we introduce a special polarisation of a completely solvable Lie algebra, called a Vergne polarisation. For the definition, we first recall the following lemma.

Lemma B.8. *Let* $\mathfrak{g}$ *be a Lie algebra and* $f \in \mathfrak{g}^*$. *Let* $\mathfrak{g}_1$ *be a Lie subalgebra of* $\mathfrak{g}$ *such that* $\dim \mathfrak{g}_1 = \dim \mathfrak{g} - 1$. *We denote the restriction of* f *to* $\mathfrak{g}_1$ *by* f_1. *Then, we have*

1. *If* $\mathrm{Ker} df \subset \mathfrak{g}_1$, *then for any* $\mathfrak{p}_1 \in \mathcal{P}_{\mathfrak{g}_1}(f_1)$, $\mathfrak{p}_1 \in \mathcal{P}_{\mathfrak{g}}(f)$.
2. *If* $\mathrm{Ker} df \not\subset \mathfrak{g}_1$, *then for any* $\mathfrak{p}_1 \in \mathcal{P}_{\mathfrak{g}_1}(f_1)$, $\mathfrak{p}_1 + \mathrm{Ker} df \in \mathcal{P}_{\mathfrak{g}}(f)$.

Proof. Case: $\mathrm{Ker} df \subset \mathfrak{g}_1$. For any $\mathfrak{p}_1 \in \mathcal{P}_{\mathfrak{g}_1}(f_1)$, $\mathfrak{p}_1$ is a totally isotropic subspace of $\mathfrak{g}$ for f. In this case, $\mathrm{Ker} df \subset \mathrm{Ker} df_1$, and thus $\dim \mathfrak{c}_{\mathfrak{g}}(f) \le \dim \mathfrak{c}_{\mathfrak{g}_1}(f_1)$. Hence, we have

$$\dim \mathfrak{p}_1 \ge \frac{1}{2}(\dim \mathfrak{g} - 1 + \dim \mathfrak{c}_{\mathfrak{g}}(f)) = \dim \mathfrak{p} - \frac{1}{2},$$

which implies

$$\dim \mathfrak{p}_1 \ge \dim \mathfrak{p}.$$

By the maximality of $\mathfrak{p}_1$, we have $\mathfrak{p}_1 \in \mathcal{P}_{\mathfrak{g}}(f)$.

Case: $\mathrm{Ker} df \not\subset \mathfrak{g}_1$. We show that

$$\mathrm{Ker} df_1 = \mathrm{Ker} df \cap \mathfrak{g}_1. \tag{B.11}$$

The inclusion ($\supset$) is clear. Hence, we prove ($\subset$). Since the codimension of $\mathfrak{g}_1$ is one and $\mathrm{Ker} df \not\subset \mathfrak{g}_1$, there exists $x \in \mathrm{Ker} df$ such that $\mathfrak{g} = \mathbb{K}x + \mathfrak{g}_1$. Since for any $y \in \mathrm{Ker} df_1 \subset \mathfrak{g}_1$,

$$df(y, \mathfrak{g}) = f([y, \mathbb{K}x]) + f([y, \mathfrak{g}_1]) = f_1([y, \mathfrak{g}_1]) = \{0\}.$$

Hence, we have $y \in \mathrm{Ker} df$, and thus ($\subset$).

(B.11) implies that $\dim \mathrm{Ker} df_1 = \dim \mathrm{Ker} df - 1$. On the other hand, for $\mathfrak{p}_1 \in \mathcal{P}_{\mathfrak{g}_1}(f_1)$, $\mathfrak{p}_1 + \mathrm{Ker} df$ is a totally isotropic subspace of $\mathfrak{g}$ for f. Since $\dim(\mathfrak{p}_1 + \mathrm{Ker} df) = \dim \mathfrak{p}_1 + 1$, $\mathfrak{p}_1 + \mathrm{Ker} df$ is a maximal totally isotropic subspace. We have completed the proof. □

Let $\mathfrak{g}$ be a completely solvable Lie algebra with a chain

$$\{0\} = \mathfrak{g}_0 \subset \mathfrak{g}_1 \subset \cdots \subset \mathfrak{g}_d = \mathfrak{g} \tag{B.12}$$

of ideals of $\mathfrak{g}$ such that $\dim \mathfrak{g}_i = i$. For $f \in \mathfrak{g}^*$, we set

$$f_i := f|_{\mathfrak{g}_i} : \text{the restriction of } f \text{ to } \mathfrak{g}_i.$$

Then, we have

Proposition B.6 *For each integer m such that $1 \leq m \leq d$, we define a subspace $\mathfrak{p}_m$ of $\mathfrak{g}_m$ by*

$$\mathfrak{p}_m := \sum_{i=1}^{m} \mathfrak{c}_{\mathfrak{g}_i}(f_i).$$

Then, we have

$$\mathfrak{p}_m \in \mathcal{P}_{\mathfrak{g}_m}(f_m). \tag{B.13}$$

In particular, for $m = d$, we have $\mathfrak{p}_d \in \mathcal{P}_{\mathfrak{g}}(f)$.

Proof. Here, we show (B.13) by induction on m. For $m = 1$, (B.13) obviously holds: Hence, we assume that $m > 1$.

Case: $\mathrm{Ker} df_m \subset \mathfrak{g}_{m-1}$. By definition, we have

$$\mathfrak{c}_{\mathfrak{g}_m}(f_m) \subset \mathfrak{c}_{\mathfrak{g}_{m-1}}(f_{m-1}).$$

Thus, we see that

$$\mathfrak{p}_m = \sum_{i=1}^{m-1} \mathfrak{c}_{\mathfrak{g}_i}(f_i) = \mathfrak{p}_{m-1}.$$

By the induction hypothesis, $\mathfrak{p}_{m-1} \in \mathcal{P}_{\mathfrak{g}_{m-1}}(f_{m-1})$. Hence, from Lemma B.8. 1, we obtain (B.13).

Case: $\mathrm{Ker} df_m \not\subset \mathfrak{g}_{m-1}$. By definition,

$$\begin{aligned} \mathfrak{p}_m &= \sum_{i=1}^{m} \mathfrak{c}_{\mathfrak{g}_i}(f_i) \\ &= \sum_{i=1}^{m-1} \mathfrak{c}_{\mathfrak{g}_i}(f_i) + \mathfrak{c}_{\mathfrak{g}_m}(f_m) \\ &= \mathfrak{p}_{m-1} + \mathrm{Ker} df_m. \end{aligned}$$

By the induction hypothesis, we have $\mathfrak{p}_{m-1} \in \mathfrak{p}_{\mathfrak{g}_{m-1}}(f_{m-1})$. Hence, (B.13) follows from Lemma B.8. 2. □

Definition B.8 $\sum_{i=1}^{d} \mathfrak{c}_{\mathfrak{g}_i}(f_i)$ *is called a* ***Vergne polarisation*** *of $\mathfrak{g}$ at f.*

Remark that a Vergne polarisation depends on the choice of a chain of ideals as in (B.12).

When we show the irreducibility of induced representations of completely solvable Lie algebras, the following lemma is helpful.

Lemma B.9. *Let $\mathfrak{g}$ be a completely solvable Lie algebra. For $f \in \mathfrak{g}^*$, let $\mathfrak{p}$ be the Vergne polarisation of $\mathfrak{g}$ at f defined by the chain (B.12). Let $\mathfrak{a}$ be a subalgebra of $\mathfrak{g}$ such that*

$$\operatorname{codim} \mathfrak{a} = 1 \quad \textit{and} \quad \mathfrak{p} \subset \mathfrak{a}.$$

Then, there uniquely exists i $(1 \le i \le d-1)$ such that

$$\mathfrak{g}_i \cap \mathfrak{a} = \mathfrak{g}_{i+1} \cap \mathfrak{a},$$

and $\mathfrak{a}$ is a completely solvable Lie algebra with the chain of ideals of $\mathfrak{a}$:

$$\{\mathfrak{g}_k \cap \mathfrak{a} | 1 \le k \le d,\ k \ne i\}. \tag{B.14}$$

Moreover, $\mathfrak{p}$ is a Vergne polarisation of $\mathfrak{a}$ at $f|_{\mathfrak{a}}$.

Proof. We fix $x \in \mathfrak{g} \backslash \{0\}$ such that $\mathfrak{g} = \mathbb{K}x \oplus \mathfrak{a}$. By considering the composition of linear maps

$$\mathfrak{g} \twoheadrightarrow \mathfrak{g}/\mathbb{K}x \simeq \mathfrak{a} \to \mathfrak{a}/(\mathfrak{g}_k \cap \mathfrak{a}),$$

we see that

$$\dim(\mathfrak{g}_k \cap \mathfrak{a}) + 1 = \dim(\mathfrak{g}_k + \mathbb{K}x).$$

Hence, $\mathfrak{a}$ is a completely solvable Lie algebra with a chain (B.14).

Next, we show that $\mathfrak{p}$ gives a Vergne polarisation of $\mathfrak{a}$. For $1 \le k \le d$, we set

$$\mathfrak{c}_k := \{x \in \mathfrak{g} | f([x,y]) = 0 \ (\forall y \in \mathfrak{g}_k)\} \subset \mathfrak{g}.$$

By definition, $\mathfrak{c}_{\mathfrak{g}_k}(f_k) = \mathfrak{c}_k \cap \mathfrak{g}_k$. Hence, for any k, we have

$$\mathfrak{c}_k \cap \mathfrak{g}_k \subset \mathfrak{p} \subset \mathfrak{a},$$

and thus,

$$(\mathfrak{g}_k \cap \mathfrak{a}) \cap \mathfrak{c}_k = (\mathfrak{g}_k \cap \mathfrak{c}_k) \cap \mathfrak{a} = \mathfrak{g}_k \cap \mathfrak{c}_k.$$

On the other hand, by setting

$$\mathfrak{d}_k := \{x \in \mathfrak{g} | f([x,y]) = \{0\} \ (\forall y \in \mathfrak{g}_k \cap \mathfrak{a})\},$$

we have $\mathfrak{c}_k \subset \mathfrak{d}_k$. Hence, we obtain

$$\mathfrak{c}_{\mathfrak{g}_k \cap \mathfrak{a}}(f|_{\mathfrak{g}_k \cap \mathfrak{a}}) = \mathfrak{g}_k \cap \mathfrak{a} \cap \mathfrak{d}_k \supset \mathfrak{g}_k \cap \mathfrak{a} \cap \mathfrak{c}_k = \mathfrak{g}_k \cap \mathfrak{c}_k = \mathfrak{c}_{\mathfrak{g}_k}(f).$$

This implies that

$$\sum_{k=1}^{d} \mathfrak{c}_{\mathfrak{g}_k \cap \mathfrak{a}}(f|_{\mathfrak{g}_k \cap \mathfrak{a}}) \supset \mathfrak{p}.$$

On the other hand, since $\mathrm{Ker}df \subset \mathfrak{p} \subset \mathfrak{a}$, by Lemma B.8. 1, the dimension of a polarisation of $\mathfrak{g}$ at f coincides with that of a polarisation of $\mathfrak{a}$ at $f|_{\mathfrak{a}}$. Hence, we have

$$\sum_{k=1}^{d} \mathfrak{c}_{\mathfrak{g}_k \cap \mathfrak{a}}(f|_{\mathfrak{g}_k \cap \mathfrak{a}}) = \mathfrak{p}$$

and thus, $\mathfrak{p}$ is a Vergne polarisation of $\mathfrak{a}$. □

B.2.3 Completely Solvable Lie p-algebras

In the sequel, we suppose that $\mathbb{K}$ is an algebraically closed field of characteristic $p > 0$.

Let $\mathfrak{g}$ be a completely solvable Lie p-algebra over $\mathbb{K}$. In this subsection, we show the following proposition:

Proposition B.7 *Let $\mathfrak{g}$ be a completely solvable Lie p-algebra. Then, there exists a chain as in (B.10) such that all $\mathfrak{g}_i$ are p-ideals of $\mathfrak{g}$.*

To show this lemma, we recall a lemma on a property of abelian Lie p-algebras.

Lemma B.10. *Let $\mathfrak{a}$ be a finite dimensional abelian Lie p-algebra. Then, there exists $x \in \mathfrak{a} \setminus \{0\}$ such that $x^{[p]} \in \mathbb{K}x$.*

Proof. We fix an element $y \in \mathfrak{a} \setminus \{0\}$. Let m be the maximal integer such that

$$\{y,\, y^{[p]},\, y^{[p^2]}, \cdots, y^{[p^m]}\}$$

are linearly independent over $\mathbb{K}$. If $m = 0$, then we may choose y as x. Hence, we assume $m > 0$.

By the assumption, $y^{[p^{m+1}]}$ is expressed as

$$y^{[p^{m+1}]} = \alpha_0 y + \alpha_1 y^{[p]} + \cdots + \alpha_m y^{[p^m]}.$$

If $\alpha_0 = 0$, then by replacing y with $y^{[p]}$, we have

$$y^{[p^m]} = \alpha_1 y + \alpha_2 y^{[p]} + \cdots + \alpha_m y^{[p^{m-1}]}.$$

Hence, we may assume that $\alpha_0 \neq 0$. We set

$$F(t) := \sum_{k=0}^{m} \alpha_{m-k}^{p^k} t^{p^{k+1}} - t.$$

Let t_0 be a zero of the equation $F(t) = 0$ such that $t_0 \neq 0$. We introduce a sequence $\{c_0, c_1, \cdots, c_m\}$ by

$$c_0 := t_0^p \alpha_0, \quad c_i := c_{i-1}^p + t_0^p \alpha_i \ (1 \leq i < m), \quad c_m := t_0,$$

and set

$$x := \sum_{i=0}^{m} c_i y^{[p^i]}.$$

Notice that $x \neq 0$, since $\alpha_0, t_0 \neq 0$.

Since $\mathfrak{a}$ is abelian, by using (B.4), (B.5) and Remark B.1, we have

$$x^{[p]} = \sum_{i=0}^{m} c_i^p y^{[p^{i+1}]} = t_0^p \alpha_0 y + \sum_{i=1}^{m} (c_{i-1}^p + t_0^p \alpha_i) y^{[p^i]}.$$

Hence,

$$\begin{aligned} c_{m-1}^p + t_0^p \alpha_m &= (c_{m-2} + t_0^p \alpha_{m-1})^p + t_0^p \alpha_m \\ &= c_{m-2}^p + t_0^{p^2} \alpha_{m-1}^p + t_0^p \alpha_m \\ &= \sum_{k=0}^{m} \alpha_{m-k}^{p^k} t_0^{p^{k+1}} \\ &= t_0 = c_m, \end{aligned}$$

and thus, $x^{[p]} = x$. Now, we have completed the proof. □

Proposition B.7 is an immediate consequence of the following lemma.

Lemma B.11. *For a completely solvable Lie p-algebra* $\mathfrak{g}$*, there exists a one-dimensional p-ideal of* $\mathfrak{g}$.

Proof. First, we suppose that $Z(\mathfrak{g}) \neq \{0\}$. By the above lemma, there exists $x \in Z(\mathfrak{g}) \setminus \{0\}$ such that $\mathbb{K}x$ is a p-ideal of $\mathfrak{g}$.

Next, we suppose that $Z(\mathfrak{g}) = \{0\}$. Since $\mathfrak{g}$ is completely solvable, there exists $x \in \mathfrak{g}$ such that $\mathbb{K}x$ is an ideal of $\mathfrak{g}$. Since $\mathrm{ad}x(\mathfrak{g}) \subset \mathbb{K}x$, we have

$$(\mathrm{ad}x)^2(\mathfrak{g}) = \{0\}.$$

In particular,

$$\mathrm{ad}x^{[p]}(\mathfrak{g}) = (\mathrm{ad}x)^p(\mathfrak{g}) = \{0\},$$

and hence, $x^{[p]} \in Z(\mathfrak{g}) = \{0\}$ by the assumption. Thus, $\mathbb{K}x$ is a p-ideal of $\mathfrak{g}$. □

Proof of Proposition B.7. We show this proposition by induction on $d := \dim \mathfrak{g}$. The above lemma ensures that there exists a one-dimensional p-ideal $\mathfrak{k}$ of $\mathfrak{g}$. By the induction hypothesis, there exists a chain

$$\{0\} \subset \bar{\mathfrak{g}}_1 \subset \cdots \subset \bar{\mathfrak{g}}_{d-1} = \mathfrak{g}/\mathfrak{k}$$

of p-ideals of $\bar{\mathfrak{g}}$ such that and $\dim \bar{\mathfrak{g}}_i = i$. For $i \geq 2$, let $\mathfrak{g}_i$ be the pre-image of $\bar{\mathfrak{g}}_{i-1}$ under the projection $\mathfrak{g} \twoheadrightarrow \mathfrak{g}/\mathfrak{k}$. Then, $\mathfrak{g}_i$ is a p-ideal of $\mathfrak{g}$ and

$$\{0\} \subset \mathfrak{k} \subset \mathfrak{g}_2 \subset \cdots \subset \mathfrak{g}_d = \mathfrak{g}$$

such that $\dim \mathfrak{g}_i = i$. □

We state a lemma on a polarisation of a Lie p-algebra for later use.

Lemma B.12. *Let $\mathfrak{g}$ be a Lie p-algebra, and let $f \in \mathfrak{g}^*$. If $\mathfrak{p}$ is a polarisation of $\mathfrak{g}$ at f, then $\mathfrak{p}$ is a p-subalgebra of $\mathfrak{g}$.*

Proof. For $x \in \mathfrak{p}$, we set $\mathfrak{p}_x := \mathfrak{p} + \mathbb{K}x^{[p]}$. We show that $\mathfrak{p}_x \subset \mathfrak{p}$ for any $x \in \mathfrak{p}$.

Since for any $y_1, y_2 \in \mathfrak{p}$ and $c_1, c_2 \in \mathbb{K}$,

$$f([c_1 x^{[p]} + y_1, c_2 x^{[p]} + y_2]) = 0,$$

$\mathfrak{p}_x$ is a totally isotropic subspace for f. By the maximality of $\mathfrak{p}$, we have $\mathfrak{p} = \mathfrak{p}_x$. □

B.3 Irreducible Representations of a Completely Solvable Lie p-algebra

In this section, we recall a theorem on the dimension of irreducible representations over a completely solvable Lie p-algebra. Throughout this section, let $\mathfrak{g}$ be a completely solvable Lie p-algebra with a chain

$$\{0\} = \mathfrak{g}_0 \subset \mathfrak{g}_1 \subset \cdots \subset \mathfrak{g}_d = \mathfrak{g}, \qquad \text{(B.15)}$$

of *p-ideals* of $\mathfrak{g}$ such that $\dim \mathfrak{g}_i = i$.

B.3.1 Simplicity of Induced Representations

First, we discuss the irreducibility of induced representations of $\mathfrak{g}$. We introduce some notation. We fix $f \in \mathfrak{g}^*$. Let $\mathfrak{p}$ be the Vergne polarisation of $\mathfrak{g}$ at f defined from the chain (B.15). Let $\mathbb{K}_f := \mathbb{K}v_f$ be the one-dimensional $\mathfrak{p}$-module defined by

$$x.v_f = f(x)v_f \quad (\forall x \in \mathfrak{p}). \qquad \text{(B.16)}$$

Since $f([\mathfrak{p}, \mathfrak{p}]) = \{0\}$ by definition, the above action of $\mathcal{P}$ is well-defined. Note that there exists $\chi \in \mathfrak{g}^*$ such that the restriction of χ to $\mathfrak{p}$ is the central character of $\mathbb{K}_f$, i.e.,

$$f(x)^p - f(x^{[p]}) = \chi(x)^p \quad (\forall x \in \mathfrak{p}).$$

The main result of this subsection is the following:

Theorem B.3 *The induced representation*

$$\mathrm{Ind}_{\mathfrak{p}}^{\mathfrak{g}}(\mathbb{K}_f;\chi)$$

is an irreducible $\mathfrak{g}$*-module.*

For the proof, a preliminary lemma is necessary. For a p-ideal $\mathfrak{a}$ of $\mathfrak{g}$, we set

$$\mathfrak{c}_{\mathfrak{g},\mathfrak{a}}(f) := \{x \in \mathfrak{g} | df(x,\mathfrak{a}) = \{0\}\}.$$

Remark that $\mathfrak{c}_{\mathfrak{g},\mathfrak{a}}(f)$ is a p-subalgebra of $\mathfrak{g}$. We have

Lemma B.13. *Suppose that* $\mathfrak{a}$ *satisfies*

1. $\mathfrak{a} \subset \mathfrak{c}_{\mathfrak{g},\mathfrak{a}}(f)$,
2. *the codimension of* $\mathfrak{c} := \mathfrak{c}_{\mathfrak{g},\mathfrak{a}}(f)$ *in* $\mathfrak{g}$ *is one.*

Moreover, let N *be a* $\mathfrak{c}$*-module whose* $\mathfrak{a}$*-action satisfies*

$$x.v = f(x)v \quad (x \in \mathfrak{a},\ v \in N).$$

We set

$$M := \mathrm{Ind}_{\mathfrak{c}}^{\mathfrak{g}}(N;\chi).$$

Then, for any non-trivial $\mathfrak{a}$*-submodule* M' *of* M,

$$M' \cap (1 \otimes N) \neq \{0\}.$$

Proof. We fix $x \in \mathfrak{g} \setminus \mathfrak{c}$. Then, $\mathfrak{g} = \mathbb{K}x + \mathfrak{c}$. Since $x \notin \mathfrak{c}$, there exists $y \in \mathfrak{a}$ such that $f([x,y]) = 1$. Let us fix such y.

By definition, we see that

$$M = \bigoplus_{i=0}^{p-1} \mathbb{K}x^i \otimes N.$$

For $j = 0, 1, \cdots, p-1$, we set

$$M_{\leq j} := \bigoplus_{i=0}^{j} \mathbb{K}x^i \otimes N.$$

It is enough to show that

$$M' \cap M_{\leq j} \neq \{0\} \ \Rightarrow \ M' \cap M_{\leq j-1} \neq \{0\} \tag{B.17}$$

for any $j \geq 1$. Recall that

$$yx^j = \sum_{k=0}^{j} \binom{j}{k} x^{j-k} \underbrace{[\cdots[y,x],x],\cdots,x]}_{k \text{ times}}.$$

Since $[\cdots[y,x],x],\cdots,x] \in \mathfrak{a}$, we have

$$\begin{aligned} yx^j \otimes v &= \sum_{k=0}^{j} \binom{j}{k} x^{j-k} [\cdots [y, x], x], \cdots, x] \otimes v \\ &= \sum_{k=0}^{j} \binom{j}{k} x^{j-k} \otimes f([\cdots [y, x], x], \cdots, x]) v \\ &\equiv x^j \otimes f(y) v - j x^{j-1} \otimes f([x, y]) v \pmod{M_{\le j-2}}. \end{aligned}$$

Hence, for any $w \in M' \cap M_{\le j} \setminus \{0\}$,

$$(y - f(y)) w \in M' \cap M_{\le j-1} \setminus \{0\}.$$

Hence, (B.17) holds and thus the lemma holds. □

PROOF OF THEOREM B.3. Let us fix a chain of ideals (B.15) and set

$$\mathfrak{p} := \sum_{i=0}^{d} \mathrm{Ker} df_i \in \mathcal{P}_{\mathfrak{g}}(f),$$

where we set $d := \dim \mathfrak{g}$ and $f_i := f|_{\mathfrak{g}_i}$. We show this theorem by induction on $\dim \mathfrak{g} - \dim \mathfrak{p}$.

Notice that

$$\mathrm{Ker} df = \mathfrak{c}_d \subset \mathfrak{c}_{d-1} \subset \cdots \subset \mathfrak{c}_0 = \mathfrak{g},$$

where we set

$$\mathfrak{c}_i := \mathfrak{c}_{\mathfrak{g}, \mathfrak{g}_i}(f_i).$$

In the case $\mathfrak{g} = \mathfrak{p}$, there is nothing to prove. Hence, in the sequel, we suppose that $\mathfrak{g} \neq \mathfrak{p}$, i.e., $\mathfrak{g} \neq \mathrm{Ker} df$.

Let j be the minimal integer such that $\mathfrak{c}_j \neq \mathfrak{g}$. The following lemma holds.

Lemma B.14. *1.* $\mathfrak{g}_j \subset \mathfrak{p} \subset \mathfrak{c}_j$.
2. $\mathfrak{c}_j$ *is a subspace of* $\mathfrak{g}$ *of codimension one.*

Proof. Let us fix $x_j \in \mathfrak{g}_j \setminus \mathfrak{g}_{j-1}$. Note that $\mathfrak{g}_j = \mathbb{K} x_j + \mathfrak{g}_{j-1}$. Since $\mathfrak{c}_{j-1} = \mathfrak{g}$, we have $\mathfrak{g}_{j-1} \subset \mathrm{Ker} df$ and

$$\mathfrak{c}_j = \{x \in \mathfrak{g} | f([x, x_j]) = 0\}.$$

Hence, the second statement of the lemma holds.

Moreover, we have

$$f([\mathfrak{g}_j, \mathfrak{g}_j]) = \{0\},$$

since $\mathfrak{g}_{j-1} \subset \ker df$ and

$$[\mathfrak{g}_j, \mathfrak{g}_j] = [\mathbb{K}x + \mathfrak{g}_{j-1}, \mathbb{K}x + \mathfrak{g}_{j-1}] \subset [\mathbb{K}x + \mathfrak{g}_{j-1}, \mathfrak{g}_{j-1}].$$

This means that $\mathfrak{g}_j \subset \mathfrak{c}_j$. Hence,

$$\mathfrak{p} = \sum_{i=1}^{d} \mathrm{Ker} df_i = \sum_{i=1}^{d} \mathfrak{g}_i \cap \mathfrak{c}_i \supset \mathfrak{g}_j \cap \mathfrak{c}_j = \mathfrak{g}_j.$$

Further, we have $f([\mathfrak{g}_j, \mathfrak{p}]) = \{0\}$, since $f([\mathfrak{p}, \mathfrak{p}]) = \{0\}$. Thus, $\mathfrak{p} \subset \mathfrak{c}_j$. Hence, the first statement has been proved. □

Combining this lemma with Lemma B.9, we see that $\mathfrak{p}$ is a Vergne polarisation of $\mathfrak{c}_j$ at $f|_{\mathfrak{c}_j}$.

Let $\mathbb{K}_f = \mathbb{K}v_f$ be the one-dimensional representation of $\mathfrak{p}$ defined as in (B.16). By the induction hypothesis,

$$\mathrm{Ind}_{\mathfrak{p}}^{\mathfrak{c}_j}(\mathbb{K}_f; \chi')$$

is an irreducible $\mathfrak{c}_j$-module, where $\chi' := \chi|_{\mathfrak{c}_j}$.

We set $r := \dim \mathfrak{c}_j - \dim \mathfrak{p}$. Let us fix $\{x_1, \cdots, x_r\} \subset \mathfrak{c}_j$ such that

$$\mathfrak{c}_j = \mathfrak{p} \oplus \mathbb{K}x_1 \oplus \cdots \oplus \mathbb{K}x_r.$$

By definition,

$$\mathrm{Ind}_{\mathfrak{p}}^{\mathfrak{c}_j}(\mathbb{K}_f; \chi') = \bigoplus_{\mathbf{n} \in I} \mathbb{K}X^{\mathbf{n}} \otimes v_f,$$

where $I := \{(n_1, \cdots, n_r) \in (\mathbb{Z}_{\geq 0})^r | n_i < p\}$ and for $\mathbf{n} = (n_1, \cdots, n_r) \in I$, we set

$$X^{\mathbf{n}} := x_1^{n_1} \cdots x_r^{n_r}.$$

For $\mathbf{m} = (m_1, \cdots, m_r) \in I$, $\mathbf{m} \leq \mathbf{n}$ means $0 \leq m_i \leq n_i$ for any i. For $\mathbf{m}, \mathbf{n} \in I$ such that $\mathbf{m} \leq \mathbf{n}$, we set

$$\mathbf{n} - \mathbf{m} := (n_1 - m_1, \cdots, n_r - m_r)$$

for simplicity.

By induction on r, one can easily show that for $y \in \mathfrak{g}$,

$$yX^{\mathbf{n}} = \sum_{\mathbf{m} \leq \mathbf{n}} \binom{\mathbf{n}}{\mathbf{m}} X^{\mathbf{n}-\mathbf{m}} Y_{\mathbf{m}},$$

where we set

$$\binom{\mathbf{n}}{\mathbf{m}} = \prod_{i=1}^{r} \binom{n_i}{m_i},$$

and

$$Y_{\mathbf{m}} := [\cdots [\underbrace{y, x_1], \cdots, x_1]}_{m_1 \text{ times}}, \cdots \underbrace{x_r], \cdots, x_r]]}_{m_r \text{ times}}.$$

Hence, for $y \in \mathfrak{g}_j$, we have

$$
\begin{aligned}
y.X^{\mathbf{n}} \otimes v_f &= \sum_{\mathbf{m}} \binom{\mathbf{n}}{\mathbf{m}} X^{\mathbf{n}-\mathbf{m}} Y_{\mathbf{m}} \otimes v_f \\
&= \sum_{\mathbf{m}} \binom{\mathbf{n}}{\mathbf{m}} X^{\mathbf{n}-\mathbf{m}} \otimes Y_{\mathbf{m}}.v_f \\
&= \sum_{\mathbf{m}} \binom{\mathbf{n}}{\mathbf{m}} X^{\mathbf{n}-\mathbf{m}} \otimes f(Y_{\mathbf{m}}) v_f \\
&= X^{\mathbf{n}} \otimes f(y) v_f,
\end{aligned}
$$

since $Y_{\mathbf{m}} \in [\mathfrak{g}_j, \mathfrak{c}_j]$ for $\mathbf{m} \neq (0, \cdots, 0)$ and $f([\mathfrak{g}_j, \mathfrak{c}_j]) = \{0\}$.

Now, we take $\mathfrak{g}_j$, $\mathrm{Ind}_{\mathfrak{p}}^{\mathfrak{c}_j}(\mathbb{K}_f; \chi')$ and

$$
\mathrm{Ind}_{\mathfrak{p}}^{\mathfrak{g}}(\mathbb{K}_f; \chi) \left(\simeq \mathrm{Ind}_{\mathfrak{c}_j}^{\mathfrak{g}}(\mathrm{Ind}_{\mathfrak{p}}^{\mathfrak{c}_j}(\mathbb{K}_f; \chi'); \chi) \right)
$$

as $\mathfrak{a}$, N and M and apply Lemma B.13. If there exists a non-trivial $\mathfrak{g}$-submodule M' of $\mathrm{Ind}_{\mathfrak{p}}^{\mathfrak{g}}(\mathbb{K}_f; \chi)$, then

$$
M' \cap \{1 \otimes \mathrm{Ind}_{\mathfrak{p}}^{\mathfrak{c}_j}(\mathbb{K}_f; \chi')\} \neq \{0\}.
$$

Since by the induction hypothesis $\mathrm{Ind}_{\mathfrak{p}}^{\mathfrak{c}_j}(\mathbb{K}_f; \chi')$ is irreducible, we have

$$
M' \supset 1 \otimes \mathrm{Ind}_{\mathfrak{p}}^{\mathfrak{c}_j}(\mathbb{K}_f; \chi'),
$$

and thus $M' = \mathrm{Ind}_{\mathfrak{p}}^{\mathfrak{g}}(\mathbb{K}_f; \chi)$. This means that $\mathrm{Ind}_{\mathfrak{p}}^{\mathfrak{g}}(\mathbb{K}_f; \chi)$ is irreducible. □

B.3.2 Dimension of Irreducible Representations over a Completely Solvable Lie Algebra

The following is a key of the proof of the main theorem (Theorem B.4).

Lemma B.15. *Let M be an irreducible $\mathfrak{g}$-module. Then, there exists $f \in \mathfrak{g}^*$ and $v \in M \setminus \{0\}$ such that $\mathbb{K}v$ is isomorphic to $\mathbb{K}_f$ as $\mathfrak{p}$-module, where $\mathfrak{p}$ is the Vergne polarisation of $\mathfrak{g}$ at f defined from the chain (B.15), and $\mathbb{K}_f$ is the one-dimensional $\mathfrak{p}$-module (B.16).*

Proof. We prove this lemma by induction on $d := \dim \mathfrak{g}$. In the case $d = 1$, any irreducible $\mathfrak{g}$-module is one-dimensional, since $\mathfrak{g}$ is abelian.

Suppose that $d > 1$. Let M' be an irreducible $\mathfrak{g}_{d-1}$-submodule of M. By the induction hypothesis, there exists $f' \in \mathfrak{g}_{d-1}^*$ and $v' \in M' \setminus \{0\}$ such that $\mathbb{K}v' \simeq \mathbb{K}_{f'}$ as $\mathfrak{p}'$-module, where $\mathfrak{p}'$ is the Vergne polarisation of $\mathfrak{g}_{d-1}$ at f' defined from $\{\mathfrak{g}_i | 1 \leq i \leq d-1\}$.

Notice that $[\mathfrak{g}, \mathfrak{g}] \subset \mathfrak{g}_{d-1}$. We consider the following cases:

Case *I*: For any $x \in \mathfrak{g} \setminus \mathfrak{g}_{d-1}$, $f'([x, \mathfrak{g}]) \neq \{0\}$.

Case II: For some $x \in \mathfrak{g} \setminus \mathfrak{g}_{d-1}$, $f'([x, \mathfrak{g}]) = \{0\}$.

Case I We take a linear map $f \in \mathfrak{g}^*$ such that $f|_{\mathfrak{g}_{d-1}} = f'$. In this case,

$$\mathrm{Ker}df \subset \mathfrak{g}_{d-1}.$$

Hence, by definition, $\mathfrak{p}'$ is a Vergne polarisation of $\mathfrak{g}$ at f. If we take the vector v' as v, then the proposition holds.

Case II Notice that for any $f \in \mathfrak{g}^*$ such that $f|_{\mathfrak{g}_{d-1}} = f'$, we have $x \in \mathrm{Ker}df$. Since $\mathrm{Ker}df \not\subset \mathfrak{g}_{d-1}$, by definition $\mathfrak{p}' + \mathrm{Ker}df = \mathfrak{p}' + \mathbb{K}x$ is a Vergne polarisation of $\mathfrak{g}$ at f. We set $\mathfrak{p} := \mathfrak{p}' + \mathbb{K}x$. Here, let us fix $\chi \in \mathfrak{g}^*$ such that

$$f(u)^p - f(u^{[p]}) = \chi(u)^p \quad (\forall u \in \mathfrak{p})$$

and set

$$V := \mathrm{Ind}_{\mathfrak{p}'}^{\mathfrak{p}}(\mathbb{K}v'; \chi).$$

Then, $\{x^i \otimes v' | i = 0, 1, \cdots, p-1\}$ forms a basis of V. Notice that $\mathfrak{p}'$ is an ideal of $\mathfrak{p}$, since

$$[\mathfrak{p}', \mathbb{K}x] \subset [\mathfrak{g}, \mathfrak{g}] \cap \mathfrak{p} \subset \mathfrak{g}_{d-1} \cap \mathfrak{p} = \mathfrak{p}'.$$

Hence, any $y \in \mathfrak{p}'$ acts on V as $f'(y)\mathrm{id}_V$, since

$$\begin{aligned}
yx^i \otimes v' &= \left\{ x^i y + \sum_{j=1}^{i} \binom{i}{j} x^{i-j} [\cdots \underbrace{[y, x] \cdots x]}_{j \text{ times}} \right\} \otimes v' \\
&= x^i \otimes yv' + \sum_{j=1}^{i} \binom{i}{j} x^{i-j} \otimes [\cdots \underbrace{[y, x] \cdots x]}_{j \text{ times}} v' \\
&= x^i \otimes yv' + \sum_{j=1}^{i} \binom{i}{j} x^{i-j} \otimes f'([\cdots \underbrace{[y, x] \cdots x]}_{j \text{ times}}) v' \\
&= f'(y) x^i \otimes v'.
\end{aligned}$$

On the other hand, there exist $v \in V \setminus \{0\}$ and $c \in \mathbb{K}$ such that $x.v = cv$, since V is finite dimensional and $\mathbb{K}$ is algebraically closed. Therefore, by taking $f \in \mathfrak{g}^*$ such that $f(x) = c$ and $f|_{\mathfrak{g}_{d-1}} = f'$, we have $\mathbb{K}v \simeq \mathbb{K}_f$ as $\mathfrak{p}$-module. □

Combining Theorem B.3 with the above key lemma, we obtain the following theorem due to B. J. Veisfeiler and V. G. Kac [VK]:

Theorem B.4 *Let M be an irreducible $\mathfrak{g}$-module and let f be an element of $\mathfrak{g}^*$ given by Lemma B.15. Then, we have*

$$\dim M = p^s,$$

where $2s := \mathrm{rank} df$.

Proof. By Lemma B.15, there exists $v \in M$ such that $\mathbb{K}v \simeq \mathbb{K}_f$ as $\mathfrak{p}$-module, where $\mathfrak{p}$ is a Vergne polarisation of $\mathfrak{g}$ at f. On the other hand, by Theorem B.3, the induced representation $\mathrm{Ind}_{\mathfrak{p}}^{\mathfrak{g}}(\mathbb{K}_f, \chi)$ is irreducible, where $\chi \in \mathfrak{g}$ is the central character of M. By Proposition B.4, we see that $M \simeq \mathrm{Ind}_{\mathfrak{p}}^{\mathfrak{g}}(\mathbb{K}_f, \chi)$. Hence,

$$\dim M = p^{\dim \mathfrak{g} - \dim \mathfrak{p}}.$$

We have completed the proof. □

Remark B.2 *For the above* $f \in \mathfrak{g}^*$ *and central character* χ *of* M,

$$f(x) - f(x^{[p]})^{\frac{1}{p}} = \chi(x) \qquad (\forall x \in \mathfrak{p}).$$

Appendix C
Vertex Operator Algebras

In this appendix, we briefly recall the definition of vertex (operator) algebras and their basic properties. In particular, the operator product expansion (OPE, for short) and the rationality are reviewed.

C.1 Basic Objects

Here, we recall the definition of a vertex (operator) algebra and summarise some useful theorems for practical computation.

C.1.1 Notation

Let R be a $\mathbb{C}$-vector space and $z_1, z_2, \cdots, z_n$ be formal variables. We denote the vector space of *R-valued formal series*

$$\sum_{i_1 \in \mathbb{Z}} \cdots \sum_{i_n \in \mathbb{Z}} A_{i_1,\cdots,i_n} z_1^{i_1} \cdots z_n^{i_n} \qquad A_{i_1,\cdots,i_n} \in R,$$

by $R[[z_1^{\pm 1}, \cdots, z_n^{\pm 1}]]$ and the vector space of *R-valued formal Taylor series*

$$\sum_{i_1 \in \mathbb{Z}_{\geq 0}} \cdots \sum_{i_n \in \mathbb{Z}_{\geq 0}} A_{i_1,\cdots,i_n} z_1^{i_1} \cdots z_n^{i_n} \qquad A_{i_1,\cdots,i_n} \in R,$$

by $R[[z_1, \cdots, z_n]]$. For a formal variable z, the space of *R-valued formal Laurent series*

$$\sum_{i \in \mathbb{Z}} A_i z^i,$$

K. Iohara, Y. Koga, *Representation Theory of the Virasoro Algebra*, Springer Monographs in Mathematics, DOI 10.1007/978-0-85729-160-8,

where $A_i \in R$ and there exists $N \in \mathbb{Z}$ such that $A_i = 0$ for all $i < N$, is denoted by $R((z))$. We also set $R((z))((w)) := S((w))$ where $S := R((z))$.

For a $\mathbb{Z}$-graded vector space $V = \bigoplus_{n\in\mathbb{Z}} V_n$, we say that a vector $v \in V$ is of **degree** n and is denoted by $\deg v = n$ if $v \in V_n$. A linear operator $\phi \in \mathrm{End}V$ is called **homogeneous of degree** m and is denoted by $\deg \phi = m$ if it satisfies $\phi(V_n) \subset V_{n+m}$ for all $n \in \mathbb{Z}$.

C.1.2 Definition of a Vertex Operator Algebra

Recall that a formal series $A(z) \in \mathrm{End}V[[z^{\pm 1}]]$ is called a **field** if, for any $v \in V$, it satisfies $A(z).v \in V((z))$.

Definition C.1 *A **vertex algebra** is a pair of data consisting of*

1. *(Space of states) a vector space V,*
2. *(Vacuum vector) a vector $|0\rangle \in V$,*
3. *(Translation Operator) a linear operator $T : V \longrightarrow V$,*
4. *(Vertex Operators) a linear operation*

$$Y(\cdot, z) : V \longrightarrow \mathrm{End}V[[z, z^{-1}]],$$

taking each $A \in V$ to a field acting on V,

$$Y(A, z) = \sum_{n\in\mathbb{Z}} A_{(n)} z^{-n-1},$$

*called the **vertex operator associated with** A.*

These data are subject to the following axioms:

1. *(Vacuum Axiom) $Y(|0\rangle, z) = \mathrm{Id}_V$ and for each $A \in V$, one has $Y(A, z)|0\rangle \in V[[z]]$. In particular, the specialisation $|_{z=0}$ is well-defined and*

$$Y(A, z)|0\rangle|_{z=0} = A.$$

2. *(Translation Axiom) $T|0\rangle = 0$. For each $A \in V$, one has*

$$[T, Y(A, z)] = \partial_z Y(A, z).$$

Here and after, we set $\partial_z = \dfrac{\partial}{\partial z}$ for simplicity.

3. *(Locality Axiom) For any $A, B \in V$, the fields $Y(A, z)$ and $Y(B, w)$ are local, i.e., there exists $N \in \mathbb{Z}_{\geq 0}$ (depending on A, B) such that*

$$(z - w)^N [Y(A, z), Y(B, w)] = 0.$$

A vertex algebra V is called **$\mathbb{Z}$-graded** *if V is a $\mathbb{Z}$-graded vector space $V = \bigoplus_{n\in\mathbb{Z}} V_n$, $|0\rangle \in V_0$, $\deg T = 1$, and for $A \in V_m$, the field $Y(A,z) = \sum_{n\in\mathbb{Z}} A_{(n)} z^{-n-1}$ satisfies $\deg A_{(n)} = -n+m-1$ for all $n \in \mathbb{Z}$. (Such a field is said to have* **conformal dimension** *m.)*

We remark that some authors also consider a $\mathbb{Q}$-graded vertex algebra (see, e.g., [DLM1]).

Now, a homomorphism of vertex algebras, a subalgebra and an ideal is defined as follows:

Definition C.2 *1. A* **homomorphism of vertex algebras**

$$\rho : (V, |0\rangle, T, Y) \longrightarrow (V', |0\rangle', T', Y')$$

is a linear map $V \longrightarrow V'$ mapping $|0\rangle$ to $|0\rangle'$, intertwining the translation operators, and satisfying

$$\rho(Y(A,z)B) = Y'(\rho(A),z)\rho(B) \qquad A, B \in V.$$

2. A T-invariant subspace $V' \subset V$ is called a **vertex subalgebra** *if it satisfies $Y(A,z)B \in V'((z))$ for any $A, B \in V'$.*

3. A T-invariant subspace $I \subset V$ is called a **vertex ideal** *if it satisfies $Y(A,z)B \in I((z))$ for any $A \in I$ and $B \in V$.*

Remark C.1 *It can be shown that a vertex ideal is in fact a two-sided ideal and V/I naturally possesses a structure of vertex algebra.*

A special class of vertex algebras, which is of our interest, is defined as follows:

Definition C.3 *A $\mathbb{Z}$-graded vertex algebra $V = \bigoplus_{n\in\mathbb{Z}} V_n$ is called a* **conformal vertex algebra,** *if it possesses a non-zero element $\omega \in V_2$ such that*

1. the Fourier modes of

$$Y(\omega, z) = \sum_{n\in\mathbb{Z}} L_n^V z^{-n-2}$$

satisfy the commutation relations of the Virasoro algebra with central charge c_V, i.e.,

$$[L_m^V, L_n^V] = (m-n)L_{m+n}^V + \frac{1}{12}(m^3 - m)\delta_{m+n,0} c_V \mathrm{Id}_V \qquad m, n \in \mathbb{Z},$$

2. $L_{-1}^V = T$ and $L_0|_{V_n} = n\mathrm{Id}_{V_n}$ for $n \in \mathbb{Z}$.

ω is called a **conformal vector** *and $c_V \in \mathbb{C}$ is called the* **central charge of** *V. In addition, if a conformal vertex algebra V satisfies $\dim V_n < \infty$ for any $n \in \mathbb{Z}$ and there exists $N \in \mathbb{Z}$ such that $V_n = \{0\}$ for all $n < N$, then V is called a* **vertex operator algebra.**

C.1.3 Strong Reconstruction Theorem

In this subsection, we recall a PBW type theorem for a vertex algebra. For detail, the reader may consult, e.g., [FB].

Let V be a vertex algebra. For $A, B \in V$, the naïve product $Y(A,z)Y(B,w)$ may not be well-defined. In order to obtain a 'well-defined product', one considers a kind of renormalisation defined as follows:

Definition C.4 *1. For two fields*

$$A(z) = \sum_{m\in\mathbb{Z}} A_{(m)} z^{-m-1}, \qquad B(w) = \sum_{n\in\mathbb{Z}} B_{(n)} z^{-n-1},$$

we define the **normally ordered product** $\overset{\circ}{\circ}A(z)B(w)\overset{\circ}{\circ}$ *by*

$$\begin{aligned}
&\overset{\circ}{\circ}A(z)B(w)\overset{\circ}{\circ} \\
:=&\sum_{n\in\mathbb{Z}}\left\{\sum_{m<0} A_{(m)}B_{(n)}z^{-m-1} + \sum_{m\geq 0} A_{(m)}B_{(n)}z^{-m-1}\right\} w^{-n-1} \\
=&A(z)_+B(w) + B(w)A(z)_-.
\end{aligned}$$

Here and after, for each formal series $f(z) = \sum_{n\in\mathbb{Z}} f_n z^n$, we set

$$f(z)_+ := \sum_{n\geq 0} f_n z^n, \qquad f(z)_- := \sum_{n<0} f_n z^n.$$

2. For fields $A_i(z_i)$ $(i = 1, \cdots, n)$, we define the normally ordered product inductively by

$$\begin{aligned}
&\overset{\circ}{\circ}A_1(z_1)\cdots A_n(z_n)\overset{\circ}{\circ} \\
:=&\overset{\circ}{\circ}A_1(z_1)(\overset{\circ}{\circ}\cdots(\overset{\circ}{\circ}A_{n-1}(z_{n-1})A_n(z_n)\overset{\circ}{\circ})\cdots\overset{\circ}{\circ})\overset{\circ}{\circ}.
\end{aligned}$$

Now, we are in a place to explain a PBW type theorem.

Suppose that we are given a vector space V, a non-zero vector $|0\rangle$ and a linear operator $T \in \mathrm{End}V$. Let S be a countable ordered set and $\{a^\alpha\}_{\alpha\in S} \subset V$ be a set of vectors in V. Assume, in addition, that we are given a field

$$a^\alpha(z) = \sum_{n\in\mathbb{Z}} a^\alpha_{(n)} z^{-n-1}$$

for each $\alpha \in S$ satisfying the following conditions:

(1) For any $\alpha \in S$, one has $a^\alpha(z)|0\rangle = a^\alpha + O(z)$.
(Here, the symbol $O(z)$ is the Landau symbol.)
(2) $T|0\rangle = 0$, and $[T, a^\alpha(z)] = \partial_z a^\alpha(z)$ for any $\alpha \in S$.
(3) All fields $a^\alpha(z)$ are mutually local.

(4) V is spanned by the vectors of the form

$$a^{\alpha_1}_{(j_1)} \cdots a^{\alpha_k}_{(j_k)}|0\rangle \qquad j_i < 0.$$

A proof of the following theorem, called the **strong reconstruction theorem**, can be found in [FB]:

Theorem C.1 *Under the above assumption,*

1. *by setting*

$$Y(a^{\alpha_1}_{(j_1)} \cdots a^{\alpha_k}_{(j_k)}|0\rangle, z) := {}^{\circ}_{\circ}\partial_z^{(-j_1-1)} a^{\alpha_1}(z) \cdots \partial_z^{(-j_k-1)} a^{\alpha_k}(z){}^{\circ}_{\circ},$$

   one defines a vertex algebra structure on V.
   Here, for $n \in \mathbb{Z}_{\geq 0}$, $\partial_z^{(n)} := \dfrac{1}{n!}\partial_z^n$ is the nth divided power of ∂_z.
2. *This is a unique vertex algebra structure satisfying (1) – (4) and that $Y(a^\alpha, z) = a^\alpha(z)$.*
3. *Moreover, if V is a $\mathbb{Z}$-graded vector space, $\deg|0\rangle = 0$, $\deg T = 1$, the vectors a^α are homogeneous, and the fields $a^\alpha(z)$ have conformal dimension $\deg a^\alpha$, then V is a $\mathbb{Z}$-graded vertex algebra.*

C.1.4 Operator Product Expansion

In this subsection, we briefly recall several useful formulae for practical computation. Here, V always stands for a vertex algebra unless otherwise stated.

The first theorem we should recall is a uniqueness theorem due to P. Goddard:

Theorem C.2 (Field-State Correspondence) *Let $A(z)$ be a field on V. Suppose that $A(z)|0\rangle \in V[[z]]$ and*

$$\partial_z A(z)|0\rangle = TA(z)|0\rangle, \qquad A(z)|0\rangle\big|_{z=0} = a$$

for some $a \in V$. If, in addition, $A(z)$ is mutually local with $Y(b,w)$ for any $b \in V$, then one has $A(z) = Y(a,z)$.

Let us recall two formulae related to the translation operator:

Lemma C.1. *For $a \in V$, one has*

$$\begin{aligned} &i) \qquad Y(a,z)|0\rangle = e^{zT}.a, \\ &ii) \qquad Y(Ta,z) = \partial_z Y(a,z). \end{aligned}$$

Notice that the second formula is a corollary of Theorem C.2. The following technical proposition is a corollary of this lemma and the locality:

Proposition C.1 (Skew-Symmetry) *For any $A, B \in V$, one has the following formula in $V((z))$:*

$$Y(A,z).B = e^{zT}Y(B,-z).A.$$

As an application of this proposition, one can prove the following important theorem:

Theorem C.3 (Associativity) *For any $A, B, C \in V$, the three elements*

$$\begin{aligned} Y(A,z)Y(B,w).C &\in V((z))((w)),\\ Y(B,w)Y(A,z).C &\in V((w))((z)), \text{ and}\\ Y(Y(A,z-w)B,w).C &\in V((w))((z-w)) \end{aligned}$$

are the expansions of the same element of

$$V[[z,w]][z^{-1},w^{-1},(z-w)^{-1}].$$

The so-called Jacobi identity, which was originally used in the definition of vertex algebras, is a corollary of this theorem stated as follows.

Let $f(z,w)$ be a rational function with only possible poles at $z=w=0$ and $z=w$. Denote the Laurent series expansion of $f(z,w)$ in the domain $|z|>|w|$ by $\iota_{z,w}f(z,w)$. The following identity follows from the Cauchy theorem:

$$\mathrm{Res}_{z-w}(\iota_{w,z-w}f(z,w)) = \mathrm{Res}_z(\iota_{z,w}f(z,w)) - \mathrm{Res}_z(\iota_{w,z}f(z,w)),$$

where one sets $\mathrm{Res}_z f(z) := a_{-1}$ for a formal Laurent series $f(z)=\sum_{n\in\mathbb{Z}} a_n z^n$. Hence, Theorem C.3 implies the following theorem:

Theorem C.4 (Jacobi identity) *For $A, B \in V$ and $m, n, l \in \mathbb{Z}$, one has*

$$\begin{aligned} &\mathrm{Res}_{z-w}(Y(Y(A,z-w)B,w)\iota_{w,z-w}F(z,w))\\ =&\mathrm{Res}_z(Y(A,z)Y(B,w)\iota_{z,w}F(z,w)) - \mathrm{Res}_z(Y(B,w)Y(A,z)\iota_{w,z}F(z,w)), \end{aligned}$$

where $F(z,w)=z^m w^n (z-w)^l$.

An important corollary of the associativity (Theorem C.3) is the operator product expansion which is explained below.

First, we recall a useful proposition which follows from the definition of the normally ordered product:

Proposition C.2 *Let $A(z), B(w)$ be fields. The following statements are equivalent:*

1. *There exist fields $C_j(w)$ $(j=0,1,\cdots,N-1)$ such that*

$$[A(z),B(w)] = \sum_{j=0}^{N-1} C_j(w)\partial_w^{(j)}\delta(z-w).$$

2. $A(z)B(w)$ (resp. $B(w)A(z)$) has an expression

$$\sum_{j=0}^{N-1} \frac{C_j(w)}{(z-w)^{j+1}} + {}^{\circ}_{\circ}A(z)B(w)^{\circ}_{\circ}.$$

Here, $\frac{1}{z-w}$ is expanded in the positive power of $\frac{w}{z}$ (resp. $\frac{z}{w}$).
3. $A(z)B(w)$ converges to the formula in 2. on the domain $|z| > |w|$ and $B(w)A(z)$ does on the domain $|w| > |z|$.

Here, $\delta(z-w)$ is the **delta function** *defined by*

$$\delta(z-w) := \iota_{z,w}\frac{1}{z-w} - \iota_{w,z}\frac{1}{z-w} = \sum_{n\in\mathbb{Z}} z^{-n-1}w^n.$$

Taking just the singular part, the second statement in this proposition is often expressed as

$$A(z)B(w) \sim \sum_{j=0}^{N-1} \frac{C_j(w)}{(z-w)^{j+1}},$$

which is called the **operator product expansion** (or OPE.for short).

This proposition together with Theorem C.3 implies the following theorem:

Theorem C.5 *For $A, B \in V$, one has the following equivalent identities:*

$$Y(A,z)Y(B,w) = \sum_{n\geq 0} \frac{Y(A_{(n)}.B,w)}{(z-w)^{n+1}} + {}^{\circ}_{\circ}Y(A,z)Y(B,w)^{\circ}_{\circ},$$

$$[Y(A,z),Y(B,w)] = \sum_{n\geq 0} Y(A_{(n)}.B,w)\partial_w^{(n)}\delta(z-w).$$

C.2 Rationality

In this section, we recall the definition of a module over a vertex operator algebra and the rationality of a vertex operator algebra. We also briefly recall some facts about rational vertex operator algebras.

C.2.1 Modules

Here, we recall several versions of the definition of a module over a vertex operator algebra (V,ω) and denote the central charge of V by c_V.

Definition C.5 *A* ***weak*** ***V-module*** *is a pair (W, Y_W), where W is a vector space and Y_W is a linear map from V to $\mathrm{End}W[[z,z^{-1}]]$ satisfying*

1. *$Y_W(|0\rangle, z) = \mathrm{Id}_W$ and the Fourier modes of $Y_W(\omega, z) = \sum_{n\in\mathbb{Z}} L_n^W z^{-n-2}$ satisfies the commutation relations of the Virasoro algebra with the central charge c_V.*
2. *For $A \in V$, one has $Y_W(TA, z) = \dfrac{d}{dz} Y_W(A, z)$.*
3. *For $A, B \in V$, there exists $N \in \mathbb{Z}_{\geq 0}$ such that*

$$(z-w)^N [Y_W(A, z), Y_W(B, w)] = 0.$$

One may impose some conditions on the above definition which are stated as follows:

Definition C.6 *1. A weak V-module (W, Y_W) is called a $\mathbb{Z}_{\geq 0}$-**graded weak** V**-module** if*

i) W admits a $\mathbb{Z}_{\geq 0}$-gradation $W = \bigoplus_{n\in\mathbb{Z}_{\geq 0}} W_n$.
ii) For any $a \in V$ and $m, n \in \mathbb{Z}$, one has

$$a_{(m)}.W_n \subset W_{\deg a + n - m - 1}.$$

*2. A $\mathbb{Z}_{\geq 0}$-graded weak V-module $(W = \bigoplus_{n\in\mathbb{Z}_{\geq 0}} W_n, Y_W)$ is called a V-**module** if*

i) L_0 acts semi-simply on W_n.
ii) For any $n \in \mathbb{Z}_{\geq 0}$, one has $\dim W_n < \infty$.

Remark C.2 *We remark that, in some literature such as [DLM2], a $\mathbb{Z}_{\geq 0}$-graded weak V-module here is called an **admissible** V**-module**.*

C.2.2 The Zhu Algebra

In this subsection, we recall the Zhu algebra associated to a vertex operator algebra. Some fundamental results of Y. Zhu [Zh] are also reviewed.

Let $(V = \bigoplus_{n\in\mathbb{Z}} V_n, \omega)$ be a vertex operator algebra.

Definition C.7 *1. For a homogeneous $a \in V$ and $b \in V$, set*

$$a * b := \mathrm{Res}_z \left(Y(a, z) \frac{(1+z)^{\deg a}}{z} b \right).$$

2. Let $O(V) \subset V$ be the vector subspace spanned by

$$\mathrm{Res}_z \left(Y(a, z) \frac{(1+z)^{\deg a}}{z^2} b \right) \qquad a \in V: \text{ homogeneous}, \quad b \in V,$$

and set $A(V) := V/O(V)$.

We denote the element of $A(V)$ represented by $a \in V$ by $[a]$.

The basic structure of $A(V)$ is described in the following theorem:

Theorem C.6 (Y. Zhu) *$O(V) \subset V$ is a two-sided ideal with respect to the multiplication $*$, hence it defines a multiplication on $A(V)$. Moreover, one has*

1. *$*$ is associative on $A(V)$.*
2. *$[|0\rangle] \in A(V)$ is the unit element, where $|0\rangle \in V$ is the vacuum vector.*
3. *$[\omega] \in A(V)$ belongs to the centre.*

The associative algebra $A(V)$ is called the **Zhu algebra** of V. By definition, one has

Proposition C.3 ([FZ]) *Let $I \subset V$ be a vertex ideal such that $\omega \notin I$. The image $A(I)$ of I in $A(V)$ is a two-sided ideal and $A(V/I)$ is isomorphic to $A(V)/A(I)$.*

Definition C.8 *For a homogeneous $a \in V$, we set*

$$o(a) := a_{(\deg a - 1)},$$

and extend this symbol linearly to any $a \in V$.

The importance of the Zhu algebra is explained by the following theorem:

Theorem C.7 (Y. Zhu) *1. Let $M = \bigoplus_{n \in \mathbb{Z}_{\geq 0}} M_n$ be a $\mathbb{Z}_{\geq 0}$-graded weak V-module. One can introduce an $A(V)$-module structure on M_0 as follows:*

$$[a]|_{M_0} := o(a) \qquad a \in V.$$

2. Conversely, for an $A(V)$-module W, there exists a $\mathbb{Z}_{\geq 0}$-graded weak V-module $M = \bigoplus_{n \in \mathbb{Z}_{\geq 0}} M_n$ such that $M_0 \cong W$ and that one has $N = \{0\}$ for any $\mathbb{Z}_{\geq 0}$-graded weak V-submodule $N = \bigoplus_{n \in \mathbb{Z}_{\geq 0}} N_n \subset M$ satisfying $N_0 = \{0\}$.

Therefore, in particular, the isomorphism classes of V-modules and those of $A(V)$-modules are in one-to-one correspondence.

C.2.3 A Theorem of Y. Zhu

Here, we first recall the definition of the rationality of a vertex operator algebra V. We also recall an important theorem of Y. Zhu [Zh] in a refined form.

Definition C.9 *A vertex operator algebra V is said to be* ***rational*** *if*

1. *every $\mathbb{Z}_{\geq 0}$-graded weak V-module is completely reducible,*

2. *V has only finitely many isomorphism classes of $\mathbb{Z}_{\geq 0}$-graded weak simple modules, and*
3. *every $\mathbb{Z}_{\geq 0}$-graded weak simple module is a V-module.*

Here and after, we assume that V is a rational vertex operator algebra with the central charge c_V. It follows from Theorem C.7 that the Zhu algebra $A(V)$ is a semi-simple associative algebra; in particular, it is of finite dimension.

Remark C.3 *In [DLM2], it is shown that if a vertex operator algebra satisfies the condition* 1. *in the above definition, then the conditions* 2. *and* 3. *are automatically satisfied.*

Let $\{M_i = \bigoplus_{n\in\mathbb{Z}_{\geq 0}}(M_i)_n | \ 1 \leq i \leq m\}$ be the complete list of simple V-modules, and let $h_i \in \mathbb{C}$ be the L_0-eigenvalue on $(M_i)_0$, i.e., $L_0|_{(M_i)_0} = h_i\mathrm{id}$. For each $1 \leq i \leq m$, set

$$\chi_i(\tau) := \mathrm{tr}_{M_i} q^{L_0 - \frac{1}{24}C} := \sum_{n\geq 0} \dim(M_i)_n q^{h_i + n - \frac{1}{24}c_V},$$

where we set $q := e^{2\pi\sqrt{-1}\tau} \in \mathbb{C}^*$. Under this setting, one has the following theorem:

Theorem C.8 (Y. Zhu) *If a rational vertex operator algebra V satisfies the C_2-**cofiniteness**, i.e., the space $C_2(V) :=$ the linear span of $\{A_{(-2)}B | A, B \in V\}$ is finite co-dimensional in V, then*

1. *$\chi_i(\tau)$ $(1 \leq i \leq m)$ is holomorphic on $\mathbb{H} := \{\tau \in \mathbb{C} | \mathrm{Im}\tau > 0\}$, and*
2. *the space $\bigoplus_{i=1}^m \mathbb{C}\chi_i(\tau)$ is $SL(2,\mathbb{Z})$-invariant, where the $SL(2,\mathbb{Z})$-action on $\mathbb{H}$ is given by the Möbius transformation.*

Notice that if V satisfies the C_2-cofiniteness, then it is shown in [DLM3] that V satisfies the condition C of Y. Zhu in [Zh].

Remark C.4 *M. Miyamoto has shown (Theorem* 5.5 *in [Miy]) that if a vertex operator algebra, not necessarily rational, satisfies the C_2-cofiniteness, then the space spanned by '**generalised characters**' is $SL(2,\mathbb{Z})$-invariant.*

C.3 Fusion Rule

In this section, we recall the definition of a fusion algebra and some of their basic properties. In particular, we assume that V is a rational vertex operator algebra.

C.3.1 Intertwining Operators

Here, we recall the definition and a basic property of intertwining operators.

Definition C.10 *Let M^i $(i = 1, 2, 3)$ be $\mathbb{Z}_{\geq 0}$-graded weak V-modules.*

1. *An **intertwining operator of type** $\begin{pmatrix} & M^3 & \\ M^1 & & M^2 \end{pmatrix}$ is a linear map*

$$I(\cdot, z): \quad v \in M^1 \quad \longmapsto \quad \sum_{k \in K} v_k z^k,$$

*(where $v_k \in \mathrm{Hom}_{\mathbb{C}}(M^2, M^3)$ and $K = \bigcup_{i=1}^n \{\alpha_i + \mathbb{Z}\}$ for some $\alpha_i \in \mathbb{C}$) satisfying (**intertwining property**):*

$$\begin{aligned} &\mathrm{Res}_{z-w} \left(I(Y(a, z-w)v, w)(z-w)^m \iota_{w,z-w}((z-w)+w)^n \right) \\ =& \mathrm{Res}_z \left(Y(a,z) I(v,w) \iota_{z,w}(z-w)^m z^n \right) \\ &- \mathrm{Res}_z \left(I(v,w) Y(a,z) \iota_{w,z}(z-w)^m z^n \right), \end{aligned}$$

for any $a \in V$, $v \in M^1$ and $m, n \in \mathbb{Z}$, and

$$I(L_{-1}.v, w) = \frac{d}{dw} I(v, w).$$

2. *Denote*

$$I\begin{pmatrix} & M^1 & \\ M^2 & & M^3 \end{pmatrix} := \left\{ \text{intertwining operators of type} \begin{pmatrix} & M^1 & \\ M^2 & & M^3 \end{pmatrix} \right\},$$

*and the dimension of this vector space is called the **fusion rule** of the corresponding type.*

By definition, one has

Proposition C.4 *Let $M^i = \bigoplus_{n \in \mathbb{Z}_{\geq 0}} M^i_n$ $(i = 1, 2, 3)$ be $\mathbb{Z}_{\geq 0}$-graded weak V-modules such that $L_0|_{M^i_n} = (h_i + n)\mathrm{id}$ for some $h_i \in \mathbb{C}$. An intertwining operator $I(\cdot, z)$ of type $\begin{pmatrix} & M^3 & \\ M^1 & & M^2 \end{pmatrix}$ has the following Fourier expansion:*

$$I(v, z) = \sum_{n \in \mathbb{Z}} v_{[n]} z^{-n-1} \cdot z^{-h_1 - h_2 + h_3}.$$

In particular, if $v \in M^1$ is homogeneous, then the following inclusion holds for $m \in \mathbb{Z}_{\geq 0}$:

$$v_{[n]} M^2_m \subset M^3_{m + \deg v - n - 1}.$$

C.3.2 $A(V)$-Bimodule associated to a V-Module

In this subsection, we recall an $A(V)$-bimodule associated to a V-module.

Definition C.11 *Let M be a $\mathbb{Z}_{\geq 0}$-graded weak V-module, and $a \in V$ be a homogeneous element.*

1. For $v \in M$, set

$$a * v := \mathrm{Res}_z \left(Y(a,z) \frac{(1+z)^{\deg a}}{z} v \right).$$

2. For $v \in M$, set

$$v * a := \mathrm{Res}_z \left(Y(a,z) \frac{(1+z)^{\deg a - 1}}{z} v \right).$$

3. Set

$$O(M) := \left\{ \mathrm{Res}_z \left(Y(a,z) \frac{(1+z)^{\deg a}}{z^2} v \right) \middle| \ a \in V : \text{ homogeneous}, \ v \in M \right\}.$$

We extend the operations $*$ linearly to any $a \in V$ and $v \in M$.

The next structure theorem is proved by I. Frenkel and Y. Zhu [FZ]:

Theorem C.9 *Let M be a $\mathbb{Z}_{\geq 0}$-graded weak V-module and set $A(M) := M/O(M)$.*

1. *$A(M)$ has a left $A(V)$-module structure by the operation $*$.*
2. *$A(M)$ has a right $A(V)$-module structure by the operation $*$.*
3. *The left action and the right action of $A(V)$ on $A(M)$ commute.*

*$A(M)$ is called the $A(V)$-**bimodule associated to a $\mathbb{Z}_{\geq 0}$-graded weak V-module** M.*

C.3.3 Fusion Rule

In this subsection, we describe the space of intertwining operators of a type corresponding to simple modules in terms of $A(V)$-bimodule recalled in the previous subsection. Here, we assume that V is a rational vertex operator algebra.

Let $M^i = \bigoplus_{n \in \mathbb{Z}_{\geq 0}} M^i_n$ $(i = 1,2,3)$ be $\mathbb{Z}_{\geq 0}$-graded weak V-modules such that $L_0|_{M^i_n} = (h_i + n)\mathrm{id}$ for some $h_i \in \mathbb{C}$ and consider an intertwining operator of type $\begin{pmatrix} & M^3 & \\ M^1 & & M^2 \end{pmatrix}$. By Proposition C.4, one has $v_{[\deg v - 1]}.M^2_0 \subset M^3_0$ for a homogeneous $v \in M^1$.

Definition C.12 *For a homogeneous $v \in M^1$, set*

$$o'(v) := v_{[\deg v - 1]}$$

and extend this symbol linearly.

By direct calculation, one can show the following lemma:

Lemma C.2. *For $a \in V$, $v \in M^1$ and $v' \in O(M^1)$, the following identities hold:*

$$o(a)o'(v) = o'(a * v),$$
$$o'(v)o(a) = o'(v * a),$$
$$o'(v') = 0.$$

Let $\langle \cdot, \cdot \rangle$ be the dual pairing between $(M_0^3)^*$ and M_0^3. We regard $(M_0^3)^*$ as a right $A(V)$-module via the adjoint of the left $A(V)$-module structure on M_0^3 defined in Theorem C.7. Looking at the zero-mode of an intertwining operator, one obtains the next proposition (see [FZ]):

Proposition C.5 *Under the above setting, the pairing $\langle v_3, o'(v_1)v_2 \rangle$ ($v_1 \in A(M^1), v_2 \in M_0^2, v_3 \in (M_0^3)^*$) defines a functional f_I on $(M_0^3)^* \otimes_{A(V)} A(M^1) \otimes_{A(V)} M_0^2$.*

We remark that the pairing in this proposition is well-defined by Lemma C.2. Hence, by this proposition, one obtains a linear map,

$$\pi : I \begin{pmatrix} & M^3 & \\ M^1 & & M^2 \end{pmatrix} \longrightarrow (M_0^3)^* \otimes_{A(V)} A(M^1) \otimes_{A(V)} M_0^2. \tag{C.1}$$

As a special case, one has the next theorem due to I. Frenkel and Y. Zhu [FZ]:

Theorem C.10 *Suppose that M^i ($i = 1, 2, 3$) are simple V-modules. Then, the linear map (C.1) is an isomorphism.*

For a more general case, see [Li].

C.4 Vertex Superalgebras

Here, we briefly recall the definition of a vertex superalgebra and some properties of them analogous to those recalled in § C.1. For details, see, e.g., [Kac5].

C.4.1 Notations

A vector space V is called a **superspace** if it is $\mathbb{Z}/2\mathbb{Z}$-graded

$$V = V^{\bar{0}} \oplus V^{\bar{1}},$$

where $\bar{0}$ and $\bar{1}$ stand for the coset in $\mathbb{Z}/2\mathbb{Z}$ of 0 and 1. If V is a superspace, then End$V$ naturally inherits the superspace structure

$$\phi \in (\mathrm{End}V)^{\bar{j}} \quad \Longleftrightarrow \quad \phi(V^{\bar{i}}) \subset V^{\bar{i}+\bar{j}} \quad \forall\, \bar{i} \in \mathbb{Z}/2\mathbb{Z}.$$

We say that an element $a \in V$ has **parity** $|a| \in \mathbb{Z}/2\mathbb{Z}$ if $a \in V^{|a|}$. An element of $V^{\bar{0}} \setminus \{0\}$ is called **even** and that of $V^{\bar{1}} \setminus \{0\}$ is called **odd**.

Let R be an associative superalgebra, i.e., $R = R^{\bar{0}} \oplus R^{\bar{1}}$ satisfying $R^{\bar{i}} \cdot R^{\bar{j}} \subset R^{\bar{i}+\bar{j}}$ for $\bar{i}, \bar{j} \in \mathbb{Z}/2\mathbb{Z}$. One defines a bracket operation $[\cdot,\cdot]$ on R by letting

$$[a,b] := ab - (-1)^{|a||b|}ba,$$

for homogeneous $a, b \in R$. Via this bracket, R becomes a Lie superalgebra. We recall that a **Lie superalgebra** $\mathfrak{g}$ is a vector superspace $\mathfrak{g} = \mathfrak{g}^{\bar{0}} \oplus \mathfrak{g}^{\bar{1}}$ equipped with a bilinear operation $[\cdot,\cdot] : \mathfrak{g} \times \mathfrak{g} \longrightarrow \mathfrak{g}$, called a **Lie superbracket**, satisfying

1. $[\mathfrak{g}^{\bar{i}}, \mathfrak{g}^{\bar{j}}] \subset \mathfrak{g}^{\bar{i}+\bar{j}}$ for $\bar{i}, \bar{j} \in \mathbb{Z}/2\mathbb{Z}$,
2. for homogeneous $x, y \in \mathfrak{g}$,
$$[x,y] = -(-1)^{|x||y|}[y,x],$$
3. for homogeneous $x, y, z \in \mathfrak{g}$,
$$[x,[y,z]] = [[x,y],z] + (-1)^{|x||y|}[y,[x,z]].$$

The second property is called the **skew-symmetry** and the third property is called the **super Jacobi identity**.

C.4.2 Definition of a Vertex Superalgebra

A field $A(z) = \sum_j A_j z^{-j} \in \mathrm{End}V[[z^{\pm 1}]]$ is said to have **parity** $|A| \in \mathbb{Z}/2\mathbb{Z}$ if $A_j \in (\mathrm{End}V)^{|A|}$ for all $j \in \mathbb{Z}$.

Definition C.13 *A* ***vertex superalgebra*** *is a pair of data consisting of*

1. *(Space of states) a vector superspace* $V = V^{\bar{0}} \oplus V^{\bar{1}}$,
2. *(Vacuum vector) a vector* $|0\rangle \in V^{\bar{0}}$,
3. *(Translation Operator) an even linear operator* $T : V \longrightarrow V$,
4. *(Vertex Operators) a linear operation*
$$Y(\cdot, z) : V \longrightarrow \mathrm{End}V[[z, z^{-1}]],$$
preserving the parity, and taking each $A \in V$ *to a field acting on* V*, which is called the* ***vertex operator associated with*** A.

These data are subject to the following axioms:

1. *(Vacuum Axiom)* $Y(|0\rangle, z) = \mathrm{Id}_V$ *and for each* $A \in V$*, one has* $Y(A,z)|0\rangle \in V[[z]]$*. In particular, the specialisation* $|_{z=0}$ *is well-defined and*

$$Y(A,z)|0\rangle|_{z=0} = A.$$

2. *(Translation Axiom)* $T|0\rangle = 0$. *For each* $A \in V$, *one has*

$$[T, Y(A,z)] = \partial_z Y(A,z).$$

3. *(Locality Axiom) For any* $A, B \in V$, *the fields* $Y(A,z)$ *and* $Y(B,w)$ *are local, i.e., there exists* $N \in \mathbb{Z}_{\geq 0}$ *(depending on* A, B*) such that*

$$(z-w)^N [Y(A,z), Y(B,w)] = 0.$$

Here, the bracket is defined as in the previous subsection.

C.4.3 Operator Product Expansion

As one can imagine, most of the statements recalled in § C.1.4 are still valuable for a vertex superalgebra without any change. Here, we recall the only statements that have to be modified in the supersetting and we assume that V is a vertex superalgebra.

Proposition C.1 should be read as follows:

Proposition C.6 (Skew-Symmetry) *For any homogeneous* $A, B \in V$ *with respect to the* $\mathbb{Z}/2\mathbb{Z}$*-gradation, one has the following formula in* $V((z))$*:*

$$Y(A,z).B = (-1)^{|A||B|} e^{zT} Y(B,-z).A.$$

For homogeneous fields $A(z), B(w)$ acting on V, the **normally ordered product** ${}^\circ_\circ A(z)B(w)^\circ_\circ$ is defined by

$${}^\circ_\circ A(z)B(w)^\circ_\circ := A(z)_+ B(w) + (-1)^{|A||B|} B(w) A(z)_-.$$

Proposition C.2 should be read as follows:

Proposition C.7 *Let* $A(z), B(w)$ *be homogeneous fields with respect to the* $\mathbb{Z}/2\mathbb{Z}$*-gradation. The following statements are equivalent:*

1. *There exist fields* $C_j(w)$ $(j = 0, 1, \cdots, N-1)$ *such that*

$$[A(z), B(w)] = \sum_{j=0}^{N-1} C_j(w) \partial_w^{(j)} \delta(z-w).$$

2. $A(z)B(w)$ *(resp.* $(-1)^{|A||B|} B(w)A(z)$ *) has an expression*

$$\sum_{j=0}^{N-1} \frac{C_j(w)}{(z-w)^{j+1}} + {}^{\circ}_{\circ}A(z)B(w)^{\circ}_{\circ}.$$

Here, $\frac{1}{z-w}$ *is expanded in the positive power of* $\frac{w}{z}$ *(resp.* $\frac{z}{w}$*).*

3. $A(z)B(w)$ *converges to the formula in* 2. *on the domain* $|z| > |w|$ *and* $(-1)^{|A||B|}B(w)A(z)$ *does so on the domain* $|w| > |z|$.

Hence, Theorem C.5 also holds for a vertex superalgebra by interpreting the bracket in the theorem as a Lie superbracket.

Further Topics

Here, we provide a quick guide for further reading on some topics that are not treated in this book.

Combinatorics The famous Rogers–Ramanujan identities is related to the so-called Lee–Yang model (cf. § 5.1.5), i.e., $(2,5)$-series, and its generalisation due to G. E. Andrews (see, e.g., [And] for several related topics) is related to the $(2,2k+1)$-series which was discovered by B. Feigin and E. Frenkel [FeFr4]. These identites can be regarded as equalities relating *bosonic expressions* and *fermionic expressions*. The former, namely, the factorisation problem was systematically treated by A. G. Bytsko and A. Fring [BF] and an interesting generalisation was obtained by E. Mukhin [Mu]. The latter was first considered by R. Kedem *et al.* [KKMM] in the context of the regime II of the RSOS model (or Andrews–Baxter–Forrester model) of statistical mechanics. This approach was generalised by T. A. Welsh [Wel] for all of the BPZ series representations.

Group of diffeomorphisms of the circle The group $\mathrm{Diff}_+(S^1)$ of orientation preserving diffeomorphisms of the circle, whose commutator subgroup was proved to be simple by D. B. A. Epstein [E], is known to be a Fréchet Lie group (see, e.g., [Ham], [Miln]) which appears in several contexts such as dynamical systems (see, e.g., [Gh]) and Teichmüller spaces (see [Kon]). A non-trivial 2-cocycle of the Witt algebra, which defines the universal central extention, can be extend to a 2-cocycle of its smooth completion and the corresponding 2-cocycle of $\mathrm{Diff}_+(S^1)$ is explicitly given by R. Bott [Bott]. A. Kirillov [Kir1] showed that the infinite dimensional manifold $M := \mathrm{Diff}_+(S^1)/\mathrm{Rot}(S^1)$, where $\mathrm{Rot}(S^1)$ signifies the group of rotations of the circle, admits a Kähler structure and L. Lempert [Lem] showed that the central extension of $\mathrm{Diff}_+(S^1)$ by $\mathbb{R}$ is a $\mathbb{C}^*$-bundle over M, in particular, it admits the Kähler structure inherited form on M. Some discrete series representations are shown to be realisable [Kir2] in terms of M. See, e.g., [GR] for more information in this direction.

Logarithmic conformal field theory An interesting example of conformal field theories, which can be regarded as a non-semisimple theory, called the logarithmic theories has been studied recently. For example, B. Feigin

K. Iohara, Y. Koga, *Representation Theory of the Virasoro Algebra*, Springer Monographs in Mathematics, DOI 10.1007/978-0-85729-160-8,

et al. have studied the logarithmic $(1,p)$ conformal field theory [FGST1], [FGST2] and the logarithmic extension of the (p,q) minimal models [FGST3], [FGST4]. It was shown by D. Adamović and A. Milas [AdM2] that the triplet vertex operator algebra $\mathcal{W}(p)$ associated to $(1,p)$-series provides an example of a vertex operator algebra which is C_2-cofinite but irrational.

Semi-infinite cohomology In 1976, C. Becci *et al.* [BRS] and indenpendently I. Tyutin introduced a new method to treat quantum field theory with gauge invariance. In 1984, B. Feigin [Fe] introduced a new cohomology theory, which nowadays is called a semi-infinite cohomology to explain the critical dimension 26 of the string theory. This theory was formalised and generalised by I. Frenkel *et al.* [FGZ] to a certain class of $\mathbb{Z}$-graded Lie algebras. In particular, they showed what is called the no-ghost theorem. B. Kostant and S. Sternberg [KS] showed that this theory can be regarded as a quantisation of Hamiltonian reduction (cf. [MW]). In 1993, A. Voronov [Vor1] initiated semi-infinite homological algebra. See, e.g., [FeFr1], [FeFr3] and [Vor2], for further topics.

Super Virasoro algebras There are several super-extensions of the Virasoro algebra, among which we briefly discuss some known facts for $N = 1$ and $N = 2$ super Virasoro algebras. The $N = 1$ super Virasoro algebras have two classes, the Neveu–Schwarz and Ramond algebras. The structure of Verma modules were studied by the authors [IK2], [IK5] and that of Fock modules in [IK3]. The fusion algebras associated to minimal series representations were also determined by D. Adamović [Ad1] and the authors [IK1], [IK7]. The necessary condition for a highest weight module to be unitarisable was proved by F. Sauvageot [Sau]. The $N = 2$ super Virasoro algebras have several variants among which we mention the twisted and untwisted sectors. The untwisted sector has been extensively studied by B. Feigin and his colleagues [FST], [FSST], [SeF], [ST]. See also [Ad2] and [Ad3]. For the twisted sector, the structure of Verma modules and Fock modules was determined by the authors [IK4] and the classification of unitarisable highest weight modules was given in [Io]. See also [BFK].

$\mathcal{W}$-algebras V. Drinfeld and V. Sokolov [DS1], [DS2] showed that a certain Hamiltonian structure, called the second Gelfand–Dickey structure, of generalised mKdV hierarchies can be obtained from a more simple structure by a Hamiltonian reduction. A quantisation of such structure has been studied by several authors, e.g., [FaZ], [FaLu], [Lu] and [LuFa] but the quantisation in the esprit of BRST cohomology was first proposed by B. Feigin and E. Frenkel [FeFr2] and the algebra obtained in this way is called a quantum $\mathcal{W}$-algebra, which contains the Virasoro algebra as the simplest example. A conjectural character formula of it was proposed by E. Frenkel, V. Kac and M. Wakimoto [FKW] which was proved by T. Arakawa [Ara].

Now, the reader might be impressed by the artistic achievements of B. Feigin, one of the most important founders of the representation theory of the Virasoro algebra.

References

[AdM1] D. Adamović and A. Milas, *Vertex Operator Algebras Associated to Modular Invariant Representations for $A_1^{(1)}$*, Math. Res. Lett. **2**, (1995), 563–575.

[AdM2] D. Adamović and A. Milas, *On the triplet vertex algebra $\mathcal{W}(p)$*, Adv. Math. **217**, (2008), 2664–2699.

[Ad1] D. Adamović, *Rationality of Neveu–Schwarz Vertex Operator Superalgebras*, Int. Math. Res. Not., (1997), 865–874.

[Ad2] D. Adamović, *Representations of the $N = 2$ Superconformal Vertex Algebra*, Int. Math. Res. Not., (1999), 61–79.

[Ad3] D. Adamović, *Vertex Algebra Approach to Fusioin Rules for $N = 2$ Superconformal Minimal Models*, Jour. Alg. **239**, (2001), 549–572.

[AAR] G. E. Andrews, R. Askey and R. Roy, *Special Functions*, Encyclopedia Math. and Appl. **71**, Cambridge Univ. Press, Cambridge, 1999.

[AK] K. Aomoto and M. Kita, *Theory of Hypergeometric Functions, translated from the Japanese original by K. Iohara*, Springer Monographs in Mathematics, Springer, to appear.

[And] G. E. Andrews, *Theory of Partitions*, Reprint of the 1976 original. Cambr. Math. Library, Cambr. Univ. Press, Cambridge, 1998.

[Ara] T. Arakawa, *Representation theory of $\mathcal{W}$-algebras*, Invent. Math. **169**, (2007), 219–320.

[ADKP] E. Arbarello, C. De Concini, V. G. Kac and C. Procesi, *Moduli spaces of curves and representation theory*, Comm. Math. Phys. **117**, no. 1, 1–36, 1988.

[Ark] S. Arkhipov, *Semi-infinite cohomology of associative algebras and bar duality*, Internat. Math. Res. Notices, no. **17**, (1997), 833–863.

[As] A. Astashkevich, *On the structure of Verma modules over Virasoro and Neveu–Schwarz algebras*, Comm. Math. Phys. **186**, (1997), 531–562.

[AsFu] A. Astashkevich and D. B. Fuchs, *Asymptotics for singular vectors in Verma modules over the Virasoro algebra*, Pacific J. Math., **177**, No. 2, (1997), 201–209.

[AtM] M. F. Atiyah and I. G. Macdonald, *Introduction to Commutative Algebra*, Addison–Wesley, Reading, Massachusetts, (1969).

[BEG] M. Blume, V. J. Emery and R. B. Griffiths, *Ising Model for the λ Transition and Phase Separation in $He^3 - He^4$ Mixtures*, Phys. Rev. A **4**, (1971), 1071–1077.

[BRS] C. Becchi, A. Rouet and R. Stora, *Renormalization of Gauge Theories*, Ann. Phys. **98**, (1976), 287–321.

K. Iohara, Y. Koga, *Representation Theory of the Virasoro Algebra*,
Springer Monographs in Mathematics, DOI 10.1007/978-0-85729-160-8,

[BFM] A. Beilinson, B. Feigin and B. Mazur, *Note on Conformal Field Theory*, in preparation.

[BMS] A. A. Beilinson, Yu. I. Manin and V. V. Schechtman, *Sheaves of the Virasoro and Neveu–Schwarz algebras*, in Lect. Notes in Math. **1289**, (1987), 52–66.

[BPZ1] A. A. Belavin, A. M. Polyakov and A. B. Zamoldchikov, *Infinite Conformal Symmetry of Chiral Fluctuations in Two Dimensions*, Jour. Stat. Phys. **34**, (1984), 763–774.

[BPZ2] A. A. Belavin, A. M. Polyakov and A. B. Zamolodchikov, *Infinite Conformal Symmetry in Two-Dimensional Quantum Field Theory*, Nucl. Phys. B **241**, (1984), 333–380.

[BS] L. Benoit and Y. Sain-Aubin, *A proof of the explicit formula for certain singular vectors of the Virasoro algebra*, in Lie Theory, differential equations and representation theory, Univ. Montréal, Montréal, PQ, (1990), 77–84.

[BeSch] A. A. Beilinson and V. V. Schechtman, *Determinant Bundles and Virasoro algebras*, Comm. Math. Phys. **118**, (1988), 651–701.

[BGG1] I. N. Bernstein, I. M. Gelfand and S. I. Gelfand, *On a category of $\mathfrak{g}$-modules*, Funct. Anal. and Appl. **10**:2, (1976), 87–92.

[BGG2] I. N. Bernstein, I. M. Gelfand and S. I. Gelfand, *Differential Operators on the Base Affine Space and a Study of $\mathfrak{g}$-modules*, in Lie Groups and Their Representations, Proceedings of summer school of the Bolyai János Math. Soc., Budapest, I. M. Gelfand ed., 21–64, Hilger, London, 1975.

[Bl] R. E. Block, *On the Mills–Seligman Axioms for Lie Algebras of Classical Type*, Trans. Amer. Math. Soc. **121**, (1966), 378–392.

[BFK] W. Boucher, D. Friedan and A. Kent, *Determinant Formulae and Unitarity for the $N = 2$ Superconformal Algebras in Two Dimensions or Exact Results on String Compactification*, Phys. Lett. B **172**, (1986), 316–322.

[BGR] W. Borho, P. Gabriel and R. Rentschler, *Primideale in Einhüllenden auflösbarer Lie-Algebren*, Lect. Notes in Math. **357**, (1973).

[Bott] R. Bott, *On the characteristic classes of groups of diffeomorphisms*, L'Enseign. Math. **23**, (1977), 209–220.

[BF] A. G. Bytsko and A. Fring, *Factorized Combinatorics of Virasoro Characters*, Comm. Math. Phys. **209**, (2000), 179–205.

[C] E. Cartan, *Les groupes des transformations continus, infinis, simples*, Ann. Sci. Ec. Norm. Sup. 3e sér. **26**, (1909), 93–161.

[Chan] K. Chandrasekharan, *Elliptic Functions*, Grund. math. Wiss. **281**, Springer-Verlag Berlin Heidelberg (1985).

[Ch] H. J. Chang, *Über Wittsche Lieringe*, Abh. Math. Sem. Univ. Hamburg **14**, (1941), 151–184.

[Char] V. Chari, *Integrable representations of affine Lie algebras*, Invent. Math. **83**, (1986), 317–335.

[CP1] V. Chari and A. Pressley, *Integrable representations of twisted affine Lie algebras*, Jour. Alg. **113**, (1988), 438–464.

[CP2] V. Chari and A. Pressley, *Unitary representations of the Virasoro algebra and a conjecture of Kac*, Compo. Math. **67**, (1988), 315–342.

[CE] C. Chevally and S. Eilenberg, *Cohomology theory of Lie groups and Lie algebras*, Trans. Amer. Math. Sci. **63**, (1948), 85–124.

[DJKM] E. Date, M. Jimbo, M. Kashiwara and T. Miwa, *Transformation groups for soliton equations. Euclidean Lie algebras and reduction of the KP hierarchy*, Publ. Res. Inst. Math. Sci., **18** (1982), no. 3, 1077–1110.

[DGK] V. V. Deodhar, O. Gabber and V. Kac, *Structure of some categories of representations of infinite-dimensional Lie algebras*, Adv. Math. **45**, (1982), 92–116.

[Di] J. Dixmier, *Enveloping Algebras*, Graduate studies in mathematics **11**, Amer. Math. Soc., (1996).

[Do] C. Dong, *Vertex algebras associated with even lattices*, Jour. Alg. **161**, (1993), 245–265.

[DLM1] C. Dong, H. Li and G. Mason, *Vertex Operator Algebras Associated to Admissible Representations of* $\widehat{\mathfrak{sl}}_2$, Comm. Math. Phys. **184**, (1997), 65–93.

[DLM2] C. Dong, H. Li and G. Mason, *Twisted representations of vertex operator algebras*, Math. Ann. **310**, (1998), 571–600.

[DLM3] C. Dong, H. Li and G. Mason, *Modular-Invariance of Trace Functions in Orbifold Theory and Generalized Moonshine*, Comm. Math. Phys. **241**, (2000), 1–56.

[DS1] V. G. Drinfel'd and V. V. Sokolov, *Equations of Kortweg-de Vries Type and Simple Lie Algebras*, Soviet Math. Dokl. Vol. **23**, (1981), 457–462.

[DS2] V. G. Drinfel'd and V. V. Sokolov, *Lie Algebras and Equations of Kortweg-de Vries Type*, Jour. Sov. Math. **30**, (1985), 1975–2036.

[E] D. B. A. Epstein, *The simplicity of certain groups of homeomorphisms*, Compos. Math. **22**, (1970), 165–173.

[FaLu] V. A. Fateev and S. L. Lukyanov, *The Models of Two-Dimensional Conformal Quantum Field Theory with* $\mathbb{Z}_n$ *Symmetry*, Int. Jour. Mod. Phys. A **3**, (1988), 507–520.

[FaZ] V. A. Fateev and A. B. Zamolodchikov, *Conformal Quantum Field Theory of Models in Two Dimensions Having* $\mathbb{Z}_3$ *Symmetry*, Nucl. Phys. B **280**, (1987), 644–660.

[Fe] B. L. Feigin, *Semi-infinite homology of Lie, Kac–Moody and Virasoro algebras*, Uspekhi Mat. Nauk **39**, (1984), 195–196.

[FeFr1] B. L. Feigin and E. Frenkel, *Bosonic ghost system and the Virasoro algebra*, Phys. Lett. B **246**, (1990), 71–74.

[FeFr2] B. L. Feigin and E. Frenkel, *Quantization of the Drinfeld–Sokolov reduction*, Phys. Lett. B **246**, (1990), 75–81.

[FeFr3] B. L. Feigin and E. Frenkel, *Semi-infinite Weil complex and the Virasoro algebra*, Comm. Math. Phys. **137**, (1991), 617–639, *Erratum* in **147**, (1992), 647–648.

[FeFr4] B. L. Feigin and E. Frenkel, *Cohomology of nilpotent subalgebras of the Virasoro algebra and partitions identities*, in *I. M. Gelfand Seminar*, Adv. Soviet Math. **16** Part I, (1993), 139–148.

[FeFu1] B. L. Feigin and D. B. Fuchs, *Skew-Symmetric Invariant Differential Operators on the Line and Verma Modules over the Virasoro Algebra*, Funct. Anal. Appl., **16**, (1982), 47–63.

[FeFu2] B. L. Feigin and D. B. Fuchs, *Verma Modules over the Virasoro Algebra*, Funct. Anal. Appl., **17**, (1983), 91–92.

[FeFu3] B. L. Feigin and D. B. Fuchs, *Verma Modules over the Virasoro Algebras*, Lect. Notes in Math. **1060**, (1984), 230–245.

[FeFu4] B. L. Feigin and D. B. Fuchs, *Representations of the Virasoro algebra*, Adv. Stud. Contemp. Math. **7**, 465–554, Gordon and Breach Science Publ. New York, 1990.

[FeFu5] B. L. Feigin and D. B. Fuchs, *Cohomology of some nilpotent subalgebras of the Virasoro and Kac–Moody Lie algebras*, Jour. Geom. Phys. **5**, (1988), 209–235.

[FGST1] B. L. Feigin, A. M. Gainutdinov, A. M. Semikhatov and I. Yu. Tipunin, *Modular Group Representations and Fusion in Logarithmic Conformal Field Theories and in the Quantum Group Center*, Comm. Math. **265**, (2006), 47–93.

[FGST2] B. L. Feigin, A. M. Gainutdinov, A. M. Semikhatov and I. Yu. Tipunin, *Kazhdan–Lusztig Correspondence for the Representation Category of the Triplet W-algebra in Logarithmic CFT*, Theor. Math. Phys. **148**, (2006), 1210–1235.

[FGST3] B. L. Feigin, A. M. Gainutdinov, A. M. Semikhatov and I. Yu. Tipunin, *Logarithmic extensions of minimal models: Characters and modular transformations*, Nucl. Phys. B **757**, (2006), 303–343.

[FGST4] B. L. Feigin, A. M. Gainutdinov, A. M. Semikhatov and I. Yu. Tipunin, *Kazhdan–Lusztig-dual quantum group for logarithmic extension of Virasoro minimal models*, Jour. Math. Phys. **48**, 2007, 46 pp.

[FKW] E. Frenkel, V. Kac and M. Wakimoto, *Characters and Fusion Rules for W-algebras via Quantized Drinfeld–Sokolov Reduction*, Comm. Math. Phys. **147**, (1992), 295–328.

[FeMa] B. Feigin and F. Malikov, *Modular functor and representation theory of $\hat{sl}_2$ at a rational level, Operads: Proceedings of Renaissance Conference*, Contemp. Math. **202**, 357–405, Amer. Math. Soc. Providence, RI, 1997.

[FST] B. L. Feigin, A. M. Semikhatov and I. Yu. Tipunin, *Equivalence between chain categories of representations of affine $sl(2)$ and $N = 2$ superconformal algebra*, Jour. Math. Phys. **39**, (1998), 3865–3905.

[FSST] B. L. Feigin, A. M. Semikhatov, V. A. Sirota and I. Yu. Tipunin, *Resolutions and characters of irreducible representations of the $N = 2$ superconformal algebra*, Nucl. Phys. B **536**, (1999), 617–656.

[Fr] E. Frenkel, *Determinant Formulas for the Free Field Representations of the Virasoro and Kac–Moody algebras*, Phys. Lett. B **286**, (1992), 71–77.

[FB] E. Frenkel and D. Ben'zvi, *Vertex Algebras and Algebraic Curves, 2nd edn.*, Math. Survey and Monographs **88**, Amer. Math. Soc., Providence, 2004.

[FGZ] I. Frenkel, H. Garland and G. J. Zuckerman, *Semi-infinite cohomology and string theory*, Proc. Natl. Acad. Sci. USA **83**, (1986), 8442–8446.

[FZ] I. Frenkel and Y. Zhu, *Vertex operator algebras associated to representations of affine and Virasoro algebras*, Duke Math. Jour. **66**, (1992), 123–168.

[FQS1] D. Friedan, Z. Qui and S. Shenker, *Conformal Invariance, Unitarity, and Critical Exponents in Two Dimensions*, Phys. Rev. Lett. **52**, (1984), 1575–1578.

[FQS2] D. Friedan, Z. Qiu and S. Shenker, *Details of the nonunitarity proof for highest weight representations of the Virasoro algebra*, Comm. Math. Phys. **107**, (1986), 535–542.

[Gab] O. Gabber, *The Integrability of the Characteristic Variety*, Amer. Jour. Math. **103**, (1981), 445–468.

[Gar] H. Garland, *The arithmetic theory of loop groups*, Publ. Math. IHES **52**, (1980), 5–136.

[Ger] J. Germoni, *On the classification of admissible representations of the Virasoro algebra*, Lett. Math. Phys. **55**, (2001), 169–177. Erratum **59**, (2002), 277.

[GF] I. M. Gelfand and D. B. Fuchs, *Cohomology of Lie algebras of the vector fields on the circle*, Funct. Anal. and Appl. **2**, (1968), 92–93.

[Gh] E. Ghys, *Groups acting on the circle*, L'Enseign. Math. **47**, (2001), 329–407.

[GKO1] P. Goddard, A. Kent and D. Olive, *Virasoro algebras and coset space models*, Phys. Lett. B **152**, (1985), 88–92.

[GKO2] P. Goddard, A. Kent and D. Olive, *Unitary Representations of the Virasoro and Super-Virasoro Algebras*, Comm. Math. Phys. **103**, (1986), 105–119.

[GW1] R. Goodman and N. R. Wallach, *Structure and unitary cocycle representations of loop groups and the group of diffeomorphisms of the circle*, Jour. f. Reine u. Angew. Math. **347**, (1984), 69–133, *Erratum* **352**, (1984), 220.

[GW2] R. Goodman and N. R. Wallach, *Projective unitary positive-energy representations of* $\mathrm{Diff}(S^1)$, Jour. Funct. Anal. **63**, (1985), 299–321.

[Gro] A. Grothendieck, *Sur quelques points d'algèbre homologique*, Tohoku Math. J. **9**, (1957).

[GR] L. Guieu and C. Roger, *L'algèbre et le groupe de Virasoro (with an appendix by V. Sergiescu)*, Publ. CRM, Montreal, 2007.

[GSW] M. Green, J. Schwarz and E. Witten, *Superstring theory*, Vol. **1**, Vol. **2**, Second Edition, Cambridge University, 1988.

[Ham] R. Hamilton, *The inverse function theorem of Nash and Moser*, Bull. Amer. Math. Soc. (New Ser.) **7**, (1982), 65–222.

[HiSt] P. J. Hilton and U. Stammbach, *A Course in Homological Algebras*, Second Edition, Graduate Texts in Mathematics **4**, Springer-Verlag, 1996.

[Hu1] Y.-Z. Huang, *Vertex operator algebras, the Verlinde conjecture, and modular tensor categories*, Proc. Natl. Acad. Sci. USA **102**, (2005), 5352–5356.

[Hu2] Y.-Z. Huang, *Vertex operator algebras and the Verlinde conjecture*, Commun. Contemp. Math. **10**, (2008), 103–154.

[HHK] L. A. Hugel, D. Happel and H. Krause, *Handbook of Tilting Theory*, London Math. Soc. Lect. Notes Ser. **332**, Cambr. Univ. Press, 2007.

[ID] C. Itzykson and J.-M. Drouffe, *Statistical field theory* **1,2**, Cambr. Monogr. Math. Phys., Cambr. Univ. Press, 1989.

[In] A. E. Ingham, *A Tauberian theorem for partitions*, Ann. Math. **42**, (1941), 1075–1090.

[Io] K. Iohara, *Modules de plus haut poids unitarisables sur la super-algèbre de Virasoro $N = 2$ tordue*, Ann. Inst. Fourier, **58**, (2008), 733–754.

[IK1] K. Iohara and Y. Koga, *Fusion Algebras for $N = 1$ Superconformal Field Theories through Coinvariants,* II*: $N = 1$ Super-Virasoro-Symmetry*, Jour. Lie Th. **11**, (2001), 305–337.

[IK2] K. Iohara and Y. Koga, *Representation Theory of Neveu–Schwarz and Ramond Algebras* I*: Verma Modules*, Adv. in Math., **178**, (2003), 1–65.

[IK3] K. Iohara and Y. Koga, *Representation Theory of Neveu–Schwarz and Ramond Algebras* II*: Fock Modules*, Ann. Inst. Fourier, **53**, (2003), 1755–1818.

[IK4] K. Iohara and Y. Koga, *Representation Theory of $N = 2$ super-Virasoro algebra: Twisted sector*, Jour. Funct. Anal, **214**, (2004), 450–518.

[IK5] K. Iohara and Y. Koga, *The structure of pre-Verma modules over the $N = 1$ Ramond algebra*, Lett. Math. Phys., **78**, (2006), 89–96.

[IK6] K. Iohara and Y. Koga, *Tilting Equivalence for Superconformal Algebras*, Math. Scand. **99**, (2006), 17–52.

[IK7] K. Iohara and Y. Koga, *Fusion Algebras for $N = 1$ Superconformal Field Theories through Coinvariants,* II*: Ramond sectors*, Int. Math. Res. Not., IMRN 2009, no. 13, Art ID rnp020, 2374–2416.

[IK8] K. Iohara and Y. Koga, *Note on spin modules associated to $\mathbb{Z}$-graded Lie superalgebras*, J. Math. Phys., **50**, (2009), 103508, 9 pp.

[Is] E. Ising, *Beitrag zur Theorie des Ferromagnetismus*, Zeits. f. Physik **31**, (1925), 253–258.

[IZ] C. Itzykson and J. B. Zuber, *Two-dimensional conformal invariant theories on a torus*, Nucl. Phys. B **275**, (1986), 580–616.

[Jac] N. Jacobson, *Lie Algebras*, Dover Publications, New York, 1979.

[Ja1] J. C. Jantzen, *Moduln mit einem höchsten Gewicht*, Lect. Notes in Math. **750** Springer-Verlag, 1979.

[Ja2] J. C. Jantzen, *Einhüllende Algebren halbeinfacher Lie-Algebren*, Ergebnisse der Math. und ihrer Grenzgebiete **3** Springer-Verlag, 1983.

[Ja3] J. C. Jantzen, *Representations of Algebraic Groups*, Second edn., Math. Surv. Monogr. **107**, Amer. Math. Soc., Providence, 2003.

[Kac1] V. G. Kac, *Highest Weight Representations of Infinite-Dimensional Lie Algebras*, Proc. Inter. Congr. Math. Helsinki, (1978), 299–304.

[Kac2] V. G. Kac, *Contravariant form for infinite dimensional Lie algebras and superalgebras*, Lect. Notes in Phys. **94**, (1979), 441–445.

[Kac3] V. G. Kac, *Some Problems on Infinite Dimensional Lie Algebras and Their Representations*, Lect. Notes in Math. **933**, (1981), 117–126.

[Kac4] V. G. Kac, *Infinite dimensional Lie algebras*, Third Edition, Cambridge University Press, 1990.

[Kac5] V. G. Kac, *Vertex Algebras for Beginners*, 2nd edn., Univ. Lect. Ser. **10**, Amer. Math. Soc., Providence, 1998.

[Kac6] V. G. Kac, *Superconformal Algebras and Transitive Group Actions on Quadrics*, Comm. Math. Phys. **186**, (1997), 233–252, *Erratum*, **217**, (2001), 697–698.

[KR] V. G. Kac and A. K. Raina, *Bombay lectures on highest weight representations of infinite-dimensional Lie algebras*, Adv. Ser. Math. Phys. **2**, World Scientific, 1987.

[KW1] V. G. Kac and M. Wakimoto, *Unitarizable highest weight representation of the Virasoro, Neveu–Schwarz and Ramond algebras*, Lect. Notes in Phys. **261**, (1986), 345–372.

[KW2] V. G. Kac and M. Wakimoto, *Modular invariant representations of infinite-dimensional Lie algebras and superalgebras*, Proc. Natl. Acad. Sci. USA **85**, (1988), 4956–4960.

[Ka] I. Kaplansky, *The Virasoro algebra*, Comm. Math. Phys. **86**, (1982), 49–54.

[KaSa] I. Kaplansky and L. J. Santharoubane, *Harish–Chandra modules over the Virasoro algebra*, in Infinite-dimensional groups with applications, Math. Sci. Res. Inst. Publ. **4**, Springer, New York-Berlin, (1985), 217–231.

[KT] M. Kashiwara and T. Tanisaki, *Kazhdan–Lusztig conjecture for symmetrizable Kac–Moody Lie algebras. III – Positive rational case*, Asian J. Math. **2**, (1998), 779–832.

[KKMM] R. Kedem, T. R. Klassen, B. M. McCoy and E. Melzer, *Fermionic sum representations for conformal field theory characters*, Phys. Lett. B **307**, (1993), 68–76.

[Kir1] A. Kirillov, *Kähler structures on K-orbits of the group of diffeomorphisms of a circle*, Funct. Anal. Appl. **21**, (1987), 122–125.

[Kir2] A. A. Kirillov, *Geometric approach to discrete series of unireps for* Vir, Jour. Math. Pure Appl. **77**, (1998), 735–746.

[Kon] M. L. Konstevich, *Virasoro algebra and Teichmüller spaces*, Funct. Anal. Appl. **21**, (1987), 156–157.

[KS] B. Kostant and S. Sternberg, *Symplectic Reduction, BRS Cohomology, and Infinite-Dimensional Clifford Algebras*, Ann. Phys. **176**, (1987), 49–113.

[Kostr] I. A. Kostrikin, *Irreducible Graded Representations of Lie algebras of Cartan Type*, Soviet Math. Dokl. **19**, (1978), 1369–1371.

[Kosz1] J. L. Koszul, *Homologie et Cohomologie des Algèbres de Lie*, Bull. Soc. Math. France **78**, (1950), 65–127.

[Kosz2] J. L. Koszul, *Sur un type d'algèbres différentielles en rapport avec la transgression*, in Colloq. de Top., Bruxelles, (1950), 73–81.

[La] R. P. Langlands, *On Unitary Representations of the Virasoro Algebra*, in Infinite Dimensional Lie Algebras and Their Applications, World Sci., Teaneck, NJ, (1988), 141–159.

[Lem] L. Lempert, *The Virasoro group as a complex manifold*, Math. Res. Lett. **2**, (1995), 479–495.

[Len] W. Lenz, *Beitrag zum Verständnis der magnetischen Erscheinungen in festen Körpern*, Zeists. f. Phys. **21**, (1920), 613–615.

[Li] H. Li, *Determining Fusion Rules by $A(V)$-Modules and Bimodules*, Jour. Alg. **212**, (1999), 515–556.

[Liu] L. Liu, *Kostant's Formula for Kac–Moody Lie algebras*, Jour. Alg. **149**, (1992), 155–178.

[Lu] S. L. Lukyanov, *Quantization of the Gelfand–Dikii Brackets*, Funkts. Anal. Prilozhen., **22**, (1988), 1–10.

[LuFa] S. L. Lukyanov and V. A. Fateev, *Conformally invariant models of two-dimensional quantum field theory with $\mathbb{Z}_n$-symmetry*, Sov. Phys. JETP **67**, (1988), 447–454.

[Mac1] I. G. Macdonald, *Symmetric functions and Hall polynomials*, Second edition, Oxford Mathematical Monographs. Oxford University Press, New York, (1995).

[Mac2] I. G. Macdonald, *Commuting differential operators and zonal spherical functions*, Algebraic groups Utrecht 1986, 189–200, Lecture Notes in Math., 1271, Springer, Berlin, (1987).

[Mal] F. G. Malikov, *Verma modules over Kac–Moody algebras of rank* 2, Leningrad Math. J., **2**, No. 2, (1991), 269–286.

[Mand] S. Mandelstam, *Dual Resonance Models*, Phys. Rep. **13**, (1974), 259–353.

[MW] J. Marsden and A. Weinstein, *Reduction of Symplectic Manifolds with Symmetry*, Rep. Math. Phys. **5**, (1974), 121–130.

[MaP] C. Martin and A. Piard, *Indecomposable Modules Over the Virasoro Algebra and a Conjecture of V. Kac*, Comm. Math. Phys. **137**, (1991), 109–132.

[Mat1] O. Mathieu, *Classification des algèbres de Lie graduées simples de croisance* ≤ 1, Invent. Math. **86**, (1986), 371–426.

[Mat2] O. Mathieu, *Classification of Harish–Chandra modules over the Virasoro Lie algebra*, Invent. Math. **107**, (1992), 225–234.

[Mat3] O. Mathieu, *Classification of simple graded Lie algebras of finite growth*, Invent. Math. **108**, (1992), 455–519.

[Matsu] H. Matsumura, *Commutative ring theory*, Cambridge University Press, (1986).

[Mill] D. V. Millionschikov, *Algebra of Formal Vector Fields on the Line and Buchstaber's Conjecture*, Funct. Anal. and Appl. **43**, (2009), 264–278.

[Miln] J. Milnor, *Remarks on infinite-dimensional Lie groups*, Proc. Summer School on Quantum Gravity, B. DeWitt ed., Les Houches, (1983).

[MY] K. Mimachi and Y. Yamada, *Singular vectors of the Virasoro algebra in terms of Jack symmetric polynomials*, Comm. Math. Phys. **174**, (1995), 447–455.

[Mit] B. Mitchell, *The Full Imbedding Theorem*, Amer. Jour. Math. **86**, (1964), 619–637.

[Miy] M. Miyamoto, *Modular Invariance of Vertex Operator Algebras Satisfying C_2-cofiniteness*, Duke Math. Jour. **122**, (2004), 51–92.

[MP] R. V. Moody and A. Pianzola, *Lie Algebras With Triangular Decompositions*, Can. Math. Soc., John Wiley and Sons, (1995).

[Mu] E. Mukhin, *Factorization of alternating sums of Virasoro chracters*, Jour. Comb. Theory, Ser. A **114**, (2007), 1165–1181.

[Mum] D. Mumford, *Stability of projective varieties*, Monograph **24**, L'Enseign. Math., Genève, 1977.

[NBRS] B. Nijenhuis, A. N. Berker, E. K. Riedel and M. Schick, *First- and Second-Order Phase Transitions in Potts Models: Renormalization-Group Solution*, Phys. Rev. Lett. **43**, (1979), 737–740.

[OT] P. Orlik and H. Terao, *Arrangements and hypergeometric integrals*, Second edition. MSJ Memoirs **9**, Mathematical Society of Japan, Tokyo, (2007).

[Pol] J. Polchinski, *String Theory*, Vol. **1**, Vol. **2**, Cambridge Monographs on Mathematical Physics, Cambridge University Press, (1998).

[P] R. B. Potts, *Some generalized order-disorder transformations*, Proc. Cambr. Philos. Soc. **48**, (1952), 106–109.

[Ro] A. Rocha-Caridi, *On highest weight and Fock space representations of the Virasoro algebras*, Proc. Sympo. Pure math. **49**, Part 1, Amer. Math. Soc. Providence, RI, (1989), 259–272.

[RW1] A. Rocha-Caridi and N. R. Wallach, *Projective modules over Graded Lie algebras. I*, Math. Z. **180**, (1982), 151–177.

[RW2] A. Rocha-Caridi and N. R. Wallach, *Highest weight modules over graded Lie algebras: resolutions, filtrations and character formulas*, Trans. Amer. Math. Soc. **20**, (1983), 133–162.

[RW3] A. Rocha-Caridi and N. R. Wallach, *Characters of irreducible representations of the Lie algebra of vector fields on the circle*, Invent. Math. **72**, (1983), 57–75.

[RW4] A. Rocha-Caridi and N. R. Wallach, *Characters of irreducible representations of the Virasoro algebra*, Math. Z. **185**, (1984), 1–21.

[Sau] F. Sauvageot, *Représentations unitaires des super-algèbres de Ramond et de Neveu–Schwartz*, Comm. Math. Phys. **121**, (1989), 639–657.

[Sel] A. Selberg, *Bemerkninger om et multipelt integral*, Norske Math. Tidsskr. **26**, (1944), 71–78.

[SeF] A. M. Semikhatov and B. L. Feigin, *Free-Field Resolutions of Unitary Representations of the $N = 2$ Virasoro Algebra:* II *The Butterfly Resolution*, Thoer. Math. Phys. **121**, (1999), 1462–1472.

[ST] A. M. Semikhatov and I. Yu. Tipunin, *The structure of Verma Modules over the $N = 2$ Superconformal Algebra*, Comm. Math. Phys. **195**, (1998), 129–173.

[Sh] N. N. Shapovalov, *On bilinear form on the universal enveloping algebra of a complex semisimple Lie algebra*, Functional Anal. Appl., **6**, (1972), 307–312.

[So] W. Soergel, *Character formulas for tilting modules over Kac–Moody algebras* Represent. Theory, **2**, (1998), No. 13, 432–448.

[St] H. Strade, *Representations of the Witt Algebra*, Jour. Alg. **49**, (1977), 595–605.

[SF] H. Strade and R. Fransteiner, *Modular Lie Algebras and Their Representations*, Monographs and textbooks in pure and applied mathematics, **116**, Marcel Dekker, New York, 1988.

[To] V. Toledano Laredo, *Integrating unitary representations of infinite dimensional Lie groups*, Jour. Funct. Anal. **161**, (1999), 478–508.

[TK1] A. Tsuchiya and Y. Kanie, *Unitary representations of the Virasoro algebra*, Duke Math. J. **53**, (1986), 1013–1046.

[TK2] A. Tsuchiya and Y. Kanie, *Fock Space Representations of the Virasoro Algebra – Intertwining Operators –*, Publ. RIMS Kyoto Univ. **22**, (1986), 259–327.

[VK] B. J. Veisfeiler and V. G. Kac, *Irreducible Representations of Lie p-algebras*, Funct. Anal. Appl. **5**, (1971), 111–117.

[Ver] E. Verlinde, *Fusion Rules and Modular Transformations in 2D Conformal Field Theory*, Nucl. Phys. B **300**, (1988), 360–376.

[Vir] M. A. Virasoro, *Subsidiary Conditions and Ghosts in Dual Resonance Models*, Phys. Rev. D **1**, (1970), 2933–2936.

[Vor1] A. Voronov, *Semi-infinite homological algebra*, Invent. Math. **113**, (1993), 103–146.

[Vor2] A. Voronov, *Semi-infinite Induction and Wakimoto Modules*, Amer. Jour. Math. **121**, (1999), 1079–1094.

[Wak] M. Wakimoto, *Infinite Dimensional Lie algebras, translated from Japanese original by K. Iohara*, Transl. Math. Monogr. **195**, Amer. Math. Soc., Providence, RI, 2001.

[Wan] W. Wang, *Rationality of Virasoro Vertex Operator Algebras*, Int. Math. Res. Not., (1993), 197–211.

[WY] M. Wakimoto and H. Yamada, *The Fock representations of the Virasoro algebra and the Hirota equations of the modified KP hierarchies*, Hiroshima Math. Jour. **16**, (1986), 427–441.

[Wei] C. A. Weibel, *An introduction to homological algebra*, Cambridge studies in advanced mathematics **38**, Cambridge University Press, 1994.

[Wel] T. A. Welsh, *Fermionic expressions for minimal model Virasoro characters*, Mem. Amer. Math. Soc. **175**, 2005.

[Y] G. Yamamoto, *Algebraic structures on quasi-primary states in superconformal algebras*, Jour. Math. Soc. Japan **57**, (2005), 309–332.

[YL] C. N. Yang and T. D. Lee, *Statistical Theory of Equations of State and Phase Transitions.* I *Theory of Condensation,* II *Lattice Gas and Ising Model*, Phys. Rev. **87**, (1952), 404–409, 410–419.

[ZF1] A. B. Zamolodchikov and V. A. Fateev, *Nonlocal (parafermion) currents in two-dimensional conformal quantum field theory and self-dual critical points in $\mathbb{Z}_n$-symmetric statistical systems*, Zh. Eksp. Teor. Fiz. **89**, (1985), 380–399.

[ZF2] A. B. Zamolodchikov and V. A. Fateev, *Disorder fields in two-dimensional conformal quantum-field theory and $N = 2$ extended supersymmetry*, Zh. Eksp. Teor. Fiz. **90**, (1985), 1553–1566.

[Za] H. Zassenhaus, *Über Lie'sche Ringe mit Primzahlcharakteristik*, Abh. Math. Sem. Univ. Hamburg **12**, (1939), 1–100.

[Zh] Y. Zhu, *Modular Invariance of Characters of Vertex Operator Algebras*, Jour. Amer. Math. Soc. **9**, (1996), 237–302.

List of Symbols

K. Iohara, Y. Koga, *Representation Theory of the Virasoro Algebra*, Springer Monographs in Mathematics, DOI 10.1007/978-0-85729-160-8,

Index

K. Iohara, Y. Koga, *Representation Theory of the Virasoro Algebra*, Springer Monographs in Mathematics, DOI 10.1007/978-0-85729-160-8,